T0229514

PROCEEDINGS OF THE THIRD DIANA WORLD CONFERENCE
TOKYO/JAPAN/9–11 OCTOBER 2002

Finite Elements in Civil Engineering Applications

Edited by

Max A.N. Hendriks
TNO Building and Construction Research, Delft, The Netherlands

Jan G. Rots
Delft University of Technology, Delft, The Netherlands

A.A. BALKEMA PUBLISHERS LISSE / ABINGDON / EXTON (PA) / TOKYO

Published by: A.A. Balkema, a member of Swets & Zeitlinger Publishers
www.balkema.nl and www.szp.swets.nl

ISBN 90 5809 530 4

Printed in the Netherlands

Finite Elements in Civil Engineering Applications, Hendriks & Rots (eds.)
© 2002 Swets & Zeitlinger, Lisse, ISBN 90 5809 530 4

Table of Contents

Shear in reinforced concrete

Masonry and block mechanics

Geomechanics, steel and rubber

Reinforced concrete structures

Short papers

Finite Elements in Civil Engineering Applications, Hendriks & Rots (eds.)
© 2002 Swets & Zeitlinger, Lisse, ISBN 90 5809 530 4

Preface

The book comprises the papers presented at the Third International DIANA World Conference on Finite Elements in Civil Engineering Applications, held in Tokyo, Japan, from 9–11 October 2002. The tradition of this conference series began in 1994 in Delft where software development and research in computational mechanics were brought into contact with finite element applications in engineering practice. Subsequently, the second conference was held in 1997 in Amsterdam along the same lines, while the opportunity was taken to celebrate the 25th anniversary of DIANA with retrospective and prospective contributions.

The third conference preserves the original spirit of the series and brings together both researchers and practising engineers engaged in computational modelling. It is held in Japan, which guarantees a focus on high-level research in structural engineering, concrete mechanics and quasi-brittle materials, considering the prime concern of durability requirements and earthquake resistance of structures in Japan. This focus on concrete structures complies with the classical strength of the DIANA community. Other areas like geomechanics, geotechnics and liquefaction, masonry mechanics, and work on steel, rubber and fibre-reinforced materials are included as well in the conference and the proceedings.

We would like to express our sincere gratitude to the members of the Scientific Advisory Committee, as listed at a separate page, for their support and efforts in carefully reviewing the 80 submitted abstracts. This ensured the quality and the acceptance of 6 invited papers and 53 contributing papers and 7 short papers, coming from authors over five continents. A number of Japanese contributions has been included in this volume in English. On top of that, approx. 14 contributions in Japanese language have been accepted for publication in a separate volume. We also would like to thank Ms. Jantine van Steenbergen for her important contribution and secretarial assistance in organising this conference.

The papers have been subdivided into six categories, supplemented with a 7th chapter for the short papers:

1. Constitutive models for quasi-brittle materials.
   The chapter covers recent advances in formulations for cracking, crushing, damage, micro-plane and other models for softening materials like concrete, including implementation effects in user-supplied routines or source code.
2. Creep, shrinkage, durability and young hardening concrete.
   The chapter presents contributions on time- and temperature-dependent behaviour, both for the early age and the long-term, covering multiple scales from micro-structural level to large structures. It reflects the increasing interest of the community in durability mechanics.
3. Masonry and block mechanics.
   Papers demonstrate the possibilities of numerical modelling for the preservation and strengthening of historical structures, as well as the development of new innovative masonry applications.
4. Geomechanics, soil mechanics, steel and rubber.
   Apart from the focus on quasi-brittle materials, the past five years have witnessed an increasing DIANA related research effort for applications like gas and reservoir engineering, soil liquefaction and for materials like metals and rubber. The chapter comprises a number of contributions in these areas.
5. Shear in reinforced concrete.
   Shear failure still constitutes one of the most complex and yet incompletely understood failure mechanisms for reinforced concrete. This chapter presents the latest research results, including cyclic modelling for earthquake loading.
6. Finite element analysis of reinforced concrete structures.
   This is the most extensive chapter. It covers a variety of practice-oriented studies on buildings, bridges, tunnels, vessels, cooling towers and other reinforced, fibre-reinforced or prestressed concrete structures. The papers demonstrate the utility of computational methods and discuss the underlying choices in modelling strategies.

Also on behalf of the co-organizers of the conference, A. de Boer, M. Murakami, G.M.A. Schreppers and W.J.E. van Spanje, we express the hope that the book will serve as a reference text and guideline for those interested in finite element research for civil engineering applications.

Delft, July 2002,   Max A.N. Hendriks
                                                                                                                 Jan G. Rots

Finite Elements in Civil Engineering Applications, Hendriks & Rots (eds.)
© 2002 Swets & Zeitlinger, Lisse, ISBN 90 5809 530 4

Conference Sponsors

Ministry of Transport, Public Works
and Water Management

Finite Elements in Civil Engineering Applications, Hendriks & Rots (eds.)
© *2002 Swets & Zeitlinger, Lisse, ISBN 90 5809 530 4*

Conference Committees

Scientific Advisory Committee

R. Al-Mahaidi, *Monash Univ., Australia*
K. Andreev, *Montan Univ. Leoben, Austria*
M.M. Bakhoum, *Cairo Univ., Egypt*
B. Banholzer, *RWTH Aachen, Germany*
B.S. Bedenik, *Univ. of Maribor, Slovenia*
P.v.d. Berg, *GeoDelft, Netherlands*
N. Bićanić, *Univ. of Glasgow, Scotland*
J. Blaauwendraad, *Delft Univ., Netherlands*
I. Carol, *Univ. Politecnica de Catalunya, Spain*
A. Carpinteri, *Universita Politecnico di Torino, Italy*
R.S. Crouch, *Univ. of Sheffield, United Kingdom*
R. Eligehausen, *Univ. of Stuttgart, Germany*
B. Espion, *Vrije Universiteit Brussel, Belgium*
E.M.R. Fairbairn, *Cid. Univ.-Ilha do Fundao, Brazil*
P.H. Feenstra, *Cornell Univ., USA*
K. Fukuzawa, *Ibaraki Univ., Japan*
K. Gylltoft, *Chalmers Univ. of Techn., Sweden*
T. Hasegawa, *Shimizu Corporation, Japan*
M. Hassanzadeh, *Lund Inst. of Techn., Sweden*
J.M. Huyghe, *Eindhoven Univ., Netherlands*
K. Høiseth, *Sintef, Norway*
A.R. Ingraffea, *Cornell Univ., USA*
J. Jongedijk, *Ministry of Public Works, Netherlands*
T. Kanstad, *Norw. Univ. of Sc. and Techn., Norway*
N. Kishi, *Muroran Inst. of Techn., Japan*
K. Kosa, *Kyushu Inst. of Techn., Japan*
S. Lee, *Yonsei Univ., Korea*
P.B. Lourenço, *Univ. of Minho, Azurem, Portugal*
K. Maekawa, *The Univ. of Tokyo, Japan*
N. Masuda, *Musashi Inst. of Techn., Japan*
I.M. May, *Heriot Watt Univ., Scotland*
V. Mechtcherine, *Univ. of Karlsruhe, Germany*
C. Middleton, *Univ. of Cambridge, United Kingdom*
J.G.M.van Mier, *Delft Univ. of Techn., Netherlands*
E. Mizuno, *Chubu Univ., Japan*
A. Miyamoto, *Yamaguchi Univ., Japan*
K.M. Mosalam, *U. of California at Berkeley, USA*
S. Murata, *Kyoto Univ., Japan*

A. Mutoh, *Meijo Univ., Japan*
M. Nagai, *Nagaoka Univ. of Techn., Japan*
H. Noguchi, *Chiba Univ., Japan*
Y. Okui, *Saitama Univ., Japan*
M. Orlando, *Univ. of Florence, Italy*
C. Pearce, *Univ. of Glasgow, Scotland*
R. Pellegrini, *Enel Hydro S.p.A, Italy*
J.M. Reynouard, *INSA de Lyon, France*
K. Rokugo, *Gifu Univ., Japan*
H. Roman, *Univ. Fed. de Santa Catarina, Brazil*
Q. Ser-Tong, *Nat. Univ. of Singapore, Singapore*
T. Shimomura, *Nagaoka Univ. of Techn., Japan*
N. Shirai, *Nihon Univ., Japan*
L.J. Sluys, *Delft Univ. of Techn., Netherlands*
H. Song, *Yonsei Univ., Korea*
S.I. Sørensen, *Norw. U. of Sc. and Techn., Norway*
L. Taerwe, *Univ. of Gent, Belgium*
I. Towhata, *The Univ. of Tokyo, Japan*
Y. Uchida , *Gifu Univ., Japan*
N. Yoshida, *Oyo Corporation, Ltd, Japan*
C. Yuren, *Northern Jiaotong Univ., China*
W. Zhishen, *Ibaraki Univ., Japan*

Organizing Committee

A. De Boer, *DIANA User's Assoc., The Netherlands*
M.A.N. Hendriks, *TNO, The Netherlands*
M. Murakami, *JIPS, Tokyo, Japan,*
J.G. Rots, *Delft Univ. of Techn., The Netherlands*
G.M.A. Schreppers, *TNO, The Netherlands*
W.J.E. van Spanje, *DIANA Anal., The Netherlands*

Local Organizing Committee

M. Ishiga, *director of JIPS, Tokyo, Japan*
K. Hara, *vice director of JIPS, Tokyo, Japan*
M. Murakami, *manager of JIPS, Tokyo, Japan*
K. Akasaka, *JIPS, Tokyo, Japan*

Constitutive models for quasi-brittle materials

Finite Elements in Civil Engineering Applications, Hendriks & Rots (eds.)
© *2002 Swets & Zeitlinger, Lisse, ISBN 90 5809 530 4*

Invited paper: Enhanced microplane concrete model and multi equivalent series phase model in DIANA for fracture analysis of concrete structures

T. Hasegawa

Institute of Technology, Shimizu Corporation, Tokyo, Japan

ABSTRACT: The multi equivalent series phase model and the enhanced microplane concrete model are non-local and local constitutive models derived by treating microscopic fracture regions respectively as series phases consisting of fracture and unloading phases and as planes (microplanes) with various orientations. Both models are implemented in the finite element system DIANA, and are applied to the simulation of failures of reinforced concrete members under shear as well as bending. The analysis results give lucid explanations of the failure mechanism of the members, and bring some issues in failure analysis into sharp relief.

1 INTRODUCTION

In applying general-purpose finite element codes such as DIANA to simulate the inelastic behavior and failure of concrete structures, various concrete constitutive models based on phenomenological considerations have been introduced. These phenomenological constitutive models impose their own limitations on capability since they cannot describe the inelasticity, damage, fracture, and plasticity of concrete in a unified and transparent way with a clear physical mechanism, especially in cases of multiaxial and combined stress conditions such as tension-compression, tension-shear, or compression-shear. Unlike such constitutive models, the Enhanced Microplane Concrete Model (EMPC Model) and Multi Equivalent Series Phase Model (MESP Model) reflect clear physical mechanisms at the microscopic or mesoscopic level, and can describe macroscopic constitutive relations under multiaxial stress conditions with accuracy (Hasegawa 1992, 1995 and 1998).

In this study, as a means of demonstrating the capabilities of the nonlocal MESP Model and the local EMPC Model, they are incorporated into the Failure Analysis System for Concrete Structures (FASCOS), which is based on DIANA, to simulate the failure mechanism and size effect of reinforced concrete members under both shear and bending.

2 REVIEW OF MODELS

Brief outlines of both the MESP and EMPC Models are given in the following.

2.1 *Multi equivalent series phase model*

Since the presence of coarse aggregate particles makes concrete a heterogeneous material, fracture localization and strain softening occur at a microscopic or mesoscopic level in a relatively stable and distributed manner prior to macroscopic softening fractures. Although these types of microscopic behavior would ideally be described by appropriate micromechanics models, we in fact assume a much simpler mechanical field as shown in Figure 1(a). Distributed microscopic fracture regions are modeled by independent fibers constrained by certain conditions as a means to relate the microscopic and macroscopic levels. When a softening fracture occurs in each fiber, the microscopic fracture localizes into a fracture phase within the fiber while an elastic unloading takes place in the remainder of the fiber (the unloading phase). The result is a microscopic strain localization in which the unloading phase supplies the released elastic energy to the fracture phase once the microscopic peak stress of the fiber is reached. We assume that the stresses σ^F and σ^U in each phase depend on the corresponding strains ε^F and ε^U of the phase, and that there are unique relations between them that we call phase-constitutive laws. The equilibrium and strain compatibility conditions for stress σ^L and strain ε^L of the series phase are described by Eq. (1).

$$\sigma^L = \sigma^F = \sigma^U;$$

$$\varepsilon^L = \frac{\varepsilon^F l^F + \varepsilon^U l^U}{l^L} = \frac{\varepsilon^F l^F + \varepsilon^U \left(l^L - l^F \right)}{l^L} \qquad (1)$$

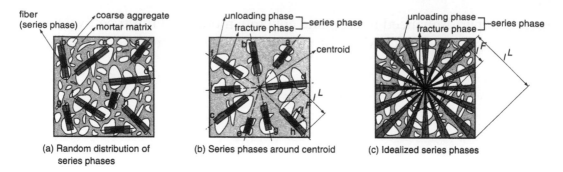

Figure 1. Fracture, unloading, and series phases in a concrete volume element.

(a) Random distribution of series phases

(b) Series phases around centroid

(c) Idealized series phases

in which superscripts F, U, and L refer to the fracture, unloading, and series phases, and l^F, l^U, and l^L are the lengths of the fracture, unloading, and series phases, respectively. This series coupling model consisting of the fracture and unloading phases is considered the basic load-carrying mechanism of concrete at the microscopic level. However, if a macroscopic constitutive model were to be formulated by deriving the softening behavior of the series phase from the fracture and unloading phase responses, it would be difficult to satisfy the conditions of Eq. (1) and calculation would be very inefficient. This problem is circumvented by introducing an equivalent series phase that is a homogenized phase combining the fracture and unloading phases, thereby taking the conditions set by Eq. (1) into account in a different form. The equilibrium and strain compatibility conditions of the series phase are satisfied by introducing a constant plastic fracture energy law (the first law of thermodynamics) to determine the stress-strain softening relations of the equivalent series phase and by utilizing relations similar to those describing the fracture phase for the equivalent series phase.

The stress-strain softening relations for the fracture, unloading, and equivalent series phases are assumed to be as follows:

For $0 \leq |\varepsilon^p| \leq |\varepsilon_0|$ (pre-peak):

$$\sigma^p = \sigma_0 \left[1 - \left(1 - \frac{\varepsilon^p}{\varepsilon_0} \right)^{C_0 \varepsilon_0 / \sigma_0} \right] \tag{2a}$$

for $|\varepsilon_0| < |\varepsilon^p|$ (post-peak):

$$\sigma^p = \sigma_0 \exp\left[-\left(\frac{\varepsilon^p - \varepsilon_0}{\varepsilon_s^p} \right) \right] \tag{2b}$$

in which σ^p and ε^p are the stress and strain of phase p; $\varepsilon_0 = \sigma_0/\zeta C_0$; and $\varepsilon_s^p = \gamma^p \varepsilon_0 = \gamma^p \sigma_0/\zeta C_0$. Superscripts $p = F$, U, and E refer to the fracture,

unloading, and equivalent series phases, but Eq. (2b) is not necessary for the unloading phase. In Eq. (2), C_0 is the initial modulus, σ_0 is the peak stress of the curve, ζ is a parameter controlling the peak strain ε_0, and γ^p is a ductility parameter that controls ε_s^p. At strain $\varepsilon^p = \varepsilon_0 + \varepsilon_s^p$, stress σ^p has decreased to σ_0/e in the softening region, i.e., the ductility in the softening region depends on the strain ε_s^p or the ductility parameter γ^p. The stress-strain relations in the pre-peak region are the same for all phases, but in the post-peak region the equivalent series phase takes a different value, γ^E, of ductility parameter from the value γ^F for the fracture phase. On the other hand, the unloading phase follows an elastic unloading path with initial modulus C_0 after the peak stress.

The constant plastic fracture energy law is written as Eq. (3):

$$V^E\left(l^E\right) g^E = V^F\left(l^F\right) g^F + V^U\left(l^U\right) w^U \tag{3a}$$

$$V^E\left(l^E\right) \int_{\varepsilon^E = 0}^{\varepsilon^E = \infty} \sigma^E d\varepsilon^E$$
$$= V^F\left(l^F\right) \int_{\varepsilon^F = 0}^{\varepsilon^F = \infty} \sigma^F d\varepsilon^F + V^U\left(l^U\right)$$
$$\times \left[\int_{\varepsilon^U = 0}^{\varepsilon^U = \varepsilon_0} \sigma^U d\varepsilon^U - \frac{\sigma_0^2}{2C_0} \right] \tag{3b}$$

in which $V^F(l^F)$, $V^U(l^U)$, and $V^E(l^E)$ are the volumes of the fracture, unloading, and equivalent series phases depending on the length of each; g^F and g^E are the plastic fracture energy densities of the fracture and equivalent series phases; w^U is the plastic energy density of the unloading phase. When parameters C_0 and ζ for the pre-peak region, the ductility parameter γ^F for the fracture phase, and the volume of each phase are given, the ductility parameter γ^E for the equivalent series phase can be determined by solving Eqs. (2) and (3). Furthermore, if the volume of the equivalent series phase is assumed to be the same as the volume of the series phase consisting of the fracture and

4

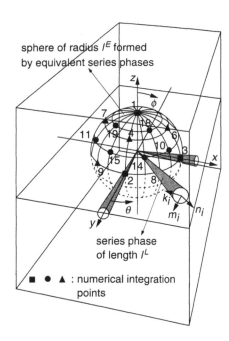

sphere of radius l^E formed
by equivalent series phases

■ ● ▲ : numerical integration
points

Figure 2. Concrete volume element and sphere.

unloading phases, i.e., Eq. (4), then the solution for γ^E is given by Eq. (5):

$$V^E\left(l^E\right) = V^L\left(l^L\right) = V^F\left(l^F\right) + V^U\left(l^U\right) \qquad (4)$$

$$\gamma^E = \frac{V^F\left(l^F\right)}{V^L\left(l^L\right)}\left(\frac{\zeta}{2} + \gamma^F\right) - \frac{\zeta}{2} \qquad (5)$$

The MESP Model is a nonlocal macroscopic constitutive model derived from the assumption that a number of equivalent series phases are distributed with various orientations in the concrete. To develop the model for the concrete volume element shown in Figure 2, the randomly oriented series phases in the element are collected around the centroid of the element as shown in Figure 1(b). It is assumed that the length l^L of each series phase is represented by the distance between the centroid and the element boundary, and differs among the series phases (Fig. 1(c)). On the other hand, the length l^F of each fracture phase is the same for all the series phases. Then every series phase between the centroid of the element and the element boundary, each having its own length l^L, is replaced with an equivalent series phase. The stress-strain softening relation of each equivalent series phase depends on the length l^L of the actual series phase, and is determined using Eq. (5).

A tensorial kinematic constraint is hypothesized to relate the macroscopic strain tensor ε_{ij} to the strains

of the equivalent series phase; i.e., the normal strain ε_N^E and shear strains ε_{TK}^E and ε_{TM}^E of the equivalent series phase are the resolved components of the macroscopic strain tensor ε_{ij}:

$$\varepsilon_N^E = n_i n_j \varepsilon_{ij};$$

$$\varepsilon_{TK}^E = \frac{1}{2}\left(k_i n_j + k_j n_i\right)\varepsilon_{ij}; \quad \varepsilon_{TM}^E = \frac{1}{2}\left(m_i n_j + m_j n_i\right)\varepsilon_{ij} \qquad (6)$$

in which n_i, k_i, and m_i are components of the unit coordinate vectors **n**, **k**, and **m** perpendicular to each other (Fig. 2). The incremental forms of the phase-constitutive relations for the equivalent series phase are written separately for the normal component and the shear components in the K and M directions:

$$d\sigma_N^E = C_N^E\, d\varepsilon_N^E - d\sigma_N^{E''} = f_{N1}^E\left(\varepsilon_N^E, \varepsilon_L^E, S_L^E\right) \qquad (7a)$$

$$d\sigma_{TK}^E = C_{TK}^E\, d\varepsilon_{TK}^E - d\sigma_{TK}^{E''} = f_{T1}^E\left(\varepsilon_{TK}^E, S_N^E\right) \qquad (7b)$$

$$d\sigma_{TM}^E = C_{TM}^E\, d\varepsilon_{TM}^E - d\sigma_{TM}^{E''} = f_{T1}^E\left(\varepsilon_{TM}^E, S_N^E\right) \qquad (7c)$$

in which C_N^E, C_{TK}^E, and C_{TM}^E are incremental elastic stiffnesses for the equivalent series phase; $d\sigma_N^{E''}$, $d\sigma_{TK}^{E''}$, and $d\sigma_{TM}^{E''}$ are inelastic stress increments for the equivalent series phase; $f_{N1}^E(\varepsilon_N^E, \varepsilon_L^E, S_L^E)$ is the normal stress increment $d\sigma_N^E$ expressed in terms of normal strain ε_N^E, the resolved lateral strain ε_L^E of the macroscopic strain tensor ε_{ij}, and the resolved lateral stress S_L^E of the macroscopic stress tensor σ_{ij} onto the phase; and $f_{T1}^E(\varepsilon_{Ts}^E, S_N^E)$ is the shear stress increment $d\sigma_{Ts}^E$ expressed in terms of shear strain ε_{Ts}^E and the resolved normal stress S_N^E of the macroscopic stress tensor σ_{ij} onto the phase ($Ts = TK, TM$).

Since a uniform state of macroscopic stress and strain is assumed in the concrete volume element, and microscopic softening localization is homogenized using the equivalent series phase, with the length of the series phase taken into account, we can use an arbitrary inner volume within the element to relate the responses of the equivalent series phases to the macroscopic behavior. Here a sphere of radius l^E formed by the equivalent series phases is adopted (Fig. 2). Using the principle of virtual work (i.e., the equality of virtual works δW^V of the macroscopic stress tensor and δW^E of the stresses in the equivalent series phases within the sphere of radius l^E), we can write

$$\delta W^V = \delta W^E \qquad (8a)$$

$$\delta W^V = \int_{r=0}^{r=l^E} \int_{\theta=0}^{\theta=2\pi} \int_{\phi=0}^{\phi=\pi} d\sigma_{ij}\,\delta\varepsilon_{ij}\, r^2 \sin\phi\, d\phi\, d\theta\, dr$$

$$= \frac{4}{3}\pi\left(l^E\right)^3 d\sigma_{ij}\,\delta\varepsilon_{ij} \qquad (8b)$$

5

$$\delta W^E = \int_{\theta=0}^{\theta=2\pi} \int_{\phi=0}^{\phi=\pi} \left(\begin{array}{c} d\sigma_N^E \, \delta\varepsilon_N^E + d\sigma_{TK}^E \, \delta\varepsilon_{TK}^E \\ + d\sigma_{TM}^E \, \delta\varepsilon_{TM}^E \end{array} \right)$$
$$\times \frac{1}{3} \left(l^E \right)^3 \sin\phi \, d\phi \, d\theta \qquad (8c)$$

in which θ and ϕ are the spherical angular coordinates (Fig. 2). Expressing $\delta\varepsilon_N^E$, $\delta\varepsilon_{TK}^E$, and $\delta\varepsilon_{TM}^E$ by Eq. (6), and substituting them into Eq. (8) along with the phase-constitutive relations Eq. (7), we obtain an incremental form of the macroscopic stress-strain relation

$$d\sigma_{ij} = C_{ijrs} \, d\varepsilon_{rs} - d\sigma_{ij}'' \qquad (9a)$$

$$C_{ijrs} = \eta \int_{\theta=0}^{\theta=2\pi} \int_{\phi=0}^{\phi=\pi/2} \Big[n_i n_j n_r n_s C_N^E$$
$$+ \frac{1}{4} \left(k_i n_j + k_j n_i \right) \times \left(k_r n_s + k_s n_r \right) C_{TK}^E$$
$$+ \frac{1}{4} \left(m_i n_j + m_j n_i \right)$$
$$\times \left(m_r n_s + m_s n_r \right) C_{TM}^E \Big] \cdot \sin\phi \, d\phi \, d\theta \qquad (9b)$$

$$d\sigma_{ij}'' = \eta \int_{\theta=0}^{\theta=2\pi} \int_{\phi=0}^{\phi=\pi/2} \Big[n_i n_j \, d\sigma_N^{E''}$$
$$+ \frac{1}{2} \left(k_i n_j + k_j n_i \right) d\sigma_{TK}^{E''}$$
$$+ \frac{1}{2} \left(m_i n_j + m_j n_i \right) d\sigma_{TM}^{E''} \Big] \cdot \sin\phi \, d\phi \, d\theta \qquad (9c)$$

in which $\eta = \eta^E$; and $\eta^E = 1/2\pi$.

2.2 Enhanced microplane concrete model

The EMPC Model is a local constitutive model derived by treating microscopic fracture regions as planes (microplanes) with various orientations. The constitutive equation for the EMPC Model results in a surface integral over a unit hemisphere with a surface consisting of microplanes, and is obtained in the form of Eq. (9) with $\eta = \eta^M = 3/2\pi$ by replacing C_N^E, C_{TK}^E, C_{TM}^E, $d\sigma_N^{E''}$ $\sigma_{TK}^{E''}$, and $d\sigma_{TM}^{E''}$ with the corresponding variables for the microplanes. The microconstitutive models for microplanes in the EMPC Model are adopted as phase-constitutive models for the fracture phases in the MESP Model, which is a very attractive approach since the EMPC Model has succeeded in describing the local constitutive relations of concrete under multiaxial stress conditions accurately. Both the MESP and EMPC Models have been implemented in the Failure Analysis System for Concrete Structures (FASCOS) based on the general-purpose finite element system DIANA for practical and structural calculations.

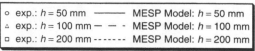

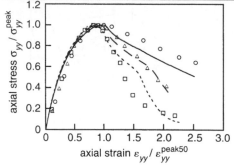

Figure 3. Uniaxial compression analysis.

2.3 Verification of constitutive relation

The orientation and size characteristics of series phases in a concrete volume element are taken into account in the resulting nonlocal macroscopic constitutive relation of the MESP Model. This yields a regularization of the MESP Model and results in the realization of size effects. In Figure 3 the calculated results of the size effect on uniaxial compressive softening are shown, compared with tests by Van Mier (Van Mier 1984) using concrete prism specimens of identical section (100×100 mm) but different heights h (50, 100, and 200 mm). The axial stress-strain relations are normalized by the peak axial stress σ_{yy}^{peak} of each specimen, and the axial strain $\varepsilon_{yy}^{peak50}$ corresponding to the peak axial stress of the specimen with $h = 50$ mm. In the analysis, the length l^F of the fracture phase is assumed to be $2l^F = 50$ mm $\cong 3d_{max} = 48$ mm, in which d_{max} is the maximum aggregate size. The MESP Model can capture the decrease in ductility with increasing specimen height.

Figure 4 shows the results of triaxial compression analysis along the compressive meridian in comparison with experiments by Kosaka et al. (Kosaka et al. 1985). In these experiments, concrete prism specimens of identical section (100×100 mm) but different heights h (100 and 200 mm) were tested under confinement pressures σ_c of 0, -3, and -6 kgf/cm^2. The length l^F of the fracture phase is taken to be $2l^F = 100$ mm $= 6.6d_{max}$ in this analysis. Although the analysis underestimates the lateral strain in the softening regime during uniaxial compression as compared with the experimental results, the MESP Model roughly predicts the size effects on triaxial compressive softening.

To examine applicability of the MESP Model under shear stress with a rotating principal direction,

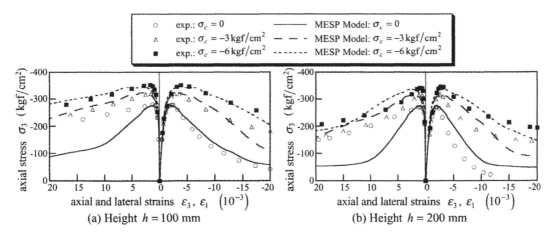

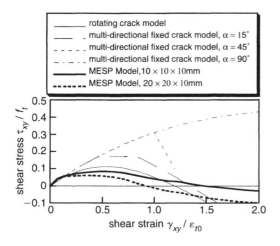

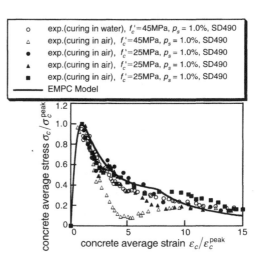

Figure 4. Triaxial compression analysis.

Figure 5. Shear responses in biaxial tension-shear analysis.

Figure 6. Tension stiffening analysis.

the biaxial tension-shear analysis of Rots (Rots 1988) is simulated. In this analysis, uniaxial tension up to uniaxial tensile strength f_t is first applied to a concrete volume element in the x-direction. The element is then immediately subjected to combined biaxial tension and shear according to $\Delta\varepsilon_{xx} : \Delta\varepsilon_{yy} : \Delta\gamma_{xy} = 0.5 : 0.75 : 1$. Two sizes of square element (10×10 mm and 20×20 mm) with thickness 10 mm are considered, and the length l^F of the fracture phase is assumed to be $2l^F = 10$ mm in the MESP Model. In Figure 5, the shear responses obtained in biaxial tension-shear analysis using the MESP Model are compared with the results calculated by Rots using the rotating crack model and the multi-directional fixed crack model (α = threshold angle). It is worth noting that the MESP Model predictions of the flexible shear

responses for both element sizes are similar to the result achieved with the rotating crack model, which has been shown capable of simulating shear-tension failures of concrete. On the other hand, the multi-directional fixed crack model, with larger values of α, results in much stiffer shear behavior. The MESP Model can simulate the size effects on shear strength and softening, as shown in Figure 5. It also incorporates the capabilities of the rotating crack model, which has previously been shown effective in its application to fracture mechanics.

The local EMPC Model can describe constitutive relations of bond concrete surrounding a reinforcing bar, which take into account the tension stiffening effect of a reinforced concrete member. Figure 6 shows the uniaxial response of the local EMPC Model for tension stiffening in bond concrete elements in comparison with experimental data (Shima et al. 1987).

3 SIZE EFFECT OF REINFORCED CONCRETE DEEP BEAMS

3.1 *Analysis models*

As a first analytical application of the models, shear tests carried out on three reinforced concrete deep beam specimens by Matsuo (Matsuo 2001) are simulated in analysis cases A, B, and C. The specimens, D200, D400, and D600, have the same $a/d = 1\ 0$, but different d of 200, 400, and 600 mm, respectively (a = shear span length; d = effective depth). In analysis

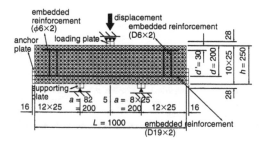

Figure 7. Analysis case B1.

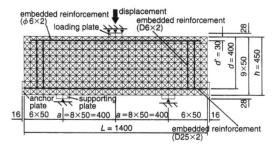

Figure 8. Analysis case B2.

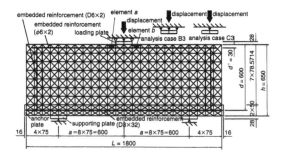

Figure 9. Analysis cases B3 and C3.

cases B, the three specimens are discretized as a full structure model without exploiting their symmetry (Figs. 7–9). The finite elements in the web concrete are enlarged similarly in each specimen to discretize the specimen into finite element meshes. On the other hand, in analysis cases A, only a half-left structure is considered, taking symmetry into account. Furthermore, in analysis case C3 on specimen D600, to examine the influence of rotation and deformation of the loading plate, the fixed displacement load is applied only to a single central node among the loading plate elements.

The MESP Model is used for plain concrete portions of the specimens in all analysis cases. Assuming length l^F of the fracture phase to be $2l^F = d_{max}$, the material parameters of the MESP Model are determined by fitting the uniaxial tension softening relation to the one provided in CEB-FIP Model Code 1990. Then regularization of the MESP Model for individual finite elements is achieved by taking into account the lengths of series phases in the element when calculating the incremental stiffness of the model. Reinforcing bars are modeled using the embedded reinforcement elements of the Von Mises elasto-plastic constitutive model taking into account the tension stiffening effect of concrete. Elements within 50 mm above and below the tension reinforcing bar are assumed to be bond concrete elements so as to take into account the tension stiffening effect of concrete. The EMPC Model, a local constitutive law, is assumed for the bond concrete.

3.2 *Analysis results and discussion*

Figures 10–12 compare the experimental shear responses of specimens D200, D400, and D600 with

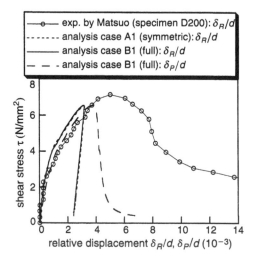

Figure 10. Relative displacement ($d = 200\,\text{mm}$).

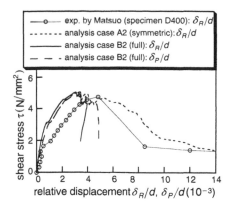

Figure 11. Relative displacement ($d = 400\,$mm).

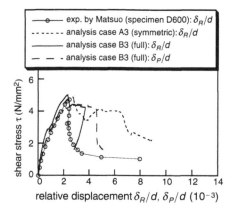

Figure 12. Relative displacement ($d = 600\,$mm).

the calculated shear responses obtained in analysis cases A (the symmetric structure model) and in analysis cases B (the full structure model). These figures indicate the relative displacement, δ_R, of the center point of the beam with respect to the supporting point at the level of the tension reinforcing bar, in which shear stress $\tau = V/bd$; $V =$ shear force; $b =$ beam width. The same figures also show the calculated shear responses in terms of relative displacement, δ_P, between the loading and supporting plates for analysis cases B. In Figures 13 and 14, the incremental deformation of analysis cases B is shown for the maximum shear load V_u and for $2V_u/3$ in the post-peak regime. Figure 15 shows the incremental deformation of analysis case A3 ($d = 600\,$mm) for V_u and for $2V_u/3$ in the post-peak regime.

It is obvious that the shear load reaches a maximum after shear-compressive failure localizes near the loading plate in all analysis cases. Since, in the post-peak regime for analysis cases A1, B (Fig. 14), and C3, shear-compressive failure localizes only near the loading plate and unloading deformation is dominant in other areas, response δ_R indicates snapback. However, in the post-peak regime for analysis cases A2 and A3 (Fig. 15), shear-compressive failure propagates downward, and shear slip deformation enlarges at the crack. This brings about relatively ductile post-peak behavior. On the other hand, snapback behavior is not observed in response δ_P for any analysis case. In the earlier stages of loading the shear response of analysis case B3, in which a uniform fixed displacement load is applied to the full structure model, is almost identical to analysis case A3 (the symmetric structure model). However, as the shear crack in the left span propagates further, a bifurcation from a symmetrical to

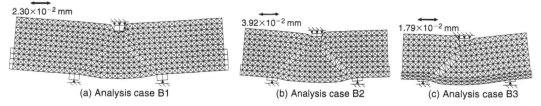

Figure 13. Incremental deformation at V_u.

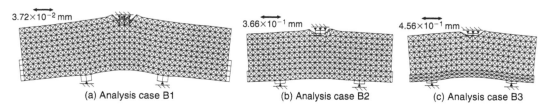

Figure 14. Incremental deformation at $2V_u/3$.

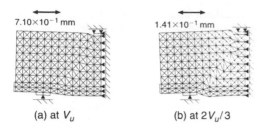

(a) at V_u (b) at $2V_u/3$

Figure 15.   Incremental deformation for analysis case A3.

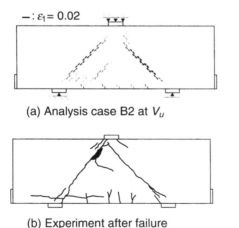

(a) Analysis case B1 at V_u

(b) Experiment after failure

Figure 16.   Cracking pattern $(d = 200\,mm)$.

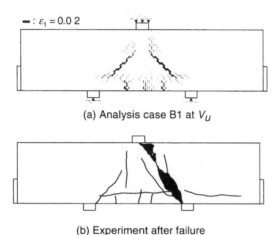

(a) Analysis case B2 at V_u

(b) Experiment after failure

Figure 17.   Cracking pattern $(d = 400\,mm)$.

an unsymmetrical fracture mode occurs in analysis case B3. Thereafter, a difference in shear response between analysis cases A3 and B3 begins to develop gradually.

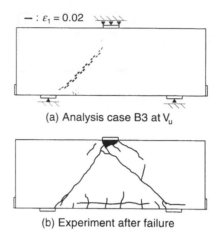

(a) Analysis case B3 at V_u

(b) Experiment after failure

Figure 18.   Cracking pattern $(d = 600\,mm)$.

In Figures 16(a), 17(a), and 18(a) is plotted the line of maximum principal strain $\varepsilon_1 \geq 5\varepsilon_{t0}$ with thickness proportional to its value. This represents crack strain and crack direction at V_u in analysis cases B, and is a good measure of crack width (ε_t0 = the tensile strain corresponding to tensile strength). The distributed crack lines near the loading and supporting plates are considered to represent shear-compressive damage of concrete. These crack strain figures are compared with the experimental cracking patterns after failure in Figures 16–18. The analysis reflects the experimental crack pattern in a relatively good manner. The incremental deformations and cracking pattern look almost symmetrical in analysis case B1 ($d = 200\,mm$), but are obviously unsymmetrical in analysis cases B2 ($d = 400\,mm$) and B3 ($d = 600\,mm$). This indicates that the shear fracture does not propagate symmetrically through the left and right spans, and that the bifurcation from symmetrical to unsymmetrical modes is likely to arise as the size of the specimen increases. This results in a complicated size effect in the fracture of reinforced concrete deep beams.

Figure 19 shows stress-strain responses of elements a and b (Fig. 9) in analysis cases B3 and C3. Very similar softening fracture responses occur in elements a and b for analysis case B3, while in analysis case C3 unloading responses take place at the maximum load in element a of the left span and softening fracture responses occur in element b of the right span. In analysis case C3, shear-compressive failure localization near the right end of the loading plate induces rotation of the loading plate, and this intensifies the unsymmetrical fracture mode (Hasegawa 2001). The loading boundary condition has an important influence on the bifurcation to unsymmetrical fracture mode and on the fracture

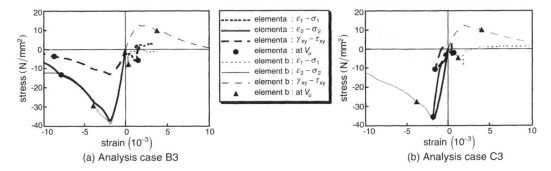

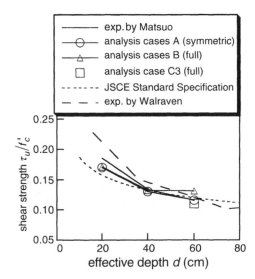

(a) Analysis case B3 (b) Analysis case C3

Figure 19. Stress-strain responses in analysis cases B3 and C3.

Figure 20. Size effect of shear strength.

localization in reinforced concrete deep beams. In the experimental study by Matsuo, details of the loading boundary conditions, such as the arrangement of spherical seats and load cells, were not made clear in the report (Matsuo 2001). These conditions relate to the rotational degree of freedom and deformation of the loading plate. For this reason, it was difficult to pursue further the issue of loading boundary conditions in the present analysis.

In Figure 20 the size effect of shear strength obtained in analysis cases A, B, and C is compared with the experimental data of Matsuo and Walraven (Walraven & Lehwalter 1994) along with the design equation given in the Japan Society of Civil Engineers Standard Specification (f_c' = compressive strength of concrete). This figure demonstrates that the analysis method using the MESP and EMPC Models can predict the size

effect of shear strength in reinforced concrete deep beams with accuracy. In a previous study (Yoshitake & Hasegawa 2000) it was clarified that tensile modeling alone, which takes into account fracture energy, is insufficient to simulate the size effect of shear strength in reinforced concrete deep beams that are geometrically similar. It is concluded that a model for the size effect of shear-compressive failure of concrete like the MESP Model is necessary for the simulation.

4 DIAGONAL TENSION FAILURE OF REINFORCED CONCRETE SLENDER BEAMS

4.1 Analysis models

As the second analytical application of the models, diagonal tension failure of reinforced concrete slender beam specimens, BN50 and BN25 (having the same $a/d = 3.0$, but different d of 450 and 225 mm, respectively) tested at the University of Toronto (Podgorniak-Stanik 1998) is simulated. In this analysis, the mechanism of diagonal tension failure is studied numerically, taking the longitudinal splitting crack at the level of the tension reinforcing bar to be the primary cause of the failure as advocated by Chana (Chana 1987). Both specimens are discretized using cross-diagonal meshes in analysis cases D (Fig. 21), and Delaunay triangulation meshes in analysis cases E and F (Fig. 22). The longitudinal splitting crack at the level of the tension reinforcing bar is caused by dowel action of the bar, which is primarily influenced by the shear stiffness of the bar. Therefore, to examine the effect of bar shear stiffness on the longitudinal splitting crack, embedded bar elements that have no shear stiffness are utilized in analysis cases D and E, and beam elements are utilized in analysis cases F. In all of these cases, the MESP Model is assumed for all concrete elements.

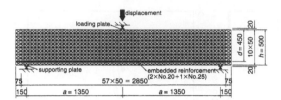

Figure 21. Analysis case D2.

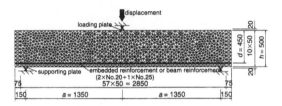

Figure 22. Analysis cases E2 and F2.

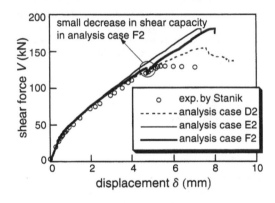

Figure 23. Shear response ($d = 450$ mm).

Figure 24. Experimental cracking pattern at failure ($d = 450$ mm).

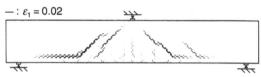

Figure 25. Cracking pattern at V_u (analysis case D2).

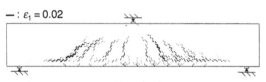

Figure 26. Cracking pattern at V_u (analysis case F2).

Figure 27. Incremental deformation at V_u (analysis case D2).

Figure 28. Incremental deformation at V_u (analysis case D1).

4.2 Analysis results and discussion

Figure 23 compares the calculated shear response in analysis cases D2, E2, and F2 ($d = 450$ mm) with the experiment. In Figures 25 and 26 the crack strain is plotted (as in Figs. 16–18) for the specimen with $d = 450$ mm in analysis cases D2 and F2; these are to be compared with the corresponding experimental cracking pattern shown in Figure 24. The incremental deformation at maximum shear load V_u in analysis cases D2 and D1 ($d = 225$ mm) is shown in Figures 27 and 28.

Analysis case D2 has difficulty in predicting exactly the flatter diagonal crack with curved shape as observed in the experiment, due to the mesh dependency of cracking in the cross-diagonal mesh. However, the analysis does reproduce the mechanism of diagonal tension failure as triggered by the longitudinal splitting crack at the level of the tension

reinforcing bar (Figs. 24, 25, and 27). In the cross-diagonal mesh, cracks can propagate easily in the direction of element alignment ($\theta = n\pi/4$; $n = 0, 1, 2, 3$), but the shear resistance due to aggregate interlock at cracked elements increases because of the large inclination angle of the cracks. This causes overestimation of shear strength in analysis case D2, as shown in Figure 23.

The lack of sufficient results in analysis cases D is due to the mesh dependency of cracking in the cross-diagonal mesh. On the other hand, relatively good representation of the cracks is achieved in analysis

4.64×10⁻¹ mm

Figure 29. Incremental deformation at a small decrease in shear capacity (analysis case F2).

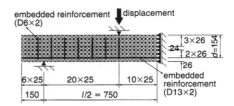

Figure 30. Analysis case G1.

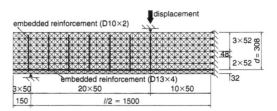

Figure 31. Analysis case G2.

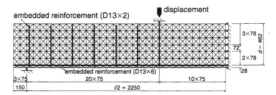

Figure 32. Analysis case G3.

cases E and F using the Delaunay triangulation meshes (Fig. 26). The results are particularly accurate in analysis case F2. When a longitudinal splitting crack at the level of the tension reinforcing bar opens widely and propagates toward the support, a small decrease in shear capacity occurs near the point where the experimental shear load reaches its maximum (Figs. 29 and 23). But since the longitudinal splitting crack does not propagate unstably thereafter, the diagonal crack connecting to the longitudinal splitting crack at the level of the tension reinforcing bar neither opens widely nor extends further up to the loading point; consequently, the cracking process fails to reduce to a diagonal tension failure, but results in higher shear capacity and ultimately bending failure. A future rational evaluation of diagonal tension failure will depend on predicting local fractures of concrete at the root of the diagonal crack due to dowel action of tension reinforcing bar as well as bond slip behavior, which result in the longitudinal splitting crack at the level of the tension reinforcing bar.

5 SIZE EFFECT OF REINFORCED CONCRETE FLEXURAL BEAMS

5.1 *Analysis models*

As the third analytical application of the models, bending tests carried out on three reinforced concrete flexural beams by Mizumachi et al. (Mizumachi et al. 1993) are simulated. Their specimens, A-S18, A-S34, and A-S49, are geometrically similar but have differing effective depths d of 154, 308, and 462 mm, respectively. Assuming structural symmetry, only half-left structures are modeled and discretized into finite elements in analysis cases G (Figs. 30–32). The finite elements in geometrically similar concrete portions are enlarged similarly to discretize the test specimens into the finite element meshes. The tension reinforcement and stirrup are modeled using the embedded reinforcement elements of the Von Mises elasto-plastic constitutive model. Each one element layer above and below a tension reinforcing bar is assumed to consist of bond concrete elements to take into account the tension stiffening effect of concrete. The EMPC Model, local constitutive law, is assumed for the bond concrete. The elasto-plastic constitutive model (yield strength) of the

tension reinforcing bar is modified taking into account the tension stiffening effect of the EMPC Model. The MESP Model is used for concrete elements other than the bond concrete elements.

5.2 *Analysis results and discussion*

In Figure 33 the calculated displacement δ/l of the center point of beam is compared with the experiment by Mizumachi et al., in which P = load; l = loading span. The predicted flexural response agrees with the experiment very well. Figure 34 shows the calculated results of yield strength f_y and ultimate strength f_u in comparison with the experiment, in which $f_y = 6M_y/bd^2$; M_y = yield moment; $f_u = 6M_u/bd^2$; M_u = ultimate (maximum) moment. The analysis is able to predict that there is no size effect of yield and ultimate strengths in these reinforced concrete flexural beams. In Figure 35, the deformation capacity is shown, i.e., yield displacement δ_y/l, ultimate displacement δ_u/l (at maximum moment), failure displacement δ_f/l, in comparison with the experimental results. Values of δ_y/l for all effective depths d as well as δ_u/l and δ_f/l for $d = 462$ mm fsare predicted with relatively good

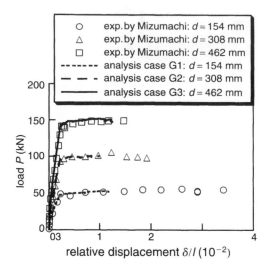

Figure 33. Bending response.

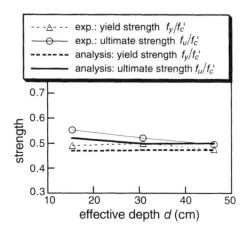

Figure 34. Yield and ultimate strengths.

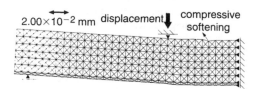

Figure 35. Deformation capacity.

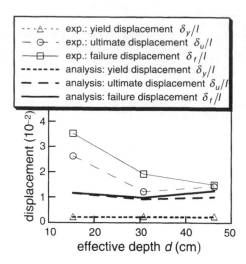

Figure 36. Incremental deformation at final step (d = 462 mm).

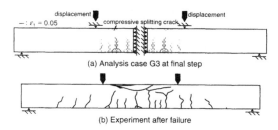

Figure 37. Cracking pattern (d = 462 mm).

accuracy, but δ_u/l and δ_f/l for $d = 154$ mm are considerably underestimated in comparison with the experiments. Thus, these simulation results indicate that there is no size effect of deformation capacity, i.e., δ_f/l, δ_u/l, and δ_f/l in reinforced concrete flexural beams, which is consistent with the experimental results obtained by Alca et al. (Alca et al. 1997), but contradicts the experiments by Mizumachi et al.

Figure 36 shows the incremental deformation of the specimen with $d = 462$ mm (analysis case G3) at the final calculation step where convergence is maintained; i.e., just before the step where numerical divergence occurs. The compressive softening fracture tends to be localized in the elements near the extreme compression fiber. In Figure 37, the crack strain is plotted (as in Figs. 16–18) for the specimen with $d = 462$ mm at the final calculation step. The result is compared with the experimental cracking pattern after failure. This figure demonstrates that the analysis provides reasonable simulation of the localization of flexural and shear cracks, and the crack spacing as well as compressive splitting cracks in the experiment.

To determine why the calculation presented here is unable to continue to large displacements and terminates at smaller values of δ_u/l and δ_f/l compared with the experiment by Mizumachi et al., details of the

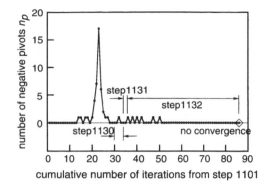

Figure 38. Number of negative pivots.

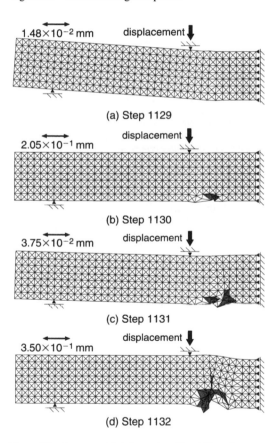

(a) Step 1129

(b) Step 1130

(c) Step 1131

(d) Step 1132

Figure 39. Incremental deformation $(d = 154\,\text{mm})$.

numerical iteration are examined for the analysis case G1 for the specimen with $d = 154\,\text{mm}$. In this case, numerical convergence is not obtained after the maximum number (fifty) of iterations at the final calculation step, 1132. Figure 38 shows the number n_p of negative pivots in the diagonal matrix **D** obtained during the $\textbf{LDL}^T$ decomposition of tangential stiffness matrix **K**. This negative pivot count, which is regarded as the number of negative eigenvalues of **K**, alternates between zero and one in steps 1130 to 1132. During these steps, the iteration process using the Newton-Raphson method with **K** passes several times through singular points, $\det(\textbf{K}) = 0$, i.e., bifurcation points. This means that perturbations with eigenmodes corresponding to eigenvalues that are very close to zero are applied to the structure in the vicinity of the bifurcation points. The perturbations cause branch-switching to bifurcation paths that are related to premature failure modes. This branch-switching is, unexpectedly, very similar to that which takes place with the scaled corrector algorithm (Noguchi & Hisada 1992) developed for post-buckling analysis.

To look into these premature failure modes, the incremental deformation at each final iteration from steps 1129 to 1132 is examined (Fig. 39). At step 1129, a normal bending mode is clearly seen, but at step 1130 a spurious mode emerges in the (shaded) elements near the extreme tension fiber. At the next step, 1131, another spurious mode appears and at the final step, 1132, these spurious modes coalesce and enlarge. The emerging modes are considered spurious kinematic modes of certain finite elements as a consequence of the strain softening constitutive model for tension (De Borst 1989). Divergence due to the appearance of these spurious kinematic modes is inevitable, and does not represent a physical failure of the structure, but a premature termination of the numerical analysis. To simulate the true limit failure state of the structure and evaluate the ductility, which is important for aseismatic design, by numerical failure analysis it will be essential to overcome these spurious kinematic modes.

6 CONCLUSIONS

(1) In the Multi Equivalent Series Phase (MESP) Model, fracture localization at the microscopic level is modeled using a series phase consisting of fracture and unloading phases. Through a simple homogenization procedure for the individual series phases, based on a constant plastic fracture energy law, the orientation and size characteristics of series phases in a concrete volume element are taken into account in the resulting nonlocal macroscopic constitutive relation in the MESP Model. This yields regularization of the MESP Model.
(2) The Enhanced Microplane Concrete (EMPC) Model is a local constitutive model derived by treating microscopic fracture regions not as series phases but as planes (microplanes) with various orientations. Both the MESP and EMPC Models have been implemented in the Failure Analysis

System for Concrete Structures (FASCOS) based on the general-purpose finite element system DIANA.

(3) The MESP Model and the EMPC Model are able to predict cracking, shear deformation, shear-compressive failure localization, unsymmetrical shear failure mode, and size effect of shear strength in reinforced concrete deep beams. However, since analysis in which structural symmetry is assumed fails to predict unsymmetrical shear failure modes with brittle response, the analysis overestimates the ductility of the beam.

(4) Diagonal tension failure of reinforced concrete slender beams was simulated, taking the longitudinal splitting crack at the level of the tension reinforcing bar to be the primary cause of failure. The results of analysis using the MESP Model show that the longitudinal splitting crack should propagate unstably, leading to widening and propagation of the diagonal cracks, in order for diagonal tension failure of the beam to occur. For rational prediction of diagonal tension failure, it will be important to simulate the longitudinal splitting crack.

(5) The size effect of failure and deformation capacity in reinforced concrete flexural beams was simulated using the MESP and EMPC Models. Analysis results show that the models are able to predict the yield and ultimate strengths, and the flexural response of the beams as well as crack localization and compressive softening failure. However, the size effect of deformation capacity does not appear in the analysis. A detailed examination of the analysis results reveals that analysis cannot be continued further into the post-peak regime because of a premature termination of the numerical calculation due to spurious kinematic modes of certain finite elements as a consequence of the strain softening constitutive model for tension.

ACKNOWLEDGMENTS

Implementation of the Enhanced Microplane Concrete Model and Multi Equivalent Series Phase Model in FASCOS based on DIANA was carried out by the author as a member of the DIANA Foundation while he was a visiting researcher at the Netherlands Organization for Applied Scientific Research (TNO). He would like to acknowledge the staff of TNO, especially Drs. Max Hendriks, Peter Feenstra, and Jan Rots for providing him with expertise in programing in DIANA.

REFERENCES

Alca, N., Alexander, S. & MacGregor, J. 1997. Effect of size on flexural behavior of high-strength concrete beams. *ACI Structural Journal* 94(1): 59–67.

Chana, P. S. 1987. Investigation of the mechanism of shear failure of reinforced concrete beams. *Magazine of Concrete Research* 39(141): 196–204.

De Borst, R. 1989. Analysis of spurious kinematic modes in finite element analysis of strain-softening solids. In J. Mazars & Z. P. Bazant (eds), *Cracking and Damage: Strain Localization and Size Effect*: 335–345. London: ELSEVIER Science Publishers.

Hasegawa, T. 1992. Multi equivalent series phase model for nonlocal constitutive relation of concrete. *Proceedings of the 47th annual conference of JSCE* 5: 18–19 (in Japanese).

Hasegawa, T. 1995. Enhanced microplane concrete model. In F. H. Wittmann (ed.), *Fracture Mechanics of Concrete Structures*: 857–870. Friburg: AEDIFICATIO Publishers.

Hasegawa, T. 1998. Multi equivalent series phase model for nonlocal constitutive relations of concrete. In H. Mihashi & K. Rokugo (eds), *Fracture Mechanics of Concrete Structures*: 1043–1054. Friburg: AEDIFICATIO Publishers.

Hasegawa, T. 2001. Size effect analysis of reinforced concrete deep beams. In R. de Borst, J. Mazars, G. Pijaudier-Cabot & J. G. M. van Mier (eds), *Fracture Mechanics of Concrete Structures*: 689–696. Lisse: Balkema.

Kosaka, Y., Tanigawa, Y. & Hatanaka, S. 1985. Evaluation of effect of confinements on compressive toughness of concrete based on triaxial compressive test data. *Proceedings of JCI* 7: 305–308 (in Japanese).

Matsuo, M. 2001. Shear tests of reinforced concrete deep beams. *JCI committee report on test method for fracture property of concrete*. Tokyo: JCI (in Japanese).

Mizumachi, M., Iwase, H., Rokugo, K. & Koyanagi, W. 1993. Size effect and localization of deformation in flexural failure of RC beams. *Proceedings of JCI* 15(2): 329–334 (in Japanese).

Noguchi, H. & Hisada, T. 1992. Development of a new branch-switching algorithm in nonlinear FEM using scaled corrector. *Transactions of JSME* (A) 58(555): 181–188 (in Japanese).

Podgorniak-Stanik, B. A. 1998. The influence of concrete strength, distribution of longitudinal reinforcement, amount of transverse reinforcement and member size on shear strength of reinforced concrete members. *M.A.S. thesis, University of Toronto*.

Rots, J. G. 1988. Computational modeling of concrete fracture. *Ph.D. thesis, Delft University of Technology*.

Shima, H., Chou, L. & Okamura, H. 1987. Micro and macro models for bond in reinforced concrete. *Journal of the Faculty of Engineering, University of Tokyo (B)* 39(2): 133–194.

Van Mier, J. G. M. 1984. Strain-softening of concrete under multiaxial loading conditions. *Ph.D. thesis, Eindhoven University of Technology*.

Walraven, J. & Lehwalter, N. 1994. Size effects in short beams loaded in shear. *ACI Structural Journal* 91(5): 585–593.

Yoshitake, K. & Hasegawa, T. 2000. Parametric study for size effect of reinforced concrete deep beam on shear strength. *Proceedings of the 55th annual conference of JSCE* 5: V-523 (in Japanese).

Finite Elements in Civil Engineering Applications, Hendriks & Rots (eds.)
© *2002 Swets & Zeitlinger, Lisse, ISBN 90 5809 530 4*

Invited paper: Comparative study of crack models

J.G. Rots

Faculty of Architecture, Delft University of Technology, Delft, The Netherlands

ABSTRACT: This paper starts with a discussion of popular crack models and describes their merits and demerits from a user's point of view. Both smeared and discrete crack models are included. Subsequently, it is demonstrated that the formation of cracks is often accompanied by snaps and jumps in the load-displacement response which complicate the analysis. The paper provides a solution by simplifying nonlinear crack models into sequentially linear saw-tooth models, either saw-tooth tension-softening for unreinforced material or saw-tooth tension-stiffening for reinforced material. The approach captures the snaps automatically and may help engineers to find practical solutions in case of complex structural cracking.

1 INTRODUCTION

For quasi-brittle materials like concrete and masonry a variety of crack models exists. For continua, the user can choose between smeared cracking with decomposed strain, total strain based models, plasticity based models or other frameworks. For cracking in discontinua, interface models based on decomposed relative displacements, total relative displacements, plasticity based approaches and other models are available. Although some overview papers exist, these often address the subject from a theoretical point of view and evaluate single-element outcomes or notched beams. Practical points of use and the way these models work out at structural level, have received less attention. The present paper first discusses a number of popular crack models and briefly describes their merits and demerits from an end-user's point of view. No completeness is claimed. For more information on continuum crack models the reader is referred to e.g. Feenstra & Rots (2001). Higher-order models and embedded discontinuity models, which are not yet available in commercial multi-purpose codes, are left out.

Structures have the ability to redistribute stresses upon cracking or crushing. A crack may initiate, propagate and snap to a free surface, whereupon stresses are redistributed and a new crack or crush may start. For large-scale structures, like unreinforced masonry facades, the balance between elastic energy stored in the structure versus fracture energy consumed in crack localization is large, which makes this redistribution process brittle, involving so-called snap-backs in the load-displacement response. This behavior is also typical of reinforced structures. In fact a local snap occurs whenever a crack crosses a reinforcing bar, giving a downward path under displacement or arc-length control followed by a new ascending path after redistribution. Tracing snap-back and snap-trough response is known to be a delicate matter in incremental-iterative nonlinear analysis. It is a source of convergence problems, independent of the constitutive crack model adopted. Sometimes jump-over techniques are tried, but these cannot guarantee equilibrium and stability. In the present paper a new somewhat unorthodox approach is proposed for snappy problems. It is a sequentially linear model with saw-tooth tension-softening or saw-tooth tension-stiffening.

2 CONTINUUM MODELS FOR CRACKING

2.1 *Decomposed-strain based smeared crack models*

Smeared crack models start from the notion of a continuum with strain, representing either the width of an individual localized crack, or an average crack strain in case of distributed cracking in reinforced concrete. It has been proven to be effective to decompose the strain into a part that belongs to the crack and a part that belongs to the material at either side of the cracks, see e.g. de Borst & Nauta (1986), Rots (1985).

An advantage of the strain decomposition is that elastic softening behavior can be described in an

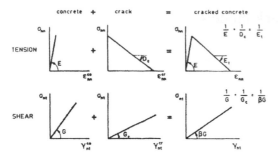

Figure 1. Smeared cracking with decomposed strain.

elegant way, both for mode I and mode II including secant unloading/reloading, see Fig. 1.

Another advantage is that the strain of the material between the cracks can be sub-decomposed into e.g. an elastic part, a creep part, a plasticity part and/or a thermal part (de Borst 1987). This allows for the combination of cracking with other nonlinear phenomena. Furthermore, the crack strain can be decomposed into the local crack strains of a number of cracks at different orientations. This makes it possible to handle non-orthogonal multi-directional cracking.

Though the model is thus conceptually attractive, the difficulty lies in the implementation and parameter choices. We mention two practical disadvantages. First, the algorithmic aspects are complicated. The model needs internal iterations to handle state changes like unloading, closing and re-opening, to handle nonlinear softening diagrams and to handle the combination with plasticity for compression strut action. When multiple cracks occur this internal iteration procedure may fail and even sub-stepping cannot always guarantee success. Secondly, the user's choices for shear retention functions and inter-crack threshold angles are not obvious, which may e.g. result in a too stiff response, stress-locking or uncontrollable principal stress rotation. These issues were addressed by Rots (1988).

2.2 Total-strain based models

The difficulties mentioned above gave rise to the development of a simplified total-strain based model by Feenstra et al. (1998). The material is described via stress-total strain relations in either fixed or rotating axes. For the fixed version, shear retention parameters are required, whereas the rotating version uses an implicit shear term to provide coaxiality between the rotating principal stress and strain. Advantages are as follows. First, the model is attractive to practicing engineers who think in terms of stress-strain relations and can plug these in directly. There is no need to input abstract yield functions or sophisticated crack laws. Secondly, the concept of a stress-strain relation cannot

only be used for softening or stiffening in tension, but also for nonlinear behavior and crushing in compression. This gives a natural combination of cracking and crushing whereby unconfined 2D cases, confined 3D cases and various coupling effects can be accounted for (Feenstra et al. 1998). Thirdly, from an implementation and algorithmic point of view, the model is attractive. It is purely explicit and does not require internal iterations. After updating a number of internal state variables, the stress is computed directly from the strain. Local convergence problems or closing/re-opening difficulties do not exist.

The disadvantages are as follows. Due to the fact that stress-strain relations are used, either in a fixed orthogonal system or in the rotating orthogonal principal stress system, only orthogonal cracks can be modeled. Non-orthogonal multi-directional cracking cannot be included. However, due to the difficulties with parameter choices for non-orthogonal cracking, and considering the fact that many engineering problems only involve orthogonal cracking, this is not very serious. Another disadvantage is that combinations with creep, shrinkage and thermal strains are not straightforward due to the description in terms of lumped total strain. Here, the decomposed model is to be preferred.

The rotating crack model can be derived as a special case of the decomposed multi-directional model assuming a zero inter-crack threshold angle, so that a new crack under slightly different angle is initiated in each step while the previous cracks unload elastically. Regarding the choice of fixed versus rotating, it can be said that for localized cracking the rotating model is to be preferred. It provides less stress-locking than the fixed version. For smeared cracking in reinforced structures, explicit shear across fixed cracks can be relevant. If so, the fixed model provides more possibilities to model that, although the choices for shear retention functions are not obvious considering their interrelation with possible uncontrollable principal stress rotations.

2.3 Plasticity based crack models

A unified approach for tension and compression can also be achieved by taking a plasticity based approach instead of a direct stress-strain approach for the two domains. Rankine-Von Mises and Rankine-Drucker Prager crack-crush models including softening for the two domains were developed by Feenstra (1993). The advantage is that the additive strain decomposition and general theory of elasto-plasticity can be used. The implementation and algorithmic aspects follow a concise format, and it has been demonstrated that quadratic convergence can be achieved for a number of localized fracture problems. However, other experiences exist with spontaneous divergence, probably

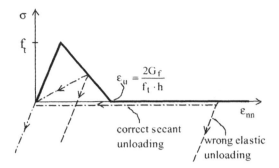

Figure 2. Secant and elastic unloading options.

due to local internal convergence problems or problems with maintaining the global consistent (negative) tangent stiffness.

An advantage of the plasticity based models is that they can be extended to orthotropic materials, still using the same principles. This was developed and carefully evaluated by Lourenco (1996) for masonry, with Rankine-Hill criteria. Also the framework of the decomposed smeared crack model allows for extension to orthotropic elastic and fracture behavior (Rots 1997). Another advantage of a plasticity formulation is that it allows for a transparent extension to creep, shrinkage and thermal effects. Van Zijl (2000) combined Rankine-Hill with visco-elasticity, shrinkage and rate-effects. Practical examples were shown for both concrete and masonry.

A disadvantage of plasticity based models is the inherent elastic unloading/reloading. A secant approach back to the origin, as included in the decomposed as well as total-strain based smeared crack models, is a better approximation of the real unloading behavior for quasi-brittle materials. Figure 2 gives an example in tension.

On purpose, in Fig. 2 the unloading behavior for a fully open crack is included, i.e. a crack for which the strain is far beyond the ultimate strain of the tensile softening diagram. This ultimate strain for tension-softening is based on the crack band model by Bazant and Oh (1983) and reads for a linear diagram: $(2\,G_f)/(f_t\,h)$ with G_f the fracture energy, f_t the tensile strength and h the crack band width which is related to finite element size and lay-out. Often, researchers evaluate models for small-scale notched beams. Then, h is small, and consequently ε_u is large, so that one does not pass the ultimate point of the diagram easily. For large-scale structures, however, h is large and ε_u relatively small). The Rankine plasticity model with elastic unloading then gives a response which is far from the real unloading behavior that passes almost along the origin. For cyclic problems the consequences thereof have been discussed by Feenstra & Rots (2001). It is stressed

here that also for monotonic non-proportional loadings erroneous results may be found with the elastic unloading option. The author experienced this when analyzing buildings that cracked initially under dead load and were subsequently subjected to a different load, e.g. wind load or imposed settlements. Upon adding this second load, the initial cracks unloaded and closed which resulted in a too stiff behavior when Rankine plasticity was used. Plasticity models should be extended with damage options to improve this behavior.

2.4 Other continuum approaches and future directions

At present, the above three models are popular under DIANA users. Other continuum crack models, however, exist and users can activate their personal models via user-supplied subroutines.

In this Volume Feenstra (2002) shows that isotropic damage models are attractive for simulating localized fracture. Damage models describe the degradation process by some scalar damage variable. The unloading is of secant type, and it was demonstrated that stress-locking does not occur and that curved mode-I fractures can be simulated accurately. A disadvantage of isotropic damage models is that also the compression strut action parallel to the crack degrades to zero. This makes these models impractical for reinforced concrete where the compression strut action is crucial. Extensions to anisotropic damage are required. These models come closer to smeared crack models that are inherently anisotropic as only the material in the direction normal to a crack degrades.

Hasegawa (2002) and Hoehler & Ozbolt (2002) discuss micro-plane models in this Volume. Here, normal and shear stresses across a fixed set of microplanes are monitored. These models also unify tension and compression. The multiple planes suggest a similarity with the non-orthogonal smeared crack model. However, in the non-orthogonal smeared crack model a set of planes is not predefined, but the cracks/planes emerge during the process, in the direction normal to the principal stress when this principal stress violates a tension cut-off. There is room for further studies to compare the pro's and contra's of the various formulations.

A model geared towards use for heavily reinforced structures is the four-way fixed crack model by Maekawa and co-workers, see e.g. Pimanmas & Maekawa (2001). Here, two sets of orthogonal cracks can be present, with a non-orthogonal angle between them. Good results were reported for non-proportional loading and path-dependence in case of pre-cracked reinforced structures, also including shear and compression non-linearity.

In the author's opinion, there is room for future research to make the decomposed multi-directional model of section 2.1 more robust, by allowing only a limited number of cracks under fixed angles from the first arisen crack. Taking 45 degrees, a 4-plane crack model for 2D situations results, which is suffcient for many engineering RC situations. It allows for non-orthogonal cracking, but does not loose the conceptual attractiveness of the decomposed formulation providing for coupling with creep, thermal and other effects. The implementation becomes compact and internal iterations and algorithmic aspects are under better control as the maximum number of cracks is constrained and "hesitating" cracks under small angles do not occur. For 3D cases and shells, the 45 degree pre-assumption results in a 9-plane crack model. Here, positive elements from smeared cracking, micro-plane models and Maekawa's model are combined.

3 DISCONTINUUM INTERFACE MODELS

3.1 *Total-relative displacement based models*

Predefined interface elements can be used as potential planes for discrete cracks, discrete crush zones, shear slip planes or bond shear zones, see e.g. Rots (1988, 1994). The interface elements define a relation between stresses and relative displacements across the interface. These have a normal and a shear component. In analogy to the total-strain based models of section 2.2, direct explicit relationships can be used. Such traction-displacement models have been implemented in user-supplied routines, for the tension, compression and shear domains, see Fig. 3.

The stresses are computed directly from the total relative displacements, taking internal states like loading/unloading and possible explicit coupling effects between the three domains into account. Secant unloading is included. An example of a coupling effect is that upon arising of a discrete tensile crack, the shear stress is reduced to zero suddenly or gradually. For tension-shear or compression-shear also "isotropic" degradation of the resultant normal-shear traction can be used, still in an explicit formulation.

This model is robust as it does not require internal iterations. The discrete cracking version for tension is also available in the current standard version of DIANA. For compression, the lumping of crushed zones into an interface that "implodes" is less known, but has been proven to be effective for modeling crushing in the toe of piers or shear panels (Frissen & Rots 1999). In fact, a compression cut-off is used in the same way as a tension cut-off. Frissen & Rots (1999) also employed explicit relations between shear stresses and shear displacements for joints, either dry or cohesive, in pier-wall connections and complete prefab buildings. These shear relations can be directly

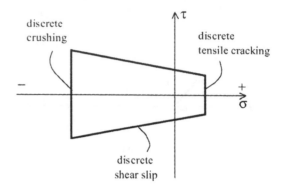

Figure 3. Three domains with direct relations between stress and total-relative displacement.

placed in the context of design codes that prescribe required shear capacities for joint.

3.2 *Discrete cracking based on decomposed relative displacements*

In early days, a decomposed discrete crack model was developed (Rots 1988). The relative displacement was split into an elastic part and a crack part. Internal iterations were required to compute these parts from the total relative displacement. In many cases, the elastic part refers to a dummy penalty stiffness that is given to the interface elements in order to keep them initially inactive in the stage before the tension cut-off is reached. It is no worth to exactly keep track of the elastic dummy part in the cracked stage. An exception might be the cases where the interface elements represent real joints that have a stiffness with a physical meaning, like mortar joints in masonry. However, the model has not been further developed. In fact, the use of interface elements is already a "decomposition" as such, as the elastic bulk behavior at either side of the discontinuity is decoupled from this discontinuity.

3.3 *Plasticity based interface models*

Similar to the plasticity based continuum models of section 2.3, plasticity based interface models can be formulated. With a view to masonry, Lourenco (1996) developed a composite yield criterion which combines tensile cracking and compression crushing with non-associated Coulomb friction for the shear domain. The correct inclusion of Coulomb friction for the shear domain is a major advantage. Other advantages and disadvantages are along the same lines as for the plasticity based continuum approaches, see section 2.3. The problem with the inherent elastic unloading option, as described in 2.3, also holds here, which calls for future extensions with damage and secant unloading. Following section 3.2, the method may be cumbersome as it involves a decomposition of the interface displacements into an elastic and plastic part, on top of the overall "decomposition" in bulk material and interface. In fact, only a rigid plastic model would be required. When the elastic part of the relative displacements represents a high dummy stiffness, strong demands are posed on the internal iteration procedure, also considering the corners in the yield surface. However, these issues were tackled successfully by Lourenco (1996). For masonry real stiffness values derived from zero-thickness mortar joints between "blown-up" bricks (Rots 1994) are used and this issue is less relevant. Regarding the Coulomb friction component, studies on confined masonry shear walls demonstrate that best results are achieved for a zero or low dilatancy angle, or with dilatancy softening (Van Zijl 2000).

3.4 Discussion and pitfalls with interface elements

The major advantage of interface elements is that the behavior of the bulk material is fully decoupled from the behavior of the discontinuity. This allows for a pure representation of elastic-softening behavior. The softening is not spread over continua, which circumvents objectivity problems and convergence problems for zigzag bands. When the bulk material undergoes creep or thermal effects, visco-elasticity or temperature-dependence can be easily plugged in for the bulk without interference with the crack model. The counterpart of these pro's is that the interfaces have to be pre-assumed in the mesh on the basis of engineering judgement and anticipated softening lines and softening hinges. Developments in remeshing and embedded discontinuities are upcoming.

Finally, we mention two pitfalls in practical interface modelling. First, the choice for the initial dummy stiffness should not be set naively. Users sometimes just take some very high value, e.g. 10^{+20} N/mm^3. A better philosophy is that in the initial uncracked stage, the interface deformations should remain below, say 0.1% of the elastic deformation of the body modeled. Figure 4 illustrates that the normal interface stiffness k_n then should be chosen as approximately 1000 times the Young's modulus E of the body divided by a characteristic size l of the body. For the interface shear stiffness k_t, the same holds in proportion to the elastic shear stiffness G of the body. These guidelines prevent ill-conditioning of the global matrix, oscillations in traction profiles (e.g. Rots 1988) and unnecessary difficulties in local integration point procedures. Secondly, when the two shaded bodies on top of Fig. 5 have to be connected, the topology and node numbering of the interface elements should be according to the first or third choice in Fig. 5. The second and fourth choice

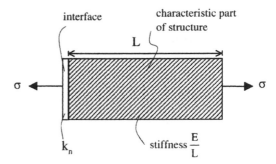

$$\text{dummy interface stiffness } k_n > \text{appr.} 1000 \times \frac{E}{L}$$

Figure 4.   Selection of dummy stiffness.

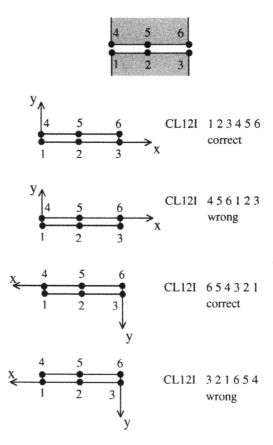

Figure 5.   Topology and numbering of quadratic six-node line interface element CL12I.

are wrong and result in a mixing up of tension with compression. For line interfaces in 3D space, an out-of-plane z-axis should be specified, in the direction that is consistent with the in-plane x- and y-axes of the first or third choice in Fig. 5. Care should be taken with automatic pre-processors that may provide inconsistent orientations.

4 SNAPS AND JUMPS

Users in FE analysis of concrete and masonry structures know that convergence problems at global level may occur due to softening and negative tangent stiffness, independent of the local constitutive model adopted. As mentioned in the introduction, a crack initiates, propagates and may either snap to a free surface or be arrested by a rebar, whereupon stresses are redistributed and a new crack or crush may start. This is accompanied by drops in the quasi-static load-displacement response. After reaching a valley, a new ascending path may be found up to a new local peak. The user cannot afford to miss these peaks as it might be the global peak for structural failure. The jumpy response is difficult to trace, although arc-length and indirect control schemes provide possibilities (e.g. de Borst 1987b).

This is not an incidental issue. We identify three examples. Figure 6 shows a reinforced tension bar (Rots 1988). Bond-slip was modeled using an explicit tau-delta relation and smeared cracking was adopted. Both the experimental and numerical result shows four local peaks associated with the sequential development of four primary cracks. Here, a subtle perturbation of strength in combination with delicate arc-length control provided a solution. Considering that such tension bar is only a minor component of a real structure, tracing the full response of structures becomes unwieldy.

Figure 7 reprints a result by Pimanmas & Maekawa (2001) for a pre-cracked reinforced structure. Indeed, many local peaks are displayed, both in the experiment and computation. This behavior is typical of many reinforced structures. The peaks cannot be simply jumped over in load control.

Figure 8 shows a result for crack propagation in a large-scale unreinforced masonry facade subjected to settlement (Rots 2000). Details are given in Rots (2001) and Boonpichetvong and Rots (2002). A global peak is visible and subsequently some sharp peaks in the valley, associated with the crack that discontinuously snapped from window to window. Only for discrete cracking, arc-length control provided a solution whereas all smeared trials failed and an unconverged jump-over was found. The difficulty here lies in the large scale (the masonry facade is 180 times larger than e.g. the SEN notched beam benchmark), which gives a large amount of stored elastic energy versus moderate consumption of fracture energy upon crack propagation, resulting in a brittle response and sharp snap-backs.

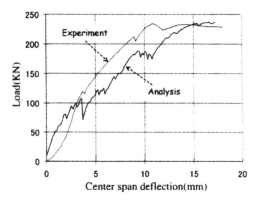

Figure 7. Response of pre-cracked RC structure (Pimanmas & Maekawa 2001).

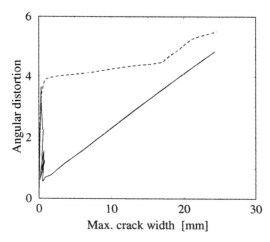

Figure 8. Response of large-scale masonry façade (Rots 2000, 2001). Smeared (dashed) and discrete (drawn).

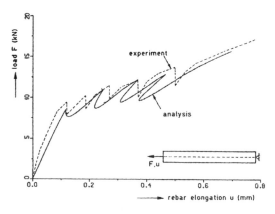

Figure 6. Response of reinforced tension bar (Rots 1988).

The irregular responses of Figs. 6–8 gave rise to the idea of sequentially linear analysis where the local peaks emerge automatically rather than delicately trying to pass them using arc-length procedures in nonlinear analysis. This model was recently pioneered and is based on a saw-tooth approximation of the tension-softening diagram (Rots 2001).

5 SEQUENTIALLY LINEAR MODEL WITH SAW-TOOTH SOFTENING

5.1 *Global procedure*

The structure is discretized using standard elastic continuum elements. Young's modulus, Poisson's ratio and a tensile strength are assigned to the elements. Subsequently, the following steps are sequentially carried out:

- Add the external load as a unit load.
- Perform a linear elastic analysis.
- Extract the critical element from the results. The critical element is the element for which the principal tensile stress is most close to its current strength. This principal tensile stress criterion is widely accepted in mode-I fracture mechanics of quasi-brittle materials.
- Calculate the critical global load as the unit load times the current strength divided by stress of the critical element.
- Extract also a corresponding global displacement measure, so that later an overall load-displacement curve can be constructed.
- Reduce the stiffness and strength, i.e. Young's modulus E and tensile strength f_t of the critical element, according to a saw-tooth tensile softening stress-strain curve as described in the next section.
- Repeat the previous steps for the new configuration, i.e. re-run a linear analysis for the structure in which E and f_t of the previous critical element are reduced.
- .... Repeat again, etc. ....
- Construct the overall load-displacement curve by connecting all load-displacement sets consecutively found in the above steps.
- Plot deformed meshes. These plots reveal the fracture localization because the series of critical weakened elements will display the largest strains, representing crack width.

5.2 *Saw-tooth softening model via stepwise reduction of Young's modulus*

The outcome of the above scheme heavily depends on the way in which the stiffness and strength of the critical elements are progressively reduced. This constitutes the essence of the model. A very rough method would be to reduce E to zero immediately after the first, initial strength is reached. This elastic perfectly brittle approach, however, is likely to be mesh dependent as it will not yield the correct energy consumption upon mesh refinement (Bazant and Cedolin 1979). In this study, the consecutive strength and stiffness reduction is based upon the concept of tensile strain softening, which is accepted in the field of fracture mechanics of concrete (Bazant and Oh 1983).

The tensile softening stress-strain curve is defined by Young's modulus E, the tensile strength f_t, the shape of the diagram, e.g. a linear or exponential diagram, and the area under the diagram. The area under the diagram represents the fracture energy G_f divided by the crack band width h, which is a discretisation parameter associated with the size, orientation and integration scheme of the finite element. Although there is some size-dependence, the fracture energy can be considered to be a material property. This softening model usually governs nonlinear constitutive behaviour in an incremental-iterative strategy. Please note that here we adopt the curve only as a "mother" or envelope curve that determines the consecutive strength reduction in sequentially linear analysis. In the present study, attention is confined to a linear softening diagram, but extension to any other shape of the diagram is possible. For a linear softening diagram, the ultimate strain ε_u of the diagram reads:

$$\varepsilon_u = (2G_f)/(f_t h) \qquad (1)$$

In a sequentially linear strategy, the softening diagram can be imitated by consecutively reducing Young's modulus as well as the strength. Young's modulus can e.g. be reduced according to:

$$E_i = E_{i-1}/a \quad \text{for } i = 1 \text{ to } n \qquad (2)$$

with i denoting the current stage in the saw-tooth diagram, $i - 1$ denoting the previous stage in the saw-tooth diagram and a being a constant. When a is taken as 2, Young's modulus of a critical element is reduced by a factor 2 compared to the previous state. n denotes the amount of reductions that is applied in total for an element. When an element has been critical n times, it is removed completely in the next step. This complete removal can be done explicitly so that "a hole in the mesh" occurs for full cracks, or it can be approximated by maintaining the element but giving it a very low residual Young's modulus for reasons of computational convenience (e.g. 10^{-6} times the initial Young's modulus). The reduced strength f_{ti} corresponding to the reduced Young's modulus E_i is taken in accordance with the envelope softening stress-strain curve:

$$f_{ti} = \varepsilon_u E_i(D/(E_i + D)) \qquad (3)$$

with

$$E_i = E/(a^i) \qquad (4)$$

and

$$D = f_t/(\varepsilon_u - (f_t/E)) \qquad (5)$$

being the tangent to the tensile stress-strain softening curve. Note that this is the softening curve in terms of stress versus *total* strain, i.e. the sum of elastic strain and crack strain of an imagined cracked continuum. The diagram includes the initial rising branch, which is steep compared to the downward slope in case of small-scale elements (small crack band width) and/or high fracture energy. As an example, Fig. 9 shows the envelope softening curve and the corresponding saw-tooth curve for an initial Young's modulus E of 38000 N/mm^2, initial tensile strength f_t of 3 N/mm^2, fracture energy G_f of 0.06 N/mm, crack band width h of 5 mm, factor a equal to 2 and number of reductions n equal to 10.

Strictly speaking, the sequentially linear approach corresponds to a set of unconnected lines of different slope, starting from the origin up to the current strength. For reasons of presentation, the lines have been connected into one discontinuous curve, by letting the stress drop vertically from the peaks to a residual level at which the new rising branch of reduced slope passes.

The model is simple. It always provides a solution, as ill-conditioning or divergence does not appear in sequentially linear analysis. A physical explanation to the model is that fracture is a gradual separation process whereby the net cross section that connects material, and thus the stiffness, is gradually reduced. An advantage of the model compared to lattice models (e.g. Schlangen & van Mier 1992, Beranek & Hobbelman 1995) is that the regular notions of fracture mechanics, like the principal tensile stress criterion, the envelope strength and fracture energy are maintained which helps in reaching realistic energy consumption and toughness as observed in experiments.

Unloading of softening zones is possible and inherently of the secant type, because of the reduction of elasticity.

5.3 Alternative definition of saw-tooth diagram

In the previous section, the starting point for defining the saw-tooth diagram was a consecutive reduction of Young's modulus via the factor a while the corresponding reduced strength was subsequently determined from the envelope curve according to (3). In first trials of the model, the starting point was taken alternatively by reducing the strength from f_t to zero in n steps, while the corresponding reduced Young's modulus was subsequently computed from the envelope curve. The decreasing line of the envelope curve is in fact split into n equidistant portions. When n is assumed to be 10 and when the element is critical for the first time, the strength is reduced to 90% of the original strength. When it is critical for the second time, the strength is reduced to 80% of the original strength, etc. The resulting saw-tooth diagram for this alternative approach is depicted in Fig. 10, using the same parameters as subject to Fig. 9. Other alternative ways of stepwise reduction might be used as well. The choice is related to the absolute dimensions of the crack band width. The notched beam studied below is of small-scale laboratory size, giving a low value of h and thus a high ultimate strain compared to the initial elastic strain limit. For large-scale structures of the same material and of the same relative mesh fineness, h becomes much larger and the pictures in Figs. 9 and 10 become different.

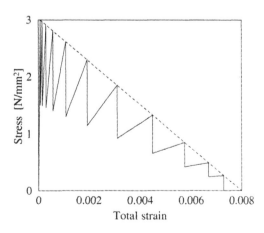

Figure 9. Envelope softening stress-strain diagram (dashed) and saw-tooth approximation (drawn).

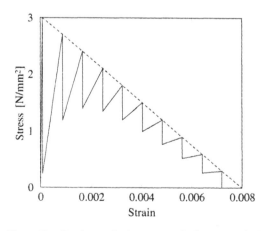

Figure 10. Envelope softening stress-strain diagram (dashed) and alternative saw-tooth approximation (drawn).

6 ANALYSIS OF NOTCHED BEAM

6.1 Geometry and meshes

A symmetric notched beam of total length 500 mm, span 450 mm, height 100 mm, thickness 50 mm and notch depth 10 mm was selected for analysis. The distance between the loading points in the symmetric four-point loading scheme is 150 mm. In the present study, on purpose a regular straight mesh was adopted, so that the solution is not affected by any disturbing effect due to zig-zag crack band paths. Two different meshes were used, referred to as coarse and fine respectively. These meshes have a symmetric center crack band of 20 mm and 5 mm width respectively. The element height for the central elements was taken the same as the element width, so that the amount of center elements over the depth of the beam amounts to 5 and 20 respectively. The ultimate strain ε_u of the envelope input curve was adapted to h according to Eq. (1). Four-node linear elements were used. These were integrated using a two by two Gaussian scheme, except for the elements in the center band which were integrated using a single center-point integration. This means that, if an integration point of an element in this center band reaches a local peak in the saw-tooth curve, Young's modulus of the entire element can be reduced. The material parameters are the same as underlying Figs. 9 and 10. Poisson's ratio was taken as 0.2. The model of Fig. 9 was adopted. The factor a in Eq. (2) was taken as 2. The factor n was taken as 10, i.e. the stiffness of an element can be reduced 10 times according to Eq. (2), assuming the element becomes 10 times critical in a global sense. Beyond that, the element is removed.

6.2 Results

Figs. 11 and 12 show the results for the course and the fine mesh in terms of the total load (sum of the two loads) versus displacement at the loading points. The curves are constructed by connecting the critical loads and the corresponding displacements for all linear analyses that have been executed sequentially. As a comparison, the reference curve (see for this beam Rots 1993) from a nonlinear softening analysis with the same parameters is included in the graphs. This reference curve appeared to be almost identical for all meshes, except for the coarse mesh where a minor deviation near peak occurred.

For both meshes, the load-displacement curve from sequentially linear analysis is irregular. This is because the process of elements becoming critical is discontinuous. If two elements have almost the same ratio of stress versus current strength, still the most critical one is selected first, but in the next step the other one will fail very soon, maybe at a lower load and lower displacement than in the previous step.

The irregular saw-tooth response at global level can be interpreted as the global pendant of the saw-tooth input at local level.

With increasing mesh fineness, the curve become smoother. Surprisingly, the envelope of the curve for the coarse mesh appears to resemble the reference curve most closely, both in terms of the maximum load and post-peak toughness. Please note that the model is especially derived with a view towards large-scale brittle fracture analysis, i.e. large element sizes, which is thus promising. With increasing mesh fineness, however, the peak in the load-displacement behavior becomes lower compared to the correct peak from the nonlinear reference analysis that is to be mimicked. This demonstrates that without special measures

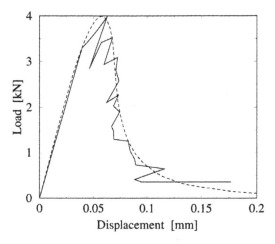

Figure 11. Load-displacement curve, coarse mesh.

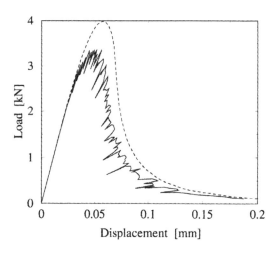

Figure 12. Load-displacement curve, fine mesh.

sequentially linear models are not objective with respect to mesh refinement. Increasing mesh fineness leads to sharper stress peaks at the crack band tip, so that the strength at the crack band tip is reached earlier than for a coarse mesh. In fact, a kind of "zip-fastener" effect is introduced. With a sudden full removal of elements, i.e. a diagram with only one saw-tooth instead of the present ten saw-teeth, this effect would be even more pronounced. That was described already by Bazant & Cedolin (1979).

6.3 Mesh-objectivity via lifting the strength of the saw-tooth diagram

A natural way to circumvent mesh-inobjective results is to enhance the strength with decreasing element size. This counteracts the zip-fastener effect. Precisely speaking, the input envelope curve and saw-tooth curve should be adapted to the mesh in two ways. First, the ultimate *strain* ε_u of the diagram is adapted to the element *width*, which is equal to the crack band width in the present case of a perfectly vertical crack in a perfectly vertically aligned mesh with constant strain in the direction normal to the crack. This is the same procedure as used for crack band modeling in nonlinear analysis. Now, we add the second adaptation, viz. the ultimate *strength* f_t is adapted to the element *height*, which is again equal to the crack band width h in the present case of a perfectly vertical crack in a mesh composed of perfectly vertical square elements.

As an example, Fig. 13 depicts the response for the fine mesh when the strength is raised by a factor 1.2 together with the previous result from Fig. 12 without strength enhancement. Note that upon lifting the saw-tooth curve by a factor not only the strength of the peaks is enhanced by this factor but also the fracture energy of the input envelope curve. The result shows

that objectivity can be reached in this way. The curve now lies precisely on the reference result.

The choice for the factor 1.2 was made here arbitrary, by fitting. Further work is required to rationalize the quantification of the strength enhancement. The challenge is to derive consistent relationships between the various length scales in a fracture problem on the one hand, e.g. the depth of the structure, the thickness of the structure, the initial notch depth, the fracture process zone length, the so-called brittleness number, and the required strength enhancement on the other hand. Also the material parameters, the type of saw-tooth approximation, the factors a and n play a role in this respect. Also for elastic-brittle lattice models, which bear some similarity with the present model, the issue of mesh-(in)objectivity should receive attention.

7 ANALYSIS OF MASONRY FAÇADE

The masonry façade underlying Fig. 8 was re-analyzed using the sequentially linear model of Fig. 9. The parameters were taken as: $E = 3000\,\text{N/mm}^2$, $f_t = 0.6\,\text{N/mm}^2$, $G_f = 0.05\,\text{N/mm}$, crack band width $h = 225\,\text{mm}$ (note: approximately 100 times larger than for the small-scale notched beam), factor $a = 2$, factor $n = 5$. The thickness is 220 mm. For geometry the reader is referred to Rots (2000) or the paper by Boonpichetvong & Rots (2002) in this Volume. A difficulty with this problem is that it involves a non-proportional loading scheme. It was outlined how this can be handled via a combination of sequentially and incrementally linear analysis (Rots 2001). Herein, a fictitious proportional load was taken as an example, viz. a vertical point load at the top of the façade, slightly off-center.

Fig. 14 shows the result in terms of the vertical point load versus displacement. The result reveals the

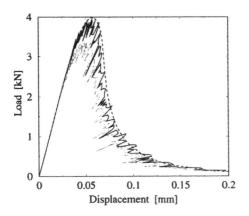

Figure 13. Load-displacement curve, fine mesh, strength and energy enhancement by a factor 1.2 (drawn line), compared with results of Fig. 12 (dashed lines).

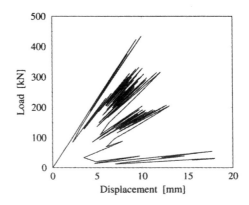

Figure 14. Load-displacement curve for sequentially linear analysis of masonry façade under point load.

very sharp snap-back behavior, which is found in the sequentially linear fashion without any numerical problems. In the post-peak behavior, we observe four nested snaps, which correspond to the subsequent jumps of the crack from window to window, starting at the bottom and ending at the top of the façade. This is an adequate alternative to the complex nonlinear analyses summarized in Fig. 8.

8 DISCUSSION AND FUTURE DIRECTIONS

Above, a sequentially linear continuum model for concrete fracture has been proposed. The model approximates an envelope softening stress-strain curve by a saw-tooth diagram. In each linear analysis, a critical element is traced by comparing element stress with current element strength, i.e. with the current peak in the saw-tooth diagram. Next the stiffness and strength of the critical element are reduced according to the subsequent tooth of the diagram, and the process is repeated.

The charm of this model is that words like iteration, ill-conditioning and divergence do not appear in the vocabulary. The stiffness of the linear analyses is always positive definite as the saw-teeth at local level are always rising. A notched beam analysis shows the ability of the method to reproduce the behavior of a nonlinear reference analysis on the same parameters. The advantage of the model appears for brittle snap-back behavior as was illustrated by analysis of a large-scale façade.

Current and future activities to extend the model include the following. We mention six points. First, the approach to achieve mesh-size objectivity should be pursued further, as described in chapter 6. Second, objectivity with regard to mesh orientation should be studied. The strategy followed for nonlinear analyses by Oliver (1989) and Jirasek & Zimmermann (2001) to accurately compute the crack band width and to use a non-local (smoothed) principal stress criterion at the "crack tip" can be adopted in the same way for sequentially linear analysis. Third, extension of the model to anisotropic degradation is possible. The present formulation is isotropic because Young's modulus is reduced for all directions. Inclusion of the model to fixed planes of orthotropy is possible, including memory of unloading/reloading and full crack closure and re-opening. Fourth, extension of the model to non-proportional loadings has been outlined (Rots 2001), which is important in building engineering as vertical dead load is often superimposed by horizontal wind load. It should be evaluated whether path-dependence can be captured sufficiently in this way. Fifth, implementation of the sequentially linear model for interface elements is underway. A cut-off criterion for tension, shear and compression, as sketched in Fig. 3, followed by saw-tooth softening for the domains is implemented.

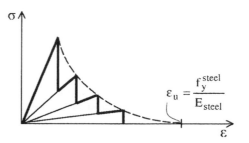

Figure 15. Saw-tooth tension-stiffening diagram for reinforced concrete.

For the continuum version, extension with saw-tooth softening in compression is also possible. Sixth, and important for engineering applications on RC structures, is the extension towards saw-tooth tension-stiffening rather than tension-softening. This approach fits the nature of tension-stiffening as the tension-stiffening curve is already a saw-tooth serrated type of curve, depending on details of reinforcing lay-out and reinforcing percentages. The response of the reinforced tension bar in Fig. 6 then becomes the global pendant of a local saw-tooth tension-stiffening input diagram as sketched in Fig. 15.

The ultimate strain in Fig. 15 is determined by the yield strength of the reinforcing steel divided by Young's modulus of the steel. When the steel yields, the concrete in between the cracks cannot contribute any longer to the tensile carrying capacity. The approach fits global practice in RC engineering, where nonlinear behavior is sometimes captured via repeating linear analyses using reduced Young's moduli at areas of anticipated cracking.

Finally, it is mentioned that the required CPU-time for sequentially linear continuum analysis can be large, depending on the number of elements and saw-teeth adopted and the number of cracks or crushes emerging. On the other hand, the approach always "converges", as the secant stiffness for the saw-teeth at local level is always positive, so is the secant stiffness for the total system. Complex solution techniques like arc-length procedures, to pass the drops and snaps as for the RC structure in Fig. 7, are no longer required and the user is protected from spending time and effort in that. The required man-hour time to steer the analyses is low.

9 CONCLUDING REMARK

Advantages and disadvantages of various existing smeared and discrete models for cracking in quasi-brittle materials have been discussed. The potential of a recently developed sequentially linear saw-tooth model for analyzing structures with drops and snaps in the response has been demonstrated.

ACKNOWLEDGEMENT

Financial support from Delft Cluster (theme 1 Soil and Structures, project Settlement damage) and from the Netherlands Technology Foundation STW (project DCT 3683 Computational Modeling) is acknowledged. The author is indebted to Dr. Peter Feenstra from Cornell University for cooperation and discussions on crack models.

REFERENCES

Bazant, Z.P. & Cedolin, L. 1979. Blunt crack band propagation in finite element analysis. ASCE *J. Engineering Mechanics Division* 105(2): 297–315.

Bazant, Z.P. & Oh, B.H. 1983. Crack band theory for fracture of concrete. RILEM *Materials and Structures* 16(93): 155–177.

Beranek, W.J. & Hobbelman, G.J. 1985. 2D and 3D-Modelling of concrete as an assemblage of spheres: revaluation of the failure criterion. In F.H. Wittmann (ed.), *Fracture mechanics of concrete structures; Proc. FRAMCOS-2*: 965–978. Freiburg: Aedificatio.

Boonpichetvong, M. & Rots, J.G. 2002. Settlement damage of masonry buildings in soft-ground tunneling. This Volume.

Borst, R. de & Nauta, P. 1986. Non-orthogonal cracks in a smeared finite element model. *Engrg. Comp.*, 2:35–46.

Borst, R. de 1987. Smeared cracking, plasticity, creep and thermal loading – A unified approach. *Comp. Meth. Appl. Mech. Engng.* 62:89–110.

Borst, R. de. 1987b. Computation of post-bifurcation and post-failure behavior of strain-softening solids. *Computers & Structures* 25(2): 211–224.

Feenstra, P.H. 1993. Computational aspects of biaxial stress in plain and reinforced concrete. PhD thesis Delft University of Technology, Delft, The Netherlands.

Feenstra, P.H., Rots, J.G., Arnesen, A., Teigen, J.G. & Hoiseth, K.V. 1998. A 3D constitutive model for concrete based on a co-rotational concept. In R. de Borst et al. (eds.), *Computational modelling of concrete structures; Proc. EURO-C 1998*: 13–22. Rotterdam: Balkema.

Feenstra, P.H. & Rots, J.G. 2001. Comparison of concrete models for cyclic loading. ASCE Committee Report *Modelling of Inelastic Behavior of RC Structures under Seismic Loads*, pp. 38–55, ASCE, USA.

Feenstra, P.H. 2002. Implementing an isotropic damage model in Diana. Use-case for the user-supplied subroutine usrmat. This Volume.

Frissen, C.N. & Rots, J.G. 1999. Numerical simulation of stability of masonry structures. Collection of working reports CUR Committee A33 Masonry Mechanics. Internal Report, TNO Building and Construction Research, Delft.

Hasegawa, T. 2002. Enhanced microplane concrete model and multi equivalent series phase model in Diana for fracture analysis of concrete structures. This Volume.

Hoehler, M. & Ozbolt, J. 2002. Application of the microplane model for three-dimensional reversed-cyclic analysis of reinforced concrete members. This Volume.

Jirasek, M. & Zimmermann, T. 2001. Embedded crack model: II. *Int. J. for Numerical Methods in Engineering* 50(6): 1269–1290.

Lourenco, P. 1996. Computational strategies for masonry structures. PhD thesis Delft University of Technology, Delft, The Netherlands.

Oliver, J. 1989. A consistent characteristic length for smeared cracking models. *Int. J. for Numerical Methods in Engineering* 28: 461–474.

Pimanmas, A. & Maekawa, K. 2001. Multi-directional fixed crack approach for highly anisotropic shear behavior in pre-cracked RC members. *J. Materials, Conc. Struct., Pavements*, JSCE, No. 669/V-50, 293–307, Japan Society of Civil Engineers.

Rots, J.G. 1985. Smeared crack approach and fracture localization in concrete. HERON 1985, No. 4, The Netherlands.

Rots, J.G. 1988. Computational modeling of concrete fracture. PhD thesis Delft University of Technology, Delft, The Netherlands.

Rots, J.G. 1993. The smeared crack model for localized mode-I tensile fracture. In F.H. Wittmann (ed.), *Numerical models for fracture mechanics*: 101–113. Rotterdam: Balkema.

Rots, J.G. 1994 (Ed.). Structural masonry – An experimental/numerical basis for practical design rules. CUR report 171, CUR, Gouda (in Dutch). English version published by Balkema, Rotterdam in 1997.

Rots, J.G. 2000. Settlement damage predictions for masonry. In L.G.W. Verhoef and F.H. Wittmann (eds.), Maintenance and restrengthening of materials and structures – Brick and brickwork, Proc. Int. Workshop on Urban heritage and building maintenance: 47–62. Freiburg: Aedificatio.

Rots, J.G. 2001. Sequentially linear continuum model for concrete fracture. Fracture Mechanics of Concrete Structures (eds. R. de Borst, J. Mazars, G. Pijaudier-Cabot and J.G.M. van Mier), Vol. 2, A.A. Balkema Publishers, 2001, pp. 831–839.

Schlangen, E. & Van Mier, J.G.M. 1992. Experimental and numerical analysis of micro-mechanisms of fracture of cement-based composites. *Cement & Concrete Composites* 14:105–118.

Zijl, G.P.A.G. van 2000. Computational modelling of masonry creep and shrinkage. PhD thesis Delft University of Technology, Delft, The Netherlands.

Finite Elements in Civil Engineering Applications, Hendriks & Rots (eds.)
© *2002 Swets & Zeitlinger, Lisse, ISBN 90 5809 530 4*

The application of Meschke's damage model in engineering practice

A. de Boer
Ministry of Transport, Public Works and Watermanagement, Utrecht, The Netherlands

E.G. Septanika
TNO Building and Construction Research, Delft, The Netherlands

ABSTRACT: The paper presents Meschke's damage model which is based on a combination between Rankine's plasticity model and Meschke's computational approach. The present damage model has been implemented into DIANA by means of a user-supplied subroutine. To demonstrate the capability of the implemented damage model two test cases will be performed, i.e. (i) tension-shear deformation of plain concrete (Willam's test) and (ii) skew slab bridge under combined loadings.

1 INTRODUCTION

The last 5 years damage models are becoming popular in research activities. Although in practice, for instance in the design department of consultancy offices, the use of a damage model in a nonlinear analysis is still not so common. To close the gap between the research activities and the civil engineering practice, the research department of the Civil Engineering Division of the Dutch Ministry of Public Works decided to implement an adequate 3-D damage model in DIANA by means of a user-supplied subroutine. A successful implementation would be an additional tool above the present library of material models for non-linear analysis in DIANA.

Within the flow theory of plasticity the state of stress in the material is monitored by means of the yield condition, which is formulated in terms of the stress vector and the so-called internal state parameters.

The relations between plastic strain rate and the state of stress, and between the internal state parameter and the state of stress, are governed by evolution equations.

The solution of the evolution equations and the updated stress and strain vectors can be obtained by means of a so-called return-mapping algorithm.

For brittle materials, such as concrete and rock, yield conditions which are based on the maximum principal stress, such as the Rankine yield criterion, can be employed to describe their tensile behavior. On the other hand, Meschke (1996) has developed an adequate numerical approach which is proven to be suitable for principal-stress-based yield conditions

such as the Rankine yield criterion. Meschke proposed a computational scheme in which singularity problems of the Rankine yield surfaces are by-passed. In the present paper, a 3-D concrete damage model will be described based on the Rankine yield criterion in combination with the Meschke numerical approach which will be further referred to as *the Meschke damage model*.

The framework of computational scheme is recapitulated in section 2, i.e. the return mapping algorithm and updating procedures.

Section 3 describes the corresponding verifications and the numerical simulation of a skew bridge is presented in section 4. Concluding remarks are notified in section 5.

2 FRAMEWORK COMPUTATIONAL SCHEME

The iteration scheme for the return mapping is formulated by employing the so-called residuals method which can be summarized as follows.

The residual functions $\boldsymbol{R}$ for the elastic strain vector ε^e and for the internal variable κ in the kth iteration step and in the $n + 1$ loading step are defined as

$$R_{\hat{\varepsilon}^e}^{(k)} = \hat{\varepsilon}_{n+1}^{e(k)} - \hat{\varepsilon}_{n+1}^{e,tr} + \lambda_{n+1}^{(k)} \left(\frac{\partial \hat{f}}{\partial \hat{\sigma}} \right)_{n+1}^{(k)} \tag{1}$$

$$R_{\kappa}^{(k)} = -\kappa_{n+1}^{(k)} + \kappa_n + \lambda_{n+1}^{(k)} \, \hat{h}_{n+1}^{(k)} \tag{2}$$

where

- the principal values of the strains and stresses are indicated by the symbol "^",
- the trial strains are indicated by the symbol "tr",
- λ is the plastic multiplier,
- f is the yield function and
- h is the hardening/softening function.

Linearization of the residuals and the yield function read

$$R_{\hat{\varepsilon}^e}^{(k)} + \Delta\hat{\varepsilon}^e + \lambda_{n+1}^{(k)}\left(\frac{\partial^2 \hat{f}}{\partial\hat{\sigma}^2}\right)_{n+1}^{(k)} \Delta\hat{\sigma} + \left(\frac{\partial\hat{f}}{\partial\hat{\sigma}}\right)_{n+1}^{(k)} \Delta\lambda \quad (3)$$

$$R_\kappa^{(k)} - \Delta\kappa + \lambda_{n+1}^{(k)}\left(\frac{\partial\hat{h}}{\partial\hat{\sigma}}\right)_{n+1}^{(k)} \Delta\hat{\sigma} + \hat{h}_{n+1}^{(k)}\Delta\lambda \quad (4)$$

Next, by employing Rankine's yield function, i.e.

$$\hat{f}_i(\hat{\sigma}_1, \hat{\sigma}_2, \hat{\sigma}_3, \kappa) = \hat{\sigma}_i - \bar{\sigma}_i(\kappa), \quad (i = 1, 2, 3) \quad (5)$$

one arrives to the following relations for the stress increment and the increment of the plastic multipliers (for more details see e.g. Septanika, 2000):

$$\hat{f} + I\Delta\hat{\sigma} + \hat{E}\Delta\Lambda = 0 \quad (6)$$

$$\Delta\Lambda = -\hat{E}^{-1}\left[\hat{f} + I\Delta\ \hat{\sigma}\right] \quad (7)$$

with

$$\hat{f} = \begin{bmatrix} \hat{f}_1 \\ \hat{f}_2 \\ \hat{f}_3 \end{bmatrix}, \quad \Lambda_{n+1}^{(k)} = \begin{bmatrix} \lambda_1 \\ \lambda_2 \\ \lambda_3 \end{bmatrix}_{n+1}^{(k)} \quad \text{and} \quad (8)$$

$$E = \sigma'\begin{bmatrix} \left(\frac{\partial\hat{f}_1}{\partial\bar{\sigma}}\right)\left(\frac{\partial h}{\partial\Lambda}\right)^T \\ \left(\frac{\partial\hat{f}_2}{\partial\bar{\sigma}}\right)\left(\frac{\partial h}{\partial\Lambda}\right)^T \\ \left(\frac{\partial\hat{f}_3}{\partial\bar{\sigma}}\right)\left(\frac{\partial h}{\partial\Lambda}\right)^T \end{bmatrix} = -\frac{\bar{\sigma}'}{h}\begin{bmatrix} \lambda_1 & \lambda_2 & \lambda_3 \\ \lambda_1 & \lambda_2 & \lambda_3 \\ \lambda_1 & \lambda_2 & \lambda_3 \end{bmatrix} \quad (9)$$

The stress increment then reads ($\hat{D}$ is the elastic modulus matrix)

$$\Delta\hat{\sigma}_{n+1}^{(k)} = -\hat{D}\left[R_{\hat{\varepsilon}^e}^{(k)} + I\Delta\Lambda_{n+1}^{(k)}\right] \quad (10)$$

implying the following relation for the plastic multipliers $\Delta\Lambda$

$$\Delta\Lambda_{n+1}^{(k)} = -\left[\hat{E} - \hat{D}\right]^{-1}\left\{\hat{f}^{*(k)} - \hat{D}\,R_{\hat{\varepsilon}^e}^{(k)}\right\}$$
$$= \left[\hat{Z}^{(k)}\right]^{-1}\hat{f}^{*(k)} \quad (11)$$

where

$$\hat{Z} = \begin{bmatrix} D_{11} + \lambda_1^* & D_{12} + \lambda_2^* & D_{13} + \lambda_3^* \\ D_{12} + \lambda_1^* & D_{22} + \lambda_2^* & D_{23} + \lambda_3^* \\ D_{13} + \lambda_1^* & D_{23} + \lambda_2^* & D_{33} + \lambda_3^* \end{bmatrix}^{(k)} \quad (12)$$

and

$$\lambda_i^* = \frac{\bar{\sigma}'}{h}\lambda_i^* \quad (13)$$

From the solution for the plastic multiplier λ, one can update the internal variable, the elastic and plastic strains and the stresses. The consistent tangent modulus can then be determined in terms of the convergence solutions for the stresses.

3 NUMERICAL VERIFICATION

To demonstrate the capability of Meschke's plasticity based damage model, two well-known numerical problems will be investigated.

3.1 Tension-shear deformation of plane model

It is a one element test, where all nodes are fixed in both directions X and Y. The element is loaded in biaxial tension and shear, which causes a continuous rotation of the principal directions after cracking typically observed in crack propagation in smeared crack finite element analysis (Willam et al., 1987).

First, the element is subjected to tensile straining in the x-direction. Immediately after cracking the element is loaded by a combined biaxial tensile strain (in the x- and y-direction) and shear strain (in the xy-direction).

The material properties are in accordance with the original proposal by Willam et al. (1987), i.e.

- Young's modulus $E = 10^4$ MPa,
- the Poisson ratio $\nu = 0.2$,
- the tensile strength $f_{ct} = 1$ Mpa and
- the fracture energy $G_f = 0.15$ J/m².

This results in an ultimate equivalent strain $\kappa_{ult} = 0.3 * 10^{-3}$ for a unit crack band width (i.e. h = 1 mm) and linear tension-softening. It implies that for $\kappa = 0 \rightarrow \sigma = 1$ and for $\kappa = \kappa_{ult} \rightarrow \sigma = 0$, which leads to

$$\sigma(\kappa) = f_{ct}(1 - \kappa/\kappa_{ult}) \quad \text{for } (0 \leq \kappa \leq \kappa_{ult})$$

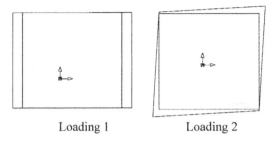

Loading 1   Loading 2

Figure 1.   Loading types plane-strain test.

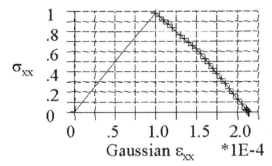

Figure 2.   Relation stress-strain in x-direction.

with

$$\kappa_{ult} = 2G_f/(hf_{ct}) = 3.0 * 10^{-4}$$

For this elementary problem the finite element analysis using both the shell element and the solid element will be performed.

Since the solid element is not constrained in the direction perpendicular to the tension-shear plane (i.e. the xy-plane), the analysis with the solid element should also give similar results with respect to the plane-stress analysis.

The model and the total deformations are shown already in Figure 1 the plane strain element.

For the shell or plane strain model one uses the plane-strain DIANA element Q8EPS.

To verify the proper responses for the present linear softening case, the element is loaded by an uniaxial straining. Up to the stress level f_{ct} the material should deform according to the linear elastic response. If the crack has developed, the increase of uniaxial strain will cause the stress decay in the material according to the linear-softening law. The tensile behavior under uniaxial straining is shown in Figure 2. The shear strain responses in the xy-direction (Figure 3) show the typical behavior of the rotating-crack type.

Thus, for a linear softening case the maximum principal stresses criterion of the Meschke damage model predicts a crack-growth mechanism comparable to the rotating-crack model. The corresponding stresses and in Figures 2 and 3, which are in resemblance with the results shown in Meschke, Mang and Lackner (1997).

In the same way the relation between the xy-direction of a plane stress test can be shown.

The last figure in this field is the stress-strain relation in y-direction.

All type of stress-strain relations show us a rather smooth distribution, what is nice. Further on we see that the σ_{yy} is starting at 0.2 N/mm^2, which is dependent of the Poisson ratio $\nu = 0.2$.

3.2   Tension-shear deformation test on solid model

The second elementary test is the same test like loading and material properties, but then with a solid element,

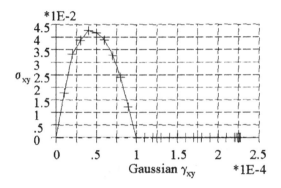

Figure 3.   Stress-strain relation in xy-direction.

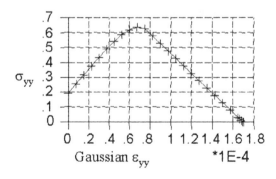

Figure 4.   Stress-strain relation in y-direction.

so a 3D test with the implemented material model of Meschke. The loading of this test can be demonstrated in the next figure.

The thickness of the plane element of 1 m is of course added to the node components in z-direction. In Figure 5 only the loading in the xy-plane is demonstrated. Also the same kind of loading can be added in yz-direction and xz-direction, to fulfill the 3D range of loadings.

Similar to the 2D example, the usual stress-strain relations can be shown in the next figures. It should

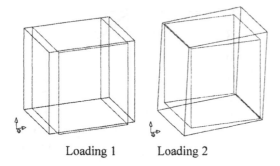

Loading 1   Loading 2

Figure 5.   Loading 3D element test.

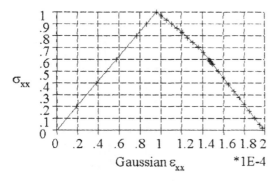

Figure 6.   Stress-strain $\sigma_{xx} - \varepsilon_{xx}$ relation 3D example.

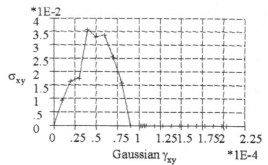

Figure 7.   Stress-strain $\sigma_{xy} - \gamma_{xy}$ relation 3D example.

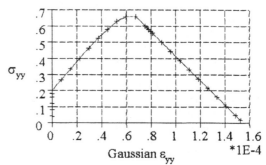

Figure 8.   Stress-strain $\sigma_{yy} - \varepsilon_{yy}$ relation 3D example.

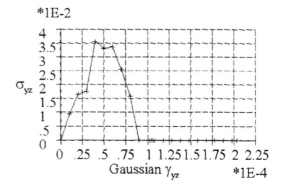

Figure 9.   Stress-strain $\sigma_{yz} - \gamma_{yz}$ relation 3D* example.

go too far on this place to show all the components of all 3 types of 3D loading cases. To minimize the pictures of these relations only the results of the loading case in yx-direction are shown.

Also these three relations between stress and strain in a 3D model under axial and shear loading are quite similar to the earlier presented relations in the 2D model. Only the stress-strain relation in the xy-direction looks a little bit different.

All figures are rather smooth, except the figure of the shear stress-strain relation. The shape and the value of the stress-strain relation is okay, but relation should be smoother. This seems to be a research point in the future.

Looking to the other type of loadings, for instance the loading in yz-direction, Figure 9 shows us the same stress-strain relation. Conclusion on this point is that the relation is independent to the way of loading. It would be deeper into the implementation of today.

4   SKEW SLAB BRIDGE UNDER COMBINED LOADINGS

The analysis of skew plate bridge has been studied extensively in the past (Feenstra, 1996). The skew slab discretization has been taken from Feenstra (1996) for both shell model and solid model.

The slab is loaded by three kinds of loading:

–   its own dead weight,
–   an uniformly distributed load and

– a number of bogies. The location of the bogies is as shown in Figure 10 by the "*".

Further, the slab is reinforced with a grid in two layers:

– the reinforcement layout at the soffit is depicted in Figure 11,
– the reinforcement layout at the top is shown in Figure 12.

The material properties of the concrete read:

Young's modulus $E = 3.13 * 10^4$ MPa,
the Poisson ratio $\nu = 0.2$,
the tensile strength $f_{ct} = 3.07$ MPa,
the fracture energy $G_f = 1.4$ J/m^2, which results in an ultimate equivalent strain $\kappa = 0.912 * 10^{-3}$ for a unit crack band width (i.e. $h = 1$ mm) and linear tension-softening.

For the steel reinforcements the following material properties have been utilized:

Young's modulus $E = 21.92 * 10^4$ MPa and a Von Mises yield criterion with a yield stress $\bar{\sigma} = 500$ MPa using the ideal plasticity assumption.

The slab bridge is first loaded with its own weight (the self-weight loading) and an uniformly distributed loading with a total load of 64 kN. Next the construction is loaded by the bogies. In the numerical simulation the self-weight loading and the uniformly distributed load is applied at the same time, divided into 2 load-steps. Then the bogies loading is applied at a number of nodes. The bogies load is incrementally increased up to approximately 40 kN (the total bogies loading is divided into 20 steps).

Conclusion of Figure 13 is that the presented Meschke model is coming to a centre displacement of 4.7 mm, instead of earlier temptations of Feenstra, who was coming till 3.5 mm.

Similarly to other examples, for the present analysis it is also found that the convergence behavior of the return-mapping algorithm (at integration-point level) is quite robust with no (noticeable) difficulty finding the convergence solution.

5 CONCLUDING REMARKS

This paper considers Meschke's damage model which is based on the Rankine yield criterion and the Meschke computational approach.

The fortran code for the present Meschke damage model has been implemented into DIANA by means of the user-supplied subroutine.

A drawback of the current implementation (through user-supplied subroutine USRMAT) is that

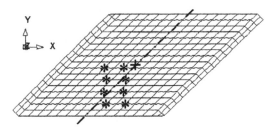

Figure 10. Upper view of the skew slab.

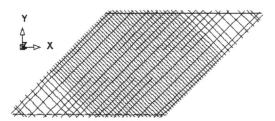

Figure 11. Soffit reinforcement.

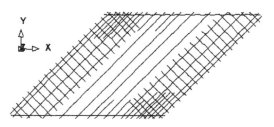

Figure 12. Top reinforcement.

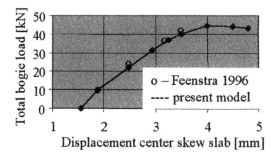

Figure 13. Displacement-bogie load relation middle of the skew slab.

plastic strains and crack directions cannot be visualized by means of a graphic, it is only available in a tabular form.

For graphical visualization, either the postprocessing features of DIANA with respect to user-supplied materials should be improved or the Meschke damage model has to be implemented inside DIANA. Based on the present numerical simulations, it can be concluded that the proposed Meschke's damage model is more stable than the earlier developed invariants approach (Feenstra, 1996).

Formulations of the return-mapping algorithm, the stress-update and the consistent tangent modulus in principal directions yield to a computationally effective implementation of multi-surface plasticity models.

In this way, problems associated with singularity of the Rankine multi yield-surfaces in the regions of the vertices are bypassed.

REFERENCES

G. Meschke (1996), *"Consideration of aging of shotcrete in the context of a 3-D viscoplastic material model"*, *Int. J. Num. Meth. Eng.*, **39**, 3123–3143.

P.H. Feenstra (1993), *"Computational aspect of biaxial stress in reinforced concrete"*, Ph.D. Thesis, TU-Delft, The Netherlands.

P.H. Feenstra and R. de Borst (1992), *"The Rankine plasticity model for concrete cracking"*, *Computational Plasticity – Proceedings of the 3rd Int. Conf.*, edited by D.R.J. Owen et al., Pineridge Press, Swansea, U.K., 657–668.

P.H. Feenstra (1996), TNO-Report 95-NM-R1562, TNO-Bouw, Delft, The Netherlands.

J.C. Simo (1992), *"Algorithms for static and dynamic multiplicative plasticity that preserve the classical return mapping schemes of the infinitesimal theory"*, *Comp. Meth. in Appl. Mech. Eng.*, **99**, 61–112.

J.C. Simo, J.G. Kennedy and S. Govindjee (1988), *"Non-smooth multisurface plasticity and viscoplasticity"*, *Int. J. Numer. Meth. Eng.*, **26**, 2161–2185.

K. Willam, E. Pramono and S. Sture (1987), *"Fundamental issues of smeared crack model"*, *Int. Conf. on Fracture of Concrete and Rock.*, *Texas*, 192–207.

Z.P. Bazant and L. Panula (1978), *"Practical prediction of time-dependent deformations of concrete"*, *Mater. Struct. Res. Testing;* **11**, 307–328 and **12**, 169–183.

CEB-FIP (1990), Lausanne, Switzerland.

Finite Elements in Civil Engineering Applications, Hendriks & Rots (eds.)
© 2002 Swets & Zeitlinger, Lisse, ISBN 90 5809 530 4

Fracture behaviour of high-performance concrete

V. Mechtcherine & H.S. Müller
University of Karlsruhe, Karlsruhe, Germany

ABSTRACT: The fracture mechanical properties, in particular uniaxial tensile strength f_t, Young's modulus E_0 and fracture energy G_F as well as the shape of the stress-strain and the stress-deformation relations were investigated for high-strength and normal strength concrete. In order to analyze failure mechanisms of these concretes, the roughness and the fractal dimension of the entire fracture surfaces and of their components (fractured aggregate, cement paste and cement paste-aggregate interface) were calculated. These data showed a clear correlation with fracture properties of the concretes investigated. Further, the mechanisms of the transfer of tensile stress across the crack were studied numerically. Hereby, a new method was applied to generate artificial concrete failure surfaces with a given fractal dimension. The results of the FE simulations confirmed the experimentally found correlations between the condition of the failure surface and the fracture mechanical properties of the investigated concretes.

1 INTRODUCTION

Due to its positive material properties and to its growing employment in the practice of construction, high-strength concrete became in the recent years an object of intensive research. With respect to the strength and deformation behaviour of high-strength concrete a better knowledge of its fracture properties is required to enable a realistic prediction of its failure.

In this study the effects of the concrete strength and the curing conditions on the formation and propagation of cracks in concrete were investigated. First, a series of deformation controlled uniaxial tension tests with different set-ups were performed for normal strength concrete (NSC) and high-strength concrete (HSC). From the test data the characteristics of the material response, i.e. uniaxial tensile strength f_t, Young's modulus E_0 and fracture energy G_F were evaluated.

As the second step, the entire fracture surfaces of the tested specimens were studied by means of optical analyses in order to understand fracture mechanisms and to explain the effect of the parameters under investigation on the fracture energy G_F and on the shape of the stress-crack opening diagrams (σ-w relations) derived from the tests on notched specimens. Furthermore, the components of the failure surfaces – fractured aggregate, cement paste and cement paste-aggregate interface – were investigated to provide additional information about the local failure mechanisms in

concrete. These data were correlated with the shape of the stress-strain curves (σ-ε relations) obtained from the experiments on unnotched prisms.

Finally, a new method was applied to generate artificial fracture surfaces of concrete with a fractal dimension as obtained from the fractological investigations. These artificial surfaces were applied for the FE analysis of the mechanisms of the stress transfer over discrete cracks.

2 FRACTURE MECHANICAL EXPERIMENTS

2.1 *Experimental program, preparation of specimens and test set-up*

The composition of the two types of concrete which were investigated is given in Table 1. In both mixtures ordinary Portland cement CEM I 32.5 R was used. As aggregate quartzite Rhine sand and gravel were applied. The mixture for the HSC had a silica fume content of 45 kg/m^3. By adjusting the dosage of a sodium naphthalene sulfonate type super plasticizer the same nominal spread (43–47 cm) for the both types of concrete could be achieved. Their compressive strength obtained from the tests on cube specimens at an age of 28 days was 96 MPa and 44 MPa, respectively.

Dog-bone shaped prisms with a gauge length of 250 mm were chosen to determine the uniaxial tensile strength f_t, the tangent modulus of elasticity E_0, as

Table 1. Composition of the concretes.

Concrete	w/c	Cement [kg/m³]	SF [kg/m³]	Aggregate 0–16 mm [kg/m³]	SP [kg/m³]	f_{cc}^{cube} [MPa]
HSC	0.33	450	45	1721	21	96
NSC	0.6	318	–	1811	1	44

SF = Silica fume. SP = Super plasticiser.

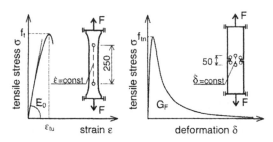

Figure 1. Schematic view of the geometry of dog-bone (left) and notched (right) specimens with typical stress-deformation relations (geometrical data in [mm]).

Table 2. Results of the uniaxial tension tests.

Concrete	f_t [MPa]	E_0 [MPa]	ε_{tu} [10⁻³]	f_{tn} [MPa]	G_F [N/m]	l_{ch} [m]
HSC	6.1	47700	0.151	5.4	162	0.21
	(0.2)	(170)	(0.005)	(0.7)	(21)	
NSC sealed	3.8	36320	0.130	2.9	135	0.34
	(0.4)	(1540)	(0.009)	(0.3)	(21)	
NSC unsealed	3.1	33060	0.128	2.7	167	0.57
	(0.2)	(750)	(0.01)	(0.4)	(24)	

Standard deviations are given in parentheses.

well as the σ-ε diagram for increasing stresses (see Fig. 1, left). Notched prisms were used to determine fracture energy G_F and the complete stress-deformation relation. Both types of specimens had the same effective cross-section 60×100 mm². A schematic view of the specimen geometries and typical stress-deformation relations are given in Figure 1 (right). Details may be found in (Mechtcherine 2000).

All specimens were cast horizontally in metal forms. After demoulding, the specimens were wrapped in a thin plastic sheet to which an aluminium foil was glued in order to protect the concrete against desiccation. For the normal strength concrete under investigation also a parallel series of specimens were stored, after demoulding, unsealed in a climatic chamber at a relative humidity of 65% and a temperature of 23°C. All specimens were tested at an age of 56 days.

To assure a stable and possibly symmetrical crack propagation in the uniaxial tension tests, stiff metal plates were glued to the specimens. Finally the metal plates were firmly connected with the bearing platens of the testing machine.

In the tests on the dog-bone shaped prisms the strain rate was controlled by means of the average signal of two LVDTs fixed to the specimens. In the experiments on the notched specimens four LVDTs with a gauge length of 25 mm were placed on the notch tips on both sides of the specimen to achieve a better deformation rate control. Two further LVDTs with a gauge length of 50 mm were placed in the middle of

the notched cross-section to provide the data for the σ-δ relation (see Fig. 1, right).

The tension tests on dog-bone shaped specimens were performed with a strain rate $\dot{\varepsilon}$ of 10^{-6} 1/s. The corresponding deformation rate $\dot{\delta}$ in the tension tests on notched specimens was $5 \cdot 10^{-5}$ mm/s.

2.2 Experimental results on strength and fracture behavior

The ascending stress-strain relation for concrete subjected to uniaxial tension has a characteristic shape as shown in the Figure 1 (left). While the relation is linear at low stresses the shape of the curve deviates from the linearity at higher stresses due to the microcrack formation, until it becomes horizontal at $\sigma = f_t$. A stable crack propagation could not be achieved with the particular test set-up used here.

According to Table 2, the tensile strength f_t, the tangent modulus of elasticity E_0 and the strain ε_{tu} (the strain at peak stress $\sigma = f_t$) of the HSC are significantly higher than the corresponding values for the NSC. The unsealed normal strength concrete showed lower f_t- and E_0-values, and slightly lower values of the strain ε_{tu} than the sealed NSC.

In the uniaxial tension tests on notched specimens also the descending branch of the σ-δ relation could be determined up to nearly complete separation of the specimens into two parts because of the localized crack due to the notches and the reduction of the gauge length for the control of deformation rate to 25 mm. The values of the net tensile strength f_{tn}, the fracture energy G_F and the characteristic length l_{ch} obtained from these experiments are listed in Table 2.

The net tensile strength f_{tn} shows a similar dependence on the concrete composition and the curing conditions as the tensile strength f_t measured on the dog-bone specimens, however f_{tn} is lower than f_t, indicating a notch sensitivity of the concretes.

The fracture energy G_F is defined as the energy per unit area needed for complete separation of a specimen into two parts. This value corresponds to the area under the σ-δ relation. The fracture energy of the high-strength concrete is higher than the corresponding value of the sealed normal strength concrete, but slightly lower than the fracture energy of the unsealed normal strength concrete.

The characteristic length l_{ch} ($l_{ch} = G_F \cdot E_0/f_t^2$) decreases with increasing strength of concrete. The specimens made of the NSC and protected from desiccation showed a smaller characteristic length than those of the unsealed concrete. The higher l_{ch}-values indicate a more ductile behavior of concrete.

3 QUANTIFICATION OF THE CONDITION OF THE FRACTURE SURFACES

3.1 Projected fringes technique

To study the effect of the concrete strength and the curing conditions on the crack propagation the fracture surfaces from the uniaxial tension test were measured using the projected fringes technique (see Fig. 3). Height differences of the surface induce a lateral displacement of the projected strip pattern. The incorporation of geometrical data of the optical configuration then allows the contour information to be detected from the phase shift of the surface strip pattern at each surface location (Wolf et al. 1996).

The measurement at intervals of 0.16 mm gives 375×625 mesh data for each failure surface. Typical contours of a fracture surface of the HSC and the NSC, sealed and unsealed, are shown in Figure 2.

3.2 Confocal microscope technique

To study the condition of the components of the fracture surfaces the confocal microscope technique was applied. In the confocal microscope a point source is imaged in the fracture surface plane. The light reflected by the surface is directed to a photo-multiplier via a small aperture (Fig. 4), which physically excludes light coming from above and below the focal plane of the microscope objective. Laser light reflected from a dichroic mirror into a scanning device moves in a defined raster in a x-y plane. The arrangement of the detector aperture ensures that only information from the focal plane reaches the detector. By coupling a step motor to the focusing unit and changing the focal

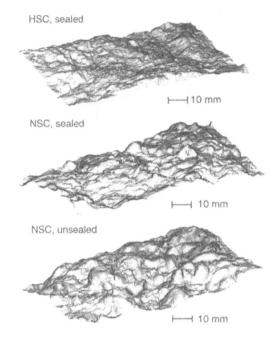

Figure 2. Typical contours of fracture concrete surfaces.

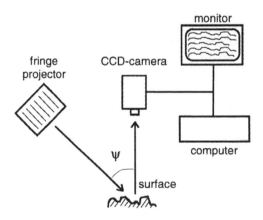

Figure 3. Principle of the projected fringes technique.

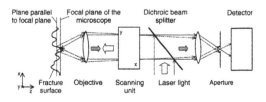

Figure 4. Basic concept of a confocal laser scanning microscope after Wilson (1990).

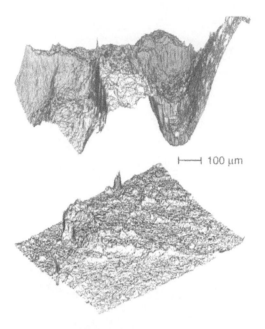

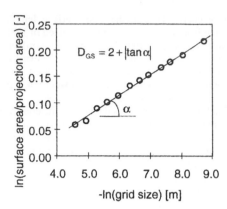

Figure 6. Plot based on the grid scaling method applied to characterize the fracture surface of a typical concrete specimen.

Figure 5. Typical contours of the fracture surfaces of cement paste (above) and cement paste-aggregate interface (below).

Table 3. Roughness and fractal dimension of concrete fracture surfaces of the concretes under investigation.

Concrete	Roughness R_S	Fractal dimension D_{GS}
HSC, sealed	1.167 (0.008)	2.029 (0.001)
NSC, sealed	1.258 (0.013)	2.044 (0.003)
NSC, unsealed	1.323 (0.040)	2.052 (0.003)

Standard deviations are given in parentheses.

plane, whole series of sectional images of the frac-ture surfaces could be performed. Three-dimensional images of the investigated surfaces were rebuild then by overlaying the images of section series.

In this study 512 × 512 mesh data have been used for each monitored spot of $0.64 \times 0.64\,mm^2$. Typical contours of the fracture surfaces of cement paste and cement paste-aggregate interface of the sealed nor-mal strength concrete are shown in Figure 5.

3.3 Determination of the roughness and the fractal dimension

From the optical measurement data the roughness and the fractal dimension of the surfaces were determined in order to quantify the condition of the entire fracture surfaces as well as of their components.

The roughness R_S of the entire fractured surface was calculated as the surface area measured with 0.16 mm mesh size and divided by the projected area. The fractal dimension was defined by the grid scaling method (Mechtcherine et al. 1995).

This method is based on the fact that the measured surface area increases as the grid size decreases. The plot of the logarithm (ln) of the measured surface area (here related to projection area) over the logarithm of the grid size gives the negative value of fractal increment = tan α (Fig. 6). The grid scaling fractal dimension D_{GS} can be calculated by adding this value to the dimension of a plane = 2.

Table 3 gives the results of the calculations. The roughness and the fractal dimension of the fracture surface increase with decreasing strength of concrete. The fracture surfaces of the unsealed normal strength concrete are rougher than those of the sealed NSC.

The roughness R_S of the components of fracture sur-faces was derived from the surface area measured with 1.252 μm mesh size. Also in this case the grid scaling method was applied to calculate the fractal dimension.

According to Table 4 the micro-roughness and frac-tal dimension of the fractured cement paste of the HSC are lower than the corresponding values of the NSC. The fractured cement paste of the unsealed NSC has the roughest surface. Also the cement paste-aggregate interface of the unsealed NSC is rougher than that of the sealed NSC. The fractured aggregates of the high-strength concrete provided rather strongly varying R_S- and D_{GS}-values. This was caused by a significant variation of the microstructure of the aggregate grains investigated.

4 DISCUSSION OF THE EXPERIMENTAL RESULTS

With increasing strength of concrete an increase of the strain ε_{tu} was observed (Table 2 and Fig. 7, above). Drying of the normal strength concrete had

Table 4. Roughness and fractal dimension of the fracture surfaces of the concrete components.

Concrete	Components	Roughness R_S [−]	Fractal dimension D_{GS} [−]
HSC,	Cement paste	1.636 (0.239)	2.087 (0.027)
sealed	Fract. aggregate	1.238–2.368	2.047–2.184
NSC,	Cement paste	1.718 (0.047)	2.097 (0.016)
sealed	CP/Agg.-Interface	1.465 (0.137)	2.080 (0.019)
NSC,	Cement paste	1.940 (0.159)	2.119 (0.011)
unsealed	CP/Agg.-Interface	1.782 (0.147)	2.111 (0.011)

Standard deviations are given in parentheses.

no significant effect on this material parameter. However, considering only the non-linear deformation component of the strain $\varepsilon_{tu,nl}$ ($\varepsilon_{tu,nl} = \varepsilon_{tu} - f_t/E_0$) the highest values are obtained for the unsealed NSC and the lowest ones for the HSC. The σ-ε relation of the NSC is more non-linear than that of the HSC due to the advanced micro-cracking which is observed for the NSC. Figure 7 (below) shows calculated non-linear strains ε_{nl} ($\varepsilon_{nl} = \varepsilon - \sigma/E_0$), i.e. strains caused by micro-cracking. The unsealed NSC provided the most pronounced non-linearity.

These results are in agreement with the observations of the development of microcracks in a normal strength and a high-strength concrete under compression using X-Ray technique (Smady & Slate 1989). Further, the non-linear strains, which represent a measure of the damage in concrete due to micro-cracking, correlate to the values of the roughness R_S and the fractal dimension D_{GS} detected for the components of the fracture concrete surfaces on the microlevel (refer to Table 4). It appears to be a consistent result, that the pronounced micro-cracking for the NSC has to cause a rougher failure surface.

The uniaxial tension tests provided for the high-strength concrete approximately 20% higher values of the fracture energy than the corresponding values of the sealed normal strength concrete. However, for the unsealed NSC a slightly higher G_F-value has been found than for the HSC. The predicted effect of the concrete strength on the fracture energy of concrete according to Model Code 1990 (CEB-FIP 1993) is much higher: for the two investigated concretes an increase of 80% is obtained. One reason may be, that the former experiments to determine the fracture energy were performed up to relatively small deformations. In this case, mainly the first, steeper part of the descending branch of the stress-deformation relation could be recorded.

Figure 8 shows the influence of the concrete strength and the curing conditions on the shape of the stress-deformation relation. For the HSC the area under the initial part of the σ-δ relation is larger than that for the NSC, because of a higher tensile strength and higher values for δ_{tu}. This indicates an increase of

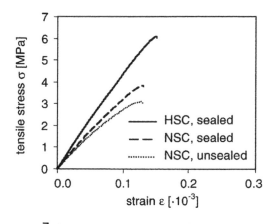

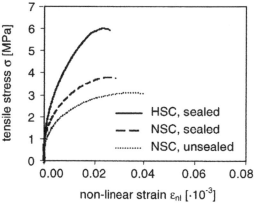

Figure 7. Measured σ-ε-relations (above) and calculated non-linear strains ε_{nl} at different stress levels (below) for the normal strength and the high-strength concretes.

energy consumption for the formation and propagation of narrow cracks with increasing strength of concrete. For larger crack widths this trend reverses, and the σ-δ relations for the NSC are above those for the HSC. This difference is small in the case of the sealed NSC, but it is significant in the case of the unsealed NSC. The condition of the fracture surfaces gives an explanation for this phenomenon: the higher roughness

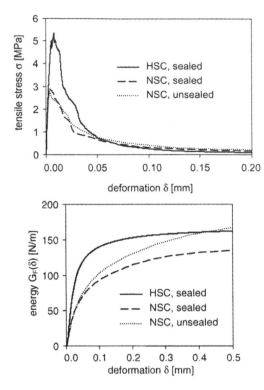

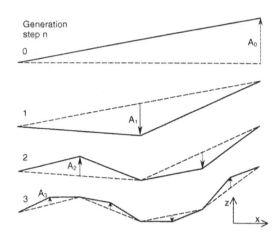

Figure 9. Principle of the midpoint displacement method.

Figure 8. Influence of the concrete strength and the curing conditions on the shape of the σ-δ relation (above) and on the energy consumption due to the fracture process (below).

and the higher fractal dimensions of the fracture surfaces of the NSC, especially of the unsealed ones (see Table 3) indicate a pronounced crack surface interlocking, which provides a better transfer of the tensile stresses across the crack.

5 NUMERICAL MODELLING

5.1 Generation of fracture surfaces

The values of the fractal dimension obtained from the fractological investigations can be used as input data for the generation of the artificial fracture surfaces. This approach was initially used for the estimation of the reliability of the values derived from fractological measurements (Mechtcherine & Müller 1999). In this study, this method was applied for the generation of FE meshes for a numerical analysis of concrete failure.

For the generation of the artificial fracture surfaces the midpoint displacement method was applied, which had been initially developed for the simulation of Brownian motion (Peitgen & Saupe 1988). Figure 9 shows schematically the first four steps in applying the one-dimensional version of the midpoint displacement

method. In the first step a straight line is generated, which one end has a zero altitude while the other end is defined as a sample of a Gaussian random variable A_0 with a mean value of zero and unit variance σ^2. In the next step the midpoint between these two end points is calculated and displaced by a random variable A_1 with a mean value of zero and variance $(\Delta_1)^2$. Further, new midpoints between the neighboring points are considered and displaced by a random value A_2, which is in average smaller than the displacements in the former steps.

The procedure can be continued until the desired resolution is achieved. The variances of the displacement variables are calculated using Equation 1, where n is the step number and H is the Hust exponent depending on the chosen fractal dimension:

$$\Delta_n^2 + \frac{\sigma^2}{(2^n)^{2H}}(1 - 2^{2H-2}) \tag{1}$$

where $H = 2 - D_L$ and D_L = fractal dimension of a line, i.e. the crack profile.

A similar procedure was applied for the generation of fracture surfaces. In this case for the calculation of the altitude of each midpoint four neighboring points in a quadratic lattice were used. The displacement variances $(\Delta_1)^2$ were derived using Equation 2 governed by the fractal dimension of a surface D_S. Details may be found in Mechtcherine 2000.

$$\Delta_n^2 = \sigma^2(\tfrac{1}{2})^{n(3-D_S)} \tag{2}$$

Figure 10 shows typical crack profiles obtained from the fracture surfaces generated using different fractal dimensions. The profile taken from the artificial surface with a fractal dimension $D_S = 2.05$, which

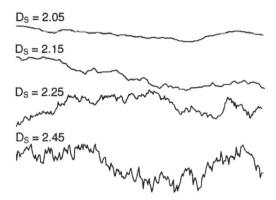

$D_S = 2.05$

$D_S = 2.15$

$D_S = 2.25$

$D_S = 2.45$

Figure 10. Typical crack profiles of generated fracture surfaces.

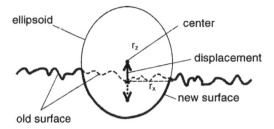

Figure 11. "Crater and hill"-method.

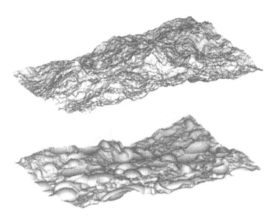

Figure 12. Example of artificial surfaces with "broken" (above) and "unbroken aggregates" (below).

approximately corresponds to the values obtained by the authors in the fractological investigations, looks rather similar to those optically measured for the high-strength concrete. The crack profiles having a considerably higher fractal dimension appear less realistic.

Figure 12 (above) shows an artificial surface, which was generated using a fractal dimension $D = 2.03$ (which corresponds approx. to the average D_{GS}-value obtained in the fractological investigations for the concrete HSC, see Table 3). Since the artificial fracture surface (Fig. 12, above) looks similar to the measured fracture surfaces of the high-strength concrete (Fig. 2, above), the midpoint displacement method seems to be quite good for the generation of fracture surfaces with dominating aggregate failure (besides high-strength concrete this is typical e.g. for lightweight aggregate concrete too). To simulate the fracture surfaces of ordinary concretes the generated surfaces have to be accomplished by some portion of unbroken aggregates.

For this purpose the "crater and hill"-method was developed on the basis of a similar technique applied for the generation of "moon landscapes" (Scholl & Pfeiffer 1991). In this method first a point on a generated fracture surface is randomly chosen. Figure 11 shows a corresponding crack profile as a sectional view of the surface in the x-z plane. In the next step this point is displaced up or down, again randomly. From now on the displaced point serves as the center of an ellipsoid with random radii r_x, r_y and r_z which are linked to the minimum and maximum aggregate size. In the following step the z-coordinates of the concerned lattice points are redefined to reproduce the shape of a "crater" or a "hill", depending on the position of the ellipsoid center with respect to the surface plane. The procedure has to be repeated so many times until the surface is provided with the desired amount of unbroken aggregates.

Figure 12 (below) gives an example of an artificial surface with a high content of "unbroken aggregates". The calculation of the fractal dimension of the generated fracture surfaces with and without "aggregate fracture" showed that the "addition" of the unbroken aggregates to the generated surfaces increased the D_{GS}-value from initially $D_{GS} = 2.03$ up to approx. $D_{GS} = 2.06$. Therefore, by means of the "crater and hill"-method the effect of the unbroken aggregate on the condition of the failure surfaces, which was observed experimentally, could be properly modeled for artificial fracture surfaces.

5.2 Numerical analysis

The generation of fracture surfaces or crack profiles is a powerful tool for a comprehensive analysis of different physical and mechanical phenomena with regard to concrete cracking (Mechtcherine 2000). In this study the artificial fracture surfaces were used for modeling the mesh in a FE analysis in order to investigate the transmission of stresses across discrete cracks in concretes with and without aggregate failure.

First a profile of a generated surface with "complete aggregate failure" was selected (Fig. 13, above).

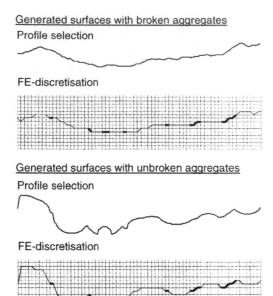

Generated surfaces with broken aggregates
Profile selection

FE-discretisation

Generated surfaces with unbroken aggregates
Profile selection

FE-discretisation

Assignment of element properties

—— mortar
----- aggregate
—— bond

Figure 13. Application of crack profiles in a FE analysis.

Table 5. Input parameters for the components of the concrete fracture surfaces.

	Input parameter			
	Model 1a	Model 2a	Model 1b	Model 2b
E_0 [GPa]	35	35	45	35
$f_{t,mortar}$ [MPa]	4	4	7	6
$G_{F,mortar}$ [N/m]	60	60	70	60
$f_{t,agg}$ [MPa]	4	–	7	–
$G_{F,agg}$ [N/m]	60	–	100	–
$f_{t,bond}$ [MPa]	–	4	–	2
$G_{F,bond}$ [N/m]	–	60	–	60

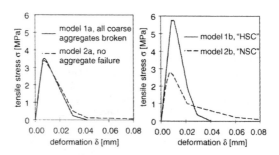

Figure 14. Effect of the crack profile geometry (left) and of the properties of the concrete components (right) on the shape of the calculated stress-deformation relations.

After having added some "craters" and "hills" to the surface a new profile of the same section was chosen again (Fig. 13, middle). Then both profiles were discretized using interface elements. Finally, the properties of the concrete components, i.e. mortar, coarse aggregate and bond were assigned to the corresponding finite elements. For this purpose both discretized profiles were compared. The common elements represented the mortar matrix. Other elements were considered to belong to the aggregates in the case of the first profile and to the bond zone in the case of the second profile (Fig. 13, below).

In the first series of calculations exactly the same fracture mechanical properties, i.e. tensile strength f_t and fracture energy G_F were assigned to all elements, see Table 5, models 1a and 2a. As a result, only the effect of the crack geometry could be studied. In the analysis a linear stress-crack opening relation was assumed. The comparison of the calculated σ-δ relations shows that the model with unbroken aggregates provided a better stress transfer across the crack than the model with aggregate failure, see Figure 14 (left).

The difference becomes much more evident if more realistic values of the tensile strength and the fracture energy for the concrete components are used, see Table 5. In this calculation series model 1b was considered to represent high-strength concrete and model 2b normal strength concrete, respectively. As a result, the obtained σ-δ relations for both concretes showed a reasonably good agreement with the corresponding curves from the experiments for small as well as for large crack openings, compare Figures 8 (above) and 14 (right).

6 SUMMARY AND CONCLUSIONS

The investigated high-strength concrete has a higher strain ε_{tu} at tensile strength f_t, and when approaching the tensile strength, it has a lower value of the non-linear strain $\varepsilon_{tu,nl}$ than a normal strength concrete. The strongest non-linearity of the σ-ε relation, indicating the most advanced micro-cracking, was observed for the unsealed normal strength concrete. These observations correlate well with the results from the optical measurements of the micro-roughness of the fracture surfaces: the high-strength concrete provided the lowest and the unsealed normal strength concrete the highest values of the micro-roughness R_S and the fractal dimension D_{GS} on the microlevel.

The higher roughness R_S and the higher fractal dimension D_{GS} of the entire fracture surfaces of the NSC, especially of the unsealed one, indicate a better transfer of tensile stress across the crack. This results in a less steep descending branch of the σ-δ relation and in a higher energy consumption at larger crack widths in comparison to the high-strength concrete.

The generation of fracture surfaces was applied for the FE analysis of mechanisms of the stress transfer over discrete cracks. The results of the FE simulations confirmed the experimentally found correlations between the condition of the failure surface and the fracture mechanical properties of the investigated concretes.

REFERENCES

CEB-FIP Model Code 1990, *CEB Buletin D'Information* 213/214, Lausanne.

Mechtcherine, V. 2000. *Investigations on crack spreading in concrete.* Doctoral thesis, University of Karlsruhe, in German.

Mechtcherine, V., Müller, H.S. 1999. Fractological and numerical investigations on the fracture mechanical behaviour of concrete. F.H. Wittmann (ed.), *Proceeding of EUROMAT 99*, 6: 119–124. Weinheim: Wiley-VCH.

Peitgen, H.-O., Saupe, D. 1988. *The Science of Fractal Images*, New York: Springer-Verlag.

Scholl, R., Pfeiffer, O. 1991. *Natur als fraktale Grafik.* München: Markt&Technik Verlag.

Smadi, M.M., Slate, F.O. 1989. Microcracking of high and normal strength concretes under short- and long-term loadings. *ACI Materials Journal* 86 (2): 117–127.

Wilson, T. 1990. *Confocal microscopy.* London: Academic Press Ltd.

Wolf, Th., Gutmann, B., Weber, H. 1996. Fuzzi logic – a new tool for 3-D displacement measurements. *International Society for Optical Engineering, SPIE-Proceedings* 2782: 363–375.

Finite Elements in Civil Engineering Applications, Hendriks & Rots (eds.)
© *2002 Swets & Zeitlinger, Lisse, ISBN 90 5809 530 4*

A coupled damage-viscoplasticity model for concrete fracture

L.J. Sluys
Faculty of Civil Engineering and Geosciences, Delft University of Technology, The Netherlands

J.F. Georgin, W. Nechnech & J.M. Reynouard
URGC-Structures, Institut National des Sciences Appliquées de Lyon, France

ABSTRACT: A coupled damage-viscoplasticity model has been developed in order to overcome shortcomings of the viscoplastic model in simulating the failure process of softening materials. The coupled model improves the fracture modeling since a stress-free crack can be simulated which grows from a localization zone of finite width into a macro-crack. This is demonstrated by means of an analytical and computational study.

1 INTRODUCTION

Viscoplasticity is known to be a good concept for the computational modelling of failure. The introduction of viscous terms in the constitutive model introduces a length scale effect and solves mesh dependence in localisation problems in statics and dynamics (Needleman 1988; Sluys 1992). However, the use of a viscoplastic model in fracture problems shows two deficiencies. Firstly, the viscous contribution to the stress causes that we can not obtain a stress-free crack at full crack opening. Secondly, the width of the localisation zone, which is set by the viscous length scale, does not converge into a true discontinuous crack. The width of the localisation zone or fracture process zone remains more or less constant. The coupling of damage to plasticity is normally performed to obtain a good behaviour under cyclic loading conditions. However, in this paper also the good characteristics of the coupled model for the description of localized processes are being demonstrated. When we couple the viscoplasticity model of Duvaut-Lions type with an isotropic damage model the abovementioned problems are solved in a natural fashion. The finite element computations with the model show that a stress-free crack can be simulated. Furthermore, the width of the localisation zone approaches zero, which implies that a true discontinuity or macro-crack is simulated. The results will be explained by means of a mathematical and algorithmic treatment of the model.

2 VISCOPLASTICITY MODEL FOR CONCRETE

2.1 *Model formulation*

The stress-strain relations for the viscoplasticity model and it's rate independent backbone are

$$\sigma = \mathbf{E}_0 \left(\varepsilon - \varepsilon^{\mathrm{vp}} \right) \qquad (1)$$

$$\bar{\sigma} = \mathbf{E}_0 \left(\varepsilon - \varepsilon^{\mathrm{p}} \right) \qquad (2)$$

in which $\bar{\sigma}$ is the projection of the current stress σ on the yield surface and ε, ε^{p} and $\varepsilon^{\mathrm{vp}}$ are the total, plastic and viscoplastic strain, respectively. Viscoplasticity is formulated by means of a Duvaut-Lions approach (Sluys 1992) according to

$$\dot{\varepsilon}^{\mathrm{vp}} = \frac{1}{\eta} \, \mathbf{E}_0^{-1} \left(\sigma - \bar{\sigma} \right) \qquad (3)$$

where η is the viscosity parameter. The rate derivative of Eq. (1) leads to

$$\dot{\sigma} = \mathbf{E}_0 : \left(\dot{\varepsilon} - \dot{\varepsilon}^{\mathrm{vp}} \right) \qquad (4)$$

Substituting Eq. (3) into Eq. (4) yields the differential equation

$$\dot{\sigma} + \frac{\sigma}{\eta} = \mathbf{E}_0 : \dot{\varepsilon} + \frac{\bar{\sigma}}{\eta} \qquad (5)$$

The hardening parameter κ is defined in a similar fashion

$$\dot{\kappa} = \frac{1}{\eta}\left(\kappa - \bar{\kappa}\right) \qquad (6)$$

in which $\bar{\kappa}$ is determined by the rate-independent plastic strain history. Eq. (6) can be solved analytically via

$$\kappa^{t+\Delta t} = \kappa^t e^{-\Delta t/\eta} + \left(1 - e^{-\Delta t/\eta}\right)\bar{\kappa}^{t+\Delta t} \qquad (7)$$

The solution of the homogeneous equation in Eq. (5) with a particular solution yield the equation for the stress update

$$\sigma^{t+\Delta t} = \sigma^t e^{-\Delta t/\eta} + \left(1 - e^{-\Delta t/\eta}\right)\bar{\sigma}$$
$$+ \frac{1 - e^{-\Delta t/\eta}}{\Delta t/\eta}\mathbf{E}_0 : \Delta\varepsilon \qquad (8)$$

The plastic response is characterized in the stress space and the yield surface is given by

$$F(\bar{\sigma}, \kappa) \leq 0 \qquad (9)$$

A non-smooth multisurface criterion is used to describe the dissymmetrical material behaviour in tension and compression of concrete (Feenstra 1993). The employed yield surfaces F_i, are function of invariants of the stress tensor $\bar{\sigma}$ and the hardening parameter κ. For tension, a Rankine yield function is used

$$F_t(\bar{\sigma}, \kappa)_t = \bar{\sigma}_t - \bar{\tau}_t(\kappa_t) \qquad (10)$$

and for compression a Drucker-Prager yield function is used

$$F_c(\bar{\sigma}, \kappa_c) = J_2(\bar{s}) + \beta_1 I_1(\bar{\sigma}) - \beta_2 \bar{\tau}_c(\kappa_c) \qquad (11)$$

where $\bar{\sigma}$ is the major principal stress, $I_1(\bar{\sigma})$ is the first invariant of the stress tensor, $J_2(\bar{s})$ is the second invariant of the deviatoric stress tensor $\bar{s}$, β_i ($i = 1, 2$) are two multiplying factors and $\bar{\tau}_x$ is the equivalent stress in tension or in compression. The cohesion capacity of the material given by the $\bar{\tau}_x$ is expressed by an analytically convenient function that is valid for tension and compression. It is consistent with the fact that experimentally observed stress-strain curves tend to attain zero-stress level asymptotically and is chosen according to

$$\tau_x = f_{x0}\left[\left(1 + a_x\right)\exp\left(-b_x\kappa_x\right) - a_x\exp\left(-2b_x\kappa_x\right)\right] \qquad (12)$$

in which a_x and b_x are material parameters and f_{x0} is the initial tensile ($x = t$) or compressive ($x = c$) strength.

At the multisurface corners in the stress space, the ambiguity of the plastic flow direction is removed using Koiter's rule (Koiter 1953; Maier 1969) by considering the contribution of each individual loading surface separately:

$$\dot{\varepsilon}^p = \sum_{i=1}^{i=2} \dot{\lambda}_i \frac{\partial F_i}{\partial\bar{\sigma}} \qquad (13)$$

where $\dot{\lambda}_i$ is the plastic multiplier associated to the plastic potential function F_i in tension or in compression.

2.2 Regularisation aspects

The introduction of rate dependence in the plasticity model prevents the model from becoming ill-posed when strain softening takes place. It introduces a length scale parameter in the problem which is dependent on η and sets a finite width to the localisation zone (the fracture process zone). The length scale effect in the Duvaut-Lions viscoplastic model is constant (Sluys 1992). For this reason, the width of the band is constant and does not narrow and finally collapse into a macro-crack of zero width when the strain reaches the ultimate strain. Furthermore, since the strain rate at ultimate strain is unequal to zero we have a viscous stress component. This viscous contribution of the stress causes that we cannot obtain a stress-free crack at ultimate strain. If the strain rate is increasing even some rehardening effects can be observed in the crack. Both the narrowing localisation zone and the stress-free crack are features that should be modelled with the coupled damage-plasticity model.

3 COUPLED DAMAGE-VISCOPLASTICITY MODEL FOR CONCRETE

3.1 Model formulation and algorithmic aspects

The damage variable, associated to concrete failure processes, can be interpreted as the surface density of materials defects (Kachanov 1986; Ju 1989), and will be defined as the ratio between the area occupied by created micro-cracks and the overall material area. This definition states that the damage variable is a non decreasing parameter, since the reduction of the effective resisting section area will continuously increase until failure occurs:

$$\tilde{S} = (1 - D)S \qquad (14)$$

where D is the damage variable.

The stress-strain relationship in a coupled damage-viscoplastic medium (4) becomes

$$\sigma = (1 - D)\mathbf{E}_0 : \varepsilon^e = \mathbf{E} : \left(\varepsilon - \varepsilon^{vp}\right)$$
$$\sigma = (1 - D)\tilde{\sigma} \tag{15}$$

where σ is the nominal stress tensor and $\tilde{\sigma}$ is the effective stress tensor. We assume an isotropic scalar damage model. The degree of brittleness of the mechanical effect of progressive micro-cracking due to external loads is described by the single internal scalar variable D which degrades the current Young's modulus of the material such as the stiffness tensor reads

$$\mathbf{E} = (1 - D)\mathbf{E}_0 \tag{16}$$

The damage evolution has an exponential form

$$1 - D_x = \exp(-c_x \kappa_x) \tag{17}$$

dependent on the cumulated viscoplastic strain (see Eq. (6)), where c_x is a material parameter (Lee 1998; Meftah, Nechnech et al. 2000; Nechnech 2000). In order to describe different behaviour under tensile loading (where subscript x = t) and compressive loading (where subscript x = c) as observed in test data, the mechanical damage variable is subdivided into two parts, one for tensile loading and one for compressive loading (Mazars 1984; Lee 1998):

$$D(\kappa, \tilde{\sigma}) = 1 - (1 - D_c)(1 - D_t) \tag{18}$$

Once micro-cracks are initiated, local stresses are redistributed to undamaged material micro-bonds over the effective area. Thus, effective stresses of undamaged material points are higher than nominal stresses. Accordingly, it appears reasonable to state that plastic flow occurs only in the undamaged material micro-bounds by means of effective quantities (Ju 1989).

The rate derivative of Eq. (15) leads to

$$\dot{\sigma} = (1 - D)\mathbf{E}_0 : \left(\dot{\varepsilon} - \dot{\varepsilon}^{vp}\right) - \dot{D}\mathbf{E}_0 : \left(\varepsilon - \varepsilon^{vp}\right) \tag{19}$$

Substituting Eq. (3) into Eq. (15) yields to the modified differential equation (Cf. Eq. (5))

$$\dot{\sigma} + \left(\frac{1}{\eta} + \frac{\dot{D}}{1-D}\right)\sigma = (1 - D)\,\mathbf{E}_0 : \dot{\varepsilon} + \frac{1}{\eta}\tilde{\sigma} \tag{20}$$

The hardening parameter κ is updated with Eq. (7). The stress update for the damage-viscoplastic model

is obtained by a Euler numerical approach where the stress rate is determined with an approximate value from Eq. (20):

$$\sigma^{t+\Delta t} = \left(1 - \frac{\Delta t}{\eta} - \frac{\Delta\dot{D}}{1 - D^{t+\Delta t}}\right)\sigma^t$$
$$+ \left(1 - D^{t+\Delta t}\right)\mathbf{E}_0 : \Delta\varepsilon + \frac{1}{\eta}\tilde{\sigma} \tag{21}$$

3.2 Regularisation aspects

If we differentiate Eq. (20) for a one-dimensional coupled damage-viscoplasticity element with respect to x and use the kinematic expression:

$$\varepsilon = \frac{\partial u}{\partial x} \tag{22}$$

and the one-dimensional equation of motion:

$$\frac{\partial\sigma}{\partial x} = \rho\frac{\partial^2 u}{\partial t^2} \tag{23}$$

with ρ the density, we obtain:

$$\eta\left(\rho\frac{\partial^3 u}{\partial t^3} - (1 - D)\mathbf{E}_0\frac{\partial^3 u}{\partial x^2\partial t}\right)$$
$$= \frac{\partial\bar{\sigma}}{\partial x} + \rho\left(1 + \frac{\eta\dot{D}}{1 - D}\right)\frac{\partial^2 u}{\partial t^2} \tag{24}$$

which is the wave equation for a coupled damage-viscoplastic element. We can distinguish three cases:

(i) Rate independence
     For this case $\eta = 0$ and the first and second term in Eq. (24) cancel. The problem is ill-posed in case of statics and dynamics.

(ii) Rate dependence-statics
     The inertia terms one and four cancel from Eq. (24) in the static case. The behaviour is set by the remaining third-order term. The problem is well-posed but approaches the ill-posed limit when the viscosity becomes zero ($\eta\rightarrow 0$) or the material is fully damaged (D$\rightarrow$1).

(iii) Rate dependence-dynamics
     All terms appear in the equation (24), but the behaviour is set by the two third-order terms. The problem remains well-posed if $\eta > 0$.

From the second case it can be concluded that for the static case the regularizing effect, which is constant for the viscoplastic model (see part. 2.2), decreases upon increasing damage. This results in a narrowing localisation zone and a stress drop to zero at full crack opening as will be explained for the bar problem in section 4.2.

4 COMPUTATIONAL RESULTS

4.1 Model response

The constitutive behaviour of the model in tension and compression is shown in Figures 1 and 2 for different values of the viscosity parameter η. Mechanical characteristics are the Young's modulus E = 35000 MPa, the compression strength fc = 40 MPa, the tensile strength ft = 4 MPa, the tension and compression fracture energies Gt = 0.035 Nmm/mm^2 and Gc = 1 Nmm/mm^2.

4.2 Bar in tension

A bar (Fig. 3) is subjected to an imposed displacement for different meshes (20, 50 and 100 elements). The damage viscoplasticity model gives the same force displacement curve for the three meshes (see Fig. 4). Furthermore, we can observe that the stress drops to zero after the failure which is in agreement with the explanation in section 3.2. A slight mesh dependence appears when the stress is almost zero. In this case the length scale (= regularizing) effect is almost zero and the corresponding width of the localization zone is close to the finite element size. From the stroboscopic evolution of the axial strain in Figures 5 and 6 we can see that the crack band narrows when the strain increases (see explanation in section 3.2). The band width even becomes smaller than the finite element size, which causes that the strain is more localized for the analysis with 100 elements (Fig. 5) than for the analysis with 20 elements (Fig. 6).

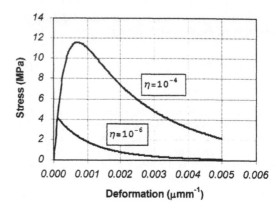

Figure 2. Tension behaviour function of η ($\dot{\varepsilon} = 3\,\mathrm{s}^{-1}$).

Figure 3. Geometry of the uniaxial bar.

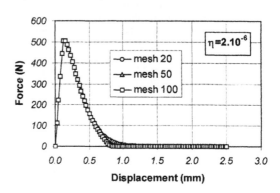

Figure 4. Uniaxial force displacement response.

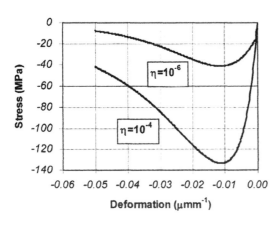

Figure 1. Compressive behaviour function of η ($\dot{\varepsilon} = 30\,\mathrm{s}^{-1}$).

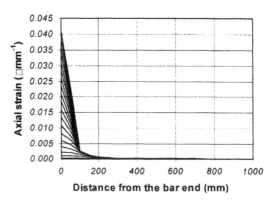

Figure 5. Axial strain gradient (mesh 20).

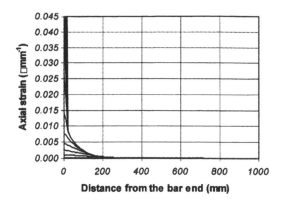

Figure 6. Axial strain gradient (mesh 100).

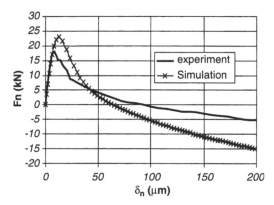

Figure 7. F_n-δ_n response for the 200 mm specimen.

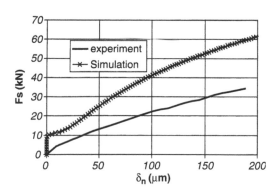

Figure 8. F_s-δ_s response for the 200 mm specimen.

4.3 Nooru-Mohamed test

In order to demonstrate the capability of this approach to predict concrete fracture, an experimental test (Nooru-Mohamed, Schlangen et al. 1993) was simulated. The Figures 7 and 8 show the force-displacement

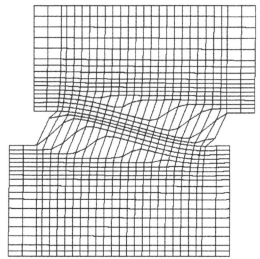

Figure 9. Simulated crack pattern.

response of the coupled model compared to the experimental results for load path 2 with δ_n/δ_s ratio equal to 1 (see (Nooru-Mohamed, Schlangen et al. 1993)). Moreover, the simulated crack patterns (Fig. 9) were found to be identical to the observed crack patterns.

5 CONCLUSION

This paper showed that a coupled damage-viscoplasticity model is a good concept for the computational modeling of failure. The two main points of this work are the ability of the model to preserve the well-posed problem and the non constant regularizing aspect. Numerical results demonstrate the narrowing localization zone and a stress drop to zero at full crack opening which is in accordance with the physics.

Future work will primarily be oriented to improve the algorithmic treatment presented in this study. To have time step independence, we practically observe that the ratio $\Delta t/\eta$ must be less than 0.5 which may be disadvantageous for simulating quasi-static problems. Furthermore, comparisons between experiments and modeling taking into account a true time scale are necessary to quantify the concrete viscosity parameter.

REFERENCES

Feenstra, P.H. Computational aspects of biaxial stress in plain and reinforced concrete Ph.D institute of technology Delft 1993.

Ju, J.W. (1989). "On Energy-Based Coupled Elastoplastic Damage Theories: Constitutive Modeling and Computational Aspects." *Int. J. Solids Struct.* **25**(7): 803–833.

Kachanov, L.M. (1986). *Introduction to continuum damage mechanics*. Dordrecht, Martinus Nijhoff.

Koiter, W.T. (1953). "Stress-strain relations, uniqueness and variational theorems for elastic-plastic materials with a singular yield surface." *Q. Appl. Math* **3**.

Lee, J. Theory and implementation of plastic-damage model for concrete structures under cyclic and dynamic loading Ph.D University of California Berkeley 1998.

Maier, G. (1969). "Linear flows-laws of elastoplasticity: a unified general approach." *Lincei-Rend. Sci. Fis. Mat. Nat.* **47**: 266–276.

Mazars, J. Application de la mécanique de l'endommagement au comportement non linéaire et a la rupture du béton de structure Ph.D Université Paris VI Cachan 1984.

Meftah, F., W. Nechnech, et al. (2000). *An elasto-plastic damage model for plain concrete subjected to combined mechanical and high temperatures loads*. 14th Engineering Mechanical Conference (A.S.C.E), Austin U.S.A.

Nechnech, W. Contribution à l'étude numérique du comportement du béton et des structures en béton armé soumises à des sollicitations thermiques et mécaniques couplées – Une approche thermo-élasto-plastique endommageable Ph.D INSA LYON 2000.

Needleman, A. (1988). "Material rate dependence and mesh sensitivity on localisation problems." *Comp. Meth. Appl. Mech. Eng.* **67**: 69–86.

Nooru-Mohamed, M.B., E. Schlangen, et al. (1993). "Experimental and Numerical Study on the Behavior of Concrete Subjected to Biaxial Tension an Shear." *Advanced Cement Based Materials* **1**: 22–37.

Sluys. L.J. Wave propagation, localisation and dispersion in softening solids. Ph.D. University of Technology Delft 1992.

Finite Elements in Civil Engineering Applications, Hendriks & Rots (eds.)
© 2002 Swets & Zeitlinger, Lisse, ISBN 90 5809 530 4

Application of the microplane model for three-dimensional reversed-cyclic analysis of reinforced concrete members

M. Hoehler & J. Ožbolt
Institute of Construction Materials, Stuttgart, Germany

ABSTRACT: This paper presents the application of a microplane based finite element approach for three-dimensional, nonlinear, reversed-cyclic analysis of reinforced concrete members. The constitutive relations and the finite element model used in the analysis are first described. The model's performance is assessed by comparison with experimental work performed previously at the University of California, Berkeley. The method is shown to accurately represent global hysteretic behavior up to failure for cases with a limited number of cycles, while simultaneously providing information about local behavior such as concrete cracking and steel to concrete bond interaction. Issues related to the modeling of steel to concrete bond and the importance of accurate representation of boundary conditions for three-dimensional cyclic analysis are addressed.

1 INTRODUCTION

The goal of this study was to investigate the performance of the program MASA for detailed three-dimensional (3D) modeling of reinforced concrete structural components subjected to reversed-cyclic loads and to target areas where further development of the program might be necessary. MASA is a Finite Element (FE) Method based computer program developed at the University of Stuttgart that is capable of 3D, fully nonlinear analysis of concrete and concrete-like materials and reinforced concrete structures. The program is based on the microplane model with relaxed kinematic constraint, which is a continuum-based model that incorporates damage and cracking phenomena (Ožbolt et al. 2001). MASA has hereto been shown to perform exceptionally well in comparisons with commercial and research finite element codes for the simulation of pre- and post-peak behavior of monotonically loaded concrete and reinforced concrete (EDF 2001, Ožbolt et al. 2000). While the microplane material formulations are capable of 3D cycling, few applications to practical problems with reversed-cyclic loading exist (Ožbolt et al. 1998).

Following a brief discussion of the material models used in the study, results from a numerical investigation of two reinforced concrete beam members that were previously tested by Ma et al. at the University of California, Berkeley (Ma et al. 1976) are presented.

2 MATERIAL MODELS

Three material models were used in this study: a microplane material model for concrete, a microplane smeared bond material model and a trilinear steel material model. A brief discussion of these material models for 3D cyclic analysis is presented in this section.

2.1 Microplane model

The microplane model is a three-dimensional, macroscopic model in which the material is characterized by uniaxial relations between the stress and strain components on planes of various orientations. At each element integration point these "microplanes" can be imagined to represent damage planes or weak planes of the microstructure (see Fig. 1a). In the model the tensorial invariance restrictions need not be directly enforced. They are automatically satisfied by superimposing the responses from all microplanes in a suitable manner.

Ožbolt et al. recently proposed an advanced version of the microplane model for concrete (Ožbolt et al. 2001). This model is based on the so-called "relaxed kinematic constraint" concept. In the model, a microplane is defined by its unit normal vector of components n_i (see Fig. 1b). Normal and shear stress and strain components (σ_N, σ_T; ε_N, ε_T) are considered on each plane. The shear stress and strain (σ_T, ε_T) are composed of two mutually perpendicular components

a)

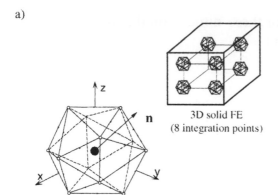

Unit-volume sphere (at integration point)

b)

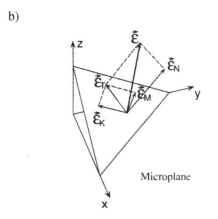

Microplane

Figure 1.   The microplane concept: a) unit-volume sphere at a finite element integration point; b) microplane stress-strain components.

σ_M, σ_K and ε_M, ε_K, respectively. These components are designated in tensor notion below by the index r. Microplane strains are assumed to be the projections of the macroscopic strain tensor ε_{ij} (kinematic constraint). Based on the virtual work approach, which represents a weak form of the equilibrium between macroscopic response and microplane responses, the macroscopic stress tensor is obtained as an integral over all possible, previously defined, microplane orientations:

$$\sigma_{ij} = \frac{3}{2\pi}\int_\Omega \sigma_N n_i n_j \, d\Omega$$
$$+ \frac{3}{2\pi}\int_\Omega \frac{\sigma_{Tr}}{2}(n_i\delta_{rj} + n_j\delta_{ri})\,d\Omega \qquad (1)$$

To realistically model concrete, the normal microplane stress and strain components must be decomposed into volumetric and deviatoric parts ($\sigma_N = \sigma_{I'} + \sigma_D$,

$\varepsilon_N = \varepsilon_{I'} + \varepsilon_D$), which leads to the following expression for the macroscopic stress tensor:

$$\sigma_{ij} = \sigma_{I'}\delta_{ij} + \frac{3}{2\pi}\int_\Omega \sigma_D n_i n_j \, d\Omega$$
$$+ \frac{3}{2\pi}\int_\Omega \frac{\sigma_{Tr}}{2}(n_i\delta_{rj} + n_j\delta_{ri})\,d\Omega \qquad (2)$$

For each microplane component the uniaxial stress-strain relations are assumed as:

$$\begin{aligned}\sigma_{I'} &= F_{I'}(\varepsilon_{I'}) \\ \sigma_D &= F_D(\varepsilon_{D,eff}) \\ \sigma_{Tr} &= F_{Tr}(\varepsilon_{Tr,eff}, \varepsilon_V)\end{aligned} \qquad (3)$$

where $F_{I'}$, F_D and F_{Tr} designate nonlinear relations between the stresses and strains. As previously mentioned the microplane strains are calculated from the macroscopic strain tensor based on the kinematic constraint approach. In Equation (3), however, only effective parts of these strains are used to calculate microplane stresses (relaxation of the kinematic constraint (Ožbolt et al. 2001)). The macroscopic stress tensor is obtained from Equation (2) in which the integration over all microplane directions (21 directions) is performed numerically.

To model unloading, reloading and cyclic loading for general triaxial stress-strain states, loading and unloading rules for each microplane stress-strain component are introduced. The "virgin loading" for each microplane strain component occurs if:

$$\varepsilon\Delta\varepsilon \geq 0 \quad \text{and} \quad (\varepsilon - \varepsilon_{max})(\varepsilon - \varepsilon_{min}) \geq 0 \qquad (4)$$

where ε_{max} and ε_{min} are the maximum and minimum values of the effective microplane strain that have occurred so far; otherwise unloading or reloading takes place. In contrast to virgin loading, for cyclic loading the stress-strain relations must be written in the incremental form:

$$d\sigma = E\,d\varepsilon \qquad (5)$$

where E represents the unloading-reloading tangent modulus, which is generally defined as:

$$E = E_0\alpha + \sigma\left(\frac{1-\alpha}{\varepsilon - \varepsilon_1}\right)$$

$$\varepsilon > \varepsilon_p; \quad \varepsilon_1 = \varepsilon_p - \frac{\sigma_p}{E_0} + \beta(\varepsilon - \varepsilon_p) \qquad (6)$$

$$\varepsilon \leq \varepsilon_p; \quad \varepsilon_1 = 0$$

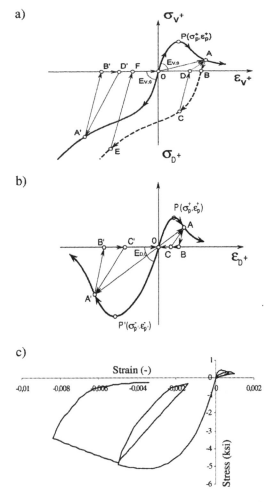

Figure 2. Cyclic stress-strain relations: a) volumetric microplane component; b) deviatoric microplane component; c) macroscopic response for a unit 3D finite element under uniaxial tension-compression load history.

In Equation (6) σ_p and ε_p denote the positive or negative peak stress and the corresponding strain for each microplane component using values σ_p^+, ε_p^+ and σ_p^-, ε_p^- to designate positive and negative peaks, respectively. The constants α and β are empirically chosen ranging between 1 and 0 and E_0 is the initial elastic modulus for the corresponding microplane component.

The loading-unloading-reloading rules for microplane components are schematically plotted in Figure 2. Figure 2a shows rules for cyclic behavior of the volumetric stress-strain component. In the compressive region the loading-unloading modulus is defined by the initial elastic modulus E_{V0}. For tension, the unloading and reloading moduli are controlled by Equation (6).

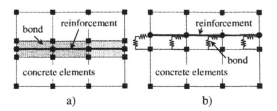

Figure 3   Modeling bond between concrete and reinforcement: a) smeared bond; b) discrete bond.

The typical load cycle for virgin loading in tension, unloading in compression and subsequent reloading in tension is: 0-P-A-B-C-D-A-B-E-F-0-A. Virgin loading in compression, unloading and subsequent loading in tension follows the path: 0-A'-B'-D'-A' or 0-A'-B'-0-P-A-B-E-F-0-A (Fig. 2a). For deviatoric compression and tension similar rules to those for the volumetric tension are employed (see Fig. 2b). The loading-unloading-reloading rules for shear are principally the same as for the deviatoric component.

The general three-dimensional cyclic response resulting from several one-dimensional microplane responses (see Fig. 2c) is an important feature of the model. The relatively simple cyclic rules allow one to better understand the macroscopic material response and to correlate it with the material structure and macroscopic stress and strain tensor. The verification of the above-mentioned cyclic rules is shown in (Ožbolt et al. 2001).

2.2   Bond model

The bond between reinforcement and concrete can be modeled using either a smeared or a discrete approach (see Fig. 3).

In the present analysis bond was represented by a layer of 3D finite elements around the reinforcement bar, i.e. a smeared approach was employed. The microplane material model was adapted for these elements such that the shear stress-strain relation yielded a realistic shear stress-slip relation. The slip was obtained by multiplying the shear strains by the element depth, i.e. by the height of the finite elements located around the reinforcement. The shear stress-slip relation used for the present studies is plotted in Figure 4. One should note that the model does not account for stiffness and strength reduction during one cycle of reversed loading, i.e. the relation is symmetric. The bond strength and stiffness are, however, reduced with repeated cycling due to accumulation of damage in the material.

2.3   Reinforcement steel

A simple trilinear cyclic material was used to represent the steel reinforcement. The skeletal curve for

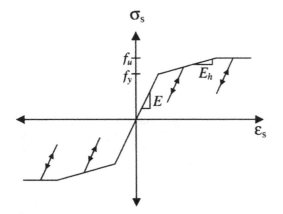

Figure 4. Shear stress-slip relation used to represent bond.

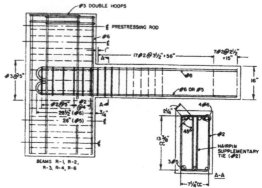

Figure 6. Reinforcement details.

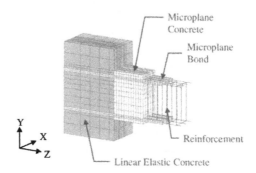

Figure 5. Trilinear cyclic steel material model.

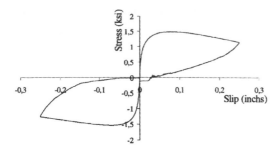

Figure 7. Section of the 3D finite element mesh for beams R-3 and R-4.

this material is shown in Figure 5. The Bauschinger effect is not accounted for in this material model.

3 TEST SPECIMENS AND FINITE ELEMENT MODEL

The specimens used for this numerical study were selected from a series of experimental tests performed at the University of California, Berkeley. Analysis of flexural beam members R-3 and R-4 investigated by Ma et al. (1976) are presented in the present paper. These beams are rectangular members adapted to represent the critical region at a beam-column interface in a lower exterior girder of a 20-story ductile moment-resisting concrete frame. The only differences between beam R-3 and R-4 were the cyclic loading history applied and a slight difference in the concrete compressive strength. Details of the beams are shown in Figure 6.

The 3D finite element mesh in Figure 7, which is shown in section to illustrate the significant features of the model, is the result of several iterations to determine the necessary mesh configuration. For the analyses, the X-axis symmetry of the structure was exploited to reduce the number of elements.

The original mesh was significantly less detailed than that shown in Figure 7. A simple fixed-end cantilever without a bond material was used. It was discovered during the study, however, that over-simplification of the model yielded a significantly different failure mechanism than was observed in the experiments. Not modeling the support block (see Fig. 7; Linear Elastic Concrete) for example led to an early onset of shearing in the member and subsequent failure.

It was found that providing a bond material for the elements surrounding the longitudinal reinforcing steel (see Fig. 7) improved the cyclic performance of the model. There are two reasons for this improved performance. The first is that in modeling bond smaller elements were provided at the steel-concrete interface and thus a better representation of cracking in this critical zone was achieved. The second reason is that the bond material somewhat decoupled the deformation of the steel from the concrete continuum.

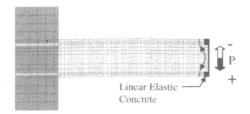

Figure 8. Shear flow at point of load application.

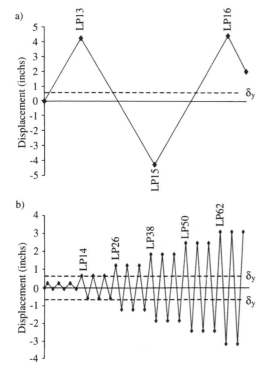

Figure 9. Applied displacement time-histories: a) beam R-4; b) beam R-3.

This allowed cracks that were opened in the tension region of the member during loading in one direction to close more easily after load reversal.

Load was applied to the member by means of a pin-type connection, i.e. free rotation, at the beam tip. Because high localized shear stresses can occur at the point of load application, a linear elastic concrete material with the geometry shown in Figure 8 was used to reduce the formation of shear cracks along a single row of elements during cycling.

All analyses were displacement controlled. Positive loading was considered to be downward bending of the beam as shown in Figure 8. The applied displacement histories shown in Figure 9 were generated based on

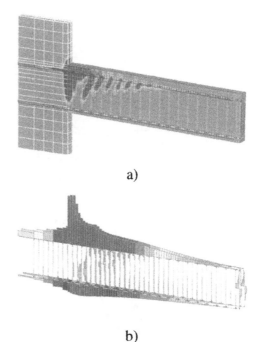

Figure 10. Virgin loading of beam R-4: a) principle strains ε_{11}; b) axial strain in reinforcement.

the information provided in Ma et al. (1976). The "LP" designations in Figure 9 are load control points used during the experimental tests and are applied in this paper as reference points. The numbering scheme is arbitrary with respect to the numerical study.

4 ANALYSIS OF RESULTS AND COMPARISON WITH EXPERIMENTAL DATA

In this section numerical results obtained using MASA are presented along with selected experimental data.

Figure 10 shows cracking and the distribution of strains in the reinforcing steel in beam R-4 shortly after yielding during the virgin loading of the member. In Figure 10b one can see the distribution of strain in the reinforcement between cracks. Unfortunately, experimentally measured steel strain distributions for the members analyzed were not available for comparison. The crack building process in the concrete is an essential feature of the initial load-displacement behavior and is captured in the analysis.

The load-displacement response is used to evaluate the ability of the program to represent the behavior of the members. Figure 11 and Figure 12 compare the load-displacement response at the member tip obtained from the numerical and experimental tests.

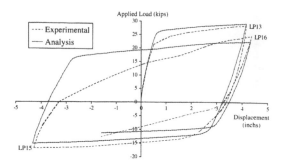

Figure 11. Load-displacement diagram for beam R-4.

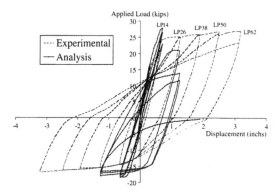

Figure 12. Load-displacement diagram for beam R-3.

Although the terminology "Applied Load" is used to coincide with the terminology in the experimental report, it is reiterated that all of the numerical analyses were carried out under displacement control.

The figures show that the hysteretic behavior is quite well represented for member R-4, however, there is a significant divergence from the experimental results for multiple reversed cycling over the yield point in member R-3. The numerical analysis of member R-3 was terminated at the fourth displacement plateau (LP38) because the member had clearly failed.

We concentrate our discussion on member R-4 because the mechanism that led to the failure of member R-3 in the numerical analysis was also present in the analysis of member R-4, however, its contribution to the ultimate failure was less pronounced for the case with a lower number of cycles.

Before addressing the failure mechanism in the members, it is noted that the sharp corners of the numerically determined load-displacement curves on the branches subsequent to the virgin loading (see Fig. 11; LP15 to LP16) can in part be attributed to the use of the trilinear material model for the reinforcing steel. A new steel material based on the Menegotto-Pinto equation, which takes the Bauschinger effect into

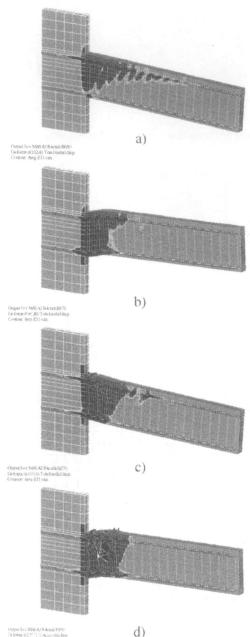

Figure 13. Principal strains in beam R-4: a) LP13; b) LP15; c) LP16; d) Failure.

account, was developed and implemented in MASA subsequent to this study (Hoehler, in press).

Figure 13 shows the computed deformations and principal strains for member R-4 at the points of load reversal. The principal strains provide a picture of cracking in the member. Figure 13a shows that the deformation up to the first load reversal is primarily flexural. At the second load reversal (Fig. 13b) one observes significant shear deformation in the member. This shear deformation is also seen in the subsequent load reversals (Figs. 13c, d).

Localized shearing was of interest during the experimental investigations and thus shear behavior was monitored even for the so-called "flexural members" R-3 and R-4. This provides the opportunity to more closely examine shearing in the members. An illustration of how an average shear angle was measured in the experiments and subsequently in the finite element analyses is shown in Figure 14.

Figure 15 shows the applied load plotted against the shear distortion γ_{av} for member R-4. While the agreement of the numerical and experimental results is quite good up to control point LP13, the shear distortion observed after the first load reversal reaches a value nearly three times that of the experimental. This confirms what is observed in Figures 13a-d. It is this shear deformation that leads to the early degradation of member R-3. The material damage caused by this shearing is more pronounced for the case with a higher number of cycles.

Throughout the study MASA was observed to be sensitive to material damage caused by shearing. In actual reinforced concrete members where simultaneous flexure and shear are present, shear forces are transferred by three mechanisms: (1) shear transfer across concrete in compression, (2) dowel action across cracks by reinforcement and (3) mechanical interlock along crack surfaces. In MASA shear is transferred primarily by the concrete in compression.

In MASA one is required to use either truss elements or fixed-end beam elements to model the reinforcing steel because no rotational degree of freedom exists in the element formulations. This method of modeling the reinforcement does not allow for shear transfer through dowel action.

The authors believe that the use of a steel material that allows for earlier re-yielding of steel through the Bauschinger effect and a bond model that more realistically represents the cyclic steel to concrete bond behavior, i.e. through an asymmetrical bond-slip relation, would reduce the amount of shear damage by allowing tensile cracks to close more easily upon load reversal. Only after the tensile cracks have closed can a compression zone for the transfer of shear loads develop.

The most significant source of the additional shear, i.e. additional in relation to the real shear that was observed in the experimental investigations (see Fig. 15), however, was likely due to the modeling of the boundary conditions at the member tip. In the presented numerical results the beam tip was allowed to displace freely in the Z-direction (longitudinal

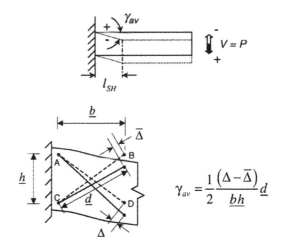

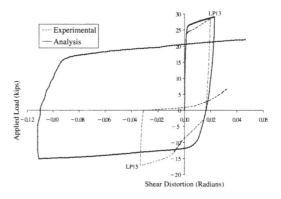

Figure 14.   Description of average shear angle γ_{av}.

Figure 15.   Load-shear distortion diagram for beam R-4.

direction). With each yielding of the longitudinal reinforcement, the beam tip was displaced in the positive Z-direction. This led to additional tensile strains at the beam-column joint (see Fig. 16). The exact boundary conditions at the beam tip in the experiment are unknown. Analysis of the members with the movement of the beam tip restrained in the Z-direction allowed for fictitiously large loads to be carried by the member and ultimately led to numerical instabilities. This issue was unresolved at the time of the writing of this article.

Extensive photographic records were maintained during the experimental investigations. A qualitative comparison of the crack patterns for members R-4 and R-3 is shown in Figure 17 and Figure 18, respectively.

In the experiments the ultimate failure mode for both members R-3 and R-4 was blowout of the concrete in compression and a buckling of the longitudinal

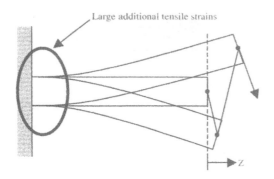

Figure 16. Mechanism resulting in additional tensile strains at the beam-column joint.

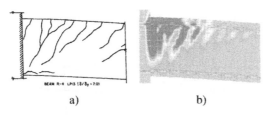

a)   b)

Figure 17. Crack patterns in beam R-4: a) experimental, $d/d_y = 7.0$; b) analysis, $d/d_y = 7.0$.

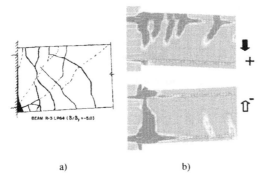

a)   b)

Figure 18. Crack patterns in beam R-3: a) experimental, $d/d_y = 5.0$; b) analysis, $d/d_y = 1.0$.

reinforcement. This failure mode was observed in the numerical investigations for member R-4 as shown in Figure 19.

5 CONCLUSIONS

In this paper the present capacity of a microplane based finite element program for the nonlinear,

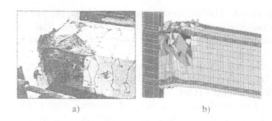

a)   b)

Figure 19. Ultimate failure mode: a) experimental (beam R-3); b) analysis (beam R-4).

three-dimensional, reversed-cyclic analysis of reinforced concrete members was presented.

A finite element model developed to study beam members R-3 and R-4 from the experiments conducted by Ma et al. was discussed. For three-dimensional cyclic modeling, accurate representation of the boundary conditions, both at the member support and at the loading point, was essential to recover the proper failure mode. Furthermore, a bond material was needed to "relax" the continuum at the steel-concrete interface.

The program was shown to represent global hysteretic behavior quite well, as long as the number of load cycles was limited. For higher numbers of cycles the damage caused by shear led to premature failure of the member. The program realistically represents cracking and provides useful qualitative and quantitative information on crack formation in the concrete and strain distribution in the reinforcement.

In the future, reinforcement elements with rotational degrees of freedom would likely reduce the shearing problem by providing a dowel effect. Furthermore, adding the Bauschinger effect to the steel model should improve the overall performance of the program for modeling of reinforced concrete. A discrete bond material with asymmetry of the stress-slip relationship dependent on the load history should improve the cyclic behavior by further separating the reinforcement elements from the continuum material formulation.

REFERENCES

Elecricité de France (EDF) 2001. *MECA benchmark – three dimensional non linear constitutive models of fractured concrete: evaluation – comparison – adaptation: comparison report #1* (draft report).

Hoehler, M. In press. Formulation and implementation of the Menegotto-Pinto cyclic steel model for the finite element program MASA. *Festschrift zum 60. Geburtstag von Prof. Eligehausen*. University of Stuttgart.

Ma, S.-Y.M. et al. 1976. *Experimental and analytical studies on the hysteretic behavior of reinforced concrete rectangular and T-beams*. Earthquake Engineering Research Center Report No. UBC/EERC 76–2. University of California, Berkeley.

Ožbolt, J. et al. 1998. 3D cyclic finite element analysis of beam-column connections. *Fracture Mechanics of Concrete Structures* 3: 1523–1536.

Ožbolt, J. et al. 2000. Compression failure of beams made of different concrete types and sizes. *Journal of Structural Engineering* 126(2): 200–209.

Ožbolt, J. et al. 2001. Microplane model for concrete with relaxed kinematic constraint. *International Journal of Solids and Structures* 38: 2683–2711.

Finite Elements in Civil Engineering Applications, Hendriks & Rots (eds.)
© 2002 Swets & Zeitlinger, Lisse, ISBN 90 5809 530 4

Application of finite element modelling to the thermo-mechanical behaviour of refractories

K. Andreev & H. Harmuth

Christian Doppler Laboratory for Building Materials with Optimised Properties,
Department of Ceramics, University of Leoben, Austria

ABSTRACT: A modelling approach for the thermo-mechanical material behaviour of bulk ceramic refractory materials has been formulated using material laws available in DIANA. This paper illustrates the application of the proposed approach for the simulation of stress fields and crack formation in refractory linings of vessels of the steel industry. Effects of thermal loading are investigated for the lining of a teeming ladle with combined monolithic and brick lining. The analysis results are used to determine mechanisms of material failure observed under service conditions. Modelling of some specific load cases, such as cyclic loading of cracked material with partially irreversible crack strains, demands extension of existing material laws.

1 INTRODUCTION

Refractories are bulk ceramic materials composed of oxide and non-oxide components with a rather coarse grain-matrix structure. Their main application is the lining of industrial furnaces, including vessels of the steel industry, where refractories are exposed to high temperatures as well as chemical and mechanical wear.

A typical microstructure of a refractory material is illustrated by a microphotograph of a magnesia carbon refractory, Figure 1. The main refractive components are magnesia and graphite. The bonding agent can be pitch or resin. The microstructure consists of rather large grains with a maximum grain size up to e.g. 5 mm imbedded in the matrix built up by the fines. The structure is porous, heterogeneous and contains defects. Usually the defect size scales with the grain size and can reach the same order of magnitude.

A teeming ladle offers a common example for a refractory lining (Fig. 2). It is used in metallurgy for

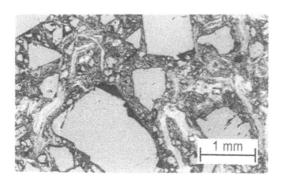

Figure 1. Typical microstructure of a magnesia-carbon brick.

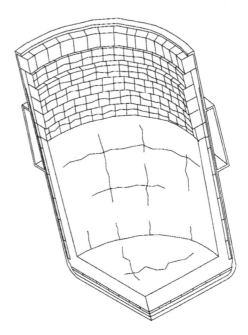

Figure 2. Teeming ladle with combined working lining with cracks caused by thermo-mechanical loads.

transport and processing of liquid steel. During the exploitation the lining is exposed to a multiple number of thermal cycles when the steel, along with the slag, is poured into the ladle, processed and tapped out of it. The slag may erode or infiltrate the lining material degrading its mechanical performance. To sustain these loads the lining of the ladle is composed of several layers of refractory products. Different elements of the lining can be composed either of shaped products such as bricks or of monolithic materials, e.g. refractory concrete.

Due to complexity and cross influence of service parameters thermo-mechanical behaviour of refractory structure is very difficult to analyse. Therefore a phenomenological design approach it still widely used for design of refractory structures and choice of service regimes. In this respect application of finite element analysis could be a more physically justified design approach. Successful implementation of FEM analysis would demand know-how in accurate representation of material properties and boundary conditions. This involves:

- choice of material laws to describe behaviour of refractory materials;
- determination of necessary material properties;
- accounting for possible change of material properties during service, e.g. infiltration, drying of monolithic products;
- representation of service loads acting on the lining, including thermal loads;
- accounting for changes in lining geometry during the service, e.g. thinning of the lining due to wear or spalling;
- reducing complexity of the lining structure for model building.

This paper reports experience obtained in modelling of thermo-mechanical behaviour of refractory lining using the commercially available FEA-code DIANA (DIANA – User's Manuel 1999).

2 REPRESENTATION OF MATERIAL PROPERTIES

2.1 Thermo-mechanical behaviour of refractory materials

The mechanical behaviour of refractory materials at room temperature has many similarities with the behaviour of concrete and rocks (Fig. 3). The characteristics of the microstructure in many cases cause deviations of the material behaviour from pure linear elastic mechanics and fracture mechanics. Failure mechanisms under tension and compression are different. Under compression micro cracking takes place and material may behave less brittle, under tension quasi-brittle failure occurs. The failure, especially

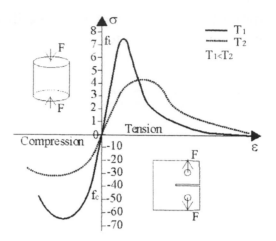

Figure 3. Failure of refractories under uniaxial tension and compression (f_t, f_c tensile and compressive strength, resp., σ – stress, ε – strain, T – temperature).

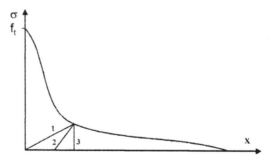

Figure 4. Strain-softening behaviour with possible unloading paths (1 – reversible, 2 – partly reversible, 3 – irreversible), σ, f_t and x are the transmitted stress, the tensile strength and the crack opening displacement, resp.

under tension, is often characterised by strain-softening, which results from the ability of the material to sustain loads even after crack initiation. During unloading irreversible displacements are observed, as the uneven crack faces do not fit together any more (Fig. 4). In the course of the crack propagation the irreversible crack opening displacement does not remain constant. It is minimal at crack initiation when faces of opening cracks are still close to each other. With further crack propagation the irreversible crack opening displacement will rise. Depending on the compressive stress state, failure may take place either due to tension or shear. Compressive strength is sensitive to the hydrostatic pressure and the dilatancy is characteristic for the plastic flow.

Under room temperatures major material properties of refractories may have following values: Young's modulus 10–70 GPa, Poison's ratio 0–0.2, uniaxial

Table 1. Parameters used for modelling the material behaviour under compression (determined at room temperature).

Material	Cohesion (MPa)	Angle (°) Friction	Angle (°) Dilatancy
Material A			
Green	12.2 ± 0.7	43.6 ± 1.0	18.4 ± 0.3
Tempered at 1000°C	9.6 ± 0.9	49.9 ± 2.4	18.1 ± 0.3
Material B			
Green	9.9 ± 0.4	36.5 ± 0.8	16.9 ± 0.3
Tempered at 1000°C	10.1 ± 0.3	36.5 ± 0.5	17.9 ± 0.4
Material C			
Fired at 1000°C	6.9 ± 3.0	43.1 ± 3.5	14.9 ± 4.2

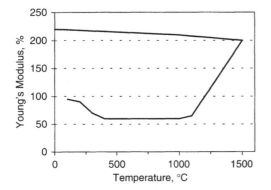

Figure 5. Normalised Young's modulus ($E/E_0*100\%$) of a high alumina castable at higher temperatures (Nonnet et al. 2001).

compressive strength 30–70 MPa, tensile strength 5–15 MPa, fracture energy in Mode I 50–350 N/m. Very little information is available on behaviour of refractories under multiaxial compression even at room temperature. The values of the angles of friction and dilatancy have been determined for only a few refractories so far. Table 1 shows some of these values. In the table Material A is a pitch bonded magnesia-carbon brick with 5% carbon, Material B is a pitch bonded magnesia-carbon brick with 9% carbon and Material C is a spinel forming castable.

Rising temperature may cause in the materials a wide range of chemical and physical processes, such as e.g. phase transformations or sintering. Due to these processes material properties will in many cases show a nonlinear change with temperature. Therefore a property is defined sufficiently only when its values are known for the whole temperature range of material application. Change of Young's modulus with temperature in a high alumina castable is given in Figure 5. The decrease of the Young's modulus taking place between the room temperature and 400°C during the first heating up procedure is due to the loss of humidity during the drying process. When the temperature rises higher than 1100°C Young's modulus rises very significantly, which is due most probably to the effects of sintering. So in this case the Young's modulus is not only dependent on the temperature itself, but also on the course of the thermal load.

Viscosity has also significant influence on material behaviour under higher temperatures. Therefore information on stress relaxation is a key property.

2.2 Material models

Due to specific material behaviour and service conditions of refractories numerical modelling of these materials and their structures is distinctive from such traditional areas of modelling as the civil or mechanical engineering. Commercially available FEM codes usually have not been developed for refractory applications. Therefore modelling of refractory structures can be enabled either by creating an own code or by using one of the general purpose commercial codes and adjusting available material laws. This investigation has been conducted by means of the commercial code DIANA. This program has been chosen after extensive bench marking as it offers a wide range of constitutive laws for modelling of brittle disordered materials.

The following material models have been implemented to describe thermo-mechanical behaviour of refractories:

– temperature dependent heat capacity and conductivity;
– temperature dependent coefficient of thermal expansion;
– temperature dependent Young's modulus;
– combined failure criterion, i.e. the criterion of Rankine for cracking under tension and the criterion of Drucker-Pragger for hydrostatic pressure sensitive plastic failure;
– temperature dependent tensile strength;
– multi-linear or exponential strain softening under tension;
– temperature dependent fracture energy;
– temperature dependent compressive strength and/or cohesion;
– non-associated plasticity;
– hardening/softening of material after plastic failure defined by cohesion and internal state variable to match experimental curves of uniaxial compression; dependence on temperature is accounted for by temperature dependent cohesion;
– temperature dependent visco-elastic material behaviour (relaxation) with discrete input of parameters of Maxwell chain;
– fully reversible or fully irreversible crack strain unloading.

Not all interesting features are readily implemented in DIANA. It may be advantageous to apply user defined subroutines to allow for partly irreversible crack strain after unloading.

A modelling approach considering a great number of non-linear material properties complicates the solution of the numerical model. In application of the approach described above for some problems convergence could be achieved only when the number of non-linear properties was reduced. The inconvergence was especially significant when non-linear analysis was combined with contact analysis.

3 REPRESENTATION OF BOUNDARY CONDITIONS

Besides the number of non-linear material properties the number of elements has a great influence on the calculation . This fact makes it very difficult to build detailed three dimensional models of refractory structures. Special modelling approaches should be developed for each problem. Such approaches will be illustrated by the analysis of the thermomechanical behaviour of a teeming ladle (Fig. 2).

The ladle consists of a steel shell lined with refractories. The lining has two layers called the working and the safety lining. The working lining is the inner layer coming into contact with liquid steel. It is made of a monolithic refractory castable in the bottom and the lower part of the wall and bricks laid without mortar in the upper part of the wall. Sometimes special expansion allowances are made between the bricks to reduce stresses due to thermal expansion. The safety lining is the layer behind the working lining. It can be made of bricks or a castable.

A detailed model of the ladle would demand description of all structural elements, including bricks and joints between them, which is almost impossible. An alternative to such a model, considering the monolithic nature of the grater part of the working lining, could be an axisymmetrical model (Fig. 6). It allows to analyse quite accurately the behaviour of the monolithic part of the ladle. Interaction between the monolithic part and the rest of the lining is modelled by contact elements. But with this model the effect of vertical joints cannot be exactly introduced. Therefore a more accurate modelling of the brick parts of the lining featuring expansion allowances in two directions is only possible by three dimensional analysis. This can be realised by a so called unit-cell modelling approach. According to this approach the complex structure of the cylindrical part of the vessel is reduced to a much smaller domain which is still representative for the whole structure. Such a model (Fig. 7) features a quarter of a refractory brick, layers of the refractory mix from the safety

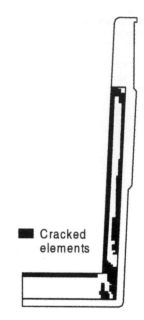

Figure 6. The axisymmetrical model showing cracks in the monolithic part of the lining after the end of the first service cycle (approx. 21.5 h. since the beginning of the heating up).

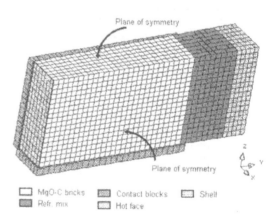

Figure 7. The unit-cell model.

lining and the steel shell. At the planes of symmetry the model is constrained perpendicular to their direction.

The mix and the shell are also constrained on both vertical faces. Two special unbreakable blocks are introduced to model the expansion allowances; they play the role of neighbouring bricks. Varying the distance between the blocks and the brick allows modelling joints of different size.

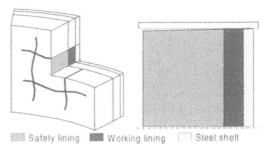

Safety lining ▮ Working lining ☐ Steel shell

Figure 9.   Model for the analysis of the plate spalling (phased analysis); the dotted and dashed line is the axis of symmetry.

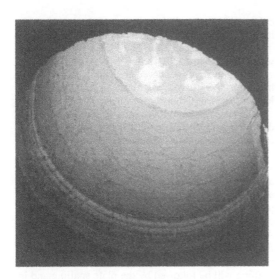

Figure 8.   Cracks on the hot face of the ladle lining.

4   ANALYSIS RESULTS

Staggered potential flow-stress analysis has been used to model thermomechanical processes taking place in the refractory lining of a teeming ladle during the initial heating of the cold ladle and the following service. Calculation of transient thermal fields in the lining was the first stage of the analysis. For this analysis the temperature of the ambient air and liquid steel were taken as 15 and 1580°C, respectively. The heat transfer due to radiation and convection was taken into account.

The thermomechanical calculation is the second stage of the analysis, in this stage previously calculated thermal fields are used as the loading. Different stress conditions develop in the lining depending on whether the temperature on the hot face is rising or falling. Heating of the lining causes thermal expansion, which is often constrained by the steel shell or elements of the lining, such as neighbouring bricks. Constrained expansion results in compressive stressing of the lining and possible failure. In the brick lining, for example, compressive failure usually causes spalling of parts of the brick lining near the hot face. During the cooling tensile stresses develop at the hot face. This may lead to formation of cracks orthogonal to the hot face. More detailed analysis of thermomechanical behaviour of refractory structures can be found elsewhere (Andreev & Harmuth 2001a, b, c).

The cyclic nature of thermal loads in the teeming ladle may lead to complex cases of lining failure. One of such cases is spalling of thin plates of material at the hot face of the monolithic lining (Fig. 8). Such plates may be 500 mm in diameter and have a thickness of 20–30 mm. During the service several plates may spall one after another causing significant thinning of the lining and the necessity of early ladle relining. Observations during service and following numerical simulations established that a net of cracks orthogonal to the hot face can develop in the monolithic lining during initial drying of the new castable after ladle relining. These cracks may extend during the phases of the service cycle when the temperature at the hot face is falling. They run parallel and perpendicular to the ladle bottom at a distance of approx. 500 mm from one another. Sometimes they cut the whole lining thickness.

The spalling of the material between cracks orthogonal to the hot face is investigated by means of a special model (Fig. 9). This model features parts of the wall lining between two cracks, the safety lining and the ladle shell. An unbreakable block is to model contact with other parts of the lining during thermal expansion. The distance between the castable and the block is equal to material shrinkage during the drying.

Thermal analysis preceded mechanical calculations. To model the initial heating up the temperature at the hot face of the lining was raised from 15 to 1260°C in 17,5 hrs., this was followed by three hour holding period at 1260°C. After that temperature was raised from 1260 to 1580°C in 400 sec. Operation cycles are modelled by simplified heating and cooling of the hot face which corresponds with filling and tapping of the ladle (Fig 10), respectively. The duration of each heating and cooling is 800 sec.

According to mechanical analysis the gap modelling the crack orthogonal to the hot face begins to close due to thermal expansion at the beginning of the heating-up period. After closing compressive stresses directed parallel to the hot face build up in the lining. The values of the stresses are not high enough to cause compressive failure even at the end of the filling of the ladle as the gap (opened crack) resulting from the shrinkage reduces the stress level. The first cooling causes formation of two crack types (Fig. 11.a). Cracks orthogonal to the hot face are caused by tensile

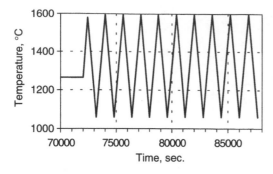

Figure 10. Thermal regime for the phased analysis.

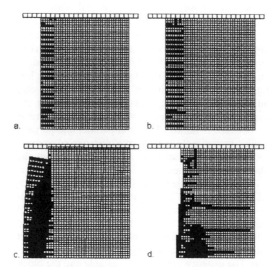

Figure 11. Crack formation in the working lining after a – the first cycle (73200 sec), b – the fourth cycle (78000 sec), c – the end of the fifth filling (78800 sec, displacements are enlarged with factor 15), d – the end of the fifth cycle (79600 sec).

stresses developing on the hot face during cooling. Cracks of the second type can occur some distance away from the hot face parallel to it. These cracks of the second type result from tensile stresses developing near the faces of the initial crack which are deformed due to rapid shrinkage of the castable in the vicinity of the hot face. The formation of these cracks is predicted 20–30 mm from the hot face.

Due to fully irreversible crack displacements defined in the model the repeated thermal shock during filling and tapping may lead to propagation of the cracks formed during the first cycle. Cracks of the two types will propagate one orthogonal and another parallel to the hot face and eventually they can intersect (Fig. 11.b). The intersection of the cracks will cause spalling of the lining corner. Under service conditions such spalling is also accelerated by the erosion due to the liquid steel bath. The spalling of the corner was modelled to happen after the end of the fourth cycle by applying the phased analysis. Thermal fields had been previously recalculated accounting for the loss of the corner after four cycles.

The loss of the corner modifies stress patterns developing in the lining. In the beginning of the fifth cycle after the loss of the corner shear stresses developed in the lining due to mismatch in thermal expansion of the parts of the lining unloaded by the spalling and those still constrained by contact with the unbreakable block. With rising temperatures shear cracks grow parallel to the hot face. Soon they span the whole height of the model and cause spalling of a plate (Fig 11.c). After this the stress distribution is similar to that of the first phase. Formation of the cracks leading to the corner spalling is observed again (Fig. 11.d). The corner spalling will be followed by the plate spalling and in this way repeated plate spalling is expected.

The proposed spalling mechanism agrees with observations. It was registered that in the ladle the major plate spalling is preceded by the spalling of smaller parts of the lining along the cracks (Fig. 8). This initial spalling along the cracks may be the corner spalling observed in the model.

5 CONCLUSIONS

Due to specific material behaviour and service conditions of refractories numerical modelling of this materials and their structures is a field distinctive from such traditional areas of modelling as the civil or mechanical engineering. An approach for thermomechanical modelling of refractory structures has been developed using the commercially available FEM program DIANA. In general DIANA has proved to be an accurate and robust tool and the developed approach has been successfully applied for the analysis of stresses and crack formation in various refractory structures. The accuracy of the analysis could probably be increased by several improvements of the constitutive laws available in DIANA. Such improvements could be heat transfer by radiation between two bodies and partially irreversible unloading of crack strains.

REFERENCES

Andreev, K. & Harmuth, H. 2001a. Calculation of crack formation in refractories – a case study. *J. of Physical Mesomechanics* 4(4): 105–111.
Andreev, K. & Harmuth, H. 2001b. Modelling of the thermo-mechanical behaviour of the lining materials of

teeming ladles. In *Proc. of 7th Unified International Technical Conference on Refractories (UNITECR), Nov. 4–8, 2001, Cancun, Mexiko* Vol. II: 830–840.

Andreev, K. & Harmuth, H. 2001c. Numerische Simulation des mechanischen und thermo-mechanischen Verhaltens der Feuerfestzustellungen von Stahlpfannen. In *Proc. of Gesteinshüttenkolloquium 2001, Okt. 19, 2001, Leoben, Österreich*: 91–99.

DIANA-User's Manuel for Release 7. 1999. TNO Building and Construction Research.

Nonnet, E. et al. 2001. In Situ X-ray and Young's Modulus Measurement during Heat Treatment of High-Alumina Cement Castables. *J. of Am. Cer. Soc.* 84(3): 583–587.

Finite Elements in Civil Engineering Applications, Hendriks & Rots (eds.)
© 2002 Swets & Zeitlinger, Lisse, ISBN 90 5809 530 4

Effect of particle density on tensile fracture properties of model concrete

G. Lilliu, A. Meda*, C. Shi
Delft University of Technology, Delft, The Netherlands

J.G.M. van Mier
ETH Hönggerberg, Zürich, Switzerland

ABSTRACT: Numerical analysis of fracture in concrete subjected to tension by means of a 3D lattice type fracture model has revealed a significant effect of the particle density on the tensile strength and the post-peak behaviour. It was found that with increasing particle density the peak-load decreases whereas the ductility increases. Experiments conducted on synthetic concrete specimens confirm the outcome from numerical simulations.

1 INTRODUCTION

Recently, a 2D lattice model used for simulating fracture processes in concrete (Schlangen & van Mier 1992) has been extended to 3D (Lilliu & van Mier 2002). The reason for this extension is two-fold: firstly, a 3D model is more appropriate than a 2D model to reproduce real material behaviour; secondly, 3D effects enhance the ductility of the lattice response, as shown with the parallel system of two 2D lattices analysed by (Vervuurt 1997). The enhanced ductility is of capital importance, since the main argument against the 2D lattice model is that its response is too brittle in comparison with the experimental results. The cause of this brittleness is that in a 2D lattice the particles are represented as cylinders, while in reality they have a nearly spherical shape. Therefore, a realistic crack surface is not cylindrical but oriented in all directions. The consequence is that the dissipated energy is larger than that predicted with a 2D model, since the area of the crack surface is larger.

Besides the 3D effects, also the internal structure of the material affects the ductility of the lattice response. This fact becomes evident when, for example, small particles are included in an initial sparse particle distribution (Schlangen 1993). These particles induce more crack branching and bridging that, as known, play a substantial role in the tail of the load-displacement diagram (van Mier 1991). Another problem related to the representation of the material structure and,

probably, to the brittle response of the lattice, was the impossibility to obtain computer-generated particle distributions as dense as in real concrete. Nowadays, this problem can be overcome by using computer programs where collision rules among the particles are adopted (Stroeven 1999).

Another possible cause of the lattice brittleness is the adopted fracture criterion, which assumes a linear-elastic behaviour of the lattice elements until failure. In order to achieve a more ductile lattice response, failure criteria based on energetic considerations have been proposed, for example (Arslan et al. 2002). According to such criterion, the matrix elements in the three-phase representation of concrete (aggregate, matrix and bond) follow a non-linear material constitutive relation. Indeed, the introduction of a softening regime at the element level can lead to a quantitative load-deformation response that better fits the experiments, but does not contribute to understanding of the physics behind fracture processes. There is no evidence of the significance of quantities like fracture energy. Even, it is doubtful whether these quantities are characteristic of the material or are just extrapolated from the global response of the structure. Although debatable, such quantities are necessary in homogeneous models, where a non-linear material constitutive relation is needed locally in order to obtain a non-linear overall response of the structure. On the contrary, structural softening may be obtained with the lattice model, despite the adopted brittle fracture criterion. Thus, it is opinion of the authors that the introduction of a softening regime at element level is a superfluous complication of the original lattice model, needed only because of the rough

---

*On leave from University of Brescia, Italy.

schematisation of the material. In fact, the only reason for schematising the matrix as homogeneous is to limit the computational time by adopting relatively coarse lattices, while in reality it is a mixture of cement paste and sand ($d_a \leq 2$ mm) and should be regarded as a heterogeneous medium.

So far, the main obstacle to 3D lattice analyses and sufficiently fine lattices, which would help to achieve a more realistic representation of the material structure, was the enormous computational time. This obstacle has been surmounted by implementing both the 2D and 3D lattice in a finite element package which uses a parallel solver (Lingen 2000).

In the first analysis conducted with the 3D lattice, a uniaxial tensile test on a concrete cube, with varying density of particle content, was simulated. This analysis showed that the particle content influences both the peak-load and the post-peak response. With increasing particle density the peak-load decreases, while the post-peak branch of the load-displacement diagram experiences an up-lift, which corresponds to an increased ductility of the numerical response. In order to verify the numerical results, a laboratory experiment was designed. Uniaxial tensile tests were conducted on notched prisms made of a synthetic concrete, with varying density of particle content. The load was applied via a cable which, having a low flexural stiffness, allowed the specimen to rotate freely at the ends. By varying the length of the cable a varying stiffness of the testing machine could be reproduced. Independently on the stiffness of the machine, stable crack propagation could be obtained by controlling the experiments with a proper feedback signal (van Mier & Shi 2002).

A direct comparison between the numerical and experimental results might seem hazardous, since the shape and size of the specimen, as well as the boundary conditions and the dimensions of the particles are different. However, the experiments conducted in the laboratory show a tendency similar to that of the numerical experiments, i.e. a decrease of the peak-load and an increase in ductility, with increasing particle density.

The capability of a simple model like the lattice model in simulating complex phenomena such as crack formation and propagation in concrete is confirmed again.

Furthermore, both numerical and laboratory experiments suggest that the brittleness of the lattice response is a consequence of the adopted sparse particle distributions, rather than of the local elastic purely brittle fracture criterion.

2 NUMERICAL EXPERIMENTS

2.1 Case study

The specimen used for the uniaxial tensile test was a cube with 24 mm length of the side. The nodes of the

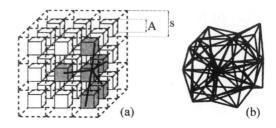

Figure 1. Generation with the Voronoi construction of the connectivities for a node at the edge (a) of a 3 × 3 × 3 3D lattice mesh with A/s = 0.5 (b).

upper and bottom face of the cube were supported in the X, Y and Z directions, and a unit vertical displacement was applied to the nodes of the upper face in Y direction. The size of the original grid used for the Voronoi construction (Fig. 1) was 0.75 mm and the randomness of the mesh was A/s = 0.001.

This value of the randomness was chosen in order to isolate the effect of the geometrical disorder, due to the random distribution of beam lengths, from the disorder induced by the presence of randomly distributed particles. The relatively small dimensions of the specimen as well as the beam length were dictated only from the desire to limit the computational time.

The longest beams of the mesh are the diagonals, with length $l_{beam} = 0.75 \cdot \sqrt{3}$ mm. The beam length and the minimum diameter of the particle that can be represented in the lattice follow the relation $l_{beam} \leq 1/3 \cdot d_{a,min}$. Since in this case all particles with diameter $d_a \geq 4$ mm would be neglected, only particle distributions with diameter varying in the range 4 to 2 mm were considered. First, a sparse particle distribution with a Fuller sieve curve and density $P_k = 34\%$ was generated in a cube with volume $72 \cdot 72 \cdot 72$ mm³. Starting from the largest size, each particle was randomly positioned at a minimum distance from all the other surrounding particles. Next, by contracting the volume and assuming some collision rules among the particles, two denser distributions were obtained, with density $P_k = 48\%$ and $P_k = 62\%$. After the lattice mesh was overlaid on top of the particle distribution, different mechanical properties were assigned to the beams falling inside each of the three phases. These properties were $f_{t,a}/f_{t,m} = 10/5$, $f_{t,b}/f_{t,m} = 1.25/5$, $E_a/E_m = 70/25$ and $E_b/E_m = 25/25$, for the tensile strengths and the elastic moduli of aggregate, matrix and bond, respectively. Due to the relatively thick interface, the effective particle density obtained after the lattice is overlaid on top of the particle distribution is $P_{k,latt} \leq P_k$. In the case which is considered here, from the original values $P_k = 34\%$, 48% and 62%, the final values $P_{k,latt} = 29\%$, 39% and 53% were obtained. With such a small ratio between grain size and dimensions of the structure (12/24), a large scatter in the

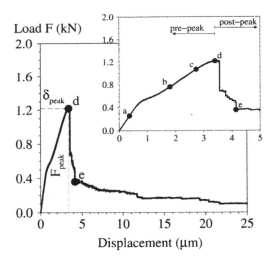

Figure 2. Load displacement obtained for $P_{k,latt} = 29\%$.

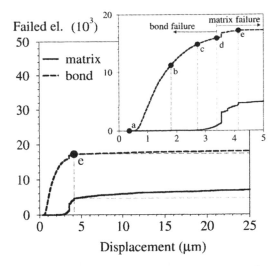

Figure 3. Cumulative number of failed beams versus displacement, obtained for $P_{k,latt} = 29\%$.

results must be expected, as a consequence of the relative position of the big grains respect to the faces of the cube. Therefore, like in the experiments, three different samples were considered for each value of the particle density. These samples were obtained cutting the specimen from three different positions in the original particle distribution, in a way that the resulting phase fractions were nearly the same, while the spatial distribution of the particles varied. Crack formation and propagation were simulated by a step-wise removal of elements from the mesh. At each loading step the element which was removed was the first to reach the critical stress $\sigma^* = 4 \cdot N/\pi \cdot h^2 = f_t$, where N is the internal axial force and h is the diameter of the circular cross section of the beam. In the case here reported $h = 0.577$ mm.

2.2 Description of the fracture process

In Figure 2 the diagram which reports the load versus the relative displacement between bottom and upper face of the prism is shown, for the case $P_{k,latt} = 29\%$.

Different stages in the fracture process, marked with different letters, can be recognised. In a the first lattice element fails. Between a and b only de-bonding occurs, i.e. only elements in the interfacial transition zone (ITZ) fail. In b failure of matrix elements starts, although de-bonding still prevails until c. In d the peak-load is reached and, until e, cracking localises in the matrix and macro-cracks form. The same phenomena appear clearly also from the diagram which shows the number of failed elements at each loading stage (Fig. 3) for the matrix and bond phase, as well as from the crack pattern (Fig. 4).

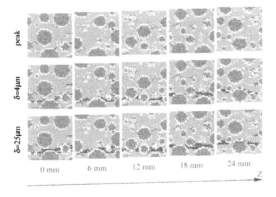

Figure 4. Exploded view of the crack pattern at different loading stages, obtained for $P_{k,latt} = 29\%$. The failed bond and matrix beams are represented in white and black respectively.

In Figure 4 an exploded view of the specimen, after it has been cut into 0.75 mm thick slices, is given. The failed beams, falling inside each slice, are projected on a plane and represented in white or black, depending whether they are interface or matrix beams, respectively.

Finally, in the tail of the softening branch, i.e. between e and the displacement 25 μm, propagation and opening of the macro-cracks already formed in the stage d-e can be observed. Essentially, a stage of micro-cracking occurs in the non-linear pre-peak branch, followed by formation and propagation of macro-cracks in the steep branch and the tail of the load-displacement curve, respectively.

Load F (kN)

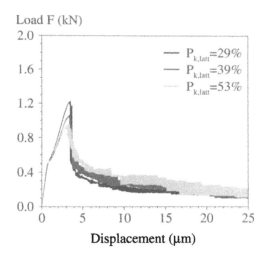

Displacement (μm)

Figure 5. Scatter in the results obtained from three different samples, for different values of particle density.

2.3 Comparison among different particle densities

For a comparison among different particle densities the load-displacement diagrams, that originally present a characteristic zig-zag, are smoothened (van Mier et al. 1997). A disadvantage of smoothening is that all information about possible snap-backs is lost and the amount of work of fracture that is computed from the area below the diagram is overestimated. In the load-displacement diagram the scatter band of the results obtained from the three different samples is shown (Fig. 5).

Note that the scatter bands develop as already found in experiments (Hordijk 1991) and 2D lattice analyses (van Mier et al. 2002). Namely, the larger scatter among the results obtained from different samples extracted from the same batch occurs in the softening of the load-displacement diagram. The 3D analyses show that such a scatter is wider when denser particle distributions are considered. The explanation of these results is quite straightforward, after the description of the fracture process made in the previous paragraph.

At the peak-load micro-cracks developed in the ITZ in the pre-peak regime start coalescing to form macro-cracks, which propagate during the tail of the load-displacement diagram. The peak-load depends on the strength and relative content of matrix and bond. Therefore, the peak-load decreases with increasing fraction of bond, which is the weakest component. Furthermore, if the samples contain approximately the same phase fractions, approximately the same peak-load can be expected. Once the macro-cracks have formed and start propagating, they may encounter particles on their way. In case of strong particles, the cracks cannot cross them but are forced to propagate

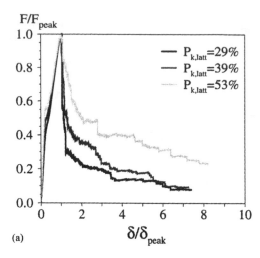

(a)

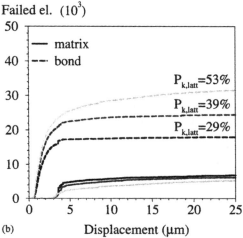

(b)

Figure 6. Dimensionless load-displacement diagrams obtained from simulations on one sample of the batch for different particle densities (a), and corresponding cumulative curves of failed elements versus displacement (b).

around them. Thus, it is evident that the crack process in the tail of the softening may be strongly affected by the spatial distribution of the particles. As a consequence, the larger scatter in the softening is closely related to the particle density.

The effect of the particle density is twofold: firstly, the peak-load decreases; secondly, the relative ductility of the lattice response increases. The latter point becomes evident from the dimensionless load-displacement diagrams (Fig. 6a), which show an up-lift in the softening regime. For more clarity in the representation, for each particle density only the results of one simulation are reported in Figure 6.

The dependence of the peak-load on the particle density is the consequence, again, of the amount of bond elements. As a matter of fact, when the particle density increases the fraction of matrix decreases, while the bond increases, producing a decrease of the peak-load. The increasing bond fraction has also another effect. If more bond elements are initially present, more elements are removed prior the peak-load is reached. This corresponds to a more pronounced non-linearity of the pre-peak branch in the load-displacement diagram.

The diagram which shows the number of failed elements during loading (Fig. 6b) may help to understand the increase of ductility which occurs when the particle density increases. This diagram shows that the total number of bond elements that are removed is larger than the matrix elements, independently on the fraction of aggregate. Furthermore, while the number of failed bond elements increases with particle density, the number of failed matrix elements decreases. This is related to the initial composition of the lattice, since more de-bonding occurs if more bond elements are initially present. In the softening branch, however, although the total number of failed bond elements stays larger than the matrix elements, the two phases have a different behaviour, depending on the aggregate fraction. In case of sparser aggregate distributions, while the number of failing matrix elements increases continuously, which corresponds to localisation of cracks in the matrix, the number of failed bond elements is practically constant. For the highest particle density, the initial matrix and bond fractions are nearly the same. In this case de-bonding prevails on matrix failure also in the post-peak region. Thus, the number of bond elements that fail increases continuously during the tail of the load-displacement diagram. However, a distinction should be made between de-bonding occurring in the pre-peak regime and in the post-peak regime. In the pre-peak, two combined effects contribute to the failure of elements falling in the ITZ: the stress concentration induced by the different stiffness between aggregate and matrix, and the lower strength of the bond elements. In the post-peak de-bonding is a consequence of crack branching and bridging. As already mentioned in the introduction, more crack branching and bridging correspond to an increased amount of dissipated energy during the fracture process, thus to an increased ductility.

3 EXPERIMENTAL TESTS

3.1 Specimens preparation

The experimental tensile tests were conducted on specimens made of a synthetic concrete. The material was a mixture of a cement mortar and glass spheres. The cement mortar was prepared using ordinary

Table 1. Mix-design of the glass spheres concrete.

Amount of glass spheres	30%	45%	60%
CEM I 52.5 R	830	652	474
Sand < 125 μm	277	217	158
Water	332	261	190
Spheres 1 mm	304	456	190
Spheres 2 mm	126	189	252
Spheres 4 mm	178	267	356
Spheres 6 mm	206	206	274

Quantities in [kg/m^3].

Portland cement (CEM I 52.5 R), sand (d_a < 125 μm) and water. The fine sand ensured a limited shrinkage of the cement paste. Next, the adopted water/cement ratio, w/c = 0.4, guaranteed a good workability. The glass spheres had diameters 1, 2, 4, 6 mm, and were distributed according to a Fuller sieve curve. The spheres with 2, 4 and 6 mm diameter were sandblasted in order to achieve a bond behaviour similar to that of natural gravel. Three different mixtures were prepared, with spheres volume fractions of 30%, 45% and 60%, respectively. Table 1 shows the three mix designs.

Firstly the cement was sieved in order to avoid lumps. Then it was mixed with the sand and the water was added. When the mortar was sufficiently homogeneous and viscous, the glass spheres were added to the mixture. A 150 mm cube was cast for every mixture and maintained sealed for 1 day, when it was demoulded. The cubes were stored in water at 20°C. After 30 days the specimens were sawn from the cube and kept in tap water for another 30 days.

The specimens were prisms with 90 mm length, 45 mm width and 10 mm thickness. A single notch with 10 mm depth was sawn with a diamond blade with squared edge.

3.2 Experimental set-up

The uniaxial tensile tests were performed by means of a 10 kN hydraulic testing machine (INSTRON 8874) with a closed-loop servo-control that permits to compensate the limited stiffness of the load system. The specimen was glued on platens that were connected to the frame through 100 mm long steel cables. In this way the exact position of the load applied to the specimen was known throughout the test. A special device was adopted to mount the specimen centrally between the platens, in order to have negligible out-of-plane eccentricities at least before onset of cracking. The cable tensile test procedure is described in (van Mier & Shi 2002). .

The displacement measured with a LVDT placed over the notch (position C in Fig. 7) was used for test-control. The measurement length was 15 mm, as shown in Figure 7. The load was applied by maintaining

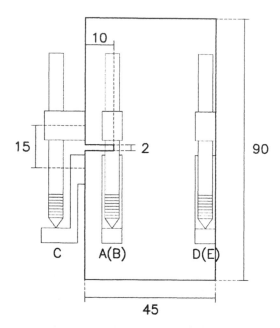

Figure 7. Specimen geometry and LVDTs location.

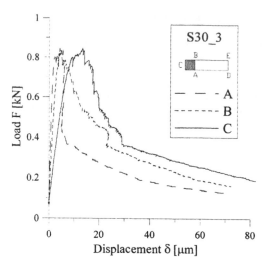

Figure 8. Load-deformation diagrams for the LVDT at the notch mouth (C) and at the notch tip (A) and (B), for a specimen with 30% particles content.

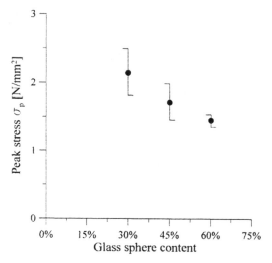

Figure 9. Average peak-stress for different glass spheres content. The maximum and minimum value for each material is also presented.

a constant deformation rate of 0.01 μm/s. Furthermore, four LVDTs were located on the surface of the specimen (as indicated in Fig. 7), in order to evaluate the relative displacement and rotation of the sections at each side of the crack surface.

A long distance optical microscope was used in some of the tensile tests in order to monitor the crack propagation. The system consists of a Questar microscope (QM100 MK-III) connected to an Ikegami black and white CCD Camera. The microscope is fixed to a cradle and subsequently to stages that can electronically move in three orthogonal directions. In the tests with the microscope, the LVDT close to the notch on the monitored specimen side (position B in Fig. 6) was removed because it obstructed the view of crack.

3.3 *Results*

For each mixture the experiments were repeated at least three times. In some cases more experiments were necessary. In fact, several specimens exhibited cracks prior to loading. These cracks, caused by shrinkage, were detected with the microscope at the beginning of the test and caused a decrease of the pre-peak stiffness. In this case the results were discarded. In other cases, the specimen failed abruptly in the post-peak branch. The results coming out from these tests were included in the computation of the average peak-stress (see Fig. 9).

In Figure 8 a typical load-deformation curve is presented. The displacements are those measured with the LVDT placed at the notch mouth (position C

in Fig. 7) and at the notch tip (positions A and B in Fig. 7). The deformations of LVDT A and LVDT B are very similar until the peak-load is reached, since the transverse eccentricity is negligible. After the crack onset an out-of-plane rotation can be observed.

The maximum nominal stress was computed as ratio between the peak-load and the area of the cross section. Thus, the contribution of the bending moment resulting from the in-plane eccentricity induced by the presence of the notch at one side was not taken into account. Figure 9 shows that the peak-stress

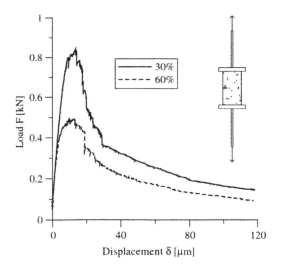

Figure 10. Load versus displacement curve for specimens with 30% and 60% particle content.

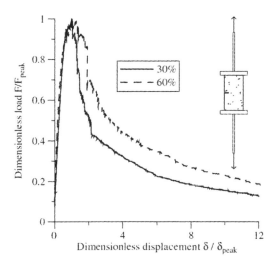

Figure 11. Dimensionless load-displacement curves for 30% and 60% particle content.

decreases with increasing content of glass spheres. From the analysis of the diagrams that represent the load versus the deformation measured by the control LVDT (position C in Fig. 7), it appears that both the initial stiffness and the deformation at the peak-load are not affected by the particle content. The results coming out from one test for 30% and 60% of particle content are plotted in Figure 10. The corresponding dimensionless diagrams, where load and

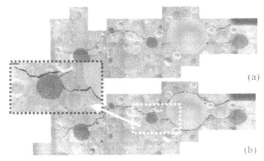

Figure 12. Example of crack pattern, filtered after (a) and before (b) creation of the mosaic of images.

deformation are divided by the values at the peak, are shown in Figure 11.

The dimensionless diagrams display a more brittle behaviour of the mix with 30% particle content, which exhibits a post-peak branch lower than the mix with 60% content of glass spheres.

The crack pattern that was detected in one of the specimens, at 65 μm deformation, is depicted in Figure 12. This figure shows a mosaic of images taken with the microscope. Each image contains 768×512 pixels. In Figure 12a the crack was filtered after generation of the mosaic, while in Figure 12b the mosaic was generated after filtering the cracks in the single images. Normally the crack became visible only after a displacement of ≈ 25 μm, which corresponds approximately to the end of the steep branch in the softening regime. At this stage a relatively long macro-crack appeared to have propagated from a corner of the notch to the opposite side of the specimen (see Fig. 12a). The crack path showed a tendency to bend around the aggregates, propagating in the plane of the specimen surface if the sawed segment of a glass sphere was bigger than a half sphere. The crack could propagate in the transversal plane if the glass segment was smaller. The two cases are shown schematically in Figure 13a and 13b, respectively. As an example, Figure 12a shows the occurrence of such phenomenon when the crack encounters a big grain near the notch. In this case, however, the initial part of the crack has two branches. The question arises whether the crack indeed branches or, rather, a second, isolated crack is present. The latter possibility seems to be confirmed by Figure 12b, which displays two cracks that are bridged by the grain. However, the opening of such cracks at the upper and lower side of the grain is negligible in comparison to the opening of the main crack(s). As a matter of fact, these cracks could be detected only after magnification of the mosaic of images. Thus, the conclusion may be drawn that only one macro-crack was present, which

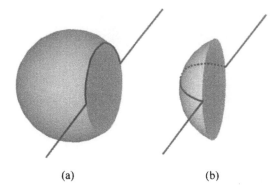

(a)   (b)

Figure 13. Crack propagation in the plane of the specimen surface (a) and in the transversal plane (b).

developed and opened not uniformly along the thickness of the specimen, due to presence of the grain. Such crack passes indeed behind the grain and the debonding cracks that appear on the surface of the specimen are a result of 3D bridging.

The images taken with the microscope show that the main crack does not propagate from a particle to an adjacent one following a straight path in the matrix, but deviates towards neighboring particles, like shown in the detail of Figure 12. In fact, the crack is attracted to the locations with lower strength (ITZ), where high stress concentrations occur due to the relatively high stiffness of the aggregate in comparison with the matrix. As a result, the crack pattern is more or less tortuous, depending on the bond and matrix strength and stiffness, as well as on aggregate dimension and content.

4 CONCLUSIONS

A 3D beam lattice model has been used for simulating uniaxial tensile tests on a relatively small prism with varying aggregate density. In order to validate the numerical results experiments were conducted using a synthetic concrete, where glass spheres with different particle densities were embedded in a cement matrix. In the experiments a long distance microscope was used for detecting the crack propagation. The numerical results predicted an influence of the particle density on the peak-load and on the post-peak behaviour. The peak-load decreases with increasing particle density whereas the ductility increases. In both cases the interfacial transition zone (ITZ) plays a major role. As a matter of fact the lowest strength of the ITZ determines the global strength of the aggregate-matrix-bond system, which is used to schematise concrete. Since the bond fraction increases with increasing aggregate fraction, the resulting bearing capacity of the structure decreases. In the post-peak regime cracks propagate through the matrix between the aggregates. In case of high particle density more crack branching and bridging occurs. The result is that the crack pattern is more tortuous and the dissipated energy and ductility increase. The experiments confirm the outcome from the numerical simulations. The images taken with the microscope show that, due to the presence of randomly distributed particles, cracks propagate in all directions. Therefore, realistic results from simulations of crack formation and propagation in heterogeneous materials such as concrete can be obtained only with a 3D model. Furthermore, it appears that the intrinsic brittleness of the lattice response is a consequence of the rough schematisation of the internal material structure, rather than of the assumed elastic brittle material behaviour.

ACKNOWLEDGEMENTS

The expert support of Mr. A. Elgersma in conducting the experiments and of Mr. F. Everdij in performing the 3D lattice simulations is gratefully acknowledged. The financial support of STW/PPM and the Ministry of Works is acknowledged.

REFERENCES

Arslan, A., Ince, R. and Karihaloo B.L. 2002. Improved lattice model for concrete fracture. *Journal of Engineering Mechanics ASCE* 128(1): 57–65.

Hordijk, D.A. 1991. *Local approach to fatigue of concrete.* PhD Thesis. Delft University of Technology.

Lilliu, G. & van Mier, J.G.M. 2002. 3D lattice type fracture model for concrete. *International Journal of Fracture Mechanics.* Submitted.

Lingen, F.J. 2000. *Design of an object oriented finite element for parallel computers.* PhD Thesis. Delft University of Technology.

Schlangen, E. & van Mier, J.G.M. 1992. Experimental and numerical analysis of micromechanism of fracture of cement-based composites. *Cement and Concrete Composites* 14 (2): 105–118.

Schlangen, E. 1993. *Experimental and numerical analysis of fracture processes in concrete.* PhD Thesis. Delft University of Technology.

Stroeven, M. 1999. *Discrete numerical modelling of composite materials.* PhD Thesis. Delft University of Technology.

van Mier, J.G.M. 1991. Crack fails bridging in normal, high strength and lytag concrete. In J.G.M. van Mier, J.G. Rots & A. Baker (eds), *Fracture Processes in Concrete Rock and Ceramics:* 27–40. Chapman & Hall/E&FN Spon.

van Mier, J.G.M., Chiaia, B. and Vervuurt, A. 1997. Numerical simulation of chaotic and self-organizing damage in brittle disordered materials. *Computer Methods in Applied Mechanics and Engineering* 142: 189–201.

van Mier, J.G.M. & Shi, C. 2002. Stability issues in uniaxial tensile tests on brittle disordered materials. *International Journal of Solids and Structures.* In press.

van Mier, J.G.M., van Vliet, M.R.A. & Wang, T.K. 2002. Fracture mechanism in particle composites: Statistical aspects in lattice type analysis. *Mechanics of Materials.* Submitted.

Vervuurt, A. 1997. *Interface fracture in concrete*. PhD Thesis. Delft University of Technology.

Finite Elements in Civil Engineering Applications, Hendriks & Rots (eds.)
© 2002 Swets & Zeitlinger, Lisse, ISBN 90 5809 530 4

Application of cohesive element approach to analysis of delamination buckling and propagation in honeycomb panels

T.-S. Han, A.R. Ingraffea & S.L. Billington
School of Civil and Environmental Engineering, Cornell University, Ithaca, NY, USA

ABSTRACT: The cohesive element approach is proposed as a tool for simulating delamination propagation between a facesheet and a core in a honeycomb core composite panel. The cohesive model has two parameters (critical energy release rate (G_c) and tensile strength (σ_c)), and the parameters are selected from the experiments. The proposed model is validated by simulating the 914 × 914 mm (36 × 36 in.) debond panel experiment under edge compression loading. It is concluded that the cohesive element approach can predict delamination propagation of a honeycomb panel with reasonable accuracy.

1 INTRODUCTION

A High-Speed Civil Transport (HSCT) has been proposed as a next generation supersonic commercial aircraft (Williams 1995; Wilhite and Shaw 1997). Realization of the aircraft has been faced with numerous environmental, economic, and technical challenges. One of the technical challenges has been the development of an airframe with light-weight, stiff and damage-tolerant structural materials, since the aircraft must withstand severe loading conditions and high temperature (about 177°C (350°F)) during a supersonic cruise (Miller et al. 1998).

A solution for weight reduction is honeycomb core sandwich panel construction (Figure 1). The high stiffness/weight ratio of the panel is obtained by using a light weight core to connect composite laminate facesheets.

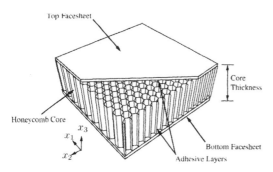

Figure 1. Honeycomb sandwich panel.

However, little is known regarding the damage tolerance of such panels not only under normal service loading conditions but also under the unexpected severe loading conditions they may encounter. Manufacturing defects, in-service mechanical loading conditions, and entrapped water in the honeycomb core under thermocyclic loading can cause debonding between a facesheet and the honeycomb core. The existing debond may lead to catastrophic delamination propagation which will eventually cause failure of the structural member.

Broad research has been done in the damage tolerance of composite materials. Chai et al. (1981) investigated buckling and post-buckling behavior; energy release rates were calculated to predict delamination propagation load for the two dimensional case. Two dimensional delamination buckling and growth in a honeycomb panel due to in-plane loading was examined by Kim et al. (1981). Interfacial crack growth in composite plates was investigated by Nilsson and Storakers (1992). A buckling induced delamination growth analysis (Nilsson et al. 1993) and a stability analysis (Nilsson and Giannakopoulos 1995) of the delamination propagation using perturbation method were performed. Using shell elements, Klug et al. (1996) applied the crack closure method to predict the delamination propagation and stability of a delaminated composite plate. Whitcomb (1981) used the three dimensional finite element approach to characterize the buckling and post-buckling behavior of homogeneous quasi-isotropic materials for through-width delamination, and later for embedded delamination

(Whitcomb 1989). He investigated the strain energy release rate along a crack front to predict delamination propagation. The effect of a contact zone was also examined (Whitcomb 1992).

Most of these previous analyses decouple the interaction between structural behavior and the fracture process, and thus can only used as a first order approximation for damage tolerance assessment. One way to couple the structural behavior and the fracture process is to use a cohesive zone model. The model was introduced to investigate the fracture process with nonlinear fracture mechanics (Barenblatt 1962; Dugdale 1960). It has been noted that, combined with the finite element method, the cohesive zone model approach can simulate the crack propagation of various types of problems (de Andres, Perez, and Ortiz 1999; Ortiz and Pandolfi 1999).

The objective of this work was to provide and to verify a computational tool for simulating delamination buckling and propagation in a honeycomb panel that can be extended to panels with fairly complicated geometry. In order to simulate the buckling-driven delamination propagation process, a computational nonlinear fracture mechanics approach was adopted. Specifically, the cohesive crack approach combined with the finite element method was used to simulate the delamination propagation.

The cohesive crack approach or cohesive zone approach is characterized by interface elements with a cohesive constitutive model. As shown in Figure 2, the interface elements are inserted between the facesheet and honeycomb core for geometrical representation of the adhesive and the debonding process, and the cohesive constitutive model is applied to reproduce traction-separation behavior between the facesheet and core. The idea of using the cohesive approach is to include a complex behavior around the crack front between the facesheet and the core into the simple cohesive model for the crack propagation analysis. To our best knowledge, application of the cohesive element model to a honeycomb sandwich panel subject to buckling-driven delamination propagation has not been reported in the literature. Thus, it is our goal to provide a systematic approach to using the cohesive model, determined from coupon tests, to predict full-scale response. Numerical studies using various material parameters of the cohesive model are carried out to assess the predictive sensitivities of structural damage tolerance.

2 COMPUTATIONAL METHODOLOGY

The cohesive crack model proposed by Dugdale (1960) and Barenblatt (1962) is applied to simulate the debond propagation. The cohesive crack model is characterized by a traction-separation law, which is a function of the fracture energy and the strength of the interface (Figure 3). According to Griffith theory, singular stresses are predicted at a crack front. However, the cohesive crack model results in a nonsingular stress at a crack front. If the cohesive zone ahead of the crack front is very small compared to other geometric dimensions of the problem, the Griffith and the cohesive crack models are equal in their prediction of fracture behavior (Rice 1968).

The tensile stress distribution on the new cracked surface is illustrated in Figure 4 which represents a

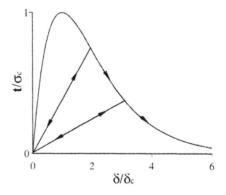

Figure 3. Cohesive traction separation model (t: Effective traction. δ: Effective displacement, σ_c/δ_c: Critical (peak) effective traction/effective displacement at peak traction).

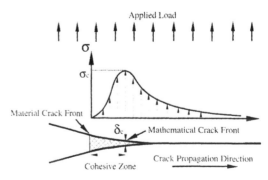

Figure 4. Cohesive crack model around crack front.

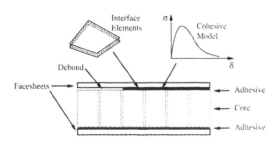

Figure 2. Modeling of honeycomb panel.

crack front at the adhesive layer with a debond in Figure 2. A true or material crack front is a point where the stress is zero after softening. A fictitious or mathematical crack front is defined as a point where the maximum stress occurs along the interface. The cohesive or fracture process zone is the region between the true crack front and the mathematical crack front. All the complicated fracture process including debonding of the adhesive and delamination between plies in a facesheet is included in the cohesive fracture model.

An interface element approach (de Andres et al. 1999; Ortiz and Pandolfi 1999) is used to model the cohesive crack propagation as opposed to a mixed boundary condition approach (Wei and Hutchinson 1997). The interface element is a zero-thickness element which is a degenerate form of a continuum element. Unlike the continuum element, the constitutive model for the interface element is formulated with a relation between traction (t) and relative displacements (δ) along the interface. In this work, an eight-noded 3-D interface element is used as shown in Figure 5. The cohesive constitutive model for the interface element is based on the formulation of Ortiz and Pandolfi (1999), and takes the form:

$$
\begin{aligned}
\mathbf{t} &= \frac{\partial \phi}{\partial \delta_n}(\delta_n, \delta_S, \mathbf{q})\mathbf{n} + \frac{\partial \phi}{\partial \delta_S}(\delta_n, \delta_S, \mathbf{q})\frac{\delta_S}{\delta_S} \\
&= t_n \mathbf{n} + t_S \frac{\delta_S}{\delta_S}
\end{aligned}
\tag{1}
$$

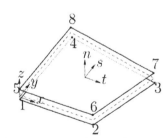

a. 3D eight-noded interface element

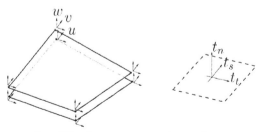

b. Displacements and tractions

Figure 5.   Schematic of interface element.

where subscript n represents a component on the n axis, and subscript S represents the resultant shear component from the t and s directions. Also, $\mathbf{t}$ is a traction at an integration point of an interface element, ϕ is the Helmholtz free potential, $\mathbf{n}$ is a unit normal of the interface element, δ_n is the relative displacement in the normal direction, δ_S is the relative displacement resultant in the shear direction ($\delta_S = \sqrt{\delta_t^2 + \delta_s^2}$), δ_S/δ_S is a unit vector in the direction of shear relative displacement resultant, and $\mathbf{q}$ is a vector of collected internal state variables. Traction t_n is a normal component of traction vector $\mathbf{t}$ as shown in Figure 5(b), and t_S is the resultant of shear tractions t_t and t_s ($t_S = \sqrt{t_t^2 + t_s^2}$).

The effective opening displacement defined for simple coupling of the normal and shear modes is:

$$
\delta = \sqrt{\delta_n^2 + \beta^2 \delta_S^2}
\tag{2}
$$

where β represents the contribution of the shear mode to the fracture process ($0 \leq \beta \leq 1$). If we further assume that 50% of the shear displacement contributes to the effective opening displacement, β reduces to $1/\sqrt{2}$. Then, the effective traction can be shown to be (Ortiz and Pandolfi 1999):

$$
t = \sqrt{t_n^2 + \beta^{-2}t_S^2}
\tag{3}
$$

The Helmholtz free potential is selected as:

$$
\phi = e\sigma_c\delta_c\left[1 - \left(1 + \frac{\delta}{\delta_c}\right)e^{-\delta/\delta_c}\right],
\tag{4}
$$

from which the fracture energy (G_c) can be expressed as:

$$
G_c = \lim_{\delta \to \infty} \phi = e\sigma_c\delta_c.
\tag{5}
$$

Assuming secant unloading/reloading (Figure 3), the relationship between the effective traction and effective displacement becomes:

$$
t = \begin{cases} e\sigma_c\dfrac{\delta}{\delta_c}e^{-\delta/\delta_c} & \text{if } \delta = \delta_{max} \text{ and } \dot{\delta} \leq 0 \\[2ex] \dfrac{t_{max}}{\delta_{max}}\delta & \text{if } \delta < \delta_{max} \text{ or } \dot{\delta} < 0 \end{cases}
\tag{6}
$$

where σ_c and δ_c are the stress and displacement at the peak of the cohesive model shown in Figure 3, and t_{max} and δ_{max} are the maximum effective traction and relative displacement throughout a loading history. In compression, a simple linear elastic behavior with a very high stiffness is assumed to model the contact between the facesheet and the core. This stress-softening, mode-coupled cohesive model was implemented via a user-supplied subroutine in a commercial finite element code.

3 DEBOND PANEL SIMULATION

3.1 Experiment

The experiment, performed at Boeing, consisted of a 914×914 mm (36×36 in.) square sandwich panel with an initial, centrally-located debond between the bag side of the facesheet and the composite honeycomb core (Figure 6). The dimension of the debond was 305×457 mm (12×18 in.), longer in the horizontal direction. The panel was loaded in edge compression in

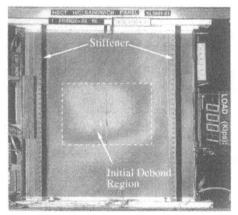

a. Before delamination propagation

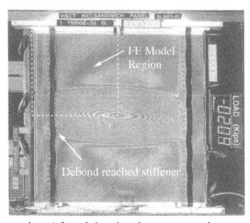

b. After delamination propagation

Figure 6. Honeycomb debond panel experiment. (*Shadow Moiré fringe patterns in (b) depict the out-of-plane displacements of the facesheet due to the facesheet buckling and delamination propagation.*)

the ribbon direction (vertical direction in Figure 6, and x_1 direction in Figure 1). More details on the experiment can be found in Boeing (1996).

The total propagation length from the initial debond front to the stiffeners was 95.3 mm (3.75 in.). Crack growth was initially stable, but it became unstable and very fast propagation was observed. The fast crack propagation was retarded when the crack front reached the stiffener. The experiment was performed under displacement control.

3.2 Modeling

The facesheets were modeled with four-noded isotropic shell elements. The equivalent elastic modulus of the facesheet was calculated based on the bending stiffness and the stacking sequence using composite laminate theory (Tsai 1987) and compared with the measured value from three point bending tests, Table 1 (Ural et al. 2001). Laminate theory predicted the elastic modulus with reasonable accuracy for the 24-ply facesheet. The equivalent elastic modulus of the facesheet from laminate theory was 4.4% larger than the measured elastic modulus (Table 1).

To reduce computational effort, eight-noded solid elements were used for the honeycomb core according to the core homogenization method proposed by Burton and Noor (1997). As shown in Table 2, the honeycomb core properties have strong orthotropy. It should be

Table 1. Material properties for facesheets.

Property	Magnitude
E	54.4 GPa (7890 ksi)[†]
	52.1 GPa (7560 ksi)[‡]
ν	0.33
Density	1580 kg/m³ (0.057 lb/in.³)
Thickness (24 layers)	3.30 mm (0.13 in.)

[†] Laminate theory.
[‡] Measured from 3-point bending test (Ural et al. 2001).

Table 2. Material properties for homogenized honey-comb core.

Property	Magnitude
E_1	689 kPa (0.1 ksi)
E_2	689 kPa (0.1 ksi)
E_3	888 MPa (128.8 ksi)
G_{12}	68.9 kPa (0.01 ksi)
G_{13}	1.47 GPa (212.75 ksi)
G_{23}	555 MPa (80.5 ksi)
ν_{12}	0.99
ν_{13}	0.00
ν_{23}	0.00
Density	96.0 kg/m³ (6 lb/ft³)
Thickness	25.4 mm (1 in.)

noted that the in-plane ($x_1 - x_2$ plane, Figure 1) properties are much lower than the out-of-plane (planes in the x_3 direction) properties since the cell walls can easily bend in the $x_1 - x_2$ plane. Fairly small values of the in-plane properties compared with the out-of-plane properties were chosen since in-plane properties have little effect on the behavior of the honeycomb panel.

The cohesive modeling approach was used to simulate the facesheet buckling and delamination propagation in the 914×914 mm (36×36 in.) debond panel. The cohesive model obtained from the DCB and FWT tests (Ural et al. 2001), and the mesh size calibrated from the DCB test simulation was used (Han et al. 2002). To simulate unstable crack growth with snapback behavior within a framework of a static analysis approach, the arc-length control method can be used. However, the snap-back behavior due to the propagation of the debond front was so severe that a stable numerical solution could not be obtained with the static analysis. Instead, a transient

analysis was performed to overcome the numerical problems of static analysis.

In the transient analysis, inertia and damping effects were considered. These effects stabilize the numerical solution process. The loading rate was calculated from the displacement loading rate in the experiment (0.318 mm/min. (0.0125 in./min.)) multiplied by the initial elastic stiffness of the debond panel. Since damping of the panel was not reported, the structural damping coefficients were assumed. Damping was selected based on the fact that the composite panels may have a substantial amount of damping and an assumption that the damping should not affect the overall response drastically.

Values of $\sigma_c = 13.8$ MPa (2000 psi), $G_c = 788$ J/m^2 (4.5 lb/in.), and 0.001% Rayleigh damping were used in a baseline Case 1, Table 3. Parametric analyses were performed on the damping coefficient, G_c, and σ_c (Cases 2, 3 and 4, Table 3). Case 2 investigated the change in the crack propagation rate due to a larger damping coefficient. Case 3 and 4 studied the sensitivity of the fracture parameters on the delamination propagation load. The analyses of Case 3 and 4 were performed with the same larger damping (1.0%) to obtain better numerical stability.

The finite element mesh is shown in Figure 7. A quarter of the whole panel was modeled considering symmetry of the panel (Figure 6(b)). Top and bottom facesheets were modeled by four-noded shell elements. Eight-noded solid elements were used for

Table 3.   Key parameters in analyses of debond panel.

Case	σ_c	G_c	Damp.
1	13.8 (2000)	788 (4.5)	0.001
2	13.8 (2000)	788 (4.5)	1.000
3	6.9 (2000)	525 (3.0)	1.000
4	6.9 (1000)	788 (4.5)	1.000

[†]Unit: MPa (psi), [‡]Unit: J/m^2 (lb/in.), [*]Unit: %

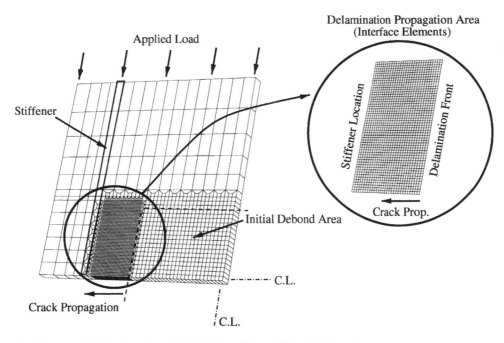

Figure 7.   Honeycomb debond panel mesh (1/4 of 914×914 mm (36×36 in.) panel).

the homogenized honeycomb core. Also, eight-noded 3D interface elements were used to represent the interface between the top facesheet and the homogenized core. To simulate the stiffeners a line of linear elastic interface elements with very high stiffness was assigned. The number of degrees of freedom for the model was 32,864. The Newmark-β implicit time integration method was used.

3.3 Results

In the 914 × 914 mm (36 × 36 in.) debond panel simulations, an eigenvalue analysis was performed before the facesheet buckling and delamination propagation analysis. The first eigenvector from the eigenvalue analysis was embedded as an initial imperfection in the panel so that a facesheet could buckle away from the honeycomb core during the delamination propagation analysis. The first two elastic buckling load and mode shapes from the eigenvalue analysis are shown in Figure 8. The first elastic buckling load was 182 kN (41 kips), which is similar to the value from

an analytical approach and the experimental result (Boeing 1996). The first mode, a half sine mode in the delamination area, was embedded into the panel geometry. The maximum magnitude of the imperfection was 0.254 mm (0.01 in.) (1% of the core thickness) at the center of the actual debond area, which is the lower right corner in the finite element mesh shown in Figure 7.

The predicted in-plane loading versus in-plane displacement is plotted in Figure 9. The in-plane loading versus out-of-plane displacement at the center of the debond area is plotted in Figure 10. The deformed shapes and the crack front in the delamination propagation area at four load levels for Case 1 are shown in Figure 11.

Figure 9 shows that the in-plane initial stiffness in the simulation is slightly larger than in the experiment. This might be due to assumptions in the simulation model, i.e. the core homogenization, the isotropy assumption of the facesheet with bending stiffness, as well as uncertainties in the experiment. As the load increased the facesheet buckled away from the honeycomb core (Figure 10). As expected, the nonlinear buckling load with imperfection was lower than the

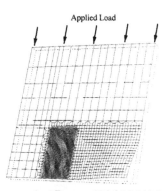

a. First mode ($P_{cr,I}$ = 182 kN (41 kips))

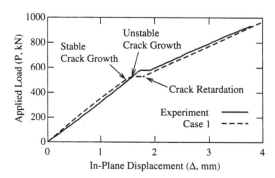

Figure 9. Applied load versus in-plane displacement (Case 1).

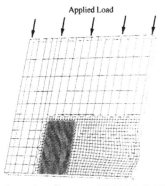

b. Second mode ($P_{cr,II}$ = 556 kN (125 kips))

Figure 8. Elastic buckling loads and mode shapes from eigenvalue analysis.

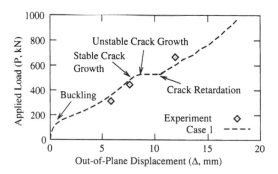

Figure 10. Applied load versus out-of-plane displacement (Case 1).

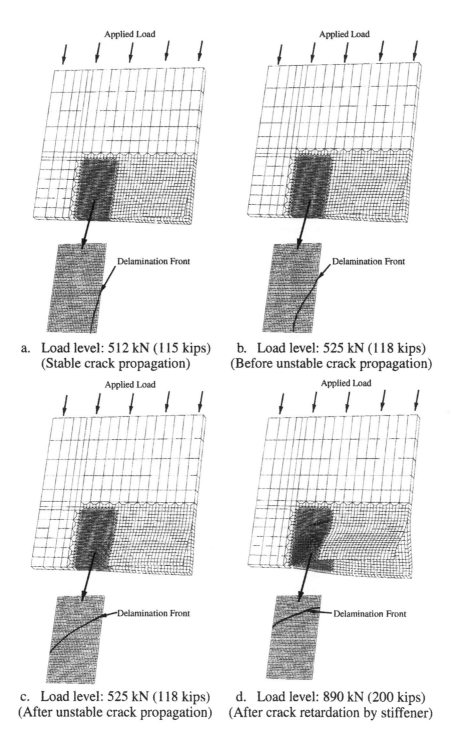

a. Load level: 512 kN (115 kips)
   (Stable crack propagation)

b. Load level: 525 kN (118 kips)
   (Before unstable crack propagation)

c. Load level: 525 kN (118 kips)
   (After unstable crack propagation)

d. Load level: 890 kN (200 kips)
   (After crack retardation by stiffener)

Figure 11. Deformed shapes of honeycomb panel and crack front in delamination area (Case 1; Magnification factor for deformed shape = 4).

ideal elastic buckling load. The predicted nonlinear buckling load (Figure 10) was slightly lower than the calculated elastic buckling load (182 kN (41 kips)). After buckling, the applied load increased by about a factor of three without delamination propagation.

The delamination first propagated in a stable fashion without much in-plane and out-of-plane stiffness change (from Figure 11(a) to Figure 11(b)). Significant in-plane and out-of-plane stiffness reduction is predicted as the crack begins to grow faster (Figure 9, Figure 10). The delamination propagation load which caused fast crack propagation was about 525 kN (118 kips). This is about 9% lower than the observed unstable delamination propagation load (578 kN (130 kips)). The discrepancy might be due to the assumptions in the simulation and uncertainties in the experiment. While the initial debond was ideal in the simulation model, the initial debond in the experiment could be behaving as a blunt notch rather than a sharp crack. The critical energy release rate for the cohesive model was determined from the DCB test results away from the initial notch to remove sawcut effects. This may have caused the under-prediction of the initial delamination propagation load when compared to the experiment.

The elapsed time during the unstable delamination process from simulation was about 9 milli-seconds. The time for unstable crack propagation from the experiment was not measured: It occurred between 2 frames of the video recording movie of the experiment. Therefore, it is certainly less than 30 milliseconds, the frame period of the video recording of the experiment. It is concluded that the period of unstable crack propagation predicted from the simulation was on the same order of magnitude as the experiment.

Unstable delamination propagation was arrested by the line of the linear elastic interface elements with very high stiffness (the left edge of the fine mesh in Figure 11(c)). The crack growth direction then changed, and the crack propagated upward, opposite to the applied load direction (Figure 11(d)). In this fashion, 95.3 mm (3.75 in.) of the crack propagation from the initial delamination front to the stiffener could be simulated. After the delamination growth retardation, the honeycomb panel partially recovers its in-plane stiffness. The recovered stiffness from the simulation was similar to the experimental result (Figure 9).

In spite of the uncertainties both in the experiment and the simulation, it is concluded that the overall simulation predictions (pre- and post-buckling behavior, delamination propagation, and behavior after crack retardation) agree well with the experimental observations.

3.4 Sensitivity

This section describes the results of parametric analyses on the damping coefficient, G_c, and σ_c. An analysis

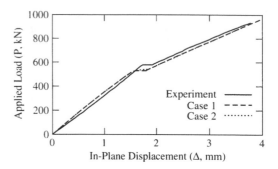

Figure 12. Applied load versus in-plane displacement. Comparison between Case 1 and Case 2 (Damping).

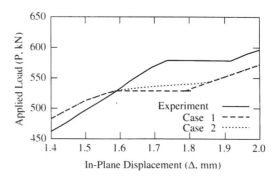

Figure 13. Applied load verses in-plane displacement. Comparison between Case 1 and Case 2 (Damping). Zoomed around unstable crack growth region.

with a different damping coefficient was performed (Case 2 in Table 3). Fracture parameters in Case 2 were the same as those in Case 1, but 1% damping was used instead of 0.001% damping. The simulation results for Case 1 and Case 2 were the same until the unstable delamination propagation load was reached. Although substantial stiffness decrease was predicted in Case 2, a finite positive slope was maintained during the delamination propagation (Figures 12 and 13). The elapsed time for the fast delamination process from Case 2 was about 5 seconds, which is significantly longer than that from Case 1 (9 milliseconds), and the upper bound time from the experiment (30 milliseconds). The crack propagation predicted from Case 2 is not a distinct unstable process as observed in the experiment. This is due to an overestimated damping coefficient. However, the difference in overall load versus displacement relationships between Case 1 and Case 2 is not significant. Therefore, 1% damping was used in subsequent parametric analyses (Case 3 and Case 4) for better numerical stability than the 0.001% damping.

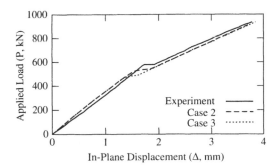

Figure 14. Applied load versus in-plane displacement. Comparison between Case 2 and Case 3 (G_c).

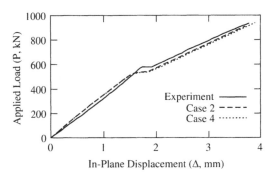

Figure 15. Applied load versus in-plane displacement. Comparison between Case 2 and Case 4 (σ_c).

To investigate the sensitivity of the stable/unstable delamination propagation load to G_c and σ_c, parametric analyses were performed. In Case 3, G_c was reduced to 525 J/m² (3.0 lb/in.) while other parameters were the same as those in Case 2. The unstable delamination propagation load was about 489 kN (110 kips) for $G_c = 525$ J/m² (3.0 lb/in.) (Figure 14). This is about 15% lower than the observed delamination propagation load (578 kN (130 kips)) from the experiment, and about 7% lower than the propagation load (525 kN (118 kips)) for $G_c = 788$ J/m² (4.5 lb/in.) (Case 2). In Case 4, σ_c was reduced to 6.9 MPa (1000 psi). The unstable delamination propagation load from Case 4 is at most 1% lower than the propagation load from Case 2 (Figure 15). Similar to the DCB simulation results, the delamination propagation load is dominated by G_c rather than σ_c.

4 CONCLUSIONS

The feasibility of applying the cohesive crack propagation model as a method to simulate delamination propagation between a facesheet and a core in a honeycomb panel was investigated. Two key parameters of the cohesive model were measured from experiments. The G_c value was determined from the DCB tests, and σ_c was selected from FWT test results. To reduce the computational cost, core homogenization was performed and both the facesheets and the core were modeled as linear elastic materials. The fracture process between the facesheet and the core was modeled using cohesive interface elements.

The overall response of the 914 × 914 mm (36 × 36 in.) debond panel simulation results captured the dominant behavior observed in the experiment in spite of the absence of precise information on the loading rate and damping of the panel. The 9% under-prediction of the unstable delamination propagation load, which is conservative compared with the experimental result, can be considered accessible for practical purposes. The stable/unstable crack growth simulation was successfully performed, and the change of delamination growth direction after crack retardation by the stiffener was also simulated.

The parametric study on σ_c for the cohesive model showed that the cohesive zone length decreases as σ_c increases, but the mathematical crack front paths remain unchanged. The parametric study on G_c showed that G_c is the sensitive parameter which determines the delamination load for crack propagation. These parametric studies suggest that the fracture-mechanics-based approach should be used in a delamination propagation analysis of a honeycomb panel as opposed to a strength-based approach.

It is concluded that the cohesive element approach allows successful simulation of a facesheet buckling and delamination propagation problem between a facesheet and a core in a realistic honeycomb panel. The current task can also be considered as a guide for simulating debonding between a facesheet and a core in a honeycomb panel using the cohesive element approach. The debonding process of a realistic honeycomb panel can be simulated by first measuring the dominant parameter, G_c and σ_c from DCB and FWT tests, respectively, and then applying the determined cohesive model to the realistic structure.

ACKNOWLEDGEMENTS

The calculations have been carried out with the finite element program DIANA of TNO Building and Construction Research, the Netherlands. This research was supported by the Boeing Commercial Airplane Company, by the Multidisciplinary Center for Earthquake Engineering Research, and by Cornell University. The authors gratefully acknowledge the contributions of Dr. Peter H. Feenstra, Prof. Alan T. Zehnder, Ms. Ani Ural, Prof. Chung-Yuen Hui, and

Dr. Paul. A. Wawrzynek at Cornell University, and Prof. Shuin-Shan Chen at National Taiwan University.

REFERENCES

Barenblatt, G. I. (1962). The mathematical theory of equilibrium cracks in brittle fracture. In *Advances in Applied Mechanics*, Volume VII, pp. 55–129. Academic Press.

Boeing (1996, August). Structural material laboratory work request. Technical report, Boeing HSCT Structures Technology. Unpublished.

Burton, W. S. and A. K. Noor (1997). Assessment of continuum models for sandwich panel honeycomb cores. *Computer Methods in Applied Mechanics and Engineering 145*, 341–360.

Chai, H., C. A. Babcock, and W. G. Knauss (1981). One dimensional modeling of failure in laminated plates by delamination buckling. *International Journal of Solids and Structures 17*(11), 403–426.

de Andres, A., J. L. Perez, and M. Ortiz (1999). Elastoplastic finite element analysis of three-dimensional fatigue crack growth in aluminum shafts subjected to axial loading. *International Journal of Solids and Structures 36*, 2231–2258.

Dugdale, D. S. (1960). Yielding in steel sheets containing slits. *Journal of the Mechanics and Physics of Solids 8*, 100–104.

Han, T.-S., A. Ural, A. T. Chen C.-S., Zehnder, A. R. Ingraffea, and S. L. Billington (2002). Delamination buckling and propagation analysis of honeycomb panels using a cohesive element approach. *International Journal of Fracture*. Accepted.

Kim, W., T. Miller, and C. Dharan (1981). Strength of composite sandwich panels containing debonds. *International Journal of Solids and Structures 30*(2), 211–223.

Klug, J., X. X. Wu, and C. T. Sun (1996). Efficient modeling of postbuckling delamination growth in composite laminates using plate elements. *AIAA Journal 34*(1), 178–184.

Miller, M., A. C. Rufin, W. N. Westre, and G. Samavedam (1998). High-speed civil transport hybrid laminate sandwich fuselage panel test. In T. Panontin and S. Sheppard (Eds.), *Fatigue and Fracture Mechanics: Vol. 29, ASTM STP 1321*, pp. 713–726. American society for Testing and Materials.

Nilsson, K.-F. and A. E. Giannakopoulos (1995). A finite element analysis of configurational stability and finite growth of buckling driven delamination. *Journal of the Mechanics and Physics of Solids 43*(12), 1983–2021.

Nilsson, K.-F. and B. Storakers (1992, September). On interface crack growth in composite plates. *Journal of Applied Mechanics 59*, 530–538.

Nilsson, K.-F., J. C. Thesken, P. Sindelar, A. E. Giannakopoulos, and B. Storakers (1993). A theoretical and experimental investigation of buckling induced delamination growth. *Journal of the Mechanics and Physics of Solids 41*(4), 749–782.

Ortiz, M. and A. Pandolfi (1999). Finite-deformation irreversible cohesive elements for three-dimensional crack-propagation analysis. *International Journal for Numerical Methods in Engineering 44*, 1267–1282.

Rice, J. R. (1968). Mathematical analysis in the mechanics of fracture. In H. Liebowitz (Ed.), *Fracture: An Advanced Treatise*, pp. 191–311. Academic Press.

Tsai, S. W. (1987). *Composites design* (Third ed.). Dayton: Think Composites.

Ural, A., A. T. Zehnder, and A. R. Ingraffea (2001). Fracture mechanics approach to facesheet delamination in honeycomb: Measurement of energy release rate of the adhesive bond. *Engineering Fracture Mechanics*. Submitted.

Wei, Y. and J. W. Hutchinson (1997). Steady-state crack growth and work of fracture for solids characterized by strain gradient plasticity. *Journal of the Mechanics and Physics of Solids 45*(8), 1253–1273.

Whitcomb, J. D. (1981). Finite element analysis of instability related delamination growth. *Journal of Composite Materials 15*, 403–426.

Whitcomb, J. D. (1989). Three-dimensional analysis of a postbuckled embedded delamination. *Journal of Composite Materials 23*, 862–889.

Whitcomb, J. D. (1992). Analysis of a laminate with a postbuckled embedded delamination. *Journal of Composite Materials 26*(10), 1523–1533.

Wilhite, A. W. and R. J. Shaw (1997, April). Hsct research picks up speed. *Aerospace America 35*(8), 24–29, 41.

Williams, L. J. (1995, April). Hsct research gathers speed. *Aerospace America 145*, 32–37.

Finite Elements in Civil Engineering Applications, Hendriks & Rots (eds.)
© 2002 Swets & Zeitlinger, Lisse, ISBN 90 5809 530 4

Implementing an isotropic damage model in Diana: Use-case for the user-supplied subroutine usrmat

Peter H. Feenstra
Computational Materials Institute, Cornell Theory Center, Cornell University

ABSTRACT: This paper discusses the usefulness of the Diana user-supplied subroutine mechanism for computational materials research at the Computational Materials Institute at Cornell University. We will present the implementation of an isotropic damage model as a test case. After a discussion of the various components of the model, we will present the implementation using the Diana usrmat routine. We show the versatility of Diana by using user-defined *input* items in the data file that allows us to develop a Diana version with a unique material model but with a standard Diana look-and-feel. This facilitates usage of the material model by students and other researchers.

1 INTRODUCTION

The range of research in computational mechanics is wide: it varies from element formulations, to linear equation solvers, to solution techniques, and to development of material models. Researchers in computational mechanics need a programming environment to develop, implement, and test these new ideas. This programming, or better, development environment has different properties: easy to understand, flexible, and with guaranteed continuity such that researchers do not have the burden of re-implementing developments if a new version of the development environment becomes available.

Basically there are two choices: the first is to develop a private development environment. The researcher/ research group controls the environment completely and has no burden of re-implementing developments. The draw-back of this option is that all features of a finite element program, such as pre-processing, element formulations, and post-processing need to be implemented too. The second option is to use a proprietary finite element code such as Diana, as the development environment but this has the draw-back that the kind of computational mechanics research is restricted to what the proprietary code allows to change and/or add through its user-supplied subroutines. But, because the vendor of the code commits to an API that is stable from one version to another, continuation of research is guaranteed since the developments can be linked with the new version without problems.

The standard development environment of Diana is limited to implementing new material models for continuum elements, interface elements, and embedded reinforcements. However, it allows for using new material models in combination with all the special features that Diana has, for instance embedded reinforcement, prestressing, and soil-structure interaction. Although there is certainly a need to add other user-supplied routines than material models to the development environment, we will concentrate on the user-supplied material model in this paper.

We present the implementation of an isotropic damage model as a test case. We also discuss how the researcher can provide a user-supplied material model with an input syntax that has the Diana look-and-feel by adding information to the SCHEMA file: the file describing the database layout. The implementation discussed in this paper is based on Diana version 7.2.

A cautionary note is necessary at this time: the information provided in the SCHEMA file is not part of the API and the vendor has the right to change the information without notice. Nevertheless, we hope to show that the combination of the standard Diana user-supplied subroutines and additional items in the SCHEMA file provide a powerful and flexible development environment for computational materials research.

2 ISOTROPIC DAMAGE MODEL

The isotropic damage model is one of the simplest material models to describe progressive damage in materials. The model has a scalar-type internal variable, α, that is a measure of the damage in the material. Assume that the free energy density in the material is given by

$$\Psi(\varepsilon, \alpha) = (1 - d(\alpha))\Psi_0(\varepsilon) \tag{1}$$

with ε the strain vector, and $\Psi_0(\varepsilon)$ the free energy density of the undamaged material

$$\Psi_0(\varepsilon) = \frac{1}{2}\varepsilon^T \mathbf{D}_e \varepsilon \tag{2}$$

The scalar d, as a function of the internal parameter α is equal to zero for the undamaged material and equal to one for the fully degraded material, i.e. $0 \leq d \leq 1$. Using standard concepts of mechanics we can derive the stress vector and the stiffness matrix from this

$$\sigma = \frac{\partial \Psi}{\partial \varepsilon} = (1 - d(\alpha))\sigma_0 \tag{3a}$$

$$\mathbf{D} = \frac{\partial^2 \Psi}{\partial \varepsilon^2} = (1 - d(\alpha))\mathbf{D}_e \tag{3b}$$

with $\sigma_0 = \mathbf{D}_e\varepsilon$. In this study, we will look at the plane stress case ($\sigma_{zz} = 0$) only. Because the damage model discussed in this paper is strain-based we take the strain in the constrained direction, the z-direction, explicitly into account as

$$\varepsilon_{zz} = -\frac{\nu}{1 - \nu}(\varepsilon_{xx} + \varepsilon_{yy}) \tag{4}$$

what allows us to use the following isotropic elastic constitutive relationship, $\mathbf{D}_e$,

$$\mathbf{D}_e = \frac{E}{(1 + \nu)(1 - 2\nu)}\begin{bmatrix} 1 - \nu & \nu & \nu & 0 \\ \nu & 1 - \nu & \nu & 0 \\ \nu & \nu & 1 - \nu & 0 \\ 0 & 0 & 0 & \frac{1 - 2\nu}{2} \end{bmatrix} \tag{5}$$

with the Young's modulus E and Poisson's ratio ν.

The damage model is determined by three assumptions: (1) the *damage criterion*; (2) the *damage norm*; and (3) the *evolution law*. We will discuss these three assumptions briefly in the following.

2.1 Damage criterion

In this study we assume the St. Venant damage criterion given by

$$f(\tilde{\varepsilon}, \alpha) = \tilde{\varepsilon} - \alpha \tag{6}$$

with $\tilde{\varepsilon}$ the damage norm, and α the damage threshold. For situations where $f < 0$, the internal state is considered elastic and damage variable does not grow, $\dot{d} = 0$. For situations where $f < 0$ and $\dot{f} < 0$, the situation is an unloading one and also here it is assumed that the damage variable is constant, $\dot{d} = 0$. Only for situations where $f = 0$ and $\dot{f} > 0$ we have a situation with increasing damage, $\dot{d} > 0$. In this study we will not consider damage recovery, nor crack closure.

2.2 Damage norm

We will assume that the damage norm is a suitable, strain-like variable, say

$$\tilde{\varepsilon} = \tilde{\varepsilon}(\varepsilon, \sigma_0) \tag{7}$$

The damage norm determines the shape of the damage surface defined by the damage criterion. The damage surface in strain space can be translated to a more traditional failure surface in stress space if we know the elasticity properties, in this study the Young's modulus E and the Poisson's ratio ν. Different formulations of the damage norm can be found in literature which are not all suited for modeling concrete behavior as we will see. In this study we will discuss four formulations: (1) the damage norm according to Mazars (Mazars and Pijaudier-Cabot 1989), (2) the energy damage norm (Simo and Ju 1987), (3) the J_2-norm, and (4) a modified Von Mises norm (De Vree, Brekelmans and Van Gils 1995).

Mazars norm
The damage norm of Mazars is given by

$$\tilde{\varepsilon} = \sqrt{\sum_{i=1}^{3}\langle\varepsilon_i\rangle^2} \tag{8}$$

with the Macaulay brackets $\langle\circ\rangle$ and ε_i the i-th principal strain. Note that even if we consider plane stress only, the strain in the constrained direction is unequal to zero and we have three principal strains with $\varepsilon_3 = -[\nu/(1-\nu)](\varepsilon_{xx} + \varepsilon_{yy})$. The damage criterion in strain space is shown in Figure 1 (top). The related failure surface is also shown in Figure 1 (bottom) and the experimental data of Kupfer and Gerstle (Kupfer and Gerstle 1973) is plotted in this figure for reference. We observe that the Mazars norm models the tension-tension region accurately, as well as the tension-compression region for moderate compressive stresses. The norm does not model the behavior of concrete under higher compressive stresses well. Note that the shape of the damage surface (and failure surface) depends on the value of the Poisson's ratio, and on the fact that we take the out-of-plane strain (ε_{zz}) into account.

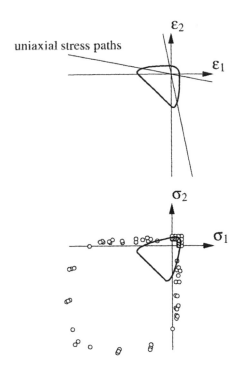

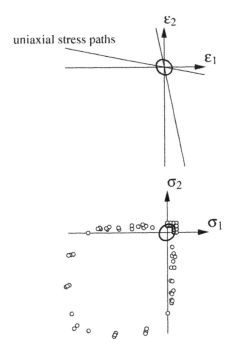

Figure 1. Failure surface of Mazars for $\nu = 0.2$ compared to experimental data of Kupfer and Gerstle.

Figure 2. Failure surface of the energy norm for $\nu = 0.2$ compared to experimental data of Kupfer and Gerstle.

Energy norm

This damage norm is defined as

$$\tilde{\varepsilon} = \sqrt{\frac{1}{E}\boldsymbol{\varepsilon}^{\mathrm{T}}\boldsymbol{\sigma}_0} = \sqrt{\frac{1}{E}2\Psi_0} \qquad (9)$$

which is a scaled energy norm of the undamaged material. The damage and failure surfaces are shown in Figure 2, with the failure surface again compared to the data of Kupfer and Gerstle. It is clear that this norm would work well in analyses where the structure is primarily loaded in tension-tension, but does not work if there are compressive stresses in the structure.

J_2-norm

The J_2-norm is given by

$$\tilde{\varepsilon} = \sqrt{J_2} \qquad (10)$$

with the second invariant of the strain vector given by

$$J_2 = (\varepsilon_1 - \varepsilon_2)^2 + (\varepsilon_2 - \varepsilon_3)^2 + (\varepsilon_3 - \varepsilon_1)^2$$

$$= (\varepsilon_{xx} - \varepsilon_{yy})^2 + (\varepsilon_{yy} - \varepsilon_{zz})^2 + (\varepsilon_{zz} - \varepsilon_{xx})^2 + \frac{6}{4}\gamma_{xy}^2 \qquad (11)$$

The damage and failure surfaces are shown in Figure 3, with the failure surface again compared to the data of Kupfer and Gerstle. The observation for this model is

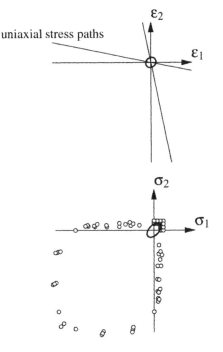

Figure 3. Failure surface of the J_2-norm for $\nu = 0.2$ compared to experimental data of Kupfer and Gerstle.

91

the same as for the energy norm: it does not model the biaxial behavior of concrete.

The energy norm and the J_2-norm are better suited for materials that have approximately equal strength in tension and compression, such as metals, and are less applicable for materials like concrete.

Modified Von Mises norm

The modified Von Mises norm (De Vree, Brekelmans and Van Gils 1995), takes the different strength in tension and compression into account and reduces the sensitivity of the analysis to compressive strains. The norm is defined as

$$\tilde{\varepsilon} = \frac{n-1}{2n(1-2\nu)} I_1$$

$$+ \frac{1}{2n} \sqrt{\left(\frac{n-1}{1-2\nu}\right)^2 I_1^2 + \frac{2n}{(1+\nu)^2} J_2} \qquad (12)$$

with I_1 the first invariant of the strain vector $I_1 = \varepsilon_{xx} + \varepsilon_{yy} + \varepsilon_{zz}$, and J_2 the second invariant of the strain vector defined in (11). The factor n reflects the ratio between the compressive and tensile strength of the material; for concrete approximately equal to 10. For a value of n equal to one, the modified Von Mises norm reduces to a scaled version of the J_2-norm. See Figure 4 for the damage surface and the failure surface. We observe that this norm models the uniaxial tensile and compressive strength properly, but overestimates the tension-compression region and the compression-compression region.

2.3 Evolution law

The damage threshold and the damage norm are defined according to the following evolution laws,

$$\begin{cases} \dot{\alpha} = \lambda \\ \dot{d} = \lambda \dfrac{\partial G}{\partial \alpha} \end{cases} \qquad (13)$$

with the damage consistency parameter λ which should fulfill the Kuhn-Tucker conditions:

$$\lambda \geq 0 \quad \lambda f(\tilde{\varepsilon}, \alpha) = 0 \quad \lambda \dot{f} = 0 \quad \text{for } f(\tilde{\varepsilon}, \alpha) = 0 \quad (14)$$

The *damage evolution function G* is a function of the internal variable α. Consequently, the damage scalar d is defined as

$$d = \int_0^t \frac{\partial G}{\partial \alpha} d\alpha = G(\alpha) \qquad (15)$$

2.4 Mapping of material data

Mapping of material data to model parameters is an important aspect of all material models. The damage

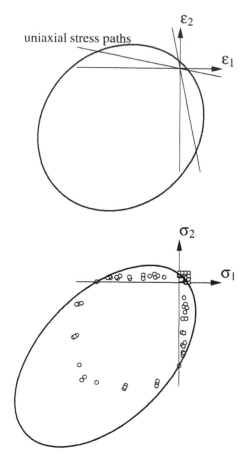

uniaxial stress paths

Figure 4. Failure surface of the modified Von Mises norm for $\nu = 0.2$ compared to experimental data of Kupfer and Gerstle.

evolution function $G(\alpha)$ determines the shape of the stress–strain relationship, or reversely, if we know the stress–strain curve we can determine the damage evolution function. The usual approach is to use an experimental uniaxial stress–strain curve and calculate the parameters by fitting this curve. So, we have the uniaxial strain, ε which translates to the internal parameter α by the definition of the damage norm. If we know the internal variable as a function of the strain, we can determine the damage evolution function as follows

$$\begin{cases} \varepsilon \mapsto \alpha = f(\varepsilon) \\ \sigma \mapsto G(\alpha) = d = 1 - \dfrac{\sigma}{E\varepsilon} \end{cases} \qquad (16)$$

With the relationship $\alpha = \dot{\lambda} = \tilde{\varepsilon}$ it follows that $\alpha = \tilde{\varepsilon}$ for continuous loading. Given a uniaxial stress–strain relationship with

$$\boldsymbol{\sigma}^T = \{\sigma, 0, 0, 0\}$$
$$\boldsymbol{\varepsilon}^T = \{\varepsilon, -v\varepsilon, -v\varepsilon, 0\}$$

the damage norms considered in this study result in

1. Mazars: $\alpha = \sqrt{\varepsilon_1^2} = \sqrt{\varepsilon^2} = \varepsilon$

2. Energy: $\alpha = \sqrt{\dfrac{1}{E}\varepsilon\sigma} = \sqrt{\dfrac{1}{E}\varepsilon E\varepsilon} = \varepsilon$

3. J_2-norm: $\alpha = \sqrt{2\varepsilon^2(1 + v)^2} = \varepsilon(1 + v)\sqrt{2}$

4. modified von Mises norm:

$$\alpha = \frac{n-1}{2n}\varepsilon + \frac{1}{2n}\sqrt{(n-1)^2\varepsilon^2 + 4n\varepsilon^2} = \varepsilon$$

With the formulations of the damage norms in this study, the mapping from uniaxial stress–strain relationship to the damage evolution function is straight-forward except for the J_2-norm. But the mapping as defined in (16) reduces to $\alpha = s\varepsilon$ with $s = (1 + v)\sqrt{2}$ in case of a J_2 damage norm. In the other cases the scalar s is equal to one.

In this study we implement two different softening curves, a linear softening curve, and an exponential softening curve.

Linear softening
For linear softening the uniaxial stress–strain relationship is given by

$$\sigma = \begin{cases} E\varepsilon & \text{if } \varepsilon \le \varepsilon_0 \\ f_t\left(1 - \dfrac{\varepsilon - \varepsilon_0}{\varepsilon_u - \varepsilon_0}\right) & \text{if } \varepsilon_0 < \varepsilon \le \varepsilon_u \\ 0 & \text{if } \varepsilon < \varepsilon_u \end{cases} \quad (17)$$

with ε_0 the strain at the maximum tensile strength, f_t, and $\varepsilon_u = 2G_F/(l_c f_t)$, the ultimate strain at which the stress is equal to zero. This can be mapped to

$$G(\alpha) = \begin{cases} 0 & \text{if } \alpha \le \alpha_0 \\ \dfrac{\alpha_u}{\alpha}\dfrac{\alpha - \alpha_0}{\alpha_u - \alpha_0} & \text{if } \alpha_0 < \alpha \le \alpha_u \\ 1 & \text{if } \alpha > \alpha_u \end{cases} \quad (18)$$

with $\alpha_0 = sf_t/E$ the initial damage threshold and $\alpha_u = s2G_F/(l_c f_t)$ the ultimate damage threshold at which the damage is equal to one. The parameter l_c is the equivalent length, or crack bandwidth, which is used to regularize the results with respect to mesh refinement. If we refine the finite element discretization, the response should stay approximately the same. One way to regularize the response, is to make the post-peak response a function of the element size, for instance relate the equivalent length to the area of the element (Rots 1988). This method works well if the elements are approximately square and the cracks

are aligned with the edges. A more refined formulation is proposed by Oliver (Oliver 1989) that uses the element geometry (the nodal connectivity and coordinates) and the direction of the crack to calculate the equivalent length.

Exponential softening
The exponential softening curve is given by

$$\sigma = \begin{cases} E\varepsilon & \text{if } \varepsilon \le \varepsilon_0 \\ f_t \exp\left(-\dfrac{\varepsilon - \varepsilon_0}{\gamma}\right) & \text{if } \varepsilon_0 < \varepsilon \le \infty \end{cases} \quad (19)$$

with $\gamma = G_F/(l_c f_t) - (1/2)\varepsilon_0$ a material parameter which determines the decaying rate of the strength. This can be mapped to

$$G(\alpha) = \begin{cases} 0 & \text{if } \alpha \le \alpha_0 \\ 1 - \dfrac{\alpha_0}{\alpha}\exp\left(-\dfrac{\alpha - \alpha_0}{\gamma}\right) & \text{if } \alpha_0 < \alpha \le \infty \end{cases}$$

$$(20)$$

with $\gamma = sG_F/(l_c f_t) - \frac{1}{2}\alpha_0$.

3 IMPLEMENTATION

We implemented the model described above using the user-supplied subroutine `usrmat`. For the current implementation we have the following model choices:

Model entity	Possible input
damage criterion	venant
damage norm	mazars
	energy
	j2
	modivm
softening curve	linear
	expone

The usual implementation of a user-defined material model only allows to define input through the user-indicators USRIND, and the user-values USRVAL. For instance, the choice of the damage norm should be input as an integer indicator in USRIND, and the parameters of the model, such as fracture energy and tensile strength, with the USRVAL input parameters. This could look like this:

```
'MATERI'
1      USRMAT     DAMAGE
       USRIND     0 0 1 1 2
       USRVAL     3.1 0.06
```

Novice users will have to know what the meaning is of these integer indicators and real numbers. To facilitate

93

the user we would like to have an input syntax that is close, or similar, to the standard Diana input, for instance like

```
'MATERI'
  1     USRMAT    DAMAGE
        NORM      MAZARS
        CRITER    VENANT
        CURVE     LINEAR
        FCT       3.1
        GF        0.06
```

The flexibility of Diana allows us actually to do this. The layout of the database is fully determined by the entries in the SCHEMA file. We can add entries to a file named SCHEMA.add, that contains the additional entries in the database. For instance, in case of the above mentioned input syntax we have to add:

```
/materi()/NORM.c*6
/materi()/CRITER.c*6
/materi()/CURVE.c*6
```

We can retrieve information, such as the damage norm, from the database, using Diana IO routines. See Diana Users Manual, *User-supplied Subroutines*, for IO routines for database access. For instance

```
CALL GTCH ( '../MATERI/NORM', NORM )
```

The model only has one user-defined status variable, the internal variable α, and we have two user-defined indicators, one to indicate the initialization and one to indicate the status of the sampling point (elastic, cracked, unloading). So we have to define the following two items in the data file

```
USRSTA 0.0
USRIND 0 0
```

4   APPLICATIONS

The applications we present here are only to show how the material model can be applied to different structures. First we show the basic behavior of the model with a uniaxial tension test and with the tension-shear model problem. These tests allow us to compare the model on sampling point level. For testing with Diana we use a single-element test. After that we present a Single-Edge-Notched beam (SEN beam) under four-point tension-shear loading.

4.1   *Uniaxial tension problem*

The uniaxial tension problem is a single element test under displacement control, see Figure 5. The displacements in the y-direction are not constrained. The stress–strain response is shown in Figure 6 for the four damage norms discussed earlier. As expected, the responses of the four norms are the same.

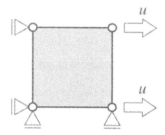

Figure 5.   Uniaxial tension problem. $E = 30000$ [N/mm^2], $\nu = 0.15$ [–], $f_t = 3.0$ [N/mm^2], $G_F = 0.075$ [Nmm/mm^2], $l_c = 100.0$ [mm].

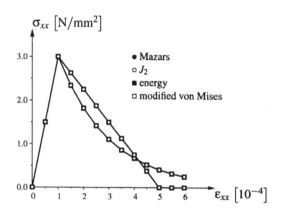

Figure 6.   Uniaxial tension problem. Stress–strain response (all four damage norms result in the same curves).

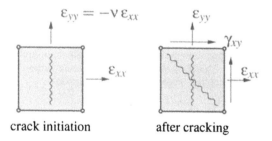

Figure 7.   Tension-shear model problem. $E = 10000$ [N/mm^2], $\nu = 0.2$ [–], $f_t = 1.0$ [N/mm^2], $G_F = 0.15 \cdot 10^{-3}$ [Nmm/mm^2], $l_c = 1.0$ [mm].

4.2   *Tension-shear model problem*

The tension-shear model problem (TSMP) is a single element test for comparing the basic behavior of constitutive models for cracking. The test, proposed by Willam et al. (Willam, Pramono and Sture 1987), loads an element up to the failure stress in the x-direction and subsequently applies biaxial tension and shear in a ratio of $\Delta\varepsilon_{xx} : \Delta\varepsilon_{yy} : \Delta\gamma_{xy} = 0.5 : 0.75 : 1.0$.

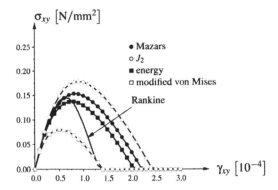

Figure 8. Tension-shear model problem. Shear stress–strain response).

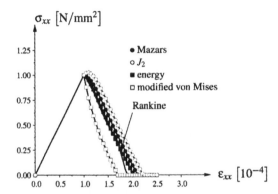

Figure 9. Tension-shear model problem. Normal stress–strain response).

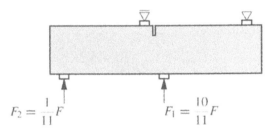

Figure 10. Tension-shear model problem. Normal stress–strain response).

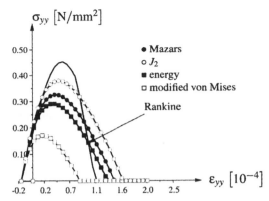

Figure 11. SEN beam. $E = 35000\,[\text{N/mm}^2]$, $\nu = 0.15\,[-]$, $f_t = 2.8\,[\text{N/mm}^2]$, $G_F = 0.07\,[\text{Nmm/mm}^2]$.

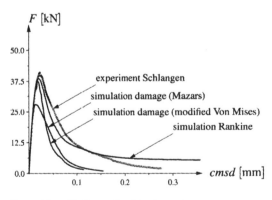

Figure 12. SEN beam. Total load–cmsd response.

We show the response of the different stress–strain components and compare the responses to the response of the Rankine principal stress model (Feenstra 1993). See Figures 8, 9, and 10. We observe clear differences between the four damage norms; especially the modified von Mises norm shows a more brittle response.

4.3 SEN beam

The Single-Edge-Notched beam has been used to study crack propagation under tension-shear loading experimentally and numerically for many years (Arrea and Ingraffea 1982; De Borst 1986; Rots 1988; Feenstra 1993; Schlangen 1993). We will simulate one of the experiments of Schlangen (Schlangen 1993), a SEN-specimen of $440 \times 100 \times 100\,[\text{mm}^3]$ with a notch of $5 \times 20\,[\text{mm}^2]$. The distance between the inner loading platens is $40\,[\text{mm}]$, and the distance between the outer loading platens is equal to $400\,[\text{mm}]$, see Figure 11. The structure is analyzed using an automatic load-incrementing method and a constrained Newton-Raphson iteration method. The load–cmsd history (load–crack mouth sliding displacement) for the damage model with a modified Von Mises norm and a Mazars damage norm are compared to results obtained with a Rankine principal stress criterion; see Figure 12. In Figure 13 the deformed shape with the distribution of the damage parameter is shown for the damage model with a modified Von Mises norm, and

95

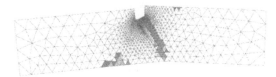

Figure 13. SEN beam. Distribution of damage parameter at final load. Damage model with a modified Von Mises norm.

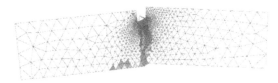

Figure 14. SEN beam. Distribution of damage parameter at final load. Damage model with a Mazars norm (the anomaly at the notch is a spurious displacement due to zero stiffness at the node).

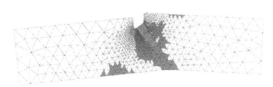

Figure 15. SEN beam. Distribution of damage parameter at final load. Rankine principal stress model.

in Figure 14 the damage distribution for the damage model with the Mazars norm. Compare the results of the Rankine model in Figure 15. It is clear that the damage models results in a more localized crack with a curved shape starting at the lower-right-hand side of the notch and growing towards the right-hand side of the bottom-center loading platen in case of the modified Von Mises norm. The localization in the case of the Rankine model is more straight with a considerable spreading of the damage. The distribution of the damage for the Mazars norm shows a different failure pattern where the crack propagates almost vertically through the specimen. Comparing the load–cmsd history of the three analyses we observe that the damage model with the Mazars norm gives a failure load that is too low compared to the experimental result. Although the other two models have comparable failure patterns, the load–cmsd curves are considerably different. The simulation with the damage model results in an even larger crack length than for the Rankine model, but the load–cmsd curve of the damage model is more brittle than the Rankine model. This result has been obtained before, see for instance (Simone 2000), and can be understood from the results of the tension-shear model problem, but it is

certainly puzzling and needs further study. One possible explanation is that the behavior of the specimen is more plane strain ($\varepsilon_{zz} = 0$) than plane stress considering the thickness of the specimen (100 [mm]). This would influence the failure pattern of the damage model with the Mazars norm, for instance.

Note that the post-processing of user-defined status variables is *not* standard in Diana. Currently (Diana version 7.2) it is not possible to output USRSTA to the Femview data base. We mimicked the post-processing of the standard available total strain-based models in Diana to post-process the user-defined variables.

5   CONCLUDING REMARKS

The run-time version of Diana can be used as the development environment for computational materials research. We have used the user-supplied subroutine usrmat to implement a new material model for continuum elements. We showed that the new model can be applied to a variety of problems, to single element tests but also to simulate physical experiments, using different element types and using standard features of Diana such as the solution procedures (indirect displacement control with full Newton-Raphson). There are, however, two major limitations: the post-processing of the user-defined status variables is rather poor; and the input of parameters of the user-defined models is not user-friendly.

The limited post-processing capabilities are avoided in this paper by using knowledge of the implementation details of the post-processing for the standard crack models in Diana. Cornell is a member of the Diana Foundation and has the source of Diana available what makes it possible to mimic the post-processing for standard crack models. The privilege of access to the source code is not granted to a large number of (potential) users of user-supplied subroutines and therefore the post-processing has to be improved. At least scalar-type post-processing of user-defined status variables should become available in Femview.

The input can be made custom-designed by adding items to the SCHEMA file so that users can input parameters in application-domain specific terms. The contents of the SCHEMA file is subject to (continuous) change, and as such, it is important to realize that the user-defined models should not depend heavily on the SCHEMA file. But if the developer only *adds* entries to the SCHEMA file, the only check for each new release is that the added entries do not conflict with possibly new entries with the same name. This is a relatively easy check, much easier than re-implementing a complete material model.

In summary, Diana has unique assets that make the program a potential development environment for materials research. But the current development environment should be improved by: (1) adding post-processing of user-defined status variables; (2) making the SCHEMA file part of the API; and (3) providing more information about the working and content of the SCHEMA file. With the last two improvements the development environment becomes more stable and could be an alternative to the development environment provided by the Diana Foundation.

ACKNOWLEDGEMENTS

This study is partially funded by the Multidisciplinary Center for Earthquake Engineering (MCEER) through grant 02-4301, and supported by the Cornell Fracture Group. The simulations are performed using the standard version of Diana 7.2.

REFERENCES

Arrea, M. and A. R. Ingraffea (1982). Mixed-mode crack propagation in mortar and concrete. Technical Report 81-13, Cornell University.

De Borst, R. (1986). *Non-linear Analysis of Frictional Materials*. Ph.D. thesis, Delft University of Technology.

De Vree, J. H. P., W. A. M. Brekelmans, and M. A. J. Van Gils (1995). Comparison of non-local approaches in continuum damage mechanics. *Comp. Struct.* 55, 581–588.

Feenstra, P. H. (1993, November). *Computational aspects of biaxial stress in plain and reinforced concrete*. Ph.D. thesis, Delft University of Technology.

Kupfer, H. B. and K. H. Gerstle (1973). Behavior of concrete under biaxial stresses. *J. Engrg. Mech., ASCE* 99(4), 853–866.

Mazars, J. and G. Pijaudier-Cabot (1989). Continuum damage theory – Application to concrete. *J. Engrg. Mech., ASCE* 115(2), 345–365.

Oliver, J. (1989). A consistent characteristic length for smeared cracking models. *Int. J. Numer. Meth. Engrg.* 28, 461–474.

Rots, J. G. (1988). *Computational modelling of concrete fracture*. Ph.D. thesis, Delft University of Technology.

Schlangen, E. (1993). *Experimental and numerical analysis of fracture processes in concrete*. Ph.D. thesis, Delft University of Technology.

Simo, J. C. and J. W. Ju (1987). Strain- and stress-based continuum damage models I. Formulations. *Int. J. Solids Struct.* 7(23), 821–840.

Simone, A. (2000). Assessment of a gradient-enhanced continuum damage model and its implementation. Technical Report CM2000.009, Delft University of Technology.

Willam, K. J., E. Pramono, and S. Sture (1987). Fundamental issues of smeared crack models. In S. P. Shah and S. E. Swartz (Eds.), *Proc. SEM/RILEM Int. Conf. Fracture of Concrete and Rock*, pp. 142–157. Springer-Verlag.

Creep, shrinkage, durability and young hardening concrete

Finite Elements in Civil Engineering Applications, Hendriks & Rots (eds.)
© 2002 Swets & Zeitlinger, Lisse, ISBN 90 5809 530 4

Prediction of column shortening due to creep and shrinkage in tall buildings

Hitoshi Kumagai
Shimizu Corporation, Tokyo, Japan

ABSTRACT: Axial shortening due to creep and shrinkage in tall buildings should be taken into consideration. When different types of structure such as steel columns and reinforced concrete shear walls are used together, vertical displacements resulted from accumulation of the axial shortening may become unequal, and cause floor inclination. This paper describes a prediction of axial shortening of a 300 m-high model structure by visco-elastic analysis.

1 INTRODUCTION

Recently, reinforced concrete structure (RC) or composite structure of steel and reinforced concrete (SRC) are being widely used for tall buildings, because high-strength concrete that withstands stress more than 60 N/mm^2 can be used. In conventional buildings, axial shortening due to creep and shrinkage is insignificant, but in tall buildings over 300 m in height, axial shortening should be taken into consideration. When different types of structure such as steel columns and reinforced concrete shear walls are used together, vertical displacements resulted from accumulation of the axial shortening may become unequal, and cause floor inclination.

While the creep test has been often conducted on plain concrete, the important thing to predict the magnitude of axial shortening in an actual tall building is to identify how steel reinforcement has influence on that. For this reason, we have conducted a series of creep test on RC and SRC columns for about 2 years and analysis for the test results with visco-elasticity model. Since the analytical results are in good agreement with the test results, we have also attempted to predict axial shortening in a tall building.

This report is to summarize the result of these studies.

2 CREEP TEST ON COLUMNS

2.1 *Specimen*

In this test, 2 kinds of creep specimens, RC and SRC, were applied. In addition to this, non-loaded specimens were also fabricated to identify drying and autogenous shrinkage, so that the total 4 different kinds of specimens were provided. Figure 1 illustrates the shape and the dimension of a specimen. The specimen represents a story of column and part of the beams above and underneath. This is a 3/10 scale model of the model structure mentioned later including the column in sectional dimension of 300 mm in width, 360 mm in depth and 1210 mm in height. The longitudinal bars 16-D13 (D13 represents a deformed steel bar with a diameter of 13 mm, 1.9% of longitudinal bar reinforcement ratio) and the rectangle hoops of D6 with spacings of 60 mm were used in common. The SRC

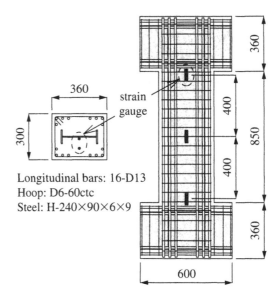

Longitudinal bars: 16-D13
Hoop: D6-60ctc
Steel: H-240×90×6×9

Figure 1. Shape and dimension of the SRC specimen.

Table 1. Mechanical properties.

(a) Concrete

Age (days)	Compressive strength (N/mm²)	Modulus of elasticity (N/mm²)
28	65.0	32920
280	80.6	35470
560	80.2	37200

Note: Cylindrical specimens 100 mm in diameter and 200 mm high were used, curing under same condition as SRC and RC specimen. Modulus of elasticity is a secant modulus at the stress of 1/3 of compressive strength.

(b) Steel

Test piece	Yield strength (N/mm²)	Tensile strength (N/mm²)	Elongation (%)
D6	359	445	17
D13	345	499	21
6 mm (web)	371	450	23
9 mm (flange)	323	457	30

specimen also contains a steel frame of H-240 × 90 × 6 × 9 (2.7% of steel reinforcement ratio) in addition to this.

2.2 Material properties

The mechanical properties of materials used for the specimens are listed in Table 1.

Cylindrical specimens 150 mm in diameter and 300 mm in height were sampled simultaneously, and creep test on plain concrete has also been carried out. 2 kinds of curing conditions, standard curing and heat curing, were applied. Standard curing conditions were 98% relative humidity moisture curing for 7 days and 60% atmospheric curing after the 7th day at a constant 20°C. Heat curing conditions were accelerated curing in a hot water tank simulating the heat of hydration of a column for 7 days (approximately 70°C at the maximum) and standard curing conditions after the 7th day. Loadings started from the 28th day where the axial stress was maintained at 0.3 times of compressive strength. The test results on plain concrete are shown in Figure 2. It revealed that the creep strain under heat curing conditions, which gained strength faster, is less than that under standard curing conditions.

2.3 Testing method

Figure 3 shows axial loading over age of specimens. Loadings started on the 28th day, applied in 10 equal increments, until reaching the predetermined value on the 280th day. Retaining this value for another 280 days,

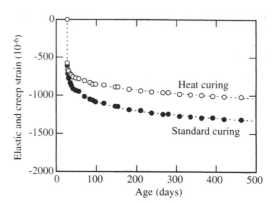

Figure 2. Test results on plain concrete.

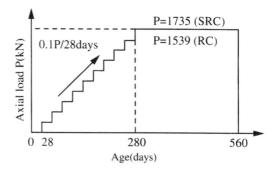

Figure 3. Axial loading diagram.

the load was released. The axial force on the 280th day shall be calculated as: the equivalent sectional area considering steel reinforcement is multiplied by the value of stress which is 0.2 times of the compressive strength on the 28th day. Axial strains of the specimen were measured by strain gauges embedded in the concrete (80 mm in gauge length) and also strain gauges attached to steel frames at the top, the mid-height and the bottom of the column respectively. Strain described below is the mean value at these three points. The specimens and the testing setup were placed under normal laboratory conditions with an average temperature of approximately 15°C and an average relative humidity of approximately 60%.

2.4 Test results

Creep strain was led by a calculation: shrinkage strain derived from the non-loaded specimen and elastic strain that would change immediately after applying a load increment are deducted from the measured total strain of the specimen. Each component of strains is shown in Figure 4. The shrinkage strain from the 28th day to the completion of the test was about 80×10^{-6} and

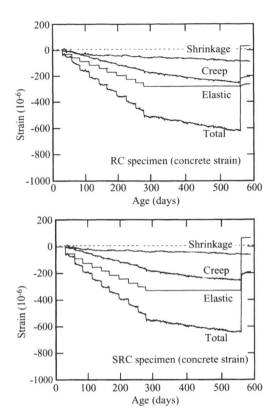

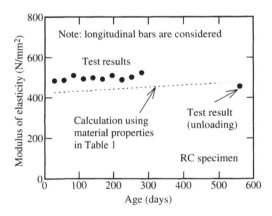

Figure 4. Test results on RC and SRC specimens.

Figure 5. Moduli of elasticity at load increments.

60×10^{-6} for RC and SRC specimens respectively, which are found to be very low. The creep strain of RC and SRC specimens was 250×10^{-6}, which shows little difference.

Figure 5 shows the modulus of elasticity obtained by dividing stress increment with elastic strain measured

at each loading increment. The modulus of elasticity after the 28th day remained almost unchanged. This is thought to be because the specimen achieved certain strength faster after being exposed to high temperature due to the heat of hydration in the early stage. Hence, the aging influence on the modulus of elasticity was negligible in the following analysis.

3 ANALYSIS WITH VISCO-ELASTIC MODEL

3.1 Analysis for creep test

Creep analysis of actual tall buildings comprised of a large number of structural members with the conventional methods is very complicated. Accordingly, visco-elastic analysis of FEM code "DIANA" has been used. Creep characteristics of plain concrete were represented by the Kelvin-chain model, one of the visco-elastic theoretical models.

Total strain $\varepsilon(t)$ and creep function $J(t, \tau)$ at age t when loading started at τ was induced from the principle of superposition below:

$$\varepsilon(t) = \int_{-\infty}^{t} J(t, \tau) \, \dot{\sigma}(\tau) d\tau$$

$$J(t, \tau) = \sum_{i=0}^{n} \frac{1}{E_i} \left\{ 1 - \exp\left(-\frac{t - \tau}{\lambda_i} \right) \right\}$$

In the creep test on plain concrete, stress was constant $\sigma = \sigma_0$ for the range of $t > 28$ ($\tau = 28$):

$$\varepsilon(t) = \sigma_0 \, J(t)$$

$$J(t) = \frac{1}{E_0} + \sum_{i=1}^{n} \frac{1}{E_i} \left\{ 1 - \exp\left(-\frac{t - 28}{\lambda_i} \right) \right\}$$

where $J(t)$ is creep compliance and $\lambda_i = \eta_i / E_i$ is the retardation time (unit: days). In this equation, the number of Kelvin-chain n was specified as 3, which is sufficient for practical use. As for the 3 types of retardation time, 1, 50 and 500, the modulus of elasticity of ith spring E_i were estimated with the least square method. The estimated Kelvin-chain characteristics are listed in Table 2. Figure 2 shows the results of the creep test on plain concrete and the estimation with the Kelvin-chain (dotted line). The approximation was highly accurate.

Next, an elastic spring, which was arranged with a Kelvin-chain in parallel, was used to represent steel reinforcement for the analysis of the RC and SRC specimen. Figure 6 compares the result of experiment and analysis on creep strain. For both RC and SRC specimen, analytical results plotted closely to the measured test results.

103

Table 2. Kelvin-chain characteristics of the creep tests on plain concrete.

i	λ_i (days)	E_i (N/mm^2)	η_i (N/mm^2 days)
0	–	29620	–
1	1	139519	1.4×10^5
2	50	116854	58.4×10^5
3	500	55972	279.9×10^5

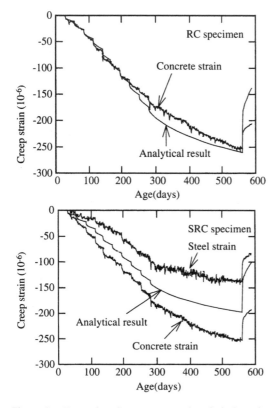

Figure 6. Comparison between test and analytical results of RC and SRC specimen.

3.2 Prediction of long-term displacements

The next analysis was to predict long-term vertical displacements of a model structure (75 floors, 300 m in height) as shown in Figure 7. This is the "Tube-in-tube" structure with a center-core surrounded by RC shear walls and peripheral framed tube with columns at 4 m intervals. 2 types of peripheral columns, SRC and square tube steel columns were used for comparison. Table 3 shows the data used for the model structure (cross sectional and material data). The following assumptions were applied to the analysis.

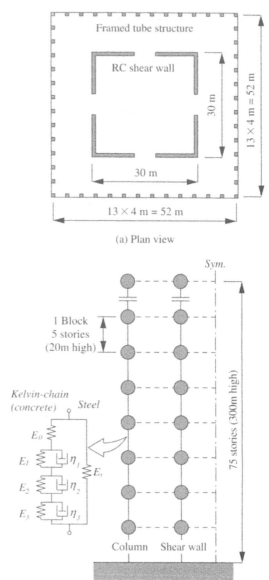

(a) Plan view

(b) Schematic view of the analytical model

Figure 7. Outline of the model structure.

– For the Kelvin-chain model, the approximation of the creep test for plain concrete (heat curing) is applied.
– Shrinkage strain was calculated by the ACI 209 code assuming the average temperature 20°C and humidity 60%.
– Loadings were applied on 1 block containing 5 stories simultaneously, to simplify the analysis.

Table 3. The data used for the model structure.

Story	B_s, D_s	t_s	B	D	t_w	F_c	E_0
71–75	600	20	750	900	400	30	25075
66–70	600	20	750	900	400	30	25075
61–65	600	20	750	900	400	30	25075
56–60	650	25	750	1000	400	30	25075
51–55	650	25	750	1000	400	36	26800
46–50	700	30	800	1000	500	36	26800
41–45	700	30	800	1000	500	42	28386
36–40	750	30	800	1100	500	42	28386
31–35	750	35	900	1100	500	42	28386
26–30	750	35	900	1100	500	48	29863
21–25	750	35	900	1100	600	48	29863
16–20	750	35	900	1100	600	48	29863
11–15	750	40	900	1100	600	60	32562
6–10	750	40	900	1100	600	60	32562
1–5	800	40	1000	1100	700	60	32562

Note: B_s, D_s: width and depth of square tube steel column (mm); t_s: thickness of square tube steel column (mm); B, D: width and depth of SRC column (mm); t_w: thickness of RC shear wall (mm); F_c: compressive strength of concrete (N/mm^2); E_0: modulus of elasticity of concrete (N/mm^2); Kelvin-chain characteristics in Table 2 are revised in proportion to modulus of elasticity of concrete E_0.

- Displacement is equal to zero on the completion of each floor.
- For the steel column, creep strain was not considered (only elastic strain).
- Bar reinforcement ratio of the RC shear wall was 1.8% and steel reinforcement ratio (including steel frame and bar reinforcement) of the SRC column was 5.0% in the entire blocks.
- No flexural stiffness was assumed as the floor beams (pinned at both the ends).
- Dead load for structure was applied at a rate of 5 stories per month, and followed by dead load for finishing in 2 months.
- Live load was applied to the entire block in 6 months upon completion of the top block.
- It was applied with dead load of 3.0 kN/m^2 for structure, 0.7 kN/m^2 for finishing, and live load of 1.2 kN/m^2.

Figure 8 shows the strain histories for each block and Figure 9 shows the distribution of vertical displacements for each block. This chart shows the displacements of each block from the completion to 1 year after live loads were applied. It showed that vertical displacements would reach the maximum value when the height was the level about 2/3 of the total height of the structure. This is because the creep and shrinkage strain at the bottom layers would have been converted enough to get strength by the time when the top layers of the structure is constructed, even though the vertical displacements increases with the increase of structure height due to the cumulative axial shortening.

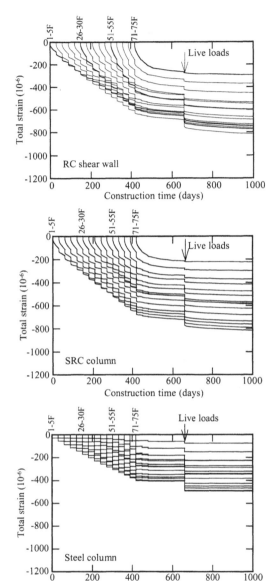

Figure 8. Strain histories for each block.

Figure 10 shows relative displacements based on RC shear wall (note that the displacement shows a positive number since RC shear wall with less amount of steel in is more settled). The maximum relative displacements are about 15 mm and 35 mm for SRC and steel column respectively. These are equivalent to 1/730 and 1/310 in terms of inclination angle. The SRC model structure in question had less floor inclination. However, care still needs to be taken against the floor inclination if steel columns are used.

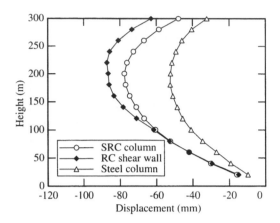

Figure 9. Distribution of vertical displacements.

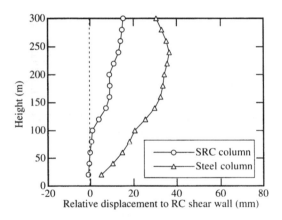

Figure 10. Relative displacements based on RC shear wall.

4 CONCLUSION

Creep strain of RC and SRC columns can be evaluated by the Kelvin-chain model based on the creep test results for plain concrete and by an elastic spring for steel reinforcement. In this case, it is more appropriate to use the result of creep test under heat curing conditions that simulate the heat of hydration.

Prediction of axial shortening of a 300 m-high building with RC shear wall and SRC framed tubes was conducted in consideration of the creep and shrinkage. This analysis revealed that floor inclination related to the vertical displacement of RC shear walls and peripheral SRC framed tube was insignificant. However, caution still should be applied if peripheral steel framed tube is used.

REFERENCES

Fintel, M., Ghosh, S.K. & Iyenger, H. 1987. *Column Shortening in Tall Structures*. Skokie: Portland Cement Association.
Colaco, J.P. 1985. 75-Story Texas Commerce Plaza, Houston–the Use of High-Strength Concrete. *American Concrete Institute* SP-87(1): 1–8.
Bažant, Z.P. & Wittmann, F.H. 1982. *Creep and Shrinkage in Concrete Structures*. New York: John Wiley & Sons.
ACI 209 Committee. 1992. *Prediction of Creep, Shrinkage, and Temperature Effects in Concrete Structures*. Farmington Hills: American Concrete Institute.
TNO Building and Construction Research. 1998. *DIANA User's Manual Release 7*. Nonlinear Analysis: 302–310.

Finite Elements in Civil Engineering Applications, Hendriks & Rots (eds.)
© 2002 Swets & Zeitlinger, Lisse, ISBN 90 5809 530 4

Field measurements versus calculated results for Norwegian high performance concrete structures

T. Kanstad, P.F. Takács & D. Bosnjak
Department of Structural Engineering, The Norwegian University of Science and Technology, Trondheim, Norway

ABSTRACT: Within several Norwegian research projects, extensive field testing of high performance concrete structures has been carried out, and calculation methods and materials models for the hardening phase and the serviceability limit states in general have been verified. Within a recent Brite-Euram project, IPACS, a field instrumentation programme related to the hardening period and assessment of early age cracking has been carried out. Temperatures and strains have been measured in more than 70 points in several sections of a large culvert structure. In the instrumented sections the temperature history, and the corresponding stresses and strains were calculated by Diana. The second part of this paper considers the long-time behaviour of three concrete cantilever bridges with main span width ranging from 220 to 301 metre. Regarding the oldest bridge, built in 1987, deflections and strains have been recorded for a time period of 13 years. The bridges have been analysed by Diana utilising the phased analysis method, where the construction history is sub-divided into phases representing the real construction history as closely as possible.

1 INTRODUCTION

Two types of structures are considered in this paper; (1) Large culvert structures, which are used as parts of highway crossings or as submerged concrete tunnels, and (2) long span concrete cantilever bridges, which frequently are used to bridge Norwegian straits and fjords. Some of the world's longest spans of this type have been built in Norway. An important design aspect for the culvert structures is to obtain a crack free structure during the hardening phase. Through cracks may occur in this phase, and the reasons to avoid them are due to water leakage, durability and aesthetics. The work reported here is partly from a recent doctoral study (Bosnjak 2000) and from the Brite Euram project IPACS (i.e. Heimdal et al. 2001), for which a major objective was to evaluate and extend the existing knowledge about early age crack prediction in engineering practice. A specific objective by the field tests is verification of theoretical modelling on real concrete structures. Considering long and slender cantilever bridges, the deformation prediction is an important part of the design, and significant deviation from the expected deflection may result in serviceability and esthetical problems. The second author completed his doctoral study recently (Takács 2002) which aimed to contribute to improved deformation prediction in this type of structure.

2 EARLY AGE CONCRETE FIELD TEST

2.1 *Field test overview*

The Maridal Culvert, which is a part of the Ring Road around Oslo, is a 340 meters long box culvert constructed with 2 by 2 traffic lanes. Two significantly different sections, as illustrated in Figure 1, have been used; either with a full bottom slab or with a strip foundation. A comprehensive field test programme has been carried out, and the temperature and strain development were recorded in 70 points distributed on 3 foundations, 5 walls and 3 top slabs. All points were located in the middle of the thickness. Figure 2 presents how the measurement points were located for one of the culvert walls. In general the highest tensile stresses in the hardening phase occur in the lower parts of the wall close to the middle section. Due to the combination of high degree of restraint and high temperatures, the critical position for through cracking is approximately one wall-thickness above the foundation slab. For further details regarding geometry, location of gauges, and type of instrumentation, etc. (see Heimdal et al. 2001).

Although 4 different types of strain gauges were used in the first parts of the field test, only the two following were appropriate with respect to reliability and price. (1) Strain gauges welded to the reinforcement

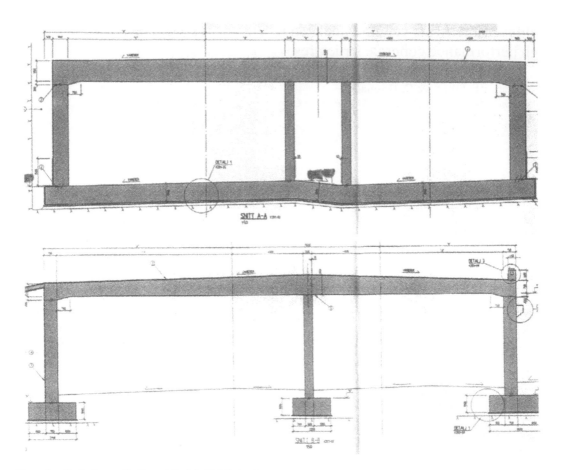

Figure 1. Typical cross Sections at the Maridal Culvert.

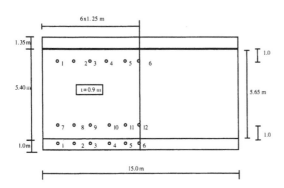

Figure 2. Overview of instrumented points in one of the culvert walls.

bars, type TML AWC-8B (gauge length 8 mm). And (2) Geokon vibrating wire type VCE-4200 directly embedded in the concrete (gauge length 100 mm). These types were verified towards the permanent instrumentation systems (LVDT) for the Dilation rig and the Stress rig in the laboratory at the Department of Structural Engineering at NTNU. Both these rigs are temperature controlled, and while the free temperature movement and the autogenous deformation is measured in the Dilation rig, the specimen in the Stress rig is fully or partly restrained, and consequently the stress development caused by the volume changes is recorded (Bjøntegaard 1998). For both methods very good agreement with the LVDT-systems was achieved. This holds both for temperature and strain development, and thus a sound basis for the subsequent field test was established. This is illustrated in Figure 3 which presents results from Geokon vibrating wire in the free dilation rig. It is worth noting, that the strain gauges record temperature compensated strains, while it always is the total strains that are most interesting regarding the structural behaviour:

$$\varepsilon_{Total} = \varepsilon_{recorded} + \alpha_T \times \Delta T \qquad (1)$$

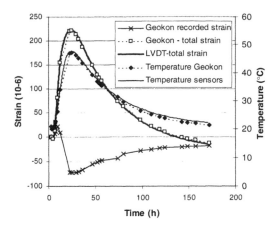

Figure 3. Verification of the Geocon vibrating wires and the permanent instrumentation systems in the dilation rig.

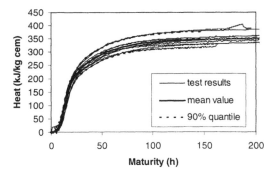

Figure 4. Hydration heat release converted to adiabatic conditions from semi-adiabtic test at different Norwegian laboratories.

in which α_T should be the thermal dilation coefficient of the strain gauge itself.

In the first part of the field tests all the instrumentation type investigation was continued and for the type (1) and (2) gauges it was concluded that the results agree well, i.e. within $\pm 20 \, \mu\varepsilon$.

2.2 Structural and materials modelling

In the FE analyses, in which a modified version of the Diana 7.2 release (Bosnjak 2002) has been used, the thermal problem is solved first, and these results are used as input for the subsequent stress calculation. To identify model parameters, a comprehensive laboratory test program has been performed. Both thermal and mechanical properties have been tested. The material properties used in the temperature calculation are the hydration heat release, the heat capacity, the activation energy and the thermal conductivity. To determine the heat release, measured concrete temperatures from semi-adiabatic tests are converted to produced heat by compensating for the heat loss from the concrete to the environment. A total of 11 tests have been performed in different Norwegian laboratories and the mean value of the hydration heat development has been used in the temperature calculations (Helland et al. 2001).

The thermal dilation and the autogenous deformation, which represent the driving forces in the cracking process, were determined from tests in the free dilation rig (Bjøntegaard et al. 2001). The major problem is that it is only the sum of the two properties that can be directly measured, while they have to be separated to achieve practical materials modelling. This separation is complicated both experimentally and theoretically due to the dependencies of temperature level and

rates. In this project the problem has been solved rather simply by assuming that the thermal dilation coefficient is a constant ($\alpha_T = 8.0\mathrm{E}-6/°\mathrm{C}$). Furthermore an autogenous deformation curve relevant for the average temperature history of the structure has been used. For the particular concrete used in the culvert, which has a relatively small autogenous deformation ($30\mathrm{E}-6$ at 7 days when the maximum temperature was 45°C), the source of error due to this simplification is probably not significant. The modified version of the CEB-FIP MC1990 equation (Kanstad et al. 2002) was used to describe the development of the compressive strength, tensile strength, and modulus of elasticity:

$$f_c(t_e) = f_c(28) \cdot \exp\left[s \cdot \left(1 - \sqrt{\frac{28}{t_e - t_0}}\right)\right] \tag{2}$$

$$f_t(t_e) = f_t(28) \cdot \left\{\exp\left[s \cdot \left(1 - \sqrt{\frac{28}{t_e - t_0}}\right)\right]\right\}^{n_t} \tag{3}$$

$$E_c(t_e) = E_c(28) \cdot \left\{\exp\left[s \cdot \left(1 - \sqrt{\frac{28}{t_e - t_0}}\right)\right]\right\}^{n_E} \tag{4}$$

in which the parameters s and t_0, which are common for all three properties were determined from the compressive strength tests, whereas parameters n_t and n_E were determined from the tensile strength and E-modulus tests, respectively (Bosnjak 2001, Kanstad et al. 2002).

Creep was described by the theory of linear viscoelasticity for ageing materials, although it is obvious that stress and strain conditions far beyond the

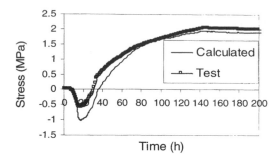

Figure 5. Calculated and measured stress development in one of the stress rig tests.

general validity range of this theory occur in the hardening phase. The compliance function, expressed by the well known double power law $(J(t, t') = (1 + \phi_0 t'^{-d} (t - t')^p)/E(t'))$, was supplemented by an additional term representing the coupling between the thermal dilation and the stress (i.e. Atrushi et al. 2001). The creep model parameters were determined by standard compressive creep tests, an approach, which completely defines the compliance function within the linearity range. The model parameters were verified by comparison to results from Stress rig tests. Typical agreement between measured and calculated results is as illustrated by the results presented in Figure 5.

The culvert was cast in sections of 15 meters length. The temperature, stress and strain versus time were calculated in the foundation, the wall and the top slab in several sections. Some typical 3D FE models made with solid elements are shown in Figure 6. Because of the symmetry only one half of the wall and one fourth of the top slab has been modelled. The structure was cast on an approximately 200 mm thick compacted gravel layer lying on rock, and in the simulations the subgrade reaction modulus method was used to model the deformability of the ground. Practically, this was done by means of interface elements, which do not transfer tensile stresses. The exchange of temperature with the ambient environment was modelled by the conduction coefficient of boundary elements. Since a major objective by this work was to verify the structural and the material modelling against measurements, the actual conditions on site were used as input to the analyses (air temperature, fresh concrete temperature, temperature of the previously cast slab, type of formwork and time of striking, etc.). The wind velocity, however, was estimated.

2.3 General experience and important results

In Figure 3 calculated and measured temperatures and strains in one of the top slabs are compared. Considering the temperatures (Figure 3a and 3b), the results

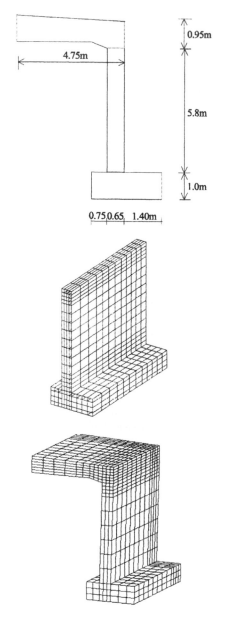

Figure 6. Typical cross section of the culvert and corresponding FE models.

illustrate the general observation that measured and calculated temperatures agree well

In a few points, however, the deviations are as large as ±5°C. This can partly be explained by the variations in the external conditions (i.e. fresh concrete temperature, wind velocity, formwork stripping details, etc.) which all have proven to have large influence on the results.

a)

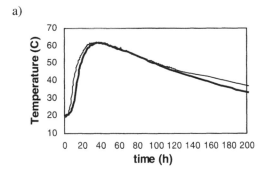

b)

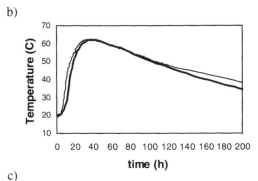

c)

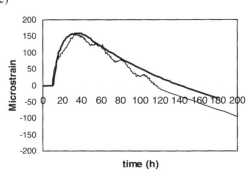

d)

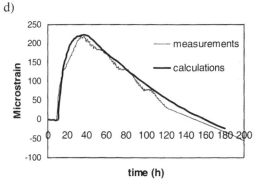

Figure 7. Calculated and measured temperatures and strains in the middle of a 1.2 m thick top slab [a, c] 1.4 m *from the wall, [b, d] 5.5 m from the wall.*

Regarding the strains, the general experience is that in areas with low or medium degree of restraint (<60%) measured and calculated strains agree well as illustrated in Figure 3c and 3d. In the most restrained areas, however, the deviations can be as large as $\pm 50\,\mu\varepsilon$. The most probable reasons for the deviations are due to: (1) High stress non-linearity and cracking in the zones with largest tensile stresses, (2) slip (or cracking) in the casting joints between newly cast and hardened concrete, and (3) the simplifications made in the materials and the subgrade modelling. To identify the effect of (1) and (2) is rather difficult due to the long calculation times needed for the analyses.

Crack observations were also done, and all walls with calculated stresses in the range 3.0–3.5 MPa cracked (1–4 cracks (<0.25 mm) on each 15 metre long wall). This corresponds to a stress ratio in the range 0.75–0.88, when the uniaxial tensile strength is estimated to 4.0 MPa.

3 LONG SPAN CONCRETE CANTILEVER BRIDGES

3.1 *Project overview*

Deformation prediction is an important part of the design calculation for segmentally cast concrete cantilever bridges (Figure 1). Deflections in the superstructure must be properly compensated in the construction stages in order to achieve the intended smooth camber in the completed bridge deck. Since the deflection development continues virtually through the entire life span of the bridges, long-term deformations must also be taken into account during the erection of the cantilevers. Deformation prediction in prestressed concrete bridges, however, is not as reliable as it would be desirable. The main reason is the inherent uncertainty in the prediction of creep and shrinkage. Significant deviation from the expected deflection may result in serviceability and esthetical problems. The long and slender free concrete span enhances the concern over the accuracy of the deformation prediction. As a consequence of several ill-fated bridges, the problem received considerable attention lately in Norway as well as worldwide. The second author recently finished his doctoral study (Takács 2002), which aimed to contribute to improved deformation prediction in this type of structure. The main objective was to establish a reliable methodology for deformation analysis taking advantage of advanced numerical methods.

One important part of the study was to establish a database on observed deformations which contains deflection and strain measurements in three, long span concrete cantilever bridges in Norway (see Table 1). The measurements cover the construction stages and the service life of the bridges until 2001. The numerical

Table 1. Key data of the investigated bridges.

Bridge	Span length [m]	Height of box girder at the piers [m]	Concrete grade*	Year of completion	Instrumentation
Norddalsfjord Bridge	98 + 230.5 + 68.5	13.0	C45	1987	Deflections and strains
Støvset Bridge	100 + 220 + 100	12.0	C55, LC55	1993	Deflections and strains
Stolma Bridge	94 + 301 + 72	15.0	C65, LC60	1998	Deflections

* In the Norwegian Standard, the concrete grade refers to the characteristic cube strength.

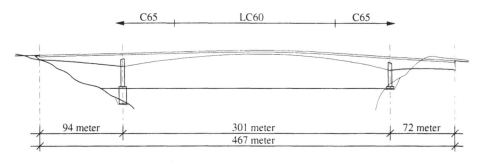

Figure 8. Stolma Bridge (1998), world record long free concrete girder.

studies were organised mainly around these three monitored bridges. For Norddalsfjord bridge, which is the oldest bridge, strain measurements are successfully carried out in 15 years. The strain gauges were vibrating wires welded to the reinforcement (P300/NGI, Norway) one type which frequently has been used in field instrumentation projects related to Norwegian concrete oil platforms and bridges. The deflections were measured by levelling, carried out regularly in the construction phase and annually as part of the inspection programme.

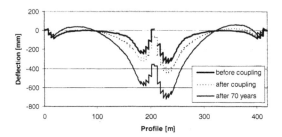

Figure 9. Calculated deflection in the construction period and at the end of the lifetime.

3.2 Numerical model and analysis

The deformation analysis, carried out by the Diana 7.2 release, is based on a two-dimensional beam model, which has proved to be an effective, yet economical solution. It is suitable for large-scale practical applications. The beam model has been verified against a more robust two-and-a-half dimensional shell model to test its general performance and some specific issues. In particular, the effect of non-uniform creep and shrinkage characteristics across the height of the box-girder has been studied and found to be very small. The prestressing tendons are individually modelled with their true profile as linear elements embedded in the beam elements. Concrete is considered as an ageing viscoelastic material and it is modelled with the Kelvin Chain model. The materials model, which is mainly used in the study, is the CEB-FIP Model Code 1990 and its recent extensions.

The numerical simulation is a non-linear phased structural analysis, which is a series of quasi independent calculation phases. Each phase models the construction of one segment and spans a typical time period of one week. The last phase covers the life span of the completed bridge up to 70 years. The main objective of the analysis is to determine the long-time deflection diagram of the superstructure (Figure 9).

3.3 Results

Available Norwegian and Swedish creep and shrinkage test results on lightweight aggregate concrete and high strength concrete (Hynne 2000, Persson 1998) were used to establish a data base, and to evaluate the existing theoretical formulations. Figure 10 compares measured and calculated creep compliance for lightweight

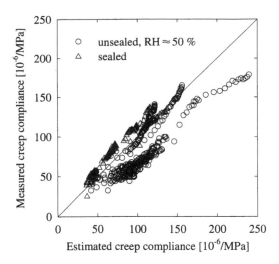

Figure 10. Measured versus calculated creep compliance for lightweight aggregate concrete.

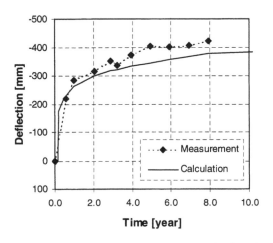

Figure 11. Calculated versus measured deflections in Støvset bridge.

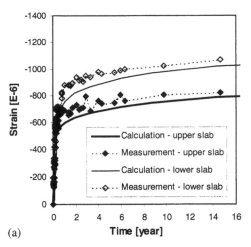

(a)

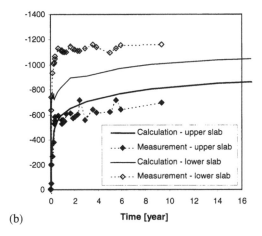

(b)

Figure 12. Calculated versus measured strains – Norddalsfjord bridge. (a) At a cross section near the columns, (b) In the middle of the main span.

aggregate concrete exposed to drying, and as for ordinary concrete it is seen that the scatter is relatively large. Because nearly all creep tests on lightweight aggregate concrete have been carried out for relatively short periods of time, it is important that the drying process acts differently in lightweight aggregate, and consequently, the long-time behaviour cannot be determined by the same time-functions as for ordinary concrete. Obviously, more long-time experiments should be conducted before a refined creep model can be made.

In Figure 11 calculated and measured deflections for Støvset bridge are compared, and it is seen that the agreement is very good. Regarding all the three

bridges, the general conclusion is that calculated and observed deformations were in reasonable agreement, and that no indication of a general tendency toward either over- or underestimation was observed.

Furthermore Figure 12a and b compares calculated and measured strain development for two different cross sections at Norddalsfjord bridge. It is seen that very good agreement is obtained for three of the four curves in the figures.

The deterministic analysis incorporates several theoretical models, which are marked with significant statistical uncertainty. Most notably the creep and shrinkage prediction models can be rather inaccurate in any particular case. The room for improvement is limited even with very sophisticated material models. It is therefore necessary to take into account and prepare for a potential deviation from the expected deformation in order to minimise the risk that intolerable deformations will compromise the serviceability and the aesthetics of the bridge. A robust probabilistic model based on a Monte Carlo simulation was therefore presented (Takács 2002). The main objective was to estimate the coefficient of variation of the deformation responses.

ACKNOWLEDGEMENT

The early age field test at the Maridal culvert was financially by the Brite-Euram project IPACS (BRPR-CT97-0437), The construction company Selmer Skanska and the Norwegian Public Roads Directorate, while the doctoral study of Daniela Bosnjak was funded by the Norwegian research council. The financial support of the doctoral project of Peter F. Takács was provided by the Norwegian Research Council through the research program *Computational Mechanics in Civil Engineering*, carried out at SINTEF in the period 1999–2002.

REFERENCES

Atrushi, Bjøntegaard, Ø., Kanstad, T. & Sellevold, E.J. 2001. Creep deformations due to self-stresses in hardening concrete, effect of temperature, 6th International Conference on Creep, Shrinkage & Durability Mechanics of Concrete and other Quasi-Brittle Materials, August, 20–22, M.I.T., Cambridge, USA.

Bjøntegaard, Ø. 1999. Thermal dilation and autogenous deformation as driving forces to self-induced stresses in high performance concrete. Doctoral thesis, Department of Structural Engineering, Norwegian University of Science and Technology, Trondheim, Norway.

Bjøntegaard, Ø. & Sellevold E.J. 2001. Interaction between Thermal Dilation and Autogenous Deformation in HPC, Materials and structures (RILEM). Vol. 34, June, pp. 266–272.

Bosnjak, D. 2000. Self-induced cracking problems in hardening concrete structures. Doctoral thesis, Department of Structural Engineering, Norwegian University of Science and Technology, Trondheim, Norway.

Heimdal, E., Kanstad, T. & Kompen, R. 2001. Maridal culvert Norway, Field test I and Field test II, IPACS reports 2001:73–7 and 2001:74–5. TU Luleå, Sweden.

Helland, S. & Smeplass, S. (2001) Concrete for the Maridal culvert, round robin test. IPACS-report 2001:75–3. TU Luleå, Sweden.

Hynne, T. 2000. Creep properties of LWAC, Euro-lightcon report, Sintef civil and environmental engineering, Norway.

Kanstad, T. 2001. The Maridal culvert Field test, Evaluation of field test results towards theoretical investigations. IPACS-report TU Luleå, Sweden.

Kanstad, T., Hammer, T.A., Bjøntegård, Ø. & Sellevold, E.J. 2001. Mechanical properties of young concrete: Experimental results and determination of model parameters. Paper accepted for publications in Materials and Structures (RILEM, Paris). Part I and II.

Persson, B. 1998. Quasi-instantaneous and longterm deformations of high-performance concrete with some related properties, Ph.D. thesis, Div. of building materials, Lund University, Sweden.

Takács, P.F. 2002. Deformations in Concrete Cantilever Bridges: Observations and Theoretical Modelling, Doctoral thesis, Department of Structural Engineering, Norwegian University of Science and Technology, Trondheim, Norway.

Finite Elements in Civil Engineering Applications, Hendriks & Rots (eds.)
© 2002 Swets & Zeitlinger, Lisse, ISBN 90 5809 530 4

Heat and stress development during chemical hydration in concrete

J.A. Øverli
SINTEF Civil and Environmental Engineering, Trondheim, Norway

K. Høiseth
MARINTEK Structural Engineering, Trondheim, Norway

D. Bosnjak
Norconsult AS, Oslo, Norway

ABSTRACT: During the hardening process of concrete energy is released as heat. Due to temperature differences the concrete will tend to expand or contract and there is a risk of cracking of the concrete due to restrained volume changes from the temperature and shrinkage. When casting takes place upon an existing concrete structure, the relative stiffness will have an influence on whether or not cracks are formed. On basis of the heat of hydration in the hardening concrete and the thermal boundary condition, the temperature distribution as a function of time can be calculated in a finite element analysis. In order to determine the stress situation, it is necessary to describe the modulus of elasticity and creep as function of maturity or degree of reaction. This paper presents material models that are capable of describing both the heat production and the stress development during the hydration process of concrete.

1 INTRODUCTION

During the hardening process of concrete energy is released as heat. Due to temperature differences the concrete will tend to expand or contract and there is a risk of cracking of the concrete due to restrained volume changes from the temperature and shrinkage. When casting takes place upon an existing concrete structure, the relative stiffness will have an influence on whether or not cracks are formed. A typical problem is the casting of a wall on an existing deck, as illustrated in Figure 1. Keeping control of the cracking in early age

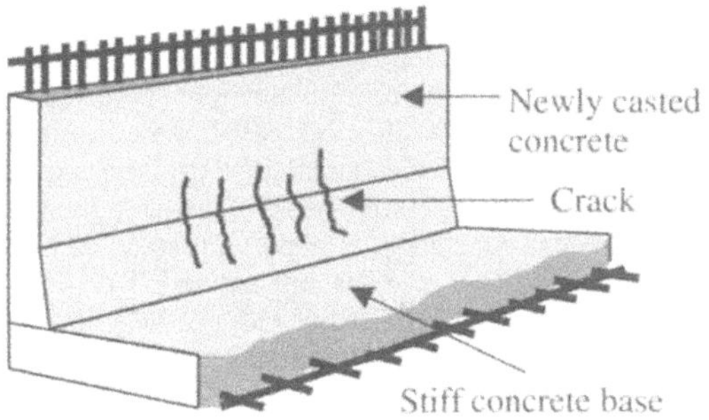

Figure 1. A typical problem with cracking in concrete.

concrete is important with respect to durability and therefore also on the lifetime of a concrete structure.

On basis of the heat of hydration in the hardening concrete and the thermal boundary condition, the temperature distribution as a function of time can be calculated in a finite element analysis. In general this distribution is 3-dimensional which results in a 3-dimensional stress distribution in the structure. To obtain reliable results, a simulation of hardening concrete structures has to take into account temperature development due to hydration, development of mechanical properties, creep and shrinkage, and restraint conditions of the particular structure. This calls for well-documented material models. Within the Brite-Euram project IPACS (1997) and the Norwegian NOR-IPACS (1998) project a comprehensive laboratory test programme has been performed to identify parameters for models used in such calculations. Both thermal and mechanical properties have been investigated.

2 FEATURES OF YOUNG CONCRETE PROPERTIES

2.1 *General*

Concrete is a mixture of cement, water, aggregate (fine and coarse) and admixtures. The main compounds in cement are calcium silicates. In presence of water, a chemical reaction takes place between the cement and the water (hydration process), which in time produces a firm and hard concrete. With focus on numerical simulations, this paper describes the hydration process and the subsequent response on concrete structures. In the following a brief description is given of the most important properties of young concrete. A more detailed description can be found in Bosnjak (2000). In practice all concrete structures are reinforced with steel bars. However, when estimating the risk for cracking during the hardening process, the influence from the reinforcement is minor.

2.2 *Temperature development*

The temperature development in a newly cast structure is determined by the balance between the development of heat due to hydration and the exchange of heat with the surroundings. The most important factors are:

- Thermal properties of concrete
- Geometry and size of structure
- Boundary conditions
- Initial conditions

Heat of hydration is the most important property affecting temperature development in hardening concrete. The rate and magnitude of heat generation depends on the type of cement, water to binder ratio, type and amount of chemical additives and temperature. The total quantity of heat generated in concrete is directly proportional to the amount of cement. The thermal conductivity of concrete depends on the temperature, moisture conditions, type and content of aggregate, porosity, density and age. Normally it varies between 1.2 and 3.0 W/mK. The specific heat of concrete varies between 0.85 and 1.15 kJ/kgK. For a given concrete mix the specific heat depends on the water content and temperature.

The heat transfer between a surface and the environment can be described by the coefficient of heat transfer and depends on the formwork and insulation, the time of the formwork, removal, the ambient temperature, wind velocity, and solar radiation.

Geometry and size of the structure influence temperature history in hardening concrete. In massive concrete structures the generated heat does not pass easily to the surroundings and hardening conditions are almost adiabatic. Hence, high maximal temperature can be reached and large temperature differences between the internal and the surface can occur. In slender concrete structures the exchange of heat with the surroundings can result in through cracking.

Fresh concrete temperature influences the hardening concrete temperature. High casting temperature accelerates the cement hydration and results in higher and earlier maximal temperature.

2.3 *Volume changes*

Volume changes in hardening concrete are induced by two mechanisms, thermal dilation and shrinkage.

Thermal strains caused by temperature changes are predicted by the thermal dilation coefficient. For hardened concrete the value varies from $5 \cdot 10^{-6}$ to $15 \cdot 10^{-6}/°C$. It is influenced by the concrete mix design, especially type and amount of aggregate, water content and temperature.

Shrinkage describes the volume changes of unloaded concrete due to the moisture exchange with the surroundings, cement hydration or carbonation. A thorough description of different types of shrinkage can be found in Bjøntegaard (1999).

2.4 *Mechanical properties*

Mechanical properties important in the analysis of hardening concrete are:

- Strength
- Young's modulus of elasticity
- Creep/relaxation

In young concrete the development of these properties as function of time and temperature history is of great concern. The compressive strength is not itself very important because of low compressive stresses. However, it is of interest because of its correlation with other mechanical properties. Experimental determination of the compressive strength is simple and can therefore be employed to estimate other mechanical properties. The tensile strength of concrete is an important parameter to estimate the risk of cracking in early age concrete. Typical, the compressive strength varies from 20 MPa to 100 MPa and the tensile strength is approximately 10% of the compressive strength.

Time dependent behaviour of concrete is described by means of the basic phenomena, creep and relaxation. Creep is time dependent deformation of concrete subjected to sustained load and relaxation is a decrease in stress with time in concrete subjected to constant deformation. In most cases, concrete in a structure is both loaded and restrained to some extent, and its behaviour is a combination of creep and relaxation. Creep in concrete is an important factor but also difficult to treat in a consistent manner. It depends both on intrinsic factors (dictated by the concrete mix) and external factors (temperature and moisture, age of loading, level of loading, etc.). The effect of varying temperature and humidity complicates it even more.

3 MATERIAL MODELING OF EARLY AGE CONCRETE

3.1 General

In order to perform numerical simulations of concrete during the hardening process, some simplifications are necessary. Assuming the mechanical behaviour of concrete does not influence the thermal and hydration processes, the thermomechanical problem in young concrete behaviour may be decoupled. The hydration process and temperature developments are assumed to be independent of stresses. Another assumption is that the influence of the moisture conditions on the hydration process and mechanical behaviour may be neglected, and only the influence of temperature is considered. These assumptions significantly simplify the numerical simulation of concrete behaviour at early ages.

3.2 Temperature calculation

The temperature development in hardening concrete due to hydration may be described by the Fourier equation for heat conduction in a homogenous and isotropic body:

$$k \cdot \left(\frac{\partial^2 T}{\partial x^2} + \frac{\partial^2 T}{\partial y^2} + \frac{\partial^2 T}{\partial z^2} \right) + Q = \rho \cdot c \cdot \frac{\partial T}{\partial t} \qquad (1)$$

where k is the thermal conductivity, T is the temperature, ρ is the mass density, c is the specific heat, Q is the rate of internal heat generation per unit volume, t is the time, and x, y, z are coordinates.

Boundary conditions are described by convection heat transfer:

$$q = h \cdot (T_f - T) \qquad (2)$$

where q is the heat flux, h is the heat transfer coefficient, T is the boundary temperature, and T_f is the temperature of the surroundings.

The mathematical modelling of the rate of heat generation caused by hydration process Q is a central part in temperature calculation of early age concrete. The parameters degree of reaction and equivalent age are employed to describe the process. Degree of reaction is defined as the ratio of the amount of heat liberated at time t, $Q(t)$, to the amount of heat at complete hydration Q_{max}:

$$r(t) = \frac{Q(t)}{Q_{max}} \qquad (3)$$

Temperature influences the hydration process. Hence, the maturity concept is introduced. It is based on the assumption that the rate of hydration at a given degree of reaction is a function of temperature only. Mathematically it can be expressed as:

$$\frac{dr}{dt} = g(r) \cdot f(T) \qquad (4)$$

where $g(r)$ is the function of the degree of hydration and $f(T)$ is the function of the temperature. Freiesleben Hasen & Pedersen (1977) proposed the following function for $f(T)$, based on Arrhenius' equation:

$$f(T) = k \cdot \exp\left(\frac{E_a}{RT} \right) \qquad (5)$$

where k is a proportionality constant, E_a activation energy (J/mol), R universal gas constant, and T concrete temperature. The activation energy describes the effect of temperature on the rate of hydration and is defined as:

$$E_a = \begin{cases} A & \text{for } T > 20°C \\ A + B \cdot (20 - T) & \text{for } T \leq 20°C \end{cases} \qquad (6)$$

where A and B are constants which may be determined by compressive strength tests. By using Equation 3, the hydration process at an arbitrary temperature history may be compared to the process at a reference temperature (usually 20°C). First the rate factor is defined as a relation between speed of hydration at reference temperature T_{ref}, and the speed at T:

$$H(T) = \frac{\text{speed at } T}{\text{speed at } T_{ref}} = \exp\left[\frac{E_a}{R} \cdot \left(\frac{1}{T_{ref}} - \frac{1}{T} \right) \right] \qquad (7)$$

Then comparison between two processes is done by calculating the equivalent age of the concrete. The equivalent age of a concrete is the age at which hydration at the reference temperature has reached the same state. It is determined as:

$$t_{eq} = \int_0^t H(\theta) \, dt \qquad (8)$$

The concept of equivalent age is in this work employed to describe all thermal and mechanical material properties.

Different approaches can be taken to model the hydration heat development in early age concrete. The fact that the hydration process is a self-accelerating process (the produced heat increase the rate of hydration) makes the modeling difficult. A simple and commonly used approach in engineering is the three-parameter equation proposed by Freiesleben Hasen & Pedersen (1977):

$$Q(t_{eq}) = Q_\infty \cdot \exp\left[-\left(\frac{\tau}{t_{eq}}\right)^\alpha\right] \tag{9}$$

where Q_∞, τ and α are model parameters which can be determined from adiabatic curves.

3.3 Stress calculation

Description of the constitutive behaviour of hardening concrete is a very complex task, since it must include all phenomena important for stress generation at early ages. To simplify, it is assumed that the strain rate may be decomposed as follows:

$$\dot{\varepsilon} = \dot{\varepsilon}_e + \dot{\varepsilon}_c + \dot{\varepsilon}_{th} + \dot{\varepsilon}_{th} + \dot{\varepsilon}_{crack} \tag{10}$$

where ε_e is the elastic strain, ε_c the creep strain, ε_{th} the thermal strain, ε_{sh} the shrinkage strain and ε_{crack} is the cracking strain. In this way, the constitutive law for each component may be defined independently. A brief description of the properties of the most important components will be given.

At present there are no generally accepted models for thermal dilation coefficient and shrinkage based on equivalent age Bjøntegaard (1999). Thus, in this study a constant thermal dilation coefficient is employed. Shrinkage is modelled using predefined curves from experiments. However, this is a simplification since the curves are valid only for a specific temperature history, which is very inconvenient having in mind the temperature variation existing in a structure.

In this work modulus of elasticity and strength are modelled by means of the following expressions from Kanstad et al. (1999):

$$f_c(t_{eq}) = f_c(28) \cdot \exp\left[s \cdot \left(1 - \sqrt{\frac{28}{t_{eq} - t_0}}\right)\right] \tag{11}$$

$$f_t(t_{eq}) = f_t(28) \cdot \left\{\exp\left[s \cdot \left(1 - \sqrt{\frac{28}{t_{eq} - t_0}}\right)\right]\right\}^{n_t} \tag{12}$$

$$E_c(t_{eq}) = E_c(28) \cdot \left\{\exp\left[s \cdot \left(1 - \sqrt{\frac{28}{t_{eq} - t_0}}\right)\right]\right\}^{n_E} \tag{13}$$

$f_c(28)$, $f_t(28)$ and $E_c(28)$ are compressive strength, tensile strength and Young's modulus at 28 days, s, t_0, n_t and n_E are model parameters.

Time dependent behaviour of concrete is described by the theory of linear viscoelasticity for aging materials. According to the linear theory, the strain is proportional to the applied stress. Hence, the strain at time t caused by a constant uniaxial stress σ applied at time t' is given by:

$$\varepsilon(t) = \sigma \cdot J(t, t') \tag{14}$$

where $J(t,t')$ is the compliance function (or creep function). An important property of the compliance function of concrete is that it is a function of two variables, the current age and age at loading t', which takes age dependencies into account. For a variable stress history, the principle of superposition is assumed to be valid. A great number of creep functions for concrete is proposed in the literature. In the numerical example presented in this paper, the Double Power Law, Bažant (1975), is employed. This function is defined as:

$$J(t,t') = \frac{1}{E(t')}(1 + \phi_0 t'^{-d}(t - t')^p) \tag{15}$$

where $E(t')$ is modulus of elasticity, and ϕ_0, p and d are creep model parameters determined from tests in compression.

Increase of creep of hardened concrete at changing temperature, so-called transient creep, can be taken into account in a simplified way by modifying the thermal expansion coefficient according to Jonasson (1994). A transient creep strain term is added to Equation 10 and defined as:

$$\Delta\varepsilon_{trc} = \alpha_T \cdot \rho \cdot \frac{\sigma}{f_t(t_{eq})} \text{sign}(\Delta T) \tag{16}$$

where α_T is the linear thermal expansion coefficient, ρ a material parameter, σ normal stress component, sign(ΔT) is the sign for the temperature difference and f_t is the maturity dependent tensile strength as defined in Equation 12.

4 FINITE ELEMENT ANALYSIS

The influence of the mechanical responses on the temperature development in hardening concrete is negligible, and the thermal and mechanical problem may therefore be separated. The thermal problem is solved first, and results are used as input for the subsequent stress calculation. This is a so-called staggered analysis.

The finite element method is adopted to solve the governing set of differential equations for the thermal and mechanical response of early age concrete.

Since the thermal and mechanical problem is decoupled, the finite element discretisation may be done separately. However, to be efficient the same element mesh is normally used in both analyses. To avoid spatial oscillation of stresses, elements in the stress analysis then must have an order higher interpolation polynomial than element in temperature analysis, Cook et al. (1989).

5 NUMERICAL ANALYSIS OF A CULVERT

To determine the risk for cracking in concrete during the hardening process of is a relevant problem for different types of structures. Typical structures are foundations, culverts in roadfillings or subsea tunnels like parts of the Øresund Link between Denmark and Sweden or the Bjørvika project outside Oslo in Norway. This section presents a numerical investigation of temperature and stress distribution in a real concrete structure. Figure 2 shows the cross section of the culvert under consideration. The finite element program Diana (1999) is employed to implement the material models described in Section 3, Bosnjak (2000), and to model and analyse the culvert. In this paper one wall with thickness 1.0 m and height 9.0 m, and a top slab with thickness 1.5 m is considered. The wall is cast on a previously cast foundation (10 × 1.5 m). Temperature, stress and strain development during the first 10 days of the wall are calculated and presented here.

Different ways of modelling the structure and the restraint conditions is possible. In a culvert it would normally be sufficient with a two-dimensional temperature distribution. However, the stress distribution is three-dimensional. Therefore both the thermal and stress analyses must be performed with 3D models. The most important stress component is in the longitudinal direction of the culvert.

If the complete culvert is modelled in the longitudinal direction, the subgrade reactions can be taken into account directly. Modelling of the subgrade stiffness is done by means of interface elements, which also can account for a no-tension bedding.

Instead of employing a full 3D model, the analyses presented in this section employs a simplified model which is a "slice" of the wall with unit thickness. The basic assumption for the simplified model is that the wall can be treated like an infinitely long structure with constant temperatures and stresses in the longitudinal direction. Different assumptions of the subgrade restraint conditions can be made for the simplified model representative for an infinitely long structure: fixed, or alternatively free rotation of the cross section of the wall. The simplifications represent two extreme cases of the boundary conditions of the real structure. The first case corresponds to infinitely stiff subgrade, while in the second case the subgrade stiffness is totally neglected. In both cases the horizontal frictional forces between foundation and subgrade are neglected. A thorough discussion for the validity of the simplifications can be found in Bosnjak (2000). The analyses in this work assume free rotation of the cross section of the wall.

Figure 3 shows the finite element model used in the analysis. Due to symmetry only one half of the wall is modelled. The same element model is used both in the temperature and stress analysis. Solid elements describing linear variation of temperature and quadratic variation of stresses are used in this case. The exchange of temperature with surroundings is modelled by the

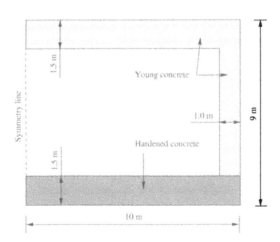

Figure 2. Cross-section geometry of the culvert.

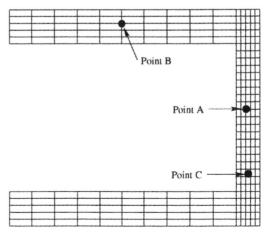

Figure 3. Finite element mesh of the culvert.

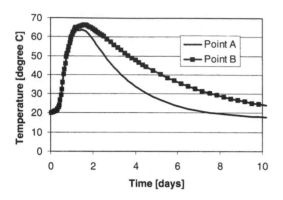

Figure 4. Temperature development in the young concrete.

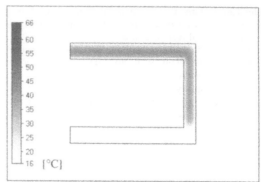

Figure 5. Temperature distribution in the cross-section after 3 days.

conduction coefficient of boundary elements. Quadrilateral elements were used at the faces of three-dimensional model.

The main material properties are:

Cement content	418	kg/m^3
Density	2394	kg/m^3
Heat capacity	1.04	kJ/kg° K
Total heat Q_∞	313	kJ/kg cement
Conductivity	2.22	J/ms° K
Conv. coeff. formwork	$3.6 \cdot 10^{-3}$	kJ/m^{2}s° K
Conv. coeff. air	0.017	kJ/m^{2}s° K
Modulus of elasticity	36200	MPa
Concrete strength	65	MPa

The air temperature is constant 17°C. The hardened and young concrete has initial temperatures of 17°C and 20°C respectively. Creep in the young concrete is taken into account according to Equation 15 with $\phi_0 = 0.8$ and p = d = 0.8. Creep in foundation are calculated according to Model Code 1990 (1993). The formwork on the wall and the underside of the slab is removed after 36 hours. To describe the effect of the casting history for wall and slab, they could be modelled in two phases with difference in starting time of the hydration process. This is not taken into consideration in this analysis.

The culvert is analysed for a time period of 10 days. Figure 4 illustrates results from the thermal analysis. The temperature development in two points, point A and B in Figure 3, in the middle of the cross-sections is given. The variation in development is due to the difference in thickness of the cross-section. Compared to the wall with thickness 1.0 m, the slab with thickness 1.5 m reaches a higher maximum temperature of 66°C at a later time. The difference in geometry also explains why the slab maintains a higher temperature for a longer period. If the analysis has been continued after 10 days, the complete structure would reach the air temperature of 17°C. In practical engineering, cooling pipes with water would be introduced in the

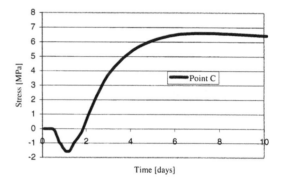

Figure 6. Development of out-of-plane stress in the wall.

cross-section if the temperature or the temperature differences were too high.

Figure 5 shows the temperature distribution in the cross-section after 3 days. As expected there is no significant heat exchange with the foundation slab.

The main objective for performing numerical analysis of young concrete is to estimate the risk for cracking due to volume changes in the hydration process. For the culvert in this example, stresses in the longitudinal direction can form vertical cracks. Figure 6 presents the development of the longitudinal stress in the wall at a point 0.6 m above the foundation, point C in Figure 3. During the expansion phase, when the concrete is heated, compressive stresses are introduced in the young concrete. In the contraction phase, rather large tensile stresses develop in the cross-section. A maximum stress of 6.7 MPa is reached after 7 days. Typical tensile strength of concrete is in the range 2–5 MPa. Hence, vertical cracks will most likely occur in this culvert.

Figure 7 presents the distribution of the stresses after 9 days in the longitudinal direction. As expected the largest tensile stresses are in the wall above the

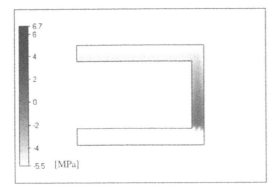

Figure 7. Distribution of out-of-plane stress after 9 days.

foundation. Since the foundation with hardened concrete is much stiffer than the wall with young concrete, the effect of restraining is higher in this area. The top slab can expand and contract more easily. Therefore the stress level is lower. From Figure 7 it can be seen that in the area above the foundation, the complete thickness of the wall have high tensile stresses. Consequently cracks can form through the cross-section, which results in increased durability.

6 CONCLUSIONS

This paper deals with numerical simulation of the heat and stress development in concrete during the chemical hydration process in concrete. The work has focused on the numerical aspects of predicting early age cracking. The crucial factors in the prediction are thermal dilatation, shrinkage, mechanical properties and restraint conditions. The behaviour of early age concrete is considered as a thermomechanical problem. Because of the small influence of stresses on the hydration process, the problem may be decoupled. The thermal problem is solved first, and the results are used in the subsequent stress analysis. The hydration process is described within the framework of degree of reaction and equivalent age. All material properties are described using equivalent age. A short mathematical description for the mechanical response is given in form of constitutive equations for different strain components. The finite element method is used to solve the differential equations for both the thermal

and the mechanical response. A culvert is analysed to exemplify the assumptions, difficulties and possibilities of employing a numerical analysis to estimate the risk of cracking at early age of a concrete structure. However, it must be emphasised that a difficult task and a major challenge in such analyses are to identify all the necessary material properties.

REFERENCES

Bažant, Z.P., 1975. Double Power Law for basic creep of concrete, *Materials and Structures*, 9(49).

Bjøntegaard, Ø., 1999. *Thermal Dilation and Autogenous Deformation as Driving Forces to Self-Induced Stresses in High Performance Concrete*. Dr.ing. thesis, Norwegian University of Science and Technology, Trondheim, Norway.

Bosnjak, D., 2000. *Self-induced cracking problems in hardening concrete structures*. Dr.ing. thesis, Norwegian University of Science and Technology, Trondheim, Norway.

CEB-FIP Model Code 1990, 1993. Comité Euro-International du Béton.

Cook, R.D., Malkus, D.S. & Pleisha, M.E., 1989. *Concepts and applications of finite element analysis*. John Wiley and Sons.

Diana 7.2, 1999. *User's Manual*, TNO Building and Construction Research, The Netherlands.

Freiesleben Hansen, P. & Pedersen, E.J., 1977. *Måleinstrument til kontrol af betons hærdning*. Nordisk Betong 1, Stockholm (in Danish).

IPACS, 1997–2001. Brite-EuRam project BRPR-CT97-0437, *Improved Production of Advanced Concrete Structures*.

Jonasson, J.E., 1994. *Modelling of temperature moisture and stresses in young concrete*. Ph.D. thesis, Division of Structural Engineering, Luleå University of Technology, Sweden.

Kanstad,T. Hammer, T.A., Bjøntegaard, Ø. & Sellevold, E. J., 1999. *Mechanical properties of young concrete: Evaluation of test methods for tensile strength and modulus of elasticity. Determination of model parameters*. NOR-IPACS report STF22 A99762, ISBN 82-14-01062-4.

NOR-IPACS 1998-2001. A project supported by the Research council of Norway (NFR), *Improved Production of Advanced Concrete Structures*.

Finite Elements in Civil Engineering Applications, Hendriks & Rots (eds.)
© 2002 Swets & Zeitlinger, Lisse, ISBN 90 5809 530 4

Deformation behaviour of concrete highway pavements

Sam Foos, Viktor Mechtcherine & Harald S. Müller
University of Karlsruhe, Germany

ABSTRACT: In this paper, results of experimental and numerical investigations on deformations and stresses in concrete pavements subjected to practical weather conditions are presented. The numerical simulations of large concrete slabs under critical weather conditions and with boundary conditions corresponding to different field situations were performed using the finite element programme DIANA. In order to verify and to improve the numerical model and to calibrate some parameters of the model a series of large-size experiments on concrete slabs was carried out with the same loading conditions as used in the numerical study. The major purpose of these investigations is to develop a new design method taking into account the effect of weather conditions in addition to static and dynamic live loads. This will improve the durability of concrete pavements while reducing the costs of their production.

1 INTRODUCTION

Concrete pavements are widely used in highway, airport as well as in industrial construction works because of their high resistance against abrasive stresses and high loadings. However, considerable damages in terms of both development of cracks and flaking of the edges in the joint zones have been observed in the practice (see Figure 1). These problems may result in high repair costs.

Besides live loads, concrete pavements are normally exposed to severe weather conditions. As a consequence of great changes of the ambient temperature as well as the relative humidity, the concrete is subjected to significant temperature and moisture gradients. Furthermore, the moisture gradients are increased by bedewing the slabs from underneath. These gradients cause large deformations. Today, the resulting stresses and strains as well as the combination of thermal and hygral loads with the live loads are not sufficiently considered in the design process of concrete slabs.

Extensive experimental investigations were carried out by Eisenmann (1979) and Springenschmid (1987), which substantially contributed to the development of design rules for concrete pavement slabs. Numerical analyses concerning the deformation and cracking behaviour of concrete pavement slabs were performed by Müller et al. (2002), however no corresponding experimental investigations were carried out. Within the current research programme both experimental and numerical investigations have been performed.

Using the FE programme DIANA an advanced numerical model has been developed, which describes the deformation, cracking and stress behaviour of concrete slabs, Foos (in prep.). In order to verify and calibrate this model, experimental investigations were carried out on large concrete slabs, which were

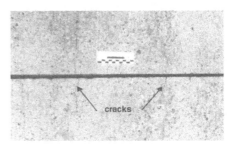

Figure 1. Cracks in the joint zone and flaking of the corners of concrete pavement slabs.

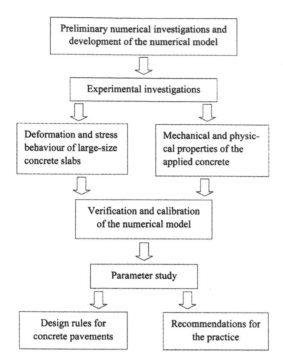

Figure 2.   Draft of the research programme.

Figure 3.   Set-up of the tests on large-size concrete slabs.

exposed to temperature and moisture conditions similar to real life applications, for example wetting of the slab underneath, solar radiation with following thermal shock by rain and hail, wind as well as deformation restrain at the ends of the slab. Furthermore, various mechanical and physical properties of the applied concrete were determined.

Based on the gained experimental data, the numerical model is presently being verified, calibrated and refined before conducting an extended parameter study. The objective of this research programme is to develop a new design method for concrete pavement slabs exposed to extreme weather conditions as well as recommendations for the practice. Figure 2 shows schematically the various phases of the research programme.

2   EXPERIMENTAL INVESTIGATIONS

2.1   *Description of the experimental programme*

The simulation of different, typical cases of the hygral and thermal loading including the modeling of the corresponding boundary conditions was carried out on four large-size concrete slabs measuring 5000 × 1000 × 260 mm³ (length × width × thickness).   In order to enable a controlled wetting on the bottom side, the slabs were placed in sealed metal tanks seven days after the casting of the concrete (see Figure 3). These tanks were equipped with transverse beams, which allowed a continual wetting underneath the slabs.

The investigation programme on these concrete slabs involved the following loadings, boundary conditions and testing parameters:

– drying of the upper side
– wetting of the bottom side
– hygral alternating loading (2 h wetting of the upper side with following 22 h drying)
– influence of wind and curing conditions
– thermal loading with thermal shock (thundershower)
– deformation restrain at the ends of the slabs, while wetting of the bottom side.

Slab 1 was exposed to wetting on the bottom side combined with drying on the upper side. The deformations resulting from the drying of the upper side only were determined on slab 2, which was sealed on the bottom side. The influence of wind and no curing of the concrete were simulated on slab 3. The boundary conditions of slab 1 were also kept for slab 4, which however was additionally subjected to a deformation restrain at the ends of the slab. Further, the thermal loading using infrared lamps was simulated. Thereby, the upper side of the slabs was heated. Subsequently it was rapidly cooled using water with various temperatures.

In order to register the vertical deformation of the slabs LVDTs were installed along the longitudinal axis of the slabs. Additionally strain gauges were applied over the height of the slab.

During thermal loading, the temperature changes were recorded with temperature sensors, which were fixed at short intervals near the upper side of the slabs. Further, so called Multi Ring Electrodes were installed on the upper and the bottom side in order to

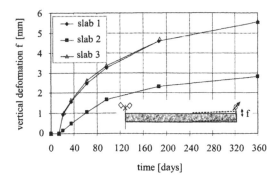

Figure 4. Average vertical deformations at the ends of the slabs 1, 2 and 3.

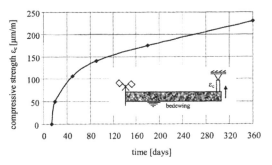

Figure 5. Average compressive strain in the restrain construction at the ends of slab 4.

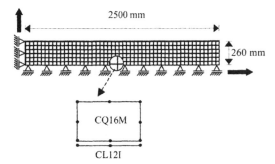

Figure 6. Schematic view of the FE model of a concrete pavement slab.

specify the moisture distribution in the vicinity of the surface of the concrete slabs.

In order to provide the data for the numerical model various mechanical and physical properties of the applied concrete were determined, for example compressive strength, uniaxial tensile strength, modulus of elasticity, creep and shrinkage functions, sorption isotherms, hygral expansion function, moisture conductivity, capillary suction, coefficient of capillary suction, diffusion coefficient, pore size distribution, thermal expansion coefficient and thermal conductivity. Furthermore, temperature as well as moisture distributions for different ages were measured.

2.2 *Results of the experimental investigations*

In this paper only the results concerning the hygral loading are presented.

Figure 4 shows the average vertical deformations at the ends of slabs 1, 2 and 3 (the deformations obtained on slab 4 due to the deformation restraint were too small to register in this diagram).

The vertical deformations of the ends of slab 1 (drying on the upper side while wetting on the bottom side) and slab 3 (accelerated drying and no curing) were approximately 5.6 mm after one year, whereas the ends of slab 2 exhibited a vertical deformation of 2.9 mm only resulting from drying on the upper side. This means that the share of the deformation due to drying on the upper side is approx. 50% of that resulting from the wetting on the bottom side combined with the drying on the upper side. During the first 100 days this ratio was even smaller than 50%. This may be caused by the rapid absorption of water on the bottom side followed by large deformations due to hygral expansion.

The effect of the deformation restrain at the ends of the slab was investigated on slab 4. Figure 5 shows the average compressive strain in the restrain construction indicating that the determined strains continue to

increase till the end of the experiments meaning no formation of through-cracks, in spite of the almost absolute deformation restrain.

3 NUMERICAL INVESTIGATIONS

3.1 *Description of the numerical model and applied parameters*

Using the finite element programme DIANA a two dimensional FE model of a concrete pavement slab (length × thickness = 5000 × 260 mm²) has been developed in order to simulate its deformation and stress behaviour due to thermal and hygral loadings. For symmetry reasons and in order to increase the efficiency of the calculations only one half of the slab has been discretized. According to the experiments, the vertical bedding on the bottom side of the slab was rigid with a shear stiffness of zero, which permitted free horizontal deformation of the slab. On the symmetry axis the slab was horizontally fixed. The support conditions are presented in Figure 6. For the discretisation isoparametric 8-node elements of the type CQ16M were applied. In order to simulate

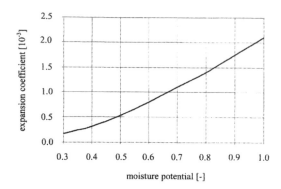

Figure 7. Hygral expansion coefficient as a function of the moisture potential.

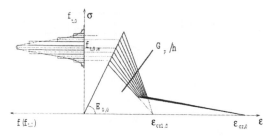

E_0	modulus of elasticity
$f_{t,0,m}$	mean tensile strength
G_F	fracture energy
$\varepsilon_{cr1,0}$, $\varepsilon_{cr,0}$	critical strains
h	characteristic finite element length

Figure 8. Crack band model for the description of the behaviour of concrete subjected to tension with the variation of tensile strength of concrete.

the rigid supports on the bottom side 3-node interface elements of the type CL12I were chosen. Boundary elements were used on the upper and the bottom side of the slab in order to simulate the moisture exchange due to convection between the concrete slab and the surrounding atmosphere.

Based on the assumptions of Alvaredo (1994) the convective boundary conditions in Equation (3.1) were assumed with a film coefficient $H_F = 5$ mm/day.

$$J_\Gamma = H_F \, (h_\Gamma - h_a) \qquad (3.1)$$

where J_Γ = moisture flux normal to the exposed surface; H_F = coefficient of surface hygral transfer (film coefficient); h_Γ = relative humidity at the surface; h_a = relative humidity of air.

By using the sorption isotherms of the applied concrete, the diffusion coefficient has been applied as a function of the moisture potential based on Kießl (1983). The initial potential of moisture was assumed to be 1.0. The hygral expansion coefficient (see Figure 7) and the creep function were specified on the basis of CEB-FIP Model Code 1990 (1993).

In order to determine the influence of the formation and propagation of cracks on the deformation behaviour of concrete slabs under critical loadings, the crack band model by Bažant & Oh (1983) was used, which allows the description of the behaviour of concrete subjected to tension. In the actual calculation the descending branch of the stress-strain relation was assumed to be bilinear. In order to obtain information on the quantity, periodicity and development of cracks, the heterogeneity of concrete was considered using statistical approach described in Mechtcherine & Müller (2000). The tensile strength $f_{t,0}$ was introduced to be an independent random variable following a Gaussian distribution (see Figure 8). The modulus of elasticity $E_{0,0}$ as well as the critical strains $\varepsilon_{cr,0}$ and $\varepsilon_{cr1,0}$ were kept constant for all

Table 1. Characteristic values of concrete and boundary conditions as used in the calculations.

Compressive strength (f_{cm})	30 MPa
Mean tensile strength ($f_{t,0,m}$)	3 MPa
Modulus of elasticity (E_{28})	30 GPa
Poisson's ratio	0.2
Fracture energy (G_F)	80 N/m
Relative humidity	35%
Age at beginning of drying	7 d
Ambient temperature	20°C
Hygral transient factor	5 mm²/d

finite elements. The elements were divided into 9 groups according to their tensile strength, which varied between 2.7 and 3.3 MPa. To each group its own material law was assigned.

Table 1 gives some input values for the applied concrete as well as the corresponding boundary conditions.

3.2 Results of the analysis of hygral deformation

The vertical deformations of the ends of the slab as calculated by the numerical model for slab 2 (drying on the upper side) agree very well with the results of the experiments (see Figure 9).

Figure 10 shows the calculated moisture potential distributions in slab 2 (drying on the upper side). The drying level with a potential of 0.9 reaches a level of 65 mm from the top of the slab after 90 days and a maximum depth of the drying front of 120 mm after 360 days. The ambient humidity of 35% was assumed to be constant in the investigated period of 360 days, which is in accordance to the experiments.

126

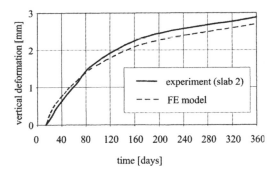

Figure 9. Development of the vertical deformation at the ends of slab 2 as a function of time.

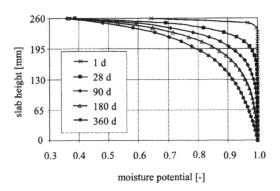

Figure 10. Moisture distribution over the cross-section of the slab as a function of moisture potential.

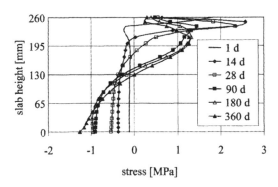

Figure 11. Stress distribution over the cross-section of a concrete pavement slab subjected to drying on the upper side.

The distribution of the stresses over the cross-section resulting from the hygral gradients due to drying of the upper side are shown in Figure 11. It can be seen, that a few days after beginning of the hygral loading the tensile stresses reach 2.5 MPa causing the formation of micro-cracks on the upper side of the slab. The peak of the tensile stresses moves toward the interior of the slab together with the drying front. The analysis showed that the cracks on the upper side reached a depth of approx. 30 mm after 360 days. This numerical model will be used in the parameter study to determine the stress and deformation behaviour of concrete slabs in various ambient boundary conditions. The results gained, contribute in the development of the new design method.

4 SUMMARY AND CONCLUSIONS

Using the finite element programme DIANA the experiments on large-size concrete slabs were simulated in order to investigate the deformation and cracking behaviour of concrete pavements, which were exposed to severe weather conditions. A numerical model has been developed using realistic static and dynamic material laws. The major target of these investigations is to develop a new design method considering the interaction of the thermal and hygral loads with live loads, taking into account the non-linear behaviour of concrete, in contrast to conventional methods. Further, specified investigations concerning the joint zones and the connecting anchors promise improvements with regard to the construction of more durable concrete pavement slabs.

REFERENCES

Alvaredo, A.M. 1994. Drying shrinkage and crack formation. Building Materials Reports No. 5. ETH Zürich.
Bažant, Z.P. & Oh, B.H. 1983. Crack band theory for fracture of concrete. Matériaux et Constructions, Vol. 16, No. 93.
CEB-FIP Model Code 1990, 1993. Bulletin D'Information No. 213/214, Comité Euro International du Béton. Lausanne.
DIANA Finite Element Analysis, User's Manuals release 7.2. TNO Building and Construction Research, Delft.
Eisenmann, J. 1979. Betonfahrbahnen Handbuch für Beton-, Stahlbeton- und Spannbetonbau. Verlag Wilhelm Ernst & Sohn.
Foos, S. (in prep.). Nichtlineare Bemessung von befahrbaren Betonplatten. Dissertation, University of Karlsruhe.
Kießl, K. 1983. Kapillarer und dampfförmiger Feuchtetransport in mehrschichtigen Bauteilen – Rechnerische Erfassung und bauphysikalische Anwendung. Dissertation, University of Essen.
Mechtcherine, V., Müller, H.S. 2000. A stochastic type continuum model for analysis of fracture behaviour of cementitious materials. Proceedings of the Sixth International Symposium on Brittle Matrix Composites,

127A.M. Brandt, V.C. Li and I.H. Marshall (eds.), Woodhead Publishing Ltd./Cambridge – ZTUREK/ Warsaw, pp. 241–250.

Müller, H.S., Hörenbaum, W., Maliha, R. 2002. Numerische Untersuchungen zur Rissbildung von Fahrbahndecken aus Beton. Final Report, FE-Nr.: 08.123 R 93 L, Institute of Concrete Structures and Building Materials, University of Karlsruhe.

Springenschmid, R. 1987. Neue Erkenntnisse beim Straßenbeton. Straße und Autobahn 12/87.

Finite Elements in Civil Engineering Applications, Hendriks & Rots (eds.)
© 2002 Swets & Zeitlinger, Lisse, ISBN 90 5809 530 4

Finite element modeling of early age concrete behavior using DIANA

R. Witasse & M.A.N. Hendriks

TNO Building and Construction Research, Delft, The Netherlands

ABSTRACT: The main nonlinear phenomena that govern the deformational behavior of early-age concrete are the evolution of the stiffness properties, the development of thermal and shrinkage strains, creep and cracking. In this article, a set of particularly useful features provided by DIANA for this kind of application is presented thanks to the analysis of a purification wall construction. Calculation of temperature and stress evolutions is carried out in a quite complex situation for which a one way coupled thermo-mechanical analysis has been set in combination with a phased analysis. Both the background theory and the construction of the finite element model is presented with special emphasis to the Japanese design code.

1 INTRODUCTION

The chemical processes that occur during hardening of young concrete are accompanied by significant volume and temperature changes. Indeed, hydration of cement is a highly exothermic and thermally activated reaction. The exothermic nature of the chemical reactions leads to heat generation which may result in high temperature rises of up to 50°C. As the rate of hydration slows down, the temperature decreases resulting in a thermal shrinkage which induces significant tensile stresses. Furthermore, this hydration process is accompanied by a volume reduction usually called autogenous shrinkage which will also result in the development of tensile stresses inside the constitutive materials. The damage induced at this moment has major consequences for the long-term structural performance of concrete structures and on its durability and serviceability.

At the TNO Building and Construction Research, investigations on concrete structures that are sensible to early age cracking dates back from the late eighties (Bogert et al. 1987, de Borst & van den Boogaard, 1994). Intensive work has been made on numerical modeling of young hardening concrete behavior for the prediction of deformation and cracking of young concrete.

The aim of this article is to describe the analysis of a purification wall under early temperature load with special emphasis to Japanese design code. It presents a set of particularly useful features provided by DIANA for this kind of application.

2 PRESENTATION OF THE STRUCTURE

The structure which has been modeled for this article is a concrete specimen that has been subjected to experimental and numerical investigations by the Japanese Concrete Institute for a study on temperature stresses in mass concrete (JCI 1998).

It consists of a base slab, with a thickness of 0.8 m and a width of 5.3 m, on which has been cast 36 days later a water-purification wall with a thickness of 0.7 m and a height of 2.3 m. This structure is 28.6 m long and is laying on a soil foundation. The casts employed both for the base slab and the wall were removed at the concrete age of 7 days. The configuration of the structure along with measurement points are illustrated in figure 1.

3 MODELING STRATEGY

The starting point for a calculation is to decide which kind of coupling between temperature and stress has to be taken into account. In the present case, it is widely accepted that it is accurate enough to calculate the fields of temperature and degree of reaction in a first step, and then use the results from the thermal analysis as input for the computation of stress.

To perform the complete analysis of the structure, two separate analysis have been performed. Within each analysis the two construction phases, i.e. the casting of the slab and the casting of the wall, are

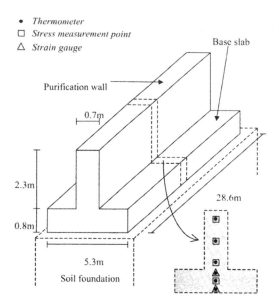

Figure 1. Presentation of the specimen.

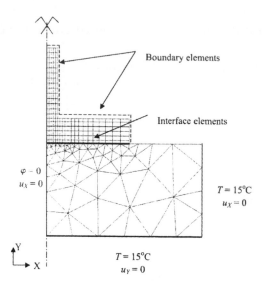

Figure 2. Mesh definition and boundary conditions.

distinguished. The first analysis is a phased potential analysis for the calculation and the storage of temperature and degree of reaction and the second analysis is a phased structural analysis with time-dependent material properties.

An interface application is used to map the temperatures and any other state variables (in our case the degree of reaction) that are calculated in the discretization used for the heat flow analysis onto the discretization that is used for structural analysis.

The mesh that is employed for our study is represented in figure 2. Due to the large length of the structure, plain strain elements are used with the same discretization in space for the thermal and the stress analysis. These isoparametric elements are quadratically interpolated with 8 nodes for the concrete parts and 6 nodes for the soil foundation. However, within the thermal analysis, DIANA converts them to linear elements with respectively 4 and 3 nodes in order to ensure compatibility between the temperature in the thermal analysis and the thermal strains in the structural analysis.

4 THERMAL ANALYSIS

4.1 *Background theory*

The temperature development in hardening concrete due to hydration can be described by:

$$\rho C \frac{\partial T}{\partial t} = \operatorname{div}\left(k \ \overrightarrow{\operatorname{grad}} \ T\right) + q_{V}, \tag{1}$$

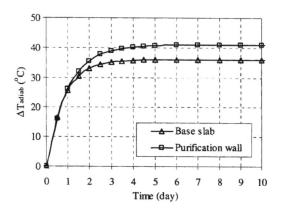

Figure 3. Adiabatic temperature rises.

where ρ is the mass density, C is the specific heat, k is the thermal conductivity, T is the temperature and q_V is the heat production rate defined as a function of the adiabatic temperature rise T_{adiab}:

$$q_{V} = \rho C \frac{\partial T_{\text{adiab}}}{\partial t}. \tag{2}$$

The evolutions of T_{adiab} with time are given by JCI (1998) and are represented in figure 3.

The development of the concrete maturity is described by the degree of reaction which is defined as:

Table 1. Thermal material properties.

	Foundation	Base slab	Wall
Thermal conductivity ($W \cdot m^{-1} \cdot K^{-1}$)	2.21	3.60	3.60
Mass density ($kg \cdot m^{-3}$)	1700	2301	2286
Specific heat ($kJ \cdot kg^{-1} \cdot K^{-1}$)	1.88	1.17	1.17
Convection coefficient ($W \cdot m^{-2} \cdot K^{-1}$)	Adiabatic condition	With cast: 17.4 Without cast: 8.1	

$$r(t) = \frac{\int_0^t q_V(\tau)d\tau}{\int_0^\infty q_V(\tau)d\tau}. \quad (3)$$

which is saved after each time step for further use in the stress analysis.

4.2 Main results

The material properties that are used for the thermal analysis are reported in table 1.

For this analysis, we just present a comparison of numerical values against available experimental data recorded during the whole construction process as shown in figures 4(a) and 4(b) respectively for the base slab and the purification wall.

Numerical results are represented by continuous lines while symbols (squares, circles and triangles) correspond to experimental results. They show really close agreement to each other.

5 STRESS ANALYSIS

5.1 Concrete creep modeling

The model that is used to deal with concrete behavior has been originally developed by de Borst & van den Boogaard (1994).

As point of departure for the creep formulation we take:

$$\boldsymbol{\varepsilon}(t) = \int_0^t J(t,\tau)\boldsymbol{C}\dot{\boldsymbol{\sigma}}(\tau)d\tau \quad (4)$$

in which σ is the stress matrix, $J(t, \tau)$ is the creep function and $\boldsymbol{C}$ is the matrix representation of the fourth order tensor defined by:

$$\boldsymbol{C} = \frac{1}{2}(1+v)\left(\delta_{ik}\delta_{jl} + \delta_{il}\delta_{jk}\right) - v\delta_{ij}\delta_{kl}, \quad (5)$$

wherein v is the Poisson's ratio which is assumed to be independent of the time of load application $\tau = t$.

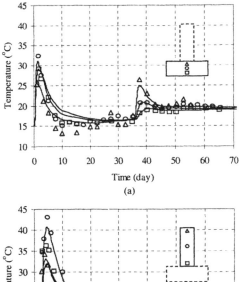

(a)

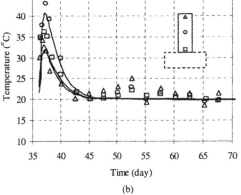

(b)

Figure 4. Temperature evolution inside the structure (a) for the base slab and (b) for the purification wall.

We now consider the time interval $\Delta t = t_{i+1} - t_i$ and define the corresponding increments of stress $\Delta\boldsymbol{\sigma} = \boldsymbol{\sigma}(t_{i+1}) - \boldsymbol{\sigma}(t_i)$ and strain $\Delta\boldsymbol{\varepsilon} = \boldsymbol{\varepsilon}(t_{i+1}) - \boldsymbol{\varepsilon}(t_i)$, so that we can write from (4) the following expression of strain increment:

$$\Delta\boldsymbol{\varepsilon} = \int_0^{t_i}\left[J(t_{i+1},\tau) - J(t_i,\tau)\right]\boldsymbol{C}\dot{\boldsymbol{\sigma}}(\tau)d\tau$$
$$+ \int_{t_i}^{t_{i+1}} J(t_{i+1},\tau)\ \boldsymbol{C}\dot{\boldsymbol{\sigma}}(\tau)d\tau. \quad (6)$$

Within displacement-method-based finite elements, it is more convenient to have constitutive relations in which the stress increment is explicitly written as a function of the strain increment. For this reason, (6) is rearranged such that the stress increment $\Delta\boldsymbol{\sigma}$ is a function of strain increment and strain history. Assuming the stress varies linearly over the time increment and introducing the dimensionless matrix $\boldsymbol{D} = \boldsymbol{C}^{-1}$, we obtain the following general expression for computing the stress increment in linear-viscoelastic, aging materials:

$$\Delta\boldsymbol{\sigma} = \tilde{E}(t^*)\boldsymbol{D}\Delta\boldsymbol{\varepsilon} + \tilde{\boldsymbol{\sigma}}(t_i), \tag{7}$$

with:

$$\tilde{E}^{-1}(t^*) = \frac{1}{\Delta t}\int_{t_i}^{t_{i+1}} J(t_{i+1},\tau)\,\mathrm{d}\tau, \tag{8}$$

and

$$\tilde{\boldsymbol{\sigma}}(t_i) = -\tilde{E}(t^*)\int_0^{t_i}\left[J(t_{i+1},\tau) - J(t_i,\tau)\right]\dot{\boldsymbol{\sigma}}(\tau)\,\mathrm{d}\tau. \tag{9}$$

To avoid the storage of the entire load history as it is implied by the supplied algorithm, the creep function $J(t,\tau)$ is commonly expanded into a series of negative exponential powers (Dirichlet series) or polynomials (Taylor series). For creep in early age concrete relatively short time spans have to be analyzed. Moreover, the stress fluctuations during this period may be more pronounced than in long-term processes. For these reasons, it may be advantageous to develop $J(t,\tau)$ in a Taylor series instead of in a Dirichlet series.

For study of basic creep in young hardening concrete, the double power law is often used. It is expressed as:

$$J(t,\tau) = \frac{1}{E(\tau)}\left[1 + \alpha\tau^{-d}(t - \tau)^p\right], \tag{10}$$

where α, p and q are material parameters. The part $g(t - \tau) = (t - \tau)^p$ of the double power law is expanded in a Taylor series around $\tau = t_d$ which leads to:

$$g(t - \tau) = \sum_{\mu=0}^{5} h_\mu(t - t_d)\tau^\mu, \tag{11}$$

wherein h_μ is a function of $t - t_d$ and is dependent on the power p and the development point t_d. Using (11), equations (8) and (9) can finally be rewritten as:

$$\tilde{E}^{-1}(t^*) = E^{-1}(t^*)$$
$$\times\left[1 + \frac{\alpha}{\Delta t}\sum_{\mu=0}^{5}\frac{(t_{i+1}^{\mu-d+1} - t_i^{\mu-d+1})}{(\mu - d + 1)}h_\mu(t_{i+1} - t_d)\right], \tag{12}$$

and:

$$\tilde{\boldsymbol{\sigma}}(t_i) = -\tilde{E}(t^*)\sum_{\mu=0}^{5}\left[h_\mu(t_{i+1} - t_d)\right.$$
$$\left. - h_\mu(t_i - t_d)\right]\tilde{\boldsymbol{\varepsilon}}_\mu(t_i). \tag{13}$$

where the internal variable $\tilde{\boldsymbol{\varepsilon}}_\mu$ is updated after each time step according to:

$$\tilde{\boldsymbol{\varepsilon}}_\mu(t_{i+1}) = \tilde{\boldsymbol{\varepsilon}}_\mu(t_i) + \frac{\alpha(t_{i+1}^{\mu-d+1} - t_i^{\mu-d+1})}{E(t^*)(\mu - d + 1)\Delta t}\Delta\boldsymbol{\sigma}. \tag{14}$$

Thanks to experimental data, de Borst & van den Boogaard (1994) suggested the following material parameters $\alpha = 4.0$, $p = 0.25$ and $q = 0.35$. The evolution of the Young modulus of concrete is given by the JSCE Design Code (1999):

$$E(t) = 4700\phi(t)\sqrt{f_c(t)}, \tag{15}$$

with:

$$\phi(t) = \begin{cases} 0.73 & \text{for } 0 \le t < 3 \text{ days} \\ 0.135t + 0.325 & \text{for } 3 \text{ days} \le t < 5 \text{ days} \\ 1.00 & \text{for } t \ge 5 \text{ days} \end{cases} \tag{16}$$

and for slowly hardening cement:

$$f_c(t) = \frac{t}{6.2 + 0.93t}f_c(91), \tag{17}$$

where $f_c(91)$ is the compressive strength at the age of 91 days and equals to 29 MPa (JCI 1998).

5.2 Shrinkage deformations

When we additionally include a shrinkage strain increment $\Delta\boldsymbol{\varepsilon}^{sh}$ in the constitutive model, we obtain the following general formalism:

$$\Delta\boldsymbol{\sigma} = \tilde{E}(t^*)\boldsymbol{D}(\Delta\boldsymbol{\varepsilon} - \Delta\boldsymbol{\varepsilon}^{sh}) + \tilde{\boldsymbol{\sigma}}. \tag{18}$$

In our case, the shrinkage strain increment is the sum of thermal and autogenous strain increments.

Thermal shrinkage strain increment is expressed as:

$$\Delta\boldsymbol{\varepsilon}^{th} = \beta\boldsymbol{I}\Delta T, \tag{19}$$

with $\boldsymbol{I} = \mathrm{diag}\{1, 1, 1, 0\}$ for plane strain elements and where β is the thermal expansion coefficient equals to $7.4\cdot 10^{-6}\text{K}^{-1}$ for the soil foundation and $10\cdot 10^{-6}\text{K}^{-1}$ for concrete parts.

Autogenous shrinkage strain increment is given by the following formula:

$$\Delta\boldsymbol{\varepsilon}^{au} = 2\boldsymbol{\varepsilon}^{au,\infty}(1 - r)\boldsymbol{I}\,\mathrm{d}r; \tag{20}$$

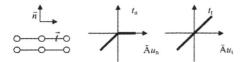

Figure 5.   Constitutive model for interface element.

wherein r is the degree of reaction and $\varepsilon^{au,\infty}$ is the ultimate autogenous strain defined according JCI (1996) and equals to $50 \cdot 10^{-6}$.

5.3   Modeling of the soil-concrete interface

To model the frictional interaction between base slab and soil, we make use of interface elements as shown in figure 5 (Bosnjak 2000). These elements are quadratically interpolated with lower nodes which belong to the soil discretization and upper nodes which are shared with the concrete structure. The material model which is used sets a relation between the stresses and the relative displacement across the interface.

If the vertical settlements of the structure lead to compression, we assume a linear relationship between the normal stress t_n and the normal relative displacement Δu_n such as:

$$t_n = k_n \Delta u_n,   \tag{21}$$

with $k_n = 63 \cdot 10^6 \, \text{MN·m}^{-3}$. As the Young modulus of soil equals 630 MPa, the normal stiffness k_n represents a thickness of 10^{-5} m and is thus considered under compression as stiff as the soil. If the structure lifts up, k_n is set to 0 (no tension bedding).

The horizontal restraining forces from the ground are caused by friction between the ground and the base slab. For the sake of simplicity, the normal stress t_t in the contact area is always proportional to the shear slip Δu_t (relative tangential displacement) with a modulus $k_t = 25 \cdot 10^6 \, \text{MN·m}^{-3}$.

5.4   Crack index output

To judge whether the crack arises or not the following index is used (JSCE 1999):

$$I_{cr} = \frac{f_t(t)}{\sigma_1(t)},   \tag{22}$$

in which $\sigma_1(t)$ is the maximal principal stress and $f_t(t)$ is the tensile strength of concrete which is defined as (JCI 1998):

$$f_t(t) = 0.35\sqrt{f_c(t)},   \tag{23}$$

where $f_c(t)$ has been previously defined in (17).

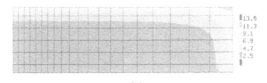

(a)

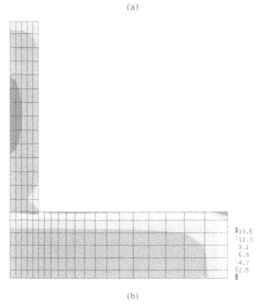

(b)

Figure 6.   Crack index contour levels (a) after 36 days and (b) after 66 days.

Contour levels of crack index values are represented in figures 6(a) and 6(b) at the end of each construction stage for which the minimum value of crack index is observed ($I_{cr} = 2.5$). We can conclude that with the chosen material parameters, no cracks will appear.

5.5   Comparison with experimental results

We now present a comparison of stress values calculated by DIANA against available experimental results in figures 7(a) and 7(b). Symbols correspond to experimental values with the same meaning as for the presentation of the thermal analysis. Lines correspond to numerical values obtained respectively for the bottom part (dotted line), the middle part (bold line) and the top part (ordinary line) of concrete structure parts.

Qualitatively, our results are satisfactory. Physical phenomena are correctly reproduced by the finite-element model: the temperature increase in the heating phase at low Young, modulus values generates small compressive stresses whereas the temperature decrease in the cooling phase at higher Young, modulus values leads to more significant stresses in tension.

Quantitatively, we observe a quite reasonable agreement between the model simulation and the

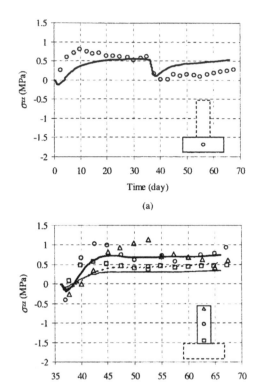

(a)

(b)

Figure 7.   Stress evolution inside the structure (a) for the base slab and (b) for the purification wall.

experimental results even if our calculation leads to rather slow evolution of longitudinal stresses compared to the experiment.

6 CONCLUSION

We demonstrate that DIANA offers advanced features to handle with studying young hardening concrete behavior. Indeed, calculations of temperature and stress evolutions have been carried out with a one-way-coupled thermo-mechanical analysis. Both the thermal and the structural computations comprise a phased analysis simulating the two construction phases. Realistic simulations can be obtained with the model although the predictive values of the computations can be further enhanced by improved constitutive models for the stiffness evolution, the creep behavior and the shrinkage deformations of early-age concrete.

ACKNOWLEDGEMENT

This work is the result of an intensive collaboration between TNO Building and Construction Research Department and Japan Information Processing Service Company. The authors would like to express their grateful appreciation to those who have collaborated in this project at JIPS and particularly to Dr Shengbin Gao and Mr Kazuhiro Kawaguchi for their scientific support and English translations of Japanese reports.

REFERENCES

Bosnjak, D. 2000. *Self-Induced Cracking Problems in Hardening Concrete Structures.* PhD Dissertation, Department of Structural Engineering, The Norwegian University of Science and Technology. Trondheim.

De Borst, R. & van den Boogaard, A.H. 1994. Finite-Element Modeling of Deformation and Cracking in Early-Age Concrete. *Journal of Engineering Mechanics,* American Society of Civil Engineers, 120(12): 2519–2534.

JCI (ed.) 1996. *Report by Research Committee of Autogenous Shrinkage,* Japan Concrete Institute, 11: 117–118 (in Japanese).

JCI (ed.) 1998. A Reconsideration of External Constraint Coefficient and an Expansion of Application Scope for Compensation Plane Method, *Report by Research Committee of Temperature Stress of Mass Concrete,* Japan Concrete Institute, 4: 141–144 (in Japanese).

JSCE (ed.) 1999. *Japan Concrete Specification.* Technical report. Japan Society of Civil Engineers, Tokyo (in Japanese).

Van den Bogert, P.A.J., de Borst, R. & Nauta, P. 1987. Simulation of the mechanical behavior of young concrete. IASBE report, 54: 339–347.

Finite Elements in Civil Engineering Applications, Hendriks & Rots (eds.)
© 2002 Swets & Zeitlinger, Lisse, ISBN 90 5809 530 4

Numerically simulated pore structure and its potential application in high performance concrete

Guang Ye, K. van Breugel & Jing Hu
Faculty of Civil Engineering and Geosciences, Delft University of Technology, Delft, The Netherlands

ABSTRACT: High Performance Concrete (HPC) has been defined as concrete with improved overall performance compared to conventional concrete. The high density of the microstructure is responsible for high strength and improved durability. However, the dense microstructure may have an adverse effect on the fire resistance. The formation of vapor goes along with expansion and the built-up of large stresses in the pore system, since the dense microstructure does hardly permit any pressure relief through the pores. The capillary pore structure, particularly the connectivity of the pores, plays a dominant role in this respect. In this contribution, a 3-D pore structure of cement paste, simulated with the numerical simulation program HYMOSTRUC, will be presented. The specific geometrical parameters, including the pore size distribution and the connectivity of capillary pores, can be deduced from the program. The potential of the program for analyzing the susceptibility of high performance concrete to spalling if subjected to fire loads will be discussed.

1 INTRODUCTION

Fire remains one of the main threads to modern buildings. The increasingly intensive use of concrete in construction industry makes it necessary to fully understand the effects of fire on concrete. Generally concrete is thought to have good fire resistance. However, further development and use of high performance concrete (HPC) give rise to some doubt on the fire resistance of this extremely dense material.

Concrete is, in essence, a non-combustible material. If exposed to fires, however, the properties of concrete will change due to chemical and physical processes and phenomena that are caused by exposure to elevated temperatures. The fire resistance could be evaluated by residual compressive strength of concrete after exposure to high temperature. Extensive investigations (Xu 2001, Chan 1999, Castillo et al. 1990, Rostasy et al. 1980, Bentz et al. 2000, Consolazio et al. 1998, Atlassi 1993) were undertaken in this field. It is now commonly accepted that the behavior of concrete subjected to high temperature depends on many factors, including both environmental conditions and the constituents of the concrete. As regards the physical response of the concrete to elevated temperatures, the thermal incompatibility between the aggregates and cement paste and the pore pressure within the cement paste have to be mentioned

as the main detrimental factors (Chan 1999). Furthermore, at high temperature, the C-S-H gel may start to decompose. From previous research work, HPC has been found to be prone to spalling under high temperature in some cases (Castillo et al. 1990). As investigated by Bentz et al. (2000), the failure mechanism of the explosive spalling is due to the buildup of very high pore pressures within the HPC, due to the liquid-vapor transition of the capillary pore water as well as that bound in the cement paste component of the concrete. If this water vapor cannot escape from the specimen, significant pressures will develop and may eventually cause spalling of the concrete. In this respect concretes made with silica fume, due to their very dense microstructure, are very susceptible to the built-up of high vapour pressure in the pore system.

According to Xu (2001), concrete suffered from a loss of integrity when exposed to 250°C. This could be explained, partly, by the coarsening of the pore structure in hardened cement paste. Therefore, it is of great importance to acquire a better understanding of the pore structure of cement paste in the assessment of fire resistance of high performance concrete (HPC). In Xu's work, mercury intrusion porosimetry (MIP) technique was employed to determine the pore size distribution in cement paste. From these experimental data temperature-induced changes of the

135

microstructure could be obtained. By comparing these microstructural changes with the changes in the macroscopic properties, like the compressive strength, the judgement of the response of concrete to elevated temperatures could be given a more fundamental basis. However, the MIP-technique for analyzing pore structure characteristics and of changes thereof has also serious limitations. Moreover, the tests are tedious and time-consuming.

With rapid development of computer technology, numerical simulations have been widely used in research of cement-based materials, including the research of microstructural features of these materials. It is possible and realistic to accomplish a complete 3D presentation of concrete microstructures. As mentioned before, the capillary pore structure, particularly the connectivity of the pores, plays a dominant role in the fire resistance and durability of HPC. In this paper, 3D numerical simulation of concrete with the numerical simulation program HYMOSTRUC will be presented. After that, a method will be proposed to characterize the pore structure of cement paste. The specific geometrical parameters, including the pore size distribution and the connectivity of capillary pores, can be deduced from the HYMOSTRUC program. The potential of this simulation program for analyzing the susceptibility of HPC to spalling when subjected to fire loads will be discussed as well.

2 PORE STRUCTURE OF CONCRETE SUBJECTED TO HIGH TEMPERATURE

Analyses of concrete that had been subjected to fire exhibit a reduction of the mechanical strength, spalling and cracking. As regards the microstructure of fire-exposed concrete, a change in pore structure has been observed, i.e. "pore structure coarsening" (Rostasy et al. 1980). This could account for the increase in concrete permeability and loss of the durability, as had been confirmed in experimental tests.

The pore structure has been recognized as an important characteristic of the concrete, influencing the properties such as the strength, durability and permeability. It has been demonstrated that the total porosity of hardened cement paste increased by only about 1–2% after exposure to 250°C with an insignificant change of average pore diameter (Xu 2001). The coarsening of hardened cement paste mainly occurred after concretes have been exposed to 450°C and 650°C. The porosity in hardened cement paste increased by about 4% after exposure to 450°C and by around 10% after exposure to 650°C. It has also been observed that the average pore diameter dramatically increases, even by a factor two or three, after the concrete has been exposed to 450°C and 650°C.

Chan's research (Chan 1999) also confirmed the coarsening of the pore structure at high temperatures. The cumulative volume of pores in the range greater than 1.3 μm, which are supposed to be responsible for the permeability of HCP, increased dramatically after exposure to 600°C (Neville 1981). This finding allows us to draw to the conclusion that high temperatures have reduced the permeability-related durability of HPC, as well as its mechanical strength. It was proved that, compared with conventional concrete, HPC suffered a smaller reduction of its mechanical strength but a greater reduction of the permeability-related durability (Chan 1999).

The mechanism causing the deterioration of fire-exposed concrete is quite complex because of the complicated chemical and physical changes in hardened cement paste, aggregate, and at the matrix-aggregate interfaces. Decomposition of the major hydrate, known as C-S-H, is inevitable as the exposure temperature reaches up to 650°C, which causes severe coarsening of the microstructure and a loss of binder property.

3 NUMERICAL SIMULATION OF MICROSTRUCTURE IN CEMENTITIOUS MATERIAL

3.1 Simulation of the hydration in cementitious material

It has been widely accepted that the mechanical properties and also the long-term behavior of concrete structures are strongly related to concrete's microstructure and the changes thereof during the concrete's lifetime. The microstructure originates from the hydration process, and in fact most the concrete's future is determined in the first period of the hydration process. Accurate simulation of the cement hydration process requires the knowledge of complicated chemical reactions between different components of cementitious materials and of the physical interaction between cement particles and of physical solid-moisture interactions. Moreover, percolation processes – percolation of solid matter and percolation of pores – play a very important role in the performance of cementitious materials and help to explain the dependency of transport properties, like ionic diffusivity and fluid permeability, on the microstructure (Garboczi et al. 2000). In the last 20 years research in this field has become known as Computational Material Science of Concrete (CMSC). The microstructural models ideally incorporate much of the basic physics and chemistry of the evolution of the microstructure.

In the HYMOSTRUC model (van Breugel et al. 1997), the cement hydration is simulated on the basis of a "growth mechanism" with the particle size distribution, the chemical composition of cement, the water/cement ratio and the reaction temperature as important model parameters. The simulation starts from unhydrated cement particles distributed in a 3D representative volume. With progress of the hydration process, cement particles and hydration products form a porous structure.

An example of a simulated cement paste and its pore structure are shown in Figure 1. In the upper part the cement paste model is shown, whereas in the bottom part the accompanying pore structure is presented.

It is interesting to compare the simulated pore structure with a realistic pore structure. For example, a cylindrical pore shape is commonly assumed to quantify the pore size distribution of cement paste when mercury intrusion porosimetry (MIP) technique is used. However, the simulated pore structure of extreme irregularity illustrated in Figure 1b will not agree to this assumption.

3.2 Geometrical and topological properties of capillary porosity

Quantitative description of the geometrical and topological properties of a simulated microstructure is of utmost importance for prediction of the transport properties. A program with the name of HYMOSTRUC3D (Ye et al. 2002) was developed for this purpose. In this program a serial sectioning algorithm associated with overlap criteria was utilised as shown in Figure 2. With this technique, that will be explained in the more detail in the following, it is possible to analyse the connectivity of the pores in a simulated microstructure and hence of the pressure relief paths for water vapour in case the materials is exposed to elevated temperatures.

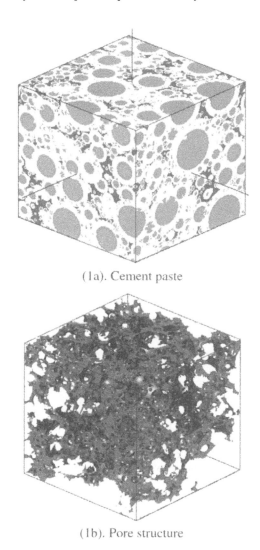

(1a). Cement paste

(1b). Pore structure

Figure 1. Simulated cement paste and its pore structure with w/c = 0.4, at degree of hydration 65%. The physical size of specimen is $100 \times 100 \times 100\ \mu m^3$.

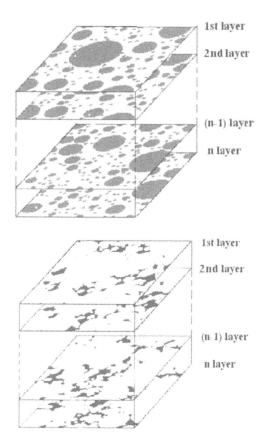

Figure 2. Determination of pore topology by using serial sectioning method and overlap criteria.

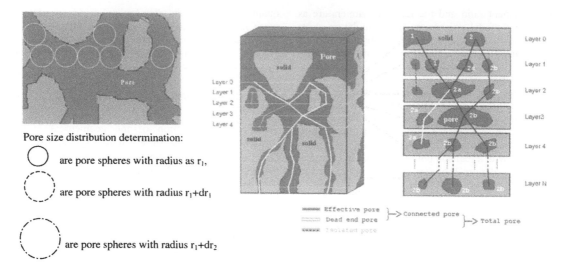

Pore size distribution determination:

○ are pore spheres with radius as r_1,

⟨ ⟩ are pore spheres with radius r_1+dr_1

⟨ ⟩ are pore spheres with radius r_1+dr_2

Figure 3. The pore network structure was determined by using an overlapping criterion. The effective pores, dead end pores and isolated pores were distinguished. The pore size distribution was determined by calculating the pore which could be covered by fiction spheres of radius r.

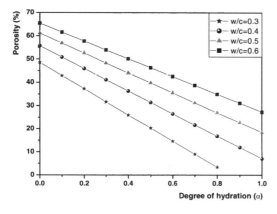

Figure 4. The development of porosity with hydration time.

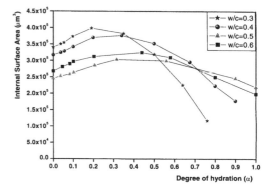

Figure 5. The development of total pore internal surface area as function of degree of hydration.

The serial sectioning method starts with scanning the 3D microstructure layer by layer. The features of all points in each 2D layer were determined by checking whether an arbitrary point belonged to the subset of the solid phase or to the subset of the capillary pore phase. Secondly, the neighbor points were checked to see whether they are of the same feature, renumber each set with points of the same feature as an individual object (pore or solid). The information of each object such as coordinate, area and perimeter were recorded. The connectivity of the object between neighbor layers was calculated by an "overlap criterion". The overlap criterion checked each object on its upper-layer and lower-layer to examine whether this object was connected with its neighbors of the same feature. The topological parameter genus was calculated. The volume fractions of the connected solid/capillary pore phase and the total internal pore surface area were derived. The reconstructed network structure is showing in Figure 3.

Figure 4 shows example where the total porosity and connected porosity are as function of degree of hydration, the sample with water/cement ratio from 0.3–0.6, at curing temperature 20°C. The simulation results demonstrate that the water/cement ratio is the most important factor that influences the total porosity and connected porosity. Figure 5 shows total internal pore surface area as function of degree of hydration.

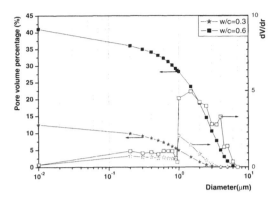

Figure 6. Pore size distribution of simulated cement paste with w/c ratio 0.3 and 0.6 at degree of hydration 0.64. Left axis: pore volume percentage. Right axis: derivative of pore size distribution.

The characteristic pore size distribution curve can be determined in the following way. Consider the sub-volumes of the system accessible to spheres of different radii. Let $\rho_{(r)}$ be the volume fraction of the pore space "coverable" by spheres of radius r. We call these spheres as pore spheres. $\rho_{(r)}$ is a monotonically decreasing function of r and can easily be compared with the "cumulative pore volume" curves measured by mercury intrusion porosimetry. The derivative $-d\rho_{(r)}/dr$ is the fraction of volume coverable by pore spheres of radius r but not by spheres of radius $r + dr$ and is a direct definition of the pore size distribution. The $\rho_{(r)}$ of the pore spheres with radius r can be computed by densely packing pore spheres in the free space. An example of pore size distribution for cement paste simulated by HYMOSTRUC model is displayed in Figure 6.

4 POTENTIAL OF HYMOSTRUC FOR ANALYZING THE FIRE RESISTANCE OF HPC

Built on earlier research work and extensive experiments investigations, this paper contributes to advancing the insight into the numerical simulation of cementitious materials subjected to high temperature and the pore structure characterization. It is pursued to present an outlook on possible future application of HYMOSTRUC for simulating and analyzing the susceptibility of HPC to fire load.

4.1 Simulation of HPC structure and pore space characterization

The pores structure of HPC could be simulated with HYMOSTRUC, taking into account the particle size distribution of the cement, the aggregate content, the water-cement ratio and the temperature. Different phases in the hardened cement paste could be represented by different color as shown in Figure 1. Pore structure of the simulated HPC could be visualized on the basis of Constructive Solid Geometry (CSG) as indicated in Figure 1. Pore size distribution, mean pore size, pore surface area and pore connectivity could be calculated according to algorithms introduced in section 3. Therefore, a complete and comprehensive characterization of the pore structure in a simulated HPC sample is available. Based on this, the permeability of HPC under normal state could be estimated as a function of porosity, pore surface area and tortuosity of the pore structure.

4.2 Pore space changes and degradation when subjected to high temperatures

As we know, when subjected to high temperature, hardened cement paste suffered from a "pore structure coarsening" effect. The porosity of the paste increases to some extent depending on the environment conditions such as heating rate and peak temperature. A concurrent increase in average pore diameter is to be expected as well. The pores structure, particularly the connectivity of the pores, and the amount of water in the pore system form the starting point for a micromechanical analysis of the built-up of temperature-induced pressures due to an external thermal load. Due to the built-up of vapour pressure in the pore system spalling of the surface layer of the concrete may occur, which directly leads to damage of the concrete.

In the past a variety of high temperature conditions tested and a variety of constituents in different concrete mixtures have been tested in view of their performance under thermal loads. This data is of utmost significance, since it provides us with valuable material for validating the reliability of numerical micromechanical models for analyzing concrete under fire loads. The present model, described in this paper, provides us with a microstructure that can be modified as a function of imposed thermal loads and the exerted stresses caused by the vapour pressure in the pore system. These temperature-induced changes in the pore systems will change the "degree of connectivity" of the pore system and hence of the stresses that will occur in the pore system. These temperature-induced changes of the pore system can in principle be evaluated with the HYMOSTRUC model.

Changes of the pore system will cause a change of the permeability. This means that after exposure of the concrete to a temperature load the resistance of the concrete against ingress of water and/or aggressive substances has decreased. With the help of numerical models, like the HYMOSTRUC, it becomes possible of evaluate the remaining durability of the concrete after exposure to a fire load.

5 CONCLUDING COMMENTS

In performance of a cement-based material, like concrete, depends on its microstructure, i.e. its pore structure. This holds for all mechanical and physical properties, including transport properties. Predictive models for analyzing transport processes require a reliable description of the pore structure, particularly of the pore size distribution and the connectivity of the pores. In the contribution the potential was shown of the HYMOSTRUC model to quantify the evolution of the pores structure, including the connectivity of the pores. This connectivity is crucial for the permeability of the material. If subjected to extreme thermal loads the connectivity of the pores determine the possibility of relief of vapour pressure and hence the mitigation of the risk of spalling of the concrete. At the moment only the potential of the model for in-depth thermal analysis is shown. For validating the model the numerically obtained pore systems must be compared with experimentally obtained pore structures and permeability studies. These studies are now in progress.

Once the applicability of microstructural models for quantitative analysis of the performance on cement-based systems is established, these models can be used for designing property-defined materials. This is in fact the major challenge for the future. The aim of this contribution was to stimulate the development of this type of these computer-based models.

ACKNOWLEDGEMENTS

The research was financially supported by the Dutch Technology Foundation (STW), which is gratefully acknowledged.

REFERENCES

Xu, Y. 2001. Impact of high temperature on PFA concrete, *Cement and Concrete Research* Vol. 31: 1065–1073.

Chan, Y.N. 1999. Residual Strength and Pore Structure of High-strength Concrete and normal strength concrete after exposure to high temperatures, *Cement and Concrete Composites* Vol. 21: 23–27.

Castillo, C. & Durrani, A.J. 1990. Effect of transient high temperature on high-strength concrete, *ACI Materials Journal* Jan-Feb: 47–53.

Rostasy, R.S., Weiss, R. & Wiedemann, G. 1980. Changes of pore structure of cement mortars due to temperatures, *Cement and Concrete Research* 10:157–164.

Bentz, D.P. 2000. Fibes, percolation, and spalling of high performance concrete, *ACI Materials Journal* Vol. 97(3): 351–359.

Consolazio, G.R., McVay, M.C. & Rish, J.W. 1998. Measurement and prediction of pore pressures in saturated cement mortar subjected to radiant heating. *ACI Materials Journal* Vol. 95 No. 5, 525–536.

Atlassi, E. 1993. *A quantitative thermogravimetric study on the nonevaporable water in mature silica fume concrete.* PhD thesis, Chalmers University of Technology, Goteborg, Sweden.

Neville, A.M. 1981. *Properties of Concrete.* London: Pitman Publishing Ltd.

Garboczi, E.J., Bentz, D.P. & Frohnsdroff, G.J. 2000. The paste. present, and future of the computational materials science of concrete. *Proceedings of the J. Francis Young Symposium* (Materials Science of Concrete Workshop), Lake Shelbyville, IL, April 27–29.

van Breugel, K. 1991. *Simulation of Hydration and Formation of Structure in Hardening Cement-based Materials,* PhD thesis, Delft University of Technology.

Ye, G., van Breugel, K. & Fraaij, A.L.A. 2002. Three-dimensional microstructure analysis of numerically simulated cementitious materials. Accepted to publish in *Cement and Concrete Research.*

Breugel, K. van, Koenders, E.A.B., Lokhorst, S.J. & van der Veen, C. 1997. Numerical prediction of the evolution of material properties in hardening concrete structures. *Proc. DIANA Computation Mechanics '97:* 425–436. Amsterdam.

Finite Elements in Civil Engineering Applications, Hendriks & Rots (eds.)
© 2002 Swets & Zeitlinger, Lisse, ISBN 90 5809 530 4

Numerical modelling of microstructure development and strength of cementitious materials

K. van Breugel, Ye Guang & V. Smilauer
Delft University of Technology,
Faculty of Civil Engineering and Geosciences, The Netherlands

ABSTRACT: For reliable predictions of strength, stresses and the risk of cracking in hardening concrete structures the evolution of the material properties should be known. In the past a couple of expressions have been launched which correlate the evolution of materials properties to the maturity or the degree of hydration. A more fundamental approach, however, would be to correlate the material properties to specific microstructural parameters. In this contribution it will be shown how a numerical microstructure can be generated with the simulation program HYMOSTRUC and how specific microstructural properties can be used for quantifying the evolution of material properties, like compressive and tensile strength. Examples of the potential of microstructure-based strength models will be presented. It will be shown how information of the microstructure can be made operational in finite element packages. Finally, the potential of microstructure-based material models will be mentioned.

1 INTRODUCTION

For numerical analyses of temperature fields, strength development and estimations of the risk of cracking in hardening concrete structures, the evolution of the material properties must be known. Although temperature predictions have been carried out since 1920, mainly for analyses of temperature fields in concrete dams, the credits for reliable temperature predictions should be given to the more recent computer programs.

The main difference between old and new programs concern the accuracy with which the reaction of the cement with water is modelled. For the determination of the temperature distribution in hardening concrete the differential equation of Fourier is normally used, viz.

$$\frac{\partial T}{\partial t} = a_c \cdot \nabla^2 T + q(x, y, z, \alpha, T) \tag{1}$$

In this equation $q(x, y, z, \alpha, T)$ is the heat source, which is a function of the degree of hydration α and the temperature T at the co-ordinate x, y, z. The degree of hydration indicates the percentage of the cement, or binder, that has been converted into hydration product, or gel.

For the effect of the temperature T on the rate of hydration the Arrhenius function is generally adopted, viz.:

$$\frac{\partial \alpha}{\partial T} = A \cdot e^{-\frac{E_A(\alpha)}{R \cdot T}} \tag{2}$$

in which E_A [kJ/mol] is the apparent activation energy, R is the universal gas constant, i.e. 8.361 kJ/mol K, and T [K] the reaction temperature.

In this contribution it will be outlined how the degree of hydration can be used to define a number of microstructural parameters and how these parameters can be used for predicting the evolution of the materials properties in a consistent way. The advantage of an approach where the materials properties are directly related to microstructural parameters will be discussed. First an overview of currently used strength concepts will be discussed.

2 MODELLING OF MATERIALS PROPERTIES

2.1 *Maturity and equivalent age*

For quantification of the evolution of material properties in hardening concrete, the engineering practice often uses the maturity concept. This concept is widely used for mathematical description of the development of strength gained at an arbitrary temperature

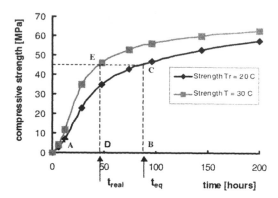

Figure 1. Principle of maturity concept for strength development.

history by relating this strength development to the strength development measured at a reference temperature of 20°C. The basic idea behind the maturity concept is:

same maturity = same strength

Maturity has often been expressed in "degree-Celsius-hours". It represents the area below the temperature curve, measured from a certain reference temperature. Schematically this is shown in Figure 1. In that figure the strength development is shown of two concrete specimens, cured isothermally at 20°C and 30°C, respectively. In the figure the reference temperature is taken at 0°C. If the areas ABC and ADE are equal, i.e. the concrete specimens hydrated at 20°C and at 30°C have the same maturity, the strength of the two concrete specimens should be the same.

In the last two decades a number of maturity "laws" have been proposed. The basis of the maturity concept is, in essence, phenomenological, and does not deal directly with the physical cause of the evolution of material properties. An exception must be made for the maturity concepts discussed by Powers (1960) and Hansen (1970). In their maturity concepts maturity is identical to degree of hydration.

Strongly associated with the maturity concept is the term "equivalent age". The equivalent age t_e is defined as the time that a concrete, that hardens at a temperature history T(t), needs to reach a certain value of the material property of interest. The equivalent time can be calculated on the basis of the Arrhenius function:

$$t_e = \int_0^t \exp\left[\frac{E_A}{R}\left(\frac{1}{273 + T_r} - \frac{1}{273 + T(t)}\right)\right] dt \quad (3)$$

The reference temperature T_r is generally taken 20°C. In case the actual hydration process proceeds at the reference temperature we get $t_e = t_{real}$, where t_{real} is the real time. If the curing temperature deviates from the reference temperature T_r, the equivalent time is longer or shorter than the reference time, depending on whether the actual temperature is higher or lower, respectively. Schematically this is also shown in Figure 1.

In eq. (3) the temperature sensitivity is determined by one parameter, i.e. the apparent activation energy E_A. The value of E_A is generally assumed constant throughout the hydration process. Experiments have shown, however, that the activation energy is a function of the degree of hydration. Moreover, a distinct temperature dependency has been found as well, even though this contradicts in fact with the principle that the activation energy should be a temperature independent quantity. The fact that the apparent activation energy has often been found to exhibit a temperature dependency indicates that we are dealing with a very complex process, of which the temperature dependency does not obey the rules of relatively simple chemical processes.

2.2 Porosity and gel-space ratio concept

In essence, the evolution of strength and of all other material properties is determined by the microstructure of the material. The microstructure of cement-based systems, however, is extremely complex. Models which represent these complex microstructures are, therefore, scarce. Precursors of microstructural models for strength development are the strength-porosity models and the gel-space ratio concept. In the porosity models the strength is related to the capillary porosity of the cement paste. With increasing hydration the capillary porosity decreases and the strength increases.

Examples of strength-porosity relationships proposed by several authors are presented in Figure 2. It is noticed that the capillary porosity P_{cap} can be represented as a function of the degree of hydration according to:

$$P_{cap}(\alpha) = \frac{\left(\dfrac{\omega - 0.3375\alpha}{\rho_w}\right)C + V_{air}}{\left(\dfrac{\rho_w}{\rho_{ce}} + \omega\right)C + V_{air}} \quad (4)$$

with ω the w/c ratio, C [gr] the cement content, ρ_w and ρ_{ce} the specific mass [gr/cm³] of the water and the cement, respectively, and V_{air} the initial air content (v/v). From eq. (4) it can easily been seen that the porosity is a linear function of the degree of hydration α.

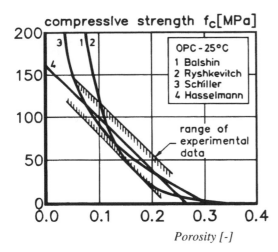

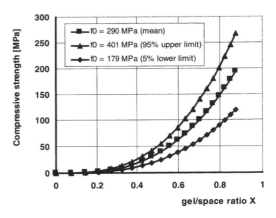

Figure 4. Strength f_c versus gel-space ratio X for different values of the intrinsic strength f_0.

Figure 2. Strength-porosity relationship as proposed by different authors (after Rossler 1985).

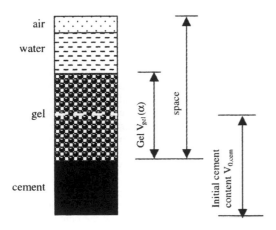

Figure 3. Schematic representation of the gel-space ratio concept.

According to Powers (1946/47) the strength of cement paste, mortar and concrete can be related to the gel-space ratio $X(\alpha)$. The gel-space ratio is defined as the quotient of the volume of the gel and the total volume of the paste *minus* the volume of the still unhydrated cement (Fig. 3). In formula form:

$$X(\alpha) = \frac{V_{gel}(\alpha)}{V_{air}(\alpha) + V_{capw}(\alpha) + V_{gel}(\alpha)} \qquad (5)$$

In the gel-space ratio concept the strength f_c is related to the gel-space ratio $X(\alpha)$:

$$f_c = f_0 * X(\alpha)^3 \qquad (6)$$

where f_0 is the intrinsic strength of the paste. The intrinsic strength f_0, i.e. the strength of the capillary pore-free cement paste, varies from about 200 MPa for ordinary cement (Locher 1976) up to 600 MPa for hot pressed cement pastes. For the intrinsic strength of mortar a value of $f_0 = 180 \ldots 342$ MPa is applicable with a mean value of 290 MPa and a standard deviation of 68 MPa (Fagerlund 1987). For three values of the intrinsic strength of the cement paste the f_c/X relationship according eq. (6) is presented in Figure 4.

2.3 Strength versus degree of hydration

In both the strength-porosity concept and the gel-space ratio concept the parameters to which the strength is related, i.e. the capillary porosity and the gel space-ratio, are functions of the degree of hydration α. In the introduction it has been noticed that in numerical predictions of temperature fields in hardening concrete the degree of hydration plays an important role. In those calculations the degree of hydration is considered to be identical with the amount of liberated heat at a certain stage of the hydration process divided by the potential amount of heat liberated in case all the cement present in the mixture has reacted. In formula form:

$$\alpha(t) = \frac{Q(t)}{Q_{pot}} \qquad (7)$$

A distinct relationship between strength and degree of hydration has been observed by many authors. Examples of f/α relationships are shown in Figure 5.

143

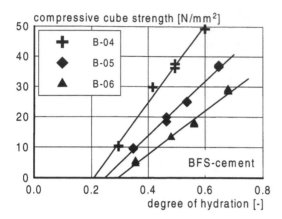

Figure 5. Compressive strength versus degree of hydration for concrete mixtures made with blast furnace slag cement with w/c ratio 0.4, 0.5 and 0.6 (Lokhorst 1998).

The relationship between strength and degree of hydration follows the general format:

$$f_c(t) = f_{max} \frac{\alpha(t) - \alpha_0}{1 - \alpha_0} \tag{8}$$

in which f_{max} is the fictitious maximum strength in case of complete hydration ($\alpha = 1.0$) and α_0 the critical degree of hydration below which the strength is zero. Figure 5 clearly shows that beyond the critical degree of hydration α_0 the f/α-relationship is almost linear and that α_0 increases with increasing water-cement ratio.

3 MICROSTRUCTURAL MODELS

3.1 Evolution of microstructural models

The strength concepts presented in section 2 are all based on changes in volume concentrations of individual components in the cement paste. In essence, however, the strength is determined by the strength of contacts between hydrating cement particles. In the early seventies Bache (1970) was among the first who tried to relate the strength of materials to the number of interparticle contacts. This is what microstructural models are actually aiming for. Microstructural models that are of interest in this respect are the NIST model (Bentz et al. 1997), the model developed by Navi (1996), the DUCOM system launched by researchers from Tokyo University (Meakawa et al, 1999) and the HYMOSTRUC model (van Breugel 1991). In the following only the HYMOSTRUC model will be discussed in more detail.

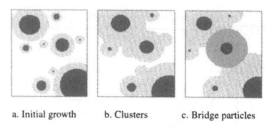

a. Initial growth   b. Clusters   c. Bridge particles

Figure 6. Stages in the formation of the microstructure – schematic.

3.2 Simulation model HYMOSTRUC

In the HYMOSTRUC program cement particles are considered as spheres. The hydration process is modelled as a process of concentric inward penetration of the reaction front and simultaneous outward growth of the shell of reaction products. Individual chemical components of which the cement particles consist are assumed to hydrate at equal rates. The initial rate of penetration depends on the clinker composition, mainly the calciumtrisilicate and calciumdisilicate content. At first the reaction starts as a phase-boundary reaction. With progress of the hydration process the contacts between the cement particles become more intense and the rate of the reaction process will gradually become diffusion-controlled. Figure 6 shows three stage of the process of formation of interparticle contacts and hence of the development of strength and stiffness.

Let us start now from the hypothesis that strength is a function of the number of contact points or contact areas between hydrating cement grains. A closer look on the results of a numerically simulated microstructure teaches that not all the contact points can be considered equally effective in yielding strength and stiffness. This is explained further with the help of Figure 6. The process starts with a homogeneous or random distribution of particles in the paste. Until a particle becomes embedded in the outer shell of a bigger particle it is called a "free" particle. The system consisting of a free particle with smaller particles embedded in its outer shell is called a cluster (Figure 6b). The distances between clusters are bridged by particles that are not embedded completely in a cluster. These particles, denoted bridging particles, exceed the shell thickness of the free particle of a cluster and can also be partly embedded in another neighbouring cluster (Figure 6c). The clusters and the bridging particles are considered as the two basic "structural elements" which determine the stiffness and strength of the microstructure.

With HYMOSTRUC the interparticle contacts and the bridge volume can be calculated as a function of

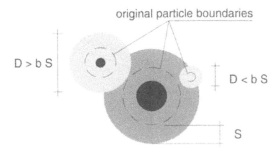

original particle boundaries

embedding & bridging

Figure 7.   Schematic presentation of bridge criterion principle. A particle can act as a bridge in case its diameter D is
larger than the product of the bridge length factor b and the
thickness of the outer shell of the free particle (Lokhorst
1998).

the degree of hydration and the water/cement ratio.
For the volume of the hydrating clusters, including
the bridge volume, it holds:

$$v_{clus} = (1 - \alpha) \cdot v_{0,cem} + \alpha \cdot \nu(T) \cdot v_{0,cem} \qquad (9)$$

with $v_{0,cem}$ is the initial volume of the anhydrous
cement, ν the ratio of the volumes of the cement
gel and the anhydrous cement and α the degree of
hydration.

3.3 Bridge criterion

In order to act as a bridge, a bridge particle should
exceed a certain minimum size, which size is related
to the thickness of the outer shell of the free particle
(Fig. 7). This minimum particle size can be formulated in a *bridging criterion*. This criterion states that
the diameter D of the bridging particle must be b
times the thickness S of the outer shell of a free particle. The factor b is called the *bridge length factor*. A
bridge length factor b = 0 means that all embedded
cement particles are equally effective in the formation
of interparticle contacts. Based on an extensive
parameter study Lokhorst (1998) found that a bridge
particle should have a diameter of about 1 to 3 times
the shell thickness of the free particle i.e. a bridge
length factor in the range of 1 to 3.

3.4 Modelling of temperature effect

Higher reaction temperatures will increase the rate of
the hydration process, but will also result in a coarse
capillary pore structure. The latter phenomenon can be
simulated by adopting a temperature dependent growth
factor $\nu(T)$. Based on experiments by Bentur et al.

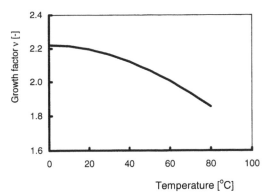

Figure 8.   Temperature dependency of growth factor $\nu(T)$
(Van Breugel 1991).

(1979), the temperature dependency of $\nu(T)$ is
approximated by the expression:

$$\nu(T) = 2.22 \cdot e^{-28 \, 10^{-6} \, T^2} \qquad (10)$$

The temperature dependency of the growth factor
$\nu(T)$ is schematically shown in Figure 8.

4   MECHANICAL PROPERTIES AND
MICROSTRUCTURE

4.1 Effect of water-cement ratio

For a bridge length factor b = 0 all the embedded
cement is assumed to be equally effective in making
interparticle contacts. As a first approximation it is
assumed that the volume of embedded cement represents the intensity with which growing cement particles become connected and hence represents the gain
in strength. For concrete mixtures made with Portland
cement and with water-cement ratii from 0.37 to 0.70
the cylinder compressive strength is shown in Figure 9
as a function of the degree of hydration. Similar to
Figure 5, also here the strength exhibits a linear relationship with the degree of hydration. Even though the
strength and the degree of hydration appear to be
strongly correlated, it is also clear that the water-
cement ratio plays a dominant role in this relationship.

Figure 10 shows the relationship between the compressive strength and the embedded cement volume
V_{emb}. It appears that the relationship between the
compressive strength and the embedded cement volume is less dependent on the water-cement ratio than
the relationship between strength and degree of
hydration. This suggests that the embedded cement
volume is a more unique, i.e. fundamental, strength
parameter than the degree of hydration.

145

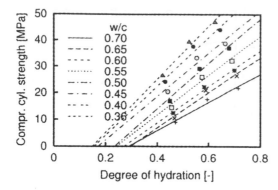

Figure 9. Linear f/α relationship for concrete mixtures with different w/c ratio (Smilauer 2001).

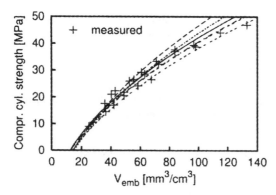

Figure 10. Compressive strength versus calculated embedded cement volume. Bridge length factor b = 0. Concrete mixtures with different w/c ratio (Smilauer et al. 2001).

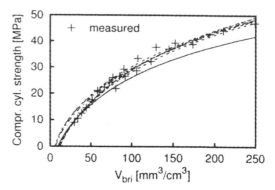

Figure 11. Compressive strength versus calculated bridge volume. Bridge length factor b_{bri} = 1.5. Concrete mixtures with different w/c ratio (Smilauer et al. 2001).

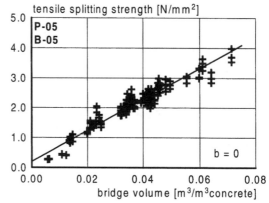

Figure 12. Tensile splitting strength as function of the bridge volume b = 0 (Lokhorst 1998).

In Figure 11 the compressive strength of the same concretes as meant in Figure 9 and 10 is presented as a function of the calculated bridge volume. In the calculation of the bridge volume a length factor b_{bri} = 1.5 was adopted. Figure 11 shows that the correlation between the compressive strength and the bridge volume with b_{bri} = 1.5 is stronger than that between strength and the embedded cement volume with b_{emb} = 0 shown in Figure 10. This is to be expected, since in the embedded cement volume all particles, including the very small ones, are assumed to be equally effective in the built up of the strength. In the model, however, the bridge particles dominate the development of the strength.

4.2 *Tensile strength of concrete as function of the bridge volume*

A typical result of the correlation between the splitting tensile strength and bridge volume is shown

in Figures 12 and 13. Mixtures made with both Portland cement and blast furnace slag cement were considered, denoted as mixtures P-05 and B-05, with w/c = 0.5. The fineness of the cements was 300 and 450 m²/kg, respectively. In Figure 12 the bridge volume was calculated for a bridge length factor b = 0, whereas in Figure 13 a value b = 2 was used. A comparison of these two figures shows that for a bridge volume up to 0.04 m³/m³ the figures are almost identical. At higher values the scatter around the mean value is less for a bridge length factor b = 2 than for b = 0. This result supports the concept of the bridge length factor, i.e. the idea that particularly those particles are relevant for the development of the strength which are able to connect hydrating clusters with each other.

Another detail of the comparison of the Figures 12 and 13 concerns the strength at the zero-value of the bridge volume. The trendline in Figure 12 suggests

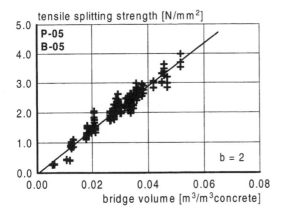

Figure 13. Tensile splitting strength as a function the bridge volume b = 2 (Lokhorst 1998).

that for a bridge volume $v_{br} = 0$ the concrete would have some strength already. Within the concept of the model this is difficult to understand, since strength cannot be expected as long as hydrating clusters are not connected with each other. Figure 13 shows that for a bridge length factor b = 2 this inconsistency has disappeared.

4.3 *Effect of temperature*

As explained in section 3.4, the effect of elevated curing temperatures on the final strength is simulated by adopting a temperature dependent growth factor. Figure 14 shows the result of a simulated microstructure of a cement paste, w/c = 0.5, obtained at two isothermal temperatures, viz. 20°C and 50°C. The degree of hydration of the two pastes is the same.

A coarser pore structure implies less interparticle contacts and hence a lower strength (Kjellsen et al. 1991). A lower strength gain at elevated temperatures has been observed in many experiments. At reduction of the 28-strength up to 20% is conceivable if hydration takes place at curing temperatures in the early stage of the reaction process of about 60°C.

In Figure 15 the relative compressive strength at 28 days $f_{c,28}(T)/f_{c,28}(22°C)$ is presented as function of the calculated relative amount of embedded cement $V_{emb,28}(T)/V_{emb,28}(22°C)$. Curing took place isothermally at 22, 30, 40 and 50°C. At 28 days the degree of hydration varied from 0.73 for the specimens cured at 22°C to 0.75 for curing at 50°C. The strength and the calculated embedded cement volume at 28 days were fixed at 100%.

Figure 15 shows that a lower strength is accompanied with a lower amount of embedded cement. This gives support to the assumed correlation between the evolution of the strength and of the interparticle

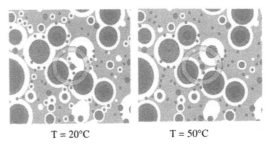

Figure 14. Simulated microstructures, obtained at different temperatures. w/c = 0.5. Degree of hydration the same in both cases.

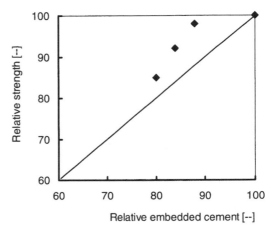

Figure 15. Relative strength versus relative amount of embedded cement after 28 days hydration at different temperatures (deduced from van Breugel 1991).

contacts, of which the amount of embedded cement is an indicative parameter.

5 DISCUSSION AND CONCLUSIONS

For reliable predictions of the risk of cracking in hardening concrete structures the evolution of the material properties must be known. In the past these properties have been linked with the maturity or were simply formulated as a function of time. In essence, however, the evolution of the materials properties are determined by the evolution of the microstructure. This holds for the strength and stiffness, but also for time dependent properties, like shrinkage and creep and relaxation. Moreover, transport properties depend on the microstructure as well, more in particular on the pore structure.

In this contribution it has been explained how the evolution of materials properties of cement-based systems can be described as a function of microstructural parameters. By relating the evolution of materials properties to microstructural parameters, a basis for a consistent quantitative evaluation of the performance of cement-based materials is obtained.

Multi-scale modelling is the most practical way to make microstructural models operational for engineering applications. At the moment multi-scale modelling is still in a process of development. For the time being the degree of hydration is a suitable quantity by the help of which FE-packages can get access to microstructural models. The evolution of the degree of hydration is obtained directly from temperature calculations of hardening concrete. For a certain concrete mixture, microstructural data and pore structure data can be linked to the actual degree of hydration. In this way finite element packages, like DIANA, have access to microstructural data. This data can then be used, in essence, in numerical analyses of the overall performance of concrete structures, for example for the determination of the evolution of strength, stiffness, creep and relaxation in hardening concrete. With the generated microstructure as a basis and with mathematical "monitoring" of the state of water in the system, the model can also been used to quantify the self-desiccation process, through which process the microstructure is put in compression by the forces exerted by the water in the pore system (Koenders et al. 1998). The simulated pore structure can also be used for numerical analysis of transport phenomena and in durability studies.

Although the potential of microstructural models is great, it has to be emphasized that we are still at the beginning of the evolution of numerical modeling of cement-based systems. However, the development of numerical microstructural models goes very fast. The increasing number of this kind of models make it conceivable that in the near future the use of them in the engineering practice will increase significantly. The authors hope that this contribution might stimulate the development of these models in view of the enhancement of the quality of concrete structures.

REFERENCES

Bache, H.H. 1970. Model for strength of brittle materials built up of particles joined at points of contact, *J. of The Am. Ceram. Soc.* 53 (12). 654–658.

Bentur, A. 1979, Berger, R.L. & Kung, J.H. & Milestone, N.B. & Young, J.F. Structural properties of calcium silicate pates – Effect of curing temperature. *Journal of American Ceramic Society*, Vol. 62, 362–366.

Bentz, D.P. 1997. Three-Dimensional computer simulation of Portland cement hydration and microstructure development", *J. Am. Ceram. Soc.* 80 (1) 3–21.

Breugel, K. van 1991. Simulation of hydration and formation of structure in hardening cement based materials. Delft: Delft University Press.

Fagerlund, G. 1987. Relations between the strength and the degree of hydration or porosity of cement paste, cement mortar and concrete. *Proc. Seminar of Hydration of cement*, Copenhagen.

Hansen, T.C. 1970. Physical composition of hardened Portland cement paste. *Journal Americal Concrete Institute*, 404–407.

Kjellsen, K.O. (1991), Detwiler, R.J. & Gjørv, O.E. Development of microstructure in plain cement pastes hydrated at different temperature. *Cement and Concrete Research*. Vol. 21, 179–189.

Koenders, E.A.B. & Breugel, K. van, 1998. Volume changes in hardening cement paste, in *Heron*, 43 (2) 99–118.

Locher, F.W. 1976. Die Festigkeit des Zements. *Beton* (7) 247–249, (8) 283–285.

Lokhorst, S.J. 1998. *Deformational behaviour of concrete influenced by hydration-related changes of the microstructure.* Research Report TU Delft.

Navi, P. & Pignat, C. 1996. Simulation of effects of small Inert grains on cement hydration and its contact surfaces. In *The modelling of microstructure and its potential for studying transport properties and durability*, Eds. Jennings et al., NATO ASI Series E, 304. 227–241.

Powers, T.C. 1946/1947. Studies of the physical properties of hardened Portland cement paste. *ACI Journal* Part 1–9. ACI-Journal.

Powers, T.C. 1960. Properties of cement paste and concrete. *Journal of American Concrete Institute*, Vol. 16, (4) 577–609.

Rossler, M. & Odler, I. 1985. Investigations on the relationship between porosity, structure and strength of hydrated Portland cement pastes – Effect of pore structure and of degree of hydration. *Cement and Concrete Research*, Vol. 15, 401–410.

Smilauer, V. & van Breugel, K. (2001). *Evolution of hydration process and microstructure – Relation between microstructure and strength.* Research Report, Delft University of Technology. 30 p.

Finite Elements in Civil Engineering Applications, Hendriks & Rots (eds.)
© 2002 Swets & Zeitlinger, Lisse, ISBN 90 5809 530 4

Development of bond between reinforcement steel and early-age concrete

K. van Breugel & M.S. Sule

Delft University of Technology, Faculty of Civil Engineering and Geosciences

ABSTRACT: Bond properties between steel and concrete form essential input for crack width calculations. For hardened concrete these properties are well known. There is, however, hardly any information about the bond properties of young concrete. As a consequence of this the reliability of crack width predictions in hardening concrete is not very high. In order to improve our knowledge about the bond between steel and hardening concrete, pull-out tests have been performed on normal strength and high strength concrete according to RILEM Recommendations. Reinforcement bars (Ø16 mm) were cast in cubes, with rib length 200 mm. The specimens were cured at different curing temperatures (20°C, 30°C, 40°C isothermal and semi-adiabatic). This was realised by using temperature-controlled moulds. In order to find a correlation between the development of cube compressive strength and pull-out strength the test specimens were cured under the same conditions and tested at the same age (after 8 h, 24 h, 30 h, 48 h and 28 days).

From the experimental results the development of bond stress is formulated in an alternative bond-slip relation. This formulation could be implemented as a user-defined model in DIANA for a more accurate prediction of the cracking behaviour of reinforced concrete at early ages.

1 INTRODUCTION

The durability of concrete structures largely depends on the quality of the concrete cover on the reinforcing steel and on the crack width. Cracks in concrete structures have, therefore, to be controlled at any stage. The basis for crack width control is the force transmission between steel and concrete. In order to be able to calculate crack widths at early-age, more must be known about bond mechanisms between reinforcing bars and young concrete. Stresses between steel and concrete start to develop when strains in concrete and steel differ from each other. The local bond behaviour is generally characterised by local bond stress τ and the related slip value Δ. The bond stress is often described with a power function (Noakowski 1978)

$$\tau = a * \Delta^b \qquad (1)$$

The parameters a and b depend on the geometry of the steel surface and the concrete compressive strength. The parameter a is often described as a constant which depends on the cube compressive strength f_{ccm} and has the form:

$$a = k * f_{ccm} \qquad (2)$$

With eq. (2) the aforementioned power function (1) becomes

$$\tau = k * f_{ccm} * \Delta^b \qquad (3)$$

For hardened concrete König et al. (1996) suggests for the parameter k the value 0.31 and for b the value 0.3. For these value of k and b in eq. (3) we get:

$$\zeta(\Delta) = 0.31 * f_{ccm} * \Delta^{0.3} \qquad (4)$$

In the present investigation the bond strength in normal strength concrete (NSC) and high strength concrete (HSC) is studied at very early age. At the start of the research project it was still unclear whether the parameters k and b in eq. (4) would have the same values in case of young concrete. In hardening concrete the bond strength between steel and concrete may change due to a change of the dominating bond mechanisms in subsequent stages of the hardening process. This may result in factors k and b, which are functions of the evolution of the concrete properties, i.e. the concrete strength. For consistent modelling of the properties of hardening concrete it would be advantageous if the parameters k and b could be formulated as a function of the degree of hydration. The degree of hydration can easily be obtained in a temperature calculation of a hardening concrete structure (van Breugel, 2001). Alternatively the degree of reaction is calculated, which parameter can be used in the same way as the degree of hydration as a key parameter for describing the evolution of the material properties.

2 BOND MECHANISMS

Between hardened NSC and reinforcing bars three bond mechanisms can be distinguished (Fig. 1):

- Adhesion
- Friction
- Bearing of the lugs against the concrete

After failure of adhesive bond, two mechanisms are triggered in concrete with rebars: friction and bearing of the lugs against the concrete. Friction is the resistance against a parallel displacement between two surfaces that are kept in contact by a compressive force perpendicular to the contact plane (Eligehausen et al. 1998). Bond of rebars is mainly governed by bearing of the lugs against the concrete. The concentrated bearing forces in front of the lugs radiate out into surrounding concrete with a certain inclination α. These forces can be decomposed into directions parallel and perpendicular to the bar axis. The parallel components equal the bond force, whereas the radial components induce circumferential tensile stresses in the surrounding concrete. Unless sufficient confinement is available, these tensile stresses may result in splitting bond failure. With sufficient confinement no splitting failure will occur, but instead of that pull-out of the bar will take place.

As there is a significant difference between NSC and HSC in microstructure and internal microcracking (Han 1996), bond behaviour differs also substantially. About this difference in bond performance Azizinamini et al. (1993) developed the following hypothesis.

In NSC the ultimate bond stress distribution is assumed to be uniform, which implies that all lugs bear against concrete at the ultimate stage (Fig. 1c). Their test observation gives reason to the conclusion that this is not the case for HSC. Here the first lugs at the loaded side resist more than the other ones and consequently create higher bearing forces. The latter in turn result in higher radial tensile forces which cause splitting of concrete surrounding the rebars in the vicinity of these lugs. This type of failure takes place before all lugs participate in resisting the applied axial force.

3 EXPERIMENTS

3.1 Equipment

The standard test for determination of the bond strength between steel and concrete is the pull-out test

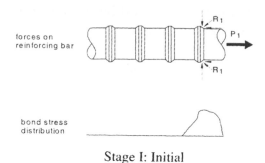

forces on reinforcing bar

bond stress distribution

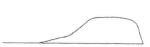

Stage I: Initial

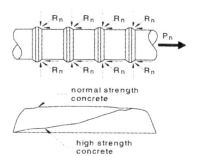

Stage II: Middle stage

normal strength concrete

high strength concrete

Stage III: Final stage

Figure 1. Schematic representation of bond mechanisms between concrete and reinforcing steel subjected to tension. Idealization of behaviour of deformed reinforcing bars at several loading stages (Azizinamini et al. 1993).

according to RILEM (1970). The reinforcement bars with 16 mm diameter were cast in cubes with a side length of 200 mm. The rebars had an embedment length of 5 times in NSC ($l_b = 80$ mm) and 3 times the bar diameter in HSC ($l_b = 48$ mm). The casting direction was parallel to the orientation of the rebar in loading direction.

The moulds for the cubes and the pull-out specimens were made of 10 mm thick steel (Fig. 2). As the

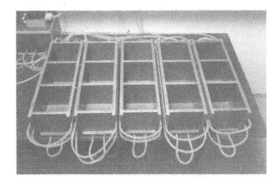

Figure 2. Temperature controlled moulds for concrete cubes and pull out strength specimens.

Table 1. Concrete compositions.

Component	HSC	NSC
Water	125.4 kg	175.0 kg
CEM III/B 42.5 LH HS	237.0 kg	–
CEM I 52.5 R	238.0 kg	–
CEM I 32.5 R	–	350.0 kg
Microsilica	50.0 kg	–
Admixture BV1	1.0 kg	–
Admixture FM951	9.5 kg	–
Gravel 4–16 mm	973.5 kg	–
Gravel 4–8 mm	–	827.6 kg
Sand 0–4 mm	796.5 kg	1011.5 kg
Cube compressive strength after 28 days of series I	98.4 MPa	34.4 Mpa

As indicated in the previous section, in most experiments isothermal curing conditions were used. Experimental results obtained in semi-adiabatic tests will be presented as function of the degree of hydration. This makes it possible to evaluate them in the same way as isothermal tests.

The reinforcing steel, with diameter Ø16 mm, was made of hot rolled steel FeB 500 HWL with a yield strength of 568 MPa and a projected rib area f_R of 0.086.

The cube compressive strength tests and pull-out tests were carried out at different concrete ages, viz. 8 h, 24 h and 31 h. In fact two test series were performed, series I and II. In these two series the same types of concrete were used.

The pullout tests have been conducted displacement-controlled. The test velocity was 0.05 mm/s. Due to the limits of the test rig (100 kN) the top of pullout strength could not be reached in all tests, particularly not in the case of the high strength concretes.

development of the hydration process and the associated reaction temperature is of major importance for the development of the microstructure, and consequently for the material properties, all specimens have been cast in temperature-controlled moulds. For that purpose the exterior of the moulds were foreseen with copper pipes, which are connected with silicone hoses to a cryostat. In one concrete specimen the temperature of the hardening concrete is measured. This temperature is sent to a PC, which regulates the temperature of all the moulds via the water temperature of the cryostat.

In fact any chosen curing temperature can be imposed onto the concrete, either semi-adiabatic or isothermal. In this paper mainly results of isothermal tests will be considered.

3.2 Mixture composition and test variables

In the project two types of concrete were considered, viz. a traditional normal strength concrete (NSC) and a more advanced high strength concrete (HSC). The mixture compositions of the two concretes are presented in Table 1.

4 TEST RESULTS

From the experimental program the following information was obtained:

- Cube compressive strength of the concrete at the age of testing pullout
- Pull-out force versus free end slip (at a certain deformation)
- Visual appearance of interface after failure

4.1 Cube compressive strength

The mean value of the cube compressive strength of the two concrete mixtures is shown in Figure 3 as a function of time. In that figure the strength development in the first 48 hours is shown.

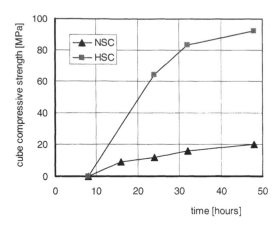

Figure 3. Concrete cube compressive strength of NSC and HSC as a function of time (Series I).

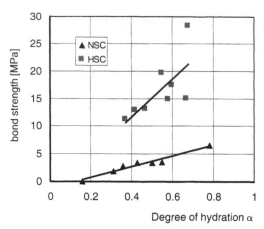

Figure 5. Development of bond stress at a slip of 0.1 mm as a function of the degree of hydration. Mean values of two test series.

4.2 *Bond strength*

The bond strength observed in the pull out tests is shown in Figure 5 as a function of the degree of hydration at a free end slip of 0.1 mm. The bond stress is calculated by dividing the measured pull out force by the bond surface ($A_b = 2\pi r l_b$). For NSC this gives the bond stresses which are distributed uniformly. Presuming that bond stress distribution in HSC is nearly triangular, the calculated value is a mean value.

In the figure linear trend lines show that the bond strength develops in the same way as the cube compressive strength (compare with Fig. 4).

4.3 *Bond – slip relation*

In the Figures 6 and 7 two examples are presented of bond – slip relations in young concrete. In Figure 6 the bond force – slip relations is shown at an age of the concrete of 8 hours for both NSC and HSC. After 24 and 31 hours the bond strength has increased significantly. The bond – slip relations are shown in Figure 7. The latter figure clearly shows that in HSC the free end slip is much smaller than in NSC. This shows that the bond strength and the bond stiffness in HSC are higher than in NSC.

For HSC mixtures Figure 8 shows the evolution of the bond stress – slip relations in the first 48 hours after casting of the concrete. Peak stresses occur systematically at a slip of about 1 mm.

In HSC small slip values, from 0 to 0.2 mm, yielded bond stresses as presented in Figure 9. The figure shows the bond stress – slip values for subsequent values of the degree of hydration, i.e. from

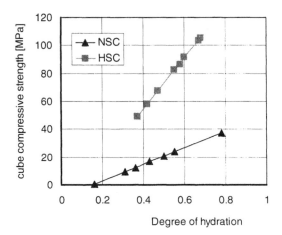

Figure 4. Concrete cube compressive strength as function of the degree of hydration α (mean values of two series).

In Figure 4 the cube compressive strength is shown as a function of the degree of hydration. In this figure strength data cover a longer curing period. In the latter case the degree of hydration was determined analytically with a specially developed software package, called UCON (Van Beek, 1991). Figure 4 confirms the often observed trend that, after a certain initial period with hardly any strength development, the strength is a linear function of the degree of hydration. This observation is very important in case the strength development in hardening concrete structures has to be "monitored" by numerical simulation.

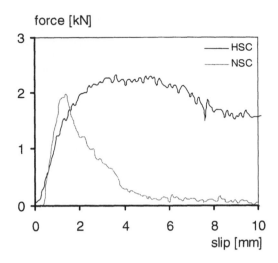

Figure 6. Pull-out force versus free end slip at a concrete age of 8 hours (Series I).

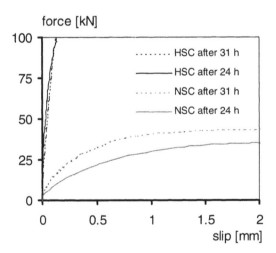

Figure 7. Pull-out force versus free end slip after 24 and 31 hours (Series I).

$\alpha = 0.37$ to $\alpha = 0.68$. In Figure 10 the ratio τ/f_{ccm} is plotted against the slip observed at different values of the degree of hydration. As will be shown in section 5, the presentation of the results shown in Figure 10 constitute a suitable starting point for developing a computer-friendly equation for the bond stress – slip relationship for hardening concrete.

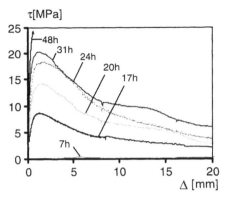

Figure 8. Bond stress – slip relation of HSC cured isothermally at 20°C from 7 hours to 48 hours.

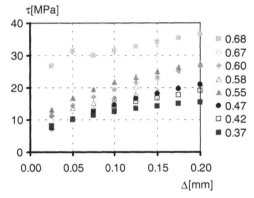

Figure 9. Bond stress as function of the slip for different values of the degree of hydration of HSC.

4.4 Visualisation of failure modes

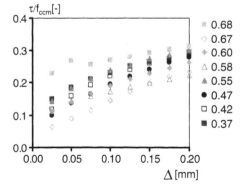

Figure 10. Relationship between the ratio of bond stress and compressive strength as function of the slip of HSC.

Figure 11. Rebar embedded in NSC, pulled out after 24 hours.

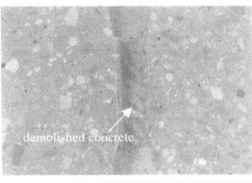

Figure 12. Rebar embedded in HSC pulled out after 8 hours.

As the test load was limited to 100 kN not all the specimens failed. In case failure did occur, pull out failure was the dominating failure mode. Obviously the size of the concrete specimens, i.e. $200 \times 200 \times 200 \, mm^3$, provided sufficient confinement for the steel bar of 16 mm to prevent splitting failure. The photos in Figures 11 and 12 show NSC and HSC specimens tested after 24 h and 8 h, respectively. Characteristic differences between pull out failure in NSC and HSC are illustrated in these figures. In NSC the concrete between the lugs shears from the confining concrete. In HSC the lugs demolish the concrete console.

5 EVALUATION

In the introduction the bond stress – slip relationship was formulated with equation (3):

$$\tau = k * f_{ccm} * \Delta^b$$

In hardening concrete the cube compressive strength $f_{ccm}(\alpha)$ and the bond strength $\tau(\alpha)$ are functions of the degree of hydration. In the introduction it was suggested that the parameters k and b might be functions of the degree of hydration. An extensive evaluation of the dependency of the parameters k and b on the degree of hydration revealed that such a relationship did exist. However, in the same study it was found that, if written as function of the degree of hydration, these k and b values did not lead to satisfactory results in crack width calculations.

At the present state of knowledge it was found that reasonable crack widths could be found with the bond stress – slip equation:

$$\tau(\alpha) = 4.8\alpha_0 f_{ccm}(\alpha)\Delta^{3.6\alpha_0} \qquad (5)$$

in which α_0 is the critical degree of hydration beyond which the strength and the stiffness of the concrete start to develop. This critical value depends on the water/cement ratio (w/c) of the concrete was found to be:

$$\alpha_0 = 0.3 \, w/c \qquad (6)$$

According to our experimental data the value of α_0 is 0.15 and 0.1 for the NSC and HSC, respectively.

In the Figures 13 and 14 the ratio between bond strength and compressive strength, τ/f_{ccm}, is represented as function of the slip, together with the analytical bond stress curves according to eq. (4) and (5). In order to ease the comparison between experimental data and analytical curves the eq. (4) and (5) were reshaped into the form $\tau/f_{ccm} = k \, \Delta^b$.

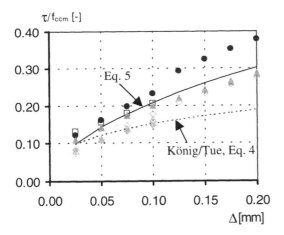

Figure 13. Ratio τ/f_{ccm} versus slip Δ of NSC. Experimental data for different degrees of hydration and analytical bond stress curves.

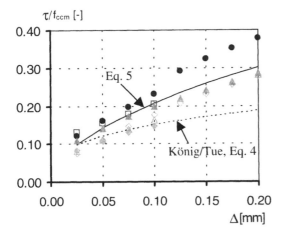

Figure 14. Ratio τ/f_{ccm} versus slip Δ of HSC. Experimental data for different degrees of hydration (see also Fig. 10) and analytical bond stress curves.

6 DISCUSSION AND CONCLUSIONS

Reliable predictions of the cracking process in hardening concrete structures require information about the evolution of the bond behaviour between steel and concrete. Data on bond between steel and young concrete is scarce and new experimental data is badly needed to feed computer packages with which hardening concrete structures can be analysed. For consistent modelling of hardening concrete structures it would be an advantage, if not a prerequisite, to formulate

all material properties, including the bond properties, as a function of one material parameter. The degree of hydration of hardening concrete is the most obvious parameter that can fulfil this requirement. The advantage of adopting the degree of hydration as a key-parameter is that this parameter is easily obtained from well-accepted numerical temperature analysis of hardening concrete.

In this study bond properties of hardening concrete were determined for both a normal strength concrete and a high strength concrete mixture. In the experiments it was found that the age of testing and the concrete composition strongly influence the bond performance. After 8 hours, when there is almost no compressive strength, bond between steel and concrete relies on adhesion. With elapse of time the compressive strength and the pull out strength increase. In HSC this development is faster than in NSC (Fig. 4 and 5).

It was investigated whether traditional bond relationships could also be used for describing the bond properties of hardening concrete. The starting point was a power function to describe the bond stress – slip relation. As a result of an extensive evaluation of the experimental data it was found that the bond stress – slip relation of young concrete could be described with the expression:

$$\tau(\alpha) = 4.8\alpha_0 \, f_{ccm}(\alpha) \, \Delta^{3.6\alpha_0}$$

The expression shows that the evolution of the bond properties can be related to the evolution of the cube compressive strength $f_{ccm}(\alpha)$. The parameters k and b in the original eq. (3) could be replaced by factors, which remain constant throughout the hydration process. Their magnitude was found to be a function of the critical degree of hydration α_0, i.e. the degree of hydration at which the strength and stiffness of the concrete in view starts to develop. The critical degree of hydration depends on the water/cement ratio of the concrete mixture as indicated in eq. (6).

The fact that the detailed analysis has resulted in a rather simple correlation between the bond strength and the cube compressive strength is confirmed by the experimental data presented in the Figures 4 and 5. In both figures the trend lines which approximate the compressive strength and the bond strength are linear curves, which suggests a linear relationship between these two properties.

In conclusion it can be said that the bond properties of hardening concrete can be described as a function of the cube compressive strength of the concrete. This facilitates the implementation in currently used software packages for analysing the behaviour of concrete structures at early ages. In the numerical analysis of hardening concrete structures the degree of hydration is a key-parameter to which all materials properties can be linked.

ACKNOWLEDGEMENTS

The assistance of Mr. A. van Rhijn in carrying out the experiments is highly appreciated. The financial support of the Dutch Technology Foundation (STW), the Ministry of Transport, Public Works and Water Management (RWS) is gratefully acknowledged.

REFERENCES

Azizinamini, A. et al., 1993. Bond Performance of Reinforcing Bars Embedded in High Strength Concrete. *ACI Structural Journal*, Vol.90, 5, 554–561.

Beek, A. van, 1991. Determination of strength of young concrete. Msc-Thesis, TU Delft.

Breugel, K. van, 1991. Simulation of hydration and formation of structure in hardening cement-based materials. PhD Thesis, TU Delft.

Eligehausen, R. & Bigaj-van Vliet, A.J., 1998. Anchorage and Detailing Principles. *fib International Course on Advanced Design of Concrete Structures*, Treviso, Italy.

Han, N., 1996. Time Dependent Behaviour of High Strength Concrete, Thesis, TU Delft.

König, G. & Tue, N.V., 1996. Grundlagen und Bemessungshilfen für die Rißbreitenbeschränkung im Stahlbeton und Spannbeton. *DafStb (Deutscher Ausschuss für Stahlbetonbau)*. Heft 466.

Noakowski, P., 1978. Die Bewehrung von Stahlbetonbauteilen bei Zwangsbeanspruchung infolge Temperatur. DafStb, Heft 296.

RILEM/CEB/FIP committee, 1970. Tests and specifications of reinforcements for reinforced and prestressed concrete. Bond test for reinforcing steel, Pullout test. *Materials and Structures, Research and Testing 3*, 15, 175–178.

Rehm, G. 1961. Über die Grundlagen des Verbundes zwischen Stahl und Beton. *DafStb*, Heft 138.

Sule, M.S. & Breugel, K. van, 2000. The Effect of Reinforcement on Stress Development in Early Age High Strength Concrete. *Int. Conf. Onf High Performance Concrete*. Eds. Leung, et al. Vol.I, 985–992. Hong Kong & Shenzhen.

Shear in reinforced concrete

Finite Elements in Civil Engineering Applications, Hendriks & Rots (eds.)
© 2002 Swets & Zeitlinger, Lisse, ISBN 90 5809 530 4

Fracture analysis of bar pull-out in beam-column joints

D. Jankovic
Civil Engineering and Geo-Sciences Department, Delft University of Technology, Delft, The Netherlands

S.K. Kunnath
Department of Civil and Environmental Engineering, University of California at Davis, Davis, USA

M.B. Chopra
Department of Civil and Environmental Engineering, University of Central Florida, Orlando, USA

ABSTRACT: In order to predict the bond-slip behavior of reinforcing bars in a typical interior beam-column joint of a reinforced concrete frame structure, experiments and numerical analyses were conducted. A concrete cylinder with a centrally embedded deformed rebar served to represent the pull-out condition. The bar was subjected to a monotonic tensile axial load at the free end while the other end was embedded in concrete. One of the primary variables was the content of propylene fibers in the concrete since one of the objectives of the study was to investigate the bond characteristics of polypropylene fiber concrete. The resulting bond-slip mechanism was studied as an axisymmetric 2D problem using the Finite Element Analysis software, DIANA. Input data for the numerical analysis were derived from the experiments. The influence of the variations in the bar embedment length as well as the addition of different percentage of polypropylene fibers in fiber reinforced concrete on bond-slip, bond stress, deformation and cracking patterns were examined. The results from the numerical analysis are presented.

1 INTRODUCTION

The design of interior joints in a reinforced concrete frame structure under seismic loads points to a problem of proper force transmission through the joints since large shear forces develop in beam-column joints (Paulay et al. 1978). Bars, that go through the interior joints, are exposed to "pushing" and "pulling" of the adjacent beams in order to transfer the force from steel to the surrounding concrete. Shear force is transmitted in the interface zone between concrete and steel by bond stresses, which can be extremely large and exceed material bond strength.

Once a bar is loaded, the strain in the interface between the concrete and steel in the joint does not remain constant over a bar length. The difference in the strain in concrete and steel interface invokes the mechanism of bond slip (Lutz & Gergely 1967) due to insufficient bond capacity. A "development length" (the accepted term from ACI rather than a bond stress) in the interior joint for a given size of a straight beam bar is usually larger than the depth of the adjacent column but also insufficient to develop adequate bond capacity. Depending on the size of the development length (shortest bar length in which steel stress increases to the yield strength f_y), steel yielding or concrete cracking may occur.

Slipping might be allowed to some extent in the frame joints where no seismic load is expected (Mindess 1989). However, reinforced concrete frame structures in seismic zones are designed to be ductile in order to resist seismic forces. Bond stresses that exceed the bond strength, developed at the bar-concrete interface, could result in bond-slip and significantly influence the hysteretic response of ductile frames. A 15% loss in bond strength may cause a loss of about 30% of the energy dissipation capacity in a typical beam-column joint (Pauley & Priestley 1992).

1.1 *How to improve bond strength?*

Much effort in the past has been devoted to the improvement of the bond characteristics in the interface zone between bar and concrete using deformed instead of plain bars (Abrams 1913) in order to resist slipping with a mechanical interlock. The succeeding research concentrated mostly on deformed reinforcement re-modeling taking into account different shapes of ribs and rib face angle, height of ribs, rib-spacing etc. (Brown et al. 1993, Goto 1971, Lutz & Gergely 1967, Clark 1946, 1949). Less has been done on the improvement of bond strength of concrete although shear and bond strength are affected by the tensile strength of concrete. The tensile strength develops more quickly than the compressive strength and influences quick development of shear and bond strength (MacGregor 1997). Although first cracks appear in the interface zone, most likely due to the break of concrete compression strength, the concrete core collapses due to concrete splitting. ACI Committee 544 (1982, 1994) recommends the use of fiber reinforced concrete (FRC) for seismic resistance to increase the concrete tensile strength. Regarding compressive strength, Yin et al. (1989, 1990) presented the results of different compression tests (uniaxial and biaxial test) with plain concrete and FRC. The compressive strength of FRC specimens increased by as much as 35% in biaxial tests, while the gain in uniaxial compression was found to be negligible. Studies by Soroushian et al. (1994) proved that bond strength increased with the fiber addition.

In order to gain a better insight into the bond characteristics of concrete with polypropylene fibers, Moradi (1994) carried out an experimental study in which reinforcing bars with different embedment lengths and configurations were subjected to pull-out tests. A concrete cylinder, made of ordinary and fiber reinforced concrete, centrally reinforced with a single embedded deformed bar, was used in the pull-out tests. In the fiber reinforced concrete mix, polypropylene fibers of two fiber lengths: "short" fibers with lengths of 19 mm (0.75 in.) and "long" fibers with a length of 50 mm (2 in.) were added to the cement matrix. During the experiments, only mixes with 0.5% and 1% fibers per volume were used due to unworkablility of the mix and bundling of fibers when a higher fiber percentage was added.

A deformed steel bar $\phi22$ (# 7, Grade 60, nominal strength $f_y = 414$ MPa) was embedded in the concrete cylinders using different embedment lengths (Moradi 1994). The bar had lateral lugs and the total bar length was 300 mm (12 in.). The ratio of the diameter of the cylinder 300 mm $\times$ 250 mm (12 in. $\times$ 10 in.) to the rebar $\phi22$ was approximately 11. The bar embedment length in some samples was 150 mm (6 in.) to represent normal bond across the complete development length. For inducing local bond, the remaining samples had shorter development lengths of 50 mm (2 in.) and 75 mm (3 in.). In order to simulate local bond conditions, debonding of rebars outside the embedment length was provided by two 50 mm PVC tubing around the rebar (Moradi 1994) which were sealed at the ends with grout to prevent penetration of concrete into the tubing. The reason for using local bond and normal full embedment was to investigate the variation of bond across the development length.

The inverted cylindrical specimens were mounted in a Universal Testing machine so that pull-out forces could be applied at one end of the reinforcing bar (Moradi 1994). Load measurements were obtained through direct readings from the machine's pressure transducer while deformations were computed from a precision gage mounted to a fixed location in the bar.

2 FINITE ELEMENT ANALYSIS

The numerical analysis was first verified using data from the pull-out tests by Gijsbers et al. (1978). In the current numerical analysis considering the experimental data set from Moradi, only half the cylinder (Fig. 1) was modeled due to axial symmetry of the problem (Jankovic 1998). Linear 2D analysis was followed by nonlinear fracture analysis using DIANA (Version 6.2, 1996). Numerical load-tests were performed using incremental displacement control applied at the loaded end of the rebar.

2.1 Finite element mesh and material modeling

Concrete was modeled by eight noded quadratic solid ring isoparametric elements (100 elements); six three-noded truss elements were used for the modeling of a deformed bar. The nonlinearity was introduced in the interface concrete-steel zone, with interface (bond) elements. These elements were not provided in the first 50 mm of the bar length where the bar enters the concrete and where no bond was expected (Gijsbers et al. 1978). The line interface elements modeled the relationship between debonding and slipping behavior (displacement in x and y direction) and bond stress (tractions in the tangential direction t_t and the radial direction t_n). For that purpose, dual nodes at the concrete-steel interface were applied. In order to calibrate the FEM model with the experimental results, the shear and normal stiffness of the interface (bond) elements varied between 100 and 300 N/mm^3. The chosen stiffness value was assumed and kept constant

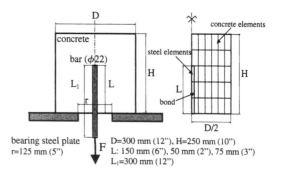

Figure 1. Experimental configuration and schematic drawing of the element mesh used in the numerical analysis.

throughout the linear and the nonlinear analysis. The value for the maximum bond stress was accepted as $5\,\mathrm{N/mm^2}$ in the linear zone (Gijsbers et al. 1978, De Groot et al. 1981).

For the reinforcement, an elastic-plastic model was used both in tension and compression, with Von Mises yield criterion. In general, concrete has been numerically modeled using two different models based upon its behavior under tension and compression: a crack model valid for the load combinations with at least one tensile component involved and a plasticity model for describing the concrete behavior under multiaxial compressive stress. The concrete was modeled as a linear material before cracks occurred. The Young's modulus (E) and the Poisson's ratio (ν) were specified and assumed to have constant values throughout the analyses. For concrete in compression, an elastic-plastic model has been chosen. Calculated E value for the plain concrete varied from $19786\,\mathrm{MPa}$ to $23223\,\mathrm{MPa}$ but was lowered to $10000\,\mathrm{MPa}$ (Gijsbers et al. 1978). A multiaxial state of stress was applied. For concrete as a brittle material the Mohr-Coulomb yield criterion (in 2D) was selected as well as Mohr-Coulomb friction model. It was assumed that the shear stress τ on the fracture plane was induced by friction.

In the analysis of FRC specimens, the Young's modulus (E) and Poisson's ratio (ν) depended on the additional percentage of fibers. The Young's modulus was increased to $30000\,\mathrm{MPa}$ with Poisson's ratio of 0.4. Regarding Coulomb failure criterion, Nielsen et al. (1978) suggested that a rigid-plastic constitutive model be assumed for the FRC with zero tensile cut-off. The same procedure was attempted in the numerical analysis but the results obtained were not as expected. The value for tensile cut-off was calculated as a fraction of the compression strength, which was assumed to be the same as in the case of plain concrete. Yield criterion for concrete was kept as Mohr-Coulomb with cohesion. The value for the friction angle was decreased from $55°$ to $30°$.

2.2 *Fracture analysis*

The crack model based on strength criterion was used for crack initiation. The fracture mechanics approach implemented in DIANA through material modeling was applied for the crack propagation. In fracture mechanics theory, the combination of two criteria, yield and fracture criterion provides a suitable way to find a safe region in the material.

In the case of bond-slip, due to the brittleness and splitting nature of concrete (loss of bond results from the splitting of concrete, Brown et al. 1993), fracture mechanics can be applied. According to Tepfers (1979), Reinhardt & Van der Veen (1992) compressive forces develop due to the bar pull-out and grow to the circumferential forces forming a hydrostatic pressure on

the inner ring. The cracked inner concrete part has a conical shape while the rest of the concrete stays elastic. Cracks develop at a very early period in the loading, long before the maximum pull-out force is reached.

The application of fracture mechanics to concrete structures has been discussed for many years due to the heterogeneity and nonlinearity of concrete since it was inappropriate to apply linear fracture mechanics (LEFM) to concrete (LEFM being a theory only valid for linear-elastic and macroscopic homogeneous materials). Since cracking in heterogeneous material such as concrete is a very complex phenomenon, fracture process mechanisms can be quantified by distribution of energy during the fracture process (Shah et al. 1995). Energy criterion can be applied for crack propagation. The criterion states that the crack propagates when the released energy is equal or greater than the absorbed energy.

Fracture energy release rate G_f (Griffith 1921) is used as a fracture parameter and a material property and the criterion for the crack propagation. RILEM (1985) established a method for the experimental determination of G_f as the energy required for breaking a notched three-point bending specimen into two pieces. In plain concrete, the fracture energy G_f was $0.09\,\mathrm{N/mm}$ ($90.5\,\mathrm{J/m^2}$) with linear softening. This value was obtained by curve fitting (Wittmann et al. 1987, 1988). RILEM (1985) suggests a higher value of $0.112\,\mathrm{N/mm}$ ($112.5\,\mathrm{J/m^2}$). For FRC, G_f was increased to $0.15\,\mathrm{N/mm}$ ($150\,\mathrm{J/m^2}$) with linear and exponential softening. Mode I and mode II were both included in the nonlinear analysis since both opening and sliding mode contribute to the shear failure.

Cracking in concrete was modeled combining a tension cut-off, tension softening and shear retention. Tension cut-off has been taken as linear i.e. crack arises if the major principal tensile stress exceeds the value for f_t and $f_t(1 + \sigma_{lateral}/f_c)$ where $\sigma_{lateral}$ is the lateral principal stress. The important parameter related to the concrete cracking is the shear retention factor β which takes into account a reduction of the shear stress that is transferred through the cracked concrete and it also models aggregate interlock. The β value is based on experience and it varies in the range $0 < \beta < 1$ (Reinhardt 1989). The value for β has been taken as 0.2 as it leads to similar results compared to the most of the experiments. This showed that the influence of the β value was significant in a way that the inclination of the diagonal crack decreased with decreasing the β value. In the present study, β value was kept constant.

3 EFFECT OF FIBER CONTENT AND EMBEDMENT LENGTH

The effect of fiber content on the load capacity, bond stress and crack pattern was investigated for the cases of plain concrete and FRC cylinders. Short fibers of 19 mm (0.75 in.) length and long fibers of 50 mm (2 in.) length were added to the plain concrete at 0.5% and 1% by volume. The embedment lengths varied from: the "full" bond length of 150 mm (6 in.) to local bond lengths of 50 (2 in.) and 75 mm (3 in.).

3.1 Full embedment length

The "full" embedment length of the reinforcing bar in the cylindrical specimens of 300 mm × 250 mm (12 in. × 10 in.) was kept constant at 150 mm (6 in.), in the first group of tests. In this test-group (1A, 6B, 3C), only the percentage of fibers (0%, 0.5% and 1%) has been varied.

Experimental and numerical results were compared (Fig. 2). Both results confirmed an increase in force and bond stress when the fiber percentage was increased. Linear concrete behavior was assumed before cracks occurred and softening behavior was simulated after the peak load. The peak load increase in the specimen 3C (1% fibers) was more emphasized in the numerical analysis (Fig. 2a) than in the experiments. Numerical analysis predicted a minimum peak load value of 96 kN for the samples without fibers. The maximum reached 125 kN (30% increase) with

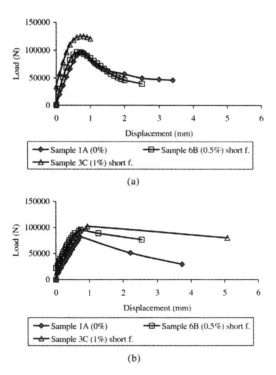

(a)

(b)

Figure 2. The effect of varying fiber content in (a) FEM analysis and (b) experiments based on an embedment length of 150 mm (6 in.).

1% of fiber addition. The experimentally observed maximum peak load was 102 kN for 1% of fibers and 82 kN (20% less) for plain concrete samples (Fig. 2b). The results did not show considerable variations in the peak load with the increase of fiber volume from 0% to 1%.

The experimental bond stress was the highest for 1% fibers with a measured stress of 9.6 MPa.

The bond stress for 0% and 0.5% fiber content decreased by about 17%. The maximum bond stress in the FEM simulation was 17.6 MPa (1% fibers). The stress value was lower by 20% in plain concrete and FRC with 0.5% fibers. Almost the same displacement values were recorded both in the numerical analysis and in the experiments. The increase in the fiber content has not influenced the slip displacements. All peak displacement values varied in the range from 0.6 to 0.8 mm. In the case of the numerical model with 1% fibers, the maximum displacement was approximately 1 mm since the results did not converge for larger displacements. The difference in the experimental and numerical results can be attributed to the difference in the assumed Young's modulus since the computed results gave always stiffer behavior than the experimentally observed behavior.

3.2 Local bond

The load-displacement curves for the cases with full and local bond are introduced in Figures 3 and 4. The samples with no fiber addition (Fig. 3a) showed a load increase of 14% with the increase of the embedment length from 50 mm (71 kN) to 150 mm (83 kN). The FEM displacements at peak values were more or less the same for all bond lengths and varied from 0.7 mm to 0.8 mm. Bond stress had the opposite effect: the maximum (20 MPa) was achieved with 50 mm embedment. The lowest stress was 13.5 MPa in the case of 150 mm (6 in.). Comparisons with the experiments (Fig. 3b) showed an increase in load capacity of 13% with 150 mm (6 in.) embedment length. The minimum force was measured with a bond length of 50 mm (2 in.), which is 38% less than calculated. The displacements, at which these peak values were reached, increased with the increase in bond length. The maximum bond stress was 11 MPa for 50 mm (2 in.) and the minimum was 7.8 MPa.

In the Figure 4 all samples had the same fiber content (0.5%) but different embedment lengths. An increase in the load capacity is observed with an increase in the embedment length (Fig. 4a). The maximum load capacity of 97 kN was reached with 150 mm (6 in.) and the minimum was 64 kN for 50 mm (2 in.), which is less than the case of specimens without fibers. The analysis for the specimen 3D stopped after 2.1 mm slippage without converging. After the peak force has been reached, the resisting load became the same for

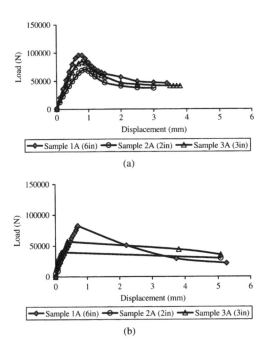

(a)

(b)

Figure 3. Load vs. slip response in the samples with no fiber addition; comparison of full and local bond (a) FEM results and (b) experimental results.

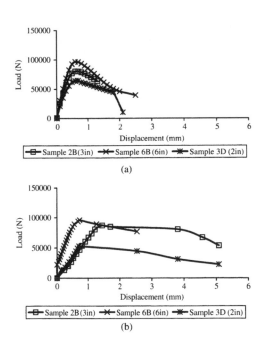

(a)

(b)

Figure 4. Load vs. slip response in FRC (0.5% fibers) samples with the full bond (150 mm or 6 in.) and local bond 50 mm (2 in.) and 75 mm (3 in.); (a) FEM results and (b) experiments.

all three samples. Bond stress showed rather different values for local and full embedment length. The highest value was 18.2 MPa for specimens with 50 mm (2 in.) embedment. It decreased to 15.1 MPa with 75 mm (3 in.). For the full embedment of 150 mm (6 in.), the bond stress value was 13.2 MPa. A similar trend has been seen in the experiments (Fig. 4b).

The displacement at the maximum peak load varied and the initial stiffness also varied considerably. Again, the maximum load attained for 150 mm (6 in.) embedment was 97 kN. The lowest force was for 50 mm (53 kN). The highest value of bond stress was 16 MPa for 75 mm and 14.7 MPa for 50 mm.

3.3 Bond stress distribution

With respect to the interface elements in the FEM model, the bond stress distribution for both components, radial and shear, was calculated. Shear bond stress is presented in Figure 5 along the specimen without/with fibers. For specimens with local bond, the stress distribution had two concentration peaks at the nodes where the no-bond zone ended (Fig. 5). The difference of only 25 mm (1 in.) in local bond samples gave a visible difference in the values of shear stress in the case of no fiber addition (Fig. 5a). In the case of full embedment length (150 mm), the stress was

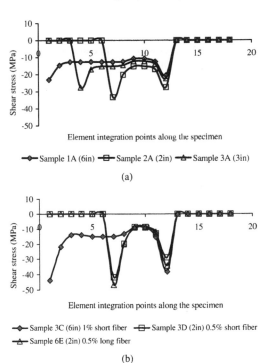

(a)

(b)

Figure 5. FEM results of shear stress distribution; (a) specimen without fibers, and (b) with short and long fibers.

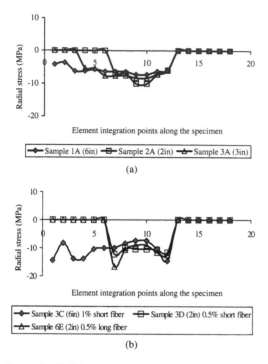

(a)

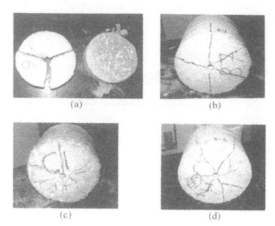

(b)

Figure 6. Radial stresses along the sample (a) without fibers and (b) with fibers.

distributed more evenly along the specimen length. Bond stress did not show any change in sign as in case of the pull-out test applied on both sides of the specimen (Gijsbers et al. 1978).

A comparison of the shear stress between 0% and 0.5% fiber addition with 150 mm (6 in.) embedment length, showed an increase in the shear stress with a max of 10 MPa in the concentration areas. The same was valid for the local bond (Fig. 5b). The addition of longer fibers (50 mm) slightly increased the bond stress. The average stress did not show any significant variation. The distribution of the radial bond stress component is presented in Figure 6 for no fiber addition. The values for the radial stress (Fig. 6) were much lower compared to the shear stress with almost similar peak values, though slightly higher for specimens with 50 mm (2 in.) of local bond.

Fiber addition (Fig. 6b) increased the stress values at the concentration points but not drastically as in the case of shear stress. The shear component was more critical at their concentration points and could contribute to the bond breakdown. The size of the fibers did not make a big influence on the results.

3.4 Cracks

The experimental observations of the cracks in the concrete cylinder (Fig. 7, Moradi 1994) after pulling

Figure 7. State of the specimens after the pull-out test (a) no fibers, (b) 0.5% short fibers, (c) 1% short fibers, (d) 0.5% long fibers (after Moradi 1994).

of the rebar indicated the following. A cylinder made of plain concrete (Fig. 7a) was ruptured into three almost equal parts due to three radial splitting cracks. The fiber content (0.5%, 1% short fibers and 0.5% long fibers) in Figures 7b–d, retarded this development significantly. Many radial cracks were visible on the surface but they did not cause splitting of the cylinder. Hence, it can be inferred that the addition of only 0.5% of synthetic fibers to the mix could improve the ductility and the energy dissipation capacity compared to conventional plain concrete samples. One possible reason is that in FRC the presence of fibers retards crack growth and improves the composite toughness.

In the finite element model, all samples showed identical initiation of cracks in the most softened concrete zone around the bond zone while in the no bond zone only some symbolic cracking was noticed. The crack pattern was notified after the first "load" step of 0.0025 mm in the numerical analysis (Fig. 8a).

The first cracks were inclined and perpendicular to the maximum stress direction. As the slip displacement reached 0.4 mm (Fig. 8b), the crack pattern began to assume a conical shape. Crack zones formed in the middle of the cylinder while the rest of concrete stayed uncracked. During loading, cracks changed their status from open (fully or partially) to closed. The first set of cracks occurred primarily in the vicinity of the rebar (interface bond zone) but closed in the following load steps.

Along with the inclined (shear) cracks, radial cracking occurred until whole specimen was totally cracked (Fig. 8c). No experimental observations (microscopic or injection of ink, Goto 1971 etc.) were performed regarding the internal crack initiation. No exact conclusions can therefore, be drawn at this point regarding the crack propagation.

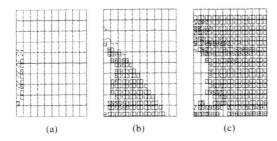

Figure 8. Cracking pattern in the plain concrete cylinder after slip displacements of (a) 0.0025 mm, (b) 0.4 mm, (c) 3.5 mm.

4 DISCUSSION AND CONCLUSIONS

In order to investigate the bond-slip characteristics of a reinforcing bar due to the presence of polypropylene fibers in a concrete mix, a series of pull-out experiments were carried out on a representative specimen. Two fiber types, different embedment lengths and full vs. local bond were studied.

Nonlinear finite element analysis was employed on an axisymmetric model to simulate the experiments. A fracture mechanics approach was applied to study the bond mechanisms with due consideration to material modeling of concrete, steel and the interface zone. Fracture mechanics overcomes several drawbacks from conventional methods: ultimate-limit analysis that considers only a single critical section for stress-strain analysis gives a different picture of the fracture process. The continuous debonding and slipping of the steel-concrete in the interfacial zone in the vicinity of the cracked matrix can influence the service performance and the load capacity of reinforced concrete structures but it is difficult to consider such effects in conventional methods.

The bond-slip of the bar can be treated as independent of the concrete cover since the bar diameter was small compared to the cylinder diameter ($d_c/d_b = 11$). This ratio was kept constant throughout the analysis. The effect of the fiber content on the bond stress including the decrease of the development length from the full bond 150 mm (6 in.) to local bond of 50 mm and 75 mm (2 and 3 in.) was observed. An increase in bond stress was evident for specimens with local bond. Results were also influenced by the fiber addition. A change in the local bond length from 50 to 75 mm made a significant difference in shear and radial bond stress results. Increasing the percentage of fibers from 0% to 0.5% to 1% resulted in an increase in the load capacity. The displacements of the bar (slipping) at the peak load was not influenced by the fiber addition.

Concrete was considered to be an elastically isotropic material before cracks occurred. Numerically,

the crack origin was in the interface zone (bond elements). The explanation as to why cracks developed in the interface zone probably had to do with the different E (Young's modulus) of concrete and steel which "confront" each other in the transition interface zone. The samples without fibers were completely cracked. The other samples with 0.5% and 1% were cracked on the surface and did not rupture completely. Slipping of rebars along the cylinder increased with the increase of radial cracks, which was observed in both experiments and numerical 2D analysis.

The stiffness of the bond elements was kept constant during the loading as well as other parameters, which is not the realistic situation. The analysis was constrained by the limitations of the finite element analysis software. That might be one of the reasons for the difference in the results between the finite element model and the experiments. The results presented here, both experiments and FEM analyses, are for monotonic loading only. Extrapolation to cyclic loading conditions is not valid.

The addition of fibers helped the increase of the fracture energy. By bridging the cracks, fibers prevented the applied stress to concentrate at the crack tip. Additional experiments should be conducted with a smaller sized cylinder (in the bigger one) with higher fiber percentage in FRC around the rebar. Assuming a higher tensile and compressive strength of FRC, the higher resistance can be expected in the mechanical interlock when chemical adhesion and friction are overruled. That could decrease crack propagation such that the rest of the cylinder could still be made of the plain concrete. The maximum value of the stress in this smaller cylindrical member would depend on the elasticity of the fibers, the number of the fibers and possibly their inclination, depending on the fiber material. These variables could be applied in the numerical model.

As previously mentioned, the addition of fibers changed the structure of concrete. When fibers are added to the reinforced cementitious matrix, the interfaces that are formed between the fiber and matrix are significant and can be helpful in determining the effective stress transfer between them. Additional research is needed to study these effects.

REFERENCES

Paulay, T., Park, R. & Priestley, M.J.N. 1978. Reinforced Concrete Beam-Column Joints under Seismic Actions. *ACI Journal, Proc.* Nov. 75 (11): 585–593.

Lutz, L.A. & Gergely, P. 1967. Mechanics of Bond and Slip of Deformed Bars in Concrete. *ACI Journal, Proc.* Nov. 64 (11): 711–721.

Mindess, S. 1989. Interfaces in Concrete. In J. Skalny (ed.), *Material Science of Concrete I*: 163–180. Westerville, OH: The American Ceramic Society, Inc.

Paulay, T. & Priestley, M.J.N. 1992. *Seismic Design of Reinforced Concrete and Masonry Buildings.* New York: John Wiley & Sons, Inc.

Abrams, D.A. 1913. Tests of Bond between Concrete and Steel, *Bulletin 71, Engineering Experiment Station:* 105. Urbana, IL. University of Illinois.

Brown, C.J., Darwin, D. & McCabe, S.L. 1993. Finite Element Fracture Analysis of Steel-Concrete Bond. *Structural Engineering and Engineering Materials, SM Report* 36, Lawrence, Kansas: University of Kansas Center for Research, Inc.

Goto, Y. 1971. Cracks Formed in Concrete around Deformed Tension Bars. *ACI Journal, Proc.* April 68 (4): 244–251.

Clark, A.P. 1946. Comparative Bond Efficiency of Deformed Concrete Reinforcing Bars. *ACI Journal, Proc.* Dec. 43 (4): 381–400.

Clark, A.P. 1949. Bond of Concrete Reinforcing Bars. *ACI Journal, Proc.* Nov. 46 (3): 161–184.

MacGregor, J.G. 1997. *Reinforced Concrete-Mechanics and Design.* Third Edition. New Jersey: Prentice Hall.

ACI Committee 544. 1982. State of the Art on Fiber Reinforced Concrete, *Concrete International* 4 (5): 9–25.

ACI Committee 544. 1994. Fiber Reinforced Concrete, Developments and Innovations. In J.I. Daniel & S.P. Shah (eds) *American Concrete Institute.* SP-142. Detroit.

Yin, W.S., Su, Eric C.M., Mansur, M.A. & Hsu, T. 1989. Biaxial Tests of Plain and Fiber Concrete, *ACI Materials Journal*, May–June 86 (3): 236–243.

Yin, W.S., Su, Eric C.M., Mansur, M.A. & Hsu, T. 1990. Fiber–Reinforced Concrete Under Biaxial Compression. *Engineering Fracture Mechanics* 35 (1/2/3): 261–268.

Soroushian, P., Mirza, F. & Alhozaimy, A. 1994. Bonding of Confined Steel Fiber Reinforced Concrete to Deformed Bars. *ACI Journal* March–April 91 (2).

Moradi, M. 1994. *Bond Characteristics of Polypropylene Fiber Reinforced Concrete.* M.S. Thesis, University of Central Florida. USA, FL.

Gijsbers, F.B.J., De Groot, A.K. & Kusters, G.M.A. 1978. A Numerical Model for Bond-Slip Problems: 37–48. *IASS.* Darmstadt.

Jankovic, D. 1998. *Numerical Modeling of Bond-Slip in Plain and Fiber Reinforced Concrete using Fracture Mechanics.* M.S. Thesis. University of Central Florida. USA, FL.

TNO Building and Construction Research. 1996. *DIANA, Finite Element Analysis, User's Manual, release 6.1.* (eds) de Witte, F.C., Nauta, P.

De Groot, A.K., Kusters G.M.A. & Monnier Th. 1981. A Numerical Model for Bond-Slip Problems. *HERON* 26 (1B).

Nielsen, M.P., Braestrup, M.W. & Bach, F. 1978. Rational Analysis of Shear in Reinforced Concrete Beams, *Int. Assoc. Bridge and Structural Engineering, Proc.* May, P-15178.

Tepfers, R. 1979. Cracking of Concrete along Anchored Deformed Reinforcing Bars. *Magazine of Concrete Research* 31: 3–12.

Reinhardt, H.W. & Van der Veen, C. 1992. Splitting Failure of a Strain Softening Material due to Bond Stresses. In A. Carpinteri (ed.), *Application of Fracture Mechanics to Reinforced Concrete, Proc. Int. Workshop, Turin, Italy 1990:* 333–346. London: Elsevier Applied Science Publishers.

Shah, S.P., Swartz, S.E. & Ouyang, C. 1995. *Fracture Mechanics of Concrete, Applications of Fracture Mechanics to Concrete, Rock, and Quasi-Brittle Materials.* New York: John Wiley & Sons, Inc.

Griffith, A.A. 1921. The Phenomena of Rupture and Flows in Solids. *Phil. Trans. Roy. Soc.:* 163–198. London, UK.

RILEM Draft Recommendations. Determination of the Fracture Energy of Mortar and Concrete by Means of Three-Point Bend Tests on Notched Beams. *Materials and Structures, Research and Testing* 18 (106): 285–290.

Wittmann, F.H, Roelfstra, P.E., Mihashi, H., Huang, Y.Y., Zhang, X.H. & Nomura, N. 1987. Influence of Age of Loading, Water-Cement Ratio and Rate of Loading on Fracture Energy of Concrete. *Materials and Structures* 20: 103–110. Paris: RILEM Publications.

Wittmann, F.H., Rokugo, K., Bruhwiler, E., Mihashi, H. & Simonin, P. 1988. *Materials and Structures.* 21 (21). Paris: RILEM Publications.

Reinhardt, H.W. 1989. Shear. In L. Elfgren (ed.) *RILEM: Fracture Mechanics of Concrete Structures; From Theory to Applications.* London: Chapman and Hall.

Finite Elements in Civil Engineering Applications, Hendriks & Rots (eds.)
© 2002 Swets & Zeitlinger, Lisse, ISBN 90 5809 530 4

FEM study on the shear behavior of RC beam by the use of discrete model

Kenta Hibino
Graduate School of Science and Engineering, Ritsumeikan University, Kusatu, Shiga, Japan

Takayuki Kojima & Nobuaki Takagi
Department of Civil Engineering, Ritsumeikan University, Kusatu, Shiga, Japan

ABSTRACT: In this study, the FEM analysis using discrete crack and bond elements is developed to apply to the RC beam without shear reinforcement subjected to bending and shear, and the analytical study was carried out on the occurrence of flexural and diagonal cracks, their developments and the shear bearing capacity of the beam. In this analysis, the characteristics of crack and bond elements are determined individually by plain concrete test and bond test respectively. By using this FEM analysis, it is possible to represent the development of flexural crack and diagonal crack, and the decrease of stiffness of beam after cracking. It is also possible to clarify the shear bearing mechanism of RC beam without shear reinforcement by this FEM analysis.

1 INTRODUCTION

It is basically important how to introduce cracking and bond slip into analysis when analyzing the mechanical properties of RC structures. In the finite element method (FEM) analysis, there are two approaches, (1) to represent cracking and bond slip as the change of element characteristics of concrete, "smeared crack idea", and (2) to introduce special element such as bond links element, "discrete crack idea". The approach (1) is widely used. On the other hand, the approach (2) is easy to analyze a structure, which is already cracked. However it is complicated when analyzing developing cracks, because it is necessary to have a re-meshing program due to developing cracks.

The authors have already developed a computer program to analyze developing cracks by introducing isoparametric crack elements and isoparametric bond elements. These elements are expressed as the isoparametric quadratic linear element, and both sides of crack or bond element has same nodes of an adjacent concrete element or a re-bar element. By this program it is possible to analyze crack development in RC beams.

In this study, the FEM analysis with discrete crack and bond linkage elements is developed to apply to the RC beam without shear reinforcement subjected to bending and shear. Shear resistance of RC beam without shear reinforcement is consisted of compressive concrete V_c, aggregate interlock V_{ay} and dowel

force of re-bar V_d. In this analysis, the characteristics of crack and bond elements are determined individually by plain concrete test and bond test respectively.

Shear failure of RC beam without shear reinforcement is affected the shear span effective depth ratio (a/d). In this study, loading tests of RC beams were carried out under the flexure-shear condition, i.e. the influence of a/d (2.0, 3.0 and 3.6) on the shear failure mode of RC beam without shear reinforcement. The adaptability of these analytical methods for experimental results was examined.

This FEM analysis makes it possible to represent the development of flexural crack and diagonal crack, and the decrease of stiffness of beam after cracking. It can also express the bond along crack (aggregate interlock), and the dowel action of longitudinal steel at crack. It is then, possible to make clear the shear bearing mechanism of RC beam without shear reinforcement by this analysis.

2 CHARACTERISTICS OF ELEMENTS

2.1 *Discrete crack element*

A crack element is inserted at the point, where the maximum principal tensile stress occurs, and to the direction perpendicular to it. Generally normal stress σ_n and shear stress τ_t are acting across a crack plane, and normal and tangential displacements δ_n and δ_t occur

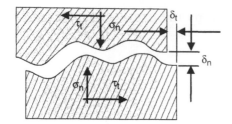

Figure 1. Transfer of stress across crack plane.

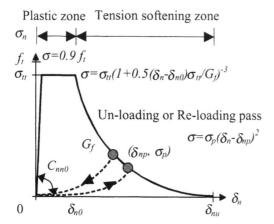

Figure 2. Relationship between normal stress and crack width.

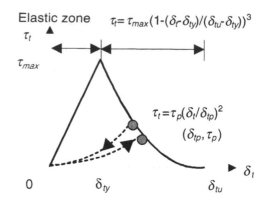

Figure 3(a). Relationship between shear stress and tangential displacement.

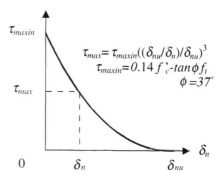

Figure 3(b). Relationship between shear stress and crack width.

as shown in Fig. 1. The relationship between these stresses and displacements can be expressed as Eq. (1):

$$\{\sigma\} = \begin{Bmatrix} \sigma_n \\ \tau_t \end{Bmatrix} = \begin{bmatrix} C_{nn} & C_{nt} \\ C_{tn} & C_{tt} \end{bmatrix} \begin{Bmatrix} \delta_n \\ \delta_t \end{Bmatrix} = [D_c]\{\delta\} \qquad (1)$$

The relationship between σ_n and δ_n in the crack element was assumed as that shown in Fig. 2. This relationship is referred by tension softening curve of concrete. On the other hand, the relationship between τ_t and δ_t was assumed as that shown in Fig. 3(a) and (b). Figure 3(a) is divided into two regions defined as "elastic stage" and "softening stage". The δ_{ty} is the transferring displacement from "elastic stage" to the "softening stage", and the τ_{max} is the maximum shear stress. In addition, the coefficient τ_{max} can be determined from the relationship between τ_{max} and δ_n in the crack element (See Fig. 3 (b)). The relationship between σ_n-δ_t is not so important. In the analysis a constant value of $C_{nt} = C_{tn} = 0$ was assumed.

The proposed crack element is a quadratic linear element with 6 nodes as shown in Fig. 4. Both sides of crack element have same nodes of adjacent concrete

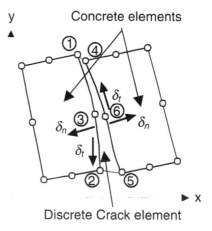

Figure 4. Discrete crack element.

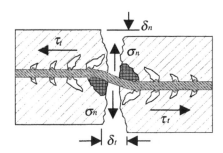

Figure 5. Transfer of stress across concrete and re-bar.

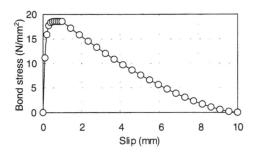

Figure 6(a). Relationship between bond stress and slip of bond element.

elements. When the element is expressed in local coordinate, the stiffness matrix $[K_c]$ of the crack element can be obtained by the numerical integration of the following Eq. (2):

$$[K_c] = \int_s [B]^T [D_c][B]t\, ds$$
$$= \int_{-1}^{1} [B]^T [D\hat{c}][B]t\sqrt{(dx/d\xi)^2 + (dy/d\xi)^2}\, d\xi \quad (2)$$

where $[B]$ is matrix between displacement and relative displacement; t is thickness of concrete where crack occurs.

2.2 Discrete bond element

Generally dowel stress σ_n and bond stress τ_t are defined as surface tractions acting on a surface between concrete and re-bar, and normal and tangential displacements δ_n and δ_t are defined as shown in Fig. 5. The relationship between these stresses and displacements can be expressed as Eq. (3):

$$\{\sigma\} = \begin{Bmatrix} \sigma_n \\ \tau_t \end{Bmatrix} = \begin{bmatrix} B_{nn} & B_{nt} \\ B_{tn} & B_{tt} \end{bmatrix}\begin{Bmatrix} \delta_n \\ \delta_t \end{Bmatrix} = [D_B]\{\delta\} \quad (3)$$

In order to estimate the bond stress-slip relationship $(\tau_t\text{-}\delta_t)$ between re-bar and concrete, the pull-out bond test and RILEM beam type bond test were carried out. By trial FEM analyses to fit the analytical results to the experimental results, the most suitable bond stress-slip relationship was obtained in each test. On the other hand, in order to estimate the dowel stress-displacement relationship $(\sigma_n\text{-}\delta_n)$, by trial FEM analyses to fit the analytical results to the experimental results carried by Suzuki et al., the most suitable dowel stress-displacement relationship was obtained.

The relationship between τ_t and δ_t in the bond element was assumed as that shown in Fig. 6(a). This relationship was obtained by experiment of RILEM beam type bond test. On the other hand, the relationship between σ_n and δ_n was assumed as that shown in

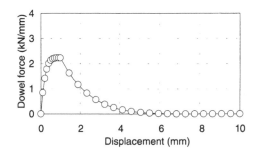

Figure 6(b). Relationship between dowel force and displacement of bond element (Cs = 25.05 mm).

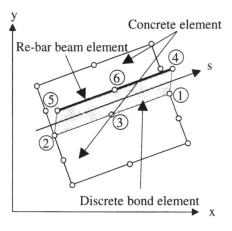

Figure 7. Discrete bond element.

Fig. 6(b). The relationship between τ_t-δ_n and σ_n-δ_t is not so important. In the analysis a constant value of $B_{nt} = B_{tn} = 0$ was assumed.

The proposed bond linkage element is a quadratic linear element with 6 nodes as shown in Fig. 7. Both sides of bond linkage element have same nodes of concrete element and re-bar element respectively.

When the element is expressed in local coordinate, the stiffness matrix $[K\hat{B}]$ of the bond linkage element can be obtained by the numerical integration of the following Eq. (4):

$$[K_B] = \int_s [B]^T [D_B][B] t \, ds$$
$$= \int_{-1}^1 [B]^T [D_B][B] t \sqrt{(dx/d\xi)^2 + (dy/d\xi)^2} \, d\xi$$

(4)

where $[B]$ is matrix between displacement and relative displacement; t is periphery of re-bar.

2.3 Characteristics of other elements

(1) Concrete element

In the FEM analysis, a concrete specimen was divided into finite element mesh as a two dimensional plane stress problem. Quadratic quadrilateral elements with 8 nodes were used for concrete. The relationship between stress and strain of concrete was assumed as shown in Fig.8 (a). Mohr-Coulomb criterion was used as the failure criteria of concrete element as shown in Fig.8 (b).

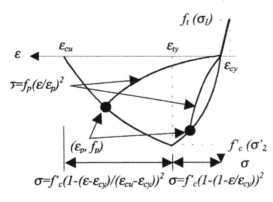

Figure 8 (a). Relationship between stress and strain of concrete.

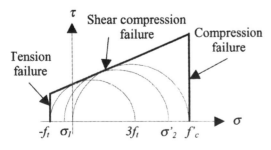

Figure 8 (b). Failure criteria of concrete.

(2) Re-bar element

Quadratic beam elements with 3 nodes were used for re-bar. The relationship between stress and strain of re-bar was assumed as shown in Fig. 9. To consider the relationship between axial force and flexural moment, fiber method was used for calculating the stiffness of re-bar.

3 CRACK PROPAGATION AND ANALYTICAL METHOD

In this analysis, the condition of cracking is defined as the maximum principal tensile stress that is calculated in the concrete elements reaches the tensile strength of concrete. One crack element is inserted at the location perpendicularly to the direction of the principal tensile stress. The FEM mesh is modified to fit the crack element newly introduced. Re-bar and bond elements are re-arranged according to the transformation of concrete elements.

The location of stress and strain calculation is designed on the base of the original mesh, in order to avoid the effect that the element size and location may change by re-meshing. Proposed re-meshing program can express the flexural crack, diagonal crack, and branch or combined cracks.

Figure 10 shows the flow chart of the analysis. The flow of the analysis is as follows.

(1) First stage

First of all, the analysis is carried out on the specimen with an arbitrary load. Then the location where the maximum principal tensile stress occurs and its direction are calculated. The load is adjusted as the stress becomes the tensile strength of the concrete.

(2) Second stage

One crack element is inserted at the location perpendicularly to the direction calculated above. The FEM mesh is modified to fit the crack element newly introduced.

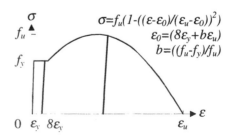

Figure 9. Relationship between stress and strain of re-bar.

(3) Third stage

For each increment by introducing one crack element, the direct iteration is carried out. In each step of iteration, the load is adjusted as the maximum principal tensile stress in concrete element becomes its tensile strength. The iteration is terminated when the adjusted load satisfies the condition of Eq. 5:

$$(P_n - P_{n-1})/P_N \leq er \tag{5}$$

where f_n is load, n is number of iteration; er is a constant for convergence $\{er = 1 \times 10^{-2}\ (n < 10),\ er = 5 \times 10^{-2}\ (n > 11)\}$.

Then the location where the maximum principal tensile stress occurs and its direction are calculated.

(4) Fourth stage

Second stage and Third stage are repeated.

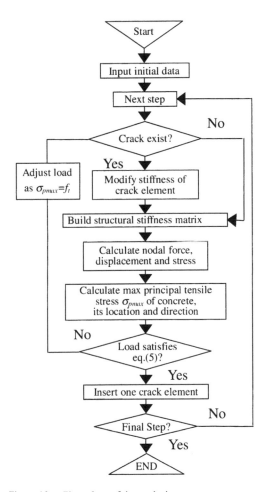

Figure 10. Flow chart of the analysis.

This analysis is based on the crack element increasing method in which one crack element is inserted in the mesh in each step. By this method, the snap back can be analyzed properly, thought it is very difficult both in experiment and in other analysis.

4 OUTLINE OF THE LOADING TESTS AND THE ANALYSIS

4.1 The loading test

Loading tests of RC beams were carried out. In the test, three levels of shear span effective depth ratio (a/d) were chosen, 2.0, 3.0 and 3.6, in order to obtain the experimental data of shear compression failure, shear tension failure and diagonal tension failure.

Target compressive strength of concrete was 24 N/mm² in all series. The size of RC beam and loading condition are shown in Fig. 11. The size of beam was 150 × 240 × 2000 mm and the effective depth was 200 mm. Its span length was 1800 mm. In the experiment, the beams were assumed to be simply supported and subjected to two-point concentrated loading. The deflection at mid span was measured.

The mechanical properties of concrete at loading test of RC beams are shown in Table 1. The mechanical properties of re-bar are shown in Table 2. Test

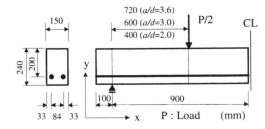

Figure 11. Size of RC beam and loading condition.

Table 1. Mechanical properties of concrete.

f_c' (N/mm²)	E (N/mm²)	f_t (N/mm²)	f_b (N/mm²)
26.7	35500	2.88	5.05

Table 2. Mechanical properties of Re-bar (D16).

Nominal diameter (mm)	Nominal cross-section area (mm²)	f_y (N/mm²)	f_u (N/mm²)	E (N/mm²)
15.9	198.6	396	601	206000

Table 3. Test results of RC beams.

a/d	F_{cr}^{*1} load (kN)	S_{cr}^{*2} load (kN)	Ultimate shear load (kN)	Failure mode
2.0	34.3	73.5	95.7	shear tension
3.0	29.4	71.1	88.2	shear tension
3.6	22.1	68.6	71.1	diagonal tension

$*^1$ Flexural crack, $*^2$ diagonal crack.

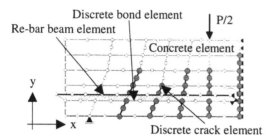

Figure 12. FEM mesh.

results of RC beams without shear reinforcement are shown in Table 3.

4.2 Outline of the analysis

The analytical results are compared with experimental results, in order to examine the appropriations and problems of this analytical method.

In this analysis, in order to decrease the number of the element, the half of the specimen was analyzed by taking the advantage of symmetry (see Fig. 12).

5 RESULTS AND DISCUSSIONS

5.1 Relationship between load and deflection and crack propagation

Load and mid span deflection diagrams obtained by the analyses and the experiments are shown in Fig. 13. In this figure load I and II correspond to the loads at the occurrence of flexural crack and flexural shear crack respectively. The load level III and IV correspond respectively to the maximum load and the load at which unstable state occurred in the analysis.

Up to the load level II, good agreement between the analysis and the test can be obtained.

The shear failure modes were affected by the a/d ratio, after the load level II. In a/d = 3.6, shear capacity could be assumed appropriately by the FEM analysis. The load and mid span deflection relationship in the analytical result fitted to the experimental

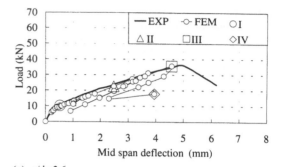

(a) a/d=3.6

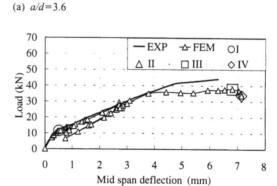

(b) a/d=3.0

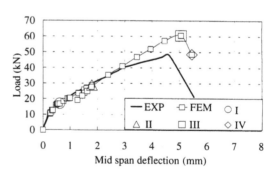

(c) a/d=2.0

Figure 13. Load and mid span deflection diagram.

result in the maximum load (III). The shear resistance of RC beam decreased suddenly after the maximum load. Diagonal crack, which developed into lower side of beam, occurred. Finally, the calculation was finished when it was difficult to stabilize the iterative process of calculation. In a/d = 30, stiffness of the beam was decreased before the maximum load (III). Load became the maximum load (III) when diagonal crack occurred. The shear resistance of RC beam decreased suddenly after the maximum load. The calculation was finished when it was difficult to stabilize

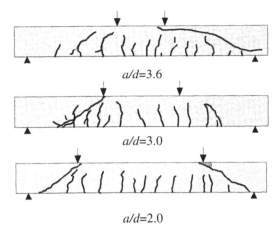

a/d=3.6

a/d=3.0

a/d=2.0

Figure 14.   Crack patterns (Experiments).

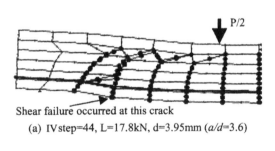

Shear failure occurred at this crack

(a) IV step=44, L=17.8kN, d=3.95mm (*a/d*=3.6)

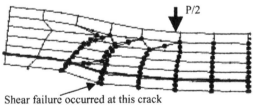

Shear failure occurred at this crack

(b) IV step=49, L=33.8kN, d=7.15mm (*a/d*=3.0)

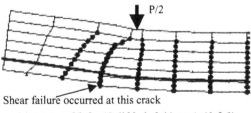

Shear failure occurred at this crack

(c) IV step=38, L=48.6kN, d=5.46mm (*a/d*=2.0)

Figure 15.   Crack Pattern (FEM).

the iterative process of calculation. In a/d = 20, load and mid span deflection relationship in experiment result was reached third slope after the load level II. However, load and mid span deflection relationship in analytical result reached the maximum load (III) with maintaining the stiffness of second slope. Finally, the calculation was reached the load level (IV) when concrete element near the loading point was failed by compressive stress. These results show that experimental result assume shear compression failure, on the other hand analytical results assume shear tension failure.

Crack patterns obtained by the experiments are shown in Fig. 14. Crack patterns obtained by the analyses are shown in Fig. 15.

When comparing crack patterns obtained in the experiment and the analysis, all cracks by the analysis seem to develop higher than the observed cracks. It is mainly because the width of the tips of the analytical cracks is very small and they are invisible.

5.2 *Shear bearing capacity of each element*

Shear resistance of RC beam without shear reinforcement is consisted of contributions of compressive concrete V_c, aggregate interlock V_{ay} and dowel force of re-bar V_d. In this analysis, the characteristics of crack and bond elements are determined individually by plain concrete test and bond test respectively. The contributions of shear resistance of each element are calculated at each step.

Contributions for shear resistance and mid span deflection diagrams obtained by the analyses are shown in Fig. 16(a), (b) and (c) at the position of the failure.

In all specimens, V_c consisted all the shear force in the stage when flexural crack develops.

Shear resisting mechanisms are affected by the shear failure mode, after the stage when flexural crack develops. In this stage (a/d = 3.6, deflection = 3.2 mm, a/d = 3.0, deflection = 2.8 mm), flexural shear crack, which developed into upper side of beam, occurred. V_{cd} decreased according to the development of the flexural shear crack, while V_{ay} increased. In the cases of a/d = 3.6 (diagonal tension failure) and a/d = 3.0 (shear tension failure), the contribution of aggregate interlock, V_{ay} became greater up to the maximum load (III). On the other hand, in the case of a/d = 2.0 (shear compression failure), the contribution of V_{ay} was small and the contributions of compressive concrete and dowel action, V_{cd} and V_d became greater up to the maximum load (III).

From this analysis, the contribution of each factors to the shear resistance could be appropriately evaluated.

173

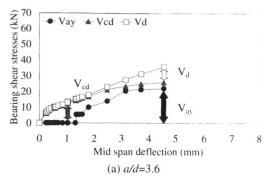

(a) *a/d*=3.6

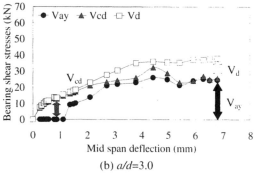

(b) *a/d*=3.0

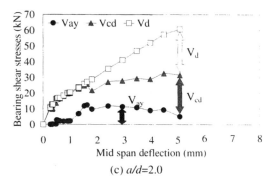

(c) *a/d*=2.0

Figure 16. Contributions for shear resistance and mid span deflection diagram.

6 CONCLUSIONS

In this study, the following conclusions were obtained.

1. Through the use of crack element and bond element, the characteristics of shear failure in RC beam without shear reinforcement can be simulated by the FEM analysis.
2. By applying the re-meshing program, the crack propagation can be represented and the shear failure mode can be estimated.
3. The contributions of compressive concrete V_c, aggregate interlock V_{ay} and dowel force of re-bar V_d to shear resistance can be calculated at each step of loading.

REFERENCES

CEB-FIP. 1990. *CEB-FIP Model Code.*

Maki Matsuo, Kenta Hbino, Nobuaki Takagi & Takayuki Kojima. 2000. Discrete Model for Cracking Concrete and its Fundamental Characteristics. *Journal of Materials, concrete structures and pavmentss (JSCE).* No.556/V-48, pp. 1–11.

Motoyuki Suzuki, Taisuke Nakamura, Makoto Horiuchi & Yoshio Ozaka. 1991. Experimental Study on Influence of Tension Force on Dowel Effect of Axial Bars. *Journal of Materials, concrete structures and pavmentss (JSCE).* No.426/V-14, pp. 159–166.

Ngo. D. & Scordelis, A.C. 1967. Finite Element Analysis of Reinforced Concrete Beams. *ACI Journal.* No.64, pp. 152–163.

Nilson. A.R. 1968. Nonliner Analysis of Reinforced Concrete by the Finite Element Method. *ACI journal.* No.65. pp. 757–766.

Niwa. J. 1993. Size Effect Analysis for Flexural Strength of Concrete Beams Using Nonlinear Rod Element. *Proc. of JCI.* Vol.15. No.2, pp. 75–80.

Rots. J.G. Nauta, P. Kusters, G.M.A. & Blaauwendraad, J. 1985. Smeared Crack Approach and Fracture Localization in Concrete. HERON. Vol.30, No.1.

Saito. S. & Hikosaka, H. 1999. Numerical analyses of models. J. of Materials. *Journal of Materials, concrete structures and pavmentss (JSCE).* V-44, pp. 289–303.

Scordelis, A.C. 1972. Finite Element Analysis of Reinforced Concrete Structures (survey paper). *Proc. Spec. Conf. Finite Element Method Civ. Eng.* Montreal. pp. 71–113.

Takayuki Kojima, Nobuaki Takagi & Kenta Hibino. 2001. Development of a Discrete Bond Linkage Element Between Concrete and Reinforcing Bar. *2001 Second International Conference on Engineering Materials.* Vol.I. pp. 315–326.

Wawrzynek. P.A. & Ingraffeam, A.R. 1987. Interactive Finite Element Analysis of Fracture Processes. An Integrated Approach. Theoretical and Applied Fracture Mechanics. V-8, pp. 137–150.

Finite Elements in Civil Engineering Applications, Hendriks & Rots (eds.)
© 2002 Swets & Zeitlinger, Lisse, ISBN 90 5809 530 4

Finite element modeling of RC members subjected to shear

V.T. Hristovski
Post-doctoral Researcher, Chiba University, Chiba, Japan

H. Noguchi
Professor, Chiba University, Chiba, Japan

ABSTRACT: A hypo-elastic FEM based model has been proposed, taking into account the shear-transfer mechanism across the smeared cracks. The induced shear stress on the crack surface generates a slippage resulting in unbalanced forces that can be further redistributed throughout the system. Also, the discrete-crack approach has been used for simulation of the experimentally observed slippage on the contact between the portions with different thickness, as well as for simulation of the uplift and sinking of tested members relatively to the basement. As a result, original software for non-linear analysis of reinforced concrete structures, based on tangent stiffness approach has been developed. Characteristic specimens (columns and shear-walls) have been used for comparison of the analytically obtained results, showing a good agreement with the experimental ones.

1 INTRODUCTION

Constitutive modeling of reinforced concrete and its finite element implementation for non-linear analysis of RC members and structures subjected to shear remains one of the most interesting research fields for structural engineers. Especially, the modeling of cracking, as major source of concrete non-linearity, still offers inexhaustible perspectives for investigation.

Within this research, cracking of concrete has been considered from different aspects, using appropriate physical models and hypo-elastic incremental numerical algorithms. Perhaps the most common treatment of a crack is to consider it as a product of principal tensile stresses acting perpendicular to the crack plane, defining the so called *normal cracking mode*. However, as many experiments have shown, *tangential cracking mode* resulting from local shear stresses acting along crack planes could have significant influence on overall structural response, particularly for RC members with predominant shear behavior, like shear walls. Despite of the discontinuity nature of cracking, very often the above-mentioned phenomena have been modeled using smeared cracking approach, as also proposed in this paper. However, the limitation of this approach is that it can be used under the assumption of small crack widths distributed in smeared manner without emphasizing any individual cracks. If crack widths become large, then the usage of *discrete crack models* would be more appropriate.

However, we are not able often to predict the location of these large cracks in advance, so that the usage of discrete crack models via joint, link or interface elements becomes less attractive in usual finite element analyses. Still, there is one specific case when the discrete crack model could be successfully applied – the case of abrupt changes in stiffness occurring at joint planes connecting two structural components of different thickness, so that the phenomena like pullout of steel bars, slippage along the joint plane, uplift and sinking of the components relatively to the basement, etc. could be taken into account in the analyses.

In order to deal with all described aspects of cracking, a rotating-crack smeared definition for the normal crack mode, a smeared shear-slip definition including the frictional effects (aggregate-interlock) for the tangential crack mode and a discrete crack approach for modeling the local discontinuities have been discussed, emphasizing the particular influence of each contributors.

2 SMEARED-CRACK CONSTITUTIVE MODELS FOR CONCRETE

2.1 *Uniaxial constitutive models for tension and compression*

The proposed physical model for the normal cracking mode is based on the rotating-crack concept. This concept can grasp the experimentally observed phenomenon of formation of new cracks in rotated direction, while previously formed ones close. Practically, it means that the rotating crack model is treated as a special case of fixed non-orthogonal crack model storing the information only for the latest formed cracks, so that an optimal finite element implementation is provided.

The mathematical model is based on hypo-elasticity (Noguchi 1985). The principal stresses and strains are allowed to rotate in coaxial directions during the loading process that is necessary condition for satisfying the form-invariance of the crack-induced orthotropic material behavior. From this condition, the tangent shear stiffness of the cracked material can be determined, as follows:

$$G_T = \frac{\sigma_1 - \sigma_2}{2(\varepsilon_1 - \varepsilon_2)} \quad (1)$$

In Equation 1, σ_1 and σ_2 are principal stresses, and ε_1 and ε_2 are principal strains. In the cases where a large rotation is expected, its control becomes necessary in order to prevent numerical instabilities during calculation of the *equivalent uniaxial strains*. Therefore, according to Noguchi (1985), the improved accuracy is provided by transformation of the principal axes in case when the rotation relating to the original state becomes greater than 45 degrees. This can be expressed, as follows:

$$\alpha_{b0} = |\theta_{new} - \theta_0| \quad (2a)$$

$$\alpha_{b1} = |\theta_{new} - \theta_{old}| \quad (2b)$$

In Equations 2a and 2b, θ_0, θ_{old}, θ_{new} are initial, previous and current angle of the principal axis 1, related to global x-axis, respectively. If $\alpha_{b0} > 45°$ or $\alpha_{b1} > 45°$ then a transformation of axes is needed in such a way that the stresses, strains and stiffness regarding the axis 1 are substituted to the corresponding quantities related to axis 2, and vise versa.

The ultimate surface adopted in the model is based on the Kupfer's yield curve for bi-axial stresses (Kupfer & Gerstle 1973), taking into account the influence of the strength reduction of cracked concrete in compression (Noguchi et al. 1989). Hence, particularly, for the combination of stresses "tension-compression", when tension strain is greater than cracking strain, the following expressions are used:

$$\sigma_{c2} = \frac{f_c}{0.27 + 0.96\left(\dfrac{\varepsilon_{u1}}{\varepsilon_{cu}}\right)^{0.167}} \geq f_c \quad (3a)$$

$$\sigma_{c1} = \sigma_{c2} \quad (3b)$$

In Equations 3a and 3b, σ_{c1} and σ_{c2} are ultimate stresses in principal directions, $f_c < 0$ is uniaxial concrete strength in compression, $\varepsilon_{u1} > 0$ and $\varepsilon_{cu} < 0$ are equivalent uniaxial strain function in principal direction 1 (in tension) and uniaxial strain in compression for the corresponding strength f_c, respectively. The constitutive relation of concrete for compression is based on the equation of Saenz (1964), as follows:

$$\sigma_{u2} = \frac{E_0 \varepsilon_{u2}}{1 + \left(E_0 \dfrac{\varepsilon_{c2}}{\sigma_{c2}} - 2\right)\dfrac{\varepsilon_{u2}}{\varepsilon_{c2}} + \left(\dfrac{\varepsilon_{u2}}{\varepsilon_{c2}}\right)^2} \quad (4)$$

Taking into account the influence of confinement (Kent & Park 1971 and Scott et al. 1982), the descending branch of the stress-strain uniaxial relation for concrete (Fig. 1) is expressed as follows:

$$\sigma_{u2} = k\sigma_{c2}\Big[1 - z\big(|\varepsilon_{u2}| - k|\varepsilon_{cu}|\big)\Big] \leq 0.2k\sigma_{c2} \quad (5a)$$

where:

$$k = 1 + \frac{\rho_s f_{yh}}{|f_c|} \quad (5b)$$

and

$$z = \frac{0.5}{\dfrac{3 + 0.29|\sigma_{c2}|}{145|\sigma_{c2}| - 1000} + 0.75\rho_s\sqrt{\dfrac{h_c}{e}} - |\varepsilon_{cu}|\,k} \quad (5c)$$

In Equations 5, σ_{c2} is given in MPa, $\sigma_{u2} < 0$ is compressive stress in concrete and $\varepsilon_{u2} < 0$ is corresponding strain. ρ_s is the ratio of the volume of confining reinforcement to the volume of concrete core, h_c is the width of concrete core, e is the spacing of confining reinforcement and f_{yh} is its yield strength. The influence and contribution of the confinement to the

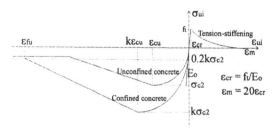

Figure 1. Adopted constitutive uniaxial model for concrete.

ductility and capacity of RC members have been demonstrated in the numerical example of shear-wall specimen SW25, tested by Lefas, et al. (1990).

Shirai & Sato (1984) investigated the effects of bond-slip between steel and concrete, aggregate-interlock between cracked surfaces and dowel action of steel bars crossing the cracks, predicting the crack widths and spacing, factors that determine the tension-stiffening effect. For the sake of simplicity in this mathematical formulation, however having the mentioned factors in mind, the tension-stiffening effect has been taken into account by the empirical Shirai's function, as follows:

$$\sigma_{ul} = f_t\left(1 - 2.748\xi + 2.654\xi^2 - 0.906\xi^3\right) \quad (6a)$$

$$\xi_{ul} = \frac{\varepsilon_{ul} - \varepsilon_{cr}}{\varepsilon_m - \varepsilon_{cr}} \quad (6b)$$

where f_t is the concrete uniaxial tension strength, ε_{cr} is a concrete cracking strain and ε_m is a tension strain for zero stresses, as shown in Figure 1 (adopted in the analyses: $\varepsilon_m = 20\,\varepsilon_{cr}$).

In this model, the smeared reinforcement approach has been adopted for modeling of reinforcing bars using an elastic-plastic bi-linear constitutive model, assuming perfect bond between steel and concrete.

2.2 Smeared constitutive model for shear-slip

The physical model for the tangential cracking mode (shear-slip) is generally based on the Disturbed Stress Field Model, DSFM (Vecchio 2000), however slightly modified and with following assumptions (Fig. 2): 1. The stress-strain states for both concrete and steel are described in average manner within the finite elements integration points; 2. Local variations of the stresses in steel-bars crossing the cracks are taken into account, so that their tensile stresses become higher than average at crack locations; 3. The local normal stresses in concrete at the crack locations approach zero, but they are greater than zero between cracks due to tension-stiffening effect. The magnitude of the average normal tensile stresses that can be transmitted by the concrete is limited by the capacity of the reinforcement crossing the cracks. 4. Rotating crack model is adopted, resulting in equivalence of directions of principal stresses and material axes. As a consequence, average shear stresses in crack directions are equivalent to zero. However, due to the local stress condition in the reinforcement crossing the crack, as well as the friction between the crack surfaces (aggregate-interlock), the equilibrium requirement at crack locations leads to development of local shear stresses along the crack surfaces. 5. Proposed rotation lag in the original DSFM model has not been considered herein, keeping pure rotating-crack formulation. 6. Perfect bond between steel and concrete has been assumed within this smeared crack and smeared-steel formulation.

The shear-slip formulation is based on the stiffness portion of Walraven's (1981) relationship, as follows:

$$\delta = \frac{\tau_c}{1.8w^{-0.8} + \left(0.234w^{-0.707} - 0.2\right)f_{cc}} \quad (7)$$

In Equation 7, δ is tangential slip along the crack, w is crack width, f_{cc} is cube concrete compressive strength and τ_c is local shear stress acting on the crack. Satisfying the equilibrium conditions along the crack surface, this local shear stress can be calculated, as follows:

$$\tau_c = \sum_{i=1}^{n} \rho_i\left(f_{s,cr,i} - f_{s,i}\right)\cos\theta_i \sin\theta_i \quad (8)$$

In Equation 8, ρ_i is the reinforcement ratio of the i-th reinforcement (with the direction α_i related to the global x-axis) crossing the crack, while θ_i is the difference between the angle of the principal direction 1 and the angle of the reinforcement direction α_i. Functions $f_{s,cr,i}$ and $f_{s,i}$ are local steel stress at the actual crack and the average steel stress in the finite element's integration point, respectively. The idea of the proposed numerical approach is to calculate the incremental slip strains $d\varepsilon_{slip}$ from the slippage δ and consequently to find the additional unbalanced stresses $d\sigma_{slip}$ induced by these strains (with C_T being a tangent elasticity matrix for concrete), as follows:

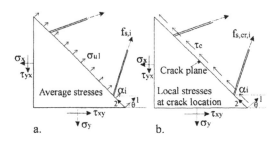

Figure 2. Smeared shear-slip approach based on Vecchio's (2000) physical model: a. equilibrium conditions via average stresses; b. local stresses at crack location.

$$d\sigma_{slip} = C_T\,d\varepsilon_{slip} \quad (9)$$

As previously mentioned, the average concrete normal tensile stress is limited by the capacity of the reinforcement, and it should be checked, as follows:

$$\sigma_{ul} \leq \sum_{i=1}^{n} \rho_i \left(f_{s,y} - f_{s,i} \right) \cos^2 \theta_i \qquad (10)$$

where $f_{s,y}$ is yield strength of the reinforcement.

From numerical point of view, compared to the *basic model* (which takes into account only the normal crack mode, as described in the previous section), the *integral model* (with included shear-slip effects) demands more iterations due to the additional unbalanced forces resulting from the shear-slip deformations. This comes out also due to the nature of the algorithm that does not require any modification of the stiffness matrix. Decreasing the incremental loads, stable solutions with monotonic convergence rate have been obtained.

3 DISCRETE MODEL FOR CONTACT PROBLEMS

3.1 Link element

Studying the behavior of reinforced concrete structural members consisting of more components with different thickness leads to the contact (or interface) problems arising due to stiffness discontinuity and local phenomena, like sinking of thinner components into thicker ones and uplifting of upper components related to the basement, as well as shear slippage along components boundaries. Since these phenomena are essentially three-dimensional, a discrete model becomes indispensable in order to capture the additional non-linear sources resulting from the contact areas. Hence, in addition to the proposed smeared-based constitutive relations, a discrete model via a link element has been herein proposed.

The proposed link element is a standard one (Fig. 3), with zero length and consisting of two points, each having two degrees of freedom (translations in horizontal and vertical direction). Therefore, the basic problem is formulation of the appropriate stiffness coefficients K_s and K_n, i.e. the incremental relationships between the normal force N and dilatancy w, as well as tangential force T and slippage s:

$$\begin{Bmatrix} dT \\ dN \end{Bmatrix} = \begin{bmatrix} K_s & 0 \\ 0 & K_n \end{bmatrix} \begin{Bmatrix} ds \\ dw \end{Bmatrix} \qquad (11)$$

For simplicity of the model, it is assumed that the coefficients in the stiffness matrix are uncoupled. However, as described in the following section, their formulation is based on coupled relations between w

and s. Constitutive relations in the model are based on simple bi-linear elastic-plastic laws including linear strain hardening for relation T–s and linear strain softening for relation N–w in tension.

3.2 Basic assumptions of the model

The determination of the stiffness coefficients K_s and K_n is based on the proposed physical model shown in Figure 4. The model takes into account the effect of bond-slip between the reinforcement crossing the contact area and the surrounding concrete, dowel action effect and frictional effects.

According to the equilibrium conditions in the contact zone, the following relations can be obtained:

$$N = \sigma_s \left(\varepsilon_s \right) A_s - \sigma_{st} A_c \qquad (12a)$$

$$T = \tau_{st} A_c + D_w \qquad (12b)$$

where $\sigma_s(\varepsilon_s)$ is stress in reinforcement, τ_{st} is frictional tangential stress, σ_{st} is accompanying normal reaction stress in concrete, A_c is considered effective concrete area and D_w is force due to dowel action of reinforcement. In order to obtain the stress σ_s in reinforcement, determination of the strain ε_s is needed. To this end, the relationship between bond-slip (equivalent to dilatancy w) and strain in reinforcement, with $w_o = w/d_b$

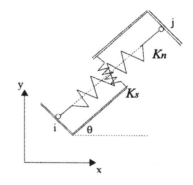

Figure 3. Formulation of link element.

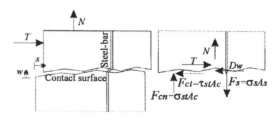

Figure 4. Physical model for contact surface.

$(|f_c|/20\,\text{MPa})^{2/3}$, proposed by Okamura & Maekawa (1991) is herein adopted, as follows:

$$w_o = w_y = \varepsilon_y\left(2 + 3500\varepsilon_y\right) \quad \text{for } \varepsilon_s = \varepsilon_y \qquad (13a)$$

$$w_o = \varepsilon_s\left(2 + 3500\varepsilon_s\right) \quad \text{for } \varepsilon_s \leq \varepsilon_{sy} \qquad (13b)$$

$$w_0 = w_y + 0.047\left(f_{su} - f_{sy}\right)\left(\varepsilon_s - \varepsilon_{sh}\right) \quad \text{for } \varepsilon_s \geq \varepsilon_{sh} \qquad (13c)$$

where d_b, f_{su} and f_{sy} are diameter, ultimate and yielding strengths of the reinforcement crossing the contact zone respectively, and ε_y and ε_{sh} are yielding strain and strain at onset of strain hardening, respectively. The frictional tangential stress τ_{st} and accompanying normal reaction stress σ_{st} can be obtained using the contact density model, described by Okamura & Maekawa (1991), as follows:

$$\beta = \frac{s}{w} \qquad (14a)$$

$$\tau_{st} = f_{st}\frac{\beta^2}{1 + \beta^2} \qquad (14b)$$

$$f_{st} = 3.8|f_c|^{1/3} \quad \text{in MPa} \qquad (14c)$$

$$\sigma_{st} = f_{st}\left[\frac{\pi}{2} - \arctan\left(\frac{1}{\beta}\right) - \frac{\beta}{1 + \beta^2}\right] \qquad (14d)$$

The force D_w due to dowel action can be calculated using the relationship proposed by Vintzeleou & Tassios (1982), as follows:

$$D_w = \begin{cases} 3D_u s & \text{for } s \leq 0.1\,\text{mm} \\ 0.7D_u\sqrt[4]{s} & \text{for } s \geq 0.1\,\text{mm} \end{cases} \qquad (15a)$$

$$D_u = 1.3d_b^2\sqrt{f_{cc}f_{sy}} \qquad (15b)$$

where s is given in mm, f_{cc} and f_{sy} in kPa, and forces D_w and D_u in kN.

3.3 Determination of the stiffness coefficients

Determination of the stiffness coefficients K_s and K_n can not be performed explicitly from the relations of the described physical model. In order to obtain them in a form suitable for implementation into the described link element, a separate implicit procedure, independent on the finite element solution, has been developed.

The idea is to develop parametric functional relations N-w and T-s for several ratios $\beta = s/w$. First, for parametrically given constant slip s, using Equations 13 and 14, normal force functions depending on the dilatancy w can be developed (Fig. 5). Also, for parametrically given constant dilatancy w, using Equations 14 and 15, tangential force functions depending on the slip s can be developed (Fig. 6). From the trend of these developed functions it is evident that the assumed bi-linear elastic-plastic laws are correct approximation for the physical problem. Also, it can

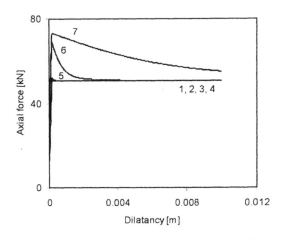

Figure 5. Typical diagrams normal force–dilatancy (N–w) for parametrically given slips from $s = 1 \times 10^{-8}$m (curve 1) to $s = 1 \times 10^{-2}$m (curve 7), specimen W4NF-N0 for the contact column–base structure.

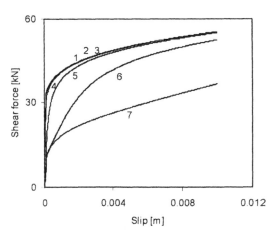

Figure 6. Typical diagrams tangential force–slip (T–s) for parametrically given dilatancies from $w = 1 \times 10^{-8}$m (curve 1) to $w = 1 \times 10^{-2}$m (curve 7), specimen W4NF-N0 for the contact column–base structure.

179

be concluded that the yielding deformations are almost independent on the ratio β, for small values of s and w up to 1×10^{-5} m. Having developed these functions, initial and yielding stiffness coefficients can be quantified, depending on the expected values of s and w during the loading process.

Deficiency of this approach is the fact that in reality for higher load intensities, the ratio β does not remain constant. Therefore, iterative finite element analyses are necessary in order that the differences between assumed and calculated values of s and w become as small as possible.

4 MODEL VERIFICATION

4.1 Finite element implementation

The above described constitutive models for concrete have been implemented into the original, general-purpose computer program FELISA/3M for linear and non-linear static and dynamic analysis of structures. The smeared crack and smeared reinforcement algorithms have been incorporated within plane-stress isoparametric finite elements with 6 nodes (triangles), as well as with 4, 8 and 9 nodes (quadrilaterals). The discrete contact model has been implemented within non-linear link element formulation, as described in the previous chapter. The non-linear equations solver is based on tangent-stiffness Newton-Raphson incremental-iterative scheme using incremental-strains update approach.

For the sake of model verification extensive comparative analytical tests have been performed. Within this paper, three examples are presented: shear wall specimen SW25 tested by Lefas et al. (1990), and shear-wall specimens W4BF-N0 and W4BF-N1 tested by Shibata et al. (1990). The results of the comparative analyses with discussion about the failure modes, as well as the effectiveness and limitations of the proposed constitutive models to simulate the observed members behavior are presented in the following section.

4.2 Numerical examples and discussion

The shear-wall specimen SW25, according to the experimental observations, failed in high-compression ductile mode as a result of confinement provided by additional transverse reinforcement in the concealed columns. The basic constitutive parameters for plane concrete (see Tables 1 and 2) have been calculated according to Vecchio (1992), using the following equations:

$$f_t' = 0.33\sqrt{f_c'} \quad [\text{in MPa}] \tag{16a}$$

Table 1. Steel material properties for shear-wall specimen SW25.

Zone	t [m]	$\rho x/\rho y/\rho z$ [%]	fyx/fyy/fyz [MPa]*
1	0.065	0.82/2.09/0	520/470/0
2	0.065	1.156/3.312/0.9	490/470/420
3	0.20	0.818/1.013/0.27	520/460/420
4	0.20	1.675/1.013/0.135	520/460/420

* Notation: t – zone thickness, $\rho x/\rho y/\rho z$ – reinforcement ratios for directions x, y and z, fyx/fyy/fzz – steel yielding strengths for directions x, y and z, Esx/Esy/Esz – steel initial elasticity moduli for directions x, y and z. All initial steel moduli of elasticity are taken by default Es = 210 GPa.

Table 2. Concrete material properties for shear-wall specimen SW25.

f_c' [MPa]	ε_c'	E_c [GPa]	f_t' [MPa]*
38.3	0.00247	30.95	2.04

* Notation: f_c' – concrete compressive strength, ε_c' – corresponding strain for peak concrete compressive strength f_c', E_c – initial concrete modulus of elasticity, f_t' – concrete tensile strength.

$$E_c = 5000\sqrt{f_c'} \quad [\text{in MPa}] \tag{16b}$$

$$\varepsilon_c' = \frac{2f_c'}{E_c} \tag{16c}$$

The geometry and obtained results via force-displacement diagrams are comparatively presented in Figure 7. The effectiveness of confinement is obvious, comparing the results obtained without (see curve Bo) and with confinement effects (see curves B and I). According to the experimentally observed failure mode, shear-slip in the wall cracks did not participate significantly in the member's overall response. This has been analytically observed, too, obtaining almost identical responses for both discussed models: B (basic) and I (integral). It should be noted that the discrepancy between the analytically obtained maximum shear force of 143.58 kN and the experimental one of 150 kN probably appears due to the concrete expansion effect (via Poisson's ratio), recognized by Vecchio (1992) analyzing the same specimen. This effect has not been taken into account within the proposed models.

The experimental tests of shear-wall specimens W4BF-N0 and W4BF-N1 (see material properties in Tables 3 and 4 and geometry in Figs 8 and 9) were conducted using cyclic loading. These specimens failed in limited ductile mode due to slippage of the upper structure relative to the basement (especially the specimen W4BF-N0 which was not subjected to initial

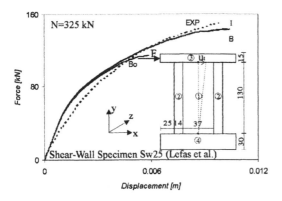

Figure 7. Comparison of the obtained diagrams shear force – displacement for specimen SW25: dashed curve (EXP) – experiment, curve (Bo) – basic model without confinement, curve (B) – basic model with confinement, curve (I) – integral model with confinement.

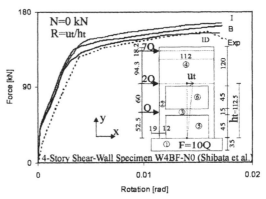

Figure 8. Comparison of the obtained diagrams shear force – rotation for specimen W4BF-N0: dashed curve (Exp) – experiment, curve (B) – basic model, curve (I) – integral model, curve (ID) – integral model plus discrete crack model.

Table 3. Steel material properties for 4-story shear-wall specimens W4BF-N0 and W4BF-N1.

Zone	t[m]	$\rho x/\rho y$[%]	fyx/fyy [MPa]	Esx/Esy [GPa]*
1	0.35	1.85/0.81	575/381	181/175
2	0.15	0.33/2.82	387/385	167/182
3	0.15	1.75/0.33	385/387	182/167
4	0.15	0.45/0.33	381/387	175/167
5	0.05	0.43/0.43	375/375	177/177
6	0.04	0.54/0.54	375/375	177/177

* Notation: t – zone thickness, $\rho x/\rho y$ – reinforcement ratios for directions x and y, fyx/fyy – steel yielding strengths for directions x and y, Esx/Esy – steel initial elasticity moduli for directions x and y.

Table 4. Concrete material properties for 4-story shear-wall specimens W4BF-N0 and W4BF-N1.

Specimen	f_c' [MPa]	ε_c'	E_c [GPa]	f_t' [MPa]*
W4BF-N0	32.7	0.00249	22.9	1.887
W4BF-N1	36.5	0.00277	24.5	1.994

* Notation the same as for Table 2.

axial load), also developing bending and shear-cracks, and with large deformations in reinforcing bars.

In the analyses, the values of ε_c' and E_c were taken from the experiments, and f_t' was calculated according to Equation 16a. Comparative push-over analyses including all 3 models (B-basic one, I-integral one and ID − integral + discrete crack model) have been

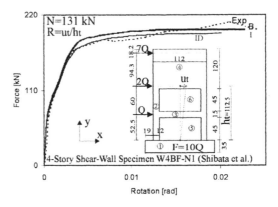

Figure 9. Comparison of the obtained diagrams shear force – rotation for specimen W4BF-N1: dashed curve (Exp) – experiment, curve (B) – basic model, curve (I) – integral model, curve (ID) – integral model plus discrete crack model.

performed, as shown in Figures 8 and 9. The ID model has proved to be most successful in predicting the structural response, since it has provided simulation of the contact phenomena between walls and base structures. Although the slip-page has not contributed in rotation, it has affected the final capacity. On the other hand, the dilatancy has contributed in both rotation and capacity. In Table 5, some of the calculated stiffness coefficients and yield points for slippage and opening phenomena are shown, based on the methodology described in the previous chapter (also see diagrams in Figs. 5 and 6). It should be noted that the phenomenon of "sinking", i.e. compression of the contact surfaces, has been assumed as linear-elastic

Table 5. Obtained stiffness coefficients for LINK elements (specimen W4BF-N0, contact zone column-base structure).

Kso	Ksy	Sy	Kno	Kny	wy
2×10^6	1.8×10^4	2.5×10^{-4}	2.3×10^7	-3.4×10^4	1.9×10^{-6}

Notation: Kso – Initial tangential stiffness, Ksy – Yielding tangential stiffness, Sy – Yielding slip, Kno – Initial normal stiffness, Kny – Yielding normal stiffness, wy – Yielding dilatancy. The stiffness coefficients are given in kN/m, slippage Sy and dilatancy wy in m.

having identical stiffness coefficient as for tension-opening.

In Figure 8, it can be observed that the integral model (I) has given somewhat "softer" response in the loading stage before yielding point compared to the basic one (B), owning to the influence of shear-slip deformations. However, the mobilized aggregate-interlock, taken into account by Walraven's (1981) Equation 7, has led to increased overall shear capacity after the yielding point. Since these two models have not captured the contact phenomena, the obtained shear-capacity has been greater than experimentally observed. The improvement of the shear-capacity prediction has been achieved using the integral plus discrete crack models (ID), as can be seen in Figure 8. The similar discussion could be done for the specimen W4BF-N1, however, due to the influence of axial load, contact phenomena have shown to be less significant than the previous case. In this case the specimen showed greater ductility and the shear-slip along cracks in the wall has not affected significantly the response (see B and I in Fig. 9). However, finally, as prediction using integral plus discrete crack model (ID) has shown, the shear capacity has been reduced owning to the contact effects. Also, it can be observed that for rotation greater than 0.015 radians the prediction using ID model is in disagreement with the experimental result, probably due to the strain-hardening effect in the reinforcement, since the proposed model for steel is based on bi-linear elastic-plastic approximation. This example shows that for cases of large deformations in reinforcement, the proper modeling of steel constitutive law could be very significant in order that more successful prediction of the structural response be performed.

Despite of discussed limitations of the proposed models, significant accordance of experimental and analytical results has been generally obtained. Comparative results for the progressive failure process of specimens W4BF-N0 and W4BF-N1 are presented in Tables 6 and 7.

A good agreement with respect to characteristic points (bending cracking, yielding and ultimate points) has been obtained for both specimens. In order to have insight into the failure progress, graphic presentations of deformation states including crack patterns

Table 6. Progressive failure process – comparison between experimental and analytical results, specimen W4BF-N0.

	Qbc	Qsc	Qy	Qu*
(EXP)	15.4	63.18	125.08	155.76
(B)	18.4	87.20	135.12	163.12
(I)	18.4	81.62	143.05	166.25
(ID)	18.4	80.86	127.63	154.64

* Notation: Qbc – shear force at bending cracking, Qsc – shear force at shear cracking, Qy – shear force at steel yielding, Qu – maximum shear force, (EXP) – experimental results, (B) – basic analytical model, (I) – integral analytical model, (ID) – integral plus discrete crack analytical models. Shear forces are given in [kN].

Table 7. Progressive failure process – comparison between experimental and analytical results, specimen W4BF-N1.

	Qbc	Qsc	Qy	Qu*
(EXP)	39.24	156.86	166.57	209.64
(B)	37.24	162.47	180.90	199.25
(I)	37.40	160.16	180.90	198.12
(ID)	37.40	156.05	172.01	192.86

* Notation the same as in Table 3.

obtained using ID model are shown in Figure 10 for the specimen W4BF-N0, showing a good agreement with experimentally observed damage progress described in Shibata et al. (1990).

5 CONCLUSIONS

Within hypo-elastic formulation, smeared-crack models for normal and tangential cracking mode have been discussed, emphasizing the influence of each separate mode to the structural response. Also, within the uniaxial constitutive models for concrete in compression, the influence of confinement has been considered. In addition, a discrete formulation for contact problems arising on the boundaries between the members components has been proposed via link element. The comparative analyses using the separate

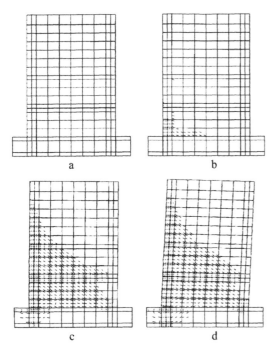

Figure 10. Progressive failure process simulation, specimen W4BF-N0, integral model plus discrete crack model: a) at first bending cracking, b) at first shear cracking, c) at steel yielding onset, d) at maximum load.

models have shown the following: 1. For members failing in bending mode (as the case of specimen SW25 failed in high compression), the influence of shear-slip does not affect significantly the response; 2. For members failing in mixed and shear mode, shear-slip has significant influence; 3. Slip-page and opening on the boundary surfaces, especially between upper structure and basement could affect the response, depending on the axial force level. Having pointed out the model's limitations (cases of concrete expansion due to Poisson's ratio effects and cases of large steel strains), the future refinement of the model should be directed toward these problems.

ACKNOWLEDGMENTS

The authors would like to express their appreciations to the Japan Society for the Promotion of Science (JSPS) that supported this research through the Post-doctoral Fellowship Program for Foreign Researchers.

REFERENCES

Kent, D. C. & Park, R. 1971. Flexural members with confined concrete. In *ASCE Journ. of structural division* Vol. 97, No. 97, No. ST7, July, 1971: 1969–1990.

Kupfer, H. & Gerstle, K. H. 1973. Behavior of concrete under biaxial stresses. In *ASCE Journ. of engineering mechanics division* Vol. 99, No. EM4, Proc. Paper 9917, Aug. 1973: 852–866.

Lefas, I. D., Kotsovos, M. D. & Ambraseys, N. N. 1990. Behavior of reinforced concrete structural walls: strength, deformation characteristics and failure mechanism. In *ACI Journ.* Vol. 87, No. 1: 23–31.

Noguchi, H. 1985. Analytical models for reinforced concrete members subjected to reversed cyclic loading. In *Seminar on finite element analysis of reinforced concrete structures: Proc. intern. seminar, Tokyo 21–24 May 1985.* Vol. 2: 93–112. JSPS, Tokyo.

Noguchi, H., Ohkubo, M. & Hamada, S. 1989. Basic experiments on the degradation of cracked concrete under biaxial tension and compression. In *JCI Proc.* Vol. 11, No. 2, 1989: 323–326 (In Japanese).

Okamura, H. & Maekawa, K. 1991. *Nonlinear analysis and constitutive models of reinforced concrete.* Tokyo: University of Tokyo.

Saenz, L. P. 1964. Discussion of equation for the stress-strain curve of concrete by Desayi and Krishman. In *ACI Journ.* Vol. 61, September 1964: 1229–1235.

Scott, B. S., Park, R. & Priestley, M. J. N. 1982. Stress strain behavior of concrete confined by overlapping hoops at low and high strain rates. In *ACI Structural Journ.* Vol. 79, No. 1, Jan.–Feb., 1982: 13–27.

Shibata, T., Hirabuki, M., Goto, Y. & Joh, O. 1990. Performances at various deformations in RC multi-story structural walls. In *Summaries of technical papers of annual meeting, Architectural institute of Japan,* Structures II, October 1990: 573–578 (In Japanese).

Shirai, N. & Sato, T. 1985. Bond cracking model for reinforced concrete. In *Trans. of the JCI,* Vol. 6, 1984: 457–468.

Vecchio, F. J. 1992. Finite element modeling of concrete expansion and confinement. In *ASCE Journ. of structural engineering* Vol. 118, No. 9, September, 1992: 2390–2406.

Vecchio, F. J. 2000. Disturbed stress field model for reinforced concrete: formulation. In *ASCE Journ. of structural engineering* Vol. 126, No. 9, September, 2000: 1070–1077.

Vintzeleou, E. N. & Tassios, T. P. 1982. Dowel action under cyclic loading. In *7th Symposium on earthquake engineering, Proc. Roorkee, 1982.*

Walraven, J. C. 1981. Fundamental analysis of aggregate interlock. In *ASCE Journ. of structural engineering* Vol. 107, No. 11, 1981: 2245–2270.

Finite Elements in Civil Engineering Applications, Hendriks & Rots (eds.)
© 2002 Swets & Zeitlinger, Lisse, ISBN 90 5809 530 4

Size effect analysis for shear strength of reinforced concrete deep beams

K. Yoshitake & T. Hasegawa
Institute of Technology, Shimizu Corporation, Tokyo, Japan

ABSTRACT: Failure analysis of reinforced concrete deep beams is presented using the general purpose finite element system DIANA, with attention towards size effect and effects of bending span to depth ratio on shear behavior. Effects of shape and size of finite elements and boundary conditions on the failure of the beams are also investigated. It is shown that the maximum shear strength and failure process of the deep beams can be well simulated by the present analysis method. It has been proved that the element shape can influence on the pre-peak shear response and tensile modeling alone, which takes into account fracture energy, is not sufficient to simulate the size effect of shear strength in reinforced concrete deep beams. Furthermore, the bending span to depth ratio has a significant influence on the shear strength, same as the shear span to depth ratio.

1 INTRODUCTION

In contrast to reinforced concrete slender beams, reinforced concrete deep beams increase the load after the formation of diagonal cracks due to tied arch action. Failure of deep beams dominated by either the yielding of reinforcement, or shear-compressive failure of web concrete, and the localized shear-compressive failure influences size effect of reinforced concrete deep beams on shear strength. According to the design equation of the Japan Society of Civil Engineers Standard Specification, shear strength decreases inversely proportional to the fourth root of the effective depth due to size effect. Therefore, it is very important to evaluate the complicated failure process of reinforced concrete deep beams for safety and rational design.

This paper describes the simulation of reinforced concrete deep beams performed by Matsuo (Matsuo 2001) and Niwa (Niwa 1983) using the general purpose finite element system DIANA (Yoshitake & Hasegawa 2000, 2001) with attention towards size

effect and the effect of bending span to depth ratio on the shear behavior. Moreover, the effects of the shape and size of finite elements and boundary conditions on the failure of the beams are investigated.

2 ANALYSIS MODELING

2.1 *Analysis cases 1, 2, and 3*

In analysis cases 1, 2, and 3, a shear test of a reinforced concrete deep beam specimen performed by Matsuo is simulated to investigate the effect of the shape and size of finite element mesh and loading boundary conditions as shown in Table 1. Figure 1 shows the finite element mesh for the specimen called D400 which has the effective depth of 400 mm and the same shear span to depth ratio $a/d = 1.0$ (a = shear span length; d = effective depth). A three-node triangular isoparametric plane stress element for analysis case 1, and a four-node quadrilateral isoparametric plane stress

Table 1. Analysis cases and parameters for analysis cases 1–3.

Analysis case	Loading condition	Element shape	Crack band width (mm)
1C	Central upper node	Three-node triangular element	25
1A	All of the upper nodes	Three-node triangular element	25
2C	Central upper node	Four-node quadrilateral element	25
2A	All of the upper nodes	Four-node quadrilateral element	25
3C	Central upper node	Four-node quadrilateral element	50
3A	All of the upper nodes	Four-node quadrilateral element	50

element for analysis cases 2 and 3 are used. Crack band width h is assumed as shown in Figure 1. For each analysis case, a fixed displacement load is applied to either all of the upper nodes or just the central upper node of the loading plate.

2.2 Analysis cases A–H

In analysis cases A–E, a shear test of three reinforced concrete deep beam specimens performed by Niwa is simulated with particular attention to the relationship between shear strength and the effective depth. The effective depths of specimens called SI0, SI1 and LR0 are 300, 600, and 900 mm, respectively. The specimens have the same $a/d = 0.5$, but are not geometrically similar. The effects of several parameters, such as the constitutive model of concrete and the finite element size are investigated (Table 2). Compressive elasto-plastic model with yield plateau (perfectly plastic model) and tensile perfectly plastic model are used for some analysis cases to reduce the cause of the analytical size effect. In analysis cases F–H, hypothetical specimens which are completely geometrically similar to specimen SI0 are calculated

to investigate the effect of the bending span to depth ratio.

The finite element meshes used in analysis cases A–E and F–H are shown in Figures 2–4, and Figure 5, respectively. The web concrete portion used in analysis case A is discretized into the same size triangular elements as specimen SI0. On the other hand, the web concrete portion for specimens SI1 and LR0 in analysis cases B–E is discretized into enlarged similar triangular elements that are used for specimen SI0. In analysis cases F–H, hypothetical specimens (d = 300, 600, and 900 mm) with completely similar meshes to the one for specimen SI0 are analyzed.

2.3 Material model

The compressive elasto-plastic model and tensile fracture models for concrete are determined based on CEB-FIP Model Code 1990 as shown in Figure 6.

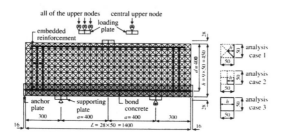

Figure 1. Finite element model of specimen D400.

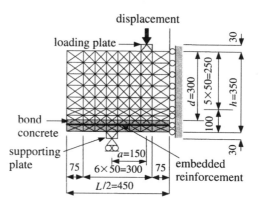

Figure 2. Finite element model of specimen SI0.

Table 2. Analysis cases and parameters for analysis cases A–H.

Analysis case	Tensile modeling		Compressive modeling	Notes
	Web concrete	Bond concrete		
A	Tension softening model	Tension stiffening model	Elasto-plastic model	Niwa's specimen (identical and fine mesh)
B	Tension softening model	Tension stiffening model	Elasto-plastic model	Niwa's specimen (coarse mesh with similar shape)
C	Perfectly plastic model	Perfectly plastic model	Elasto-plastic model	
D	Tension softening model	Tension stiffening model	Elasto-plastic model with plateau	
E	Perfectly plastic model	Perfectly plastic model	Elasto-plastic model with plateau	
F	Tension softening model	Tension stiffening model	Elasto-plastic model	Hypothetical specimen (completely similar to specimen SI0)
G	Tension softening model	Tension stiffening model	Elasto-plastic model with plateau	
H	Perfectly plastic model	Perfectly plastic model	Elasto-plastic model with plateau	

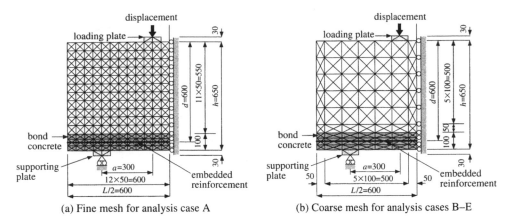

(a) Fine mesh for analysis case A

(b) Coarse mesh for analysis cases B–E

Figure 3.   Finite element model of specimen SI1.

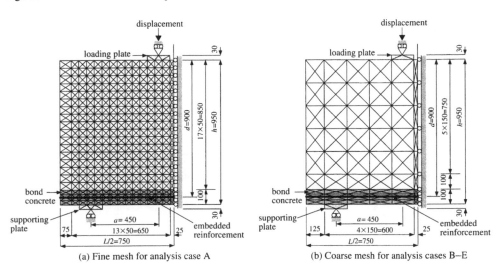

(a) Fine mesh for analysis case A

(b) Coarse mesh for analysis cases B–E

Figure 4.   Finite element model of specimen LR0.

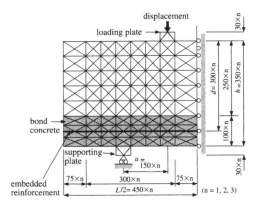

Figure 5.   Hypothetical completely similar specimen for analysis cases F–H.

The compressive yield conditions of Drucker-Prager and Von Mises, which are derived from the experimental data (Kuper and Gerstle 1973), are used in analysis cases 1–3, and analysis cases A–H, respectively. In all analysis cases, reinforcing bars are modeled with the embedded reinforcement elements of Von Mises elasto-plastic constitutive model, taking into account tension stiffening effect of concrete. The concrete around the main reinforcement is modeled as a tension stiffening region for the bond concrete. In analysis cases C, E, and H, a perfectly plastic model under tension is used for both the web concrete and bond concrete. After cracking, the shear stiffness is assumed to be reduced to 1% of the elastic stiffness. The strength and elastic moduli values for concrete and steel reinforcement are as measured during the experiments.

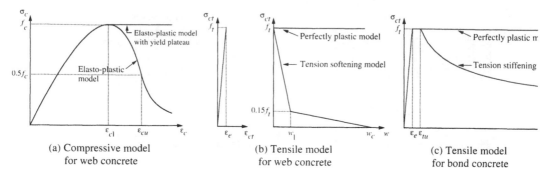

(a) Compressive model
for web concrete

(b) Tensile model
for web concrete

(c) Tensile model
for bond concrete

Figure 6.   Concrete models.

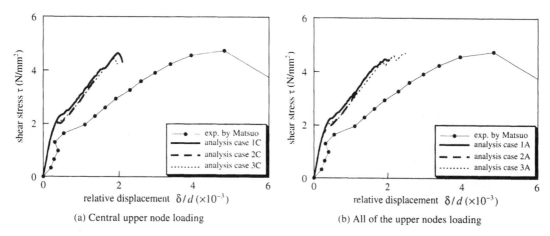

(a) Central upper node loading

(b) All of the upper nodes loading

Figure 7.   Shear response.

3   ANALYSIS RESULTS AND DISCUSSION

3.1   *Analysis cases 1, 2, and 3*

Figure 7 shows the shear response in each analysis case compared with the experiment, where $\tau = V/bd$ is shear stress; V = shear load; b = beam width; δ = relative displacement between the loading and supporting plates. The shear strength is well predicted for each loading boundary condition, although those of analysis cases 2C and 2A are significantly underestimated. And the analytical shear responses are stiffer than the experiment due to a discrepancy in initial stiffness. In all analysis cases, the main reinforcement does not yield. The cracking patterns and the incremental displacements of analysis cases 1A, 2A, 3A and 1C at maximum load are shown in Figures 8 and 9.

In analysis cases 1 and 3, the diagonal shear cracks propagate from the supporting plate to the loading plate after flexural cracks develop near the midspan in the early loading stage. Finally, the shear-compressive failure localizes near the loading plate. The numerically predicted fracture process agrees well with the experiment.

On the other hand, in analysis cases 2C and 2A the calculation terminates due to divergence before diagonal shear cracks propagate. The equivalent plastic strain, the total displacement, and the stress-strain responses of elements a and b in analysis case 2A are shown in Figures 10, 11, and 12, respectively. It is observed that a plastic compressive region forms locally near the loading plate. Since compressive softening fracture responses do not occur in element b, the maximum load is not obtained after the shear-compressive failure near the loading plate. Figure 13 compares profiles for an incremental displacement at the extreme tension fiber and the third nodal row above the extreme tension fiber near the maximum load. The incremental displacement of only one node at each shear span is

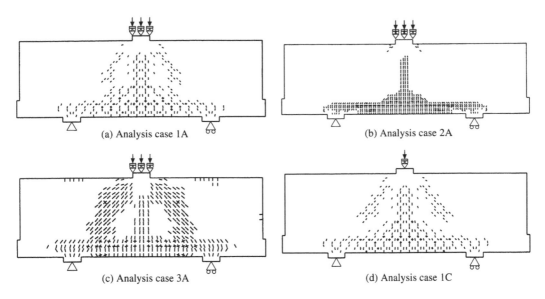

(a) Analysis case 1A

(b) Analysis case 2A

(c) Analysis case 3A

(d) Analysis case 1C

Figure 8.   Cracking pattern at maximum load.

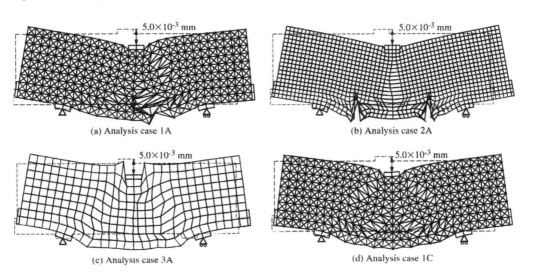

5.0×10^{-3} mm

5.0×10^{-3} mm

(a) Analysis case 1A

(b) Analysis case 2A

5.0×10^{-3} mm

5.0×10^{-3} mm

(c) Analysis case 3A

(d) Analysis case 1C

Figure 9.   Incremental displacement at maximum load.

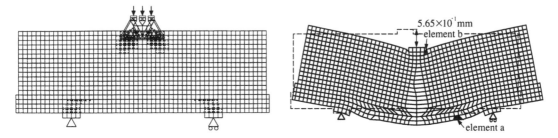

5.65×10^{-1} mm
element b

element a

Figure 10.   Equivalent plastic strain in analysis case 2A.

Figure 11.   Total displacement in analysis case 2A.

189

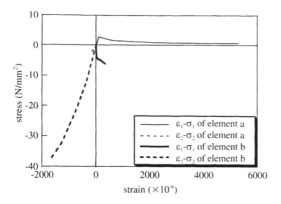

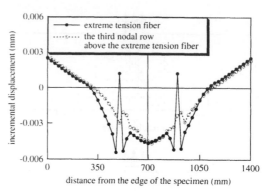

Figure 12. Stress-strain responses in analysis case 2A.

Figure 13. Incremental displacement in analysis case 2A.

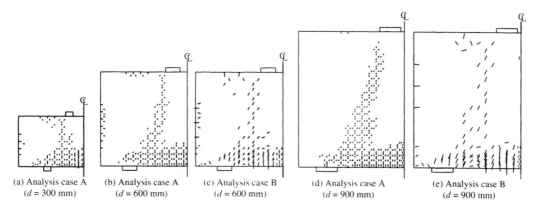

(a) Analysis case A
(d = 300 mm)

(b) Analysis case A
(d = 600 mm)

(c) Analysis case B
(d = 600 mm)

(d) Analysis case A
(d = 900 mm)

(e) Analysis case B
(d = 900 mm)

Figure 14. Analytical cracking pattern at maximum load.

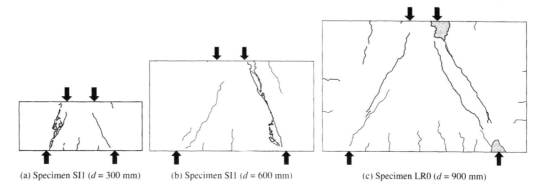

(a) Specimen SI1 (d = 300 mm)

(b) Specimen SI1 (d = 600 mm)

(c) Specimen LR0 (d = 900 mm)

Figure 15. Experimental cracking pattern at maximum load.

longitudinally increased at the extreme tension fiber, however, it relieves at the third nodal row above the extreme tension fiber. Therefore, it is concluded that the calculation might terminate because of the spurious kinematic mode as shown in Figures 9(b) and 13.

The maximum loads for different boundary conditions are almost the same, but the special case is observed, in which the incremental displacement mode differs between them due to the effect of the rotation of the loading plate.

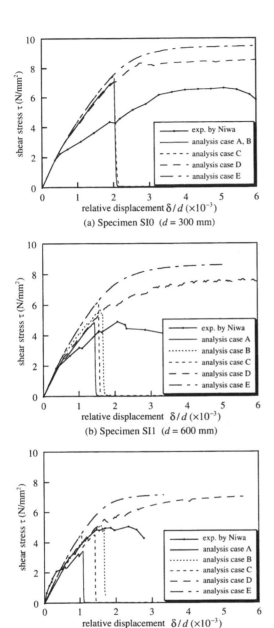

(a) Specimen SI0 ($d = 300$ mm)

(b) Specimen SI1 ($d = 600$ mm)

(c) Specimen LR0 ($d = 900$ mm)

Figure 16. Shear response.

3.2 Analysis cases A–H

The cracking patterns of analysis cases A and B at maximum load are shown in Figures 14 and 15 along with the experimental ones. The numerically predicted cracking patterns agree well with the experiment in the analysis A with a fine mesh and analysis B with a coarse mesh.

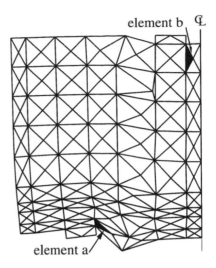

Figure 17. Incremental displacement in analysis case B ($d = 600$ mm).

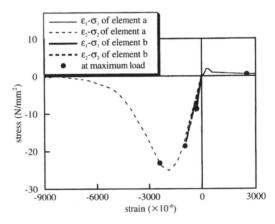

Figure 18. Stress-strain responses in analysis case B ($d = 600$ mm).

Figure 16 shows the shear responses in analysis cases A–E compared with the experiment. In analysis cases A and B, the shear strength is well predicted and maximum difference is about 15%, except in analysis case A for specimen LR0, although the analytical shear responses after cracking are stiffer than in the experiment. The incremental displacement for specimen SI1 at the maximum load in analysis case B and the stress-strain responses of elements a and b are shown in Figures 17 and 18, respectively. The shear-compressive softening failure occurs after cracking at element a. On the other hand, the unloading takes place after maximum load at element b in the biaxial compressive state.

Figure 19 shows the strain distribution of the main reinforcement at the maximum load. In analysis case B for specimen SI1, the main reinforcement does not yield. In common with this analysis case, the shear-compressive softening failure occurs in the all specimens for analysis cases A and B. A difference of deformation capacity between the analysis and the experiment is considered to be caused, since the region in which the analytical shear-compressive softening failure localizes is smaller than the experiment.

The maximum load and deformation capacity in analysis case A using the fine finite element mesh, which takes into account fracture energy, are underestimated. Comparison between analysis cases B, C, and D shows that the tensile modeling of concrete

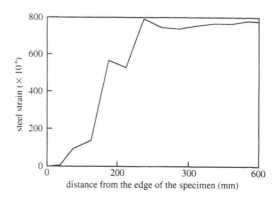

Figure 19. Steel strain in analysis case B ($d = 600 \, \text{mm}$).

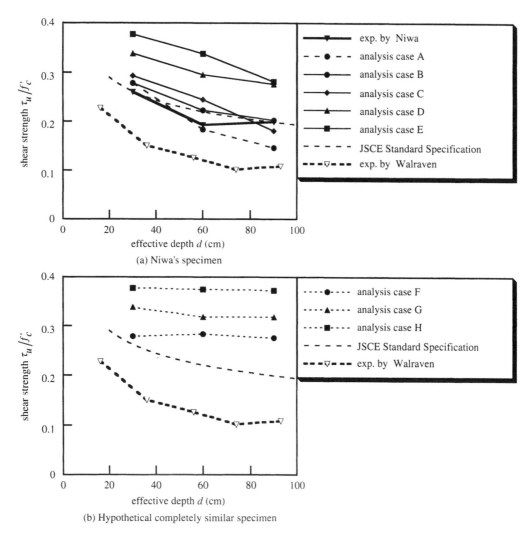

(a) Niwa's specimen

(b) Hypothetical completely similar specimen

Figure 20. Size effect on shear strength.

does not influence much on the shear response, compared with compressive modeling.

The shear strength obtained in each analysis case is shown in Figure 20 with experimental test data of Niwa and Walraven (Walraven & Lehwalter 1994), in which $\tau_u = V_u/bd$ = shear strength; V_u = maximum shear load; b = beam width; f_c' = concrete compressive strength. The design equation of the Japan Society of Civil Engineers Standard Specification is compared with the size effect in the same figure. The experimental shear strengths of specimens SI1 and LR0 decreases about 25% compared with that of specimen SI0. The size effect of shear strength is well predicted for analysis cases A and B. Analysis case E which reduces the cause of the analytical size effect by using perfectly a plastic model for both compressive and tensile of concrete, also predicts the size effect. On the other hand, analysis case H, using the hypothetical specimens which are completely similar to specimen SI0 and a perfectly plastic model, cannot predict the decrease of the shear strength as the effective depth increases. It is concluded that the experimental size effect might be influenced by the geometrical unsimilarity in beam shape.

Analysis case F, using the hypothetical specimens and the crack band model, cannot predict the size effect. Since the deep beams collapse due to the shear-compressive failure as described above, tensile modeling alone, which takes into account fracture energy, is not sufficient to simulate the size effect of the shear strength of the reinforced concrete deep beams. Therefore, it is necessary to consider the size effect on compressive softening failure of concrete.

4 CONCLUSIONS

While failure analysis of reinforced concrete deep beams described here demonstrate the potential of the present analysis method to predict the shear strength, shear-compressive softening failure process and cracking patterns. It has been proved that the element shape can influence on the pre-peak shear responses because of the spurious kinematic mode. The different boundary conditions predict almost the same shear strength but different incremental displacement mode due to the effect of the rotation of the loading plate. Moreover, tensile modeling alone, which takes into account fracture energy, is not sufficient to simulate the size effect on the shear strength of reinforced concrete deep beams, and the bending span to depth ratio has a significant influence on the shear strength, same as the shear span to depth ratio.

REFERENCES

Matsuo, M. 2001. Shear tests of reinforced concrete deep beams. *JCI committee report on test method for fracture property of concrete*. Tokyo: JCI.
Niwa, J. 1983. *Shear load carrying mechanism of reinforced concrete members like deep beam*, Ph.D. thesis, The University of Tokyo. Tokyo.
Yoshitake, K. & Hasegawa, T. 2000. Parametric study for size effect of reinforced concrete deep beam on shear strength. *Proceedings of the 55th annual conference of JSCE 5*: V-523.
Yoshitake, K. & Hasegawa, T. 2001. Effects of shape and size of finite element on shear behavior of reinforced concrete deep beam. *Proceedings of the 56th annual conference of JSCE 5*: V-452.
Kupfer, H.B. & Gerstle, K.H. 1973. Behavior of concrete under biaxial stresses. *Journal of Engineering mechanics, ASCE* 99, 4: 853–866.
Walraven, J. & Lehwalter, N. 1994. Size effects in short beams loaded in shear. *ACI Structural Journal* 91(5): 585–593.

Finite Elements in Civil Engineering Applications, Hendriks & Rots (eds.)
© 2002 Swets & Zeitlinger, Lisse, ISBN 90 5809 530 4

Shear behavior of reinforced ultra light weight concrete beams with shear reinforcement

K. Fukuzawa
Department of Urban and Civil Engineering, Ibaraki University, Hitachi, Japan

M. Mitsui
Satellite Venture Business Laboratory, Ibaraki University, Hitachi, Japan

S. Soh
Construction Deptartment Building Engineering Section, Takenaka Corporation, Osaka, Japan

ABSTRACT: This study concerns the shear behavior of beams with shear reinforcement using the ultra light weight aggregate, comparing with that of beams using normal weight aggregate. The results of this study can be summarized as follows: (1) The ultimate shear strength of an ultra light weight concrete beam with shear reinforcement is calculated as the sum of the concrete component and the shear reinforcement component based on the modified truss theory, using 0.57 as the reduction factor of the concrete component. (2) The proposed nonlinear FEM, that modelizes the longitudinal and transverse reinforcements as embedded rebar elements, can very adequately simulate the structural behavior such as the development of cracks, the strain of rebar/concrete and deformation under the load and the ultimate strength, irrespective of the kinds of concrete.

1 INTRODUCTION

Recently, with the enlargement of the civil engineering structure and the multi-storization of the building, the dead weight reduction of the concrete has been required, and more studies on light weight concrete is actively being carried out. Furthermore, the artificial light weight aggregate made from waste glass or perlite has been developed. These aggregates are called the ultra light weight artificial aggregates, since they are stronger and more impermeable than ordinary light weight aggregates due to their solid surface and the internal independent bubbles of the aggregate. It is well known that the concrete using ultra light weight artificial aggregate (the ultra light weight concrete) is highly brittle since the concrete has a larger brittleness coefficient (ratio of compressive strength to tensile strength) and a lower fracture energy compared with those of the concrete using a normal weight aggregate (the normal weight concrete). It is considered that such mechanical properties are very influential with the shear behavior of structural members using the ultra light weight concrete. This present paper experimentally and analytically discusses the shear behavior of the RC beam with shear reinforcement using an ultra light weight concrete, whose strength is about 35 N/mm² and whose

unit weight is about 1300 kg/m³, compared with the behavior of the RC beam using a normal weight concrete that has the same strength. Furthermore, the structural behavior is simulated using a nonlinear finite element method (FEM) that modelizes the rebar as an embedded element.

2 OUTLINE OF THE EXPERIMENTS

2.1 *Test beams*

An example of the details of the beam and the arrangement of the rebar are shown in Figure 1. The effective height of beam d is 250 mm, and the shear span ratio a/d is 2.4. The U-shaped stirrups were used as the shear reinforcement as shown in Figure 1. The wire strain gages, whose gage lengths were 5 mm, were affixed to the rebar as shown in Figure 1 in order to measure the strain of the stirrups. The list of test beams is shown in Table 1. The type of concrete and the stirrup ratio p_w are taken as the factors of the experiments as shown in the table. The deformed bar nominal diameters of 22 mm ($f_{ty} = 432$ N/mm²) or 25 mm ($f_{ty} = 435$ N/mm²) were used for the longitudinal reinforcement, and a steel bar nominal diameter of 9 mm ($f_{wy} = 500$ N/mm²) was used for the stirrups.

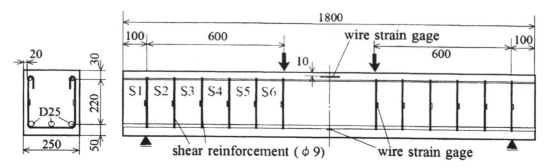

Figure 1.   Cross section and longitudinal section of N-4 or L-4 beam.

Table 1.   List of test beams.

No.	Type of concrete	Unit weight [kg/m³]	Comp. strength [N/mm²]	Elastic modulus [N/mm²]	Tensile strength [N/mm²]	Diameter, number of bars	Reinforcement ratio p_t [%]	Diameter	Reinforcement ratio p_w [%]
		Concrete				**Longitudinal rebar**		**Shear rebar**	
N-1	NWC	2290	29.0	23900	2.58	D22, 2	1.24	φ9, 0	0.00
N-2								φ9, 3	0.17
N-3		2270	27.6	25412	2.07	D25, 3	2.43	φ9, 6	0.42
N-4								φ9, 8	0.59
L-1	LWC	1370	34.2	12568	1.77	D22, 2	1.24	φ9, 0	0.00
L-2								φ9, 3	0.17
L-3		1260	36.2	11670	1.48	D25, 3	2.43	φ9, 6	0.42
L-4								φ9, 8	0.59

Table 2.   Materials using for test beams.

	Material		Specification
Cement	Ordinary portland cement	C	Density: 3.16 g/cm³
Fine aggregate	Artificial ultra light weight aggregate	S_G	Density: 0.72 g/cm³, made from glass
	Crushed sand	S_C	Density: 2.65 g/cm³, F.M.: 2.79
Coarse aggregate	Artificial ultra light weight aggregate	G_A	Density: 1.20 g/cm³, made from perlite
	Crushed stone	G_C	Density: 2.60 g/cm³, G_{max}: 20 mm
Admixture	AE agent	AE	Anionic detergent
	High-performance AE water reducing agent	SP	Polycarboxylic acid type

Table 3.   Concrete mixture.

Concrete	W/C [%]	s/a [%]	W	C	S_G	S_C	G_A	G_C	AE	SP
			Unit weight [kg/m³]							
NWC	55.0	43	191	346	–	759	–	953	17.3	–
LWC	35.0	49	214	612	199	–	255	–	–	1.22

2.2   Materials and concrete mixes

The materials used for making the beams are shown in Table 2. The pelletized ultra light weight aggregate made from the perlite was used for the coarse aggregate. The pelletized ultra light weight aggregate made from waste glass was used as the fine aggregate. The concrete mixes are shown in Table 3. A high-performance AE water reducing agent was used

Table 4. Summary of experimental details.

Concrete	No.	Ultimate shear strength [kN]				$1 - V_{u,exp}/$ $V_{u,cal}$	$1 - V_{u,exp}/$ $V'_{u,cal}$	$1 - V_{u,exp}/$ $V_{u,FEM}$	Failure type
		$V_{u,exp}$	$V'_{u,cal}$	$V'_{u,cal}$	$V_{u,FEM}$				
NWC	N-1	109	78	85	108	0.39	0.28	0.00	Shear, diagonal tension
	N-2	129	124	131	123	0.04	0.02	0.05	Shear, diagonal tension
	N-3	229	213	221	210	0.08	0.04	0.09	Shear, diagonal tension
	N-4	211	–	–	–	–	–	–	Flexural
LWC	L-1	44	56	46	47	0.21	0.03	0.05	Shear, diagonal tension
	L-2	94	102	92	103	0.07	0.03	0.08	Shear, diagonal tension
	L-3	157	188	176	156	0.16	0.08	0.01	Shear, diagonal tension
	L-4	196	234	222	206	0.16	0.12	0.05	Shear, diagonal tension

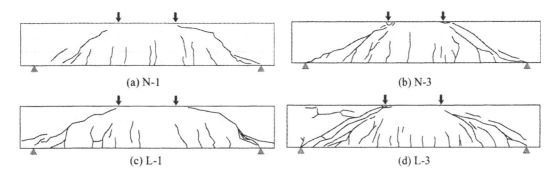

(a) N-1   (b) N-3

(c) L-1   (d) L-3

Figure 2. Crack patterns at failure of beam.

in the ultra light weight concrete in order to control the slump and to prevent the bleeding.

3 TEST RESULTS

3.1 Cracking at failure

Table 4 gives the summary of the experimental results from the shear tests. The failure mode of all test beams, except the N-4 beam, was a diagonal tension mode which failed by the yielding of shear reinforcement. The mode of the N-4 beam was a flexural mode which failed by the yielding of longitudinal rebar. The mode of the N-3 beam is decided by the failure of shear, since the yielding of the stirrup preceded the yielding of the longitudinal rebar. Examples of the cracking patterns failure are shown in Figure 2.

3.2 Calculation method of the ultimate shear strength

The experimental ultimate shear strength $V_{u,exp}$ and the calculated ultimate shear strength $V_{u,cal}$ are shown in Table 4. The ultimate shear strength V_u is calculated by the modified truss theory that is expressed as the sum of the concrete component V_c and the shear

reinforcement component V_s as shown in Equation 1. The concrete component V_c is calculated using Equation 2 proposed by Niwa et al. [1]. The shear reinforcement component V_s is calculated using Equation 3, which is written in the recommendation authorized by the Japan Society of Civil Engineers

$$V_u = V_c + V_s \ (= V_{u,cal}) \tag{1}$$

$$V_c = 0.20 f_c'^{1/3} (d/1000)^{-1/4} p_t^{1/3} b_w d \\ (0.75 + 1.4 (a/d)) \tag{2}$$

$$V_s = A_w f_{wy} d (\sin \alpha_s + \cos \alpha_s) / s_s \tag{3}$$

where

V_c = concrete component;
V_s = shear reinforcement component;
a/d = shear span-depth ratio;
A_w = area of shear reinforcement within spacing s_s;
f_{wy} = yield strength of shear reinforcement;
α_s = angle of inclined stirrup to the longitudinal axis;
s_s = stirrup spacing.

The ratio between the difference of the experimental and the calculated value ($V_{u,exp} - V_{u,cal}$) and the calculated value $V_{u,cal}$ is shown in Table 4. The reduction factor for the calculation of concrete component V_c of the ultra light weight concrete is 0.7 which is given in the recommendation for the light weight concrete (about 1800 kg/m³ over of the unit weight). The average value of the rate of beams, except N-4, is 16% which is comparatively large. The ratio of the ultra light weight concrete beams enlarges the average of the ratio. This means that the ultimate shear strength of the beam is overestimated at 0.7 as the reduction factor. This fact shows that it is not suitable to use the same reduction factor for the beams using the ultra light weight concrete, whose mechanical properties are different from that of the light weight concrete. Maeda et al. [2]. proposed a calculation formula (Equation 4) which evaluates the reduction factor η as a function of the unit volume mass ρ (kN/m³):

$$\eta = 0.053\rho - 0.12 \qquad (4)$$

The ultimate shear strength $V'_{u,cal}$ using the reduction factor η is also shown in Table 4. It is confirmed that the difference between the calculated value $V'_{u,cal}$ and the experimental value $V_{u,exp}$ to the ultra light weight concrete is decreased by using the reduction factor proposed by Maeda et al. This tendency is also confirmed in the case of the normal weight concrete beams.

4 THE SIMULATION USING A NONLINEAR FINITE ELEMENT METHOD

4.1 The outline of the analysis

Stress-strain relationships under the compression of normal weight or ultra light weight concrete used in the FE analysis is shown in Figure 3. The stress of the normal weight concrete is supposed to increase according to a parabolic curve with the increase of the strain until the stress attained is $0.85 f_c'$. The ultimate compressive strain supposed to be 0.35%. The stress strain relationship of the ultra light weight concrete is supposed to be expressed as a straight line, since no softening portion is observed from the compressive test using high stiffness compressive machine. Both concretes under tension are supposed as elastic materials till the stress attains the tensile strength. The softening diagrams of the both concretes are determined based on the compact tension test results. Tension softening diagrams are shown in Figure 4. It is observed from the figure that the starting point of tensile stress and the ultimate crack strain of the ultra light weight concrete are smaller than those of the normal weight concrete. Therefore, the fracture energy G_f of the ultra light weight concrete is remarkably smaller than that of the normal weight concrete. Shear rigidity G_c after cracking is decreased by multiplying the shear rigidity reduction coefficient β ($0 < \beta < 1$), which is inherent to each concrete, by the shear rigidity before cracking. Shear rigidity G_c after cracking is

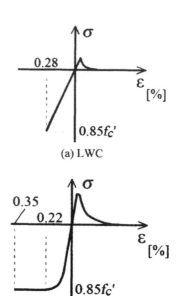

(a) LWC

(b) NWC

Figure 3. Stress-strain relationships in the analysis.

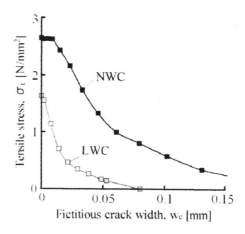

Figure 4. Tension softening diagrams in the analysis.

affected by the roughness of the cracked surfaces. It is natural to consider that the shear rigidity after cracking of the ultra light weight concrete is smaller than that of the normal weight concrete, since the roughness of the cracked surface of the ultra light weight concrete is smoother than that of the cracked surface of the normal weight concrete due to the low strength of the ultra light weight coarse aggregate. In the present study, the reduction coefficient β was decided to be 3.0×10^{-2} for normal weight concrete and 4.6×10^{-3} for the ultra light weight concrete based on the sensitivity analysis. Two-dimensional plane-stress elements with a smeared crack model, consisting of 4 node quadrilateral elements, were used to model the concrete as shown Figure 5. The longitudinal rebar and the stirrups were modeled as the embedded rebar elements. The embedded rebar element that can be freely placed on the concrete element increases the stiffness of the concrete element in the direction of the rebar. It is assumed that the concrete and the rebar are bonded to each other

completely. DIANA Ver.7.2 (TNO (1993)) is used in this analysis.

4.2 *Analytical results and discussion*

The comparisons between the experimental and analytical load-mid span deflection, load-compressive fiber strain of concrete, load-longitudinal steel strain and load-stirrup strain response of beam L-3 are shown in Figure 6. It is shown that the proposed analysis simulates the deformation and strains with the increase of the load very well. Especially, the analysis reflects the change of the position of stirrups and the presence of the crack. The ultimate shear strength by the FEM $V_{u,FEM}$ is shown in Table 4. It is shown that the difference between the experimental and analytical value of the ultimate shear strength by FEM is less than that by the modified truss theory. The comparison between the experiment and the analysis on the development of cracking with the increase of load is shown in Figure 7. It was

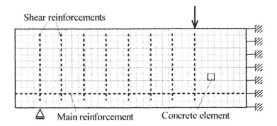

Figure 5. FE analytical model and meshing division.

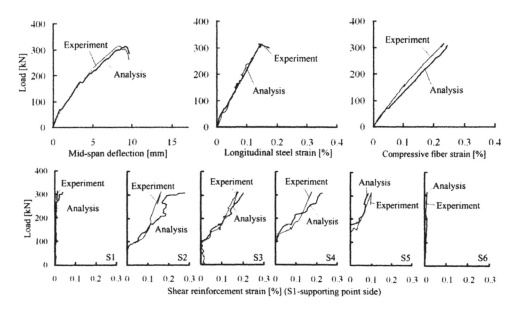

Figure 6. Comparison between experiment and analysis of L-3 beam.

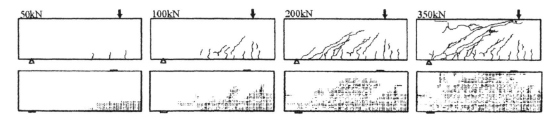

Figure 7. Comparison between experiment and analysis on the cracking with the increase of load.

confirmed that the proposed FEM simulated the experimental behavior of other beams including normal concrete very well.

5 CONCLUSIONS

The shear behavior of the RC beam with shear reinforcement using ultra light weight concrete, whose strength is about $35 \, N/mm^2$ and the unit weight is about $1300 \, kg/m^3$, was compared with the RC beam using the normal weight concrete that has the same strength. The findings obtained from this paper can be concluded as follows:

(1) The ultimate shear strength of an ultra light weight concrete beam with shear reinforcement is calculated as the sum of the concrete component and the shear reinforcement component based on the modified truss theory, using 0.57 as the reduction factor of the concrete component.

(2) The proposed nonlinear FEM, that modelizes the longitudinal and transverse reinforcements as embedded rebar elements, can very adequately simulate the structural behavior such as the development of cracks, the strain of rebar/concrete and deformation under the load and the ultimate strength, irrespective of the kinds of concrete.

REFERENCES

1. J. Niwa, K. Yamada et al. Revaluation of the equation for shear strength of reinforced concrete beams, Journal of material, concrete structural and pavement, No.372, pp.167–176, 1986 (in Japanese).
2. T. Maeda, K. Hibino et al. Shear capacity of reinforced concrete beam with high-performance lightweight aggregate, Proc. of the Japan Concrete Institute, Vol.23, No.3, pp.913–918, 2001 (in Japanese).

Finite Elements in Civil Engineering Applications, Hendriks & Rots (eds.)
© 2002 Swets & Zeitlinger, Lisse, ISBN 90 5809 530 4

Non-linear analysis of prestressed concrete beams with a total strain based model: FEM model and full-scale testing

P.F. Takács, K.V. Høiseth, S.I. Sørensen, T. Kanstad
Norwegian University of Science and Technology, Trondheim, Norway

J.A. Øverli & E. Thorenfeldt
SINTEF Civil and Environmental Engineering, Trondheim, Norway

ABSTRACT: The load-deformation characteristics and the capacity of prestressed concrete T and double T elements were analysed with the finite element method using DIANA's total strain based constitutive model. The numerical simulations accompanied two experimental programs. The first program concerned old bridge beam elements strengthened with carbon fiber reinforced polymer plates. The second program concerned beam elements without shear reinforcement. The non-linear behaviour of concrete was considered in both tension and compression including the influence of lateral cracking on the compressive strength. The elastoplastic material behaviour of the prestressing steel was also taken into account. Predicted and measured load-deformation responses showed good agreement.

1 INTRODUCTION

Two experimental programs were carried out recently at the Norwegian University of Science and Technology and SINTEF, which involved the testing of full-scale prefabricated prestressed concrete beam elements. In the first test program, beam elements of an old demolished bridge were tested in laboratory with and without additional carbon fiber reinforced polymer (CFRP) strengthening. The second test program dealt with DT elements which did not contain shear reinforcement.

The experimental programs were accompanied by the numerical simulation of the tests. Beyond the specific objectives of the programs, the experiments furnished valuable test data that could be used for the evaluation of numerical models. The present paper deals with the finite element modelling of the experiments with the use of DIANA's total strain based constitutive model for concrete. The constitutive model incorporates features of the modified compression field theory.

2 TOTAL STRAIN BASED CONSTITUTIVE MODEL IN DIANA

The total strain based constitutive model in DIANA is based on the well-known modified compression field

theory, which was developed by Vecchio and Collins (1986), originally for two dimensional loadings of reinforced concrete panels. In DIANA the model is implemented in a three-dimensional version according to a formulation proposed by Selby and Vecchio (1993).

The modified compression field theory became widely accepted in the structural engineering community, as a general and reliable approach, during the design of offshore concrete structures of the last decades. It was for this purpose actually that the total strain model was developed in 1997 (Feenstra et al. 1998).

The original total strain model in DIANA relies on the approximation that the principal stress and principal strain directions coincide for all states of stress, also for stress levels exceeding the elastic limit, that is, after the onset of cracking or crushing. The principle is known as the coaxial stress-strain concept, which implies that crack-planes under non-proportional loading conditions will not remain in a fixed position. Therefore, the total strain model in DIANA is frequently also known as the rotating crack model.

The benefit of the coaxial stress-strain concept is that the constitutive modelling needs to be handled in the principal directions only. In an incremental/iterative FEM-procedure, this means that the strain tensor is always transformed to the principal directions before the material description is entered and the belonging principal stresses are determined.

Although the coaxial model operates in the principal stress-strain space, the mechanical behaviour in the principal directions are not independent. Hence, compressive strength is reduced by lateral cracking and increased by lateral confinement. The total strain model accounts for these effects, either according to design code regulations, such as CEB-FIP Model Code 1990, or by user specifications. This is especially attractive, as material behaviour is described by means of terms and notations, which are not only recognized in the computational mechanics world, but familiar also to structural engineers dealing with traditional design of concrete structures.

DIANA offers now also a total strain model where the reference frame of the constitutive model does not follow the directions of the principal strains, but is being kept in a fixed position after cracking (de Witte and Kikstra 1999). In this so-called fixed crack concept, shear resistance in crack-planes are accounted for by a shear retention factor, which permits a reduced shear stiffness in crack-planes.

3 EXPERIMENTAL PROGRAMS

3.1 Program 1: Concrete beams strengthened with carbon fiber reinforced polymer plates

When Isakveien Bridge in Oslo was demolished in 2000 after 35 years of service, two prefabricated prestressed concrete DT (double T) elements with in-situ concrete overlay were preserved for testing. The objective of the test program was to investigate the application of CFRP plates for strengthening old prestressed concrete beams (Takács and Kanstad 2000). The two DT elements were cut half lengthwise, so eventually four T elements were obtained. Two T elements were used to determine the flexural moment and shear capacity of the original beams for reference. The other two beams were strengthened for bending with longitudinal CFRP plates in two different quantities.

3.1.1 Description of the beams

The elements were 11.20 meter long. The cross-section is shown in Figure 1. The design concrete grade was C55 for the prefabricated element and C25 for the in-situ overlay. It should be noted that the concrete grade refers to the characteristic cube strength in accordance with the Norwegian Standard. The actual concrete strength was determined in laboratory on six cylinder samples taken from the prefabricated element and six cylinder samples taken from the overlay. The mean compressive cylinder strength was found to be 77 MPa and 55 MPa, respectively.

The prefabricated element was prestressed with 54 $\phi 4$ St 1600/1800 prestressing wires, 40 in the tension zone and 14 distributed in the flange (with respect to one half of the DT element). The applied initial

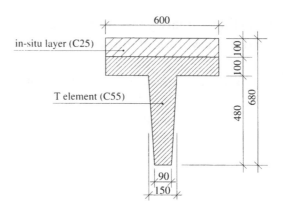

Figure 1. Main dimensions of the cross-section of the prefabricated T element with the in-situ overlay.

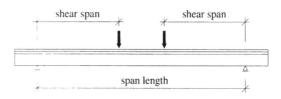

Figure 2. Test arrangement (see Table 1 for actual values).

prestressing force was 12.6 kN in each wire. That gave a total axial compression force of 680 kN with an eccentricity of 193 mm with respect to the neutral axis of the prefabricated element alone. The shear reinforcement consisted of $\phi 8$ mm stirrups in the web with 300 mm spacing along the entire length of the beam.

The bond between the prefabricated element and the concrete overlay was estimated by simple pull-off tests to check that the shear stress over the contact surface would not exceed the failure stress. In the finite element model perfect bond was assumed between the prefabricated element and the overlay.

The two strengthened beams were reinforced with longitudinal CFRP plates at the bottom of the web with quantities of 108 mm^2 and 216 mm^2 in cross-section area, respectively. The CFRP was SIKA's CarboDur S which had the following material properties: modulus of elasticity, $E = 165000$ MPa, mean tensile failure strength, $f_{tm} = 3050$ MPa and strain at failure, $\varepsilon_u = 1.85\%$. The plates were attached to the concrete element with epoxy-based adhesive.

3.1.2 Test arrangement

The beams were loaded symmetrically with two concentrated loads as it is illustrated in Figure 2. The span length and the shear span varied from test to test (see Table 1) in accordance with the intended or assumed failure mode.

Table 1. Test arrangement.

Test	Type	Span length (m)	Shear span (m)
No. 1	Original beam, assumed moment failure	10.80	3.60
No. 2	Original beam, assumed shear failure	9.20	1.80
No. 3	Strengthened beam, with 108 mm² CFRP	10.20	4.10
No. 4	Strengthened beam, with 216 mm² CFRP	9.20	3.60

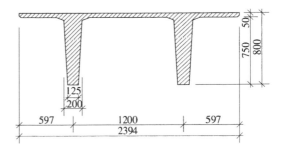

Figure 3. Main dimensions of the cross-section of the DT element.

The beams were instrumented at mid-span with strain gauges in the compression zone, on the top surface of the flange and linear variable displacement transducers (LVDTs) on a basis length of 200 mm in the tension zone, 50 mm above the bottom of the web. Deflection measurements were carried out at mid-span and 1.8 m from mid-span on both sides.

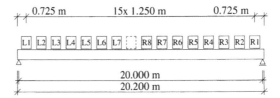

Figure 4. Test arrangement and loading scheme at the field test of the DT element (see also Table 2).

3.2 Program 2: Prestressed concrete DT element without shear reinforcement

The other test program dealt with the investigation of prefabricated prestressed concrete beam elements without shear reinforcement. It has been long claimed by the Norwegian concrete element industry that the requirement for a minimum amount of shear reinforcement is a nuisance. At present the Norwegian Standard requires the obligatory use of stirrups along the entire length of the beam irrespective of the magnitude of the acting shear forces. If the stirrups can be avoided in beams where the prestressed concrete alone has the sufficient shear capacity, that would reduce the production cost of the elements considerably. With the lower cost the market potential of these prefabricated elements would increase.

The experimental program involved the laboratory test of six prefabricated T elements and the field test of two prefabricated DT elements. The beams that were tested in laboratory were 9.8 meter long and were loaded symmetrically with two concentrated loads. The shear span length and the amount of prestressing varied. The DT elements at the field test were 20.2 and 15.2 meter long. The beams were loaded gradually with concrete blocks.

The present paper concerns only the field test of the 20.2 meter long DT element because that is the only beam which was thoroughly analysed with the finite element method.

3.2.1 Description of the beam

The cross-section of the DT element is shown in Figure 3. The design concrete grade was C55. The actual concrete strength was measured in laboratory

on three drilled cylinder samples with the concrete age of 54 days. The average strength was 53 MPa.

The element was prestressed with 18 Y1860S7 prestressing strands which had a nominal diameter of 12.7 mm. Eight strands were in the tension zone and one strand was in the compression zone in each web. The applied initial prestressing force was 100 kN in each strand. That gave a total axial compression force of 1800 kN in the DT element. The actual material properties of the prestressing steel were determined in laboratory: the tensile strength, $f_{pt} = 1900$ MPa, the elongation at the maximum load, $\varepsilon_{pu} = 7.1\%$, the yield stress (at a plastic strain of 0.2%), $f_{p0.2} = 1740$ MPa and the elastic modulus, $E_p = 195000$ MPa.

Shear reinforcement were placed only in a length of 0.85 m at both ends.

3.2.2 Test arrangement

The arrangement of the field test is shown in Figure 4. The span length was 20.0 meter. The beam was loaded with concrete blocks which weighted 32.5 kN each. In the first five load steps, the blocks were placed on the beam symmetrically, two at a time, beginning from the supports (see Table 2). In the subsequent steps the loading was continued with adding one block at a time, switching side in each step. After having the beam loaded with block "R7" the load-deformation diagrams and the crack pattern indicated that failure might occur by any further loading. It was decided to continue with smaller increments. Therefore the weight of block "R8" was shifted onto the beam

Table 2. Loading scheme.

Load step	Block no.	Load increment (kN)	Sum load load (kN)
1	L1, R1	65.0	65.0
2	L2, R2	65.0	130.0
3	L3, R3	65.0	195.0
4	L4, R4	65.0	260.0
5	L5, R5	65.0	325.0
6	L6	32.5	357.5
7	R6	32.5	390.0
8	L7	32.5	422.5
9	R7	32.5	455.0
10	R8 (10 kN)	10.0	465.0
11	R8 (15 kN)	5.0	470.0
12	R8 (20 kN)	5.0	475.0
13	R8 (25 kN)	5.0	480.0
14	R8	Failure occurred	

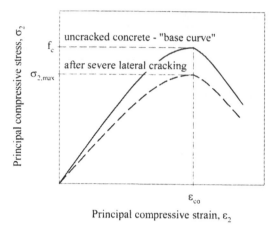

Figure 5. Stress-strain diagram in compression for laterally cracked concrete.

gradually by measuring the acting weight on the crane. The shear failure occurred when the weight of the block was almost entirely supported by the beam.

The beam was instrumented at mid-span with four strain gauges on the top surface of the flange and two linear variable displacement transducers on a basis length of 350 mm in the tension zone, 50 mm above the bottom of the web. Deflection was also measured at mid-span.

4 FINITE ELEMENT ANALYSIS

4.1 General

The finite element analysis of the experiments was carried out with DIANA. The non-linear mechanisms which were considered in the models were cracking and crushing of the concrete and yielding of the prestressing and the reinforcing steel. The behaviour of the concrete was modelled with DIANA's total strain based constitutive model. The rotating crack approach was used to model the beams in the first test program while the fixed crack approach was used to model the DT element without the shear reinforcement.

The finite element models were two dimensional consisting of plane stress elements. The T elements of the first experimental program were modelled with only their left half due to the symmetry of the structure and the loading scheme. No such symmetry could be utilized for the DT element in the second program because the loading scheme was asymmetric. On the other hand it was sufficient to model only one half of the DT cross-section.

In the first test program, the simulation of the experiments was preceded by a phased analysis in order to calculate the proper initial stress and strain conditions in the elements. It was necessary since the

concrete overlay was cast on the prefabricated element which had already been prestressed and loaded with its own dead weight and the weight of the fresh concrete of the overlay. Similar considerations were made in regard to the later installed CFRP polymer plates.

4.2 Material models

4.2.1 Concrete in compression

Vecchio and Collins (1986) found that the principal compressive stress in concrete is a function not only of the principal compressive strain, but also of the coexisting principal tensile strain. In cracked concrete, the compressive strength is reduced by large perpendicular tensile strains. Vecchio and Collins (1993) suggested to reduce the peak stress value of the compressive stress-strain diagram according to Equation 1. It is assumed that the stress-strain diagram is fully determined with its peak stress value and the corresponding strain value for a given base curve (see Figure 5):

$$\sigma_{2,max} = f_c \cdot \frac{1}{1 + K_c} \leq f_c \qquad (1)$$

with

$$K_c = 0.27 \cdot \left(-\frac{\varepsilon_1}{\varepsilon_{co}} - 0.37 \right) \qquad (2)$$

where $\sigma_{2,max}$ = peak stress value in compression; f_c = uniaxial compressive strength; ε_1 = principal tensile strain; and ε_{co} = compressive strain corresponding to the maximum compressive stress.

In the DIANA implementation, internal damage variables are incorporated in the model which record the maximum tensile and compressive strains that occurred during the entire course of loading. This is necessary to prevent damage recovery after unloading and also to govern reloading. With respect to Equation 1, the damage variables ensure that the compressive strength will not recover after deloading takes place in the lateral direction.

In the present analyses the serpentine curve was used as the base curve in compression. The serpentine curve was first used for concrete by Popovics in 1973, and later adapted to high strength concrete by Thorenfeldt et al. (1987):

$$\sigma_2 = -\sigma_{2,max} \frac{\varepsilon_2}{\varepsilon_{co}} n \left(n - 1 + \left(\frac{\varepsilon_2}{\varepsilon_{co}} \right)^{nk} \right)^{-1} \quad (3)$$

where ε_2 = principal compressive strain; n and k are model parameters which can be adjusted based on test results. If no such data is available, n and k may be estimated from the following formulas:

$$n = 0.8 + \frac{f_c}{17} \quad (4)$$

$$k = \begin{cases} 1 & if \ 0 > \varepsilon_2 > \varepsilon_{co} \\ 0.67 + \frac{f_c}{62} & if \ \varepsilon_2 \leq \varepsilon_{co} \end{cases} \quad (5)$$

where the uniaxial compressive strength, f_c must be given in MPa.

4.2.2 Concrete in tension

The stress-strain diagram in tension was defined with a linear part until the tensile strength and an exponential softening branch (see Figure 6). The softening curve is determined from the tensile strength, f_t, the mode-I fracture energy, G_F^I and the numerical crack band width, h_{cr}. The numerical crack band width is related to the dimensions of the finite element. For higher order two dimensional elements, DIANA estimates its value as the square root of the element area. The fracture energy in the present analysis was estimated from the CEB-FIP Model Code 1990 (CEB-FIP 1991) formula:

$$G_F^I = \alpha_F \cdot (f_{cm}/10)^{0.7} \quad (6)$$

where α_F = coefficient which depends on the maximum aggregate size, d_{max} ($\alpha_F = 0.03$ Nmm/mm^2 for $d_{max} = 16$ mm); and f_{cm} = mean cylinder strength in MPa.

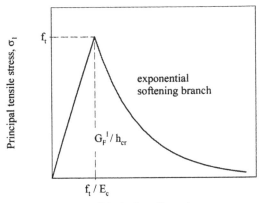

Figure 6.   Stress-strain diagram in tension.

4.2.3 Prestressing steel and reinforcement

The stress-strain relationship for the prestressing steel was defined with a multilinear diagram. The diagram consisted of three linear parts with the breaking points being at the stress that is 80% of the tensile strength and the stress corresponding to the plastic strain of 0.2%.

The stress-strain relationship for the ordinary reinforcement was defined with a bilinear diagram which consisted of an elastic part and a yield part with no hardening.

5   EVALUATION AND DISCUSSION OF THE TEST AND ANALYSIS RESULTS

5.1   Experimental program 1

Figure 7 shows the comparisons of the predicted and measured load-deflection responses at mid-span. Two predicted responses are seen for each beam. The two diagrams came from two different sets of material parameter values of the prestressing steel. The mean values of the material properties were not known for the prestressing steel, unlike for the concrete and the carbon fiber reinforced polymer. The magnitude of the yield stress, however, has a significant influence on the load-deformation response of the element. Therefore two analyses were carried out. In analysis "A" the material properties of the prestressing steel were taken with the known characteristic values. In analysis "B" the mean values were estimated to be ten percent higher than the characteristic values in respect to both the yield stress, $f_{p0.2}$ and the failure tensile strength, $f_{p,max}$. Such an arbitrarily chosen value can only be seen as a rough estimate.

The first beam was used to determine the flexural moment capacity of the unstrengthened beam. The

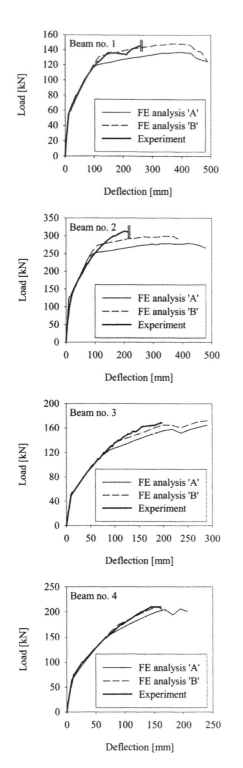

Figure 7. Predicted and measured load-deflection responses at mid-span.

test was terminated without failure due to the large deflection and the geometric limitations of the test rig. Nevertheless the structural responses suggested that the maximum load level was almost reached and the failure would have happened according to the expected failure mode, i.e. moment failure due to the crushing of the concrete in the compression zone. The maximum moment at the termination of the test was considered as the flexural moment capacity.

The intention with the second beam was to determine its shear capacity. This test was also terminated without reaching failure over the normal course of loading. The structural responses, however, indicated that the maximum load level was reached and failure was near. At the last load step the beam was already severely damaged by wide flexural-shear cracks.

The failure of the third beam occurred in a way that the CFRP plate split apart along a horizontal surface which originated from a wide flexural-shear crack. It can be explained with the combination of high shear stress between the concrete and the CFRP plate, high normal stress concentration in the plate at the crack opening and the coexisting transverse shear stress in the plate due to the vertical movement at the crack opening. The failure was very brittle having a rather explosive character. Failure mechanism of this kind was not incorporated in the present FE model. It would require a more sophisticated model of the CFRP plate, the realistic representation of the interface of the concrete surface and the CFRP plate and the realistic representation of crack opening (i.e. discrete model).

The failure of the fourth beam was a typical shear failure. That was expected since the moment capacity of the element was significantly increased with the CFRP plates while its shear capacity remained unchanged.

The finite element analysis furnished very accurate predictions of the load-deflection responses in the elastic regime and the postcracking regime until the prestressing steel began yielding. Small deviation can only be seen for the second beam.

Figure 8 compares the predicted and the measured flexural moment and shear capacity of the beams. Beside the FE analysis, they were also calculated according to the Norwegian Standard, NS 3473 (NBR 1998) under a common set of assumptions and with idealised stress-strain relationships. The conservative nature of the shear prediction formula is well apparent.

5.2 Experimental program 2

The field test of the DT element provided a good opportunity to compare the predicted and the measured load-deformation response of a prestressed concrete beam under an almost uniformly distributed load. Most laboratory tests deal with beams which are

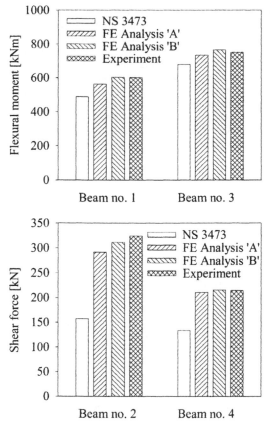

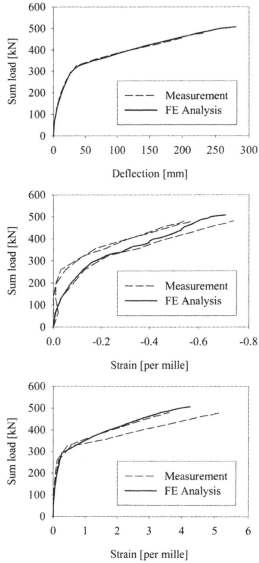

Figure 8. Predicted and measured shear and moment capacity.

loaded with two concentrated loads. Previous experience with such experiments shows that if the failure mode is shear failure, the maximum load may significantly differ among identical elements that were tested under the same conditions, depending on the formation of the critical shear crack. Besides, a distributed load is more conform to the actual conditions of loading.

Figure 9 compares the predicted and measured load-deformation responses of the beam at mid-span. The three curves that belong to the measured strain in the extreme compression fiber came from three strain gauges while the fourth gauge furnished unreasonable readings. The two curves of the measured strain in the tension zone came from one measurement in each web of the DT element.

In the experiment the failure occurred after the loading block "R8" (see Figure 4) was placed onto the beam with its full weight. In the FE analysis failure was not indicated in that stage, therefore the model was incrementally loaded further in the position where the block "L8" would have been placed

Figure 9. Measured and predicted load-deformation responses of the DT beam at mid-span: (a) deflection, (b) strain in the extreme compression fiber and (c) strain at 50 mm above the extreme tension fiber.

(marked with dashed line in Figure 4). About half weight of the block was loaded onto the beam when the analysis was stopped because convergence was not reached. Since the numerical analysis involved fixed load increments, going through the peak point was not possible.

The agreement between the predicted and measured load-deflection response is very good. Similar

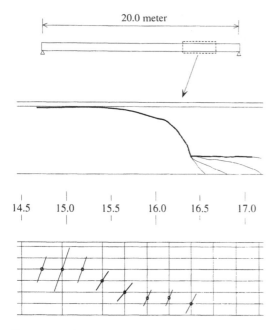

Figure 10. The shear crack that caused the failure of the beam in the field test and the vector representation of the largest values of the principal tensile strain in the FE model in the last load step.

good agreement was found in the first test program where deviations occurred only after the prestressing steel started yielding. In the present experiment failure happened before the strands reached the yield point. Yielding of the prestressing steel would have started at a corresponding strain of approximately 4.5‰ at 50 mm above the bottom of the web (see Figure 9c). The maximum strain in the extreme compression fiber is also small which indicates that the beam failed by shear well before the bending capacity was utilized.

Figure 10 shows the shear crack which caused the failure of the beam. Similar crack formation was observed in both webs of the DT element. Figure 10 also illustrates the largest values of the principal tensile strain in the last load step of the numerical analysis. The finite element model gave a reasonably good prediction for the location of the failure.

6 CONCLUSIONS

The structural response of prestressed concrete beams was simulated with the use of DIANA's total strain based constitutive model. The material behaviour of concrete in both compression and tension was modelled with non-linear stress-strain relationship. The degradation of compressive strength due to the tensile damage in the lateral direction was taken into account. The behaviour of the prestressing steel was considered with an elastoplastic constitutive model. The predicted load-deformation characteristics of the beams were found in good agreement with the test results. Also the capacity of the beams was predicted with reasonable accuracy. The experience of working with the total strain based constitutive model was rather good in regard to the stability of the numerical procedure.

REFERENCES

CEB-FIP. 1991. *CEB-FIP Model Code 1990*. Comité Euro-International du Béton.

Feenstra. P.H. 1997. *Development of the INDACS material model*. TNO report. TNO Building and Construction Research. The Netherlands.

Feenstra, P.H., Rots, J.G., Arnesen, A., Teigen, J.G. & Høiseth, K.V. 1998. A 3D constitutive model for concrete based on a co-rotational concept. *Proceedings of EURO-C 1998 Conference, Badgastein March 1998*. Austria.

NBR. 1998. *Concrete structures – Design rules*. NS 3473, Norges Standardiseringsforbund.

Selby. R.G. & Vecchio, F.J. 1993. *Three-dimensional constitutive relations for reinforced concrete*. Technical Report 93-02. University of Toronto, Department of Civil Engineering. Canada.

Takács, P.F. & Kanstad, T. 2000. Strengthening prestressed concrete beams with carbon fiber reinforced polymer plates. *Nordic Concrete Research* Publication no. 25: 21–33.

Thorenfeldt, E., Tomaszewicz, A. & Jensen, J.J. 1987. Mechanical properties of high-strength concrete and applications in design. In Ivar Holand (ed.), *Utilization of High-Strength Concrete, Proc. intern. symp., Stavanger, 15–18 June 1987*. Trondheim.

Vecchio. F.J. & Collins, M.P. 1986. The modified compression field theory for reinforced concrete elements subjected to shear. *ACI Journal* 83 (22): 219–231.

Vecchio. F.J. & Collins, M.P. 1993. Compression response of cracked reinforced concrete. *Journal of Structural. Eng., ASCE* 119, 12: 3590–3610.

de Witte. F.C. & Kikstra, W.P. (ed.) 1999. *DIANA – Nonlinear Analysis*. User's manual release 7.2. TNO Building and Construction Research. The Netherlands.

Finite Elements in Civil Engineering Applications, Hendriks & Rots (eds.)
© 2002 Swets & Zeitlinger, Lisse, ISBN 90 5809 530 4

Ultimate load capacity of reinforced steel fibre concrete deep beams subjected to shear

E. Fehling & T. Bullo
Dept. of Structural Concrete, University of Kassel, Germany

ABSTRACT: A numerical model based on the concept of smeared cracking to predict the structural behaviour of reinforced steel-fibre concrete deep beams is developed. Material parameters, which are necessary to formulate a constitutive relation for randomly oriented short steel-fibres under biaxial states of stress are systematically quantified and implemented in the DIANA finite element program. The developed material model is utilised to simulate the structural behaviour of the experimentally investigated reinforced steel fibre concrete deep beams found in the literature. The model is shown to give a very good prediction of the observed ultimate load-carrying capacity and the load-displacement behaviour of the deep beams failed in shear.

1 INTRODUCTION

Investigations on the composite materials other than ordinary reinforced concrete have shown that the addition of arbitrarily oriented steel fibres to the concrete mix can significantly improve the ultimate strength, cracking characteristic and deformation behaviour of structural elements subjected to complex combined states of stress and strain (Bullo & Fehling 2000). Steel fibres are capable of controlling crack propagation, limiting crack width, creating well-distributed cracks in a large area and improving the stress/strain transfer mechanisms in the cracked zone. Furthermore, adding steel fibres to the concrete mix can also significantly improve the post-ultimate load stress-strain behaviour of the matrix. The ultimate strain or deformation capacity of steel fibre concrete under uniaxial state of stress or strain is reported to be about ten times that of a plain concrete.

Even though fibre reinforced concrete has become a promising new material in concrete technology, there are few material models available to be adopted under multiaxial states of stress. Developing material models for steel fibre concrete is relatively a very complex task. Fibres used in civil engineering industry are short, randomly oriented, non-uniformly distributed and contain only a certain part in volume of the composite. Moreover, the local bond behaviour between ordinary reinforcing bars and steel fibre concrete depends not only on the amount of fibre added to the concrete mix but also on the type of fibre used. The tension stiffening behaviour of cracked reinforced steel fibre concrete, which is important for the prediction of

the load-deformation behaviour rather than the ultimate load carrying capacity of reinforced steel fibre concrete structural elements is not yet fully investigated (Bullo & Fehling 2000).

The main difficulty in the numerical modelling of the modes of failure of reinforced steel fibre concrete is due to the different geometric scales that are involved in the initiation and propagation of damages leading to failure. Modes of failure for reinforced steel fibre concrete structures can be of a global type (Fig. 1) or local nature (Fig. 2).

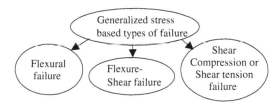

Figure 1. Generalized stress based types of failure.

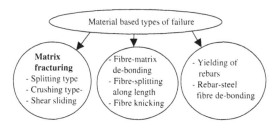

Figure 2. Material based types of failure.

In this study constitutive models are developed for steel fibre concrete and applied to numerically simulate the structural behaviour of reinforced steel fibre deep beams subjected predominantly to shear type loading. Material parameters required for constitutive modelling are identified and quantified in order to have the complete constitutive relation under all possible biaxial states of stress and strain. Since cracking is the major source of non-linearity in most cement-based composites, specific considerations are given for the modelling of matrix cracking.

2 RESEARCH SIGNIFICANCE

Very few attempts were made in the past to numerically simulate the capability of arbitrary oriented steel fibres to increase the resistance of cracked matrix to multiaxial loading in general and to enhance the biaxial compressive strength of cracked matrix subjected simultaneously to transverse tensile strains in particular. The research reported in this paper is aimed to quantify the necessary material parameters and develop a biaxial constitutive relation that may be used for the analysis of membrane type reinforced steel fibre concrete structural elements. The simulation techniques suggested in this study may help with the availability

of more experimental data, to develop efficient material models for steel fibre concrete under multiaxial loading conditions.

3 CONSTITUTIVE MODELS

3.1 *Steel fibre concrete under uniaxial tension/compression loading*

Typical stress-strain relations for steel fibre concrete under uniaxial tension and compression are given in Figures 3, 4, and 5.

3.2 *Steel fibre concrete under biaxial state of stress*

The behaviour of steel fibre concrete under proportional biaxial compressive loading depends on the magnitude of the lateral compressive stress. For a concrete matrix containing steel fibres of 2% in volume and an aspect ratio of 60 (Fig. 6), an increase in a compressive strength of about 60% over the uniaxial compressive strength is reported at a principal stress ratios of $\sigma_1/\sigma_2 = 1$ (Yin, et al. 1989).

According to the traditional flow theory of plasticity (Chen & Saleeb 1982), assuming isotropic hardening;

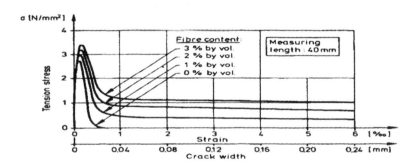

Figure 3. Influence of the volume of steel fibres on tensile stress-strain (crack width) curve.

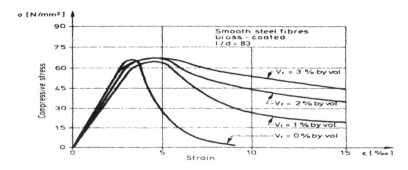

Figure 4. Influence of the volume of steel fibres on the compressive stress-strain curve.

the instantaneous yield surface in stress space is given by

$$f(\sigma_{ij}, k) = 0 \qquad (1)$$

where k is a "hardening" parameter, expressed as

$$k(\sigma_{ij}, \varepsilon_{ij}^{p}) = 0 \qquad (2)$$

For tension-compression loading the combined yield condition according to Drucker-Prager and Rankine yield surfaces is given by

$$f_{dp}(\sigma, k_{dp}) = \alpha\, \psi_{dp}^{T}\, \sigma + \sqrt{\tfrac{1}{2}\sigma^{T} P_{dp}\, \sigma} - \beta\, C(k_{dp}) \qquad (3)$$

$$f_{r}(\sigma, k_{r}) = \tfrac{1}{2}\, \psi_{r}^{T}\, \sigma + \sqrt{\tfrac{1}{2}\sigma^{T} P_{r}\, \sigma} - \sigma_{r}(k_{r})$$
$$(\text{for } \sigma_{1} > \sigma_{2} > \sigma_{3}) \qquad (4)$$

where α and $\beta * C$ are the scalar parameters that can be determined from experimental data; σ is a stress

vector; σ_1, σ_2 and σ_3 are principal stresses; and $\beta * C$ and σ_r are yield stresses for Drucker-Prager and Rankine yield criteria respectively.

For granular materials, the use of cohesion C and internal friction ϕ as material parameters is usually recommended to numerically simulate the behaviour of structural elements investigated experimentally. If the Drucker-Prager yield surface is made to coincide with the outer corners of the Mohr-Coulomb hexagonal surface under triaxial loading, the scalar Parameters α and $\beta * C$ are obtained to be

$$\alpha = \frac{2\sin\phi}{3 - \sin\phi} \qquad (5a)$$

$$\beta * C = \frac{6\,C\cos\phi}{3 - \sin\phi} \qquad (5b)$$

The scalar parameters for the Drucker-Prager yield surface are calibrated based on the biaxial strength

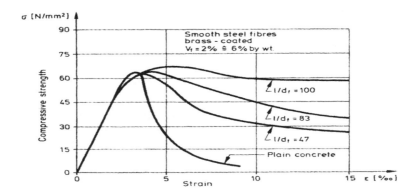

Figure 5. Influence of the aspect ratio of steel fibres on compressive stress-stain curve.

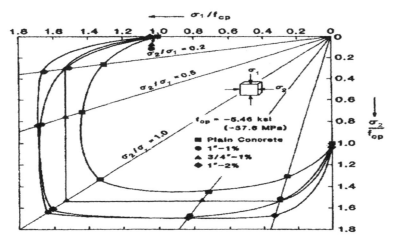

Figure 6. Biaxial strength of plain and steel fibre concrete (Yin, et al. 1989).

211

Table 1. Scalar parameters for the Drucker-Prager yield surface.

Parameters	α	$\beta * C$	ϕ
$\sigma_1/\sigma_2 = 0.5$	0.2917	$0.7083\,f_{cp}$	22.45^0
$\sigma_1/\sigma_2 = 1.0$	0.2727	$0.7272\,f_{cp}$	21.09^0
Average	0.2822	$0.7178\,f_{cp}$	21.77^0

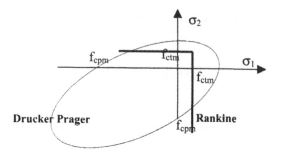

Figure 7. Systematic representation of the combined yield curve.

curve given in Figure 6, for 2% in volume of steel fibres at principal stress ratios of $\sigma_1/\sigma_2 = 0.5$ and $\sigma_1/\sigma_2 = 1$. For a given cylinder compressive strength of steel fibre concrete $f_{cpm,}$ the determined scalar parameters are given in Table 1.

For a plane stress state of stress, the projection matrix P_{dp} for the Drucker-Prager loading surface and P_r for Rankine principal stress criterion are given by

$$
P_{dp} = \begin{bmatrix} 2 & -1 & 0 \\ -1 & 2 & 0 \\ 0 & 0 & 6 \end{bmatrix} \quad P_r = \begin{bmatrix} \frac{1}{2} & -\frac{1}{2} & 0 \\ -\frac{1}{2} & \frac{1}{2} & 0 \\ 0 & 0 & 2 \end{bmatrix} \quad (6a)
$$

The projection vector ψ for both yield conditions is expressed as

$$
\psi_{dp} = \psi_r = \begin{Bmatrix} 1 \\ 1 \\ 0 \end{Bmatrix} \quad (6b)
$$

The tensile cracking under biaxial state of stress is governed by maximum principal tensile stress criteria (tension cut-off) as given in Figure 7. The post-cracking behaviour is modelled by using constitutive relations consistent with the state of cracking and element size.

In a three-dimensional Cartesian coordinate system, the infinitesimal total strain for a time-independent case can be generally be decomposed into elastic and plastic strains. Which is expressed as

$$
d\varepsilon_{ij} = d\varepsilon_{ij}^e + d\varepsilon_{ij}^p \quad (i = 1, 2, 3 \ \& \ j = 1, 2, 3) \quad (7)
$$

The infinitesimal elastic stress strain relation is given by

$$
d\sigma_{ij} = C_{ijkl}^e\, d\varepsilon_{kl}^e = C_{ijkl}^e(d\varepsilon_{kl} - d\varepsilon_{kl}^p) \quad (8)
$$

According to the Normality principle (flow-rule) suggested by von Mises and assuming an associated plasticity, the infinitesimal plastic strain is related to the loading surface by

$$
d\varepsilon_{ij}^p = d\lambda \frac{\partial f}{\partial \sigma_{ij}} \quad (9)
$$

Where $d\lambda$ is a non-negative proportionality constant or the so-called plastic multiplier that has to be determined in terms of the infinitesimal plastic work or strain. For combined loading surfaces f_{dp} and f_r, the normality condition at the common edge can be obtained according to the work of Koiter (1953). It states that the flow at the common edge is a linear combination of the flow directly to the left and to the right of the edge and is expressed as

$$
d\varepsilon_{ij}^p = d\lambda_1 \frac{\partial f_{dp}}{\partial \sigma_{ij}} + d\lambda_2 \frac{\partial f_r}{\partial \sigma_{ij}} \quad (10)
$$

Where the proportionality constants $d\lambda_1$ and $d\lambda_2$ are obtained from the two yield surfaces. If these proportionality constants are known, then the plastic flow at the common edge can be determined both in direction and magnitude. The introduction of these two arbitrary proportionality constants $d\lambda_1$ and $d\lambda_2$ helps to satisfy the strain compatibility condition at the common edge. It is generally advantages for fixed principal directions, to formulate the plastic flow in the space of principal stresses as given by

$$
d\varepsilon_i^p = d\lambda \frac{\partial f}{\partial \sigma_i} \quad (i = 1, 2, 3) \quad (11)
$$

Rewriting Equation 8 yields the total strain rate for a single yield surface as given by

$$
d\varepsilon_{ij} = [C_{ijkl}^e]^{-1}\, d\sigma_{kl} + d\lambda \frac{\partial f}{\partial \sigma_{ij}} \quad (12)
$$

At plastic yielding, the stresses are on the loading surface. Differentiating Equation 1 yields the continuity or consistency condition

$$
df = \frac{\partial f^T}{\partial \sigma_{ij}} d\sigma_{ij} - A' d\lambda = 0 \quad (13)
$$

212

In which the plastic hardening parameter A' is

$$A' = -\frac{\partial f}{\partial k} dk \frac{1}{d\lambda} = \frac{\partial f}{\partial \sigma_{ij}} C^e_{ijkl} d\varepsilon_{kl} \frac{1}{d\lambda} \qquad (14)$$

According to the effective (equivalent or generalized) stress σ_{ef} and strain $\varepsilon_{ef,p}$ concept, it is assumed that the effective stress-effective plastic strain relation for multiaxial loading reduces to a simple uniaxial stress-plastic strain curve under a uniaxial loading condition. The effective stress-effective plastic strain relation is given by

$$d\sigma_{ef} = H\left(\sigma_{ef}\right) d\varepsilon_{ef,p} \qquad (15)$$

In which the effective plastic hardening modulus H is determined from the slope of the uniaxial stress-plastic strain curve at the current value of σ_{ef}. For a work hardening material the internal state parameter $dk = dw$ is given by

$$dk = dw = \sigma^T_{ij}\, d\varepsilon^p_{ij} = \sigma_{ef}(\varepsilon_{ef,p}) d\varepsilon_{ef,p} \qquad (16)$$

Substituting from Equation 9, the expression for plastic strain into Equation 16 yields for the internal state parameter dk

$$dk = d\lambda\, \sigma^T_{ij} \frac{\partial f}{\partial \sigma_{ij}} \qquad (17)$$

Moreover, substituting the expression for dk in to Equation 14 yields

$$A' = -\frac{\partial f}{\partial k} \sigma^T_{ij} \frac{\partial f}{\partial \sigma_{ij}} = -\frac{H}{\sigma_{ef}} \sigma^T_{ij} \frac{\partial f}{\partial \sigma_{ij}} \qquad (18)$$

According to Equation 18, if an explicit relationship between f and k is determined, the plastic hardening modulus A', for both strain- and work-hardening material can be obtained for each instantaneous loading surface. It is worth noting to mention that for hydrostatic pressure independent loading surfaces, such as von Mises yield condition, $A' = H$. For elastic-perfectly plastic material $H = 0$. In case of strain hardening, for pressure independent yield surfaces; the internal parameter dk can be derived using the incompressibility condition under plastic deformation given by

$$d\varepsilon^p_1 + d\varepsilon^p_2 + d\varepsilon^p_3 = 0 \qquad (19)$$

But for pressure dependent yield surfaces, the internal parameter dk can only be determined, if the ratio between the infinitesimal plastic strains is known.

If the internal parameter dk is taken as the effective plastic strain $d\varepsilon_{ef,p}$, then

$$dk = d\varepsilon_{ef,p} = D\sqrt{d\varepsilon^p_{ij}\, d\varepsilon^p_{ij}} \qquad (20)$$

In which D is a loading surface related constant. The loading surface related constant D, the effective stress σ_{ef} and effective plastic strain $d\varepsilon_{ef,p}$ for both Drucker-prager and Rankine yield conditions are given respectively as

$$\sigma_{ef} = \frac{\alpha I_1 + \sqrt{3}\sqrt{J_2}}{1 + \alpha}$$

$$dk = d\varepsilon_{ef,p} = \frac{\alpha + 1}{\sqrt{3\alpha^2 + \frac{3}{2}}}\sqrt{d\varepsilon^p_{ij} d\varepsilon^p_{ij}} \qquad (21)$$

and

$$\sigma_e = \sigma_1 \quad if \ \sigma_1 > \sigma_2 > \sigma_3 \quad dk = \sqrt{\frac{2}{3}}\sqrt{d\varepsilon^p_{ij} d\varepsilon^p_{ij}} \qquad (22)$$

Based on the Drucker-Prager yield criterion for the uniaxial and biaxial stress conditions, the relation between the equivalent cohesion C and the uniaxial strength f_{cp} for a constant friction angle ϕ can be expressed as

$$C = f_{cp}\frac{1 - \alpha}{\beta} = f_{cp}\frac{1 - \sin\phi}{2\cos\phi} \qquad (23a)$$

$$C = \omega f_{cp}\frac{(-1 - \zeta)\alpha + \sqrt{(1 + 3^2)/2}}{\beta} \qquad (23b)$$

where ω = strength multiplication factors under biaxial compressive loading; ζ = principal stress ratio (σ_1/σ_2). If a principal stress ratio of $\sigma_1/\sigma_2 = 1$ is assumed, Equation 23b reduces to

$$C = \omega f_{cp}\frac{1 - 2\alpha}{\beta} = \omega f_{cp}\frac{3 - 5\sin\phi}{6\cos\phi} \qquad (23c)$$

It is to be mentioned that the strength multiplication factor ω depends on the percent in volume, aspect ratio and type of steel fibres used. A typical plastic hardening/softening curve in compression is derived for steel fibres concrete containing fibres having an aspect ratio of 60 (Fig. 8).

Substituting the expression for $d\lambda$ from Equation 13 into Equation 12 and applying the Sherman-Morrison-Wood-bury formula for the inversion of matrices; the

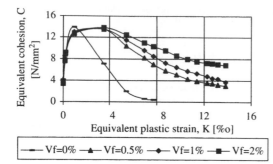

Figure 8. Typical compression hardening/softening diagram for steel fibre concrete.

infinitesimal stress in terms of an imposed strain changes can be expressed as

$$d\sigma_{ij} = C_{ijkl}^{ep}\, d\varepsilon_{kl} \qquad (24)$$

In which the elastic-plastic tangential modulus for associated plasticity is given by

$$\left[C_{ep}\right] = \left[C_e\right] - \frac{\left[C_e\right]\left\{\dfrac{\partial f}{\partial \sigma}\right\}\left\{\dfrac{\partial f}{\partial \sigma}\right\}^T\left[C_e\right]}{A' + \left\{\dfrac{\partial f}{\partial \sigma}\right\}^T\left[C_e\right]\left\{\dfrac{\partial f}{\partial \sigma}\right\}} \qquad (25)$$

For a case of ideal plasticity with no hardening the plastic hardening parameter $A' = 0$. For a plane-stress condition, the isotropic linear-elastic stiffness matrix for steel fibre concrete is given by

$$\left[C_e\right] = \frac{E}{1-\nu^2}\begin{bmatrix} 1 & \nu & 0 \\ \nu & 1 & 0 \\ 0 & 0 & (1-\nu)/2 \end{bmatrix} \qquad (26)$$

where E = the elastic modulus and ν = the Poisson's ratio. According to the simple law of mixture for composite materials

$$E = V_m E_m + V_f \eta_\theta \eta_l E_f = (1 - V_f)E_m + V_f \eta_\theta \eta_l E_f \qquad (27a)$$

$$\nu = V_m \nu_m + V_f \eta_\theta \eta_l \nu_f = (1 - V_f)\nu_m + V_f \eta_\theta \eta_l \nu_f \qquad (27b)$$

where E_m = elastic modulus of the matrix; E_f = elastic modulus of the fibre; V_m = volume fraction of the matrix; V_f = volume fraction of the fibre; ν_m = Poisson's ratio of the matrix; ν_f = Poisson's ratio of the fibre; η_θ = fibre orientation coefficient and η_l = fibre

bond efficiency parameter. According to Schnütgen (1975), for a two dimensional orientation, the fibre orientation coefficient η_θ is given by

$$\eta_{\theta,2D} = 1 - \frac{2}{15}(1-V_f)\left(\frac{6(n-1)}{n} + \nu_m\right) \qquad (28)$$

where $n = E_f/E_m$ and ν_m = Poisson's ratio of the matrix.

The interfacial properties of a composite are of primary importance for the prediction of the composite overall behaviour. A composite does not necessarily fail once the matrix has cracked if the bond at the interface is able to transfer the forces that can longer be taken by the cracked part of the matrix into the fibres. The transfer of the bond stress from the matrix to the fibres and the bond failure mechanism, namely fibre pullout or fracture are governed by the fibre aspect ratio of fibres, anchorage length, tensile strength of fibre and the strength of the matrix. Increase in the aspect ratio of steel fibres to be used, generally improves the effectiveness of fibres, but impedes the workability of the freshly mixed concrete. The bond efficiency parameter is related to the aspect ratio of fibres by

$$\eta_l = 0.5 \quad \text{for}\ \left(\frac{l}{d}\right)_{fibre} \leq \left(\frac{l}{d}\right)_{crit} \qquad (29a)$$

$$\eta_l = 1 - \frac{\left(\dfrac{l}{d}\right)_{crit}}{2\left(\dfrac{l}{d}\right)_{fibre}} \quad \text{for}\ \left(\frac{l}{d}\right)_{fibre} > \left(\frac{l}{d}\right)_{crit} \qquad (29b)$$

where the critical aspect ratio $(l/d)_{crit}$ can be determined from

$$\left(\frac{l}{d}\right)_{crit} = \frac{f_{yd}}{2\tau} \qquad (30)$$

where f_{yd} = fibre tensile strength; τ = mean bond stress between fibre and matrix; l = length of fibre; and d = diameter of fibres.

3.3 Cracked steel fibre concrete

The cracked steel fibre concrete can also be modelled using both the multi-directional smeared crack model suggested by Litton (1974) and the traditional and well developed smeared fixed or/and "rotating" crack model based on total strain (Bicanic et al. 1991).

Due to strain localization, both the mean tensile strength and the non-linear post peak load-deformation response of structures made of cement-based materials, which are characterized by softening, exhibit mesh objectivity. It has to be noted that due to the non-homogeneous strain field in the specimen, the stress-strain relation will not be unique for geometrically similar specimens of different size subjected to external loading. For Mode-I or splitting type of fracture, the mesh objectivity due to the so-called size effect could be alleviated using the energy-based criterion for the formation and propagation of fracture rather than the principal stress based criterion (Bazant 1984). For Mode-I fracture, it has been suggested to use some equivalent strain softening modulus and equivalent tensile strength depending on the selected mesh size in order to obtain consistent results. According to crack band approach proposed by Bazant (1984), for a finite ultimate crack strain $\varepsilon^{cr}_{nn,ult}$, the Mode-I fracture energy G^I_f per unit length of the crack bandwidth for a unit thickness is given by

$$G^I_f = h_{cb} \int_{\varepsilon^{cr}_{nn}=0}^{\varepsilon^{cr}_{nn}=\varepsilon^{cr}_{nn,ult}} \sigma^{cr}_{nn}(\varepsilon^{cr}_{nn}) \, d\varepsilon^{cr}_{nn} \qquad (31)$$

Based on the crack band theory, the softening material modulus after cracking is modified depending on the mesh size to be used. It is assumed that the fracture energy G^I_f is independent of the mesh size used and is a material property. In this study the crack bandwidth h_{cb} is approximated using an effective width of the element $h_e = \sqrt{A}$. In which A is the area of the element used.

The capability of the intact steel fibre concrete between the cracks to carry the tensile stresses or strains and its ability to share with the reinforcements in the vicinity of the cracks the so called tension stiffness behaviour is influenced generally by the tensile strength of steel fibre concrete, volume and distribution of steel fibres in the matrix, crack spacing, reinforcement bar sizes and arrangements with respect to the crack direction. In this study both tension softening and tension stiffening effects are modelled using a multi-linear descending stress-strain function (Fig. 9). According to the tension-softening or/and tension-stiffening model used, active cracks follow the multi-linear descending branch, whereas arrested cracks follow a path back to the origin. The tension-stiffening curve is developed based on the uniaxial tension softening curve and the ultimate strain of steel fibre concrete ε_{cu}. The modified ultimate strain $\varepsilon_{cu,t}$ for cracked reinforced steel fibre concrete can obtained using

$$\varepsilon_{cu,t} = \gamma(V_f, \eta_\theta, \eta_l) \varepsilon_{cu} \qquad (32)$$

where γ is a tension stiffening parameter, which depends mainly on the orientation of the steel fibres

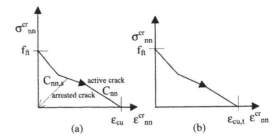

Figure 9. Multi-linear tension softening and tension stiffening models for cracked steel fibre concrete (Loading/Unloading).

with respect to the crack direction. In the absence of an experimentally accurate data, a tension stiffening parameter γ varying from 5 to 10 can be adopted for reinforced steel fibre concrete. For reinforced concrete without steel fibres, values varying from 1 to 2 may be used. For the fixed crack model a constant value of a shear retention factor $\beta_c = 0.2$ is usually recommended for reinforced concrete. Similarly, the shear retention factor given in the past for reinforced concrete can be modified and adopted for steel fibre concrete, as given by

$$\beta_{s,f} = \psi(V_f, \eta_\theta, \eta_l) \beta_c \qquad (33)$$

where ψ is a shear retention parameter. A shear retention parameter ψ varying from 2 to 3 can be assumed for steel fibre concrete depending on the magnitude of principal strain perpendicular to the crack directions. For "rotating" crack model the shear retention factor depends on the values of the principal stresses and strains parallel and perpendicular to the crack direction.

Moreover, the reinforcements are assumed to be embedded in the mother element. This technique helps to generate the mesh for the complete structural model regardless of the locations, distributions and directions of the reinforcements.

The stiffness contribution from the reinforcement bars to the composite is obtained independently from the surrounding mother element. The strain-displacement matrix of the surrounding element helps to determine the strains in the reinforcement. This assumes that the strains in the embedded reinforcement bars are coupled to the degrees of freedom of the surrounding element. This implies that perfect bond is assumed even though it is fully clear that the shear strength is affected by local cracking, bond stress and the relative displacement (slip) between re-bars and concrete. The plastic material behaviour of the reinforcements is modelled using von Mises yield criterion. The uniaxial stress-strain diagram for the reinforcement bar is assumed to be elastic-plastic without hardening.

4 FINITE ELEMENT SOLUTION PROCEDURES

The background theory on the weak form of the governing differential equations of equilibrium in non-linear structural mechanics, the associated numerical techniques applied for discretization, the applied incremental-iterative procedures to solve the system of the assembled incremental non-linear equations and the convergence criteria for the approximate solutions can be found elsewhere (Crisfield 1991, 1997). In this work only the structural model applied, the stress and strain measures adopted and the numerical solution procedures used will be briefly discussed.

In this study the total Lagrangian procedure is adopted. The Cauchy stress (true stress) and Green-Lagrange strain are used as stress and strain measures respectively. Both low-order linear four-node (Q8MEM) and high-order nine-node bi-quadratic (CQ18M) Lagrange isoparametric plane stress elements are used to discretize the deep beam. For the low-order element a 2 × 2 Gauß-Legendre (quadrature) and for the high-order a (3 × 3) Gauss-Legendre integration scheme is applied. The increase in the area of cracked zones, as well as the load-deformation behaviour of the steel fibre concrete beam is numerically monitored using DIANA (2000) for each load step at element integration points.

5 NUMERICAL INVESTIGATION

The capability of the proposed material model is illustrated using reinforced steel fibre concrete deep-beams, which are experimentally investigated by Tan et al. (1993). The tested beams without web reinforcement had an effective depth of $d = 340$ mm and were subjected to four point loading (third point loading). Typical experimental set-up for the deep beams tested is given in Figure 10.

The yield strength of the reinforcement used was 460 N/mm². The effective depth d was kept constant during the experimental investigation but the shear-span a is assumed to vary. Linear variable differential transformers (LVDTs) were used to determine beam deflection (Figure 10, T1–T4). The deflections were measured at 5 KN load intervals. Due to symmetry in geometry, load and material properties only one half of the beam is discretised.

6 RESULTS AND DISCUSSIONS

For both reinforced concrete and steel fibre concrete deep beams having variable shear-span to effective depth ratios; the ultimate load, displacement,

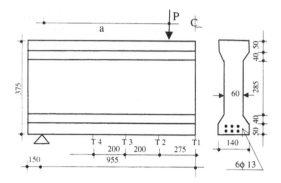

Figure 10. Typical load arrangement and dimensions for the investigated deep beams.

Table 2. Geometry, material parameters and predicted ultimate load of the deep beams

Beam	a/d	V_f %	Fctm N/mm²	Fcpm N/mm²	Ult. Load Exp.	Pred.	Pred. Exp.
1	2.0	0.0	3.08	32.8	108	83	0.77
2	2.0	0.5	3.06	35.0	177	173	0.98
3	2.0	1.0	3.71	36.0	210	198	0.94
4	2.5	1.0	3.71	36.0	147	145	0.99
5	1.5	1.0	3.71	36.0	300	270	0.90

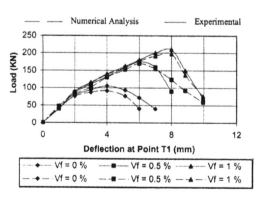

Figure 11. Influence of the volume of steel fibres on the load-displacement curve of reinforced steel fibre deep beam (a/d = 2).

stress/strain fields and crack pattern corresponding to the ultimate load are predicted. Some of the predicted results using a low-order linear four-node (Q8MEM) finite element and the developed elastic-plasticity theory based material model, with independent hardening/softening curve for tension and compression loading are given in Table 2 and in Figures 11, 12, 13 and 14.

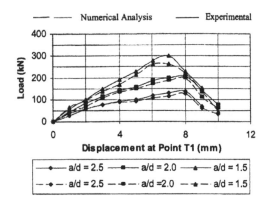

a) Crack pattern from experimental investigation (Tan et al)

Figure 12. Influence of the shear-span to effective depth ratio on the load-displacement curve of reinforced steel fibre deep-beams.

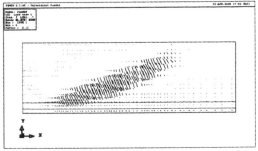

b) Crack strains from non-linear finite element analysis

Figure 14. Typical crack pattern for reinforced steel fibre concrete deep beam (a/d = 2 and V_f = 0.5%).

compression loadings have helped to simulate the increase in ductility (deformation capacity) due to steel fibres.

– Using the developed models, ultimate load carrying capacity and post ultimate load-deformation behaviour were predicted, which reasonably fit the experimental results.

– The increase in ultimate shear capacity due to steel fibres and the typical failure mode of deep beams were reasonably predicted using the developed model.

– The crack patterns at ultimate load observed during the experimental investigation for both reinforced concrete and reinforced steel fibre concrete deep beam were well predicted using the material model proposed in this study.

– With the availability of more experimental data on the behaviour of steel fibre concrete, it would be a relatively an easy to task to extend the constitutive models developed for biaxial states of stress to multiaxial state of stress or strain including time dependent and ambient conditions.

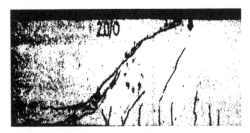

a) Crack pattern from experimental investigation (Tan et al)

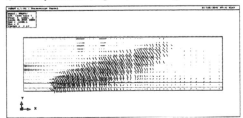

b) Crack strains from non-linear finite element analysis

Figure 13. Typical crack pattern for reinforced steel concrete deep beam (a/d = 2 and; V_f = 0%).

7 CONCLUSIONS

Based on the scope of the present theoretically study, the following conclusions are drawn:

– The systematically quantified material parameters such as fibre volume content, fibre orientation, fibre bond efficiency has helped to formulate material models for steel fibre concrete composite material under biaxial state of stress.

– The hardening/softening curves developed in this study for steel fibre concrete under both tensile and

ACKNOWLEDGEMENT

The theoretical research presented in this paper is a part of the experimental research programme on the compressive strength of cracked reinforced steel fibre concrete membrane elements subjected to tension-compression loading supported by Deutsche Forschungsgemeinschaft (DFG). The support is gratefully acknowledged.

REFERENCES

Bullo, T. & Fehling, E. 2000. The Behaviour of Steel Fibre Reinforced Concrete Panels subjected to Biaxial Tension-Compression Loading, Proceedings of the International PhD Symposium in Civil Engineering. Vol. 1, Vienna, Austria.

Yin, W.S., et al. 1989. Biaxial Tests of Plain and Fibre Concrete. ACI Mat. J., 86(3): 236–243.

Chen, W.-F. & Saleeb, A.F. 1982: Constitutive Equations for Engineering Materials. Vol. I: Elasticity and Modelling. New York. John Wiley & Sons.

Koiter, W.T. 1953. Stress-Strain Relations, Uniqueness and Variational Theorems for Elastic-plastic Materials with a Singular Yield Surface. Quart. Appl. Math 11(3): 350–354.

Schnütgen, B. 1975: Das Festigkeitsverhalten von mit Stahlfasern bewehrtem Beton unter Zugbeanspruchung. Dissertation, Ruhr-Univerisität Bochum, Bochum.

Litton, R.W. 1974. A Contribution to the analysis of concrete structures under cyclic loading. PhD thesis, University of California, Berkeley.

Bicanic, N. 1991. Computational Aspects. C. Meyer & J. Isenberg (eds.), Finite Element Analysis of Reinforced Concrete Structures II, Proceedings of the International Workshop: 367–489. New York.

Bazant, Z.P. 1984. Mechanics of Fracture and Progressive cracking in concrete Structures. G.C. Sih & A.D. Thomas (eds). Fracture Mechanics of Concrete, Dordrecht, Martins Nijhoff: 1–94.

Crisfield, M.A. 1991. Non-linear finite element analysis of Solids and structures (1). Essentials. John Wiley & Sons.

Crisfield, M.A. 1997. Non-linear finite element analysis of Solids and structures. Advanced Topics. (2). John Wiley & Sons.

DIANA 2000. Finite element Analysis, user's manual, release 7.2.

Tan, K.H., et al. 1993. Shear Behaviour of Steel Fiber Reinforced Concrete Beams. ACI Str. J., 89(6): 3–11.

Finite Elements in Civil Engineering Applications, Hendriks & Rots (eds.)
© *2002 Swets & Zeitlinger, Lisse, ISBN 90 5809 530 4*

Finite element analysis of shear wall specimens made of ductile fiber reinforced cementitious composites subjected to lateral loading

N. Shirai
Professor, Department of Architecture, College of Science and Technology, Nihon University, Tokyo, Japan

K. Watanabe, T. Oh-oka, S. Hakuto
Institute of Technology, Tokyu Construction Co., Ltd., Tokyo, Japan

T. Fujita
Tokyo Gas Urban Development Co., Ltd., Tokyo, Japan

ABSTRACT: The objective of this study is to investigate capability of the finite element analysis code; DIANA, for simulating elasto-plastic behavior of shear walls made of the ductile fiber reinforced cementitious composites (DFRCC) subjected to monotonic or cyclic lateral loading. To model ductile material properties of DFRCC, the fracture mechanics parameters were evaluated and incorporated into the finite element material models. Then, the shear wall specimens were analyzed by the FE code and applicability of the code was verified through the comparison between the test and analytical results.

1 INTRODUCTION

In recent years, research and development of the ductile fiber reinforced cementitious composites (DFRCC) superior to the existing ones have been actively conducted (JCI 2001). DFRCC is a cementitious composite material reinforced with fibers, and it represents crack dispersing properties under tensile, flexural or compressive stress. Consequently, DFRCC is a material with highly improved ductility at the flexural, tensile or compressive failure (JCI 2001).

For applying DFRCC to RC structures as an effective seismic element, it is required to understand its basic behavior by conducting the experimental test at least on a limited number of specimens. Furthermore, it is an important task to provide proper physical explanation to experimental findings and also to construct an objective mechanical model.

The numerical procedure is one of the candidates for aiding in accomplishing the above tasks and complementing test data; especially, the finite element method (FEM) is a promising tool because it can simulate stress distribution and fracture state in RC members in detail.

In the present study, the shear wall specimens made of DFRCC tested under lateral loading were analyzed by the FE procedure, and capability of the procedure for evaluating ductile performance of the specimens

was verified through the comparison between the test and analytical results. Note that a general purpose computer code, DIANA-7.2 (TNO 1999) was utilized as an analysis code.

2 LATERAL LOADING TEST ON SHEAR WALL SPECIMENS

Structural details of the shear wall specimens to be analyzed are listed in Table 1 (Hakuto et al. 2001). Configuration and reinforcement layout of the specimens are shown in Fig. 1. Size of the specimens is roughly 1/3 of the actual one; i.e., section of the column (B × D) is 240 × 240 mm and distance from the bottom of the column to the center line of the top stub (h) is 1400 mm. Normal strength concrete (N1) was used for the wall panel in the RC specimen, and DFRCC was used for the wall panel in the NMV specimen.

Figure 2 shows the loading setup. First, a constant axial load was applied to each column and this amounts to 1/6 of the compressive strength of concrete. Then, while keeping the axial load constant, cyclic lateral load was applied to the specimen in an incremental fashion. The lateral loading was controlled by the lateral displacement (δ_L) at the measuring point in the top stub; that is, the drift angle ($R = \delta_L/h$).

Table 1. Outline of specimens.

Specimen	RC	NMV
Column		
Type of concrete	N1	N2
Section B × D (mm)	240 × 240	
Rebar	SD345 12-D13 Pg = 2.65%	
Hoop	SD295A D6@50 Pw = 0.53%	
Wall		
Type of concrete	N1	DFRCC
Section B × D (mm)	60 × 1760	
Vertical and horizontal wall reinforcing bar	SD295A D6@200 Ps = 0.27%	
Loading schedule		
Drift angle R (rad.) = Lateral displacement/1400 mm	±1/6400(1), ±1/3200(1), ±1/1600(1), ±1/800(1), ±1/400(1), ± 1/200(1),	±1/800(1), ±1/400(1), ±1/200(2), ±1/100(2), ±1/50(2)

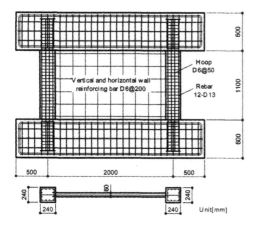

Figure 1. Reinforcement layout of specimens.

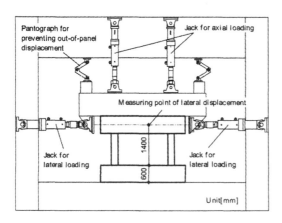

Figure 2. Loading setup.

Mix proportions for concrete and DFRCC are listed in Table 2. Material properties of concrete and DFRCC and reinforcing bars are listed in Tables 3 and 4, respectively.

3 SOFTENING BEHAVIOR OF CONCRETE AND DFRCC

Tension softening behavior of concrete and DFRCC are evaluated by applying the inverse analysis procedure (JCI 2002). For this purpose, 3-point bend tests on the concrete and DFRCC specimens with notch were carried out. Then, cracking analysis on the specimens was conducted, and a poly-linear tension softening curve was identified so that the predicted load-displacement curve gives the best fit to the observed one. Then, for use of the calculation purpose, the identified curve was approximated by a tri-linear curve (Oh-oka et al. 2000) for N1 and a poly-linear curve for DFRCC as shown in Fig. 3. Note that an initial cohesive stress for N1 is defined as the splitting tensile strength and that for DFRCC is defined as an initial cracking strength by the splitting tensile test.

Fracture mechanics parameters characterizing compression softening behaviors of concrete and DFRCC are evaluated as shown in Fig. 4 (Watanabe et al. 2000). At first, the measured compressive load (P_c) versus longitudinal displacement (δ_c) relation by the test is converted to the compressive stress (σ_c) versus plastic deformation (δ_c') relation (see Fig. 4(a) and (c)). Then, the area under the σ_c–δ_c' curve up to $\delta_c' = 3.0$ mm is calculated. This area is referred to as the compressive fracture energy (G_F^c) and it is regarded as a fracture mechanics parameter. Finally, for use of the calculation purpose, the observed σ_c–δ_c' relation was simply modeled by a bi-linear curve as shown in Fig. 4(c).

Table 2.   Mix proportions for concrete and DFRCC.

	Name of mixing type		
	N1	N2	DFRCC
(a) Mix proportion			
Water-cement ratio W/C (wt. %)	54.3	58.4	60.0
Sand percentage s/a (vol. %)	48.0	49.9	100
Sand-cement ratio S/C (wt. %)	226	276	191
Fiber volume fraction (vol. %)	–	–	2
Water content per unit (kg/m^3)	195	185	334
(b) Materials used			
Cement	Rapid-hardening portland cement		Ordinary portland cement
Sand	Pit sand: crushed sand = 3:7 (maximum size: 5 mm)		Pit sand (maximum size: 2.5 mm)
Gravel	Crushed stone (maximum size: 10 mm)		–
Fiber	–		Vinylon (diameter: 0.2 mm, length: 24 mm, Young's modulus: 25 GPa, tensile strength: 912 MPa)

Table 3.   Material properties of concrete and DFRCC.

Specimen			Compressive strength (MPa)	Splitting tensile strength (MPa)	Young's modulus (GPa)	Weight per unit volume (t/m^3)
Type	Region	Mixing type				
RC	Column, wall and stub	N1	28.6	2.21	21.9	2.25
NMV	Column and stub wall	N2	27.7	2.51	24.6	2.33
		DFRCC	26.2	2.25	14.1	2.06

Note: Splitting tensile strength: Cracking strength for DFRCC.

Table 4.   Material properties of reinforcing bars.

Specimen		Type of reinforcing bars	Yield point (Mpa)	Tensile strength (MPa)	Young's modulus (GPa)	Elongation (%)
Type	Region					
RC	Rebar in column	D13(SD345)	384	541	206	27.4
	Reinforcing bar in column or wall	D6(SD295A)	335	631	201	23.5
NMV	Rebar in column	D13(SD345)	374	516	190	19.3
	Reinforcing bar in column or wall	D6(SD295A)	306	484	193	19.3

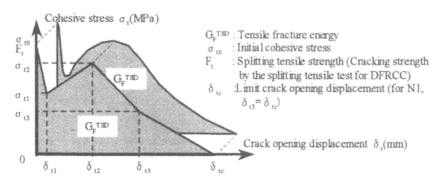

Figure 3.   Determination of tension softening behavior.

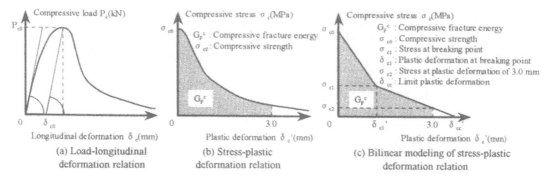

(a) Load-longitudinal
deformation relation

(b) Stress-plastic
deformation relation

(c) Bilinear modeling of stress-plastic
deformation relation

Figure 4.   Determination of tension softening behavior.

4   FINITE ELEMENT ANALYSIS

4.1   *Analytical modeling of test specimens*

Under assumption of the plane stress, the specimens were discretized as shown in Fig. 5. Concrete and DFRCC were modeled by a 4-node iso-parametric element. Reinforcing rebar in the column was modeled by a truss element and bond slip between reinforcement and concrete was taken into account. Hoop in the column and shear reinforcement in the wall panel were modeled by an embedded reinforcement element and perfect bond was assumed between concrete or DFRCC and hoop or shear reinforcing bars.

4.2   *Analysis parameters*

Analysis parameters are listed in Table 5. The following parameters are considered; i.e., effects of (1) cracking due to shrinkage, (2) compressive deterioration of cracked concrete, and (3) characteristic length. The characteristic length (Lc) is a length of fractured zone and is assumed to be equivalent to a representative element length to be used to convert plastic deformation after the peak to plastic strain. Five different cases of analysis were conducted to investigate effect of the parameters. The reference case of analysis shall be referred to as the "Analysis-I".

4.3   *Constitutive law for constituent materials*

The top and bottom concrete stubs are assumed to be elastic body, and concrete and DFRRC in the column or the wall panel are assumed to be elasto-plastic material. Von Mises yield criterion is applied to concrete and DFRCC in compression, and the tension cut-off criterion is used as a cracking criterion. Also, von Mises yielding criterion is applied for reinforcing bars. Constitutive laws of concrete and reinforcing bars used for the Analysis-I are described below.

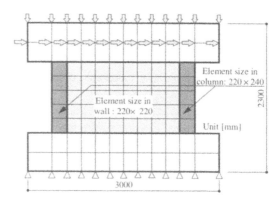

Figure 5.   Finite element discretization of shear wall specimen.

4.3.1   *Concrete and DFRCC*

The tension and compression softening curves obtained by the inverse analysis procedure stated in Chapter 3 are shown in Fig. 6. The calculated fracture energies are also listed in Table 6.

An ascending branch of the uni-axial stress versus strain relationship in tension is assumed to be linear elastic up to the tensile strength. The uni-axial stress versus strain relationship for a descending branch was modeled by a tri-linear curve (Oh-oak et al. 2000) for normal strength concrete and modeled by a poly-linear curve for DFRCC as shown in Fig.6(a).

An ascending branch of the uni-axial stress versus strain relationship in compression was modeled by a bi-linear curve with a breaking point at 1/3 of the compressive strength. The softening curve after the peak was modeled by a bi-linear curve as shown in Fig. 6(b).

The smeared crack model was used to represent cracking behavior in concrete or DFRCC. Total strain in the cracked concrete is assumed to be the sum of strain in the solid part and strain in the cracking part.

Table 5. Analysis parameters considered.

| Analytical model | Specimen to be analyzed | Concrete and DFRCC | | | |
| | | Shrinkage cracking (Initial strain) | Compressive deterioration of cracked concrete | Characteristic element length $L_{c(mm)}$ | |
				In tension	In compression
Analysis I	RC, NMV	×	×	311	
Analysis II	RC, NMV			311	
Analysis III	RC	○	○	311	311
Analysis IV	NMV		×	56.6	
Analysis V			○	56.6	

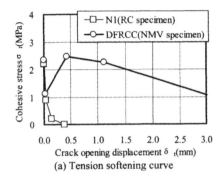
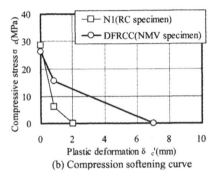

(a) Tension softening curve   (b) Compression softening curve

Figure 6. Assumed tension and compression softening curves.

Table 6. Fracture energy.

| Specimen | | | Tensile fracture energy G_F^{TSD} (N/mm) | Compressive fracture energy G_F^C (N/mm) |
Type	Region	Mixing type		
RC	Column, wall and stub	N1	0.155	18.45
	Column and stub	N2	0.155	17.87
NMV	Wall	DFRCC	5.525	45.22

G_F^{TSD}: Tensile fracture energy obtained by inverse analysis. Note that the tensile fracture energy of N2 is the value of N1 because the test for N2 is not carried out. G_F^c: Compressive fracture energy obtained by compressive loading.

Thus, the constitutive equation for the cracking part can be expressed by (TNO 1999):

$$\begin{Bmatrix} \sigma_{nn}^{cr} \\ \tau_{nt}^{cr} \end{Bmatrix} = \begin{bmatrix} D_I & 0 \\ 0 & D_{II} \end{bmatrix} \begin{Bmatrix} \varepsilon_{nn}^{cr} \\ \gamma_{nt}^{cr} \end{Bmatrix} \quad (1)$$

where σ_{nn}^{cr} and τ_{nt}^{cr}; the stress components in the cracking part in the n–t coordinate, ε_{nn}^{cr} and γ_{nt}^{cr}: the strain components in the cracking part in the n–t coordinate, n and t: the directions normal and parallel to crack, D_I: the material stiffness normal to crack, D_{II}: the

shear stiffness in the cracking part. The shear stiffness D_{II} can be defined by:

$$\beta = \beta \cdot G/(1 - \beta) \quad (2)$$

where β: the shear retention factor of cracked concrete, G: the shear modulus of intact concrete.

For use in the smeared crack model, the crack opening displacement (δ_t) in the tension softening curve in Fig. 6(a) have to be converted to the crack strain. In this study, δ_t was converted to the crack strain (ε_{nn}^{cr}) by dividing δ_t by the characteristic length (L_c).

L_c is assumed be given in terms of the element area (A) by (TNO 1999):

$$\varepsilon_{nn}^{cr} = \delta_t \sqrt{2A} \qquad (3)$$

Also, the plastic deformation (δ_c') in the compression softening curve shown in Fig. 6(b) was converted to the plastic strain (ε_{nn}^{pc}) by:

$$\varepsilon_{nn}^{pc} = \delta_c' \sqrt{2A} \qquad (4)$$

The shear retention factor for the cracked concrete was simply assumed to be a constant value of 0.125, irrespective to a magnitude of crack widths.

4.3.2 Stress-strain relationship for reinforcing bars

The stress versus strain relationship for reinforcing bars was expressed by a bi-linear model and a second slope of the stiffness was assumed to be 1/100 of the initial stiffness. In order to simulate bond slip behavior, an interface element was inserted to the boundaries between concrete and reinforcing bars. The bond stress versus slip relationship by the CEB-FIP Model Code 1990 (CEB 1993) was applied to the interface element.

4.4 Solution procedure

After applying the constant axial load to the top stub, an equal lateral displacement increment was applied to the loading nodal points in the top stub monotonically. Newton-Raphson method was utilized as a nonlinear iterative procedure to release unbalance forces within a step.

The basic convergence criterion adopted is that the ratio of the norms of internal energies is less than 0.05 % or the maximum number of iterations within a step is 9 times for the lateral displacement increment of 0.1 mm. However, the analysis easily fails to converge if nonlinearities become noticeable in the range of large displacement. In such cases, the restart option

was utilized; i.e., the results obtained up to the step prior to the step at which solution diverges are saved and then the analysis was restarted by changing the convergence criterion and/or a magnitude of the displacement increment. For example, to overcome limit or local instability point, the ratio of the norms of internal energies was allowed to loose down to 0.10%. In addition, larger displacement increment was used in a few steps just prior to the step that solution diverges, and then the displacement increment was returned to the original value of 0.1 mm if the limit or local instability point was overcome.

5 RESULTS AND DISCUSSION

5.1 Lateral force-drift angle relationship

5.1.1 Results by reference analysis (Analysis-I)

The observed and predicted ultimate strengths are listed in Table 7. The observed and predicted lateral load (Q) versus drift angle (R) relationships are shown in Fig. 7. Note that the symbol □ in the figure indicates the observed ultimate strength and △ indicates the predicted ultimate strength. The analysis for both specimens overestimates the initial stiffness as well as the subsequent stiffness during the process that shear cracks propagate in the wall panel after flexural cracking in the column. Furthermore, the analysis overestimates the ultimate strengths of the specimens, and it cannot simulate the post-peak behavior.

5.1.2 Effect of cracking due to shrinkage (Analysis-II)

Cracks in the concrete or DFRCC wall panel due to shrinkage have been observed in both specimens before loading. In this study, the effect of shrinkage cracks is investigated simply by introducing an equivalent initial strain causing shrinkage cracks to the wall panel; i.e., the assumed initial strain is $100\,\mu$ for the RC specimen and $200\,\mu$ for the NMV specimen.

The predicted Q–R relationship (Analysis-II) is compared with the test result in Fig. 8. Note that the

Table 7. Strength, deformation and failure modes.

Specimen	Loading direction	Test result			Analytical result (analysis I)	
		Ultimate strength (kN)	Drift angle ($\times 10^{-3}$ rad.)	Failure mode	Ultimate strength (kN)	Drift angle ($\times 10^{-3}$ rad.)
RC	+	922	5.06	Shear failure with diagonal compression failure at the bottom of wall panel	1041	3.46
	−	−930	−3.97			
NMV	+	1045	6.39	Sliding shear failure at the bottom of wall panel	1202	7.11
	−	−1061	−5.00			

symbol □ in the figure indicates the observed ultimate strength and △ indicates the predicted ultimate strength. The predicted initial stiffness for both specimens well compares with the observed one. On the other hand, the subsequent stiffness during the propagating process of shear cracks in the wall panel is still overestimated.

5.1.3 Effect of compressive deterioration of cracked concrete (Analysis-III)

In order to consider effect of the compressive deterioration of cracked concrete in the Analysis-III, the reduction factor of compressive strength for cracked concrete is estimated by (Vecchio & Collins 1986):

$$\lambda = \sigma_{c0}' / \sigma_{c0} = 1/(0.8 - 0.34\varepsilon_1 / \varepsilon_c) \leqq 1.0 \qquad (5)$$

where λ: the reduction factor of compressive strength, σ_{c0}': the maximum compressive stress of cracked concrete (MPa), σ_{c0}: the compressive strength of concrete (MPa), ε_1: the average tensile principal strain in the panel, ε_c: the strain at the compressive strength of concrete.

The predicted Q–R relationship is compared with the observed one in Fig. 8(a). Now, the analytical result

is relatively in good agreement with the test result. Note that the reduction factor used in the Analysis-III is $\lambda = 0.81$.

5.1.4 Effect of characteristic length of element (Analysis-IV and V)

For evaluating ductility of the NMV specimen with the DFRCC wall panel after the peak load, it is important to incorporate the tension softening behavior properly into the FE analysis. In other words, the characteristic length of element (L_c) has to be assumed appropriately because a phenomenon characterizing localization of the fracture is cracking. For this purpose, an average crack spacing is determined from the observed cracking pattern at $R = 1/200$ rad. in the test, and this crack spacing is set equal to L_c in the Analysis-IV.

The predicted Q–R relationship of the NMV specimen by the Analysis-IV is compared with the observed one in Fig. 8(b). Although the analysis gives slightly higher maximum strength than the test, the predicted overall response including in the post-peak behavior is relatively in good agreement with the test result.

In the next place, the NMV specimen was analyzed by taking the reduction factor of compressive strength

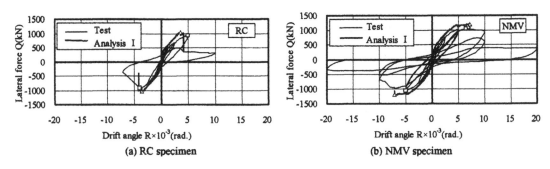

Figure 7.   Lateral force-drift angle relation (test and analysis I).

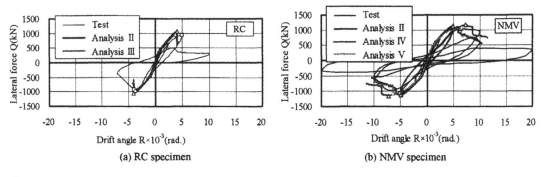

Figure 8.   Later force-drift angle relation (test and analysis II to V).

225

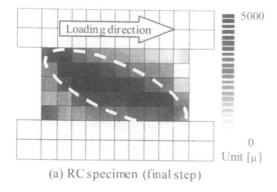

(a) RC specimen (final step)

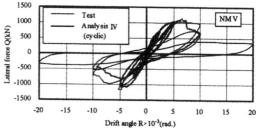

(b) NMV specimen (R=1/200rad.)

Figure 9.   Tensile principal strain distribution.

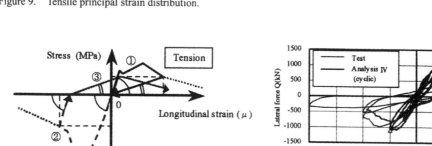

Figure 10.   Modeling of cyclic stress-strain relation of DFRCC.

Figure 11.   Lateral force-drift angle relation (test and cyclic analysis).

(Vecchio & Collins 1986) for the DFRCC wall panel into account (Analysis-V). Note that the reduction factor used in the Analysis-V is $\lambda = 0.81$. The predicted Q–R relationship is compared with the observed one in Fig. 8(b). It is seen that the analysis can predict the ultimate strength with good accuracy by considering the strength reduction factor.

5.2   *Damage distribution in specimens*

The tensile principal strain distributions for the RC (Analysis-III, R = 1/231 rad.) and NMV (Analysis-IV, R = 1/200 rad.) specimens are shown in Figs. 9(a) and (b), respectively. The color gradation in the figure distinguish a magnitude of tensile strain; darker the color, larger the tensile strain. In case of the RC specimen, larger tensile strains concentrate a narrow area circled by the broken line as shown in Fig. 9(a). In case of the NMV specimen, on the other hand, tensile strains with similar magnitude spread over a wider area in the wall panel. This indicates that cracks disperse over a wider area in the DFRCC panel, and this tendency corresponds to the observed cracking pattern. Thus, it can be said that the present analysis

procedure can explain the crack control mechanism of DFRCC.

5.3   *Cyclic alternative loading analysis*

The assumed hysteresis model of DFRCC under cyclic stresses is schematically shown in Fig. 10. Unloading and reloading in the compression range follow the initial stiffness. Unloading and reloading in the tension range are the origin-oriented type. The predicted cyclic response by the Analysis-V is compared with the observed one in Fig.11. The predicted Q–R relationship is relatively in good agreement with the test result. Therefore, it can be said that the present analysis procedure may become a promising tool for evaluating structural performance of the DFRCC shear wall structures.

6   CONCLUSIONS

In the present study, to investigate capability of the FE code for evaluating structural performance of the shear wall specimens, the RC and DFRCC shear wall specimens tested under the cyclic lateral loading were analyzed. The following conclusions are obtained:

(1) If the fracture mechanics parameters are properly evaluated and incorporated into the FE material models, the crack control mechanism of the DFRCC shear wall can be explained by the present FE procedure.
(2) The present FE analysis procedure may become a promising tool for evaluating structural performance of the DFRCC shear wall structures.
(3) The compressive deterioration due to cracking and localization at the compression failure of DFRCC have to be clarified experimentally in the future work.

ACKNOWLEDGEMENT

This research was conducted as part of the Academically Promoted Program on "Sustainable City Based on Environmental Preservation and Disaster Prevention" at Nihon University, College of Science and Technology (Head Investigator: Prof. Nagakatsu Kawahata) under A Grant from the Ministry of Education, Science, Sports and Culture and was partially supported by Japan Society for the Promotion of Science under Grant-in-Aid for Scientific Research (C).

REFERENCES

CEB 1993. CEB-FIP MODEL CODE1990, Thomas Telford.

Hakuto, S., Ozawa, J., Watanabe, K. & Nakamura, H. (2001). Test on Structural Walls using Fiber Reinforced Cement Comosites, *Summaries of Technical Papers of Annual Report, Architectural Institute of Japan, C-2, Sept. 2001*: 53–54 (in Japanese).

JCI 2001. Technical Committee on Test Method for Fracture Property of Concrete: Committee Report on Test Method for Fracture Property of Concrete, *The Japan Concrete Institute, May 2001*: 128.

JCI 2002. Technical Committee on High Performance Fiber Reinforced Cementitious Composites: Committee Report on High Performance Fiber Reinforced *Cementitious Composites, The Japan Concrete Institute*, Jan. 2002: 426 (in Japanese).

Oh-oka, T., Kitsutaka, Y. & Watanabe, K. 2000. Influence of Short Cut Fiber Mixing and Curing on The Fracture Parameters of Concrete, *Transactions of the Architectural Institute of Japan, No. 529, Mar. 2000*: 1–6 (in Japanese).

TNO 1999. DIANA Foundation Expertise Center for Computational Mechanics: *DIANA Finite Element Analysis Users Manual, TNO Building and Construction Research.*

Vecchio, F. J. & Collins, M. P. 1986. The Modified Compression- Field Theory for Reinforced Concrete Elements Subjected to Shear, *ACI Journal, March–April 1986*: 219–231.

Watanabe, K., Oh-oka, T., Shirai, N. & Moriizumi, K. 2000. Compression Softening Behavior of Various Type of Concrete, *Proceedings of the Japan Concrete Institute, Vol.22, No.2, June 2000*: 493–498 (in Japanese).

Finite Elements in Civil Engineering Applications, Hendriks & Rots (eds.)
© 2002 Swets & Zeitlinger, Lisse, ISBN 90 5809 530 4

Nonlinear FEM analysis of cap beam to evaluate shear strength against earthquake load

Kenji Kosa
Assoc. Prof. Kyushu Institute of Technology, Kitakyushu City, Japan

Jyunko Taguchi
Graduate student, Kyushu Institute of Technology, Kitakyushu City, Japan

Satoshi Yoshihara
Senior Engineer, Hanshin Expressway Public Corporation, Osaka, Japan

Katsunori Tanaka
Engineer, Yachiyo Engineering Co., Ltd, Osaka, Japan

ABSTRACT: To investigate the shear-resisting behavior of cap beams in RC rigid frame bridges under seismic loading, experiments were conducted using two seismically-strengthened pier specimens. The hoop tie ratio was used as the parameter.

According to the experiments, the beams of the specimens suffered shear damage after plastic hinges were formed at four positions. It was found that the hoop tie ratio of the beam does not affect the maximum strength of the beam, but significantly affects its deformation capacity. The FEM analysis showed that strain of hoop ties increased as the deformation of the beam increased, which suggests that shear damage to the beam was caused by the decrease of shear resisting capacity of the concrete in the beam.

1 INTRODUCTION

A significant number of RC bridge piers along the Hanshin Expressway Kobe Route suffered severe damage during the 1995 Hyogo-ken Nambu Earthquake. Though most RC single column piers were damaged, many of two-column piers of the RC rigid frame bridges survived the earthquake with small damage. One reason for this is that the latter type pier has more flexural strength than the former type, because the rigid frame structure has a higher degree of redundancy and plastic hinges are successively formed in the latter type when subjected to loading. It should be noted, however, there is not much difference in the shear strength of the two piers when subjected to loading.

After the earthquake, seismic strengthening began on all the routes of the Hanshin Expressway. The columns of single column piers and rigid frame piers are being jacketed with concrete or steel plates. In the case of rigid frame structures, there is a fear that unjacketed beams may have lost their relative strength

after the columns have been jacketed, and that shear damage may first occur at the unjacketed beam area, if subjected to seismic loading. However, seismic strengthening of the beams is rather difficult, mainly because the shear strengthening design method for the beam is not yet well-established due to design complexity and because bearings and other devices that hamper strengthening work are installed on the beam.

To date, no crucial damage to RC rigid frame piers originating from beam damage has been reported in Japan. Whereas in the U.S., shear damage to the cross beam and the column-beam joint occurred during the Loma Prieta Earthquake.

To investigate the behavior of RC rigid frame bridge beams under seismic loading, experiments were conducted using two 1/8 scale specimens modeling on a column-strengthened standard type RC rigid frame pier. The hoop tie ratio in the beam was made to normal for specimen No. 1 and reduced to one half for specimen No. 2. The shear strength, deformation capacity, and the damage mechanism of the beam were evaluated.

2 EXPERIMENTAL PROGRAM

Monotonic horizontal loading was applied to the beam end of the specimens. Table 1 shows the size and the reinforcement ratio of the two specimens and an actual pier. The actual pier means the pier that was used as the model pier for specimen construction. Used as the model pier was P-164 on the Osaka-Nishinomiya Route that has a typical configuration of rigid frame piers. According to the check by the ultimate lateral strength method specified in the Specifications for Highway Bridges (Highway Specifications), the model pier is a beam-failure preceding type. The hoop tie ratio in the beam of specimen No. 2 was reduced to 1/2 that of the model pier so that the failure mode at the beam can be evaluated.

Table 1. Actual pier and specimens.

	Actual pier	No. 1	No. 2
Pier height (m)	10.60	1.32	1.32
Column cross section (cm)	250 × 250	31 × 31	31 × 31
Beam cross section (cm)	250 × 200	31 × 25	31 × 25
Reinforcement ratio			
Column main reinforcement ratio (Axial dir.) (%)	0.49	0.89	0.89
Column main reinforcement ratio (Perp dir.) (%)	0.40	0.74	0.74
Beam main reinforcement ratio (Lower part.) (%)	0.51	0.54	0.54
Column hoop tie ratio (%)	0.71 (steel plates)	0.71	0.71
Beam hoop tie ratio (%)	0.16	0.16	0.08

In a model pier, the main reinforcement ratio of the column is 0.40% and the design flexural yielding of the column, the design flexural yielding and the design shear strength of the beam are all nearly equivalent. But, to make clear the damage mode of the beam, D13 reinforcing bars which are one size bigger were used for the main reinforcement of the column of the specimens. The column of the model pier is jacketed with 9 mm thick steel plates which is 0.71% if converted into the hoop tie ratio.

Figure 1 shows the reinforcing bar arrangement of specimen No. 1. As to the column of the specimens, the cross section is 31 × 31 cm, the main reinforcement ratio 0.74%, and the hoop tie ratio 0.71%. As to the beam of the specimens, the cross section is 31 × 25 cm and the main reinforcement ratio 0.54%. The hoop tie ratio of the beam differs between the two specimens: 0.16% (D6@130mm) for specimen No.1 and 0.08% (D4@105 mm) for specimen No. 2. As with a model pier, no hoop ties were arranged at the column-beam joint of the specimens.

Figure 2 shows a load application to the specimens. To anchor the specimens firmly, the footing and the base plate were secured with 15 reinforcing bars. Loading was applied with an increment of 30 kN up to near a maximum load. Every time a target load was reached, the load was reduced to zero and crack development was checked. Loading was applied by the load-control method up to a maximum load and

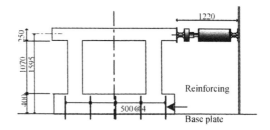

Figure 2. Load application to specimens.

Table 2. Properties of materials used.

Items		No. 1	No. 2
Concrete	Compressive strength N/mm²	27.4	36.6
Reinforcing bar D4	Yields strength N/mm²		342
	Yield strain i		1630
Reinforcing bar D6	Yield strength N/mm²	380	357
	Yield strain i	1810	1702
Reinforcing bar D10	Yield strength N/mm²	398	398
	Yield strain i	1895	1895
Reinforcing bar D13	Yield strength N/mm²	399	375
	Yield strain i	1900	1786

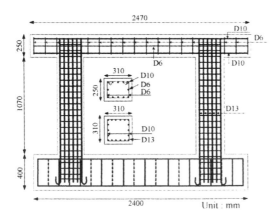

Figure 1. Reinforcement arrangement of specimen No. 1.

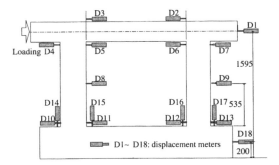

Figure 3. Positions of displacement meters.

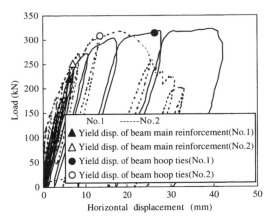

Figure 4. Load-displacement relationship.

Table 3. Experimental results.

Specimens		No. 1	No. 2
Yield displacement	äy(mm)	6.3	7.1
Ultimate displacement	äu(mm)	41.3	25.8
Yield load	Py(kN)	221.3	251.2
Maximum load	Pmax(kN)	322.0	316.5
Ductility factor	äu/äy	6.6	3.6
Pmax/Py		1.5	1.3

by the displacement-control method after reaching a maximum load. Though dead load of 1.0 N/mm^2 is acting on a model pier as the axial compressive stress, this load was neglected, considering that this small stress is not influential.

Table 2 shows the material characteristics of the concrete and the reinforcing bars used. The design strength of the concrete is 27.0 N/mm^2. Measurements included displacement of each member, strains of main reinforcement and hoop ties in the beam, strains of main reinforcement and hoop ties in the column. Figure 3 shows the positions of displacement meters installed. Displacement measurements included the horizontal displacement of each member, the rotation of the column, and displacement at the end of each member.

3 EXPERIMENTAL RESULTS

3.1 Load-displacement relationship

Figure 4 shows the load-displacement relationship and Table 3 gives other experimental results. The displacement values in the table were taken by the displacement meter D1 that was installed at the end of the beam opposite of a loading point. The yield displacement (δy) was taken when the main reinforcement in the beam reached yielding, and the ultimate displacement (δu) was taken when the load dropped to a yield load in the load-displacement envelope. The ductility factor (μ) was calculated as δu/δy.

The ductility factor thus derived was 6.7 for specimen No. 1 and 3.6 for specimen No. 2. The ductility factor of specimen No. 2 was about one half of that of specimen No. 1. But, the strength of the beam did not differ much, with a maximum load of 322.0 kN for specimen No. 1 and 316.5 kN for specimen No. 2. These experimental results suggest that the hoop tie ratio in the beam has little effect on the maximum strength of an RC rigid frame pier, but has large effect on the deformation capacity and the ductility factor of the pier.

3.2 Damage to specimens

(a) Specimen No. 1

Figures 5 and 6 respectively indicate crack development and final damage observed on the beam of specimen No. 1. Cracking started on the side of the beam when the load was 181 kN (displacement: 4.5 mm). Flexural cracking appeared on both columns of the specimen and the main reinforcement in the beam reached yielding when the load was 222 kN (displacement: 6.3 mm). Cracking continued on both upper and lower faces of the beam. The hoop ties in the beam showed a near-yield strain at the loading of 317 kN (displacement: 27.8 mm) and cracking still continued propagation at an angle of 45 degrees toward the center of the beam. The oblique cracking began to widen and a large slip was caused at a maximum load of 322 kN (displacement: 39.2 mm), resulting in shear failure.

Plastic hinges were formed at four positions after the main reinforcement in the column and the beam reached yielding and then the beam suffered shear failure. The ductility factor at the time was approximately 6, which means it was the failure with a higher deformation capacity. Though shear cracking appeared at a displacement of about 28 mm, the load did not begin to drop until reaching a maximum load of 42 mm, as shown in Figure 4. Concurrently, strain on the hoop

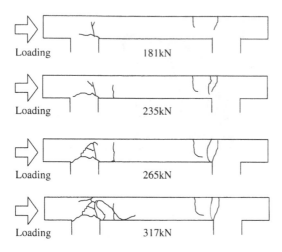

Figure 5. Crack development (No. 1).

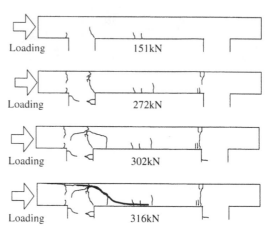

Figure 7. Crack development (No. 2).

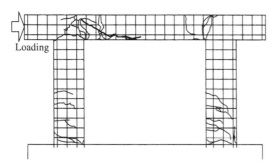

Figure 6. Damage to specimen No. 1.

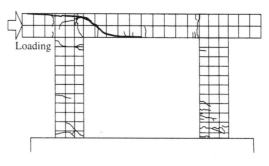

Figure 8. Damage to specimen No. 2.

tie in the beam increased up to a yield strain and then the beam failed by shear.

(b) Specimen No. 2

Figures 7 and 8 respectively show crack development and the final damage occurred to the beam of specimen No. 2. Cracking appeared on the side of the beam when the load was 150 kN (displacement: 3.0 mm). Another cracking started at the column-beam joint opposite of loading point. A strain gauge installed on the main reinforcement at that position showed a strain exceeding a yield strain when the load was 251 kN (displacement: 7.1 mm). When the load was 302 kN (displacement: 13.8 mm), hoop ties in the beam showed a strain close to a yield strain. Shear cracking that appeared at the column-beam joint on the loading side instantaneously spread at the time the load was 317 kN (displacement: 19.7 mm), causing slippage on the lower face of the beam and then a failure by shear.

Monotonic loading was continued by the displacement-control method until a maximum displacement

of 33 mm was caused. Then, the strength gradually decreased, the concrete on the lower face of the beam dropped locally, and the shear cracks extended up to the outside of the joint area.

In specimen No. 2, oblique cracks started on the beam when the displacement was about 11 mm and abruptly a shear failure came when the next load was applied, a marked difference from the results of specimen No. 1. In both specimens, numbers of cracks appeared at the bottom of the column not on the loading side but on the opposite side. Formation of four plastic hinges were confirmed in both specimens, with the fourth one formed on the left beam at a displacement of 32.7 mm in specimen No. 1 and 13.8 mm in specimen No. 2. In both specimens, shear failure occurred after the hoop ties in the beam reached yielding. It is therefore presumed that the shear failure of the beam was caused by the decrease of shear resistance provided by the concrete as deformation increased.

3.3 Sequence of reinforcement yielding

Figure 9 shows the sequence of yielding of the main reinforcement judged from the measurements using strain gauges. Yielding was assumed to have reached when the measured strain became the yield strain shown in Table 2. In both specimens, the right beam reached yielding first followed by the bottom areas of both the right and the left columns and finally the left beam, resulting in the formation of four plastic hinges. However, there was a significant difference as to the

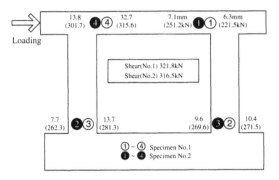

Figure 9.   Sequence of reinforcement yielding.

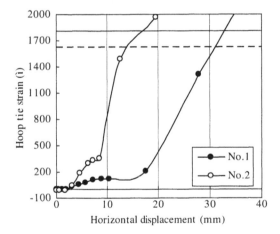

Figure 10.   Relationship between hoop tie strain and displacement.

timing of the formation of the fourth hinge. In specimen No. 2, yielding of the main reinforcement in the left beam occurred when the displacement was one-half the displacement of specimen No. 1.

3.4   Distribution of strain

Figure 10 shows the relationship between the displacement and the strain of the hoop ties in the beam. Strain was measured on the hoop ties in the beam at the position the shear damage occurred. In specimen No. 1, no visible strain appeared until displacement became 15 mm, and then strain gradually increased up to a yield strain. In specimen No. 2, strain increased little until the displacement became 10 mm and then started to increase until finally reaching a yield strain.

This suggests that most of the acting shear forces up to a displacement of 10–15 mm is carried by the concrete of the beam. And, it is after this displacement that the hoop ties start to exhibit their load-carrying capacity. The abrupt increase of strain seen in specimen No. 2 was probably because the hoop tie ratio of this specimen was rather small and had little resistance to shear.

3.5   Comparison with design values

The check method for the ultimate lateral strength of the RC rigid frame bridge piers was incorporated into the 1996 Specifications for Highway Bridges (Highway Specifications) based on the research results conducted after the 1995 earthquake. According to the specifications, check is performed by evaluating the damage mode of a pier in the same way as done with single column piers, and then comparing the strength of the pier with the response value assigned to each failure mode.

The values thus obtained are assumed as design values. Tables 4 and 5 compare the experimental and design values, and Figure 11 compares load-displacement relationships obtained by experiments and by design. It is seen from the tables and the figure that the horizontal strength of the two specimens is about 1.1 to 1.8 times the force that causes flexural yielding. The shear resistance obtained by experiment was 1.2 times the design value of specimen No. 1

Table 4.   Comparison of design and experimental values (No. 1).

No. 1	Design values		Experimental values		
Sequence	Location	Horizontal force P'_1 (kN)	Sequence	Horizontal force P_1 (kN)	P_1/P'_1
1	Right beam	170.9	1	221.5	1.3
2	Left beam	171.8	4	315.6	1.8
3	Left column	242.6	3	281.3	1.2
4	Right column	243.4	2	271.5	1.1
Shear	Right beam	269.7	Left beam	321.8	1.2

Table 5. Comparison of design and experimental values (No. 2).

No. 2	Design values		Experimental values		
Sequence	Location	Horizontal force P_2'(kN)	Sequence	Horizontal force P_2(kN)	P_2/P_2'
1	Right beam	150.6	1	251.2	1.7
2	Left beam	151.7	4	301.7	2.0
3	Left column	242.3	2	262.3	1.1
4	Right column	243.1	3	269.6	1.1
Shear	Right beam	139.4	Left beam	316.5	2.3

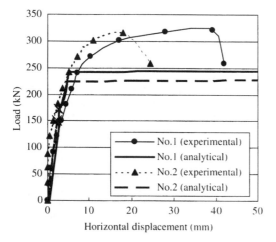

Figure 11. Comparison of load-displacement relationships.

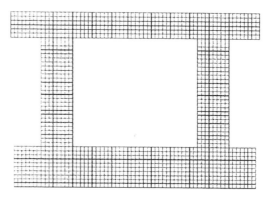

Figure 12. Elements for analysis model.

(experimental value 321.8 kN, design value 269.7 kN), but it was as high as 2.3 times in specimen No. 2 (experimental value 316.5 kN, design value 139.2 kN). This suggests that shear strength specified by the Highway Specifications has a considerable margin compared with actual behavior, depending on the hoop tie ratio.

An acting force is carried by the concrete of the beam in the range of small displacement. It is considered that the shear strength provided by the concrete is somewhat underestimated in design.

4 EVALUATION BY FEM ALALYSIS

4.1 *Analysis method*

The two-dimensional elasto-plastic FEM analysis was performed on the two specimens to investigate the reproductivity of the experimental results, the load-displacement relationship, strain development on the reinforcement, and the failure mechanism.

Figure 12 shows the elements employed for analysis. The elements used were plane stress elements for the concrete and truss elements for the reinforcement. The elements of the reinforcement and the concrete were completely bonded. As the material properties of them, the experimental values shown in Table 2 were used.

As the yield criterion for the concrete, the Drucker-Prager yield criterion and the maximum principal stress were adopted for the compressive range and the tensile range, respectively. As the yield criterion for the reinforcement, the Von-Mises yield criterion was adopted. In the shear model, a constant value of 20% was employed as the shear stress transfer value after the start of cracking.

Figure 13 shows the stress-strain model of concrete. In this model, a quadratic curve was adopted up to reaching the compressive strength in the compression-increasing range, a model with a constant stress and an increase in strain was adopted after that range, and a model with a linear stress decrease up to ε_t ($\varepsilon_t = \sigma_{sy}/E_s \fallingdotseq 2000\mu$) after reaching the tensile strength was adopted in the tensile range. For the stress-strain model of the reinforcement, a trilinear model that takes strain softening after reaching a yield strength into account was used.

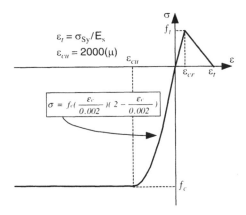

$$\varepsilon_t = \sigma_{Sy}/E_s$$
$$\varepsilon_{cu} = 2000(\mu)$$

$$\sigma = f_c\left(\frac{\varepsilon_c}{0.002}\right)\left(2 - \frac{\varepsilon_c}{0.002}\right)$$

Figure 13.   Stress-strain model for concrete.

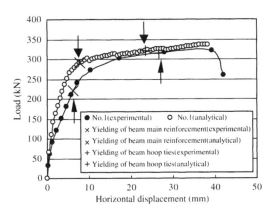

Figure 14.   Load-displacement relationship (No. 1).

Table 6.   Comparison of experimental and FEM analytical values (No. 1).

Items (mm)	Specimen No.1	
	Experimental	Analytical
Yield disp. of beam main reinforcement	6.3 (221)	7.8 (291)
Yield disp. of beam hoop ties	26.7 (313)	23.5 (321)
Yield disp. of column main reinforcement	10.4 (272)	6.0 (272)

( ) shows load: kN.

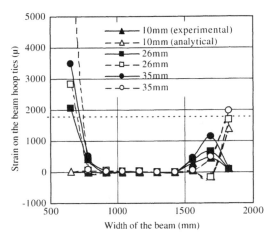

Figure 15.   Strain distribution on the beam hoop ties (No. 1).

As the loading condition, horizontal monotonic loads were applied to the beam by the displacement-control method. As the boundary condition, the bottom face of the footing was fixed completely.

4.2   *Analytical results*

(1) Results for specimen No. 1

Figure 14 shows the load-displacement relationship of specimen No. 1. The displacement shown is the displacement taken at the D1 point in both experiment and analysis. The maximum load cited in analysis is the load obtained just before convergence becomes impossible. The yield point was determined by the displacement taken when strain at each measurement point and analysis point reached to a yield strain for the first time.

From Figure 14, it is seen that both experimental and analytical inclinations are roughly similar up to near the maximum load, though the experimental inclination is slightly gentler than the analytical inclination in the initial range. The maximum load was 337 kN by analysis and 322 kN by experiment, both about 1.2 times greater than the design maximum load.

Table 6 shows the displacement and the load taken at each measurement point. As seen from the table, yielding of main reinforcements in the beam and the column occurred when the displacement was approximately below 10 mm in both experiment and analysis. In contrast, yielding of hoop ties in the beam occurred when the displacement was considerably large as 26.7 mm (load: 313 kN) for experiment and 23.5 mm (load: 321 kN) for analysis, though slight scatter was seen depending on measurement points.

Figure 15 describes the distribution of strain occurred to the hoop ties in the beam. In the experiment, the hoop tie strain increased as the main reinforcement in the both ends of the beam yielded as with an increase of displacement. In analysis, almost no strain occurred up to a displacement of 6 mm and a strain of 1000 μ occurred at a displacement of 10 mm and a strain exceeding a yield strain at a displacement of 26 mm. Figure 16 gives the relationship between the strain of the hoop ties in the beam and the displacement that was derived by comparing the hoop tie strain elements at the beam end on the loading side where shear cracking occurred. Strain did not occur up to a displacement of 15 mm in both experiment and analysis, but after this strain amount, hoop tie strain quickly increased as displacement increased.

(2) Results for specimen No. 2

Figure 17 shows the relationship between load-displacement relationship of specimen No. 2. The load-displacement relationship was compared using the displacement values, both experimental and analytical, observed at the D1 point. The initial inclination of the experimental load-displacement curve was gentler than that of analysis, as with the results of specimen No. 1. So, using displacement meters D14 to D17 installed at the column bottom of specimen No. 2, rotation at the column bottom due to a horizontal

displacement increase was corrected. About 2 mm displacement was caused due to the pullout of reinforcement at the column bottom before the main reinforcement in the beam reached yielding. If the reinforcement pullout is corrected in the experimental load-displacement relationship in Figure 14, the initial inclination will become roughly the same in both specimens.

The maximum load by analysis, 353 kN, was slightly greater than the maximum load by experiment, 317 kN, but the both values have a margin 2.3 to 2.5 times the design shear strength.

Table 7 shows the load and the displacement taken at each measurement point. Though the main reinforcement in both beam and column reached yielding at a small displacement of 7.7 mm or below, hoop ties in the beam reached yielding at a rather large displacement of 10 mm or over, specifically 13.8 mm (load: 302 kN) in experiment and 13.3 mm (load: 310 kN) in analysis.

Figure 18 shows the distribution of strain occurred to the hoop ties in the beam. The hoop tie strain at the

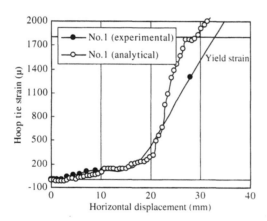

Figure 16. Relationship between displacement and hoop tie strain.

Table 7. Comparison of experimental and FEM analytical values (No. 2).

Items (mm)	Specimen No. 2	
	Experimental	Analytical
Yield disp. of beam main reinforcement	7.1 (251)	4.3 (254)
Yield disp. of beam hoop ties	13.8 (302)	13.3 (310)
Yield disp. of column main reinforcement	7.7 (262)	6.0 (287)

( ) shows load: kN.

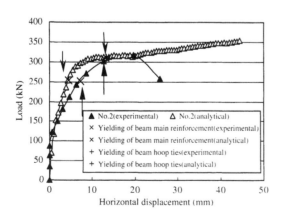

Figure 17. Load-displacement relationship (No. 2).

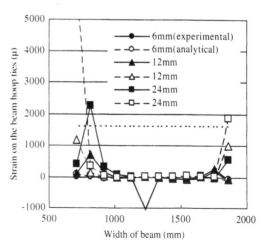

Figure 18. Strain distribution on the beam hoop ties (No. 2).

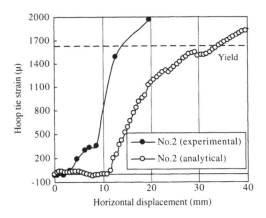

Figure 19. Relationship between displacement and hoop tie strain.

end of the beam on the loading side is below 1000 μ when the strain is 10 mm, but it exceeds a yield strain when the displacement becomes 24 mm. It means the hoop tie strain increased rather quickly as with an increase of displacement. The same tendency is found in the analytical results, in which strain is 1000 μ when the displacement is 10 mm but it exceeds a yield strain when the displacement becomes 26 mm. Figure 19 shows the relationship between the strain of the hoop ties in the beam and the displacement. Experimental and analytical results were compared using the hoop tie strain element at the beam end on the loadingside where shear cracking occurred. It is seen that strain began to increase after the displacement exceeded 10 mm in both experimental and analytical results.

As seen, the FEM analysis could follow well the behavior of the maximum load, formation of four plastic hinges, and increase of hoop tie strain with an increase of displacement in both specimens. It can be said, therefore, that the analysis is able to evaluate the behavior of specimens pretty satisfactory.

It was confirmed that shear failure of the beam of the flexural yield-preceding type occurred in both specimens, which was known through the formation of four plastic hinges, occurrence of large strain with an increase in load, and similar strain distribution patterns.

As the hoop tie strain did not increase up to a displacement of 10–15 mm, it is considered that an acting shear force was carried by the concrete of the beam in the range of smaller displacement. In contrast, according to an ultimate lateral strength check based on the Highway Specifications, a flexural yield load and a shear resistance load are balanced in specimen No. 1, but the shear strength is about 1/2 of the flexural strength and ended in shear damage in specimen No. 2.

From these investigation results, it is inferred that the beams of RC rigid frame piers have shear resistance about twice that specified by the Highway Specifications because an acting force is carried by only the concrete up to a displacement of 1.5–2.0 δy. But, after 1.5–2.0 δy shear resistance carried by the concrete begin to decrease and that carried by the hoop ties begin to increase as displacement increases, leading to a shear failure of the beam.

5 CONCLUSIONS

Horizontal monotonic loading tests were conducted using seismically-strengthened two 1/8 scale rigid frame pier specimens. Specimen No.1 was constructed to contain a standard hoop tie ratio in the beam and specimen No. 2 one half the standard hoop tie ratio. From the experiment and analysis, the following conclusions were drawn:

1. According to the experiments using the hoop tie ratio as the parameter, the positions where four plastic hinges were formed were the same in both specimens independent of hoop tie ratio in the beam. This indicates that hoop tie ratio does not have a significant effect on the failure mechanism.

2. There was no large difference in the maximum load of the specimens, with 322.0 kN for specimen No. 1 and 316.5 kN for specimen No. 2. But, difference was significant for the ductility factor, with 6.7 for specimen No. 1 and 3.5 for specimen No. 2, about 50% difference. This indicates that the hoop tie ratio in the beam does not affect the maximum strength of the beam but affects the ductility factor of the beam.

3. When the experimental results were compared with the results of the ultimate lateral strength check specified by the Highway Specifications, experimental shear strength was 1.2 times greater than the shear strength by the Specifications in specimen No. 1 and 2.3 times greater in specimen No. 2.

4. The FEM analysis could trace that shear failure occurred to the beams of the specimens was the flexural yield-preceding type. The analysis could also evaluate a tendency regarding a maximum load and formation of four plastic hinges.

5. Experimental and analytical results showed that shear damage to the beam was caused by the decrease of the shear strength of the concrete that occurred as the displacement and hoop tie strain of the beam increased. This suggests that an acting forces is carried by the shear strength of the concrete until displacement becomes 1.5–2.0 δy, and that the beam has twice the ultimate lateral strength specified by the Highway Specifications.

REFERENCES

Kosa, K., Taguchi, J., Otoguro, Y., and Tanaka, K. "Seismic Evaluation on the Beams of RC Rigid Frame Bridge Piers", Proceedings of the 26th JSCE Earthquake Engineering Symposium, 2001.

Takano, K., Mutsuyoshi, H., Sakurai, J., and Fujita, R. "Studies on the Strengthening of Beams of RC Single Layer Rigid Frame Piers", Proceeding of the Japan Concrete Institute, Vol. 22, No.3, 2000/6.

Koizumi, H., Mutsuyoshi, H., Konishi, Y., and Fujita, R. "Studies on the Seismic Strengthening of Beams of RC Single Layer Rigid Frame Piers", Proceeding of the Japan Concrete Institute, Vol. 23, No.1, 2001/6.

Adachi, Y., Unjyo, S., and Nagaya, K. "Experimental Studies on the Seismic Strengthening of RC Rigid Frame Pier Beams", Proceeding of the Japan Concrete Institute, Vol. 222, No.3, 2000/6.

Masonry and block mechanics

Finite Elements in Civil Engineering Applications, Hendriks & Rots (eds.)
© 2002 Swets & Zeitlinger, Lisse, ISBN 90 5809 530 4

Guidelines for the analysis of historical masonry structures

Paulo B. Lourenço
University of Minho, Guimarães, Portugal

ABSTRACT: The need of modeling and analysis of historical constructions is a key issue in the challenging issues of conservation and restoration. Here, the possibilities of analysis of historical constructions are addressed for two simple examples (a masonry wall loaded out-of-plane and a masonry arch. Several case studies, of different complexity levels are also presented, illustrating the application of different levels of sophistication. Finally, general recommendations for modeling are given.

1 INTRODUCTION

The analysis of historical masonry constructions is a complex task. Usually, salient aspects are:

- Geometry data is scarce or missing;
- Information about the inner core of the structural elements is missing;
- Characterization of the mechanical properties of the materials used is difficult and expensive;
- Large variability of mechanical properties, due to workmanship and use of natural materials;
- Significant changes in the core and constitution of structural elements, associated with long construction periods;
- Construction sequence is unknown;
- Existing damage in the structure is unknown;
- Regulations and codes are non-applicable.

Several methods and computational tools are available for the assessment of the mechanical behavior of historical constructions. The methods resort to different theories or approaches, resulting in: different levels of complexity (from simple graphical methods and hand calculations to complex mathematical formulations and large systems of non-linear equations), different availability for the practitioner (from readily available in any consulting engineer office to scarcely available in a few research oriented institutions and large consulting offices), different time requirements (from a few seconds of computer time to a few days of processing) and, of course, different costs. It should also be expected that results of different approaches might be also different, but this is not a sufficient reason to prefer one method over the other. In fact, a more complex analysis tool does not necessarily provide

better results than a simplified tool. Key aspects to be considered include:

- Adequacy between the analysis tool and the sought information;
- Analysis tools available to the practitioner involved in the project (it is of fundamental importance that the available engineering is compatible with the analysis tools);
- Cost, available financial resources and time requirements.

This paper addresses the possibilities of analysis of historical constructions and presents several case studies of different complexity. It is advocated that most techniques of analysis are adequate, possibly for different applications, if combined with proper engineering reasoning, see Lourenço (2001) for a more comprehensive discussion.

2 USE OF DIFFERENT ANALYSIS METHODS

The engineering assessment of ancient masonry structures requires adequate computational tools and different approaches can be used.

Common idealizations of the behavior used for analysis are elastic behavior (with or without redistribution of stresses), plastic behavior and non-linear behavior. Non-linear analysis is the most powerful method of analysis, able to trace the complete response of a structure from the elastic range, through cracking and crushing, up to complete failure. Plastic analysis or limit analysis aims at evaluating the structural load at failure. This method can be assumed as adequate for the analysis of historical masonry structures if

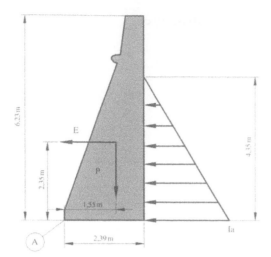

Figure 1.   Geometry and loading in masonry wall.

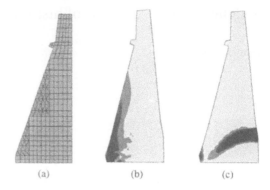

Figure 2.   Finite element analysis: (a) mesh, (b) minimum principal stresses and (c) maximum principal stresses.

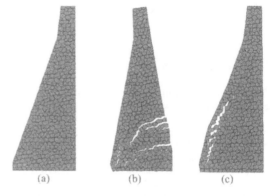

Figure 3.   Limit blocky analysis: (a) blocks, and collapses for (b) associate and (c) non-associated flow of the joints.

a zero tensile stress is assumed. The plastic analysis is either based on the static method, e.g. thrust-line analysis, or on the kinematic method, e.g. yield-line analysis. In the following, these different methods of structural analysis are applied to two simple masonry structures.

2.1   A masonry wall

Figure 1 illustrates the geometry of a cross section in a masonry wall of a Portuguese castle, together with appropriate dead loading (wall weight and soil pressure) and live loading (seismic action).

The assumed specific weight for the wall is $20\,kN/m^3$, the soil internal friction angle is 30° and the horizontal load, statically equivalent to the seismic action, is 0.22 of the weight load. With these data, the destabilizing and stabilizing design moments with respect to point A are equal to 239.5 kN.m/m and 297.7 kN.m/m, respectively. Thus, collapse of the structure occurs for a load factor equal to 1.24.

The wall is next analyzed with plane strain elements, assuming a Rankine no-tension model for tension and Von Mises for compression (uniaxial compression strength of $1.0\,N/mm^2$), see Feenstra (1993). Figure 2 illustrates the mesh and the results, where the damage in the compressed toe and the diagonal failure are clearly visible. The load factor obtained in the analysis is 1.15, being lower than the above calculated load factor due to the fact that the crack does not occur horizontally but diagonally.

Finally, the wall is analyzed with a limit analysis program with rigid blocks, Orduña & Lourenço (2002). For this purpose, three hundred rigid blocks, with dimensions similar to the stones used in the wall are used. Associated and non-associated flow of the joints was assumed, resulting in load factors of 1.03 and

0.77, respectively. In the first case, collapse occurs with a larger inclination of the diagonal crack, with respect to continuum modeling, and, in the second case, local collapse of the extrados is observed.

It is noted that the wall has collapsed recently due to a decohesion process, in a process similar to Figure 3c. An obvious recommendation of the analysis is that the masonry that constitutes the wall needs to be properly injected or tied.

It is also noteworthy to stress that the safety factor varies from 1.24 to 0.77 according to the modeling strategy adopted. This should be expected as the result of a structural analysis with a given model depends of the model itself.

The fact that different methods of analysis lead to different safety factors and different results is not a sufficient reason to select one method over the other.

2.2   A masonry arch

Figure 4 illustrates an arch with a span of 5.0 m, a rise of 2.5 m, a thickness of 0.3 m, a width of 1.0 m, a radius of 2.5 m and a backfill up to 3.0 m height.

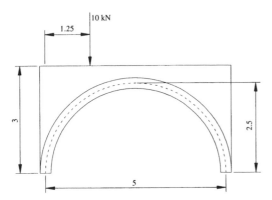

Figure 4. Geometry and loading in masonry arch.

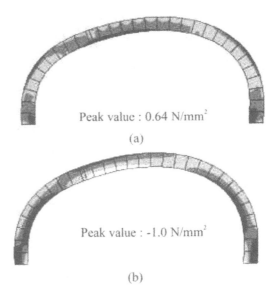

Peak value : 0.64 N/mm²

(a)

Peak value : -1.0 N/mm²

(b)

Figure 5. Results of linear elastic finite element analysis: (a) maximum and (b) minimum principal stresses.

The loads considered in the analyses include, as dead load, the weight of the arch (volumetric weight $\gamma = 20 \, kN/m^3$) and fill ($\gamma = 15 \, kN/m^3$), and, as live load, a point load of 10 kN at quarter span. The dead load is applied first, followed by the monotonic application of the live load up to failure.

The following analyses have been carried out for these structures: linear elastic finite element analysis, kinematic limit analysis, limit analysis for the calculation of the so-called geometric safety factor, non-linear physical finite element analysis and non-linear combined physical/geometrical finite element analysis.

2.2.1 Linear elastic finite element analysis

Eight-noded plane stress elements, combined with six-noded line interface elements, were adopted for the analysis. The following properties have been assumed in the analyses: Unit – Young's modulus of $10 \times 10^3 \, N/mm^2$ and Poisson ratio of 0.2; Interface – Normal stiffness $2.4 \times 10^3 \, N/mm^3$ and transverse stiffness of $1.0 \times 10^3 \, N/mm^3$. The results of the analysis are shown in Figure 5.

In order to establish the safety of the structures being considered, it is necessary to introduce the concept of "allowable maximum stress". Here, a maximum allowable tensile stress of $0.2 \, N/mm^2$ and unlimited compressive stress were assumed. For the adopted mesh discretisation, the obtained safety factor is 0.3. It is stressed that, as a general rule, the peak values of the stresses depend on the mesh discretisation and the proposed procedure is debatable.

2.2.2 Limit blocky analysis

The following properties have been assumed for the joints: zero tensile strength, unlimited compressive stress strength, friction angle of 37° and zero dilatancy. The results are given in Figure 6.

Safety of the structures is automatically evaluated when using limit analysis, resulting in a kinematic

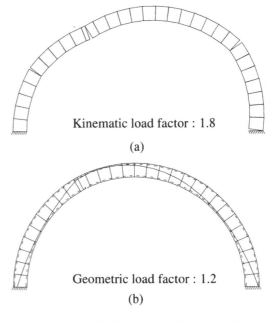

Kinematic load factor : 1.8

(a)

Geometric load factor : 1.2

(b)

Figure 6. Results of limit analysis: (a) failure mechanism and (b) thrust line for "geometric load factor (internal dotted arch).

safety factor of 1.8. Another popular concept is the so-called "geometric safety factor", Heyman (1969), which represents the ratio between the actual thickness of the arch and the minimum thickness of an internal arch with the original span and able to resist

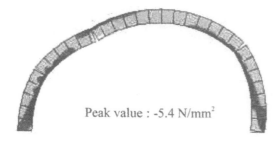

Peak value : -5.4 N/mm²

Figure 7. Results of non-linear elastic finite element analysis: minimum principal stresses and collapse mechanism.

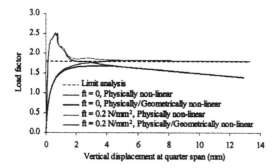

Figure 8. Load-displacements diagrams for the different non-linear analyses and limit analysis safety factor.

the original applied load. This geometric safety factors is 1.2.

2.2.3 Non-linear finite element analysis

Both physical and combined physical/geometrical non-linear behavior were analyzed. The results of the physically non-linear analyses are shown in Figure 7. Due to cracking of the joints, the compressive stresses are much higher than in Figure 5.

For the sake of completeness, a second non-linear analysis has been carried out, adopting some limited, yet non-zero, strength. For this purpose, tensile strength f_t of 0.2 N/mm², cohesion of 0.3 N/mm² and fracture energy of 0.1 N.mm/mm² were assumed in the joints. Both physical and combined physical/geometrical non-linear behavior were analyzed.

Safety of the structures is automatically evaluated when using non-linear analysis. For zero tensile strength and physically non-linear analysis, the ultimate load factor is the same as for limit analysis.

2.2.4 Comparison of results

Figure 8 presents the load-displacement diagrams for the non-linear analyses calculations together with the ultimate load factors for kinematic limit analysis and

Table 1. Safety factors for the different analyses considered.

Approach/Analysis type	Safety Factor
Allowable stresses ($f_{ta} = 0.2$ N/mm²)	0.31
Kinematic limit analysis	1.8
Geometric safety factor	1.2
$f_t = 0$, Physically non-linear	1.8
$f_t = 0$, Physically and geometrically non-linear	1.7
$f_t = 0.2$ N/mm², Physically non-linear	2.5
$f_t = 0.2$ N/mm², Phys. and geomet. non-linear	2.5

Table 1 indicates the safety factors obtained in the different analyses.

The results obtained allow to conclude that:

– Linear elastic analysis requires information on the elastic properties of the materials and maximum allowable stresses, resulting in information on the deformational behaviour and stress distribution of the structure. Limit analysis requires the strength of the materials, resulting in information on the failure mechanism of the structure. Non-linear analyses require the elastic properties, the strength of the materials and additional inelastic information (the stress-strain diagrams), resulting in information on the deformational behaviour, stress distribution *and* on the failure mechanism of the structure.

– The "safety factors" associated with a linear elastic analysis (and a maximum allowable stress) and with a static limit analysis (the so-called geometric safety factor) cannot be compared with the remaining safety factors. When such particular approaches are used, special care should be adopted if structural safety is a relevant issue;

– For the simple structures presented, physically non-linear analysis and kinematic limit analysis yield the same failure mechanisms and safety factors, if a zero tensile strength is assumed. In complex structures, when using simple hand calculations, it might be difficult to find the correct failure mechanism by using limit analysis. Additionally, if geometrically non-linear behaviour is also included in the analysis, the safety factor is reduced by around 10%, for the arch studied;

– The consideration of a non-zero, yet low and degrading, tensile strength increased the safety factors considerably. Therefore, when using non-zero tensile strength, special care might be necessary in real case applications, mainly because: (a) tensile strength is difficult to assess and (b) tensile strength might be severely reduced at critical locations. It is noteworthy to stress that different failure mechanisms might be triggered in the analyses. In the particular case of the semi-circular arch, this is a minor difference and the non-zero tensile strength

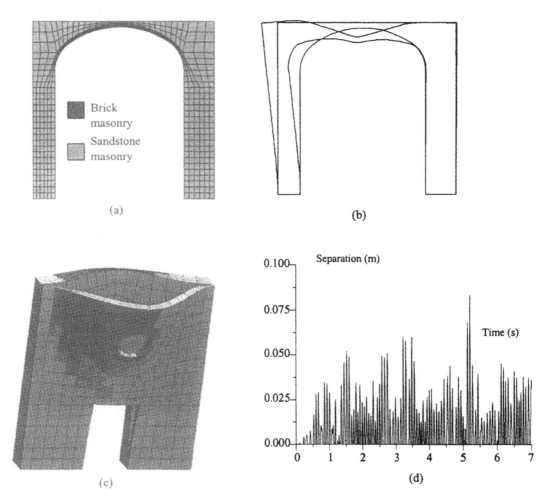

Figure 9. Examples of application of different sophistication levels to real case studies: (a,b) Lethes theatre (c. 1650): the vault is fully cracked along the key and exhibits very large abnormal vertical displacements. A plane stress model of the vault, walls and infill was considered adequate for the analysis; (c,d) Church of Saint Christ in Outeiro (1698–1738): damage of the structure is localized in the main façade and in the choir. Plane stress and shell homogeneous analyses of the façade indicated that no damage should be expected. Therefore, the external and internal leaves of the façade had to be modeled.

solution converges to the zero tensile strength solution, upon progressive tensile strength degradation. The non-zero tensile strength final solution always exhibits higher strength than the zero tensile strength solution. Finally, when using non-zero tensile strength, the consideration of geometrically non-linear behaviour seems to affect only marginally the calculation of the safety factors;

– The post-peak response obtained in a non-linear analysis is an important issue, when addressing safety factors. Indeed, brittle responses are dangerous and, from a reliability point of view, yield lower safety margins. In specific cases, it may be sensible to adopt as safety factor the residual plateau found

in the physically non-linear analyses. Of course, in the case of combined physically and geometrically non-linear analysis, no plateau will be found;

– *The fact that different methods of analysis lead to different safety factors and different completeness of results is not a sufficient reason to select one method over the other.*

3 SELECTED CASE STUDIES

Figures 9 and 10 present selected case studies with the objective of demonstrating engineering applications of different analysis tools and approaches. Table 2 indicates a

245

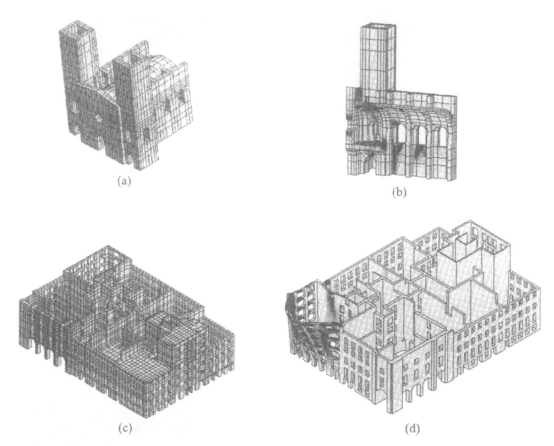

(a)

(b)

(c)

(d)

Figure 10.   Examples of application of different sophistication levels to real case studies: (a,b) Sanctuary of São Torcato (1868- to present): the structure exhibits extensive cracking in the main façade, choir and transverse walls. In order to justify the damage in the structure, a three-dimensional model of the main nave was necessary; (c,d) "Baixa Pombalina" (18th century historical center of Lisbon): safety assessment, with respect to earthquake loading. The finite element model includes shell elements, to represent (new) RC slabs, composite slabs and most masonry vaults on ground level, and volume elements, to represent masonry walls, masonry arches and vaults.

summary of the relevant aspects for the different case studies. It can be seen that a multiplicity of approaches have been adopted including: both two-dimensional and three-dimensional models; both structural details, structural parts and full constructions; both static and dynamic loading; both using and neglecting soil-structure interaction; both combined physical and geometrical non-linear behavior, or solely physical non-linear behavior; both non-linear analysis, kine-matic limit analysis and thrust-line analysis. The motto governing the decision to select one approach over the others was: *Prefer simplicity over complexity, using the analysis tool believed adequate*. Here, it must be stressed that, as pointed out above, linear elastic behavior was never considered in the analysis. This is believed to be a key issue in masonry modeling.

4   CONCLUSIONS

Constraints to be considered in the use of advanced modeling are the cost, the need of an experienced user/engineer, the need for validation and the use of the results. Cost and the need of an experienced used seem straightforward arguments. The need of validation is a key issue, meaning that the results of a complex analysis, using a complex model, might be useless if the results are not validated against in situ observations, such as cracking, crushing, displacements, flat-jack tests, etc.

Obtained results are usually valuable in understand-ing the structural behavior of the constructions. But it should not be concluded that non-linear analysis is the sole numerical tool to be used for all constructions,

Table 2. Comparison of the presented case studies.

Case study	Details of analysis	Type of analysis	Remarks
Lethes Theatre	Two-dimensional; Plane stress; Structural details	Geometrically and physically non-linear; Static loading	Damage is very localized in a barrel vault, a basically two-dimensional construction. Given the large displacements at the key of the vault and the arched shape of the construction, physically and geometrically non-linear behaviour becomes relevant.
Holy Christ	Three-dimensional; Volume elements; Structural details	Physically non-linear; Dynamic loading with time integration	Damage of the structure is localized in the main façade and choir. Previous analyses assuming a homogenous material indicated no damage, being necessary to introduce the composite constitution of the façade. The separation between the leaves of the wall (up to 0.10 m) can only be analysed by a dynamic non-linear analysis.
São Torcato	Three-dimensional; Volume elements; Half construction	Physically non-linear; Static loading with vertical actions. Soil-structure interaction	Main nave presents distributed cracking and transept presents no damage. Geotechnical survey indicated towers located in a poor quality embankment while transept located in good quality rock. A 3D model of the nave was adopted for the soil-structure interaction, while the transept was disregarded.
Downtown block	Three-dimensional; Volume and shell elements; Full construction	Physically non-linear; Static loading with vertical and horizontal actions	A 200.000 degrees of freedom example is shown to assess the seismic safety of a complete block in a historical centre. The correctness of the results should be addressed very carefully due to the size and complexity of the model. Here, a simple building was analysed first, to assess the quality of the large model.

by all engineers. As a rule, advanced modeling is a necessary means for understanding the behavior and damage of (complex) historical constructions but this requires specialized consulting engineers and it is less effective for designing strengthening. On the other hand, simplified modeling, such as limit analysis (kinematic method), is a great tool for everyday constructions, such as the buildings in historical centers.

The key message in the present paper is *"prefer simplicity over complexity and adopt an analysis tool, that can be validated and assessed by the user.* General recommendations are:

- It is better to model structural parts than complete structures;
- Do not use full structure three-dimensional models unless it is necessary;
- Avoid making linear elastic calculations for historical constructions.

REFERENCES

Feenstra, P.H. 1993. *Computational aspects of biaxial stress in plain and reinforced concrete.* Delft: Delft University of Technology.

Heyman, J. 1969. The safety of masonry arches. *Int. J. Mech. Sci.* 11: 363–385.

Lourenço, P.B. 2001. Analysis of historical constructions: From thrust-lines to advanced simulations. In Paulo B. Lourenço & Pere Roca (ed.), *Historical Constructions: Proc. intern. symp. Guimarães, 7–9 November 2001.* Guimarães: University of Minho.

Orduña, A. & Lourenço, P.B. 2002. Cap model for limit analysis and strengthening of masonry structures. *J. Struct. Engrg.*. submitted for publication.

Finite Elements in Civil Engineering Applications, Hendriks & Rots (eds.)
© 2002 Swets & Zeitlinger, Lisse, ISBN 90 5809 530 4

Analytical implications on in-plane behavior of unreinforced masonry walls

J.H. Kim
Department of Architecture, Ajou University, Suwon 442-749, South Korea

ABSTRACT: A series of unreinforced masonry (URM) walls were analytically investigated by FEM for a limited version of seismic in-plane performance. For this URM walls were assumed to be continuum and modeled as isotropic plane stress elements within which the nature of cracking was propagated. Accordingly, cracking mode of behavior in URM was modeled by smeared-crack approach. Total of 70 cases were considered for various parameters such as axial load ratio, aspect ratio and effective section area ratio due to the existence of opening. The analytical results indicate that these parameters significantly and interactively influence the ultimate strength of URM walls. Finally, it is suggested that the response modification factor for URM adopted in the current Korean Standard should be validated considering various forms of brittleness and probable failure modes in URM.

1 INTRODUCTION

1.1 *Background*

Unreinforced masonry (URM) had long been the most typical of low-rise buildings over the world due to wide availability of its constituent materials and ease in construction. In particular, URM built with reinforced (RC) tie beam-slab system as shown in Figure 1 has been popular in Korea for houses, schools and low-rise commercial buildings. This type of construction is considered quite stiff due to almost rigid diaphragm action of RC tie beam-slab system.

Historically, URM buildings have been reported to be the major victim of seismic attack by various reconnaissance report upon the occurrence of earthquakes (Mehrain 1991, Khater 1993, Bruneau 1994). As a result, the market of new construction with URM in Korea as well as other countries in the world has been continually downsized. However, it should be pointed out that despite the decrease in new construction, not a small number of URM buildings still exist over the world. Therefore, engineers and governing authorities became to concern rather seismic evaluation and retrofit of existing URM buildings than new construction.

The common modes of failure of URM buildings repeatedly observed in the earthquake-attacked area are: (1) anchorage-related failure; (2) in-plane failures; (3) out-of-plane failure; (4) combined in-plane and out-of-plane effects; and (5) diaphragm-related failure (Bruneau, 1994). The anchorage-related failure may

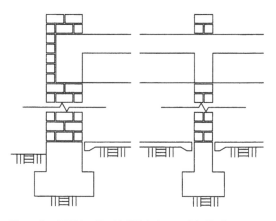

Figure 1. URM wall with RC tie beam-slab diaphragm.

result in slippage of walls on floors or foundations, or between bricks and bed mortar. The in-plane failures include flexure, shear, sliding and rocking. These modes of failure are quite dependent upon the quality of materials, workmanship and aspect ratio of URM walls. Particularly, the combined in-plane and out-of-plane effects must be considered for a real probable failure scenario. Since earthquakes include multi-directional components, URM walls subjected to an earthquake will be loaded both in-plane and out-of-plane directions at the same time. As a result, the degradation of seismic resistance will be speeded up until one of failure modes prevails. The diaphragm-related failures involve the dynamic excitation of URM wall in out-of-plane direction due to the flexibility of diaphragms. However, the RC tie beam-slab system, used in almost all URM buildings in Korea, is considered stiff enough to tie walls together, as long as the anchorage is sufficient.

1.2 *Research objectives and scope*

FEM analysis is useful to investigate the stress distribution in URM buildings with complicated configuration. Actually, there is a reported case study of historical earthquakes causing partial damages over the Roman Colosseum by utilizing the linear elastic FEM analysis (Croci & D'Ayala 1993). The linear elastic FEM analysis may be effective within the elastic range but is not enough to predict the behavior of URM walls over the ultimate state. In order to investigate the post-cracked behavior beyond the elastic limit, the nonlinear inelastic FEM analysis is required. However, it should be noted that no FEM program has been yet developed to accommodate all probable complicated failure modes mentioned previously (Bruneau 1994).

For simplicity, the scope of the present study is limited to the in-plane behavior of URM emphasizing the stress-strain relationship of URM as a governing parameter. In this notion, the purpose of present study is two-fold: (1) to investigate the appropriateness of DIANA (2000) as an analysis tool for URM walls; and (2) to evaluate the code-specified earthquake loading for URM buildings.

2 FE MODELING OF URM BUILDINGS

2.1 *Analytical model of URM walls*

URM walls considered for in-plane analysis consist of one-story 1.0B (200 mm thick) cement brick wall panel confined by 1000×450 mm RC beam-slab system at top and bottom as shown in Figure 2. The lower RC beam-slab is restrained against the horizontal and vertical translations while the upper one transfers gravity and lateral loading. The considered variables for analysis are axial load ratio ($P/f'_m A_g$), effective section area ratio (A_e/A_g) and aspect ratio (H/L), resulting in total of 35 cases as summarized in Table 1, where P = gravity load; f'_m = compressive strength of masonry; A_g = horizontal section area of URM wall; A_e = reduced section area of URM wall due to opening; H and L = height and length of URM wall, respectively.

For modeling RC beam-slab system and masonry wall, 8-noded quadrilateral plane stress element CQ16M is used. The grid pattern with mesh size of 150×150 mm is generated for convenience. This grid pattern rather than the real brick-mortar lines may not affect the accuracy of analytical results when the mesh size is small enough, since the URM is assumed as continuum.

There are two models to take the nonlinearity due to cracking into account: smeared crack model and discrete crack model. In the smeared crack model, the effect of cracking and nonlinearity are included through all elements in form of material stress-strain relationship. In the discrete crack model, the masonry layer is considered as isotropic elastic continuum,

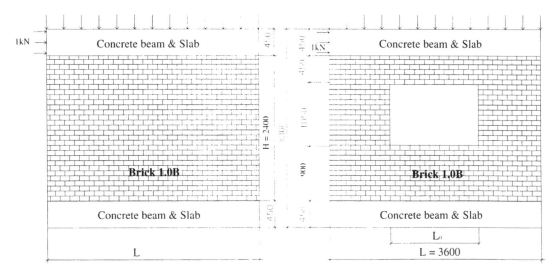

Figure 2.   URM wall models for analysis.

while the mortar is considered as the interface or gap element. When the stress exceeds the tensile strength of masonry at the cracking surface, the elements start sliding or separating against each other. In the current

Table 1.   Description of URM walls for analysis.

Model	L (m)	H/L	A_g (m^2)	A_e/A_g	$P/f'_m A_g$
S1L00	0.9	2.67	0.18	1.0	0.0
S1L025	0.9	2.67	0.18	1.0	0.025
S1L05	0.9	2.67	0.18	1.0	0.05
S1L10	0.9	2.67	0.18	1.0	0.1
S1L20	0.9	2.67	0.18	1.0	0.2
S2L00	1.8	1.33	0.36	1.0	0.0
S2L025	1.8	1.33	0.36	1.0	0.025
S2L05	1.8	1.33	0.36	1.0	0.05
S2L10	1.8	1.33	0.36	1.0	0.1
S2L20	1.8	1.33	0.36	1.0	0.2
S3L00	2.7	0.89	0.54	1.0	0.0
S3L025	2.7	0.89	0.54	1.0	0.025
S3L05	2.7	0.89	0.54	1.0	0.05
S3L10	2.7	0.89	0.54	1.0	0.1
S3L20	2.7	0.89	0.54	1.0	0.2
S4L00	3.6	0.67	0.72	1.0	0.0
S4L025	3.6	0.67	0.72	1.0	0.025
S4L05	3.6	0.67	0.72	1.0	0.05
S4L10	3.6	0.67	0.72	1.0	0.1
S4L20	3.6	0.67	0.72	1.0	0.2
O1Q00	3.6	0.67	0.54	0.75	0.0
O1Q025	3.6	0.67	0.54	0.75	0.025
O1Q05	3.6	0.67	0.54	0.75	0.05
O1Q10	3.6	0.67	0.54	0.75	0.1
O1Q20	3.6	0.67	0.54	0.75	0.2
O2Q00	3.6	0.67	0.36	0.5	0.0
O2Q025	3.6	0.67	0.36	0.5	0.025
O2Q05	3.6	0.67	0.36	0.5	0.05
O2Q10	3.6	0.67	0.36	0.5	0.1
O2Q20	3.6	0.67	0.36	0.5	0.2
O3Q00	3.6	0.67	0.18	0.25	0.0
O3Q025	3.6	0.67	0.18	0.25	0.025
O3Q05	3.6	0.67	0.18	0.25	0.05
O3Q10	3.6	0.67	0.18	0.25	0.1
O3Q20	3.6	0.67	0.18	0.25	0.2

version of DIANA, it is understood that URM wall should be idealized by either smeared crack model or discrete crack model but not by both at the same time. However, the experimental evidence (Kim & Kim 2001) indicates the possibility that the inclined cracks keep developing while sliding, which requires both models to be accommodated in an analysis segment.

The smeared crack model is adopted in this study because the discrete crack model requires the knowledge of proper location of interface element before analysis for attaining the characteristics of ultimate behavior of URM (Mosalam 1998).

2.2   Material property

Since the RC beam-slab system must remain elastic, it is assumed that $E_c = 28000$ MPa and Poisson ratio $\nu = 0.15$. For the post cracked property of URM, Rankine/Von Mises and Total strain-based models are considered as shown in Figures 3 and 4, respectively. This is to compare the biaxial stress effect (Rankine/Von Mises model) with the uniaxial stress effect (Total strain-based model). Therefore, two different material properties for 35 cases in Table 1 result in the total of 70 cases of URM walls for analysis.

The mechanical property of URM walls for analysis is: compressive strength $f'_m = 12$ MPa; modulus of elasticity $E_m = 12000$ MPa; and Poisson ratio $\nu = 0.15$. Determination of compressive strength of URM is based on the research report to the city of Seoul (2001). Since the tensile strength of URM is known to be $0.03f'_m \sim 0.09f'_m$ (Tomazevic 1999), $f'_m = 0.06f'_m$ is assumed, resulting in 0.72 MPa. The tensile fracture energy of URM is assumed $G_f = 0.15$ N·mm/mm^2 while the compressive one $G_c = 1.5$ N·mm/mm^2. The residual shear rigidity after cracking is assumed 5% of elastic one and thus $\beta = 0.05$ is used.

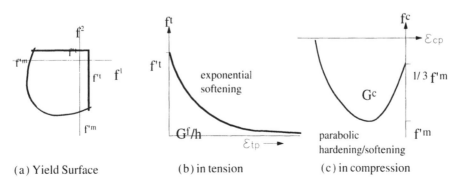

(a) Yield Surface   (b) in tension   (c) in compression

Figure 3.   Rankine/Von Mises failure criteria.

2.3 *Analysis*

One of important things in performing nonlinear analysis is the way of loading. This is because the ultimate strength of the analytical model as well as the mode of strength degradation is not known until analysis ends. In this study, the pushover analysis is performed to obtain the force-deformation curves with the loading size determined automatically at every loading step by arc-length control scheme built in DIANA. For this 1 kN of lateral loading is initially applied at the upper RC beam-slab system from left to right upon completion of gravity load analysis. In this way, the load factor itself can be interpreted as the lateral loading without a separate computational process.

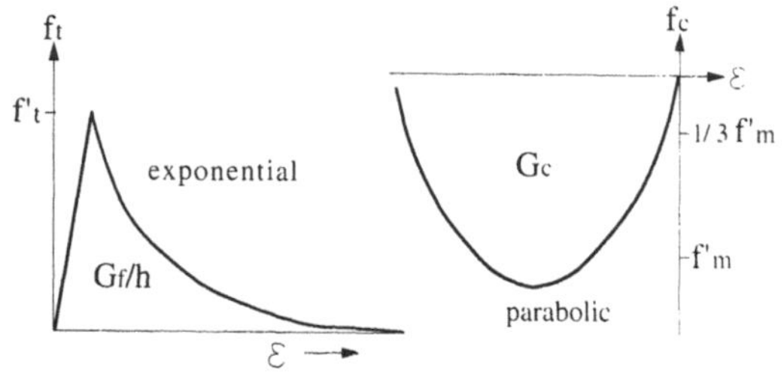
ft
f't
exponential
Gf/h
ε
fc
ε
1/3 f'm
Gc
f'm
parabolic

(a) in tension (b) in compression

Figure 4. Stress-strain relation of Total strain-based model.

3 ANALYSIS RESULTS

3.1 *Distribution of principal stresses*

Figure 5 presents the distribution of principal stresses before and after cracking for S4L10 (solid URM wall) and O2Q10 (wall with opening). The primary difference in stress distribution between before and after cracking is the influence of tensile stresses. Before cracking, the compressive stresses are balanced with the tensile stresses to form a load transferring mechanism. After cracking, the tensile strength of URM is weakened to diminish and thus the lateral

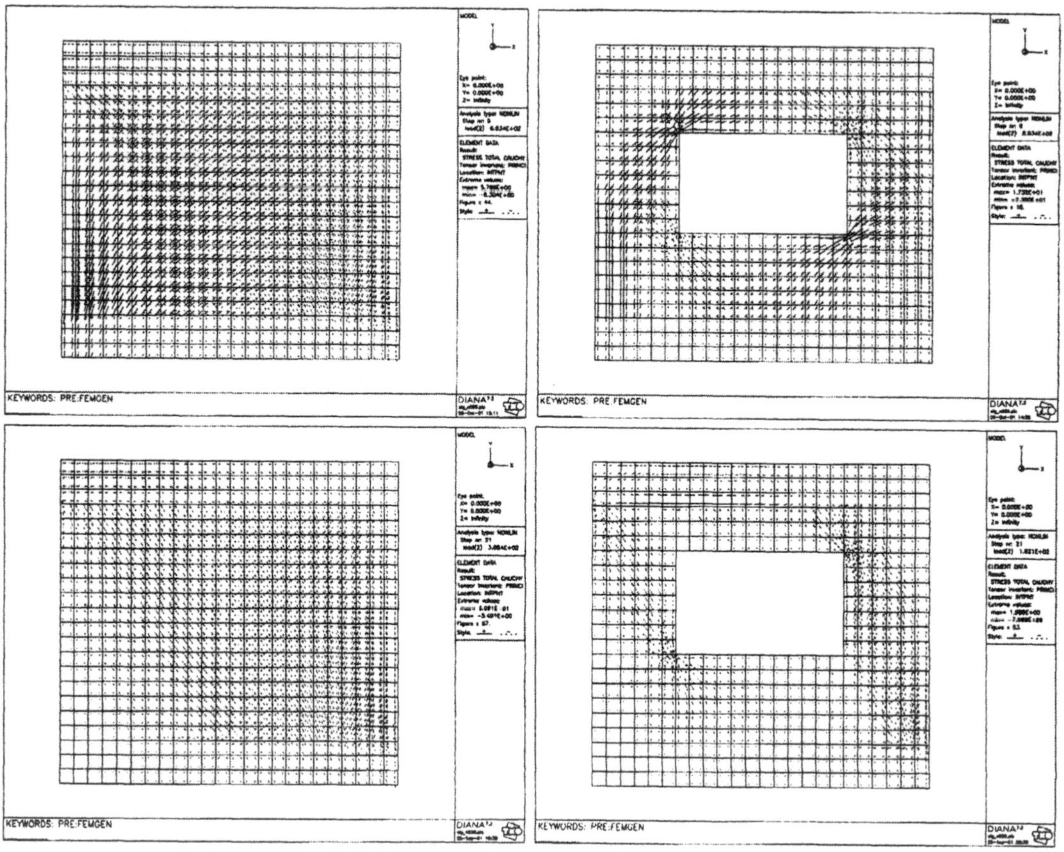

Figure 5. Distribution of principal stresses over URM walls before and after cracking.

load is transferred through the compressive stress field. The stress distribution in piers between opening after cracking resembles that of RC squat columns. The stress distribution before and after cracking indicates the possible implementation of strut-tie modeling for URM walls.

3.2 *Ultimate strength of URM walls*

Figures 6 and 7 shows the force-deformation relations obtained from the FE analysis of URM walls with various geometrical configurations. It is noted that no remarkable difference exists between Rankine/Von Mises model and Total strain-based model.

In Figure 6, solid URM walls with larger aspect ratio seem to behave as if they are ductile. But this is because of the reduced stiffness resulting from the slenderness of URM walls. The effect of axial load ratio on the base shear capacity is not so significant for slender walls, while the base shear capacity decreases with larger degree as the axial load ratio increases for squat walls. The lateral displacement required for attaining the ultimate strength increases due to the larger reduction of stiffness under a constant aspect ratio as axial load ratio increases. But this tendency becomes weak as aspect ratio decreases.

In Figure 7, the behavior of URM walls with various opening sizes is compared to that of the solid wall with the same aspect ratio. The base shear capacity

decreases with decreased effective section area but the lateral deformation to attain the ultimate strength does not make a big difference between different effective section area ratios. The elastic stiffness and the ultimate strength are maintained without reduction until the effective section area reduces to 75%. When the effective section area reduces further to 50% and 25%, both elastic stiffness and ultimate strength decrease remarkably. In particular, when axial load ratio is 0.2, about 30% of base shear capacity has been lost for $A_e/A_g = 0.75$ and the base shear capacity becomes negligible when further reduction in effective section area is made.

4 DISCUSSION

4.1 *Comparison to the code-specified demand*

Korean Standard (AIK 2000) and Eurocode 8 (1996) are considered to compare the ultimate strength of URM walls obtained by FE analysis with the code-specified demand. The design spectra corresponding to both code provisions for URM buildings on the rock are depicted in Figure 8. The difference of required base shear spectra between the two codes seems considerable and is attributed to the difference of maximum limit of dynamic factor (1.75 in Korean Standard and 2.5 in Eurocode) and the difference of

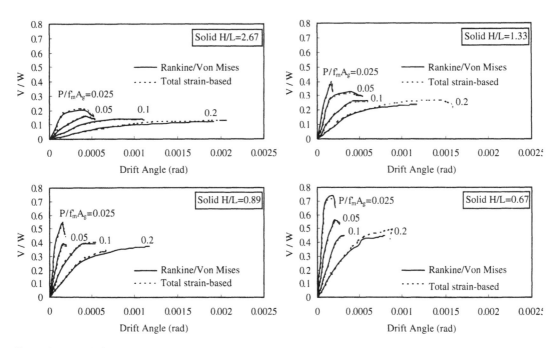

Figure 6. Force-deformation relationship of URM solid walls resulting from FE analysis.

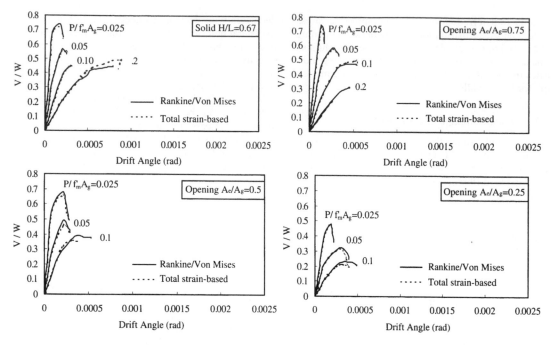

Figure 7.   Force-deformation relationship of URM walls with opening resulting from FE analysis.

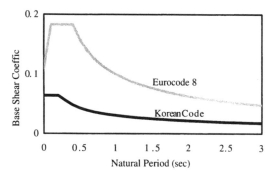

Figure 8.   Code-specified design spectrum.

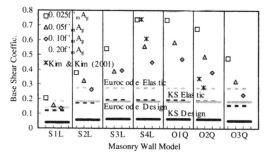

Figure 9.   Comparison of ultimate strength to demand.

response modification factor (3.0 in Korean Standard and 1.5 in Eurocode). The required base shears for various URM wall groups as per the codes can be calculated by computing natural periods of individual wall. The ultimate strengths of various URM walls obtained by FE analysis are compared to the code-specified demand in Figure 9. In the figure, the code-specified demand is expressed in thicker horizontal lines and the corresponding elastic demand in thicker dashed lines.

All solid URM walls satisfy the code requirement in strength. However, it should be noted that the response modification factor $R = 3.0$ adopted in

Korean Standard should be validated. The analysis results presented in Figures 6 and 7 indicate the approximate linearity or the limited ductility of URM walls. The presented behavioral characteristics imply the probable limitation of DIANA in capturing the descending branch of structures consisting of brittle material. Although there is an experimental report (Kim & Kim 2001) that the prototype URM walls fail in sliding mode between URM and RC beam-slab system, so that a large ductility seems attainable, it cannot represent the failure modes of URM walls in general. Therefore, the elastic demand-based comparison proceeds. Then it is found that either axial load ratio and/or aspect ratio should be limited with more or less degree upon the codes for URM walls with and without opening.

4.2 *Other experimental results*

Although the different failure mode from the one considered in this analytical study was observed, the reference to an experimental study (Kim & Kim 2001) with 4 full-scale URM walls may add some implication to the analytical results. The mentioned experimental work was performed over two solid URM walls and two URM walls with opening by pseudo-static cyclic lateral loading. The experimental results are also depicted in Figure 9. Solid walls failed in sliding show the base shear capacity close to the analytical results in the present study. Walls with opening show a considerable difference but still exceed the code-required design strength as well as the elastic demand. The comparison implies that aspect ratio, axial load ratio and effective section area ratio are the major parameters to determine the base shear capacity of URM walls.

5 CONCLUSIONS

In this study, the in-plane behavior of 70 URM walls with and without opening has been investigated for various axial load ratio, aspect ratio and effective section area ratio. These parameters are found to significantly influence the ultimate strength of URM walls. The conclusions drawn from the present study are summarized as follows:

- For solid URM walls, the effect of axial load over the base shear capacity is inversely proportional to the aspect ratio, but the effect of axial load ratio over the elastic stiffness is proportional to aspect ratio.
- For URM walls with opening, the elastic stiffness and ultimate strength are maintained without significant loss until the effective section area reduces to 75%. Further reduction of effective section area causes the significant loss of strength.
- In consideration of various forms of brittleness and the complexity of failure modes, the response modification factor of $R = 3.0$ recommended in Korean Standard should be validated.
- The present study implies the probable limitation of current version of DIANA in modeling failure of URM due to both stress-strain relation and sliding

between layers in one analytical segment, and in capturing the descending branch of structures consisting of brittle material.

ACKNOWLEDGMENTS

The present study was performed by partial financial support from the City of Seoul. The City University of Seoul granted the author the limited access to DIANA for analytical study. These aids are gratefully acknowledged.

REFERENCES

AIK 2000. *Standard deign loads for buildings* (in Korean); Architectural Institute of Korea. Seoul: Taerim.

Bruneau, M. 1994a. State-of-the-art report on seismic performance of unreinforced masonry buildings. *Journal of Structural Engineering* 120(1); ASCE: 230–251.

Bruneau, M. 1994b. Seismic evaluation of unreinforced masonry buildings – a state-of-the-art report. *Canadian Journal of Civil Engineering* 21(3): 512–539.

Croci, G. & D'Ayala, D. 1993. Recent developments in the safety assessment of the Colosseum. *IABSE Symposium – Structural Preservation of the Architectural Heritage*: 425–432.

DIANA 2000. *User's manual – Nonlinear analysis*. Delft: TNO Building and Construction Research.

Eurocode 1996. *Eurocode 8: Design provisions for earthquake resistance of structures*; DD ENV 1998-1-1. Brussels: European Committee for Standardization.

Khater, M. 1993. Reconnaissance report on the Cairo, Egypt, earthquake of October 12, 1992. *NCEER Bulletin* 7(1); National Center for Earthquake Engineering Research: 1–6.

Kim, J.H. & Kim, J.K. 2001. Experimental implications on seismic resistance of unreinforced cement brick walls. *Proc. of The 3rd Japan-Korea-Taiwan Joint Seminar on Earthquake Engineering for Building Structures*, 16–17 November. Taipei: SEEBUS.

Mehrain, M. 1991. A reconnaissance report on the Iran earthquake. *NCEER Bulletin* 5(1); National Center for Earthquake Engineering Research: 1–4.

Mosalam, K. 1998. Seismic behavior of infilled frames. *NCEER Bulletin* 12(2); National Center for Earthquake Engineering Research: 4–7.

Seoul 2001. *Seismic retrofit of masonry buildings* (in Korean); Technical report to the City of Seoul: Seoul.

Tomazevic, M. 1999. *Earthquake-resistant design of masonry buildings*. London: Imperial College Press.

Finite Elements in Civil Engineering Applications, Hendriks & Rots (eds.)
© *2002 Swets & Zeitlinger, Lisse, ISBN 90 5809 530 4*

Numerical analyses of the Bargower arch bridge

A.S. Gago, J. Alfaiate
ICIST – Instituto Superior Técnico – Lisbon, Portugal

A. Gallardo
Civil Engineer

ABSTRACT: The reported information of a full scale load test to collapse on a single span arch bridge, carried out within the aim of a research program developed by the *Transport and Road Research Laboratory* in the UK, is taken as reference to assess the ability of finite element models to reproduce the overall behaviour of stone masonry arch barrels. Two different bi-dimensional finite element models are used: (i) a simplified model where the infill behaviour is simulated with non linear springs acting in the horizontal direction and (ii) a refined model where the infill is discretized in the finite element mesh. In the latter case the non-linear behaviour of the soil is simulated with a plasticity model. In both cases the adopted masonry modelling consists of a discrete approach, in which the joints are represented by initial zero thickness interface elements and the regularly cut to shape voussoirs are represented by continuum linear-elastic elements. The numerical results are compared to the data obtained from the experimental test and the mechanical behaviour of the masonry arch bridge is discussed.

1 INTRODUCTION

The work reported in the present paper is carried out with the aim of assessing the capability and adequacy of the numerical models to trace the behaviour of a masonry arch ring throughout an entire loading range. The information collected from a set of full-scale load tests to failure undertook by the Transport and Road Research Laboratory (Hendry et al. 1986) provided a rare opportunity to validate the present method of structural analysis with real experimental data. The case studied is Bargower bridge, a semi-circular arch bridge with cut-to shape stone voussoirs.

The essential features of an arch behaviour are: (i) the fracture at the joints between stone voussoirs and (ii) the backing contribution of the arch springings, which is found to play a major role in the overall structural response of arch bridges.

The former feature is simulated in the finite element models using a discrete approach with non tensile strength interface elements. For the contribution of the backing two levels of discretization are adopted. First, as a simplified procedure, non-linear springs acting horizontally are used. Next, using Diana (Diana, 1999), a more refined finite element model is adopted, where the fill soil and the interface arch-soil are discretized in the mesh. The non-linear behaviour

of both soil and interface is modelled using plasticity Mohr-Coulomb yield criteria.

2 DESCRIPTION OF BRIDGE AND FULL-SCALE LOAD TEST

The experimental data provided by a full-scale load test to collapse conducted on a single span arch bridge is taken as reference to assess the ability of the finite element models presented herein to reproduce the overall behaviour of a stone masonry arch barrel. Present information was undertaken from a set of full-scale load tests carried out within the aim of a research program developed by the Transport and Road Research Laboratory in the UK (Hendry et al. 1986).

Bargower bridge arch ring had a semi-circular profile, built up of regularly cut to shape sandstone voussoirs. The bridge had a 16° skew angle and the arch ring courses were angled. Behind spandrel dressed stone facing walls, the presence of one meter thick inner side walls made up of rubble masonry was detected. Also, as revealed on bridge demolition, arch hauching consisted of crushed sandstone with clay traces, above which spandrel filling was constructed with a silty gravelly sand. The dimensions of the tested bridge are listed in Table 1.

Table 1.	Bargower bridge dimensions.	
Span (square) (m)	10.00	
Span (skew) (m)	10.36	
Rise at midspan (m)	5.18	
Arch thickness (m)	0.558	
Total width (m)	8.68	
Fill depth at crown (m)	1.20	

Table 2.	Interface element properties.	
Joint normal stiffness (kN/m^3)	10^{12}	
Joint tangential stiffness (kN/m^3)	10^{12}	
Tensile strength of joint (MPa)	0	

In the finite element models the adopted span is 10.36 m.

The bridge was found in moderate preservation state, presenting some defects: longitudinal cracks were observed, developing from springing to springing, away from one of the faces of the arch, probably due to frost damage; leaning outward parapets and spalling of individual stones were other observed defects.

Experimental data on sandstone specimens allowed to determine a compressive strength of 33.3 MPa and a Young modulus of 14.1 GPa. The evaluated stone self-weight is 26.8 kN/m^3 and, from available data, the fill self weight adopted in the present work is 20.0 kN/m^3.

The test procedure consisted of the incremental application of load, by means of hydraulic jacks, to a concrete strip cast in the road surface across the full width of the bridge and located at a third span, where the minimum failure load was expected. Ground anchors, set through the bridge deck, supplied the jack reactions against steel beams. Although other relevant information was obtained and studied, in the present paper attention is mainly focused on the displacement of the arch ring, measured by means of precision surveying, taken at the third span points.

In a first load test it was found that the load capacity of the jacks was insufficient to reach the maximum load that could be sustained by the bridge. The 100 tonne jacks were replaced by 200 tonne jacks and a second load test was performed. In this test the first visible sign of damage, material falling out of the longitudinal crack, occurred at 3400 kN. The maximum load applied was 5600 kN. Failure occurred by crushing of the arch beneath the load line. Prior to that, longitudinal cracks appeared in the arch ring beneath the inside edges of the spandrel walls. This indicates that the outer section of the ring, stiffened by the spandrel walls, was splitting from the (more flexible) middle section, stiffened by the fill.

3 THE SIMPLIFIED FINITE ELEMENT MODEL

The adopted masonry modelling consists of a discrete approach, in which the joints are idealized as zero-thickness interface elements and the regularly cut to shape stone voussoirs are represented by continuum linear-elastic elements. Fracture is allowed only at the masonry joints by means of a tension free model prescribed at the interfaces. Penalty functions are introduced in the constitutive relations in order to prevent overlapping at crack closure. The computational analysis of the masonry arch bridge is carried out by means of a finite element formulation developed by the authors, in which the units or blocks are represented by continuum plane stress isoparametric 4-node elements, whereas for the modelling of joints 4-node interface elements are used.

The material parameters used for the interfaces are given in Table 2.

For the linear elastic continuum elements the experimental Young modulus is used.

The arch ring is subdivided into the actual 53 equal wedge shaped masonry blocks, following the semicircular profile above mentioned, and three interface elements are used in each joint.

Backing contribution is modelled as a boundary condition, introducing horizontal spring supports at the outer faces of the arch. These springs behave under compression only with either linear or bi-linear elastic constutive laws. The horizontal subgrade coefficient is assumed to vary linearly along the depth of the arch, between 0 and 3 MNm^{-2}/m, which is considered a good estimate for the material used in the bridge construction.

Initial state of stress is contemplated applying nodal loads equivalent to the self-weight of stone voussoirs and fill and to the lateral earth pressure, which is considered due to initial steady-state ($k_0 = 0.45$).

Test loading is simulated by means of nodal forces at the extrados of the arch ring, according to the approximated elastic degradation of a line load applied at the surface of the bridge deck.

The explicit contribution of the spandrel walls is ignored and the absolute rigidity of the abutments is assumed. Geometric non-linear effects are not included in this analysis.

Five simulations are undertaken with this finite element model. In the first one, hereby designated as Case I, the backing contribution is only taken into account through the initial state of stress. The load-deflection relationship obtained beneath load line, in the vertical direction, is presented in Figure 1, together with other curves described forward in this section. The predicted relationship presents clearly a less stiff overall behaviour and an ultimate load far below the one experimentally obtained load, circa 5600 kN.

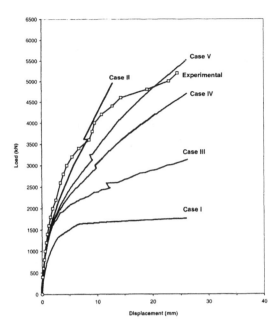

Figure 1. Vertical displacement of arch soffit beneath load line.

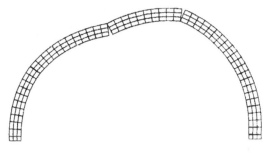

Figure 2. Case III – deformed mesh.

A plateau is achieved and an unlimited deformation is obtained, far beyond from the limits of admissibility of linear geometric analysis.

The large difference found between the experimental and numerical results indicates that the contribution of the stiffness cannot be ignored in the simulation of the arch behaviour.

In the second simulation, designated Case II, the interaction between the backing and the arch ring is attempted through the restrains provided by linear-elastic spring. Here, the fitting between numerical and experimental data improved qualitatively, showing the decisive contribution of arch haunching to the bridge behaviour. In fact, this result confirms the long known dependency of semi-circular arches from the sustaining effect of backing passive thrusts at springings (Crisfield and Page 1990). However, the global structural stiffness does not decreased significantly at the later stage of the deformation curve, giving rise to a much higher load than the maximum load obtained in the experimental test.

In the remaining three cases studied, bi-linear elastic law responses are adopted at the spring supports. The constitutive relations adopted for the springs shared the same inflection point at a predetermined contraction displacement limit and their common initial elastic stiffness is subsequently reduced to one-tenth (Case III), one-third (Case IV) and finally one-half (Case V). A better agreement with the experimental curve was achieved in the last situation, regarding all other cases.

Qualitatively, it is observed that the non-linear soil behaviour has an important contribution to approximate the actual behaviour of the arch bridge. Due to the limitations of this simplified approach no attempt is made to optimise the fitting with the experimental results, which is done in the most sophisticated model.

In all the analysis performed, the maximum compressive reaction in the springs is found to be always below the passive lateral pressure of the haunching material.

4 A MORE REFINED MODEL

As assumed in section 3, the adopted masonry modelling consists of a discrete approach, in which the joints are idealized as zero-thickness interface elements whereas the stone voussoirs and fill soil are represented by continuum elements.

Fracture is allowed only at the masonry joints and at the stone blocks-soil boundary by means of tension free and near-cohesionless frictional interfaces, respectively. The blocks are assumed to behave linear-elastically and for the fill soil a Mohr-Coulomb plasticity model is used.

The joints between voussoirs are discretized with five quadratic interface elements using a discrete crack model with a zero tensile strength Rankine criterion. For the Mode II behaviour zero shear traction and zero shear stiffness after cracking are assumed.

The joints between the arch and the fill soil are discretized with quadratic interface elements, using an associated plasticity friction model with a Mohr-Coulomb yield criterion and a near zero cohesion value.

For the continuum model the fill and the stone blocks are treated as two-dimensional elements under plane strain state. The condition of plane strain is assumed to be induced by the lateral restraint of the spandrel walls.

The stone blocks are discretized with 8-node isoparametric linear-elastic elements with a Young modulus of 14.1 GPa.

The fill is modelled as a near-cohesionless frictional material obeying the Mohr-Coulomb yield criterion with an associated flow rule. For the fill mesh triangular 6-node elements are used. A Young modulus of 40 MPa and a friction angle of 30° are adopted, which are found to be good estimations of the soil properties.

The road surface (figure 3) is modelled with triangular 6-node elements using, a Mohr-Coulomb plasticity model as well.

The material parameters used for the interfaces and for the continuum elements are given in Table 3 and result from the experimental data, when available, or from parametric tests performed with the purpose of fitting the experimental results. Large values for the interfaces normal stiffness are used to prevent overlapping at crack closure. In the voussoir joints a large value of the tangential stiffness is also used, since it is known that the joint slipping does not play an important role in the collapse of the arch (Heyman, 1982).

Similar to the simplified model, the explicit contribution of the spandrel walls is ignored.

The use of non-linear stress-strain relationships requires an incremental iterative solution procedure, i.e., the loads are applied step by step and equilibrium iterations are carried out at each increment until equilibrium is reached within acceptable limits. The arc-length method has proved to be very useful for situations where the standard Newton-Raphson method often fails, such as in the case of snap-back behaviour that is known to occur frequently in masonry structures (Rots, 1997; Almeida, 1999). In the tests presented, a monotonic increase of the main crack mouth opening displacement (CMOD) is enforced.

Four simulations are undertaken with this finite element model. In figure 4, the load-deflection relationships obtained beneath load line in the vertical direction are presented for all the performed simulations.

First the isolated arch is analyzed and the corresponding maximum load is compared with the collapse load obtained with the mechanism analysis. As expected, the collapse load obtained with this model is very far below the experimental result since, in this simulation, the backing contribution is limited to the vertical (self-weight) loads.

A second finite element model with the fill discretized in the mesh (figure 3) is used to simulate the contribution of the backing in the structural response. First a linear-elastic behavior is assumed for the fill and non-linear interfaces are used in the joints between voussoirs and between the arch and soil. Although,

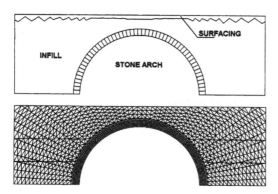

Figure 3.   Finite element mesh.

Table 3.   Finite elements properties.

Joints	Between voussoirs	Between arch and soil
Normal stiffness (kN/m^3)	0.233×10^{11}	0.233×10^{11}
Tangential stiffness (kN/m^3)	0.104×10^{11}	0.104×10^4
Tensile strength (MPa)	0	–
Cohesion (MPa)	–	1×10^{-3}
Friction angle	–	20°
Dilatancy angle	–	20°

Continuum elements	Fill	Surfacing
E (MPa)	40	5×10^3
Cohesion (MPa)	1×10^{-3}	1.443
Friction angle	30°	30°
Dilatancy angle	30°	30°

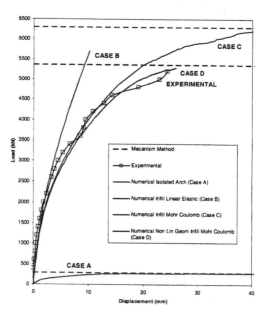

Figure 4.   Vertical displacement of arch soffit beneath load line.

in the first part of the test, a good fitting is obtained between the numerical and the experimental stiffness of the structure, the decrease of the stiffness, which is observed experimentally at a later stage, cannot be

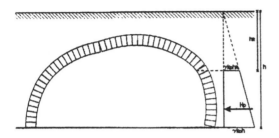

Figure 5. Mechanism model – passive impulse contribution.

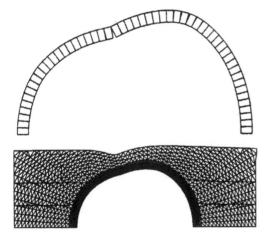

Figure 6. Deformed mesh.

reproduced with this numerical model. It is evident that a non-linear behaviour for the fill soil elements should be adopted to approximate correctly the collapse load of the structure.

A third model is adopted, in which the elements of the fill mesh obey the Mohr-Coulomb yield criterion. With this model, both geometrically linear and non-linear analyses are performed. Notwithstanding the fact that the results of these analyses do not fit exactly the experimental curve, a very good matching is obtained, specially in the case of the geometric non-linear analysis. In the latter case, the numerical obtained ultimate load is very close to the experimental one, whereas in the former a higher value of the ultimate load is found. As a consequence, it is shown that neglecting the geometric non-linear effects leads to a non-conservative estimation of the ultimate load. This difference is more pronounced as the arch span increases. For the dimensions of the Bargower bridge and for the stiffness assumed for the stone blocks and fill soil, the geometric linear model leads to an overestimation of 17% in the ultimate load.

In figure 4, the collapse loads obtained with the mechanism method are also presented. In this method, the contribution of the fill is taken into account considering the passive impulse of the soil acting only between the two hinges obtained at the right side of the arch (figure 5). Three different friction angles are used, 20°, 25° and 30°, corresponding to collapse loads of 5356 kN, 6299 kN and 7445 kN, respectively (the latter is not shown in figure 4). The mechanism analysis is a simplified and quick method for the estimation of the collapse load and it is a useful tool providing benchmarks for the numerical analyses. Given the simplicity of the assumptions of this method, non-conservative solutions are expected.

In figures 6 to 12, the numerical results obtained with geometrically non-linear effects are presented.

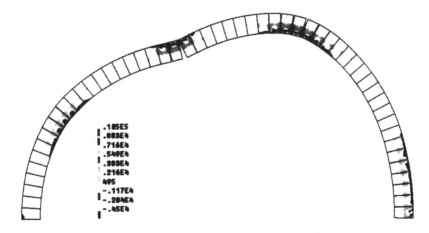

Figure 7. Principal stresses – σ_1 (Max = 0.122E05 kN/m^2 Min = -0.617E04 kN/m^2).

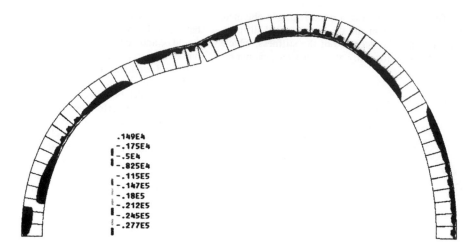

Figure 8.   Principal stresses – σ_{II} (Max = 0.474E04 kN/m² Min = −0.310E05 kN/m²).

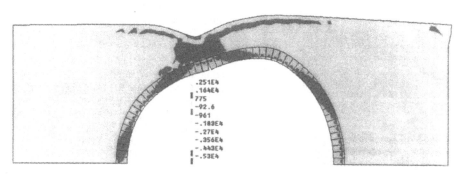

Figure 9.   Out of plane stresses – σ_{ZZ} (Max = 0.338E04 kN/m² Min = −0.617E04 kN/m²).

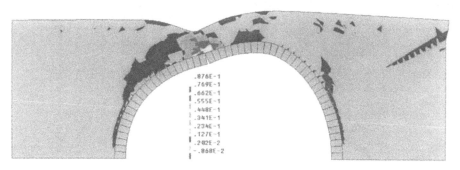

Figure 10.   Plastic deformations ε_{xx}.

In figure 6 a deformed mesh for the ultimate load is presented where the arch mechanism can be seen.

Since it is assumed a linear elastic behaviour for the blocks, it is important to check whether the compressive strength is reached. In figures 7 to 9 the principal stresses in the arch are presented and a maximum compressive stress below the compressive limit of 33.3 MPa is found.

In figures 10 to 12 the plastic deformations of the fill are presented corresponding to the expected plastification pattern of the soil.

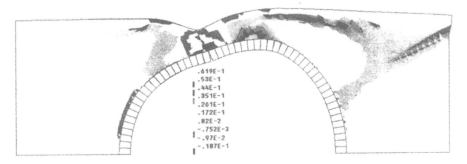

Figure 11. Plastic deformations ε_{yy}.

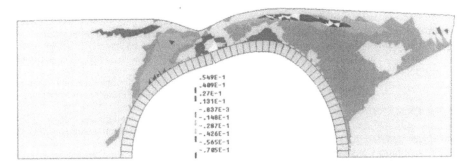

Figure 12. Plastic deformations ε_{xy}.

5 CONCLUSIONS

In this paper several numerical tests are performed sharing some basic assumptions of major relevance, since certain aspects of structural behaviour are unknown.

Although some relevant information is described in the experimental report, a correct knowledge of all the material parameters is not possible. Furthermore, the experimental tests were carried in two phases: first a monotonic loading with 100 tonne jacks and later a cyclic loading up to the collapse with 200 tonne jacks. As a consequence, and regarding both the reported information about the bridge materials and the expected damage of the materials during the cyclic loading process, some values are adopted in order to fit the experimental results.

As previously mentioned, the bridge had a skew angle, thus a three dimensional effect is expected. However, the small magnitude of the skew (16°) seems to affect only the response of the structure locally.

Experimentally it was observed that the failure followed splitting and crushing of some stone blocks under the load line. The former occurrences might have happened due to weathering and ageing effects and they are not simulated in the numerical models. The crushing of stone blocks, which is not detected in the numerical analyses, might have been due to local effects and does not seem to influence the global behaviour of the structure.

Plane stress and plane strain states are postulated in the numerical analyses and the contribution of the spandrel walls is not explicitly simulated. However, as mentioned in the report, longitudinal cracks appeared in the arch ring beneath the inside edges of the spandrel walls. This indicated that the outer section of the ring, stiffened by the spandrel walls, was splitting from the middle section stiffened by the fill. Consequently, the contribution of those walls, although relevant in intermediate stages of loading, namely inducing the plane strain state in the fill soil, may have little influence in the ultimate load. When load increases, damage occurs in the interface fill soil-spandrel walls and the ultimate load of the arch remains mainly dependent of the fill strength. In the more sophisticated model a plain strain state is assumed during the whole simulation and the spandrel walls stiffness is neglected; this contributes to some misfitting between numerical and experimental results, particularly at intermediate stages of loading.

As expected in a deep arch such as this one, the backing of the arch springings is found to play a major role in the overall structural response. The models where the backing influence is simulated strictly by vertical and horizontal self-weight loads (case I in figure 1 and case A in figure 4) lead to extremely reduced collapse loads. However, taking into account a linear elastic backing stiffness may result in ultimate overall stiffness and load which are clearly above the experimental ones. The effect of soil non-linearity is found to be most relevant. As a consequence, a model which accounts for stiffness degradation should be used to simulate the fill soil.

Although no attempt is made in the simplified numerical model to optimise the fitting with experimental results, the simplicity of mesh generation and the low computational effort of the analyses make of it an useful tool to obtain an approximate solution.

With the more sophisticated model a better simulation of the structure behaviour is achieved. In fact, good agreement is found between the numerical and experimental results, in particular when geometrically non-linear effects are included.

ACKNOWLEDGEMENTS

The work was developed within the research projects "Protecção do Património Construído da Acção dos Sismos" – PRAXIS I&D 3/3.1/CEG2606/95 and "Computational Strategies for the Protection of Historical Churches" – POCTI 33435/99 financed by the Fundação para a Ciência e Tecnologia.

REFERENCES

Alfaiate, J. and Gallardo, A., "Numerical simulations of a full scale load test on a stone masonry", Proceedings of the Historical Constructions 2001, Guimarães, Portugal (2001).

Almeida, J.R. and Alfaiate, J., "The study of confined masonry walls submitted to imposed displacements and temperature effects", Proceedings of the ECCM'99, European Conference on Computational Mechanics, Munich, Germany (1999).

Crisfield, M.A. and Page, J., "Assessment of the load carrying capacity of arch bridges", in A.M. Sowden (ed), The maintenance of brick and stone masonry structures, p. 81–113. E. & F.N. Spon (1990).

Crisfield, M.A., "Finite element and mechanism methods for the analysis of masonry and brickwork arches", Research Rep. 19, Transport and Road Research Laboratory, Crowthorne, Berks (1985).

Diana User's Manual – Release 7.2, Frits C. de Witte and Gerd-Jan Schreppers (ed), TNO Building and Construction Research, Delft, The Netherlands, (1999).

Hendry, A.W., Davies, S.R., Royles, R., Ponniah, D.A., Forde, M.C. and Komeyli-Birjandi, F., "Test on masonry arch bridge at Bargower", Contract. Rep. 26, Transport and Road Research Laboratory, Crowthorne, Berks (1986).

Heyman, J., "The masonry arch", Ellis Horwood Limited, UK (1982).

Page, J., "Load tests for assessment of in-service arch bridges", in C. Melbourne (ed), Arch bridges, p. 299–306. Thomas Telford, London (1995).

Page, J., "Load tests to collapse on masonry arch bridges", in C. Melbourne (ed), Arch bridges, p. 289–298. Thomas Telford, London (1995).

Structural masonry, an experimental/numerical basis for practical design rules, Edited by J.G. Rots, A.A. Balkema, Rotterdam (1997).

Finite Elements in Civil Engineering Applications, Hendriks & Rots (eds.)
© 2002 Swets & Zeitlinger, Lisse, ISBN 90 5809 530 4

Finite elements in the analysis of masonry structures

M. Šimunić Buršić
University of Zagreb, Faculty of Architecture, Zagreb, Croatia

Z. Žagar
University of Zagreb, Faculty of Civil Engineering, Zagreb, Croatia

ABSTRACT: In order to create virtual models of masonry structures closer to their real behavior, we used standard professional program for structural analysis (COSMOS/M2), but added specific conditions to simulate specific laws of masonry mechanics. These laws differ from those of modern structures. Being constructed of small elements (bricks, stone blocks), masonry has very low tensile strength – particularly in mortar joints. Therefore, tensile stresses cause cracking, and deactivate parts of structure. We simulated this behavior by using different finite element types for different materials. To simulate the lack of tensile strength in mortar joints of masonry structures, we used GAP finite elements not resistant to tension. In our preliminary 2D analysis, we used PLATE2D elements for modeling bricks or stone blocks, and GAP and TRUSS2D elements for modeling mortar joints. Such "virtual construction" can give us an insight into the method of building of ancient master-builders.

1 INTRODUCTION

The laws of mechanical behavior of masonry structures are more complex than those of modern materials, due to the poor resistance of masonry to tensile stresses.

The laws of the theory of elasticity were developed on the physical model of the ideal, continuous, "standard" anisotropic material, resistant equally to compressive and tensile stress, and thus can be applied only to materials with similar characteristics – such as metals. The behavior of other materials differs more or less from these laws.

2 MECHANICAL BEHAVIOR OF MASONRY

The behavior of masonry differs significantly from the behavior of a "standard" material model. First of all – masonry structures are not homogeneous and continuous. *Per definitionem*, masonry structures consist of small elements (such as bricks or stone blocks), connected with mortar. Tensile strength of bricks or stone is much lower than their compressive strength. But tensile strength of joints, made of mortar, is even much lower and may be neglected. Considering the fact that in many old buildings mortar in joints is fractured

due to the loading history, this simplification is not far from reality (Di Pasquale 1984).

On the assumption of total lack of tensile strength of masonry, Prof. Di Pasquale developed his mechanical theory of masonry structures. He proved that, due to the various actions, three states of stress may appear in masonry and similar "no-tension materials": in the region where all the principal stresses are compressive, structure acts as a structure of standard material, and all the laws of classical theory of elasticity can be applied. In the region where all three principal stresses would be tensile, which is not in accordance with the nature of material (i.e. no-tension material), fractures appear, and material is deactivated. In the region where one or two of the principal stresses are tensile, material is partly deactivated, and in this region specific laws of masonry mechanics have to be applied.

The formation and the distribution of the regions with three characteristic states of stress depend on actions on the structures. Different actions produce different distribution of states of stress. Fractured regions are deactivated, they do not participate in transferring the actions. Active structure changes according to the actions.

It means that for different actions different active structures are formed inside the visible, material

265

structure. Consequently, the law of superposition does not apply. (Di Pasquale 1992) As one of the basic laws of the theory of elasticity does not apply to masonry, then there must be other, specific mechanical laws which do. (Di Pasquale 1984)

3 EXPERIMENTAL COMPUTATIONAL PROGRAMS FOR THE MECHANICAL ANALYSIS OF MASONRY STRUCTURES

To simulate the specific mechanical behavior of masonry, several experimental computational programs, based on finite element method (FEM), were developed at the University of Florence, Italy. (Smars 1992) On the basis of standard solution (i.e. the calculation which would be a solution for a structure of standard material), these programs (CALPA, CALIPOUS, BLOC) perform iterations, eliminating at each step the deactivated part of structures, and performing next calculation with its active part. Such new structure has different value of stiffness, which means that the stiffness matrix has to be calculated at each step of iteration process. (Smars 1992) Of course, as these programs were developed at a university, without support of mighty software companies, these programs were just experimental, limited to 2D analysis (at least, at that time, ten years ago). The number of finite elements was also limited. Still, they enabled an insight into the distribution of active and deactivated regions and into the state of stress of the analyzed structures.

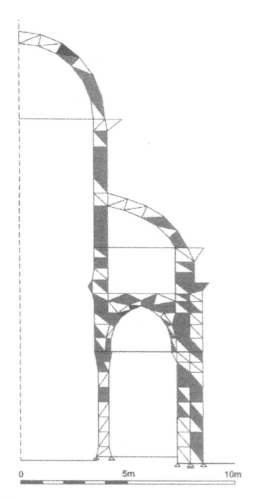

Figure 1. Structure of the Renaissance cathedral of Šibenik, analyzed with the program CALPA. Distribution of critical regions: totally deactivated (black) and partly deactivated (gray).

Figure 2. 2D finite element model of the structure of the cathedral of Šibenik Higher values of principal compressive stresses represented with darker shades (Šimunić Buršić 1995).

Prof. Dr. Salvatore Di Pasquale kindly allowed Mrs. Šimunić Buršić to use this program, to analyze the stone masonry structure of the cathedral in Šibenik, Croatia (Šimunić Buršić 1995). This Early Renaissance building, with its specific, original vault structure (its vault webs consist of large thin monolithic stone slabs, wedged into slender stone arches), was particularly suitable for the 2D analysis. In fact, as preliminary FEM analysis using ICES STRUDL system and M/STRUDL FEA proved, the thin web slabs transfer most of their load to the arches, and therefore, for dead load the simplification of 2D structure is not far away from reality (Šimunić 1989).

Of course, the finite element mesh in the CALPA FEA program was rather coarse (due to the mentioned limitation), and triangular finite elements (the only type of finite elements existing in this program at the time of the analysis), with the value of principal stresses given only in the centroids of finite elements, provide only a general idea of the state of stress field. Nevertheless, this picture is more accurate than the result of the FE calculation performed with sophisticated FEA programs, done without taking into account the specific behavior of masonry. Of course, we must be aware that we have analyzed a model, and not the real structure – but this is always true in the FEM analyses.

4 USING STANDARD FEM ANALYSIS SYSTEMS FOR THE MECHANICAL ANALYSIS OF MASONRY

To test another approach to the FEM analysis of masonry structures, we took a different path: we used existing professional programs, developed for structures of "standard" material, and used some types of finite elements (from the vast library of finite element types, now provided by software companies) to simulate the specific behavior of masonry structures. We made use of these new options to reconstruct the building process of masonry structures, in order to get the idea of its behavior closer to reality than when applying inadequate "standard" procedures.

4.1 *"Virtual construction" of masonry structures*

Nowadays, structural engineers and structural designers calculate structures by drawing first a 2D or 3D scheme of the structures (recently by using CAD), then they apply loading (actions) to the structure, and then calculate internal forces. According to these internal forces they determine dimensions of structural elements and of the whole structure, and control deformation – the state of usability. The use of CAD systems for drawing and of modern FE programs for structural analysis made this procedure a routine work, which tends to be automated. In these (routine)

calculations, the structure is usually assumed as a given fact, a completed building on which the prescribed loads and their combinations (according to regulations) are applied. Only in the recent time in the analysis of large structures (bridges, dams and similar structures, in the first place), the structures have been calculated in their phases of construction, step by step, in order to have control over the construction phases in every moment.

In reality, every building is constructed from foundation to the apex – this simple fact is ignored in usual, routine analyses. A large number of structures is still being calculated as simple assembly of structural elements (particularly in regions not subject to earthquakes). Indeed, earthquake, as the elementary action in earthquake-prone regions, forced structural engineers to consider structures as 3D unities – which should be a routine procedure nowadays, considering tools for structural analysis that we have at our disposal.

Similarly, it has been forgotten that it is possible to analyze structures during the construction, as they actually grow. It is considered a waste of time (and time is money).

In this research we explore the possibility of the analysis of construction (growth) of historic buildings, especially of ashlar. Stone blocks are real finite elements. Nevertheless, until recent time stone masonry structures were considered and analyzed as monolithic structures. Photo-elastic analyses on Plexiglas models (in which masonry structure is represented as monolithic) were carried out. Thus, the genesis of the structure itself was ignored.

Our basic assumption is that it is possible to model masonry structure better, closer to reality, if bricks or stone blocks in masonry are represented as finite elements or assemblies of finite elements, which ancient masters constructed into masonry structures.

Nowadays we are able to simulate the very construction of these historic structures, by assembling elements (stone blocks) in the same way in which they were assembled during the construction of the building in reality. We can assemble finite elements – 2D FE (PLANE) or 3D FE (SOLID or BRICK), representing stone blocks, following the same pattern as in real construction.

Joints between elements, made of mortar, can be simulated by introducing special types of connecting finite elements which can be formulated as not resistant to tension, e.g. GAP elements. Thus, real construction in time can be simulated.

4.2 *Finite element modeling of masonry*

Already in 1997 an attempt was made to create a finite element model of masonry, using standard COSMOS/M system, but simulating specific behavior of masonry by applying specific element types. Of course, nowadays one can choose from a number of FE programs.

While the above mentioned experimental programs (CALPA, CALIPOUS, BLOC), developed specially for the analysis of masonry structures, assume the characteristic of non-resistance to tension to the whole structure (i.e. masonry structure is modeled as an NTM continuum), in this research authors tried to simulate the specific behavior of masonry by introducing special types of finite elements to simulate mortar in joints.

Connecting finite elements not resistant to tension, if subject to tensile stress, "open", "break", simulating in this way opening of fractures. As fractures would indeed appear in the weakest point of structure (mortar joints), this model may be considered close to the reality.

Even though the tensile strength of bricks and stone is also much lower than their compressive strength, the tensile strength of mortar is much lower yet. Thus, if tensile stresses in masonry structure appear, mortar joints are the first to break.

Therefore, the authors simulated the specific behavior of masonry by creating a model consisting basically of two element types: for bricks or stone blocks usual plane 2D elements, such as PLANE2D or SHELL4T were used (for 2D analysis, which we considered appropriate for our initial research), and for mortar joints the GAP-FRICTION FE types, which enable the simulation of non-resistance to tension.

COSMOS/M FEA system offers GAP and TRUSS2D finite elements – their options include real constants (RC) which define resistance only to compression or only to tension, and options to define friction. GAP elements do not have material characteristics, and can be applied for formulation of contacts between geometric units (SHELL4T or PLANE2D in 2D analyses or BRICK or SOLID FE in 3D analysis). To improve the model of the finite element mesh, TRUSS2D finite elements can be used. Because of the stone arrangements within the built wall the FE should be carefully arranged to avoid overlapping or gaps and therefore accommodating the site situation and particularly the mortar layers within the stone wall.

The GAP-FRICTION element is a "flexible gap". This element may be used to model a variety of nonlinear spring or linear/nonlinear spring dampers. It can simulate:

– flexible contact surfaces (with or without forces),
– flexible one- or two-way springs, where a maximum distance is used to prevent the spring from deflecting to a length of zero or negative,
– the pre-load is useful for contact problems where an initial interference exists,
– linear/nonlinear one- or two-way spring dampers,
– nonlinear damping can be simulated,
– moving boundary problems,
– friction can be included.

In the case of regular GAP FE these elements do not affect the structural stiffness. Thus the structure stability must be supplied by soft springs. In the case where friction is not included, soft springs or TRUSS FE may be used instead (that is what we have done). Because of the limitation of the number of GAP elements to 400 in this program, at this time only structures limited in size could be analyzed.

4.3 *"Virtual experiment" on the model of a brick masonry arch*

In 1997 a "virtual experiment" was made: FE models of geometrically identical structures, but modeled in different ways, assuming different material conditions, were compared.

A brick arch was modeled first as a continuum of "standard" material (with SHELL4T FE), and then as assembly of layers of brick (PLANE2D FE) and mortar joints (simulated with GAP FE, and TRUSS2D FE used for simulating friction between layers of bricks). (Haiman & Žagar 1997)

The result of the analysis proved that these two structures, geometrically identical and loaded by same actions, behave differently. Although the displacement

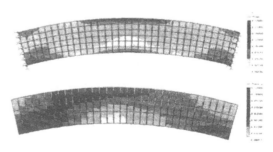

Figure 4. Principal stresses (von Mises) in brick masonry arch: modeled as standard material structure, and with evenly distributed load (above) and as no-tension resistant masonry structure, with uneven distribution of loads (below). The deformed shape displays fractures in NTM model (Haiman & Žagar 1997).

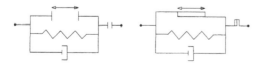

Figure 3. GAP finite element (COSMOS/M) types not resistant to tension and to compression.

pattern is similar in both structures, the stress pattern is completely different.

In the first FE model (arch simulated as being made of a "standard" material) tensile stresses appear, which is unacceptable for masonry structures.

In the second model (brick masonry with layers of mortar between brick layers) the most striking pattern was opening of some mortar joints, i.e. creation of fractures, causing discontinuity of material. This discontinuity can be closed only under another action, which would produce compressive stress in the fractured regions. In this case continuity would be regained – of course, only for compression.

Due to the fractures, caused by tensile stresses that would appear in this region, the stress pattern of the second model is completely different. Tensile stresses are almost eliminated, which proves that the modeling with GAP elements not resistant to tension approximates fairly well the behavior of masonry structures.

4.4 Examination of a model of masonry column

To examine whether the FE modeling described above matches the expected behavior and laws defined by Prof. S. Di Pasquale in his theory of masonry structures (Di Pasquale 1984), we created a very simple 2D model of masonry column, using COSMOS/M system GeoStar 2.0. modeler. Again, we used PLANE2D finite elements for modeling the bricks (stone) masonry blocks, and GAP elements not resistant to tension and TRUSS2D finite elements for modeling the mortar layers between them.

Due to this modeling blocks can transfer the actions to each other only where and when compressive stress appears. "Macro-stones" can move apart respectively – of course, according to the applied restrictions.

In this theoretical analysis we wanted to get as close to reality as possible. Therefore we modeled every single brick/stone block as a unit, and we refined the FE mesh in order to obtain more realistic model of joints and of boundary conditions. In this FE analysis GAP elements are not just orthogonal to the plane of joints – they are also disposed diagonally, in order to be able to transfer forces and stresses other than vertical or horizontal.

As we wanted to examine if there is dissipation of the stress outside the directly loaded regions in case of concentrated action, we put very simple action to the model – only one or two concentrated forces on the top of the structure, acting in oblique direction or in the horizontal direction. We did include the dead load (always present) of the structure, and so the influence of the concentrated force is less evident (Haiman & Žagar 1997).

Building our simple 2D model, we wanted also to research how the force and stress field developed during the construction. It might be called "virtual construction". Like ancient builders constructed masonry laying stone blocks, we tried to build our masonry model by laying virtual "macro-stones" one upon another. For this purpose we used step-by-step FEM analysis.

4.4.1 First step of "virtual construction"

Our first step, the starting point of a "virtual construction", was the analysis of a 2D model of a masonry column at the beginning of its building, when it consisted of only three layers of three blocks, superimposed one upon another.

Already in this early stage of construction, we examined the behavior of the structure by applying an unusual action – two concentrated forces, acting at the angle of 45° on top of the structure, in order to examine if the specific behavior of masonry (crack appearance, impossibility of dissipating stress outside the directly loaded regions) is appropriately simulated.

The FE analysis proved that our model simulates fairly well the expected behavior of masonry.

The most instructive is the figure of principal stress 3 in deformed shape. The opening of cracks in masonry joints and non-dissipation of stress in two unloaded "outer" "column" of "macro-stones" (the stress in them is due only to their own dead weight) displays clearly the expected behavior of the model – already in this early stage of construction.

4.4.2 Progression of "virtual construction"

Next step of the "virtual construction" of our 2D model of masonry column was to proceed with building of the model, by adding next three layers of "macro-stones" on top of the first model, and to apply the same action again.

As expected, the behavior of this structure is basically the same. The opening of vertical mortar joints

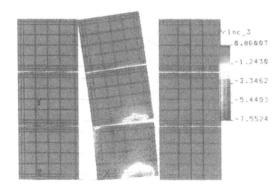

Figure 5.   First step of the "virtual construction" of masonry column. Principal stresses 3 in the deformed structure.

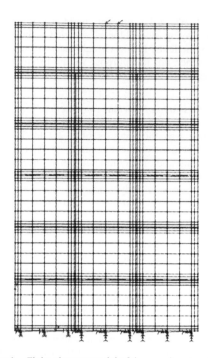

Figure 6.   Finite element model of the growing structure.

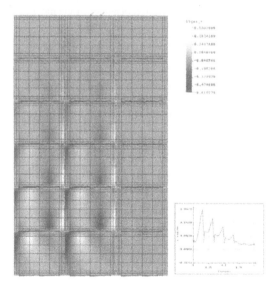

Figure 7.   Distribution of stresses in vertical direction (σ_y). The variation of σ_y stresses in vertical section A-A is presented on a diagram left.

between the middle "column of blocks" and adjoining structure, due to the concentrated actions only on top of this middle part of structure, is analogous to the behavior of the previously considered model.

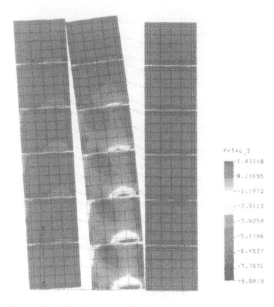

Figure 8.   Principal stress 3 in the deformed structure.

Compared to the lower column model, the deformed middle "column of macro-stones" of this taller structure has a greater effect on the adjoining "column", because middle "column" leans against the "left" part of structure, transferring in this way part of action.

4.4.3   Masonry column with connecting "macro-stone" on its top

We realized that our proposed model of the masonry structure displayed some weaknesses, i.e. its behavior differs from the behavior of the real masonry in which, of course, vertical joints are not continuous. The previous analysis of our model displayed clearly why master masons used to build in a different way, laying bricks or stone blocks of the higher layer upon the vertical joint of the lower layer.

We tried to simulate this usual solution by laying a larger block – "macro-stone" – on top of our "column". The effect is evident. The stone element on the top, connecting three "columns of blocks", diminishes the displacements by preventing the splitting of masonry structure.

If we consider the stress in the upper block, we realize that it acts nearly as a beam (with highest compressive stress σ_x on top, where action is applied, and with tensile stress in the middle of the lower part. It should be pointed out that in this case tension appears in 10 June 2002 a monolithic block, which is not a masonry structure in itself.

270

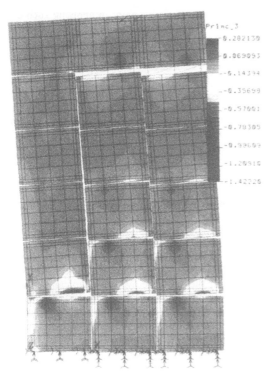

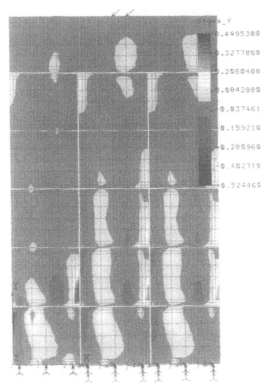

Figure 9. Masonry column modeled with connecting "macro-stone" on its top. Principal stress 3 in the deformed structure.

Figure 10. Masonry column modeled with connecting "macro-stone" on its top. Stresses in the vertical direction (σ_y).

4.4.4 *Short masonry column loaded with horizontal force on the top*

To test our model of masonry structure, we have researched also how it behaves under horizontal actions. with concentrated horizontal force applied on the top of the model. Opening of cracks in vertical mortar joints and state of stress are again explainable by the mechanical characteristics of masonry.

Of course, the modeling procedure described above can be applied to all masonry structures and structural elements. Our decision about how "deep" we want to analyze such masonry structures, constructed of stone or brick, is crucial in such analyses. The work is extensive and time-consuming, in spite of the application of various FE generators (modelers). Usually there is no willingness to research the problem in depth.

It also depends on financial questions and decisions. However, the results can be fascinating.

4.5 *Step-by-step analysis in the research of ancient structures*

In the described example we simulated the building process with step-by-step FEM analysis. Ancient

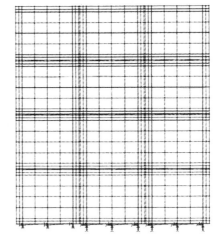

Figure 11. FE model of a short masonry column, with concentrated horizontal force acting on its top.

masters did not know theoretical mechanical laws, but based their knowledge on experience – on their own experience and on the experience of past generations. Their structures were their real-scale models.

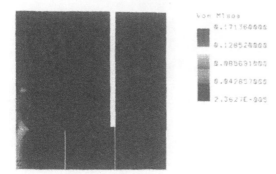

Figure 12. Principal stresses 3 (Von Mises) in the deformed structure.

During the construction, they observed carefully the joints of their masonry structures in order to detect eventual cracks, and to prevent further damage. (Mark 1982)

Nowadays we can observe our virtual models to learn about our structures. We must put in approximately the same amount of effort, but we can have results of our research in a much shorter period of time.

"Virtual construction" can also enable us to get an insight into the methods of designing and building of ancient masters. It might be said that doing a "virtual construction" is a "detective" kind of work, because when constructing a virtual structure step-by-step, we follow the ancient builders' paths and considerations – their way of thinking that resulted in fascinating forms such as Gothic cathedrals and other magnificent structures.

Of course, it should be kept in mind that the detailed survey of the analyzed structure, of each stone block, is a prerequisite for the fascinating virtual step-by-step construction. Only the profound knowledge of the examined masonry enables us to construct the proper, realistic FE mesh and to reconstruct the supposed phases of building. It requires also interdisciplinary work and collaboration with experts of other scientific fields, such as historians and art-historians.

In the virtual construction analysis, the results of the step-by-step FE analysis reveal the states of stresses and deformation encountered by ancient master-builders, all the questions they were confronted with and all the problems that they had to resolve – and thus the genesis of forms that met the requirements of a structure during the construction.

In this article only a small part of such simulation is presented. We believe that a simulation of this kind could be applied also to the larger assemblies of historic masonry structures.

Of course, there is no point in applying such an analysis if there is no need for an insight into the history of development of the state of stress and deformation during the construction (both of ancient and contemporary structures).

Perhaps in the future the FE programs with the option of automatic generation of step-by-step process of virtual construction will be developed. This is just a matter of time, willingness, decision and, of course, financing.

5 CONCLUSIONS

Appropriate modeling of structures that we analyze using FEM is essential. We should always keep on mind that model is an approximation of reality – not the reality itself. Even the most sophisticated programs give wrong results if the analyst uses an inappropriate model, which does not match the real behavior of the structure. Masonry structures, their behavior being more complex than that of modern structures, have often been modeled inappropriately, due to the lack of tools for their analysis.

Recently, new FE programs offer the possibility of modeling this type of structures too.

We used tools that we had at our disposal (COSMOS/M FE program), and modeled specific behavior of masonry (its low resistance to tension) with specific finite elements (GAP and TRUSS2D). Our simple 2D models display the behavior that can be expected in masonry structures.

Due to the opening of fractures in mortar joints, modeled with GAP and TRUSS2D finite elements, tension is eliminated. The distribution of stresses in our FE models of masonry (where mortar joints are simulated as not resistant to tension) differs fundamentally from that in geometrically identical models, simulated in usual way (i.e. as resistant both to compression and tension).

Therefore, we estimate that modeling masonry by simulating the lack of tensile strength in its mortar joints, by using appropriate types of finite elements, can be useful in further research, especially in step-by-step analysis of ancient masonry buildings. Such "virtual construction" may give us an insight into the process of designing and building of ancient master-builders – the method that enabled the construction of magnificent historic structures.

REFERENCES

Di Pasquale, S. 1992. New trends in the analysis of masonry structures. *Meccanica* 27: 173–184.
Di Pasquale, S. 1984. *Statica dei solidi murari – Teoria ed esperienze*. Firenze: Università di Firenze.

Haiman, M. & Žagar, Z. 1997. Obnova zidanih svodova i lukova, *Građevinar* 49(2): 77–85.

Mark, R. 1982. *Experiments in Gothic Structure*. Cambridge-London: The MIT Press.

Smars, P. 1992. *Etudes sur les structures en maçonneries* (M.Sc. thesis). Leuven: Katholieke Universiteit Leuven.

Šimunić, M. 1989. *Prilog istraživanju klasičnih zidanih konstrukcija na primjeru šibenske katedrale* (M.Sc. thesis). Zagreb: University of Zagreb.

Šimunić Buršić, M. 1995. Računalna analiza strukture šibenske katedrale. *Prostor* 2(10): 359–384.

Žagar, Z. 1999. *Drvene konstrukcije 2. Modeliranje drvenih konstrukcija*. Zagreb: Pretei.

Finite Elements in Civil Engineering Applications, Hendriks & Rots (eds.)
© 2002 Swets & Zeitlinger, Lisse, ISBN 90 5809 530 4

Sensitivity of masonry wall under base-restrained shrinkage

G.P.A.G. van Zijl [*], M. Boonpichetvong, J.G. Rots [†]
Faculty of Architecture, Delft University of Technology, Delft, The Netherlands

J.W. Verkleij
Research Center for Calcium Silicate Industry, Hilversum, The Netherlands

ABSTRACT: In this study, masonry strength and toughness sensitivity to the brick/block unit size is studied. Meso-analyses of masonry wall parts are performed to obtain macroscopic constitutive behaviour. Subsequently macro-analyses of base-restrained shrinking masonry walls are performed to study the sensitivity of the wall under this loading condition to the unit size. Computational models for masonry developed over the past decade and recently implemented in DIANA are employed for the analyses. The results reveal how these geometrical and material parameters affect masonry behaviour in terms of composite strength and toughness, as well as crack initiation and crack propagation.

1 INTRODUCTION

Cracking induced by shrinkage in masonry walls is an often-encountered aesthetic problem in the Netherlands. Despite the negligible impact to structural requirement, observed architectural performance can be highly impaired (de Jong 1992). Hence, there is a strong need for designers and also manufacturers to have proper guidelines and clearer insight in this matter. The omission of a movement joint has led to the cracking and eventual separation of the wall at the location of the primary crack shown in Figure 1.

Semi-analytical rules have been proposed by Copeland (1957), Hageman (1968) and Schubert (1988), to design movement joints for walls without openings. These analytical rules are based on the linear elastic stress distribution in walls, which are considered to be homogeneous continua as in Figure 2. Furthermore, these rules employ strength-based criteria, requiring that a certain average tensile stress in the wall should not exceed a strength limit. Such a criterion does not consider fracture mechanics and fails to provide the in-depth understanding of this complex phenomenon. Another serious shortcoming of these existing design rules is that no indication of crack width can be given.

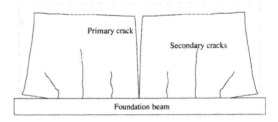

Figure 1. General crack pattern in base-restrained shinking wall.

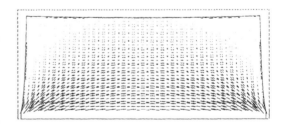

Figure 2. Principal tensile stresses for linear elastic behaviour and a homogeneous wall.

To fill in this gap, a numerical approach, based on fracture mechanics has been recently proposed and proved to be a better alternative (Rots 1994, van Zijl 1997). The strategy adopts a two-level technique of

---

[*] Also at University of Stellenbosch, South Africa.
[†] Also at TNO Building and Construction Research, Rijswijk, The Netherlands.

(1) a meso-analysis to derive the constitutive law for an equivalent macro model, and (2) a macro-analysis to predict the maximum crack-width in the structure. By such a simplified modelling strategy, crack initiation and evolution can be pragmatically captured and with a formalism of a non-linear finite element method, repetitive analyses of shrinking walls become viable within this concept. From the results rules for movement joint spacing can be derived.

Realising that different masonry types possess different fracture properties, the overview of shrinkage-resistant capability is an important concern not only for manufacturers but also designers in order to develop and choose the proper products for specific sites and environmental conditions. In this paper, the two-level computational approach is adopted to perform parametric studies of restrained shrinking masonry walls, built of certain categories of Dutch Calcium Silicate masonry units. In particular, the influence of the unit size on masonry composite strength and toughness is investigated, and in turn, its influence on the shrinkage behaviour of large walls. All the analyses are performed with the multi-purpose finite element program DIANA.

2 MODELLING APPROACH

A two-level technique is used to simulate the shrinkage response of a base-restrained masonry wall. First, a meso-analysis of a representative wall part under conditions, which reasonably imitate those in the actual area of the primary crack in a large wall, is performed. This is schematised in Figure 3. From the results a constitutive law is derived for the masonry composite, which is subsequently employed as the constitutive behaviour of an equivalent vertical crack on the macro scale. This second level schematisation is illustrated in Figure 4.

2.1 Meso-analysis

To characterise the equivalent vertical crack for the simplified macro-analysis in Figure 4, an appropriate constitutive law is sought for the interface elements, which represent the crack. In Figure 2 it is shown that for a homogeneous representation of the wall, the principal stresses are horizontal in the central area, where the primary crack is expected. Therefore, the response of a representative wall part under uniaxial tension should be derived to represent the behaviour of the wall in the vicinity of the primary crack. For this purpose, a periodic wall part is analysed by a discrete modelling approach (Rots 1994). The interface elements employed in this meso-analysis obey an interface material model capable of simulating fracture, crushing and shear-slipping, including shearing

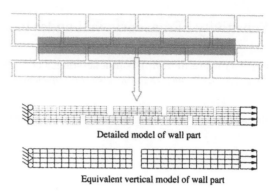

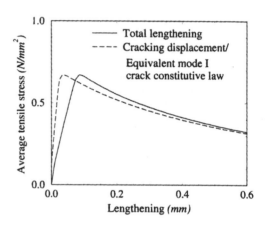

Figure 3. Wall part *meso*-analysis for the derivation of crack constitutive behaviour, to be employed as equivalent vertical crack constitutive law in *macro*-analysis.

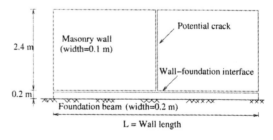

Figure 4. Simplified equivalent macro wall model.

dilatancy as described in van Zijl (1999). Figure 3 illustrates the characterisation process. The periodic part is analysed to determine its deformational response. The elastic response is then attributed to the masonry away from the crack, which is considered to be homogeneous. By subtracting the elastic deformation from the total deformation, the total cracking deformation is found, which defines the constitutive behaviour of an equivalent mode I crack in the simplified macro-model. In this manner equal global deformational behaviour can be predicted by the detailed model and the equivalent vertical crack model shown in Figure 3.

The important issues, which should be captured with reasonable accuracy in this characterisation process, are the loading conditions (uniaxial tension), the failure mechanism and the level of confinement, as is comprehensively described in van Zijl (2000).

2.2 Macro-analysis

In the macro-analyses, the equivalent vertical crack concept shown in Figure 3 is employed for the large masonry walls. The uncracked continuum is idealised to behave linear elastically and fracture is localised in a potential crack at the center of the wall. The same simplified shrinkage and thermal strain evolutions are assumed as by the analysts (Copeland 1957, Hageman 1968, Schubert 1988). This entails a combined hygral and thermal shrinkage, which is spatially uniform, isotropic and increases linearly in time. After initially activating self-weight in the model, the shrinkage is activated in the masonry wall.

In the finite element model, the elastic continua of masonry and the concrete foundation beam are represented by eight-noded quadrilateral plane-stress elements with a 3×3 Gauss integration scheme. Six-noded interface elements with a Lobatto integration scheme are adopted for the potential vertical crack, as well as for the interface between the wall and the foundation. Slipping along the wall/base interface is not included for lack of experimental data, but also to demonstrate the worst-case scenario. The continuum away from the central, primary crack is assumed to behave linear elastically. The effect of bulk creep is ignored for simplicity. As explained in van Zijl (2000), this gives a base-restrained shrinkage response, which is on the safe side. The homogenised response of the equivalent vertical crack model was confirmed to satisfactorily match with that derived by the meso-level detailed model and in the extreme situation to provide the more conservative results (van Zijl 1997). This allows the present investigation to focus directly on the effect of the vertical crack parameters on the global deformation capability of base-restrained shrinking masonry wall.

3 CRACK CONSTITUTIVE LAW: MESO-ANALYSIS

For the unit size sensitivity analyses, certain Dutch Calcium Silicate masonry products are studied. For simplicity and to concentrate on the influence of the unit geometry, it is assumed that the various units have

Table 1. Parameter for potential Element/Block/Brick crack.

Tensile strength (N/mm^2)	Tensile fracture energy (N/mm)
1.0	0.060

Table 2. Parameter for head and bed joints in meso-analysis.

Parameter	Bond type	
	Glue	Mortar
Tensile strength (N/mm^2)	0.4	0.2
Tensile fracture energy (N/mm)	0.040	0.004
Original adhesion (N/mm^2)	0.8	0.4
Shear fracture energy (N/mm)	0.04–0.03σ	0.02–0.03σ
Initial friction angle ($^\circ$)	36	36
Initial dilatancy angle ($^\circ$)	31	31

Note: σ is normal confining pressure with positive sign for tension and negative sign for compression.

equal material properties, despite evidence otherwise. The subjective classes of masonry products are (a) brick type units ($214 \times 55 \times 100$) with 10 mm mortar joints, (b) block type units ($440 \times 300 \times 100$) with 10 mm mortar joints, (c) block type units ($440 \times 300 \times 100$) with 2 mm glued bed joints and 3 mm glued head joints and (d) element type units ($900 \times 600 \times 100$) with 2 mm glued bed joints and 3 mm glued head joints. Tables 1 and 2 summarise the meso-level properties.

3.1 Meso-analysis response

Based on the meso-analysis of wall parts under tension, we obtain the average tensile stress-crack width relation. As an example, Figure 5 shows the meso-analysis results of masonry class (b). The total load deformational response is shown, together with the deformed specimen at three stages of loading. In this case all head joints crack open, followed by unit fracture. From the global load-deformation behaviour the average stress is derived by dividing the load by the total cross-sectional area. By subtracting the elastic deformation from the total lengthening to derive the total crack width, the crack constitutive law derivation is completed. In Figure 6 the crack constitutive laws for all the simulated products are shown.

3.2 Meso-analysis discussion

From the wall part responses under uniaxial tension some conclusions can be drawn regarding the influence

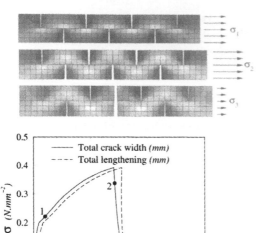

Mode I failure: head joint and unit fracture

Mode II failure: head and bed joint failure

Figure 7.   Masonry meso-level failure modes.

Figure 5.   Typical derivation of the equivalent vertical crack constitutive behaviour.

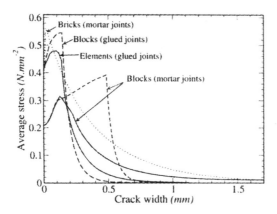

Figure 6.   Constitutive behaviour for various calcium silicate masonry types, to be employed for the equivalent central crack in simplified restrained shrinking wall analyses.

of unit size and joint type. In this regard it is informative to study the two basic failure mechanisms, namely a mode I-type failure and a mode II-type failure shown in Figure 7.

The mode I failure strength f_t^I can be approximated by integrating the limit stresses along the failing crack, giving

$$f_t^I = 1/2\left(f_{tu} + f_{tj}\right) \tag{1}$$

with f_{tu} the unit strength and f_{tj} the joint strength. This assumes elastic-perfect plastic behaviour. For the mode II failure mechanism, integration of the tensile strength along the two failing head joints and the adhesive resistance along the failing bed joint, produces an estimate of the composite mode II uniaxial strength:

$$f_t^{II} = \left(f_{tj} * H + c_j * L/2\right)/H \tag{2}$$

where H is the total height of the periodic part (equal to a unit height plus the joint thickness), L is the unit length and c_j is the adhesion between unit and mortar. Also for this case perfect plasticity is assumed.

In all cases of glued joints, i.e. for the large elements, as well as for the blocks, mode I failure occurred. The joints are so strong that unit fracture occurs rather than sliding along bed joints. However, equation (1) predicts equal tensile strength for both types of masonry ($f_t = 0.7$ MPa), while the finite element analyses, which consider the limited fracture energy, produce a lower strength for the larger unit (elements) masonry – Figure 8. This *size effect* is a well-known phenomenon. It can be concluded that larger units may produce weaker masonry. Note that this conclusion is subject to the simplifications of this study, for instance equal material for all size units, while it will be different due to different fabrication processes. This *size effect* has implications for the restrained shrinkage behaviour of walls built of large elements, as will be studied in the next section.

Mode II failure occurs in masonry with weak joints, which is mostly the case for mortar-bonded masonry. An additional condition for this failure mode is low vertical/confining pressure. Under high confining pressure, high Coulomb-Friction resistance acts along the bed joint, leading to unit fracture instead – mode I failure. For unconfined conditions, equation (2) predicts a composite tensile strength of $f_t = 0.89$ MPa for brick size masonry and $f_t = 0.48$ MPa for block masonry. The meso finite element analyses predict strengths of 0.56 MPa and 0.31 MPa

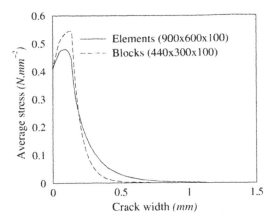

Figure 8.   Meso results for glue-bonded masonry.

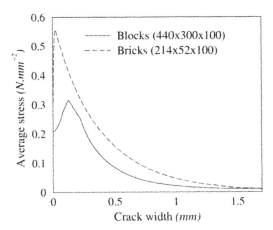

Figure 9.   Meso results for mortar-bonded masonry exhibiting mode II failure.

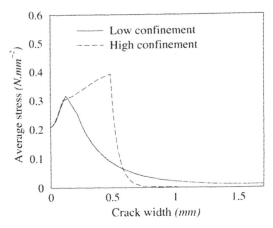

Figure 10.   Meso results for mortar-bonded block masonry with different levels of confinement.

normal/confining pressure. For low confinement, as typically occurs in the upper parts of a wall where little top load acts, mode II failure is predicted, with low composite tensile strength. For high confinement, for instance experienced by masonry low down in a wall, Coulomb-friction resistance along the bed joints may enforce mode I failure, with a higher composite strength.

For the employed set of material parameters and unit sizes, the brick masonry wall part response is insensitive to the level of confinement. This mechanism of confinement and dilatational wedging play an important role in composite behaviour. See van Zijl (2000) for a detailed discussion of this topic.

A final observation is that brick masonry has comparable composite strength to that of glued masonry, but exhibits superior, tough post-peak behaviour – Figure 6.

4   RESTRAINED SHRINKAGE: MACRO-ANALYSIS

In the previous section conclusions were drawn about the unit size effect of masonry composite strength and toughness. In this section macro analyses are performed along the lines described in section 2.2 to study the sensitivity of base-restrained wall shrinkage response to these size effects. In addition, the effect of E-modulus, wall length and foundation is studied. For these analyses equivalent primary crack constitutive laws are employed, based on the material laws obtained from the meso-analyses in the previous section. A series of macro material parameter variations have been set up, as summarised in Tables 3 and 4.

respectively for these two types of masonry – Figure 9. The fact that equation (2) overestimates the strength by 60% is due to the assumption of unlimited fracture energy (perfect plasticity). More importantly, the predictions of both equation (2) and the finite element analyses indicate another type of *size/shape effect*, which favours smaller units to larger units in terms of composite strength.

The influence of confinement is illustrated by the different responses obtained for mortar-bonded block masonry – Figure 10. The low confinement condition is modelled by allowing free vertical translation of the upper and lower edges, but constrained to remain horizontal. The high confinement is simulated by suppressing vertical translation of the upper and lower edges. Through dilatational uplift upon shearing along the bed joints, wedging occurs, causing build-up of

Table 3. Material parameters employed in macro analysis.

Component	Parameter	Value
Masonry	Young's modulus; E	See Table 4
	Poisson's ratio	0.2
	Mass density	1,800 kg/m³
Concrete beam	Young's modulus	30,000 MPa
	Poisson's ratio	0.2
	Mass density	2,400 kg/m³
Wall-foundation interface	Normal stiffness	333 N/mm³
	Shear stiffness	139 N/mm³
Potential crack	Normal stiffness	10^6 N/mm³
	Shear stiffness	10^6 N/mm³
	Tensile strength; f_t	See Table 4
	Tensile fracture Energy; G_f	See Table 4

Table 4. Outline of the material parametric studies in macro-analysis.

Test	E (MPa)	f_t (MPa)	G_f (N/mm)	Remark
Wall A	9000	0.4	0.15	Reference
Wall B	9000	0.2	0.15	Low F_t
Wall C	9000	0.6	0.15	High F_t
Wall D	9000	0.4	0.10	Low G_f
Wall E	9000	0.4	0.20	High G_f
Wall F	6000	0.4	0.15	Low E
Wall G	12000	0.4	0.15	High E

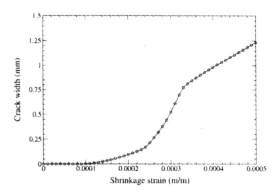

Figure 11. Shrinkage-crack width diagram of reference wall A.

4.1 Reference wall A

Firstly, wall A is analysed to serve as an average, reference case. This is to provide insight into the cracking mechanism at the macro-level. In order to delineate the shrinking wall response, the maximum crack width which arises in the central crack is plotted at each level of shrinkage strain in Figure 11. The deformation patterns, together with associated primary principal stress trajectories at certain stages are

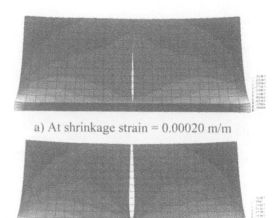

a) At shrinkage strain = 0.00020 m/m

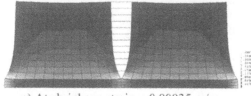

b) At shrinkage strain = 0.00025 m/m

c) At shrinkage strain = 0.00035 m/m

Figure 12. Progressive cracking in reference wall A.

plotted in Figure 12. The results indicate that cracking initiates at the bottom when the tensile stress threshold is exceeded there. Subsequently, the crack propagates towards the upper part of wall. Finally, after all fracture energy is dissipated in the central crack, the original wall is split into two separate parts, in which the stress distributions are identical to that in an elastic, base-restrained shrinking wall.

4.2 Sensitivity to material parameter variation

Variation of the tensile strength affects the maximum crack width-shrinkage strain diagram as shown in Figure 13 for walls A-C. It is seen that the higher tensile strength delays the initiation of cracking in the masonry wall, while the lower strength leads to cracking at a lower shrinkage. As the analysis is extended to a higher level of shrinkage, all the responses tend to fall on the same line (see Figure 14). This can be explained by the fact that at a specific limit point of shrinkage level, all the fracture energy is fully dissipated. After that point the wall is separated into two parts. The subsequent behaviour is elastic and, thus, similar for all cases.

The variation of toughness/fracture energy – walls A, D and E – does not affect the crack initiation, but

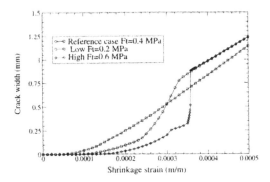

Figure 13. Influence of f_t on the shrinkage crack-width: walls A-C.

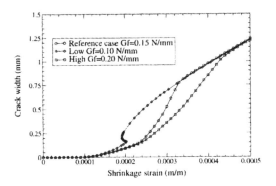

Figure 15. Influence of G_f on the shrinkage-crack width: walls A, D and E.

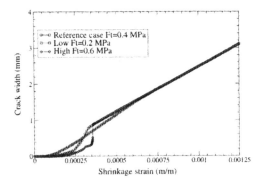

Figure 14. Influence of f_t on the shrinkage-crack width: Extended analysis.

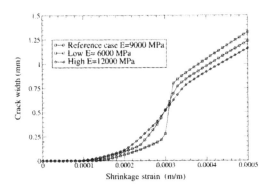

Figure 16. Influence of E on the shrinkage-crack width: walls A, F and G.

rather the cracking rate, as illustrated in Figure 15. A higher mode I fracture energy leads to ductile behaviour, hence it prolongs the crack propagation to the top part of wall. On the other hand, the lower mode I fracture energy results in a brittle response.

The effect of Young's modulus on the overall cracking resistance of the wall is displayed in Figure 16. With a higher Young's modulus, wall behaviour becomes much more stiff, causing breaching of the stress limit and subsequent cracking at lower strain. On the other hand, with a lower Young's modulus, the wall is much more flexible and deformability is improved.

4.3 Wall length and foundation restraint

Figures 16 and 17 illustrate the effect of wall length and foundation restraint to crack width-shrinkage response. Larger structures possess relatively higher elastic stored energy, hence they exhibit much more brittle response than smaller structures. For the higher degree of the foundation restraint, earlier initiation of cracking is observed.

5 REMODELLING OF MASONRY BY STANDARD SMEARED CRACK MODEL

Secondary cracking in the masonry away from the primary crack is included in this section by employing a standard smeared crack model (Rots 1988). Tensile strength and fracture energy in the continuum are assumed to be 1.5 times those in the central, primary cracks. The other material parameters are adopted as for the reference case of wall A in Table 4. The overall response, depicted in Figure 19, shows the significant change in the resisting mechanism. Allowing for such possible smeared cracking considerably affects the stress distribution and crack propagation pattern, Figure 20. Extracting the maximum tensile crack strain in masonry continuum and multiplying it with the crack-band width approximates the maximum crack-width in the continuum. For this particular wall geometry and material parameters the primary crack is arrested, while two "secondar" cracks continue to develop. This result remains to be confirmed by including cracking rate dependence

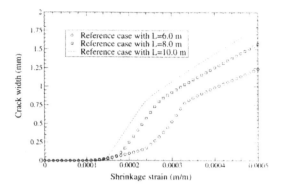

Figure 17. Influence of wall length on the shrinkage-crack width.

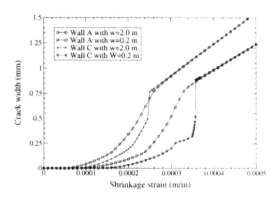

Figure 18. Influence of foundation restraint on the shrinkage-crack width.

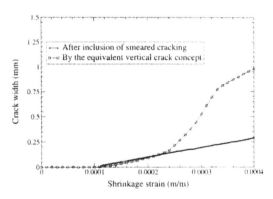

Figure 19. Changes in cracking response after inclusion of smeared cracking in the masonry continuum.

(van Zijl 2000) to regularize the crack spacing and orientation. If, for the time being, the result is accepted, it confirms the conservative character of the simplified equivalent vertical crack approach.

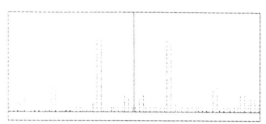

Figure 20. Crack propagation predicted by smeared cracking at shrinkage strain = 0.0004 m/m.

6 CONCLUSIONS

The response to uniaxial tension of various calcium silicate unit masonries, with both glued and mortar joints, have been studied computationally. For the employed sets of parameters and geometries, a unit size effect on masonry composite tensile strength, as well as toughness, has been detected. Smaller units produce masonry of higher strength and toughness than large unit masonry.

Subsequently, base-restrained shrinking masonry wall response has been studied in a simplified manner. The simplification to consider only the primary crack has been shown to be conservative, as secondary cracking reduces the maximum crack width. Furthermore, a sensitivity analysis has shown that a wall of lower stiffness delays cracking, as does a lower restraining foundation stiffness.

The shrinking wall response has been shown to be sensitive to the unit size effect on composite strength and toughness. Higher strength postpones cracking to higher levels of shrinkage, while higher toughness decreases the cracking rate. It must be noted that the unit size effect conclusion is based on the limited scope of material parameters employed and cannot be generalised.

ACKNOWLEDGEMENTS

The financial support by the Netherlands Research Center for Calcium Silicate Industry (RCK/CVK), the Netherlands Technology Foundation (NWO/ STW) and Delft Cluster is gratefully acknowledged.

REFERENCES

Copeland, R.E. 1957. Shrinkage and temperatures stresses in masonry. *ACI Journal*, 53: 769–780.
de Jong, P. 1992. Lessons from damage events in the building industry I (in Dutch). *Cement*, 2: 26–28.

Hageman, J.G. 1968. *Study of shrinkage cracks (in Dutch).* Research Center for Calcium Silicate Industry, Reports no.189-1-0/189-2-0, The Netherlands.

Schubert, P. 1988. About the crack-free length of non-loadbearing masonry walls (in German), *Mauerwerk-Kalendar*: 437–488.

Rots, J.G. 1988. *Computational modeling of concrete fracture*, Ph.D. Dissertation, Delft University of Technology, Delft, the Netherlands.

Rots, J.G. 1994. *Structural masonry; an experimental/ numerical basis for practical design rules* (in Dutch), CUR report 171, Chapter 6: 139–166.

van Zijl, G.P.A.G. & Rots, J.G. 1997. Towards numerical prediction of masonry walls behaviour, In H. Jongedijk, M.A.N. Hendriks, J.G. Rots & W.J.E. van Spanje (eds), *Finite elements in engineering and science*: 329–340. Rotterdam: Balkema

van Zijl, G.P.A.G. 1999. *A numerical formulation for masonry creep, shrinkage and cracking.* Series 11 Eng. Mech. 01, Delft University Press, Delft, The Netherlands.

van Zijl, G.P.A.G. 2000. *Computational modelling of masonry creep and shrinkage.* Ph.D Dissertation, Delft University of Technology, Delft, The Netherlands.

Finite Elements in Civil Engineering Applications, Hendriks & Rots (eds.)
© 2002 Swets & Zeitlinger, Lisse, ISBN 90 5809 530 4

Settlement damage of masonry buildings in soft-ground tunnelling

M. Boonpichetvong & J.G. Rots[†]
Faculty of Architecture, Delft University of Technology, Delft, The Netherlands

ABSTRACT: The application of fracture mechanics to predict the cracking damage in masonry buildings subjected to ground movement by tunnelling activity is presented in this paper. It describes the computational approach employed to capture the failure mechanism of a selected historical masonry façade. Both uncoupled and coupled soil-structure analyses are performed. Various continuum crack models are tested in large-scale fracture analyses. The results indicate the need for reliable numerical techniques for large-scale fracture analysis of highly brittle material.

1 INTRODUCTION

Bored tunnelling activities lead to surface settlements that may damage neighbouring structures. Therefore, the reliable prediction of cracking in the surrounding structures now becomes a prime issue in soft-ground tunnelling to guarantee the success of projects. This is strongly emphasized in e.g. the western part of the Netherlands where the bored tunnelling will be driven in soft soil underneath historical masonry buildings on fragile foundations.

Currently, the available practice for evaluating settlement effect to the surrounding structures due to soft-ground tunnelling can be classified into two types: uncoupled analysis and coupled analysis. In the first class, the greenfield settlement that approximated from empirical formulae is directly imposed on the building model while the second class allows for the full interaction between the above structure and the underlying soil. So far, the proper inclusion of fracture mechanics to predict the cracking damage in this large-scale fracture problem appears to be missing in the literature. Potts and Addenbrooke (1996) employed elastic beam elements to represent an overlying structure on a soil continuum. By simulating the tunnelling process, the effect of an existing surface structure i.e. axial and bending stiffness of building, on the pattern of ground movements was reported. Although such study provides useful information, the simple elastic beam model cannot simulate the realistic building response. Adopting an elastic no-tension material model to

simulate the masonry building response, coupled settlement damage analyses were carried out by Augarde et al. (1998). Although this no-tension masonry model allows for a rough estimation of damage in the masonry structures, the application of fracture mechanics models will provide a deeper insight in the performance of buildings made of quasi-brittle material. The inclusion of tensile-softening in smeared or discrete crack models has proven to give good correlation between numerical results and various fracture experiments available. To gain reliable settlement damage predictions of masonry buildings and allow for realistic fracture properties in quasi-brittle material, it is necessary to adopt such tensile-softening crack models in the numerical studies.

Recent work (Rots 2000) demonstrated the capability of fracture mechanics by both smeared and discrete crack concepts to predict the cracking damage due to settlement. That study can be considered as a semi-coupled analysis since the greenfield settlement trough, having the shape of a Gaussian distribution curve, was directly applied to the bottom of the building model while the bedding stiffness was included to represent the foundation in an average manner. In this paper, the detailed study of the uncoupled analyses is extended for the related numerical aspects i.e. performance of different crack models in large-scale fracture analysis, mesh sensitivity tests and the effect of fracture properties on the overall façade response. Furthermore, we move from uncoupled to coupled analysis by including the soil in the finite element model, to study the effect of soil-structure interaction on the cracking response of the highly brittle material. All the analyses are performed using product version of the DIANA finite element program. This is the first phase of

---

[†] Also at TNO Building and Construction Research, Rijswijk, The Netherlands.

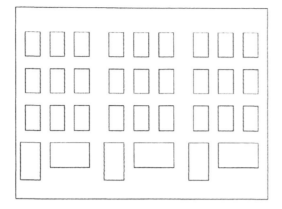

Figure 1. Selected historical masonry façade.

a research project carried out at the Faculty of Architecture, Delft University of Technology.

2 HISTORICAL MASONRY FAÇADE

An example of a historical masonry façade typical for the western part of the Netherlands is chosen for this study. This numerical study is one part of a joint research project in which the results of the numerical simulation by fracture mechanics will be interpreted and translated into the development of a damage classification system (Netzel 2000) in the near future.

The layout of the masonry façade is shown in Figure 1. This is a block of three house units. The total height of the house units is 15.5 m, starting from the foundation level at 1.5 m below green field, a ground floor level and three levels above that. The length of a house unit is 6.8 m. The three house units together have a length of 20.4 m, so that the length over height ratio is 1.3. A uniform thickness of one brick (220 mm) is adopted here for the whole façade. The opening pattern shows two large openings at ground floor and a regular pattern of three window openings at the three floors above. Above the window openings at ground floor, lintels in the form of steel beams are present to distribute the vertical load to either side of the opening.

3 NUMERICAL MODELLING OF SOIL-STRUCTURE INTERACTION

3.1 *Finite element modelling of each component*

To model the soil continuum, eight-node quadratic plane strain elements are used whereas eight-node quadratic plane stress elements represent the masonry continuum. The full, three by three integration scheme is adopted for both element types to reduce the danger

of so-called spurious kinematics modes when softening occurs. To properly connect the soil elements with masonry elements, a unit thickness is adopted for masonry continuum while a full thickness of 220 mm is taken in the uncoupled analyses. For the coupled analyses, interface elements are inserted between the underlying ground and the building structure. The inclusion of interface elements allows for the investigation of stress, strain and back-check in the assumption of the contact condition between the underlying ground and the building structure. Interface elements are also used with a different purpose i.e. as bedding elements for the uncoupled analyses.

3.2 *Material modelling of each component*

There are two major material laws involved in this numerical study i.e. the crack model for masonry and the soil model for underlying ground.

3.2.1 *Crack model for masonry*

Various smeared crack models are used to predict the settlement damage for masonry. These crack models are (a) decomposed-strain fixed smeared crack, (b) total strain fixed smeared crack, (c) total strain rotating smeared crack and (d) Rankine plasticity crack model. The structured mesh of element size 400 by 500 mm is used to represent masonry façade. Because the quadratic elements have a linear strain distribution and cracks will localise at one side of an element thus the crack bandwidth is taken by an average of 225 mm. The detailed formulations of these crack models are described in respectively Rots (1988), Feenstra et al. (1998) and Feenstra (1993). All of these crack models are now available for application in engineering practice. Each crack model has different features. Therefore, these crack models are tested in this study to check their performance when applied to large-scale fracture analysis.

3.2.2 *Soil model*

The importance of the hardening soil model able to characterize soil non-linearity at small strains in predicting ground movements adjacent to excavations is now well recognized (e.g. Gunn 1992). Adopting such non-linear soil model in the present study certainly would complicate the analysis and affect the convergence performance. As the aim is to focus on the efficiency of the crack models for masonry, it was decided to take a simpler, linear-elastic model for the soil in the coupled analysis.

3.3 *Loadings*

First, the dead load of the masonry is activated. Next, settlements have been applied and incremented. This is done in two ways: (a) increasing the magnitude of the surface settlement trough in the uncoupled analyses and (b) increasing the percentage of volume loss in the coupled analyses. By this way, the numerical results can be directly referred to the observation on site if available.

4 UNCOUPLED ANALYSES

4.1 Outline and model parameters

For the uncoupled analyses, we firstly need to evaluate the expected greenfield settlement. Based on the analysis of case records (e.g. Mair et al. 1996), the immediate transverse settlement trough (S_v) at the horizontal distance x from the tunnel center line is well described by an inverted Gaussian distribution curve as:

$$S_v = S_{max} \exp\left[\frac{-x^2}{2i^2}\right]$$

where i is the horizontal distance from the tunnel centerline to the point of inflection on the settlement trough ($i = Kz_0$). K is the trough width parameter for tunnels and is taken as 0.5 in this study. Z_0 is the depth to the tunnel axis, which is 20 m herein. The loading scheme starts with the activation of self-weight and live load of 5 kN/m at each floor. Subsequently, the settlement trough is applied incrementally. The stiffness of the no-tension bedding interface elements is taken as 0.15 N/mm³ by smearing out the stiffness of foundation system in an average manner (for such details of bedding interface, see Rots 2000).

The façade is located at the point of inflection i.e. in the hogging type situation, which is deemed as the most critical case. The elastic-softening properties of the masonry are assumed as follows. Young's modulus (E) and Poisson's ratio for the masonry are adopted as 6000 N/mm² and 0.2 respectively. Fracture properties of the masonry are taken with a tensile strength (f_t) = 0.3 N/mm², fracture energy (G_f) = 0.05 N/mm with a linear softening diagram. Mass density of masonry is 2400 kg/m³.

4.2 Standard façade response

Firstly, we employ the decomposed-strain fixed smeared crack model in which shear retention factor β is taken as 0.01 to reduce the so-called false stress locking. Figure 2 illustrates the relationship between the angular distortion (Boscardin 1989) and the maximum crack width. Due to the non-proportional loading pattern, several unloading cracks are found at the first increment of prescribed settlement. These cracks occur at the window opening at the ground-floor level during

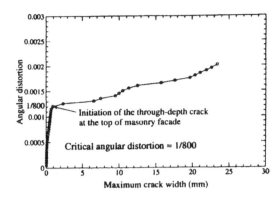

Figure 2. Standard façade response in the uncoupled analysis.

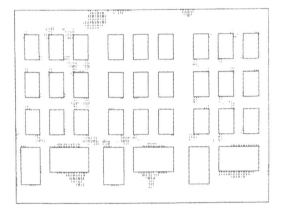

Figure 3. Predicted crack pattern at the critical angular distortion = 1/800.

the initialised state (by self-weight and live load). At each added settlement increment, new cracks propagate from the bottom part of the façade up towards the top. Along the window corners, several active cracks arise due to the high stress concentration. The response of this standard façade shows only relatively small crack widths until the angular distortion reaches the critical value, which is around 1/800. This critical angular distortion corresponds to 2.0% ground loss. The crack pattern at the critical angular distortion is shown in Figure 3. A large crack band exists at the top of the façade due to the pattern of principal tensile stress at the top prior to the onset of fully open cracks (Figure 4).

It is discovered that at the critical angular distortion, the first fully open crack emerges i.e. the crack of which the ultimate strain is beyond the end of the descending branch, the softening is completed and no more tensile stress can be transferred. Beyond this threshold, there is

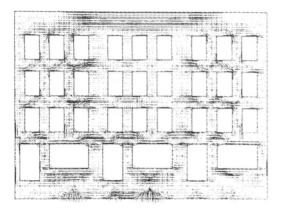

Figure 4. Principal tensile stress before the emergence of the fully open cracks.

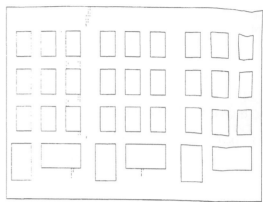

Figure 6. Fully open cracks just after the critical angular distortion.

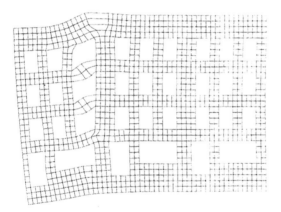

Figure 5. Localisation shown at the top of façade just after the critical angular distortion leading to unstable response.

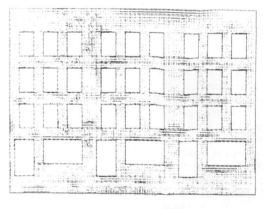

Figure 7. Stress locking just after the critical angular distortion.

a sharp localisation i.e. localised vertical crack at the top of façade as shown in Figure 5 and 6.

After this critical stage, the angular distortion is almost constant in the second branch of the façade response. Along this branch, there is a jump of the crack width in which the convergence performance becomes very poor. The difficulty to achieve a good convergence is partly due to the significant residual stress (stress locking shown in Figure 7) in the fully open crack zone. Rots (2000) demonstrated that adopting a discrete crack model together with the arc-length control technique could overcome this problem. Snap-back response could be predicted by that approach, which was not successful for the smeared crack concept. Anyway, even with the discrete crack model it was reported that much effort was required and the approach was still not attractive. Instead, herein we attempt to test

the other available crack models for the same problem with the same and more refined mesh to observe possible improvements, also in the pre-peak regime.

4.3 Performance of crack models in large-scale fracture

Crack models mentioned in section 3.2.1 are now tested in the uncoupled analyses. The overall responses of the façade simulated by each crack model are plotted in Figure 8.

All these tests are performed with the same structured mesh of quadratic elements. The same problems are encountered during testing all these crack models i.e. convergence is hardly achieved in the second branch of the façade response. A zone of the false stress locking at the top of façade is still present in

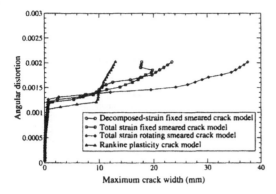

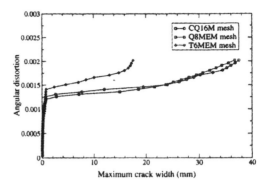

Figure 8. Façade response by various crack models.

Figure 10. Façade response in mesh refine test.

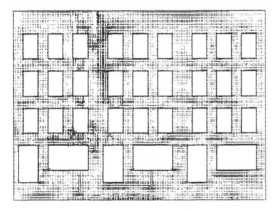

Figure 9. Reduction of stress locking zone by total strain rotating smeared crack model.

case of using the total strain fixed crack model. Stress locking is less in case of the total strain rotating crack model (see Figure 9). This feature matches with a relatively better convergence compared to the other three crack models. For the Rankine plasticity-based crack model, several undesirable negatives pivots are detected during this analysis pointing the question of the true equilibrium path. Unless branch switching techniques or indirect arc-length control techniques (de Borst 1986) are employed, the stability and uniqueness condition cannot be guaranteed. For simplicity in this pilot study, the total strain-based crack concept is selected for the rest of this investigation.

4.4 Mesh refinement test

By maintaining the density of the parental finite element mesh, mesh refinement is done by splitting each eight-node element (CQ16M) into four four-node quadrilateral elements (Q8MEM) and further dividing each four-node element into two three-node triangular (T6MEM) elements. We apply two by two integration scheme for the four-node quadrilateral elements and one-point integration for the three-node triangular elements. Element size of both elements becomes 200 by 250 mm. Since the vertical localised crack is expected to govern the final failure mode, crack bandwidth of 200 mm is adopted for both meshes.

The convergence performance is improved by using the triangular mesh whereas those of four-node and eight-node quadrilateral elements are comparable. Again, in all tests, the response in the second branch faces convergence difficulties. It was proved that problems regarding mesh dependency could not be fully solved by fracture-energy based approaches (Sluys 1992) in the smeared crack models. Nevertheless, for the real-world application, we consider this approximation reliable enough for further practical interpretation. As shown in Figure 10, the limit points of façade responses predicted by using different refined meshes are in the same magnitude and the threshold of approximately 0.0013 can be interpreted as the value that triggers notable damage for design guidelines.

4.5 Effect of fracture properties on the building performance

To understand the influence of each fracture parameter on the façade response, we reset each parameter as double and half of that of the original standard façade. Such sensitivity study is necessary for better understanding in view of the wide-range classes of various quasi-brittle materials. Variation of the fracture parameters affects the façade response as shown in Figure 11 and 12. The critical angular distortion is lower for the façade with lower tensile strength. With higher tensile strength, the initiation of concerned dynamic crack growth in the structure is delayed. The influence of the fracture energy on the façade response is pronounced. Significant crack growth is expedited in very low fracture energy material. It emphasizes the importance of tensile-softening in the smeared crack model.

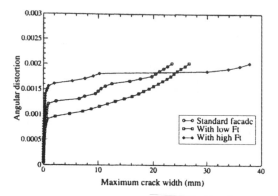

Figure 11. Influence of tensile strength on the façade response.

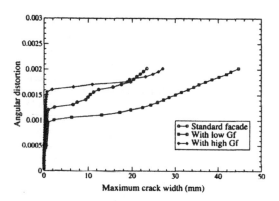

Figure 12. Influence of fracture energy on the façade response.

5 COUPLED ANALYSES

5.1 *Outline and model parameters*

The two-dimensional finite element model in Figure 13 depicts the chosen problem for the coupled analyses. Element types used for each component are as described in section 3.1. The same masonry façade as in section 4 is located at 10 m from the tunnel centerline. The tunnel axis with 6.5 m tunnel diameter is founded at 20 m below the ground surface. The segmental lining is not included in the analyses. Properties of the masonry façade are kept the same as in the uncoupled analyses. The underlying soil is divided into three strata. This is taken for simplicity, as it is not intended to incorporate all detailed characteristics of underneath soil in this pilot study. The basic elastic properties of soil are listed in Table 1. E is Young's modulus, ν Poisson's ratio, K_o coefficient of earth pressure at rest and γ unit weight. No-tension interface elements are placed along the boundary between soil

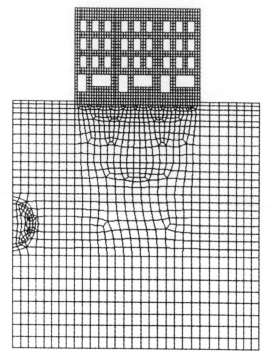

Figure 13. Finite element model for the coupled analyses.

Table 1. Elastic properties of underlying soil strata.

Depth	E (MPa)	ν	K_o	γ (kN/m³)
0–10 m	10	0.35	0.748	17
10–25 m	30	0.35	0.748	17
25–40 m	100	0.30	0.357	20

and structure. The stiffness of these elements in compression is taken very high.

Initially, the self-weight of all the components is applied in the analysis. Then, for post-processing reasons all nodal displacements are reset to zero whereas the internal variables i.e. stress and strain in crack and soil models are recorded and treated as initial conditions before tunnelling process. The excavation process is simulated in which the soil mass inside the cavity is controlled by a volume contraction procedure.

5.2 *Greenfield ground response*

The greenfield ground response at 1% volume loss is presented in Figure 14. The shape of the vertical and horizontal ground movement is identical to that of the typical empirical prediction. This justifies the application of the simple elastic soil model in this pilot soil-structure interaction analysis.

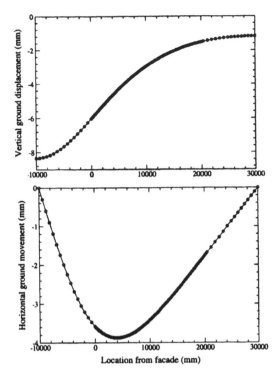

Figure 14. Greenfield movement profile at volume loss = 1%.

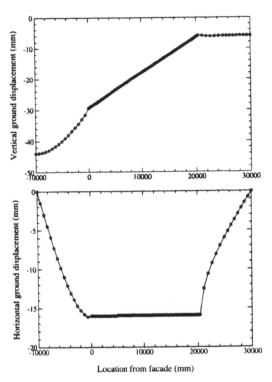

Figure 15. Influence of inclusion of the building on ground movement profile at volume loss = 5%.

5.3 Soil–façade interaction response

For the standard façade, prediction revealed that in the range of 1–5% volume loss, the façade remained free from settlement damage, no cracks from tunnelling activity were noticed. However, functionally, the observed building movements and tilt are undesirable. The pattern of predicted ground movement in Figure 15 indicates the influence of the presence of the building on the contact region. Compared to the greenfield response (Figure 14), the effect of the presence of the stiff building is clearly visible. All the interface elements are always in compressive state governed by building dead load. It is obvious that the initially selected standard masonry properties are too strong for the building to crack.

In fact, material characteristics of the historical buildings may be less than those presumed. As an example, the analysis is repeated by taking a less strong material with a reduced tensile strength $0.05 \, N/mm^2$. Two values of the fracture energy are taken, 0.05 and 0.1 N/mm (case a and b) respectively. The predicted settlement damage on the weak façade (case a) now becomes clear as shown in Figure 16. Unacceptably large crack width is detected in façade at a small amount of ground loss where full fracture occurs with

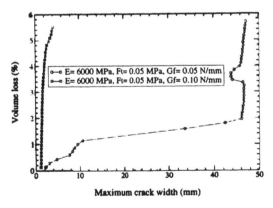

Figure 16. Façade response of the highly brittle material.

crack widths up to 45 mm. This outcome, however, is very sensitive to the precise parameters. If the fracture energy is increased from 0.05 to 0.10 N/mm (case b), the induced crack width becomes less than 5 mm.

The coupled ground response in Figure 17 shows the significant effect of horizontal ground strain on building response. This is justified for the case of shallow foundation in which the effect of foundation

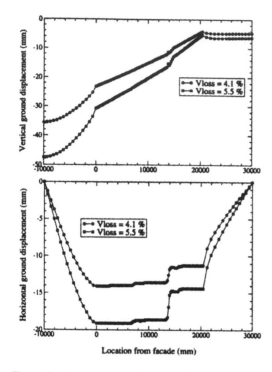

Figure 17.   Interacted ground response for case b.

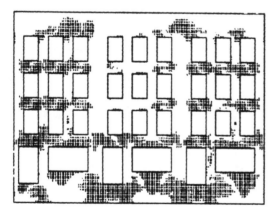

Figure 18.   Initial widespread minor cracks by dead weight action.

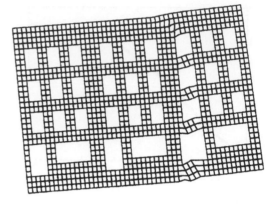

Figure 19.   Progressive localisation starting from the bottom towards the top of façade.

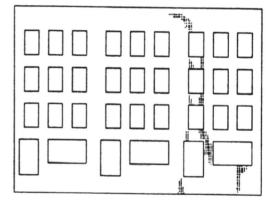

Figure 20.   Predicted localised cracking response for case a at volume loss = 1.5%.

restraint is insignificant. On the other hand, the vertical action (beam action) of this façade is relatively stiff due to the value of height over length ratio, thus the effect of differential ground settlement on the façade is hardly observed at the low ground loss.

For details of building response for case a, there are at first several negligible cracks under the action of dead weight (Figure 18). During the tunnelling process, there is a high stress-strain redistribution in which

crack growth of these initially minor cracks is progressively controlled by the horizontal ground strain. After the first appearance of the fully open crack near the ground-floor door (Figure 20), a severe damage along the window openings is rapidly induced, several localised cracks propagate from the bottom up towards the top along the same window column. The combined effect of differential settlement and horizontal ground movement is now hardly distinguished. The corresponding incremental deformation is shown in Figure 19. Serious localised cracking damage at 1.5% ground loss is illustrated in Figure 20. The response is extremely brittle, showing a large plateau in Figure 16 associated with the sudden propagation of the vertical crack. The convergence of the nonlinear analysis in this case is rarely achieved once the fully open crack arises. This emphasizes the need for future research in developing a technique for large-scale fracture analysis of the highly brittle material.

6 CONCLUSIONS

Settlement damage prediction by means of a fracture mechanics approach is presented in this paper, from which useful and practical design information can be derived. It has been demonstrated that coupled analyses may yield results that differ significantly from uncoupled analyses. The power of this method unavoidably depends on the robustness of selected crack models. We conclude that the performance of the smeared crack models in simulating settlement damage prediction yields some limitations, e.g. a proper technique to handle the highly brittle behaviour should be developed, and techniques to include the initial existing cracks should be improved. This serves as the framework for the continuation of this research.

ACKNOWLEDGEMENTS

The financial support by Delft Cluster Theme 1 "Soil and Structures" and by the Netherlands Center for Underground Structures CUR/COB Committee F220 is gratefully acknowledged.

REFERENCES

Augarde, C.E., Burd, H.J. & Houlsby, G.T. 1998. Some experiences of modelling tunnelling in soft ground using three-dimensional finite elements. In *Proc. 4th European conference on numerical methods in geotechnical engineering*: 603–612. Springer-Verlag.

Boscardin, M.D. & Cording, E.J. 1989. Building response to excavation induced settlement. *ASCE Journal of Geotechnical Engineering*, 115(1): 1–21.

de Borst, R. 1986. *Non-linear analysis of frictional materials*, Ph.D. Dissertation, Delft University of Technology, Delft, the Netherlands.

Feenstra, P.H. 1993. *Computational aspects of biaxial stress in plain and reinforced concrete*, Ph.D. Dissertation, Delft University of Technology, Delft, the Netherlands.

Feenstra, P.H., Rots, J.G., Arnesen, A., Teigen, J.G. & Høiseth, K.V. 1998. A 3D constitutive model for concrete based on a co-rotational concept, In R. de Borst et al. (eds), *Proc. Int. Conf. Computational Modeling of Concrete Structures*: 13–22. Rotterdam: Balkema.

Gunn, M.J. 1992. The prediction of surface settlement profiles due to tunnelling, In *Predictive soil mechanics*, Proc. Wroth memorial symposium: 304–316. London: Thomas Telford.

Mair, R.J., Taylor R.N. & Burland J.B. 1996. Prediction of ground movements and assessment of risk of building damage due to bored tunnelling. In R.J. Mair and R.N. Taylor (eds*), Geotechnical aspects of underground construction in soft ground*: 713–718. Rotterdam: Balkema.

Netzel, H. 2000. *Guideline on prediction of damage to buildings*, part of Guidelines for the design of bored tunnels for road and rail infrastructure, Center for Underground Structures CUR/COB, committee L500, committee L540, CUR/COB, Gouda, the Netherlands.

Potts, D.M. & Addenbrooke, T.I. 1996. The influence of an existing surface structure on the ground movements due to tunnelling, In R.J. Mair and R.N. Taylor (eds*), Geotechnical aspects of underground construction in soft ground*: 573–578. Rotterdam: Balkema.

Rots, J.G. 1988. *Computational modeling of concrete fracture*, Ph.D. Dissertation, Delft University of Technology, Delft, the Netherlands.

Rots, J.G. 2000. Settlement damage predictions for masonry, In L.G.W. Verhoef and F.H. Wittmann (eds), *Maintenance and restrengthening of materials and structures – Brick and brickwork*, Proc. Int Workshop on Urban heritage and building maintenance: 47–62.

Sluys, L.J. 1992. *Wave propagation, localisation and dispersion in softening solids*, Ph.D. Dissertation, Delft University of Technology, Delft, the Netherlands.

Finite Elements in Civil Engineering Applications, Hendriks & Rots (eds.)
© 2002 Swets & Zeitlinger, Lisse, ISBN 90 5809 530 4

Numerical modelling of masonry panels strengthened using FRPs

M. Eusebio, P. Palumbo, F. Lozza
ENEL.HYDRO – B.U. ISMES, Seriate, Italy

G. Manfredi
University of Naples Federico II (NA)

ABSTRACT: An experimental program conducted at Enel-Hydro ISMES showed that externally applied fibre reinforced polymers (FRPs) are effective in increasing the load-carrying capacity of plain ancient masonry walls that are subjected to in-plane shear loads. Twenty walls with height of 1.6 m were used to conduct 20 tests. Both plain and reinforced walls were tested. The following experimental parameters were investigated: (1) type of fibre reinforcement; (2) amount of fibre reinforcement; (3) layout of fibre reinforcement. The modelling activity has been oriented towards establishing a numerical approach, which is applicable to vulnerability assessment of real large scale structures, in support to retrofitting design.

1   INTRODUCTION

Tests on masonry shear walls and numerical simulation of the observed behaviour were carried out by Enel.Hydro-ISMES and the University of Naple, within the scope of the MURST project CIT-TANOVA. This study was mainly oriented to the evaluation of the capability of fibre reinforced polymers (FRP) in increasing the load-carrying capacity of unreinforced ancient masonry walls that are subjected to in-plane shear loads.

The modelling activity has been oriented towards establishing a numerical approach, which is applicable to vulnerability assessment of real large scale structures, in support to retrofitting design. An example of such structures will be studied in the final task of the quoted research programme.

2   EXPERIMENTAL STUDY

Twenty single-story masonry wall panels were build to reproduce the typical layout of the ancient masonry buildings of Southern Italy.

As shown in Figure 1, the specimens were 1.5 m long, 1.6 m high and 0.5 m thick and they were made of tuff blocks with dimensions $0.10 \times 0.25 \times 0.40$ m^3 and 40 mm thick mortar joints prepared with a volumetric sand-binder, with an infill rubble masonry.

The prism tests on specimens from tuff blocks and mortar showed a mean compressive strength of 3.71 MPa and 1.95 MPa respectively.

Five plain panels have been tested in order to evaluate the behaviour of masonry walls in their original condition; two of them have been tested in compression up to failure, the remaining three ones have been tested in order to evaluate the shear behaviour up to failure. The remaining 15 have been tested considering two different reinforcement layouts (grid and cross), two different type of FRP reinforcement (CST carbon sheet and GST glass sheet) and two substance of FRP (300 g/m^2 and 600 g/m^2), see Figure 1. In order to guarantee an effective anchorage of FRP to the panel's face special adhesive mortar have been used. The FRP strips, 200 mm wide, have been applied all around the panel's sides; in order to avoid any confinement effect the horizontal strips applied on the lateral short sides have been vertically cut off.

The test set-up is shown in Figure 2. Each panel was subjected to monotonically increasing load under top displacement control in a confined way, i.e. keeping the bottom and top boundaries horizontal by means of two load cells providing the counteracting forces. A horizontal actuator controlled the panel's lateral displacement. The axial vertical load, equivalent to an axial stress of 0.5 MPa, was kept constant during each test.

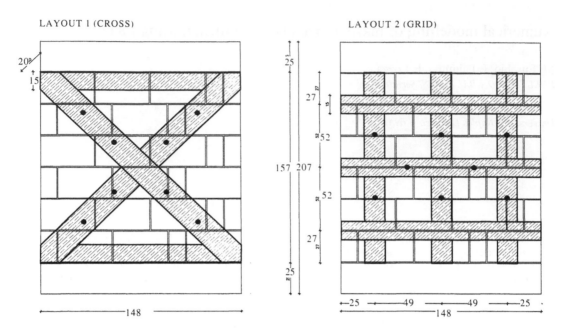

Figure 1. Masonry specimen with FRP cross layout 1 and grid layout 2.

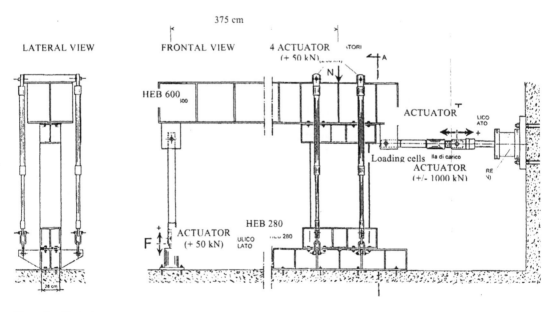

Figure 2. Test setup.

The major results of the experimental testing are briefly summarized:

– The resulting load-displacement diagram (Fig. 3) available for the plain panels showed a rather ductile behaviour. Initially, some cracks developed at the centre of the panel; upon increasing horizontal loading, cracking tended to concentrate in a large shear band spanning from one corner of the specimen to the opposite one. At the ultimate stage a well-defined failure mechanism is formed with a final shear band going from one corner of the

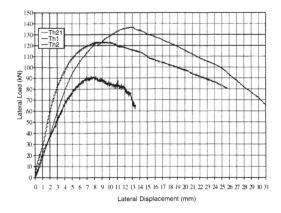

Figure 3. Experimental testing of plain panels. Lateral load versus lateral displacement.

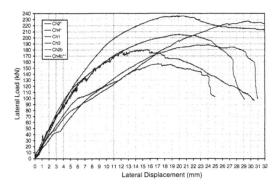

Figure 4. Experimental testing of FRP carbon reinforced panels. Lateral load versus lateral displacement.

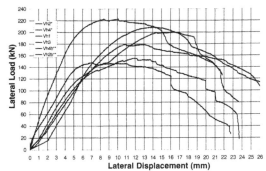

Figure 5. Experimental testing of FRP glass reinforced panels. Lateral load versus lateral displacement.

specimen to the other. Some vertical cracks formed at the panel's sides.
– The load displacement diagram (Figs 3 and 4) obtained for all 15 reinforced panels generally displayed a higher ductility with respect to the corresponding plain panel and furthermore a higher ultimate load. When the horizontal load increases, cracking tends to affect a larger area of the panel and, at the ultimate stage, the failure mechanism is accompanied by a diffused shear cracking and a vertical concentrated crack at the middle of the panel's face

Failure of the masonry is accompanied by delamination of the diagonal strips in tension for the layout 1 and by delamination with splitting of portions of masonry along the horizontal short strips for layout 2.
– Comparing the responses of plain panels and reinforced ones it is evident that the shear peak

strength increases dramatically (63%) with the introduction of fibre reinforcement to a plain panels.
– The role of the FRP reinforcement layout in increasing the shear resistance of wall panels is comparatively higher (about 20%) than substance (7%) and material properties (5%).

3 MODELLING

Modelling concerned two main aspects: representation of the plain masonry response to loads up to the ultimate state; representation of the FRP reinforced masonry system within the same range by introducing FRP action. The study has been carried out in a FE approach with the DIANA code (V.7.2).

The study has been subdivided in two steps. In the latter a finite element model of the plain masonry has been calibrated, in the former modelling fibre reinforced polymer FRP has been added.

3.1 Modelling plain masonry

Modelling criteria are constrained by the following requirements:

– To obtain a robust numerical model, capable of predicting the behaviour of the wall panel from the linear elastic stage, through cracking and degradation until total loss of strength depending on the applied load and straining level.
– To avoid heavy computational models, not justified for application to real structures.
– To allow for the calibration of response parameters of the system masonry-FRP fibres, on the grounds of laboratory test results on substructures (panels).

To this end, different options reported in the technical literature have been examined. A series of preliminary checks have been carried out to establish:

- Which finite element approach is better suited for the masonry panel.
- The necessary refinement of the mesh.
- The criteria to model the FRP strips reinforcement action and their appropriate constitutive model.
- The selection of the constitutive model for the masonry part of the system with specific focus of irreversible damage effects.

To this aim a micro-modelling strategy for masonry has been initially considered (Laurenco 1996), in which the units are described by continuum elements and the joints are reproduced with interface elements. Units (the tuff blocks) have been modelled with linear elastic isotropic elements (HX24L), bed and head joints with zero-thickness interface elements (Q24IF) responding with the Coulomb friction model criterion, with provisions for softening in tension and in compression. The above strategy is intended for predicting the joint failure mechanism, i.e. the mutual sliding of tuff blocks.

It was further analysed the actual relevance of inserting a discontinuity within the block which potentially can model the collapse of the block (Laurenco 1996); in this way, all the damage involving the unit is concentrate in a potential pure tensile cracks placed vertically in the middle of each unit. This modelling strategy was obtained by including two interface elements FRICT + DISCRA behaving in accordance to a compression yielding TRESCA criterion.

The testing of the above options has demonstrated that this very comprehensive approach cannot offer the required justification, due to:

- The scarcity of laboratory data in comparison to the amount of parameters needed for calibration.
- Heavy boundary and load conditions of the panel tests for the numerical convergence.
- The resulting finite element models are foreseen heavy to manage for real structures.

A macro-modelling approach has been therefore adopted, where blocks and mortar joints are smeared out in the continuum. The typical field of application of macro-models are large structures, subjected to loading and boundary conditions such that the state of stress and strain across a macro-length can be assumed uniform. The panel's face has been modelled as a shell: linear plane stress continuum 4 node quadrilateral elements (Q8MEM) 500 mm thick were used. This was justified by the foreseen global response of the panel given by the experiments. In fact experimental data do not separate the stiffness contribution of outer block faces and of the inner fill. The equivalent

parameters for masonry have been calibrated on the basis of the experimental results on plain panels.

Mesh refinement, which is a relevant issue in softening materials and standard continuum, has been a subject of several tests too: a regular grid of 900 elements 50×50 mm and 961 nodes was the final selection.

The tests have demonstrated that plain panels respond as an orthotropic body: in fact they showed a quite different behaviour in shear and compression respect to the tested panel, focusing on a typical situation of the orthotropic materials. An orthotropic homogeneous elastic material has been considered for masonry, where the two principal directions are coincident with horizontal X (bed mortar joints) and vertical Y (head mortar joints) model axes.

As for the modelling of irreversible strains, again some tests on panel models have been carried out to check the most effective solution among the ones available in the DIANA library. Two were specifically examined:

- Mohr-Coulomb (MOHRCO)
- Rankine-Von Mises (RANVMI)

Both constitutive models have been associated to a post-peak "smeared cracking" fracture model, where it has been assumed a mode I fracture mode in tension and compression, described by a fracture energy G_f^I. A linear tension/compression softening curve has been considered.

Even if, in principle the first criterion is more suited for confinement sensitive materials, it has the disadvantage of not providing a "compression cap", which is essential for this tuff masonry, which is comparatively weak also in compression. Moreover testing showed that the peak horizontal force and the post-peak response were poorly described.(Fig. 6).

The second criterion was better suited to catch the load-displacement experimental response (see Fig. 6) and was finally adopted.

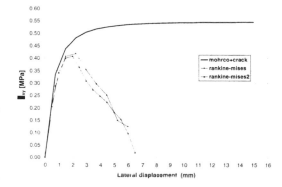

Figure 6. Plain panels. Comparison between the responses provided by Mohr-Coulomb and Rankine-Von Mises yield criteria.

The confinement effect due to the top and bottom kerbs of the specimen have been simulated using rigid cinematic boundaries constraining all the upper nodes to keep them horizontal. A rigid link has also been applied to all nodes of the upper section, to reproduce the effect of a steel beam and of the compensating load.

3.2 Addition of FRP

The study of the better justified modelling approach for the system panel + FRP. The starting point is that of keeping the results of the study for the plain panels and concentrating on the modelling of the reinforcement. The criterion is that of providing a way to capture the global load-displacement response of the panels.

Numerical simulation has been set up for FRP carbon fibres with substance of 300 and $600\,g/m^2$. Two layouts of reinforcements (grid and cross, Fig. 1) have been analysed by the finite element model.

Fibres have been represented by uni-axial finite elements (SPRING) providing a resisting action both under stretching and contraction. Their stiffness has been considered constant, whereas a limit has been established on the allowable maximum stretching and contraction at which the stiffness suddenly drops to zero (Fig. 7).

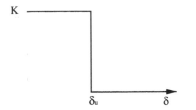

Figure 7. Axial stiffness versus stretching and contraction.

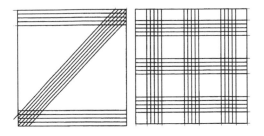

Figure 8. Modelling of FRP reinforcement with cross and grid layout.

The experimental behaviour showed that for the layout 1 (cross strips) the compressive strength contribution provided by strips was negligible and therefore it has been neglected in the analysis. For the layout 2 (grid strips), on the contrary, tests showed that strips in compression provided a significant contribution up to a limit compression strain of 0.001, evaluated from the strain measures provided by the strain gauges placed along the fibres. The maximum stretching displacement assigned has been obtained from target limit strain values available for the two different types of fibres (carbon and glass).

As to reproduce the actual contact area of the strips 5 spring elements have been used.

Full connection among nodes of the masonry FE model and the strips was considered, because the local delamination effect of the fibres on the primer was not measured.

The finite element runs were carried out with the same criteria adopted for the plain panels.

4 MODEL CALIBRATION

4.1 Loading history

Initially, the finite element model has been subjected to a vertical load, distributed over the length of the upper lines of nodes and equivalent to an axial stress of 0.5 MPa, kept constant during all the analysis. Analyses have been carried out under displacement control by specifying horizontal displacement increments at all the nodes on the top line of the model.

4.2 Plain masonry wall panels

The calibration of the material parameters has been guided by the laboratory test results, and, namely to the response of the panels to shearing. The elastic parameters required to describe the orthotropic response in DIANA are the following:

E_x = Young modulus along the X direction
E_y = Young modulus normal to X direction
v = Young modulus
G = Shear modulus

The inelastic properties required from the RAN-VMI model are the following:

f_{ct} = tensile strength
f_{cc} = compressive strength
G_f^I = fracture energy (mode I),
G_c^I = compressive fracture energy (mode I)

The compressive strength f_{cc} has been obtained using the results of the compression test up to failure of a panel from the same series (Faella et al., 1991), while f_{ct} (tensile strength), G_t (fracture energy Mode I) have been obtained directly from the experimental shear behaviour of the plain masonry wall panel. Finally, the value of G_c (fracture energy Mode I) has been calibrated starting from the results of the shear tests performed for reinforced panels.

Calibration has been carried out selecting test TH2 as the most representative of the typical response of plain panels. The best matching between numerical and experimental results in term of load-displacement diagram has been obtained using the following parameters in Table 1.

4.2.1 Sensitivity analysis

In order to evaluate the relative importance of the material parameters some sensitivity analyses have been carried out, in which some of the material parameters are varied. Specifically, keeping all the remaining parameters unchanged, the sensitivity of the load-displacement result has been investigated for the following:

– Fracture energy G_t^I (Fig. 9a)
– Compressive energy G_c^I (Fig. 9b)
– Compressive strength f_{cc} and Young moduli E_x and E_y (Fig. 9c)

The main results have been the following:

– In the post-peak regime fracture energy G_t^I is the most important parameter;
– Compressive strength f_{cc} has a small influence on the peak shear value.

Table 1. Calibrated masonry parameters.

Ex	380	MPa
E_y	400	MPa
ν	0.2	
G	70	MPa
G_t linear	0.025	N/mm
G_c linear	5	N/mm
f_c	1	MPa
f_t	0.05	MPa

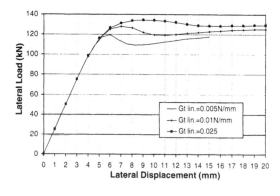

Figure 9a. Importance of fracture energy G_t^I value.

4.2.2 Numerical results

The comparison between numerical and experimental load-displacement diagrams is given in Fig. 10. As shown, the model is able to reproduce the complete

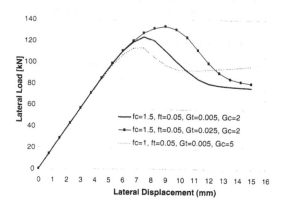

Figure 9b. Importance of compressive fracture energy G_c^I value.

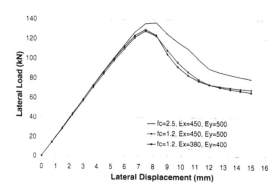

Figure 9c. Importance of compressive strength f_{cc} and Young moduli E_x and E_y.

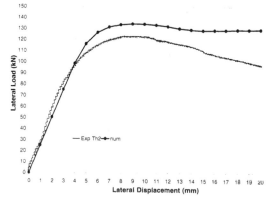

Figure 10. Plain panels. Numerical and experimental load-displacement diagrams.

path of wall panel in the pre-peak regime up to 100 kN (more than 80% of the peak). The peak predicted by the model, equal to 135 kN, is higher than the experimental one (+8%), but this discrepancy is reasonable considering the high scatter of the experimental strength values.

A significant difference in the post peak displacement path is observed, i.e. the model shows a greater ductility with respect to the wall panel. A better calibration of the softening behaviour could reduce this discrepancy, but we think that a linear softening description adopted can be an operational compromise between the requirements of a simple, unambiguous definition and accuracy of results, keeping into consideration the significant dispersion of experimental results. Furthermore, the failure mechanism observed in the experiments is well captured by the model, which is the most important validation of any model.

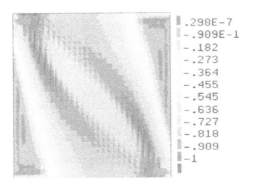

.298E−7
−.909E−1
−.182
−.273
−.364
−.455
−.545
−.636
−.727
−.818
−.909
−1

Figure 11. Results of the analysis. Plain panel. Minimum principal stresses (N/mm^2) at a displacement of 15 mm.

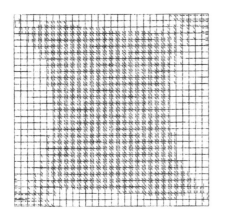

Figure 12. Results of the analysis. Plain panel. Crack pattern at a displacement corresponding to the peak load.

4.3 Reinforced wall panels

The initial values adopted for the FRP carbon strips are:

– Fibres' elastic modulus in extension: 230000 MPa,
– Maximum stretching: 1.5%,
– Mean thickness 0.165 mm.

The parameters required for the unidirectional springs, which describe the reinforcement are the axial stiffness K and the maximum displacement δ_u. The latter need to be defined along two directions.

The best matching between numerical and experimental results in terms of load-displacement diagram has been obtained using the following parameters, where tests Ch2 (cross layout, Fig. "") and Ch. 4 (grid layout, Fig. "") were taken as the most representative of reinforced panels' response (Tables 2a and 2b).

4.3.1 Numerical results

The comparison between numerical and experimental load-displacement diagrams is given in Figs 13 and 14 for cross and grid layout respectively considering the substance "a" (600 g/m^2).

As shown, a very good agreement has been reached for grid strips layout. For the cross one the model is able to reproduce the complete path of wall panel in the pre-peak regime up to 130 kN (more than 70% of the peak). The peak predicted by the model, equal to 205 kN, is higher than the experimental one (+14%), but this discrepancy is still acceptable considering the high scatter of the experimental strength values.

Table 2a. FRP carbon fibres with grid layout. Calibrated parameters.

Strips	Area [mm^2]	Length [mm]	Substance (2 faces)		δ_u [mm]
			"a" K [N/mm]	"b" K [N/mm]	
Horizontal	6.6	50	6.1E + 04	1.2E + 05	0.75
Vertical	6.6	52.5	5.8E + 04	1.2E + 05	0.79

Table 2b. FRP carbon fibres with cross layout. Calibrated parameters.

Strips	Area [mm^2]	Length [mm]	Substance (2 faces)		δ_u [mm]
			"a" K [N/mm]	"b" K [N/mm]	
Horizontal	8.7	50	8.0E + 04	1.6E + 05	0.75
Diagonal	5.1	72.5	3.3E + 04	6.5E + 04	1.09

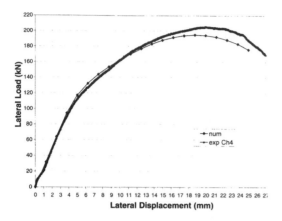

Figure 13.   Reinforced panels grid layout. Numerical and experimental load-displacement diagrams.

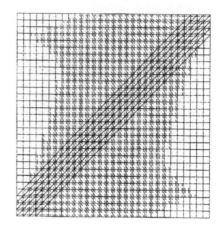

Figure 15b.   Results of the analysis. Reinforced panel cross layout. Crack pattern at a displacement corresponding to the peak load.

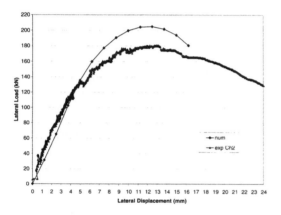

Figure 14.   Reinforced panels cross layout. Numerical and experimental load-displacement diagrams.

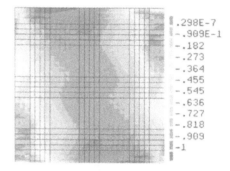

Figure 16a.   Results of the analysis. Reinforced panel grid layout. Minimum principal stresses (N/mm^2) at a displacement of 20 mm.

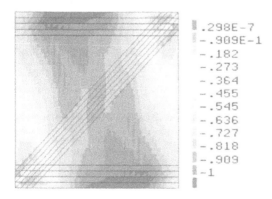

Figure 15a.   Results of the analysis. Reinforced panel cross layout. Minimum principal stresses (N/mm^2) at a displacement of 13.75 mm.

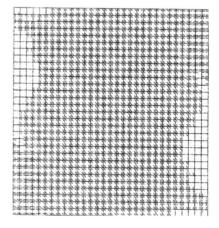

Figure 16b.   Results of the analysis. Reinforced panel grid layout. Crack pattern at a displacement corresponding to the peak load.

7 CONCLUSIONS

The study described herein is devoted to setting a modelling framework for the safety assessment and retrofitting of masonry structures with FRP technology. Models' selection and calibration has been based on literature data and on tests on plain and FRP reinforced panels, purposely carried out within the project. The resulting modelling framework is that of a macro-model describing the structure as a continuum with orthotropic elastic stiffness. Cracking and crushing is described by a "smeared-crack" non-linear model and by a fully plastic yield model respectively.

FRP are introduced explicitly in the model as spring elements of given elastic stiffness and with a threshold of elastic response based on maximum stretching.

The validation of such modelling approach is foreseen on a real case of rehabilitation of an ancient military hospital in Naples.

REFERENCES

ISMES-T131.11, Doc. RAT-ISMES-3583-2001, Rev. 0, *PROGETTO CITTANOVA Prove di laboratorio su pannelli in muratura di tufo.*

P. Benson Shing and Hamid R. Lotfi 1991, *Experimental and Finite Element analyses of single-story reinforced masonry shear walls*, Computer Methods in Structural Masonry, Books & Journals International, Swansea, UK.

Faella G., Manfredi G., Realfonzo R. 1991, *Experimental evaluation of mechanical properties of old tuff masonry subjected to axial loading*, Proc. of the 9th International Brick/Block Masonry Conference, Berlin 13–16 October 1991.

Laurenço P.B., Rots J.G., Blaauwendraad J. 1995, *Two approaches for the analysis of masonry structures: micro and macro modeling*, HERON Vol. 40, N°4.

Rots J.G. 1991, *Computer Simulation of Masonry Fracture: Continuum and Discontinuum Models*, Computer Methods in Structural Masonry, Books & Journals International, Swansea, UK.

Finite Elements in Civil Engineering Applications, Hendriks & Rots (eds.)
© 2002 Swets & Zeitlinger, Lisse, ISBN 90 5809 530 4

Strength of placed block revetments on dikes determined from field tests

C.M. Frissen
TNO Building and Construction Research, Delft, the Netherlands

H.L. Bakker
Dutch Ministry of Transport and Public Works, Delft, the Netherlands

M. Klein Breteler
Delft Hydraulics, Delft, the Netherlands

ABSTRACT: A numerical analysis on strength of placed block revetments in dike constructions is presented. A 3 dimensional finite element model is developed with physical and geometrical nonlinear characteristics for a very commonly used type of placed block revetment in the Netherlands. Earlier research showed that clamping forces can have a relevant positive contribution to the strength of block revetments in dike constructions. But adopting these clamping forces in strength-calculations of dike constructions is still not standardized because the relation between results from field-tests and wave loads is missing. To this end both field-tests and calculations of failure of the revetment caused by wave attack are numerical simulated. The general behavior of the model in both calculations is described and explained. A brief outline of parameter studies is given. The result is a relation between the strength measured during field-tests and failure in practice under wave attack.

1 INTRODUCTION

Placed block revetments are applied to protect dike constructions against wave attacks. These blocks can have several shapes. In this paper polygonal column blocks are studied which is a very commonly used type of placed block revetment. With this irregular shape, 3 dimensional effects play an important part in strength of the revetment.

Between the blocks there are joints filled with sand or gravel. Trough this the revetment is more or less permeable. When a wave attacks the dike construction, water penetrates into the permeable filter layer under the blocks. This results in an upward water pressure under the blocks when the water withdraws after the attack and the phreatic water line lies higher than the average water-level. This is shown in Figure 1.

In the past the revetments are dimensioned on the criteria that the blocks are lifted out of the revetment (resulting in failure) when an upward pressure larger than the dead weight occurs. But experiments in practice through "pull-out test" showed a higher strength of the revetment. Earlier research demonstrated that so called clamping forces bring about a positive contribution to the strength of the revetment.

The problem is that the relation between practice and theoretical research is missing. The purpose of

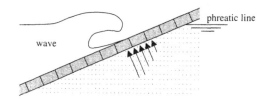

Figure 1. Upward water pressure under the blocks.

this study is to get more insight in the behavior of a placed block revetment under wave attack in relation to the pull-out test. That is why in this project both field-test and calculations of failure of the revetment under wave attack are numerical simulated. The relation between both numerical calculations can lead to more basis for adapting clamping forces in strength calculations.

This project is carried out under the authority of the Road and Hydraulic Engineering Division of the Dutch Ministry of Transport and Public Works. The work is done at TNO Building and Construction Research, department of Computational Mechanics, and advised by Delft Hydraulics.

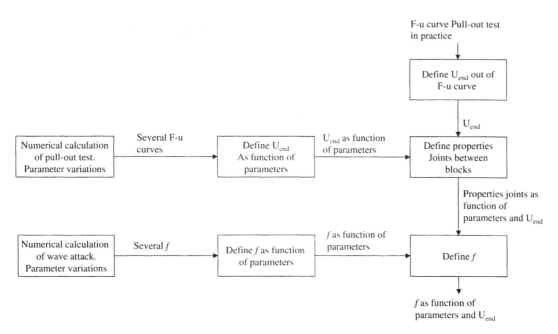

Figure 2. Process.

1.1 *Mechanisms*

The upward water pressure can cause damage to the revetment. There are two important failure mechanisms:

– Piston mechanism
  One block is lifted out of the revetment while the surrounding stay unaffected. The neighboring blocks are not lifted.
– Clamping mechanism
  The blocks are clamping to each other so that several blocks are lifted. The deformation of the revetment looks like an arch.

In this study the dike construction is so modeled that both mechanisms can occur.

1.2 *Approach*

In the current design rules the positive contribution of clamping forces is not adopted in strength-calculation because the relation between results from field-tests and wave loads is missing. In this paragraph the approach will be described how the relation between field-tests and failure under wave attack through numerical research can be made. In Figure 2 this process is visualized. The two "horizontal lines" in this process are the numerical calculations, the vertical the experiment test in practice. This path crosses one another so that a relation can be made between the experiments in practice and failure of the revetment due to wave attack.

Result of a pull-out test is a force-displacement relation (F-u curve). The calibration will be made through the result U_{end} the displacement of the block at the maximum force of 9000 N. For more detail information about these tests see §2.2.1. With the results of parameter variations of numerical pull-out tests, an empirical formula is made for U_{end} as a function of these parameters. In numerical calculations the unknown parameter is the stiffness of the joints between the blocks. This parameter can hardly be determined in practice. This stiffness can now be defined by rewriting the empirical formula of U_{end}.

The result of a numerical calculation under wave attack is a loading factor f which is related to the strength of the revetment. The loading factor f is equal to the maximum acceptable upward wave pressure divided by dead weight pressure of the blocks. Again with parameter variations an empirical formula is made for f. This formula can be combined with the formula of the stiffness of the joints. The result is an empirical relation for the loading factor f as a function of (known) parameters and U_{end}. Through this the objective is achieved namely a relation between failure in practice under wave attack (f) and field-tests (U_{end}).

2 NUMERICAL MODELING

Before numerical calculation can be made, a numerical model must be made. The model can be used for both field-tests and calculation of failure under wave attack.

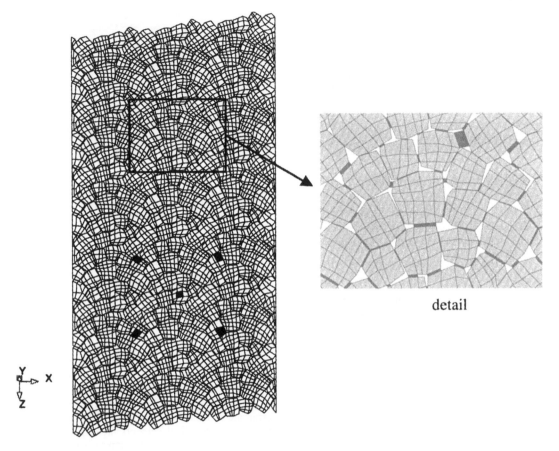

detail

Figure 3.   Element Mesh (Top view).

2.1 Geometry

A strip of a commonly used placed block revetment in dike construction is made. The model is approximately 7 meters long and 3.5 meters wide. The left and right edges are supported in horizontal x-direction (perpendicular to the edges). To simulate the blocks to be tied down at the bottom, this edge is supported in parallel direction of the slope. The top-edge is free. The slope is 16 degrees. The model consists of blocks (columns), joints between the blocks and a filter layer. The element mesh is shown in Figure 3.

The blocks are modeled by the use of 8 noded linear interpolated solid brick elements, called HX24L. The assumption is made that the blocks behave linear elastic. The blocks are placed on an elastic filter layer. This bedding is described with interface elements (Q24IF) with coulomb friction. Also the joints are modeled with interface elements with coulomb friction so that both mechanisms, piston and clamping, can occur. The elastic normal-stiffness of the joints is defined of the Young's-modulus of the filling-material E_{joint} and

the exact space between the blocks. This results in various joint stiffness in the model.

The model is characterized by both physical and geometrical nonlinear behavior.

2.2 Loads

The loading process consists of two phases. First the dead weight will be applied followed by the typical load of a pull-out test or wave load. The characteristic features of these loads are described in the next paragraphs.

2.2.1 Pull-out test

In practice the strength of the placed block revetment is measured with a sort of pull-out test. With a special car a column will be loaded centrically with a tractive force normal to the slope of the dike construction. Through the legs of the car an equal reaction force will be pressed on the blocks. The dead weight of the car will be neglected. The tractive force will be gradually

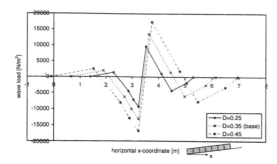

Figure 4.   Wave load during wave impact.

increased until the maximum force of 9000 N. In Figure 3 the "black" surfaces are visualized on which the pull-out force and the reaction force through the four legs is divided.

2.2.2   *Wave attack*

After the blocks are pushed into the filter layer as a result of the weight load, the wave load is applied. The wave load is coming from an experimental measurement (defined by Delft Hydraulics) and is visualized in Figure 4. In this figure the wave load is given as a function of the place x in the revetment. With x equal to zero the bottom of the slope of the revetment is meant. In this figure a negative load is an upward pressure.

For these calculations the wave load is assumed to be depended on D, the height of the blocks. With variation of D both the amplitude and the length of the wave load will be linearly scaled. Herewith the maximum upward wave load is assumed to be always on the same place. In this way the location of the wave is almost the same as the place of the traction force in the pull-out test.

The wave load is gradually increased until failure occurs. Hereby the amplitude will be increased while the wave-length stays the same. In spite of the fact that with this wave-load the water-level is at 3.3 meter, the assumptions is made that all blocks are lying under water so the dead weight of all the blocks are reduced.

3   NUMERICAL ANALYSIS AND RESULTS

To get understanding of the characteristic behavior of the placed block revetment, the results with respect to the base case are described extensively.

3.1   *Pull-out test*

The results of the pull-out test for the base case are analyzed. In this case the height of the blocks is 0.35 meter, the Young's-modulus of the joints E_{joint} is equal to $5 \cdot 10^6$ N/m^2 (assumed to be filled with sand) and the

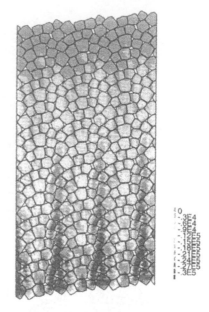

Figure 5.   Stresses parallel to the slope below in the blocks (cross-section at 1/4 D).

filling grade υ is 0.75. With υ equal to 0.75 the joints are filled with material over 75% of the block height. Due to water and wind the top 25% of the joints is empty.

For analyzing the behavior the stresses and displacements are visualized. Due to dead weight stresses occur but the (initial) displacements are suppressed. The result is that the dead weight is transported downwards through certain "vertical lines". This is visualized in Figure 5 where the stresses are shown parallel the slope in a cross-section on ¼ of the height of the blocks. The dark areas represent the high pressure zones.

After the dead weight the tractive force is applied and increased until the maximum force of 9000 N. In Figure 6 the displacements at the maximum force is visualized. This figure shows that the drawn block is slipping while lifting the neighboring blocks slightly up. A combination of the piston and clamping mechanism occurs.

These mechanisms are also recognizable in Figure 7 where the normal stresses in the filter layer are shown at the maximum force. The dark gray zone in the middle represent the blocks that came loose of the bedding. In this area the stresses are equal to zero. The white area around the drawn block represents the blocks that have a little uplift from the bedding. The four dark spots are pressure zone as a results of the reaction force of the legs of the traction-car.

The result of the pull-out test is a force-displacement relation (F-u curve) which is visualized in Figure 8. Herewith the average displacement of the surrounded blocks is used so that the influence of the appearance

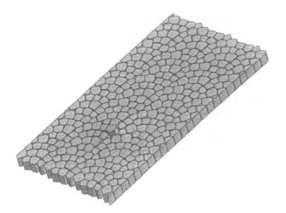

Figure 6. Displacement at maximum force of 9000 N (deformfactor = 25).

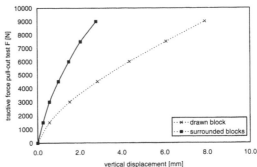

Figure 8. Force-displacement curve (F-u diagram).

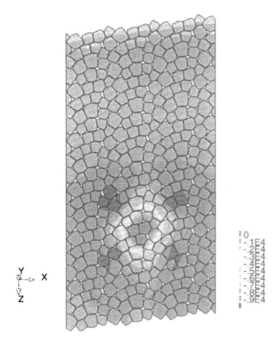

Figure 7. Normal stresses in filter layer at F = 9000 N.

of the piston-mechanism will be eliminated. This is a safe approach because the results of the pull-out test in practice will be compared with a stiffer model. Also at the experiments possible irregularities in the F-u curve (influence of appearance of "piston mechanism") will be eliminated with "engineering judgment".

With these parameters (base case) a displacement of 2.88 mm occur at the maximum force of 9000 N.

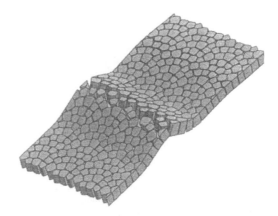

Figure 9. Displacements just before failure (deform-factor = 25).

3.2 *Wave attack*

Failure under wave attack is simulated on the same model with the same parameters (base case). After the same initial situation resulting from the dead weight (visualized in Figure 5) the wave load is applied and increased until failure. In Figure 9 the displacement at the maximum load (just before failure) is shown. In this figure the clamping mechanism is clearly visible.

The loading factor f is equal to the maximum acceptable upward wave pressure divided by dead weight pressure of the blocks. In this case this factor f is equal to 4.09.

In Figure 10 the normal stress in the filter layer is given at the maximum wave load. In this figure two large areas are identifiable where the blocks came loose of the bedding (light gray).

4 PARAMETER VARIATIONS

The influence of three model parameters is investigated. The Young's modulus of filling-material E_{joint},

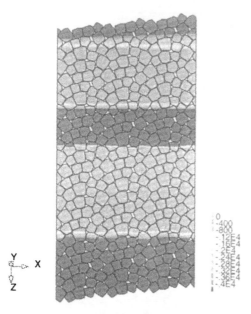

Figure 10.   Normal stresses in filter layer just before failure.

the block height D and the filling grade υ are varied sequently keeping other parameters at base case level. For every set of parameters the end displacement U_{end} as a result of the pull-out test and the loading factor f, defined with failure under wave attack is calculated. All the three parameters have the same dependency of influence to the behavior of the revetment. By increasing these parameters the revetment becomes stiffer so that U_{end} decreases and f increases.

With the results of the parameter variations, an empirical relation is made for the loading factor f as function of the block height D, the filling grade υ and U_{end} a result of a pull-out test in practice. Through this a relation is made between the pull-out test in practice and failure caused by wave attack.

5   CONCLUSIONS

This paper considers the physical and geometrical nonlinear behavior of placed block revetments.

The following conclusions can be made:

- It is possible to simulate both a pull-out test and failure under wave attack on a placed block revetment with the help of DIANA.
- This project shows that clamping forces increase the strength of placed block revetments (like earlier research).

- The numerical simulation of the pull-out test made it possible to define the unknown parameter, the stiffness of the joints. After comparison the numerical results of the pull-out test with experiment of these test in practice, a lower limit for the joint stiffness is about $0.5 \cdot 10^9$ N/m². This correspond with filling material met a Young's modulus of approximately $1.0 \cdot 10^6$ N/m².
- At the maximum tension force of 9000 N the drawn block uplifts approximately 3 millimeters. This corresponds with the revetment fails at a wave load 4 time dead weight.
- The Young's modulus of filling-material E_{joint}, the block height D and the filling state υ have a positive contribution to the strength of the revetment. By increasing these parameters the loading force f increases.
- Through this numerical research an empirical relation is made between the field test in practice and failure in practice caused by wave attack. With the help of this empirical relation the strength of the revetment can be determined. Hereby the clamping forces are automatically taken into account. The only needed input item is U_{end} that will be measured with a pull-out test in practice.

REFERENCES

Frissen, C.M. 1996. *Numerieke modellering van steenzettingen: effecten in langsrichting van de dijk, met variatiestudies (in Dutch)*, Rijswijk, TU Delft rapport nr. 03.21.1.22.26, TNO Bouw rapport nr. 96-NM-R-0994.

Frissen, C.M. 2000. *Numerieke modellering van een blokken- en zuilenbekleding, Simulaties van trekproeven en bezwijken door golfbelasting (in Dutch)*, Rijswijk, TNO Bouw rapport nr. 2000-MIT-NUM-R016.

Frissen, C.M. & Schreppers, G. 1996. *Inklemming bij steenzettingen op dijken: Numerieke analyse en praktijkaanbevelingen (in Dutch)*, Rijswijk, TNO Bouw rapport nr. 96-NM-R-1592.

Frissen, C.M. & Schreppers, G. 1998. *Numerieke modellering van steenzettingen: Zuilenbekledingen, met variatiestudies, imperfecties en praktijkaanbevelingen (in Dutch)*, Rijswijk, TNO Bouw rapport nr. 98-MIT-NM-R1762.

Frissen, C.M., Schreppers, G. & Bakker, H.L. 1997. *Strength of placed block revetment in dike constructions*. Rotterdam: Balkema.

Suiker, A.S.J. 1995. *Inklemeffecten bij steenzettingen op dijken. Eindige-elementenstudie naar geometrisch- en fysisch niet-lineair gedrag van blokkenmodellen (in Dutch)*. Rijswijk, TU Delft rapport nr. 03.21.0.31.09, TNO Bouw rapport nr. 95-NM-R0253.

Geomechanics, steel and rubber

Finite Elements in Civil Engineering Applications, Hendriks & Rots (eds.)
© 2002 Swets & Zeitlinger, Lisse, ISBN 90 5809 530 4

Invited paper: Advanced modelling to support innovative developments in tunnelling for Amsterdam North/Southline

Ir. F.J. Kaalberg
North/South LineConsultants & Witteveen + Bos Consulting Engineers Amsterdam, The Netherlands

ABSTRACT: The planning and design for the new North/South metro line in Amsterdam is a technical challenge because of the extremely difficult conditions comprising the historical environment, piled foundations and the soft subsoil. Advanced numerical modeling is required to set new standards in the understanding of soil-structure interaction regarding building response as well as TBM- and lining design and (compensation) grouting techniques. Several innovations are developed based on the principle of Value Engineering. This paper will give an overview of the key position of FE modeling in the design.

1 INTRODUCTION

1.1 *Historical background*

Previous metro construction in Amsterdam in the early 70's used cut and cover methods that required the demolition of many historic structures. This resulted in great public opposition to any further expansion of the system. However, increasing population and traffic congestion within central Amsterdam caused the city council to undertake studies to see if new methods could create a metroline without significant surface disruption and completely protect historic structures. These studies revealed that it was feasible to construct a metro in the soft-soil conditions underlying Amsterdam by using innovative mechanized shield tunneling methods. However these studies also emphasized that project success required careful and extensive subsurface investigations, intensive FE modelling to predict (settlemement) risks, innovative Tunnel Boring Machine (TBM) design and operations, and real-time monitoring of critical conditions during construction.

1.2 *Alignment and phasing*

After careful consideration of these studies and weighing alternatives, the city council of Amsterdam decided in November 1996 to permit construction of a new underground metro line, the North-South Metroline. The line is to be constructed in two phases and, when completed, will extend from the north of the greater Amsterdam region to Schiphol International Airport south of the city.

The first phase of the project will be about 9 kilometers long (Figure 1). The section north of the IJ river will run along the surface. An immersed tube construction will be used to pass under the IJ river and the Amsterdam Central Station complex. The line then traverses the historic heart of Amsterdam to reach its southern terminus at South/World Trade Center (South/WTC). When this phase is completed in 2010, it is anticipated that it will transport around 200,000 passengers per day.

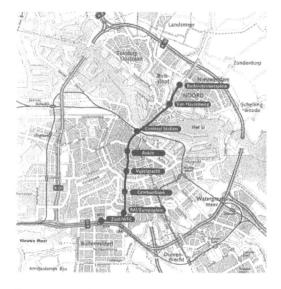

Figure 1. Map of Amsterdam including proposed alignment.

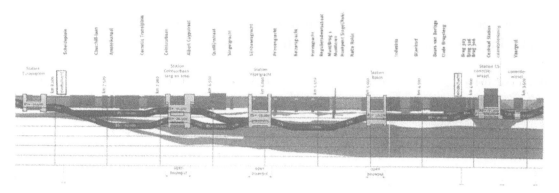

Figure 2. Vertical alignment and geological profile.

2 CONSTRUCTION CHALLENGES

Constructing the section between Central Station and South/WTC under the historic city center will be particularly challenging. At least 3.15 kilometers of the total section length of 3.8 kilometers must be constructed using advanced TBM technology in order to avoid surface disturbance and damaging historic buildings. Most adjacent buildings have a great historical value and in all cases demolishing of buildings is prohibited. Therefore potential damage has to be minimized. Two individual single-track tubes with an inner diameter of 5.82 m and a bore hole diameter of approximately 7 m must be constructed at depths between 20 m and 34 m.

Legal stipulations require the tunnels to mostly follow the present street pattern (Figure 3) and thus to avoid passing under houses or other buildings. However the narrowness of the existing streets requires the tunnels to be closely spaced, separated by only about 0.5 tunnel diameters, and along one section they even must be stacked vertically. In spite of such measures, the minimum curvature allowed for the trains (190 m radius) means that the tunnels will have to pass under a few buildings.

This section also includes three deep stations. In order to minimize surface disturbance, these stations will use covered excavations. Diaphragm walls will be placed to depths as much as 40 m, followed by placement of a roof slab. Then the street environment can be returned to its previous condition while the excavations are continued within the walls and under the roof.

This combination of TBM tunnels and station construction methods satisfy the goal of minimizing disruption of normal surface activities. However, prevention of surface deformations and damage to buildings and existing utilities requires careful design and management of construction operations based on detailed knowledge of subsurface conditions and FE modeling.

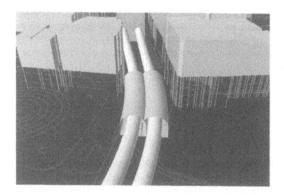

Figure 3. Typical cross section.

3 THE ROLE OF FE MODELLING IN VALUE ENGINEERING

Because it was felt that most of the critical design aspects were at the edge, or beyond the state of the art design in Civil Engineering, it was acknowledged in an early stage of the design process, to develop a design philosophy which is dominantly based on comprehensive and advanced 3D FE modeling. Although 3D FE modeling is now rapidly becoming a part of common practice in CE design this was unusual at the beginning of the design phase (1996 onwards) for the North/Southline. Therefore each revealed risk item in the design was supposed to be initially tackled by more common design methods and the implementation of a (often costly) preventative design measure, as long as it was affordable. Based on the Value Engineering philosophy each design risk aspect was later on subsequently optimized (if possible) by more advanced design methods, which were dominated by 3D FE modeling. 3D FE modelling is often considered to be too costly for Civil Engineering practice purposes and

Figure 4. Typical historical buildings along canal.

only useful in an University environment, but if the risks involved in construction are high, their contribution is certainly paying back in reduced construction costs. For a complex underground construction project like the North/Southline this principle appeared to be highly appropriate.

Because the main risk bearing design items were mostly in some way related to geotechnics, and more particularly to the even more complex interface of geotechnics and structures, in several cases even the most advanced (3D) FE modeling was considered to be not reliable enough (at this time) to asses all the risks involved properly. Therefore the Value Engineering philosophy is supported by the even more costly principle of "Design by Testing". Several major Full Scale Trials were developed and executed later on to asses the risk involved in construction in a more proper way.

Although this approach seems quiet logical, it is still not very common in Civil Engineering practice in Europe, even for Underground projects considered to be subjected to a high risk profile. The above described approach requires, above all, a well developed vision of investing in risk reducing design methods by the Client, which is not always met. This is probably not only for financial reasons but also due to limited time restrictions during the design phase which are very often dominant and the lack of economical testing opportunities available.

Regarding the time aspect the North/Southline metro project however, is characterized by an extended design phase, which was caused by program delays with either a political or a financial background. At first the design phase was extended because there was political disagreement about the optimal horizontal alignment, which was finally politically solved by a (corrective) referendum. The outcome of this referendum was that the twin bored tunnels were going to be extended with another 700 m through the Ferdinand Bolstreet, although it made the design even more challenging, because this part of the track was traversing a densely

populated area with very narrow streets and buildings with poor foundation conditions. In a later stage the overstrained situation in the Dutch market for infrastructure projects at that time, forced the Municipality to reassess their tender strategy, which caused extra delay in the program. Analogous to the design phase the subject of risk managment and -allocation dominated the discussions with the Contractors during the several tender stages.

The situation described above confirmed even more the need for Value Engineering and the principle of "Design by Testing", in order to reduce the risks and achieve a design with a risk profile which is in a good economical balance with the investments. So the above demonstrates not only time was available to optimize the design but also the will to reduce risks pushed the advanced 3D FE modeling forward as the key element in Value Engineering.

Above all the major boost in (underground) infrastructure projects in the Netherlands in the last decade created a climate for more design practice related research with the founding of the Centre for Underground Construction (COB). This government supported organization is a co-partnership between Research Institutes, Consultants and Contractors to initiate pre-competitive research. The COB played an important role in providing the conditional contractual framework in which the North/Southline organization was able to perform Full Scale Trials at the other bored tunnel projects in the Netherlands.

4 GEOTECHNICAL CHARACTERISTICS

The subsurface of Amsterdam is composed of a thick sequence of Holocene and Pleistocene sand, silt, clay and peat layers (Figure 5).

The oldest Pleistocene deposits are marine clays and fine-grained sands that currently extend far below the surface – some 250 to 350 meters deep. Tongues of Saalian glacial ice eroded a deep basin into these deposits. This basin was subsequently flooded by the sea and partially filled with marine sands and clays (the Eem Clay). During the last glacial period, the Amsterdam region was subjected to a peri-glacial climate and the basin was filled with sand (Second Sand Layer). Younger Holocene deposits consist of mainly peats and clays formed under the influence of an initially rising sea level. About 5000 years ago, a lowering of the sea level allowed rivers to create channels in the existing deposits. These channels, sometimes with widths of over 100 meters contain thin sand and soft clay layers, caused by infilling during to the continued rising sea levels of recent times.

The result of this geological evolution is a fairly consistent stratigraphy of sand clay and peat layers, as shown in Figure 2, with occasional incised river

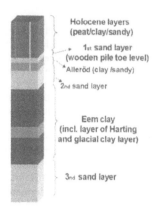

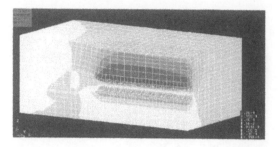

Figure 6. 3D model of subsoil and tunnel.

Figure 5.   Subsoil of Amsterdam.

channels. Historical buildings are placed on wooden piles about 13 meters long that are founded in the First Sand Layer. Larger, modern buildings are founded on concrete piles driven to the Second Sand Layer. Tunnel driving will take place almost completely in the Allerod, Second Sand, and Eem Clay units (Figure 2). The tunnels will not directly impact the older shallow pile foundations, and since the locations of the modern deeper piles are well known and not very frequent, avoidance or remedial measures can be taken during construction as required.

An important geological consideration is the presence of groundwater beginning about 1 m below the ground surface. The tunnels and station diaphragm walls are thus subjected to water pressures of around 3.5 bars. Disturbance of the groundwater regime cannot be allowed.

Dewatering would likely immediately cause unacceptable ground deformations, but also it would cause deterioration of the older wooden piles. These piles are protected as long as they are totally immersed, partial exposure by dewatering will cause them to decay.

5 SETTLEMENT RISK ASSESSMENT

5.1 General

One of the major critical design aspects are the Settlement Risk Assessment and Management issues. This includes settlement predictions, building response to settlement induced movements, pile behaviour due to tunneling, TBM performance and settlement mitigation, which will be described subsequently below. In fact all this design items interact with each other so an iterative cylclic design with mutual input is necessary. All of them required substantial 3D FE modeling in order to minimize riks.

5.2 Settlement predictions

During the design phase settlement predictions became more sophisticated. Initially the desing team started with analytical methods and 2D contraction FE models but finally complete 4D analysis were made (Figure 6). The 4D model comprises a detailed TBM (Figure 13), differentiated radial grout pressures in the annular gap around the tunnel, bentonite front pressures, and a tunnel lining. The Modified Mohr Coulomb model was used to simulate the behaviour of the subsoil after its was validated itself by means of modeling triaxial tests. The 4D model was calibrated and validated (van Dijk, Van Empel and Kaalberg, 1999) with the data from the test sites at the 2nd Heinenoordtunnel (par. 4.3), based on subsurface deformations at several depths (Figure 7).

5.3 Building response due to tunneling

5.3.1 General

Approximately 1500 historically important masonry buildings, supported by the more than 20,000 pile foundations extending to the First Sand layer, are located along the central section of the proposed North/South Metroline. In addition, the proposed tunnel passes below a number of canal bridges and quay walls. A tram line runs along the streets over the entire tunnel boring section and several underground structures are located directly over the tunnel axis.

All these are sensitive to (sub)surface deformations, so the North/South Metroline design team had to investigate the risk of settlement-induced damage. Unacceptable damage was defined to include not only any kind of functional or structural damage, but even aesthetic and repairable damage.

The adopted criteria specified that crack widths could not exceed 2 mm. A typical masonry structure within the expected influence zone of TBM-tunnelling consists of a block of several masonry houses. (Fig-ure 8) The house units are connected to one another by common masonry load bearing walls and continuous (front and back)-facade walls. The connections between the load bearing walls and the facade

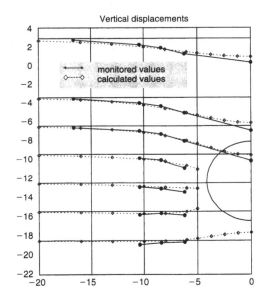

Vertical displacements

Figure 7. Comparison of monitored and calculated displacements.

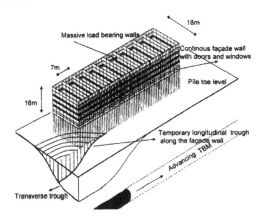

Figure 8. 3D-model of the block of ten masonry buildings (Netzel and Kaalberg, 2000).

walls consist of anchors or teeth constructions between the masonry bricks, which can transmit vertical shear forces between the construction elements.

Thus the block behaves structurally as a single construction. An individual house unit is about 7 m wide, about 16 m high and about 18 m long. The load-bearing masonry walls are founded on paired pile-groups placed under wooden cross beams. The pile groups are spaced about 1.2 m apart and the two piles in a group are about 0.5 m apart. These wooden piles are driven into the first sand layer, and are primarily end-bearing – about 80% of the total load (of about 80–100 kN is transmitted through the pile tip.

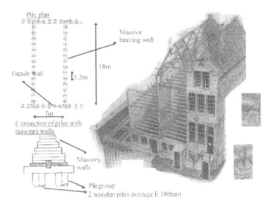

Figure 9. Open view typical historical building.

5.3.2 4D FE building response model

The risk settlement risk assessment studies are carried out in 3 stages, with an increasing level of detail. In the first stage commonly used methods like the theory of Boscarding and Cording were initially adopted, to get an idea of the risk of damage in the facades. Because it was felt these methods are conservative on the one hand, because of their elastic assumptions, and maybe to optimistic on the other hand because of the fact that they can hardly take into account the presence of windows and doors, a 3D non lineair FEM calculation was initiated, to derive Amsterdam specific criteria. Because the propagating longitudinal settlement trough along the tunnel axis was considered to be dominating the damage pattern in the front facades a staged (4D) analysis was developed.

Therefore a standardized block of ten masonry houses on pile foundations was used to model building 3D deformations and reactions to ground movements caused by passage of the TBM (Figure 7). Netzel and Kaalberg (1999) describe how the GIS capabilities assisted these 3D numerical settlement risk assessment studies. Study results were used to define the boundary conditions and the operating requirements for the tunnelling activities. The DIANA program was used to evaluate the non linear material behaviour (including crack simulation) of masonry walls, the interaction between soil and structure, and the 3D behaviour (different stiffness of structural elements) and geometric disturbances to the buildings (Netzel and Kaalberg, 2000). The calculations were carried out in a collaboration between North/Southline Consultants and TNO Building and Construction Research.

The masonry walls were modeled with 8 nodded non-linear plane stress elements, front and rear facade walls included door and window openings. The smeared crack concept was used to model the non linear material behaviour of masonry and give an indication of crack strains and crack widths. The pile foundations were

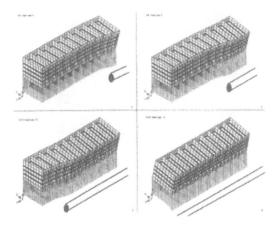

Figure 10.   3D-deformation stages during and after passage of the tunnel (Netzel and Kaalberg, 2000).

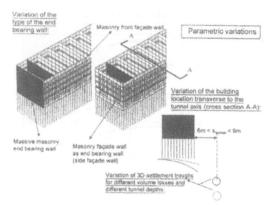

Figure 11.   Parametric variations (Netzel & Kaalberg, 2000).

modelled as non-linear springs to reflect the interaction between soil and piles when building loads are redistributed during the passage of the TBM, including both pile toe resistance and skin friction (but excluding the stress relieve effect, see par 4.3.).

The simulations began with activating the loads existing prior to the TBM passing the building (even this was quiet difficult because these loads already caused unacceptable cracks also, which was overcome by taking into account the creep effect over the centuries). Then settlements at the pile toe level imposed by the passing TBM were modelled for a large number of load steps so as to simulate the advance of the 3D-settlement wave along the building. Figure 10 shows four stages of 3D-deformation behavior of the standard building block. Passage of the settlement wave causes the building to twist because parts of the building are already undergoing settlements while other parts are still in the initial situation. Even after the passage of the TBM, the imposed settlement trough remains and the building may be deformed toward the tunnel axis.

An extensive parametric study was carried out and used to define specific damage criteria for masonry structures of Amsterdam (Figure 11). These studies evaluated:

- Material parameters of masonry walls (including E-modulus, tensile strength, brittle and plastic crack behavior);
- Settlement trough geometries caused by different volume losses and different depths of the tunnels;
- Location of the building within the settlement trough – its distance to the tunnel;
- Geometry of the end bearing walls. The load bearing walls at the ends of many blocks may have openings for doors and windows. These transverse end walls are referred to further as side facade walls.

The above discussions only consider the passage of a single tunnel. Since twin tunnels will be constructed, the modelling also included settlements caused by passage of the second tunnel. The results of these studies clearly showed that the damaging effects of the TBM are considerably reduced when the depth and distance to the tunnel is increased, or when the volume loss around the TBM excavation is decreased (Netzel and Kaalberg, 2000). Because the depth of the tunnels is restricted at the connection with the station boxes, the last aspect requires greatly improved TBM performance, and led to the third stage of research for planning the North-South Line.

5.4   Pile behaviour due to tunnelling

5.4.1   Introduction

As mentioned above the piles carry their loads mainly via the pile toes to the subsoil (80% end bearing capacity). At the start of the design phase one of the major issues was the influence of the tunnelling process on the load bearing capacity of the piles. It was suggested that piles might settle considerably more than their surrounding subsoil, because of the possible stress relieve effect in the compressed zone around the pile toe, when the distance between the tunnel and the foundations was becoming too close. Worldwide hardly any relevant practical based knowledge was available regarding this issue. At that time tests were performed in the geocentrifuge at Geodelft in the Netherlands (CUR report 177, 1995). These tests indicated that piles could indeed settle more than 10 cm. when subjected to a combination of an increasing (significant) volume loss and decreasing distance. The results of these test suggested a distance of 2 tunnel diameters between the TBM and pile toes was recommended, which caused a considerable design depth of the tunnels, which subsequently had a negative impact on the depth of the stations. This was considered to decrease the comfort in the stations and also have an increasing negative effect on the estimated costs for the project. Therefore the aim was to minimise the depth of the stations without compromising the settlement risks in the adjacent tunnel tracks.

The question however remained how close the TBM was allowed to pass in reality without causing this effect, because the tunnelling process was simulated based on the principle of concentric volume loss in these centrifuge tests. Although this schematisation, initially developed by Peck, is widely recognised and used in design calculations worldwide, it was felt this schematisation caused a pessimistic view with respect to this topic, because grout pressure distribution in the annular gap was not taken into account. It was felt that the effects caused by stress relieve could be neglected from a distance of 1D when volume loss was not exceeding 1%, which was assumed to be a situation where proper grout pressure distribution was still present.

5.4.2 Full scale pile trial

Because all were aware that great risks were involved in decreasing the depth of the tunnels, the North/South-line organisation therefore decided that this anticipation in the design was only allowed when a Full Scale Pile Trial was initiated, at the site of the first shield driven tunnel in the Netherlands, the 2nd Heinenoord-tunnel, which was just under construction at that time near Rotterdam. This Full Scale Pile trial was instigated to create more confidence in pile toe behaviour at full scale with a real TBM passing by and was also supposed to confirm above mentioned anticipation in the design. In total 38 wooden piles ad 18 concrete piles were installed and ballasted.

5.4.3 3D FEM research program

The trials were supported with an ambitious 3D FEM research program in order to use the unique opportunity to calibrate and validate 3D computer models. The 3D models were created to transfer the results of the trials from the Rotterdam subsoil conditions to the Amsterdam conditions and to be able to generate results for other (closer) configurations of tunnels and piles. The FEM program was considered to be ambitious because 3D staged calculations were used for the first time at this scale and more specific the penetrating effect of piles which were rammed into the subsoil also had to be modelled properly. Without taking into account of this effect the predictions wouldn't make any sense because the stress relieve during TBM passage was supposed to be dominating the settlements of the piles. In order to create this starting position, large strain deformation calculations were performed based on the Eulerian principle within the DIEKA program by Geodelft, supported by full scale load baring capacity tests. The DIEKA

calculations created satisfactory initial stress distribution patterns in the subsoil along the pileshaft, which was then implemented in the 3D DIANA model, before the tunnelling process was simulated.

5.4.4 Results

After the TBM had passed the trial site for the first time, (min. Distance 1.0D) the results showed that there was no evidence of any additional influence to the pile behaviour due to stress relieve, both in reality and in the 3D FE models, which confirmed the assumptions above mentioned. Becasue the FE model showed no effect at closer distances either, it was therefore decided to invest in the installation of 8 more piles for the second passage of the TBM, this time even closer to the tunnel. The distance between tunnel and piles was now reduced from 1.0D to 0.25D.

The evaluation of the results of the second passage showed that an optimum was reached, because with 0.25D distance a certain (not substantial) effect was noticeable (Figure 12). Therefore it was decided to consider 0.5D as a safe distance between piles and tunnels, which was a confirmation of the anticipation in the design. Also it was found that the settlement were steeper, than was expected based on predictions made with contraction models, which was later confirmed by the newly developed 4D model, which takes into account the grout pressure distribution in the annular gap (Figures 6 and 7).

5.5 Innovative TBM design

Initial feasibility studies for the North-South Metroline revealed that existing state-of-the-art TBM methods in soft soils produced a 1–3% volume loss around the tunnel. This would produce excessive surface deformations and building damage. Because extensive mitigation along almost the entire length of the route would be too expensive, the only alternative was to focus on improving the quality of the TBM process. Analysis of potential innovations to TBM techniques identified a compact version of the variable (vario) shield system that could be adapted to meet the project requirements. This shield could assist in reducing the degree of volume losses and subsequent surface deformations (Kaalberg and Hentschel, 1998 & 1999). The shield was developed in 3D and also implemented into the 4D DIANA model of subsoil and TBM (Figure 13).

5.6 Settlement mitigation techniques

For settlement mitigation 3 different techniques were considered; Permeation grouting, jet grouting and compensation grouting. Although there was a great need to determine the effectiveness of these methods in the design, the FE models appeared to be hardly capable of supporting the design team. Permeation grouting is

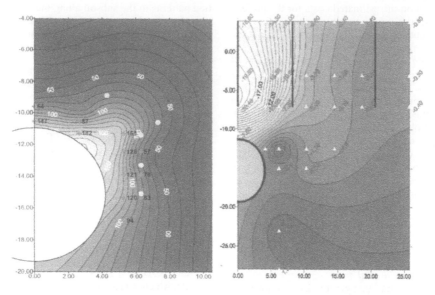

Figure 12.   Monitored vertical stress increase (left) and vertical deformations around the tunnel.

Figure 13.   3D CAD model and DIANA model of TBM.

easy to model because it is in fact not more than an increase of stiffness due to the injection fluid. However, the E modulus of this product of soil an injection material (input) is hard to determine. Unfortunately Jet grouting techniques are still very hard to model these days, because they can only be described by a flowing process. More R & D is needed in the future, to make this happen. Therefore for both techniques a Full Scale Injection trial (van der Stoel, 2001) was initiated in Amsterdam, supported by very limited FE modelling.

In case of compensation grouting (by means of horizontal fracturing) the subsoil is injected via tube a manchettes (TAM's) in order to create an on line heave to compensate settlements due to tunneling. This technique was only considered in a later stadium of the design, because it was felt hard to apply in the Amsterdam conditions and no relevant experience was available in combination with piled foundations.

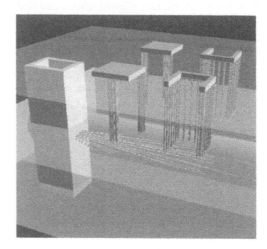

Figure 14.   Computer image of the trial set-up.

However, the advantages of the technique were clear (less hinderance, interactive control, less costs), and when the opportunity came to test this method at the Sophia railwaytunnel, which was under construction near Rotterdam, a Full Scale Compensation Grouting Trial was initiated (J.K. Haasnoot, 2001, Figure 14). The main objective of the trial was to determine the effectiveness of fracture grouting on masonry structures founded on wooden foundation piles during the passage of a tunnel boring machine.

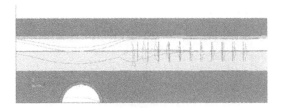

Figure 15. Horizontal and vertical discrete fractures.

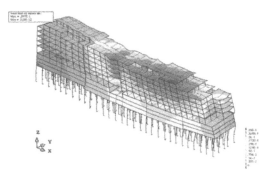

Figure 16. 3D deformed model of ABN AMRO building.

The trial learned that Compensation grouting can be effective to reduce settlements due to tunneling, although the results also showed that TBM speed has to be reduced in order give the grout some time to harden between each injections. Parallel to the trial, an FE model was developed to optimise the design for the Amsterdam situations. This was very hard because fracturing is purposely induced in the model by increasing internal pressure in the elements. It appeared to be that this was only possible when the fractures were initiated in predestined horizontal and vertical interfaces, which were integrated in the model. The goal was to make a 3D model, but so far only a 2D model could be made (van Vliet, 2002, Figure 15).

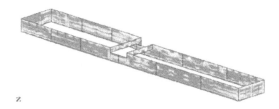

z

Figure 17. Principal stress increments in basement.

5.7 Stage 3 Risk Assessment studies

Later on in the design phase the Stage 3 SRA studies were anticipated. These are more detailed assessments of damage risk at the hot spot locations, were mostly mitigating measures were foreseen, because of the relatively high damage risk involved. If a building appears to have more deformation capacity based on a comprehensive 3D analysis than can be found by more simple means, maybe it is possible to reduce or even eliminate the anticipated mitigating measure at that location. Based on the Value Engineering philosophy very recently a 4D analysis was performed as a testcase, together with TNO Building and Construction Research for the ABN AMRO building (Figure 16).

This building is constructed by means of a reinforced concrete framework with cladding. It is supported on a 3 storey concrete basement and concrete piles. This building was chosen because it was suspected to have a greater deformation capacity than the masonry building and it does'nt have any historical value.

The 4D analysis in which the building was exposed to a progating trough showed that the building did have more deformation capacity indeed. The concrete basement plays a major role in this behaviour, but was also of special interest due to possible cracking below the watertable. The analysis showed that stress and strain increments were not significant (Figure 17). Most likely the mitigating measure will be eliminated

Figure 18. Lining concept.

at this location, showing that FE modelling contributed substantially to Value Engineering.

6 LINING DESIGN

For the lining design an innovative concept is developed which is characterized by a centrally placed elastomere profile in both the radial and circular joints (Figure 18).

The background of this innovation was to develop a lining concept that results in a defined structural behaviour in the joints in order to manage and minimize the risks regarding structural performance of the lining (especially in the Eemclay).

Comprehensive numerical 3D modeling is performed to develop the details of this concept (van Empel & Kaalberg 2002). Based on the "Design by Testing" principle full scale trials are envisaged, before the innovative concept will be implemented.

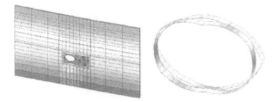

Figure 19.   Ring model with soil structure interaction.

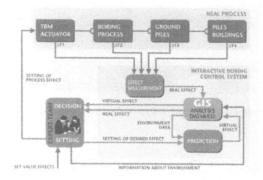

Figure 20.   Process structure with the Integrated Boring Control System (IBCS).

7   THE INTEGRATED BORING CONTROL SYSTEM (IBCS)

The Settlement Risk Assesment studies also showed that the use of a high-performance shield system would provide only a partial solution. The minimal surface deformations demanded by Amsterdam could only be achieved by precise operation of the TBM system, and this depended upon effective process control. This understanding logically led to the consideration of a process control system concept that has been named the Integrated Boring Control System (IBCS).

The IBCS design uses information concerning ground conditions as well as monitoring the operation of and the effects caused by the TBM. In this way interactions with the soil and the pile foundations can be constantly evaluated and potential damage to buildings controlled. Diverse open- and closed-loop control systems are already used by current tunnel boring systems. However, the IBCS concept is distinctive in that it integrates settlement information into the control parameters. This allows the tunnel boring process to be controlled not solely on the basis of tunnel and machine data. Rather, IBCS control and operational criteria reflect settlement information provided in real-time as the TBM advances.

Figure 20 illustrates the basic IBCS process and its relational links to data acquisition and the tunneling system operations. The initial version of IBCS contains three major components – an Effect Measurement component, a GIS Database component, and a Prediction component. These are shown in the lower right portion of Figure 20.

The Effect Measurement component accepts and integrates real-time data from the TBM (TBM and boring process parameters) and from the surroundings (monitoring systems in the ground and on piles and buildings). These data sources are shown across the top of Figure 20. The crucial point is that real effects are provided to the IBCS. Modern measuring processes, linked to the IBCS on-line, provide continuous and simultaneous data acquisition for all monitoring sites. Relevant boring process variables are extracted from TBM measurements. Together, these data streams

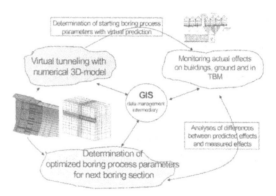

Figure 21.   Process loop of IBCS.

allow the IBCS to evaluate the real process as the tunneling progresses.

The GIS Component performs three distinct functions. First, it contains and maintains the 3D subsurface geological and geotechnical models described in the previous section. Second, the GIS component accepts information obtained and processed by the Effect Measurement component and uses this new information to up-date and improve the pre-existing 3D model (Netzel and Kaalberg, 1999). Third, the GIS provides accurate visualizations required for rapid analysis and interpretation in the Prediction component.

The Prediction component performs predictive calculations and comparisons. Prior to the actual tunnel construction, a virtual boring process is modeled using a 4D FEM numerical model (Figure 6) that incorporates determinative process parameters for the TBM and knowledge of the subsurface conditions surrounding the advancing tunnel. This numerical model evaluates the reactive forces in the bore front and annular void areas and assesses how these will be transmitted into the surrounding subsoil. This permits predictions to be made concerning soil deformation and settlement.

The Prediction component also uses the 3D subsurface model maintained within the GIS component (Netzel and Kaalberg, 1999) to combine predictions based on numerical models with current observations. Simultaneously the boring process settings are being monitored and displayed. The result is a closed, real-time information circuit for the IBCS.

In the initial development stage of IBCS, the Decision-Setting functional unit is a purely information and visualization system that monitors the TBM processes and extracts relevant process variables. Adjusting the process parameters while observing the actual effect data permits the process sequence to be successively optimized so as to minimize settlements. Decision-making and adjustment must continue to be carried out by humans, acting as an expert team. This is illustrated on the lower left portion of Figure 20. Real-time visual displays of the boring process variables, along with data defining the predicted and actual settlements, will be produced at the TBM control panel. These displays provide the basis for decision-making by the process control team. When combined with experience in TBM operations, and knowledge of the importance of various process variables and operating parameters, this initial IBCS can result in greatly reduced settlements caused by TBM operations. It allows TBM operators to make better and more rapid operational decisions, based on an extended experience base, while also reflecting both sitespecific and real-time conditions.

8 CONCLUSIONS

The overview given above shows that FE modelling plays a key role in the design of the North/Southline. The use of 3D FE models is pushed forward by the increasing needs to assess and manage risk in (underground) construction. Although is has to be considered that FE models always must be used with prudence, the great flexibility of the models to carry out extensive parametric studies, provides an excellent opportunity to assess risks during construction.

ACKNOWLEDGEMENTS

The author would like to acknowledge the NSL design team members for their contributions in the papers mentioned below, which were the basis for this overview. I also would like to thank Prof. Rots and his teammembers of TNO for his stimulating partnership in FE modelling.

REFERENCES

Kaalberg, F.J., and V. Hentschel, 1998. *Tunnelvortrieb in weichen Böden mit hohem Wasserstand – zur Entwicklung einer setzungsorientierten und setzungsminimierenden Betriebsweise der TBM. [Tunnel boring in soft soil with a high water level – development of a settlement-oriented and settlement-minimizing operating mode for the TBM.]* Forschung und Praxis 37, Neue Akzente im unterirdischen Bauen. Düsseldorf, Germany

Kaalberg F.J. and V. Hentschel 1999. *Tunnelling in Soft Soil with a High Water Level and Piled Foundations: Towards the Development of a Settlement-Orientated and Settlement-Minimizing TBM Control* – Proc. World Tunnel Congress'99, Oslo, Norway.

Teunissen, E.A.H. & Hutteman, 1998. *Pile and Surface Settlement at Full Scale Tests North/Southline Metro*, World Tunnel Congress, São Paulo.

Netzel, H. and F.J. Kaalberg 1999. *Settlement Risk Management with GIS for the Amsterdam North/South Metroline* – Proc. World Tunnel Congress'99, Oslo, Norway.

Netzel H. and F.J. Kaalberg 2000. *Numerical Damage Risk Assessment Studies on Masonary Structures due to TBM-Tunnelling in Amsterdam* – Proc. GeoEng2000, Nov. 19–24, Melbourne, Australia (Full paper on CDROM, Abstract in Extended Abstracts, Vol. 2, p.183)

Van Dijk, B. and F.J. Kaalberg 1998. *Geotechnical Model for the North/Southline in Amsterdam*. Proc. 4th European Conference on Numerical Methods in Geotechnical Engineering, Udine.

Van der Stoel, A.E.C. 2001. *Grouting to improve Pile Foundations*. Phd Thesis, Delft University of Technology.

W.H.N.C. van Empel and F.J. Kaalberg 2002. *Advanced modeling of innovative bored tunnel design Amsterdam North/Southline*. Proc. of the DIANA World Conference 2002, Oct. 9–11, Tokyo, Japan.

Haasnoot J.K., A.E.C. van der Stoel, F.J. Kaalberg 2002. *Fracture Grouting to mitigate settlements of wooden pile foundations*. Proc. of AITES-ITA World Tunneling Down Under Conference, Sydney, Australia.

Van Vliet, M., 2002. *Modelling of fracture grouting with discrete cracks.*, Msc Thesis, Delft University of Technology.

CUR report 177, *Influence of tunnel boring on piled foundations, centrifuge tests*, 1995.

van Dijk, B., W.H.N.C. van Empel and F.J. Kaalberg, 1999. *Evaluation FE models (Dutch)*, COB report K100-W-105B.

Finite Elements in Civil Engineering Applications, Hendriks & Rots (eds.)
© 2002 Swets & Zeitlinger, Lisse, ISBN 90 5809 530 4

Invited paper: Deformation of liquefied ground in shaking table tests and its prediction

I. Towhata
University of Tokyo, Japan

ABSTRACT: The present study aims at developing an analytical method to predict liquefaction-induced large deformation of ground and earth structures. This aim was firstly facilitated by examining earthquake-induced damages and, secondly, running shaking-table model tests. Consequently, an important finding was made of characteristic modes of deformation which later made the analysis simple and efficient. Furthermore, model tests as well as laboratory shear tests were conducted to find the rate-dependent nature of liquefied sand undergoing large deformation. Finally, a new numerical method was developed to conduct the deformation analysis as intended in a dynamic three-dimensional manner with large-displacement formulation. Example analyses are presented at the end of this text.

1 INTRODUCTION

The aim of the present study is to facilitate the development of performance-based geotechnical earthquake engineering. In the recent decade, the limited strength of soil cannot catch up with the recent increase in the intensity of design earthquake motion. Accordingly, the design factor of safety is not necessarily greater than unity. Liquefaction is a typical example of this situation where a satisfactory performance of facilities has to be achieved in spite of subsoil liquefaction.

To cope with this difficulty, it is important to understand the essential aspects of liquefaction-induced damage, predict what will happen during future or design strong earthquakes, and mitigate the extent of possible damage so that the serviceability of facilities will not be lost. The present text attempts to play this role in what follows.

2 ESSENCE OF LIQUEFACTION-INDUCED DAMAGE

Figure 1 shows the subsidence of a building in Dagupan City, the Philippines where the 1990 Luzon earthquake caused liquefaction in the young reclaimed fill in the new part of the city and many buildings subsided more than one meter. Figure 2 illustrates the effects of the weight of building on the magnitude of subsidence. The main part of the building was of five

Figure 1. Subsidence of building in Dagupan City.

stories and sank more than one meter. In contrast, its flat annex subsided less. It is interesting in Figures 1 and 2 that buildings were not structurally damaged by the quake; the broken state of the annex in Figure 2 was made by local people after the quake. The serviceability of buildings was lost due to the unallowable extent of vertical displacement. Significant subsidence is important in river dikes as well. Note, however, that minor subsidence does not matter, even if seismic factor of safety might be less than unity.

Floating of an embedded structure which is lighter than liquefied soil is frequently seen after seismic liquefaction (Fig. 3).

Lateral displacement is a very important feature of liquefaction-induced damage. The soil displacement

Figure 2.   Effects of building weight on extent of subsidence.

Figure 3.   Floating of embedded sewage tank in Dagupan City.

Figure 4.   Buckling of buried water pipe in Dagupan City due to lateral compression of subsoil.

induces compression or extension of ground in the horizontal direction which is a big hazard to embedded lifelines. In the case of Figure 4, the subsoil was subjected to horizontal compression and induced a buckling failure to a water pipe. Note again that this damage was caused not by shaking but by substantial displacement. Moreover, functions of quay walls have been considerably affected by large displacement; e.g., Akita Harbor in 1983, San Antonio in Chile in 1985, and Kobe in 1995.

The examples as found during past earthquakes clearly indicate that the essence of liquefaction-induced damage is the magnitude of displacement and deformation. Accordingly, the conventional approach of seismic engineering based on intensity of acceleration is not relevant. What is desired to date in mitigation of liquefaction damage is the prediction of the possible extent of displacement, reducing the displacement, and assessing the extent of reduction.

3   OBSERVATION IN SHAKING TABLE
     TESTS

Shaking table model tests have been carried out on a variety of situations, focusing on the attention on the mode of deformation of liquefied layer and the time at which the deformation ceases to develop.

The employed sand was Toyoura sand which is fine and uniform in gradation. It was deposited in a very loose state of relative density of −20 to +30% in contrast with the density of real subsoil being more than 40%. This reduced density was important in 1-g model tests where low level of effective stress made sand more dilatant. This shortcoming was cancelled by reducing the density of sand.

Figure 5 illustrates the deformed shape of a model slope. Soil moved downwards showing the important effects of gravity. The square grid of colored sand, which was embedded in the liquefiable sandy deposit, clearly indicated that the lateral displacement was greater at the upper elevation, while reduced to zero at the bottom. This mode of lateral deformation is called the *F mode* in the present text.

It is noteworthy that liquefaction in Figure 5 was triggered by a short-time impact shaking after which the liquefied model flew laterally due to its own weight. The time history of lateral displacement, which was recorded by an embedded transducer, is indicated in Figure 6. The displacement of the loosest sand ceased at 3 seconds when the displacement reached 0.24 m and the ground surface became level.

Thus, it can be pointed out that very loose liquefied sand behaves similar to liquid which attains stability

Figure 5. Liquefaction-induced deformation of sandy slope model (Towhata et al., 1995).

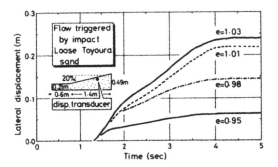

Figure 6. Time history of lateral displacement of liquefied slope after impact shaking (Towhata et al., 1995).

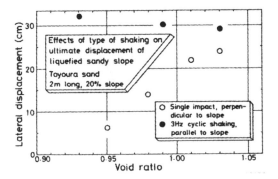

Figure 7. Effects of continued shaking on ultimate lateral displacement (Towhata et al., 1995).

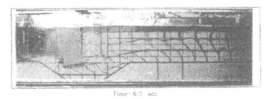

Figure 8. Deformed shape of model of gravity-type quay wall (Ghalandarzadeh et al., 1998).

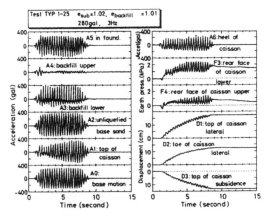

Figure 9. Time history of base acceleration (bottom left) and displacement (bottom right) in model of gravity quay wall (Ghalandarzadeh et al., 1998).

when the surface is level. Moreover, Figure 6 shows that displacement of denser sand stopped with smaller displacement and inclined surface. Later tests continued shaking longer in order to maintain the state of flow until surface became level (Figure 7).

Recently, centrifugal model tests by Okamura et al. (2002) showed that lateral displacement of a liquefied slope stops at the end of shaking when the excess pore water pressure is still high, while consolidation settlement continues over a longer period of time.

Figure 8 indicates the deformed shape of a model of gravity quay wall (Ghalandarzadeh et al., 1998). Toyoura sand of as loose as −20% relative density was employed under the wall and in the backfill. Both

parts of sand developed large strain during shaking. Similar to those in Figure 5, the lateral displacement was greater towards the upper elevation.

Figure 9 shows the time history of displacement. Although the density of sand was as low as that employed in the slope model (Figure 5), the gravity quay-wall model ceased its lateral motion at the end of shaking, while the slope model kept moving until the surface became level. This different duration of motion may be attributed to the magnitude of static shear stress. Figure 10 illustrates the stress-path diagram of foundation sand under the quay wall. This diagram was reconstructed by using the recorded response of the quay wall model and the excess pore water pressure. In the course of shaking, the stress path (relationship between shear and effective stress components) came close to the shear failure locus and large deformation developed. However, the state of zero shear stress did not occur, and, thus, the effective stress was never null. Hence, the concerned sand maintained the nature of frictional solid. This is in contrast with the behavior of sand in a slope model where the slope of 20% generated only a limited extent of static shear stress. When dynamic shear stress was superimposed, the state of zero shear stress and consequently zero effective stress were reached and sand behaved similar to liquid.

A different mode of lateral displacement was found in subsidence tests (Figure 11). The soil immediately beneath the embankment did not deform. The

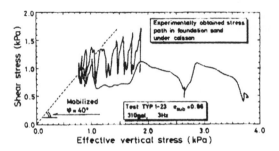

Figure 10. Stress-path diagram of foundation sand under quay wall (Ghalandarzadeh et al., 1998).

lateral displacement was null both at the top of liquefied layer (beneath the embankment) and at the bottom, while the maximum displacement was observed at the middle elevation. This mode of displacement is called the *J mode* in this text. Figure 12 shows that the subsidence was terminated when shaking was stopped. Even the greater subsidence would have occurred towards the maximum possible displacement if shaking had been continued longer. The maximum possible displacement is the one at which buoyancy and gravity are in equilibrium.

The *J mode* of lateral displacement was observed in floating tests as well. Figure 13 shows the ultimate shape of a model in which a light embedded structure came up to the surface due to liquefaction in the surrounding area. The density of Toyoura sand in this test was +20% (Towhata et al., 2003). Under the structure, liquefied sand came in and filled the space. It is evident that the lateral displacement was maximum at the middle elevation while taking small values at top and bottom. Figure 14 presents the time history of floating which stopped at the end of shaking.

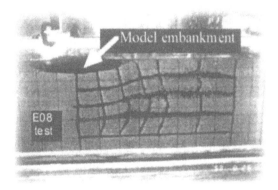

Figure 11. Deformation of liquefied foundation under embankment model (Mizutani et al., 2001).

Figure 13. Ultimate shape of embedded structure after floating (Towhata et al., 2003).

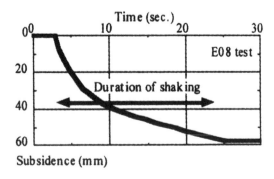

Figure 12. Time history of subsidence of embankment resting on liquefied foundation sand (Mizutani et al., 2001).

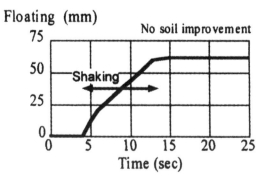

Figure 14. Time history of floating of embedded conduit model (Towhata et al., 2003).

The observation of displacement in shaking-table model tests are summarized in what follows.

- The liquefaction-induced displacement occurs towards the minimum potential energy; for example, downward motion of slope, sinking of surface structure, and floating of light embedded structure.
- There are two kinds of displacement mode which are called *F* and *J* modes.
- The *F mode* is predominant when the horizontal displacement of ground and structure is important, while the *J mode* plays a chief role when such a vertical displacement as subsidence and floating is of concern.
- The development of displacement is ceased at the end of shaking. The exception to this rule is the case in which the static shear stress is low and is overcome by cyclically generated shear stress.

4 MATERIAL PROPERTIES OF LIQUEFIED SAND

An analytical/numerical prediction of liquefaction-induced ground displacement requires the stress-strain characteristics of liquefied sand to be known. This research aim is not easy to attain mainly because liquefied sand is too soft to reasonably conduct conventional kinds of shear test. Since the effective stress is null or very low, sand does not have a rigidity to maintain a cylindrical shape of a specimen in triaxial tests. Another important problem is that the interested range of shear strain is tens of percent (see Figs 5 & 8) which is too large to be studied in conventional shear tests.

To overcome those problems, the first series of tests was conducted in shaking model tests (Towhata et al., 1999). Toyoura sand of a variety of void ratio (density) was placed and saturated with water in a container resting on a shaking table. When this model was shaken laterally, an embedded pipe model was pulled in the horizontal direction and the drag force needed for this pulling was monitored. The sand around the moving pipe developed large deformation. Moreover, monitoring the behavior of liquefied sand in the course of strong shaking was possible. The idea behind this was that the drag force is somehow related to and stands for the magnitude of shear stress in the surrounding liquefied sand, while the displacement of the pipe shows the state of shear strain. It should be borne, however, that the state of stress and strain in the model ground is never uniform, making more detailed discussion irrelevant.

The test results are illustrated in Figure 15. Note that the linear proportionality between the rate of pipe movement and the drag force. This is consistent with the theory of fluid mechanics on Newtonian viscous liquid. By thus using the theory by Lamb (1911),

the viscosity coefficient was back-calculated. The obtained values are presented in Table 1.

Conversely, the shortcomings of this model test are;

- Since the state of stress and strain is not uniform, the obtained viscosity coefficient is an overall averaged value. More detailed discussion on the viscosity is not possible.
- The overburden pressure is much lower than reality.

Consequently, an attempt was made to directly monitor the stress and strain of a specimen of liquefied sand (Nishimura et al., 2002). Although triaxial or torsional shear tests on a liquefied sand specimen were promising in this regard, the following problems had to be overcome.

- The state of null effective stress was not possible for maintaining a stable shape of a specimen.
- In case of torsional shear with very low effective stress, strain was concentrated near the top of a specimen where the effective stress was lowest. In

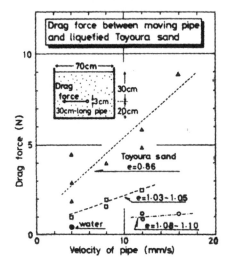

Figure 15. Rate dependency of drag force in pipe pulling tests (e = void ratio).

Table 1. Apparent viscosity of liquefied sand as back-calculated from measured drag force.

Relative density of sand (%)	Apparent viscosity (Pa sec.)
+30	100–216*
+6	45
−2	69
−20	48–66
−20	7–16*
−30	10–37.

* Observation in the course of shaking.

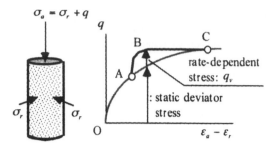

Figure 16. Monitoring rate dependent nature of sand with low effective stress.

Figure 17. Photograph of liquefaction in Niigata Airport, 1964 (JGS, 1999).

contrast, the lower part of a specimen had higher effective stress due to the own weight of sand grains.
– Undrained shear deformation was accompanied by positive dilatancy and the state of effective stress was varied. Accordingly, the change of shear stress due to dilatancy and other issues such as strain and strain rates were not separated.

Therefore, it was decided to run axial compression on a hollow cylindrical specimen in which, after isotropic consolidation under 100 kPa, high excess pore water pressure was generated by either cyclic undrained torsion or supplying high back pressure and then axial compression was made in "drained" manners. The size of a specimen was 20 cm in height, 10 cm in outer diameter, and 6 cm in inner diameter. Since the state of zero effective stress was not possible, tests were run at low stress levels of 5 to 20 kPa and the results were extrapolated to zero stress. The drained axial compression allowed the state of high pore pressure to be maintained constant during large deformation, although volume of sand changed. The state of pore pressure was monitored at the bottom of a specimen while high back pressure was supplied from the top.

The employed testing method is illustrated in Figure 16. The curve OAC designates a drained stress-strain curve under low effective stress. Suppose that a specimen is stationary at A. In the present study, the deviator stress, q, is increased quickly to B, followed by a constant value for a certain time. The strain response is small in AB and creep deformation occurs until C. Since the measured stress record, ABC, is not fully a rate-dependent behavior, it is divided into viscous (rate-dependent) and inviscid (rate-independent under the given effective stress) components. Accordingly,

viscosity coefficient = *rate-dependent*
stress/strain rate (1)

The measured viscosity coefficient is plotted against the rate of strain for several effective stress levels, σ_r'.

Since the influence of effective stress is not significant, it is reasonable to apply those values to the state of null effective stress. More important are the effects of strain rate. The decrease of viscosity coefficient with the increasing strain rate may suggest that the liquefied sand is similar to Bingham type of viscous liquid. Note that the rate of strain in reality is not known except the case in Figure 17 where a motion picture of a sinking building suggested the rate of subsidence being around 1 cm/sec. over the 10 m thickness of liquefiable sand. Hence, the rate of strain was of the order of 0.001/sec.

5 METHOD OF DISPLACEMENT ANALYSIS

This section is going to describe briefly the essence of three-dimensional dynamic analysis on liquefaction-induced displacement with large deformation taken into account. The basic idea was described by Towhata et al. (1999) in a two dimensional manner.

When the thickness of liquefied soil is H and the vertical coordinate is designated by z (positive upwards), the lateral displacement in the x direction is expressed as

$$u(x,y,z,t) = F(x,y,t)\sin\frac{\pi z}{2H} + J(x,y,t)\sin\frac{\pi z}{H} \quad (2)$$

where x, y, and z stands for three-dimensional coordinates while t designates time. F and J are unknown functions to be determined by analysis and indicate the magnitudes of F and J modes of displacement, respectively.

The two-dimensional version of the analysis was discussed in detail by Acacio et al. (2001). The vertical displacement, $w(x,y,z,t)$, is obtained by integrating a constant-volume equation:

$$\frac{\partial u}{\partial x} + \frac{\partial w}{\partial z} = 0 \qquad (3)$$

Moreover, the change of surface elevation, δH, is derived as the difference of the volume of soil flux (Figure 18):

$$\delta H = \frac{\partial}{\partial x}\left\{\int_0^H u(x,y,z,t)dz\right\} \qquad (4)$$

The effects of seepage and consolidation can be taken into account separately by adding consolidation settlement to Eqs 3 and 4. In the three-dimensional version, the displacement in the y direction, v, is formulated in a similar way as Eq.2 and added to Eqs 3 and 4.

The dynamic analysis in the time domain was developed based on an energy principle. The kinetic and potential energies of an analyzed ground, K and Q, are expressed by discretized F and J values and substituted in the Lagrangean equation of motion:

$$\begin{cases} \dfrac{d}{dt}\left\{\dfrac{\partial(K-Q)}{\partial(\partial F_i/\partial t)}\right\} - \dfrac{\partial(K-Q)}{\partial F_i} = 0 \\[2ex] \dfrac{d}{dt}\left\{\dfrac{\partial(K-Q)}{\partial(\partial J_i/\partial t)}\right\} - \dfrac{\partial(K-Q)}{\partial J_i} = 0 \end{cases} \qquad (5)$$

where F_i and J_i are F and J functions at the i-th nodal point in x-y plane. Thus, the calculated displacement is oriented towards the state of minimum potential energy. Moreover, the calculated subsidence and floating are terminated when the gravity and buoyancy due to liquefied sand become equal to each other. This is achieved by calculating the gravity potential energy

of liquefied sand by

$$\int_0^{H+\delta H} (\text{unit weight of liquefied sand})z\, dz \qquad (6)$$

Noteworthy is that this advantage cannot be achieved by popular finite element analyses based on small-displacement formulation in which the change of gravity energy is calculated by "*unit weight × vertical displacement.*" Details of the three-dimensional analysis was presented by Kobayashi (2001).

6 EXAMPLE ANALYSES

This section introduces two examples of three-dimensional dynamic analysis on liquefaction-induced displacement based on a large-displacement formulation. The first example is presented in Figure 19 where the permanent displacement of liquefied gentle slope in Noshiro City was calculated. The employed boundary condition was 1) zero displacement at the bottom of slope, and 2) open-crack boundary at the top where lateral earth pressure pushes down the moving soil mass. Low viscosity coefficient of 20 kPaSec was used due to relatively small thickness of soil. As illustrated, the surface displacement after 60 seconds of flow is a consequence of F mode, while J mode

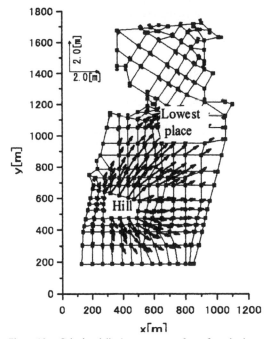

Figure 19. Calculated displacement at surface of gentle slope, Noshiro, Japan, caused by 1983 Nihonkai-Chubu earthquake.

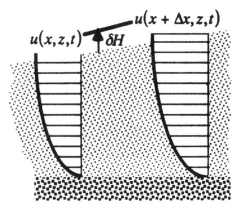

Figure 18. Simple calculation of uplift/subsidence of ground surface.

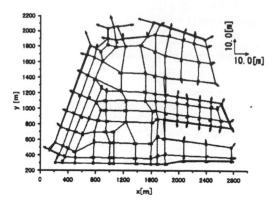

x[m]

Figure 20.   Calculated lateral displacement in Port Island of Kobe.

underneath generates more lateral soil mass movement as well. It seems that the calculated displacement at the surface is at maximum 2 m and is still smaller than the reality which was 4 m at maximum near the top of the slope. The contribution by the J mode reduced the need for F-mode flow.

The second example was made of Port Island of Kobe (1995) where the significant distortion of gravity quay walls led to lateral displacement in the backfill. With the increased 300 kPaSec of viscosity due to greater soil thickness, the displacement in Figure 20 was attained after 60 seconds of flow. Since the central part of the island avoided distortion due to soil improvement, those parts were treated as fixed area.

7 CONCLUSIONS

The present method of displacement analysis is a powerful tool which can carry out a three-dimensional dynamic analysis within several hours in PC. This goal was achieved by using experimentally detected nature of liquefied sand and displacement modes as well as eliminating unnecessary complexity in computation.

REFERENCES

Acacio, A.A., Kobayashi, Y., Towhata, I., Bautista, R. T., & Ishihara, K. (2001) Subsidence of building resting upon liquefied subsoil; case studies and assessment, *Soils and Foundations*, 41(6) 111–128.

Ghalandarzadeh, A., Orita, T., Towhata, I., & Fang, Y. (1998) Shaking table tests on seismic deformation of gravity quay walls, Special Issue on Geotechnical Aspects of the January 17 1995 Hyogoken-Nambu Earthquake, No. 2, *Soils and Foundations*, 115–132.

Japanese Geotechnical Society (1999) Air photographs of the Niigata city immediately after the earthquake in 1964, ISBN4-88644-054-1.

Kobayashi, Y. (2001) Numerical method for three-dimensional prediction of lateral flow in liquefied subsoil, Ph.D. Thesis, Univ. Tokyo.

Lamb, H. (1911) On the Uniform Motion of a Sphere through a Viscous Fluid. *Philosophical Magazine and Journal of Science*, 21(121) 112–121.

Mizutani, T., Towhata, I., Shinkawa, N., Ibi, S., Komatsu, T., & Nagai, T. (2001) Shaking table tests on mitigation of liquefaction-induced subsidence of river dikes, *Proc. 16th ICSMGE*, Vol.2 1207–1210.

Nishimura, S., Towhata, I., & Honda, T. (2002) Laboratory shear tests on viscous nature of liquefied sand, *Soils and Foundations*, 42(4).

Okamura, M., Abdoun, T.H., Dobry, R., Sharp, M.K., & Taboada, V.M. (2002) Effects of sand permeability and weak aftershocks on earthquake-induced lateral spreading, *Soils and Foundations*, 41(6) 63–77.

Towhata, I., Toyota, H., & Vargas-Monge, W. (1995) Dynamics in Lateral Flow of Liquefied Ground, *Proc. 10th Asian Regional Conf. Soil Mech. Found. Engg.*, Beijing, Vol.1 497–500.

Towhata, I., Vargas-Monge, W., Orense, R.P., & Yao, M. (1999) Shaking Table Tests on Subgrade Reaction of Pipe Embedded in Sandy Liquefied Subsoil, *Soil Dynamics and Earthquake Engineering Journal*, 18(5) 347–361.

Towhata, I., Orense, R.P., & Toyota, H. (1999) Mathematical principles in prediction of lateral ground displacement induced by seismic liquefaction, *Soils and Foundations*, 39(2) 1–19.

Towhata, I., Nakai, N., Ishida, H., Isoda, S., & Shimomura, T. (2003) Mitigation of liquefaction-induced floating of embedded structures by using underground walls, *12th Asian Regional Conf. Soil Mech. Found. Engg.*, Singapore.

Finite Elements in Civil Engineering Applications, Hendriks & Rots (eds.)
© *2002 Swets & Zeitlinger, Lisse, ISBN 90 5809 530 4*

Finite-element modelling of stress development and fault reactivation in and around a producing gas reservoir: Quantification of calculation results in DIANA

F.M.M. Mulders
Delft University of Technology

ABSTRACT: Stress changes related to the depletion of gas reservoirs in the northern Netherlands can lead to small seismic events. Field data analysis makes it very likely that these events are the result of reactivation of existing faults or joints in or near the reservoirs. DIANA is used to model 3D gas reservoirs, in order to analyse stress development and fault reactivation during gas depletion. The models described in this paper are $12,000 \times 12,000 \times 5000$ m and consist of approximately 59,380 TE12L- and T18IF-elements. Calculation results are analysed in terms of relative shear displacements (RSD), stress paths and mobilised shear capacity (MSC). Newly developed and complex formulas for MSC are successfully applied by means of the "Results calculate expression…"-option in Femview. It shows the possibility of implementing very complicated formulas in Femview. Modelling of the subsurface is a data-limited problem. The presented quantification methods are used to choose appropriate values for strength and pore pressure development of the interface elements for modelling of the fault. An application example illustrates the use of the methods and is supported by field data and earlier research work.

1 INTRODUCTION

In the northern Netherlands approximately 190 gas reservoirs are present at a depth of more or less 3000 m. The depletion of these gas fields causes changes in the effective stress field in the reservoir and its surroundings, resulting in compaction of the reservoir rock. This in turn leads in most cases to subsidence at the earth's surface and in some cases to induced seismicity. The events are with less than 3.5 on the Richter scale relatively small in magnitude (De Crook et al. 1995).

In order to get more fundamental insight in the causes of seismicity, DIANA is used to develop an understanding of how the depletion of a gas reservoir affects both a 3-dimensional tectonic stress field and the potential for reactivation of existing faults.

In general it is assumed, that seismic events are the result of reactivation of existing faults or joints in or near the reservoir (BOA 1993). Figure 1 shows this general concept in form of Mohr circles. Before gas depletion, the pore pressure in the reservoir is high and therefore the effective stresses, acting on the framework of the reservoir rock, are relatively low. When the pore pressure decreases, the effective stresses increase differentially. For a reservoir with a horizontal extension larger than the vertical extension, the effective vertical stress increase is larger than the effective horizontal

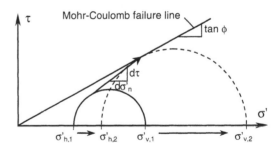

Figure 1. General concept of fault reactivation during gas depletion shown by Mohr circles for an extensional stress regime. The arrow indicates the followed stress path. Subscripts 1 and 2 denote a state of stress before and after gas depletion respectively. Subscripts v and h denote vertical and horizontal direction respectively. It is assumed that the vertical direction is a principal stress direction.

stress increase, because the horizontal reservoir contraction is more constrained than the vertical due to the elastic coupling of the reservoir rock to its surroundings. This in turn means a destabilisation of normal faults inside of a depleting reservoir in an extensional stress regime. These effects are strongly dependent on the ratio of the reservoir thickness and its lateral extension and on the elastoplastic rock properties of both reservoir and surrounding rock (Segall & Fitzgerald 1998, Rudnicki 1999).

Fault slip can be very gradually. In that case no seismic event is felt. Another possibility is that asperities on the fault might inhibit the gradual movement. If such an asperity breaks when stresses get too high, fault movement may occur very suddenly: the resulting energy release can be felt as a seismic event.

2 FINITE-ELEMENT MODEL DESCRIPTION

2.1 *Model geometry*

In order to study the effects of gas depletion on a 3-dimensional tectonic stress field and the potential for reactivation of existing faults, several 3-dimensional models of gas reservoirs are developed. Figure 2 shows the model of a disk-shaped gas reservoir. The reservoir is 150 m thick and has a radius of 3000 m. The volume surrounding the reservoir is assumed to consist of one homogeneous isotropic rock mass in

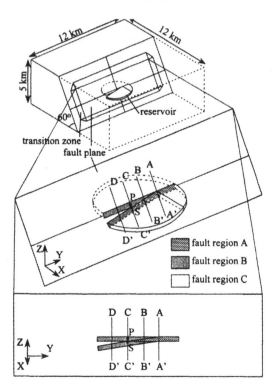

fault region A
fault region B
fault region C

Figure 2. Geometry of the finite-element model. The lowest picture shows the fault plane with view direction normal to the fault plane. For the four lines AA', BB', CC' and DD', RSD-values are plotted in Figure 8. For point P, stress paths are plotted in Figure 7. Point S is the point on line CC' where the two reservoir compartments geometrically separate from each other. Further explanations in text.

order to keep the model generic and free from any additional effects eventually produced by a layered geological structure. No gas/water contact is modelled.

A fault plane with a dip angle of 60° intersects the reservoir. It runs through the centre of the disk at 2975 m depth and divides the reservoir in two compartments. The compartment at the footwall side of the fault is horizontal with its top located at a depth of 2900 m. The compartment at the hanging wall side of the fault is partly dipping towards the negative Y-direction of the model, thus creating a normal fault geometry with a varying throw along the fault strike. The reservoir compartment dips with an angle of 7.6°. This angle is defined around an imaginary rotation axis, which runs perpendicular to the strike of the fault at 2975 m depth, 750 m in positive Y-direction from the reservoir centre. Note that a part of this reservoir compartment is not dipping, making that the two reservoir compartments at this location are located exactly opposite to each other.

Model dimensions are 12,000 × 12,000 × 5000 m. The model contains a total of 59,380 elements, consisting of 2580 T18IF elements (structural interfaces, plane triangle, 3 + 3 nodes) and 56,800 TE12L elements (structural solid elements, pyramid, 3 sides, 4 nodes).

2.2 *Creation of model geometry in Femgen*

The model is created in Femgen by first constructing the necessary shapes in form of planes, cylinders and boxes. With the Femgen-options "geometry surface intersect…" and "geometry point intersect…", the whole model geometry can be built relatively easy. This way allows the creation of various models with different sizes, fault dip angles and reservoir compartment dip angles by simply changing some co-ordinate values in a batch file, which contains all necessary commands for the model set-up.

The fault plane is created by means of the "geometry sweep…"-option. Its thickness is 0 m. Sweeping a geometry to its location requires the tolerance to be either switched off or set at 0. For the surface intersection option to function, the tolerance must be switched on again or set at its default value. Note that the fault plane does not extend until the model boundaries, but that it is surrounded by a transition zone and an outer edge of continuous surrounding rock. This prevents slipping of the entire hanging wall, which would result in divergence during iterating.

The fault plane consists of a large number of hexahedral and triangular prism bodies. This allows the generation of a regular mesh with a consistent node and element numbering, which is indispensable for the production of contour plots of relative shear displacements on the fault plane (see section 3.1). Because of the multiple bodies and thus multiple surfaces on both sides of the fault plane, reservoir and surrounding are built using general bodies. The use of general bodies generally limits the generation of bad shaped elements. A disadvantage is that besides line divisions no direct control on the element density is possible. Furthermore, the number of elements which are generated in a general body strongly depends not only on the meshing divisions of the lines bounding the general body, but also on the meshing divisions of other lines elsewhere in the model and on the sequence in which the different bodies are created.

2.3 Rock properties

Elasto-plastic stress-strain behaviour according to the Mohr-Coulomb yield criterion is taken for both reservoir- and surrounding rock. No hardening, cracking or other options such as nonlinear elasticity or cap are modelled. For simplicity in the generic model, rock properties of reservoir- and surrounding rock are equal: Young's modulus 13.0 GPa; Poisson's ratio 0.20; density 2400 kg/m^3; cohesion 5.0 MPa; friction angle 30°; dilatation angle 10°; K_0 0.4.

2.4 Fault properties

The fault interface is modelled using Coulomb friction with a very low cohesion of 10 Pa and a dilatation angle of 0°. The friction angle should be chosen with care. A too high value does not allow fault slip large enough for a good evaluation of the calculation results. On the other hand, a too low friction angle results in an unstable fault and slip on the entire fault plane before the start of gas depletion. This could lead to numerical instabilities and divergence during iterating. Assuming that the vertical stress is a principal stress, plastic fault slip occurs when

$$K_0 \leq \frac{1 - \sin \phi}{1 + \sin \phi} \qquad (1)$$

where K_0 is the effective stress ratio and ϕ is the fault friction angle. For $K_0 = 0.4$, this when $\phi \leq 25.38°$. The friction angle is chosen to be 28°. The choice of normal and shear stiffness of the fault is discussed in section 4.1.

2.5 Applied loads and boundary conditions

Before gas depletion, an initial extensional stress field is assumed with an effective stress ratio K_0 of 0.4. The pore pressure in the surrounding rock is hydrostatic and remains constant throughout gas depletion. At the start of depletion, the reservoir is overpressured with an initial pressure of 35.0 MPa. Gas depletion is modelled by decreasing the pore pressure in the reservoir linearly in 10 static depletion steps. Initial pore pressure and pore pressure development in the fault is dicussed in section 4.2. At the lateral sides of the model, horizontal translations normal to the respective sides are constrained. Vertical translations are constrained at the bottom.

3 QUANTITATIVE METHODS FOR ANALYSIS OF CALCULATION RESULTS

In order to analyse the sensitivity of a fault for reactivation in a given reservoir setting by means of numerical calculations, three quantitative methods are used to compare the calculation results on the fault plane: relative shear displacements (RSD), stress paths and mobilised shear capacity (MSC). MSC is also applied for rock volumes (MSC$_{3D}$).

3.1 Relative shear displacements

Relative shear displacement on a fault plane is defined as the differential movement in shear direction of two corresponding nodes on the opposite sides of an interface element. It is the amount of fault slip which develops during one or more steps of gas depletion.

In Femview, calculations on the attributes of result data sets stored in the database can be performed and processed to a new attribute using the option "results calculate expression...". With this option, results of different nodes or integration points of the same attribute cannot be combined. However, RSD-contour plots require this combination. A way to overcome this difficulty is to save the nodal displacements of the interface elements in a Femview ASCII file (fvi-file). Via the use of a macro in Excel, the displacement values of corresponding nodes can be subtracted and processed in order to obtain the desired results. These can then be inserted in the fvi-file, which can be imported in Femview to create the binary file for making contour plots.

Although the procedure is fairly easy, every new model geometry requires the creation of a new macro. This can be a time-consuming process for large models. Furthermore, the importation of the fvi-file requires a special format of this file, which must be maintained.

3.2 Stress paths and stress path gradients

Stress paths can be used to determine whether a state of stress on a fault plane converges to, runs parallel to or diverges from the yield criterion of the fault. When the stress path gradient is larger than the gradient of the yield criterion, i.e. when $\Delta\tau/\Delta\sigma'_n > \tan\phi$ (Fig. 1), the stress path converges and the stress development is said to be critical. For stress path gradients equal to or smaller than the gradient of the yield criterion $(\Delta\tau/\Delta\sigma'_n \leq \tan\phi)$, the stress path runs parallel to or diverges from the yield criterion, respectively. The stress development in these cases is said to be non-critical. Complications on stress paths and research results of a parameter study are described by Glab (2001).

The creation of stress path plots does not incorporate any difficulties. Contour plots in Femview of stress path gradients is in most cases not directly possible by simply dividing $\Delta\tau$ by $\Delta\sigma'_n$, since for some integration points $\Delta\sigma'_n = 0$. This difficulty can easily be overcome for the models used in this study by simply adding a scalar quantity of $1.0 \cdot 10^{-29}$ to both $\Delta\tau$ and $\Delta\sigma'_n$. Apparently, a value of $1.0 \cdot 10^{-29}$ is far below the calculation accuracy of the used models with values in the range of $1.0 \cdot 10^{+6}$. It is not sure if this trick always works.

3.3 *Mobilised shear capacity on faults*

Mobilised shear capacity (MSC) is defined as the ratio of the maximum shear stress (τ_{max}) and the shortest distance from the point of mean stress towards the yield line of the fault (d_{yield}) (Fig. 3a). An MSC-value of 1 means yielding, a value of 0 means that no shear stress is present.

MSC-values can be directly calculated in Femview using the option "results calculate expression…". The output attributes in Femview related to the state of stress of interface elements are in terms of tractions in direction of the local element co-ordinate system rather than in terms of principal stresses such as shown in Figure 3a. The definition to calculate MSC in dependence of effective normal- and shear traction is shown in Figure 3b. A is a point reflecting an arbitrary set of effective normal traction $t'_{n,A}$ and shear traction $t_{s,A}$. BC is a line through point A perpendicular to a Mohr-Coulomb yield line with arbitrary friction angle ϕ and cohesion c, where C is its intersection point with the effective normal traction axis and B is its intersection point with the yield line. $t'_{n,B}$, $t'_{n,C}$, $t_{s,B}$ and $t_{s,C}$ are effective normal- and shear traction in points B and C, respectively. MSC is then defined as:

$$MSC = AC/BC \tag{2}$$

where

$$AC = \sqrt{\left(t'_{n,C} - t'_{n,A}\right)^2 + t^2_{s,A}} \tag{3}$$

$$BC = \sqrt{\left(t'_{n,C} - t'_{n,B}\right)^2 + t^2_{s,B}} \tag{4}$$

$$t'_{n,C} = t'_{n,A} + t_{s,A}\tan\phi \tag{5}$$

$$t'_{n,B} = \frac{t_{n,A} + \left(t_{s,A} - c\right)\tan\phi}{\tan^2\phi + 1} \tag{6}$$

$$t_{s,B} = \frac{t'_{n,A}\tan\phi + \left(t_{s,A} - c\right)\tan^2\phi}{\tan^2\phi + 1} + c \tag{7}$$

3.4 *Mobilised shear capacity in rock volumes*

Figure 4 shows the Mohr-Coulomb yield criterion in a 3-dimensional stress space. Any possible state of stress $\{\sigma\}$ in a rock volume lies within or at the hexagonal pyramid and can be described by the sum of its mean stress vector $\{\bar{\sigma}\}$ and deviatoric stress vector $\{s\}$ according to:

$$\{\sigma\} = \{\bar{\sigma}\} + \{s\} \tag{8}$$

In this article the sign convention in rock mechanics is used, which states that compressive stresses are positive, with $\sigma_1 \geq \sigma_2 \geq \sigma_3$.

A state of stress in a 3-dimensional stress space can be described by the Haigh-Westergaard coordinates ξ,

(a)

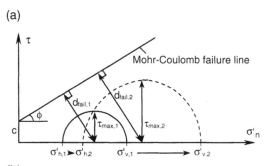

(b)

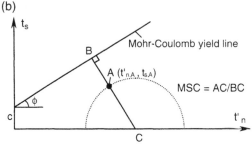

Figure 3. (a) Definition of τ_{max} and d_{fail} in an effective normal/shear stress diagram. Subscripts 1 and 2 denote a state of stress before and after gas depletion respectively. (b) Definition of MSC for an arbitrary set of effective normal and shear tractions marked by point A. AC and BC can be calculated from $t'_{n,A}$, $t_{s,A}$, ϕ and c (see formulas in text). The dashed half circle is an imaginary Mohr-circle through point A with centre point C.

ρ and θ (Chen & Han 1988). They are indicated in Figures 4 and 5. ξ is the magnitude of the mean stress vector:

$$\xi = \sqrt{3}\bar{\sigma} = I_1 / \sqrt{3} \qquad (9)$$

where I_1 is the first invariant of the stress tensor and $\bar{\sigma}$ is the mean stress (scalar quantity).

ρ is the magnitude of the deviatoric stress vector and expresses the shearing intensity of a given stress state:

$$\rho = \sqrt{\left(\sigma_1 - \bar{\sigma}\right)^2 + \left(\sigma_2 - \bar{\sigma}\right)^2 + \left(\sigma_3 - \bar{\sigma}\right)^2} = \sqrt{2J_2} \qquad (10)$$

where J_2 is the second invariant of the stress deviator.

θ is the angle between the projected vertical principal stress axis and the deviatoric stress vector:

$$\theta = \frac{1}{3}\arccos\left(\frac{3\sqrt{3}J_3}{2\left(\sqrt{J_2}\right)^3}\right) \quad \text{for } 0° \leq \theta \leq 60° \qquad (11)$$

where J_3 is the third invariant of the stress deviator.

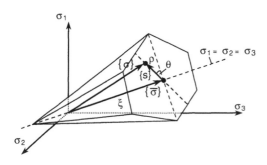

Figure 4. Graphical representation of principal, mean and deviatoric stress vectors and Haigh-Westergaard stress co-ordinates in a 3-dimensional stress space. The hexagonal pyramid is a Mohr-Coulomb yield surface.

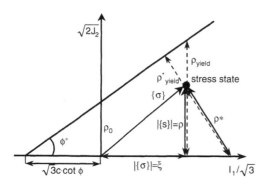

Figure 5. Rendulic plane view, showing the definitions of ρ^*, ρ^*_{yield}, ϕ^*, ξ, ρ, ρ_{yield} and ρ_0.

Figure 5 shows the rendulic plane, i.e. the plane containing all three vectors $\{\sigma\}$, $\{\bar{\sigma}\}$ and $\{s\}$. In accordance to equation (2), the mobilised shear capacity in this 3-dimensional stress space (MSC$_{3D}$) can be defined as:

$$\text{MSC}_{3D} = \frac{\rho^*}{\rho^*_{\text{fail}}} \qquad (12)$$

It is obvious that MSC$_{3D}$ depends on ξ, ρ and θ. ξ determines the size of the Mohr-Coulomb yield circumference in the deviatoric plane. ρ and θ together determine how close a stress state is located to the yield line in the rendulic plane.

In order to get to an expression for MSC$_{3D}$, one needs to express a state of stress and the Mohr-Coulomb yield surface in terms of Haigh-Westergaard co-ordinates. The principal stress vector in terms of Haigh-Westergaard coordinates follows from combination of Equations (8)–(11):

$$\begin{Bmatrix} \sigma_1 \\ \sigma_2 \\ \sigma_3 \end{Bmatrix} = \frac{1}{\sqrt{3}}\begin{Bmatrix} \xi \\ \xi \\ \xi \end{Bmatrix} + \sqrt{\frac{2}{3}}\rho\begin{Bmatrix} \cos\theta \\ \cos\left(\theta - \frac{2\pi}{3}\right) \\ \cos\left(\theta + \frac{2\pi}{3}\right) \end{Bmatrix} \qquad (13)$$

The Mohr-Coulomb yield criterion in terms of principal stresses is defined as:

$$\sigma_1 \frac{1 - \sin\phi}{2c \cdot \cos\phi} - \sigma_3 \frac{1 + \sin\phi}{2c \cdot \cos\phi} = 1 \qquad (14)$$

Substituting the expressions for σ_1 and σ_3 of equation (13) into (14) and re-arranging, one obtains the yield function f in terms of Haigh-Westergaard coordinates:

$$f = -\sqrt{2}\xi \sin\phi + \sqrt{3}\rho \sin\left(\theta + \frac{\pi}{3}\right)$$
$$- \rho \sin\phi \cos\left(\theta + \frac{\pi}{3}\right) - \sqrt{6}c \cdot \cos\phi \qquad (15)$$

where c is the cohesion of the rock. Yielding occurs when f = 0:

$$\rho_{\text{fail}} = \frac{\sqrt{6}c \cdot \cos\phi + \sqrt{2}\xi \sin\phi}{\sqrt{3}\sin\left(\theta + \frac{\pi}{3}\right) - \sin\phi \cos\left(\theta + \frac{\pi}{3}\right)} \qquad (16)$$

From goniometric relations in Figure 5, it can be easily derived that:

$$\rho^* = \frac{\rho}{\cos \phi^*} \tag{17}$$

$$\rho_{\text{fail}}^* = \cos \phi^* \left(\rho_{\text{fail}} - \rho \right) + \frac{\rho}{\cos \phi^*} \tag{18}$$

Recalling the definition of MSC_{3D} of equation (12), it then follows that:

$$MSC_{3D} = \frac{\rho \left(1 + \tan^2 \phi^* \right)}{\rho_{\text{fail}} + \rho \tan^2 \phi^*} \tag{19}$$

ρ and ρ_{fail} can be calculated with equations (10) and (16). $\tan \phi^*$ follows from Figure 5:

$$\tan \phi^* = \frac{\rho_0}{\sqrt{3} c \cdot \cot \phi} \tag{20}$$

where $\rho_0 = \rho_{\text{fail}}(\xi = 0)$:

$$\rho_0 = \frac{\sqrt{6} c \cdot \cos \phi}{\sqrt{3} \sin \left(\theta + \frac{\pi}{3} \right) - \sin \phi \cos \left(\theta + \frac{\pi}{3} \right)} \tag{21}$$

When $MSC_{3D} = 0$, the state of stress lies on the hydrostatic axis. For $MSC_{3D} = 1$, the state of stress lies on the yield surface. Note that the derivation of the formula does not take into account tension cut-off or a compression yield surface (cap).

Values for MSC_{3D} can be directly calculated in Femview using the option "results calculate expression…". Its application in Femview proved to be successful.

4 CHOICE OF APPROPRIATE FAULT PROPERTY VALUES

The amount of fault slip, which develops during gas depletion, depends on the initial state of stress and stress development on the fault plane and the strength properties of the fault. Their proper definition in the FE-models is therefore of crucial importance. The strength of the fault should be low enough to see sufficient large differences between the calculations with different parameters. On the other hand, properties should be high enough to ensure numerical stability of the model and convergence to a solution when iterating, and to avoid undesired side effects such as a large movement on the entire fault plane adding up to the

movements at reservoir level, in which we are actually interested. In the following, normal- and shear stiffness, initial pore pressure and pore pressure development are discussed.

4.1 Shear- and normal stiffness

The shear stiffness D_s of interface elements sets the relation between shear traction and elastic shear relative displacement, before the stress state on this fault reaches its yield criterion. The same applies for the normal stiffness D_n with normal traction and normal relative displacement.

Published data on stiffness properties for rock joints are limited. Summaries of data can be found in Bandis et al. (1983), Kulhawy (1975) and Rosso (1976). Values for both shear- and normal stiffness typically range from roughly 10 to 100 MPa/m for weak faults with soft clay in-filling, to over 100 GPa/m for tight joints in granite and basalt (Itasca 1995).

Appropriate values of D_s and D_n are in the range of the elastic properties of the rock formations surrounding the fault. The following rules of thumb can be used for determining appropriate values in DIANA (TNO Building and Construction Research, pers. comm.):

$$D_s = \frac{E}{2(1 + \nu)h} \tag{22}$$

$$D_n = \frac{(1 - \nu)E}{(1 + \nu)(1 - 2\nu)h} \tag{23}$$

where E and ν are the Young's modulus and Poisson's ratio of the modelled rock mass surrounding the fault and h is the fault thickness. These formulas are the basic expression of Hooke's Law for elastic stress-strain behaviour under uniaxial strain conditions, with the axis of strain normal to the interface. For the rock mass properties mentioned in section 2.3 this means $D_s = 5.4$ GPa/m and $D_n = 14.4$ GPa/m for h = 1 m.

Figure 6 shows the maximum relative shear displacement on the fault plane in dependence of shear and normal stiffness. Note that in this calculation series the shear and normal stiffness are equal. The values of other strength parameters are indicated. The location of $RSD_{\max}$ is in all calculations equal and very close to the point where the two reservoir compartments start to be geometrically separated from each other (pont S, see Fig. 2 and 8).

There exists a strong dependence of the amount of relative shear displacement on the chosen values of D_s and D_n. Values of RSD increase for decreasing values of D_s and D_n. Especially for lower stiffness values, their influence becomes very significant. Extreme high stiffness values mean glued interfaces,

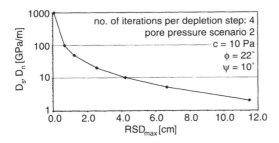

Figure 6. Maximum relative shear displacement on the fault plane as a function of shear- and normal stiffness of the interface elements.

i.e. interfaces on which any slip or separation is prevented. RSD-values are indeed as good as zero for $D_s = D_n = 1000\,GPa/m$.

For further calculations, the following values are chosen: $D_s = 5.4\,GPa/m$ and $D_n = 14.4\,GPa/m$.

4.2 Initial pore pressure and pore pressure development

There exists a direct relation between the amount of pore pressure in the modelled fault to the amount of effective normal stress on the fault plane. A higher pore pressure means a lower effective normal stress. Since shear stress is not affected by pore pressure, this always means a stress state closer to the yield criterion. Assumptions on the initial pore pressure in the interface elements and its depletion scenario should be taken with great care.

The fault plane in the model can be subdivided in three different regions (Fig. 2): region A, where the fault is bounded on both sides by overpressured reservoir rock, region B, where the fault plane is bounded on one side by overpressured reservoir rock and on the opposite side by surrounding rock with hydrostatic pore pressure, and region C, where the fault plane is bounded on both sides by surrounding rock with hydrostatic pore pressure.

Over geological times faults in the subsurface basically incorporate the same pore pressure as the neighbouring rock formation. Regions A and C are therefore assumed to have an initial pore pressure which is overpressured and hydrostatic respectively. A question arises for the initial pore pressure in region B: overpresured or hydrostatic?

Another question arises from the fact that fault region B is bounded on one side by producing reservoir rock and on the opposite side by non-producing surrounding rock formations. Does the pore pressure in fault region B remain constant at its initial value during gas depletion or does it decrease according to the (neighbouring) reservoir depletion?

Table 1. Fault pore pressure scenarios. Region C always incorporates a hydrostatic pore pressure except for scenario 1 and is never depleted.

	Region A		Region B	
Scenario	Initial pressure	Depletion	Initial pressure	Depletion
1	No pressure	–	No pressure	–
2	Overpressured	Full	Overpressured	Full
3	Overpressured	Full	Hydrostatic	No
4	Overpressured	Full	Hydrostatic	Half
5	Overpressured	Full	Hydrostatic	Full

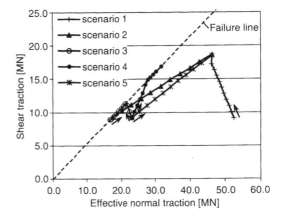

Figure 7. Stress paths in observation point P (see Fig. 2) for different fault pore pressure scenarios. Arrows indicate the initial stress state and development direction of the stress paths.

In order to answer these two questions, five different pore pressure scenarios are studied (Table 1). Figure 7 shows for each scenario the stress path in observation point P (see Fig. 2).

Scenario 1 is calculated in order to investigate whether the definition of a pore pressure in the fault can be reasonably avoided. The stress path is located very far to the right due to a high effective normal stress, which was supported by relatively low values of MSC and a restriction of relative shear displacements to a relatively small zone around point S. It can be concluded, that in order to study the effects of different parameters on fault reactivation with relative shear displacements as one of the criteria, a pore pressure has to be applied in the interface elements.

An initial pore pressure in fault region B equal to the initial (over)pressure in the reservoir rock (scenario 2) results in a state of stress in this region very close to yielding before the start of gas depletion. The stress path in Figure 7 of scenario 2 is for full depletion of region B. It diverges from the yield line, meaning that no plastic fault slip will develop. On the other hand, no or partly depletion of this fault region leads to a large amount of plastic fault slip, since the yield line is already reached after the first depletion step. It is therefore decided to take an initial pore pressure in fault region B equal to the initial (hydrostatic) pressure in the surrounding rock.

The stress paths belonging to depletion scenarios 3, 4 and 5 are for no, half and full depletion of fault zone B, respectively. The initial pore pressure for these scenarios is hydrostatic. No depletion of the fault leads to decreasing normal effective stresses (Fig. 7, scenario 3). Initially, shear stresses increase but as soon as the stress path hits the yield line they are forced to decrease. In some interface elements the normal effective stress reaches zero. Full depletion leads to a stress path diverging from the yield line, meaning that no plastic fault slip will develop (scenario 4). This would limit the estimation of the criticalness of a fault for reactivation, since this is partly based on such plastic fault slip in terms of RSD-values. Half depletion (scenario 5) causes plastic fault slip in the last five depletion steps. Scenario 5 is chosen for further calculations.

5 EXAMPLE CALCULATION

Calculation results of a calculation with property values and pore pressure scenario as mentioned in sections 2.3–2.5, 4.1, 4.2 are presented in Figures 8–11. They show the uncomplicated use of the developed quantitative methods in Femview. A constant stiffness iteration method with 20 iterations per depletion step is used.

Figure 8 shows a contour plot of RSD-values which developed on the fault plane during the 10 gas depletion steps. View direction of the plot is normal to the fault plane (see also Fig. 2). For a better visualisation, four cross sections are shown in with corresponding graphs of RSD-values.

The maximum fault slip amounts 11.4 cm and is located very close to point S. Two zones with higher RSD-values can be distinguished on either side of point S: one with a reservoir configuration corresponding to cross section BB' and one with a reservoir configuration as in cross section DD'.

Both reservoir configurations BB' and DD' are sensitive to fault slip due to differential compaction of the two reservoir compartments: the downward movement of the top of the hanging wall reservoir compartment

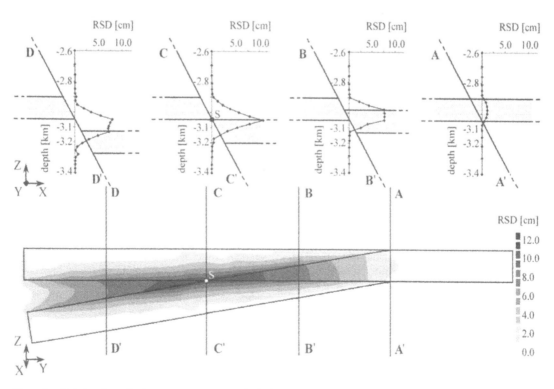

Figure 8. Relative shear displacement (RSD) on the fault plane. The maximum value is 11.4 cm, located in cross section C at a depth of −3.05 km. View direction of the contour plot is normal to the fault plane. Intersection lines of the two reservoir compartments with the fault plane are shown. View direction of the cross sections is in strike direction of the fault. Pointed areas denote the reservoir compartments.

counteracts with the upward movement of the bottom of the footwall reservoir compartment. This counter-action is at its maximum in configuration CC': an RSD peak value is observed here. In configuration BB', RSD-values are somewhat lower than in CC', but according to the contour plot high values occur over a larger area. In configuration DD', RSD-values are less than in BB' since the two reservoir compartments are geometrically not in contact to each other. A reservoir setting of type AA' has a low sensitivity to fault reactivation since the two reservoir compartments on either side of the fault compact in the same amount. RSD-values are smaller than 1.0 cm in the positive Y-direction of cross section AA'. The results are in good agreement with Roest (1994) and Roest & Mulders (2000). For a good estimation of the sensitivity of a fault to be reactivated based on fault slip, RSD-values should be integrated over the fault plane.

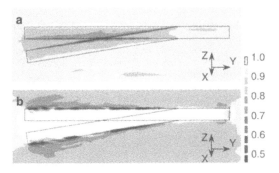

Figure 9. MSC-values on the fault plane a) before and b) after gas depletion. View is normal to the fault plane.

Figure 10. White areas denote stress paths converging towards the yield criterion during gas depletion and are therefore critical for the modelled circumstances (extensional stress regime). Grey areas denote stress paths diverging from the yield line and are stabilising. See section 3.2 for definitions.

Figure 11. Calculated MSC$_{3D}$ in the rock for cross section CC' (see Fig. 8). Reservoir edges and fault are indicated in black.

Figures 9a, b show MSC-values on the fault plane before and after gas depletion, respectively. The maximum value is exactly 1.0, according to the theory. It shows the proper implementation of the formulas in Femview. It is interesting to note that, for the given setting of an extensional stress regime, those parts of the fault plane, where most fault slip develops during gas depletion, have the lowest MSC-values prior to depletion. Almost the entire part of the fault in contact to one or both reservoir compartments is brought to yield during gas depletion for the modelled circumstances (extensional stress regime). At the top of the footwall compartment and the bottom of the hanging wall compartment, a stabilisation occurs. This can be clearly seen in Figure 10. The plot is a simplified contour plot of stress path gradients (see section 3.2 for definitions).

Figure 11 shows an example of an MSC$_{3D}$-contour plot for a cross section through the middle of the reservoir perpendicular to the fault strike. The maximum value is exactly 1.0, according to the theory, showing the proper implementation of the formulas in Femview. Zones with a stress development close to yielding occur for the modelled circumstances (relatively weak surrounding rock, extensional stress regime) outside of the reservoir at its lateral edges and close to the fault plane. Stress development inside of the reservoir is critical in an extensional stress regime, except for its lateral edges.

6 CONCLUSIONS

– In order to get more fundamental insight in the causes of gas depletion-induced fault reactivation and seismicity, several 3D models of gas reservoirs are built in DIANA. The use of intersecting shapes in Femgen allows an easy creation of various model geometries by simply changing some co-ordinate values in a batch file.
– Three quantitative methods to analyse the sensitivity of a fault for reactivation in numerical calculations are presented: relative shear displacements (RSD), stress paths and mobilised shear capacity (MSC). MSC is also applied for rock volumes (MSC$_{3D}$).
– RSD-contour plots require the combination of calculated displacements of different nodes via the use of Femview ASCII files and Excel macros. This and newly developed and complex formulas for MSC and MSC$_{3D}$ are successfully applied in Femview. It shows the possibility of implementing very complicated formulas in Femview by means of the "Results calculate expression…"-option.
– Modelling of the subsurface is a data-limited problem. The presented quantification methods are used

to choose appropriate reference values for modelling of the fault. An application example illustrates
the use of the methods and is supported by field
data and earlier research work.

ACKNOWLEDGEMENTS

The NAM is thanked in particular for providing data.
Many fruitful discussions with collegues of DUT,
NAM, SEPTAR, TNO-NITG, KNMI and SodM and
other institutions have contributed to this work.

REFERENCES

Bandis, S.C., Lumsden, A.C. & Barton, N.R. 1983.
Fundamentals of rock joint deformation. *Int. J. Rock
Mech. Min. Sci. & Geomech. Abstr.* 20(6): 249–268.

BOA (Begeleidingscommissie Onderzoek Aardbevingen)
1993. *Final report on a multidisciplinary study of the
relationship between gas production and earthquakes in
the northern part of The Netherlands.* De Bilt: Koninklijk
Nederlands Meteorologisch Instituut (KNMI).

Chen, W.F. & Han, D.J. 1988. *Plasticity for structural engineers.* Berlin: Springer.

De Crook, Th., Dost, B. & Haak, H.W. 1995. Seismisch
risico in Noord-Nederland. *Technical report TR-205,*
Koninklijk Nederlands Meteorologisch Instituut (KNMI),
De Bilt, Netherlands.

Glab, M. 2001. Localisation of sensitive reservoir settings.
*Memoirs of the Centre of Engineering Geology in the
Netherlands* 206. Delft: Delft University of Technology.

Itasca 1995. *Fast Langrangian Analysis of Continua, version
3.3.* Itasca Consulting Group Inc., Minneapolis, USA.

Kulhawy, F.H. 1975. Stress deformation properties of rock
and rock discontinuities. *Engineering Geology* 9:
327–350.

Roest, J.P.A. & Kuilman, W. 1994. Geomechanical analysis
of small earthquakes at the Eleveld gas reservoir. *Proc.
Eurock '94 Symp., Delft, 29–31 August 1994.* Rotterdam:
Balkema.

Roest, J.P.A. & Mulders, F.M.M. 2000. Overview modelling
gas production-induced seismicity mechanisms. In
Deutsche Gesellschaft für Geotechnik e.V. (ed.), *Proc.
Eurock 2000 Symp., Aachen, 27–31 March 2000.* Essen:
Verlag Glückauf.

Rosso, R.S. 1976. A comparison of joint stiffness measurements in direct shear, triaxial compression, and in situ.
Int. J. Rock Mech. Min. Sci. & Geomech. Abstr. 13:
167–172.

Rudnicki, J.W. 1999. Alteration of regional stress by reservoirs and other inhomogenities: stabilizing or destabilizing? In G. Vouille & P. Berest (eds.), *Proc. 9th Int. Congr.
on Rock Mech., General Reports, Paris, August 1999.*
Rotterdam: Balkema.

Segall, P. & Fitzgerald, S. 1998. A note on induced stress
changes in hydrocarbon and geothermal reservoirs.
Tectonophysics 289: 117–128.

Finite Elements in Civil Engineering Applications, Hendriks & Rots (eds.)
© 2002 Swets & Zeitlinger, Lisse, ISBN 90 5809 530 4

Modelling of deep subsurface for geohazard risk assessment

B. Orlic & R. van Eijs
Netherlands Institute of Applied Geoscience TNO – National Geological Survey, Utrecht, The Netherlands

ABSTRACT: Numerical simulations of the deep subsurface are increasingly used for a better understanding and more accurate predictions of subsurface behaviour in the wake of the extraction of mineral resources and storage of energy residues. Examples are described of the use of DIANA in these applications, demonstrating coupling of different modelling technologies in a workflow for integrated geomechanical modelling. The first example describes a method for construction of geological models and their conversion into unstructured finite element meshes. The second study describes an uncoupled flow-stress analysis that was used to study the relationship between reservoir depletion and the induced micro-seismicity. In the third case, a fully coupled flow-stress analysis was carried out to analyse combined load- and hydraulic-driven groundwater flow for safety assessment of a geological disposal site. The examples demonstrate that the coupled process modelling with increasingly complex three-dimensional models of the subsurface has become feasible with DIANA.

1   INTRODUCTION

The exploitation of subsurface natural resources, the underground storage of energy carriers and disposal of energy residues (e.g. CO_2) may cause deformation and compaction of the rock mass, subsidence and trigger earthquakes, usually of small magnitude. If these phenomena pose risk to human lives, property, developed land or the natural environment, then they represent man-induced geohazards.

Public authorities as well as the oil and mining industries increasingly demand fundamental insight and accurate predictions of deformation and damage related to the exploitation of subsurface natural resources and the use of subsurface for geological storage and disposal. At this moment deformation is difficult to assess and prove, although economic, environmental and societal interests are huge in terms of the granting of concessions, the failure to meet exploration targets, the resolving of damage claims and the like.

For prediction of deformation associated with the geological storage, disposal or exploitation of mineral resources we use integrated geomechanical modelling. The technologies integrated in a modelling cycle comprise the geological modelling techniques, by which "static" models of the subsurface are constructed, and the process modelling techniques, by which "dynamic" models are computed. Static models depict the structural geological relationships and, at the same time, the spatial distribution of the properties that characterise different geological materials in the subsurface. Dynamic models are used, for instance, for simulation of fluid flow and stress modelling.

The objective of this paper is to illustrate integration of different modelling technologies in recently accomplished geomechanical DIANA studies at TNO-NITG. A workflow for integrated geomechanical modelling will be introduced first. Then, a method will be described for construction of geological static models and their conversion into unstructured finite element meshes. Coupling of two process models, a flow model and a stress model, will be illustrated with two geohazard risk assessment studies. In the first study an uncoupled flow-stress analysis was used to study the relationship between reservoir depletion and induced micro-seismicity. In the second study a fully coupled flow-stress analysis was carried out for safety assessment of a geological disposal site.

2   WORKFLOW FOR INTEGRATED GEOMECHANICAL MODELLING

A workflow for integrated 3D geomechanical modelling was developed to accurately predict deformation caused by exploitation of mineral resources, geological storage of energy residues and disposal of energy residues. The workflow depicts the modelling process, which consists of three major steps (Figure 1):

– Mapping and modelling, in which the structural geological models (i.e. the framework models) are built.
– Characterisation, in which the property models are built. These models represent the spatial distribution of different (geological) materials and their properties (e.g. density, porosity, etc.) within the structural model.
– Engineering analysis, which consists of a fluid flow analysis and a stress analysis. The latter gives us predictions of surface and subsurface deformation, the assessment of which is the main objective of the geomechanical modelling.

The tools for geological modelling, fluid flow modelling and stress analysis, which are integrated in this workflow, enable efficient transfer of data between the shared earth models.

GOCAD (GOCAD, 2001) is used to model the 3D geometry of the subsurface and DIANA (DIANA, 2002) for the finite element (FE) geomechanical modelling. Different reservoir simulation packages are used to model the fluid flow.

3 CONSTRUCTING MESHES FROM STRUCTURAL MODELS

A method was developed for construction of consistent geological models of the subsurface and their conversion into unstructured finite element meshes. It

was tested on a case study of a depleting gas reservoir in order to demonstrate that it is technically possible to incorporate the complexity of the 3D geological structure of a reservoir into a geomechanical model.

The gas field under study is situated in the northeastern region of the Netherlands. This region is well known in the gas industry because of its large Permian Rotliegend fields, including the giant Slochteren Field on the Groningen High. In addition to those large fields, smaller occurrences can be found in younger deposits, such is the gas field under study. The reservoir is formed in an anticline structure, located above the Emmen Salt dome.

3.1 Method for 3D mesh generation

The method for construction of unstructured FE meshes from geological structural models preserves the main structural features of the geological setting in a FE mesh. The method starts with the construction of a geological structural model from available data. Such a model is then converted into a set of triangulated surfaces, while rearranging the position of surface nodes and reducing their number.

Following the above described procedure, the geological structural model of the reservoir is constructed solely on the basis of digital data publicly available from TNO-NITG (Figure 2). Besides public data, the inputs for the detailed structural modelling are depth surfaces of differentiated geological units and fault surfaces mapped in a mapping and contouring package (Figure 3a). These surfaces are usually available as gridded surfaces and first have to be converted to triangulated surfaces (Figure 3b). The surfaces also require some additional editing, for which GOCAD provides a versatile set of tools, in order to develop a consistent 3D structural model of the site of interest.

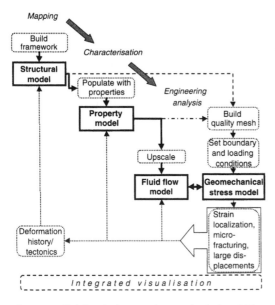

Figure 1. Workflow for integrated geomechanical modelling.

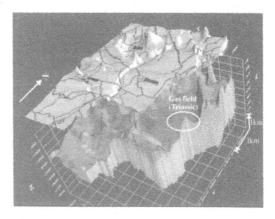

Figure 2. Geological model of the gas field under study. The model is taken from the publicly available 3D Digital Atlas of the Deep Subsurface of the Netherlands and visualised in a 3D Java viewer, developed by TNO-NITG.

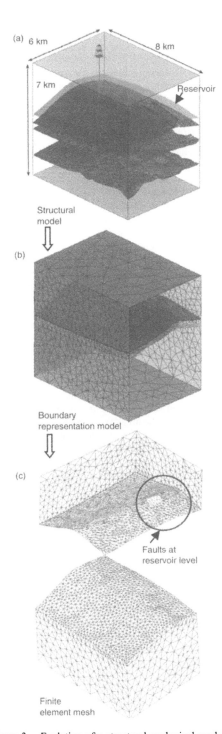

(a) 6 km

8 km

7 km

Reservoir

Structural
model

(b)

Boundary
representation model

(c)

Faults at
reservoir level

Finite
element mesh

Figure 3. Evolution of a structural geological model to a FE mesh. a) A 3D structural model of the gas field under study; b) a boundary representation model; and c) an unstructured finite element mesh, which was used to predict the subsidence due to reservoir depletion.

The mutual intersections and clipping of the surfaces may alter their regular partitioning due to incorporation of the intersection curves. Many additional vertices, and therefore triangles, are introduced into the surface along each intersection curve. As a result, some triangles and tetrahedrons in the FE mesh will have elongated and distorted shapes which are not acceptable for producing a mesh of good quality. This problem can be overcome by resampling all borderline segments that bound model surfaces at regular intervals. The interval length can be used to control the density of vertices along the line segments, which, in turn, determines the density of triangles in the surface patches and the density of the tetrahedron mesh.

The resampled keylines provide a starting point for generating resampled triangulated surfaces, which approximate the original surfaces of the structural model. The resulting surface patches marked by quality triangles define a boundary representation model that is topologically and volumetrically consistent (Figure 3b). Such a model can be meshed successfully by a 3D tetrahedral mesher, such as the one available for pre-processing in DIANA (Figure 3c). Prediction of subsidence due to reservoir depletion, calculated with this unstructured mesh, is shown in Figure 4.

4 UNCOUPLED FLOW-STRESS MODELLING

A frequently used way of combining fluid flow modelling and stress modelling in reservoir engineering is to use an uncoupled (i.e. staggered) flow-stress analysis. In this approach a reservoir simulator is used first to compute the entire time history of pressure. This time history is then used as input to the stress modelling and a transient stress solution is found. Although these two calculated time histories are independent, the procedure is often used because the conventional modelling tools for (multi-phase) fluid flow simulation and stress analysis can directly be used, without any modification.

4.1 *Modelling impacts of reservoir depletion*

Uncoupled flow-stress modelling is illustrated with a case study of man-induced seismicity due to gas extraction. The gas field under study is the same as the one described in the previous section. The analyses were performed on a 2D FE plain strain model of the field, which is shown in Figure 5.

Seismic events ranging from 1 to 3.5° on the Richter scale started to occur some 10 years after the start of gas production. According to the seismological measurements, all of the epicentres were located above the crest of the gas field. The hypocentres were at or above the reservoir depth, with an uncertainty in depth prediction of $\pm$ 500 m.

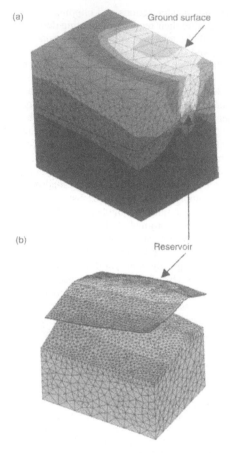

(a)   Ground surface

(b)   Reservoir

Figure 4.   Predicted subsidence due to reservoir depletion, a) at the ground surface; and b) at the reservoir level.

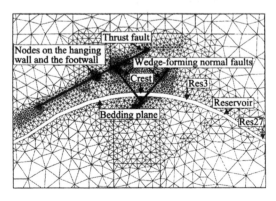

Figure 5.   Mesh showing the reservoir and the structural features incorporated into the finite element model.

Seismic events occurring in the vicinity of the gas field are believed to be caused by reactivation of faults in or above the reservoir. Following the approach of uncoupled modelling, the pore pressure changes in the reservoir during a 20-year long period of gas production were derived from reservoir simulations. These were used as input to the FE stress analyses to assess the potential of existing faults for reactivation. The influence of various tectonic in situ stress regimes on fault reactivation will briefly be discussed.

4.2   Initialising in situ tectonic stresses

A FE geomechanical model has first to be brought into the state of the in situ tectonic stresses before it can be used to simulate the effects of reservoir depletion. The tectonic stresses are known if both magnitude and orientation of the three principal stress components are determined. Magnitudes determine the lateral pressure stress ratio's (Ko), i.e. the ratio's between the magnitudes of pairs of principal stresses.

A Ko procedure in DIANA 8, now available for 3D analysis, can be used to initialise the full stress tensor of tectonic stresses at the start of a non-linear analysis. Orientation of the principal stresses, on the other hand, may in some cases influence selection of a representative cross-section for a 2D FE analysis. For example, in case of a Mohr-Coulomb plasticity, the cross-section has to be perpendicular to the intermediate principal stress (Figure 6).

The in situ state of stress in the gas field under study was not known and the prevailing stress regime had to be estimated from the current insights of the tectonic stresses of the area (World Stress Map, 2001; Van Wees & Cloetingh, 1996). The effects of a depleting reservoir were computed for a wide range of characteristic stress regimes in order to study their influence on fault reactivation.

4.3   Predicting impacts on reservoir rock

Stress evolution during reservoir depletion is computed for characteristic locations in the reservoir: in the crest, where the expected increase in the effective stresses is the highest (Crest and Res3), and in the limbs, where the smallest changes in the effective stresses are expected (Res27). The results are plotted in Figure 7 in the form of the effective stress paths on the $\sigma' - \tau$ diagrams (σ' – the normal, i.e. isotropic effective stress, and τ – the shear, i.e. deviatoric stress). The pattern, which these stress paths exhibit, is considerably different for the two stress regimes studied.

In a normal faulting stress regime, depletion of gas pressure in the reservoir causes an increase in the effective stresses throughout the reservoir. Increase in the vertical effective stresses is generally higher than increase in the horizontal effective stresses due to a Poisson ratio of $\nu < 0.5$. Consequently, the shear stresses throughout the reservoir also increase and slowly converge towards the Mohr-Coulomb failure line as illustrated by the effective stress paths plotted

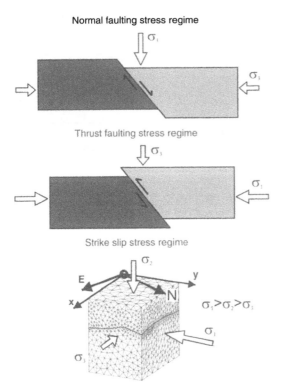

Normal faulting stress regime

Thrust faulting stress regime

Strike slip stress regime

$\sigma_1 > \sigma_2 > \sigma_3$

Figure 6. Various tectonic stress regimes determined by the principal stresses. $\sigma 1$, $\sigma 2$, $\sigma 3$ denote the major, the intermediate and the minor principal stress, respectively.

in Figure 7a. The increase in shear stress is highest in the crest of the reservoir, while the changes in the limbs are the smallest. This generally conforms to the rate of change of pore pressure in the reservoir during gas production.

In case of a thrust faulting stress regime defined by a Ko-value of 1.5, the effective stress paths are considerably different (Figure 7b). At and nearby the crest, the shear stress first decreases and, after having reached a minimum value, again increases. This can be explained by the stress rotation during depletion, i.e. the two principal stresses swap each other's directions during depletion (Orlic et al., 2001).

Associated deformation, caused by gas extraction, is the largest in the crestal part of the reservoir, as shown in Figure 8. Discontinuity of displacement contours across the two wedge-forming faults indicates that a slip on the faults at the reservoir level has occurred.

4.4 Predicting fault reactivation

Stress evolution during reservoir depletion was used to analyse the potential for fault reactivation, which is thought to be the most probable cause of induced

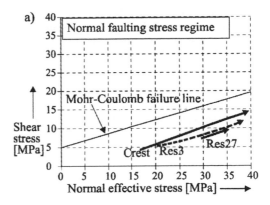

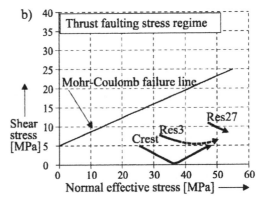

Figure 7. Stress evolution in different parts of the reservoir, for different initial tectonic stress regimes in the subsurface: a) a normal faulting stress regime; and b) a thrust faulting regime. The locations shown are the crest, a location nearby the crest (300 m off the crest; Res3) and a limb (Res27).

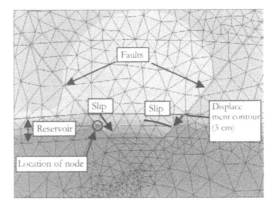

Figure 8. Displacements in the crest of the reservoir at the end of reservoir depletion.

micro-seismicity in the study area. A slip on the normal faults during reservoir depletion was obtained in a number of runs in which a normal-faulting stress regime was assumed and the friction angle of the faults was varied between $\phi = 20$ to $30°$. The graph in Figure 9a shows evolution of the normal and shear forces for a characteristic node on the normal fault in case of fault reactivation. The graphs of normal and shear displacements versus time are used to predict the time when a slip occurs, for the scenario under consideration (Figure 9b).

Stress analysis for the thrust fault, located in the overburden above the reservoir, shows a very small change in the shear stresses during reservoir depletion (Figure 10a). In a normal faulting stress regime, the shear stresses decrease during depletion (Figure 10b). This has an overall stabilising effect on the fault, which means that gas production reduces the potential for fault reactivation.

In the second case, in a thrust faulting stress regime with $Ko = 1.5$, the shear stresses slightly increase during depletion (Figure 10c). The potential for reactivation of the thrust fault is therefore also increased. In reality, a reverse slip can be triggered on a gently inclined thrust fault in case of high horizontal in situ stresses present in the subsurface. If in natural conditions, i.e. before the gas production has started, the shear forces and the resisting forces along a fault have almost reached the Limit State of equilibrium, then the reservoir depletion could cause reactivation of the thrust fault. For a given structural setting used in

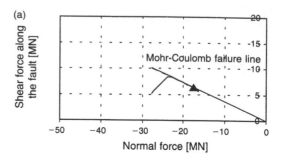

(a)

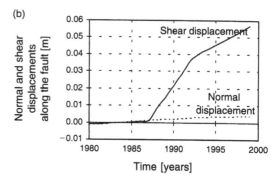

(b)

Figure 9.   a) Evolution of shear forces versus normal forces for the characteristic node, located on a wedge-forming fault at the reservoir level (i.e. the force path). b) Normal and shear displacements versus time for the same node. The graph shows that the slip occurred in 1987 when the "force path" (shown in a) touched the failure line.

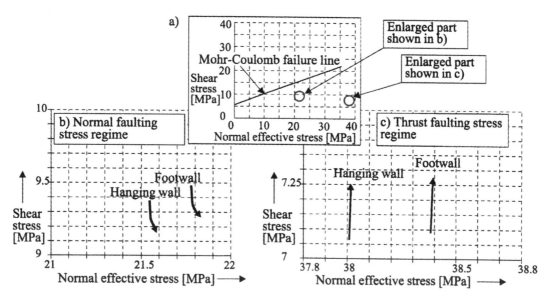

Figure 10.   a) Effective stress paths during reservoir depletion for two nodes located on the opposite sides of the thrust fault. b) Enlarged part of the graph shown above for the case of a normal faulting stress regime. c) Enlarged part of the graph shown above for the case of a thrust faulting stress regime.

this study, a slip on the thrust fault was initiated for a friction angle of $\phi = 20°$ and Ko > 2.

5 COUPLED FLOW-STRESS MODELLING

In a coupled flow-stress modelling approach the flow and stress equations are solved simultaneously. The interaction between the flow and stress is here two-directional, i.e. a mechanical load causes flow and an imposed flow causes deformation. Coupled flow-stress simulator for a single phase flow is available in DIANA and its application will be illustrated with a case study related to the safety assessment of a geological repository for hazardous waste disposal (Wildenborg et al., 2000).

5.1 Modelling glacially-driven hydromechanical processes

Deep geological disposal is generally accepted to be a viable option for long-term disposal of hazardous waste. In the Netherlands, a national research programme was launched to examine the feasibility of retrievable storage of hazardous waste in the Tertiary clay deposits. Long-term isolation – of up to 1 million years – of high-level radioactive waste under varying

conditions has to be assured. Within such a long time span, a number of geological processes can affect the clay (host rock), and accelerate the migration of radionuclides from the underground repository to the biosphere. One of the key concerns is the hydromechanical response of the clay deposits during a possible future glaciation. Climatic change leading to a glaciation is likely to occur in the Netherlands in the near geological future as it had occurred repeatedly in the recent geological past, during the Pleistocene.

The impacts of glaciation are mainly due to the mechanical loading of the subsurface by an ice sheet and due to the changes in the regional groundwater flow system. Loading causes compression and consolidation of the clays, associated with acceleration and expulsion of the pore water into adjacent aquifers (consolidation-driven flow). The changes in the groundwater flow pattern in glacial conditions are considerable: hydraulic gradients and groundwater flow velocities generally increase; direction of ground water flow in shallow aquifers changes; large quantities of melt water are released underneath the ice sheet; a permanently frozen superficial part of the ground, i.e. the permafrost, is formed, etc.

A 2D hydromechanical coupled flow-stress DIANA model was developed to assess the effects of loading of the subsurface by a thousand-metre-thick ice sheet (Figure 11). Such a model enables coupled modelling

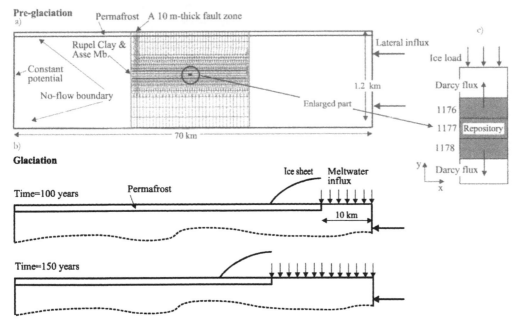

Figure 11. A coupled flow-stress finite element model for simulation of glacial loading and glacially-driven groundwater flow. a) Geohydrological boundary conditions prescribed in the initial, pre-glacial phase; b) Dynamic loading conditions, which simulate loading of the model by an advancing ice sheet, and associated geohydrological boundary conditions; c) Location of the repository for disposal of hazardous waste.

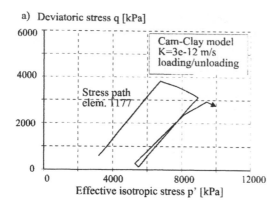
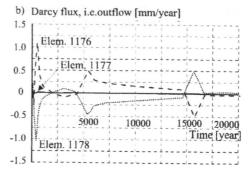

Figure 12. a) Evolution of the stress and b) the pore-water flow in the clay layer during glaciation. The clay is modelled by a Modified Cam-Clay material model. Location of elements 1176 through 1178 is shown in Figure 11.

of both load-driven flow and hydraulic-driven flow. The model was derived from a 2D schematised cross-section of the Netherlands, running approximately north to south. Geohydrogeological boundary conditions and loading scenarios for glacial period were defined for a palaeographical situation, representative for a cold climatic period in Europe during the Pleistocene. Fluxes and potentials at the boundaries of the hydromechanical model were calculated with a 3D supra-regional groundwater model of Northwest Europe for glacial conditions (Boulton & Curle, 1997; Van Weert & Leijnse, 1996).

Transient simulations were used to compute a number of ice-loading scenarios over a 20,000 year long period. Loading conditions, geohydrological boundary conditions and material properties of some differentiated model units were prescribed as time-dependent.

5.2 Predicting pore water flow

Under present-day climatic conditions, the flow of groundwater in clay deposits occurs at a very slow rate, which typically ranges between 10^{-6} and 4×10^{-7} metres a year. Diffusion is the main migration mechanism for radionuclides stored in intact clay deposits.

During glaciation, the process of consolidation has a dominant effect on the pore water flow in the clay. Loading by the ice leads to an amplified outflow of groundwater from a clay body to the aquifers above and below the clay. The outflow rate can reach a maximum value of a few millimetres per year, which is three orders of magnitude higher than the present values. Hydraulic gradients, pore pressures and flow velocities in the aquifers generally increase at least one order of magnitude with regard to the present. The duration of the time spans with high outflow rates is

very sensitive to the permeability of the clay and the dynamics of ice loading.

For example, in case of a scenario with cyclic loading by the ice, computed evolutions of the stress and the pore-water flow rate in the clay layer during glaciation show complex patterns (Figure 12). The stress path shows a cap failure and a hysteresis, which occurs due to an intermittent unloading (Figure 12a). The consolidation-driven flow rate generally follows the pattern of stress evolution (Figure 12b). During unloading, the clay swells. Swelling is accompanied by the suction-driven inflow of the pore water into the clay, which even improves the isolation capacity of the clay, as the pore water flows inwards, towards the repository.

6 CONCLUSIONS

Presented cases demonstrate that DIANA can be successfully used in geomechanical- and geological engineering applications related to the exploitation of hydrocarbons and geological storage of energy residues. Coupled process modelling with increasingly complex 3D models has become feasible. Future work in this specific application area will benefit from the advances of DIANA with respect to: (i) further development of specific process modelling tools within DIANA, their integration in a DIANA environment and the availability to the end users; (ii) further development of links to facilitate the interfacing with external geological modelling packages, which are conveniently used in the geoscience to construct complex, geologically meaningful 3D models of the subsurface; (iii) further development of DIANA specials, tailored for use in specific application areas.

REFERENCES

Boulton, G.S. & Curle, F., 1997. *Simulation of the effects of long-term climatic changes on groundwater flow and the safety of geological disposal sites.* University of Edinburgh, RIVM & RGD, European Commission, Nuclear Science and Technology, EUR 17793 EN.

DIANA 8, *Program and User's Manuals*, 2002. TNO Building and Construction Research.

GOCAD 2.0, *Program and User's Manuals*, 2001. T-Surf & National School of Geology (ENSG), Nancy.

Orlic, B., Van Eijs, R. & Scheffers, B., 2001. Geomechanical modelling of the induced seismicity for a gas field. *Proc. 63rd EAGE Conference and Technical Exhibition.* Amsterdam, Extended Abstracts, paper P 604.

Orlic, B., van Eijs, R. & Scheffers, B. 2002a. Integrated geomechanical modelling for prediction of man-induced geohazards in deep subsurface (in press). *Proc. 9th Congress of Int. Assoc. of Engineering Geology (IAEG), Theme 5*, Durban, 16–20 September 2002, South Africa.

Orlic, B., van Eijs, R., Zijl, W. & van Wees, J.D., 2002b. Building a 3D geomechanical model of a gas field for geohazard prediction. *Proc. 64th EAGE meeting*, Florence, Extended Abstracts, paper G036.

Sauter, F.J., Leijnse, A. & Beusen, A.H.W., 1993. METROPOL, User's guide. National Institute of Public Health and Environmental Protection, Bilthoven, RIVM Report 725205003.

Van Weert, F.H.A. & Leijnse, A., 1996. *Modelling the effects of the Pleistocene glaciations on the Northwest European geohydrogeological system.* National Institute of Public Health and Environment, Bilthoven, The Netherlands.

Van Wees, J.D. & Cloetingh, S., 1996. 3D flexure and intraplate compression in the North Sea Basin. *Tectonophysics* 266: 243–359.

Wildenborg, A.F.B., Orlic, B., de Lange, G., de Leeuw, C.S., Zijl, W., van Weert, F., Veling, E.J.M., de Cock, S., Thimus, J.F., Lehnen-de Rooij, C. & den Haan, E.J., 2000. *Transport of Radionuclides disposed of in Clay of Tertiary Origin (TRACTOR).* TNO-report NITG 00-223-B. Netherlands Institute of Applied Geoscience TNO – National Geological Survey.

World stress map, 2001. Internet source: http://www-wsm.physik.uni-karlsruhe.de/index.html.

Finite Elements in Civil Engineering Applications, Hendriks & Rots (eds.)
© *2002 Swets & Zeitlinger, Lisse, ISBN 90 5809 530 4*

Mechanism of remedial measures against soil liquefaction by shear deformation constraint method

N. Yoshida
Oyo corporation, Saitama, Japan

ABSTRACT: Discussion is made on how remedial measures against soil liquefaction by shear deformation constraint method works. At present, there is no relevant design method, and averaging the shear modulus in horizontal direction has been frequently used in order to evaluate shear deformation of the ground in the engineering practice. The soil element is shown to deform in the vertical direction in shear when shear deformation in horizontal direction is constrained by the stiffer underground structure, which weakens the effectiveness of remedial measures. This effect is significant when overburden stress is small, i.e., near the ground surface.

1 INTRODUCTION

It is empirically recognized that an underground wall and lattice shaped ground improvement work as remedial measures against soil liquefaction. Shear deformation of the ground is expected to become small because much shear stress is carried out by the underground wall or improved ground. This method is called remedial measures against soil liquefaction by shear strain constraint method.

In spite of the facts that this method worked in the past earthquakes and in the laboratory test, a design method to evaluate the effectiveness of this method is not established (JGS, 1993). Therefore they have been used to give additional ability to constrain the onset of soil liquefaction, but were not used in practice as only remedial measures purpose.

In this method, stiffness changes in the horizontal direction, and stiffer parts such as underground walls and improved ground are expected to constrain deformations of softer parts or ground. Therefore, interaction between stiffer parts and softer parts should be considered relevantly. A first idea as a practical engineer may be the one to use an average stiffness in evaluating the deformation of the overall ground; this method is reported as an case history in the text (JGS, 1993).

However, as pointed out in this paper, this simple assumption does not consider the interaction between the stiffer part and the softer part, and will mislead the behavior of overall ground. In this paper, we propose a mechanism to explain what happens between the stiffer part and the softer part in the ground, and to show the effect of this method as remedial measures against soil liquefaction through an example case study.

2 MECHANISM OF SHEAR STRESS PROPAGATION

Recently, a laminar box began to be used in shaking table tests and centrifugal tests, but a rigid box had been a popular test apparatus in the past and even at present in investigating the liquefaction behavior of ground. There are significant differences in behavior of the ground between a laminar box test and a rigid box test. In the test using a laminar, a surface layer may sometimes not be liquefied when liquefaction occurs at deep depths, because shear wave does not propagate through the liquefied layer. On the other hand, liquefaction occurs in the whole ground in the rigid box test because waves also propagate through the walls. This indicates that existence of the rigid underground structure cannot be said to constrain the onset of soil liquefaction; liquefaction can occur in the ground even with several tens centimeters width. Moreover, it indicates that evaluation of the ground behavior by average stiffness between stiffer parts (underground wall) and softer parts (ground) may lead bad result.

Figure 1 shows two infinitesimally small elements: the one is at the surface of the underground wall and the other is a soil element adjacent to the underground

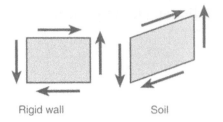

Figure 1. Shear stress transfer from rigid wall to ground.

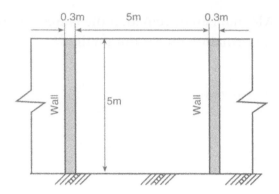

Figure 2. Ground and underground wall.

wall. Shear stress acting on each element and deformation caused by the shear stress are also drawn in the figure. Since shear stress is carried mainly by shear wall, large shear stress will be produced in the shear wall. As shown in Figure 1, if shear stress works at the edge of the wall, the same amount of shear stress should work in the soil element from the equilibrium condition. In the case of horizontal deposit, shear deformation occurs in the horizontal direction. However, since horizontal displacement is constrains by the existence of the underground wall, shear deformation occurs in the vertical direction as shown in Figure 1. Therefore liquefaction can occur in the ground by the shear waves traveling in the underground wall. The same behavior will happen in the ground improvement such as a sand compaction pile method and rod compaction method. As a result of this behavior, vertical movement appears in the ground even in the case that the ground is shaken only in the horizontal direction.

It is noted that shear stress at the surface of the wall is zero based on beam theory. This is true if the wall is placed in the air because air does not carry shear stress. However, in the ground, shear stress appears at the surface of the wall as adjacent soil can carry shear stress. This implies that shear stress at the surface of the wall can increase up to shear strength of the surrounding soil. Therefore, above discussion is valid even in beam type behavior of wall.

As shown in the preceding, a complicated interaction takes place between the underground wall and the interior ground, which affects the liquefaction characteristics of ground. It is also noted that this interaction cannot be considered by the analysis considering only horizontal displacement. Moreover, verification of this mechanism cannot be made in the total stress analysis such as FLUSH because shear stress transferred from the wall to the ground depends on the shear strength of the ground that will reduce during shaking because of excess porewater pressure generation. The effect of this mechanism to overall behavior of ground is investigated in this paper through a case study based on an effective stress dynamic response analysis.

3 MODEL

3.1 Ground and underground wall

A simplified model shown in Figure 2 is used so that mechanism can be easily recognized.

The SPT N-value of the ground is 10, and the water table coincides with the ground surface. The base layer is located at GL-5 m.

Underground walls are placed every 5.3 meters (interior width of the ground is 5 m). The 30 cm thick wall behaves in an elastic manner. The stiffness of the wall is set 100 times larger than the concrete considering the existence of the wall parallel to the shake plane. Moreover, unit weight of the underground wall is set 10 times larger than the concrete considering a weight of the superstructure that is not modeled in this study.

A repeated boundary is used in the horizontal direction; displacement at the left end of the wall is set equals to be the one at the right end of the ground. A rigid base is assumed and a half of the 1940 El-Centro NS component is applied.

3.2 Material property of soil and its modeling

A constitutive model proposed by the authors (Yoshida and Tsujino) is used under the simple two-dimensional condition. This model can simulate arbitrary defined dynamic deformation characteristics perfectly. A Ramberg-Osgood model is used for equivalent stress s versus equivalent strain e relationships for the skeleton curve. Volumetric strain caused by dilatancy is evaluated by modifying a bowl model (Fukutake and Ohtsuki, 1993) into simple two-dimensional condition. They are summarized in the followings:

Shear deformation: $\gamma = \eta(1 + \alpha\eta^{\beta - 1})$ (1)

where

$\gamma = e/e_r$: equivalent strain normalized by reference equivalent strain

$\eta = s/\tau_{max}$: dimensionless equivalent stress
$e_r = \tau_{max}/G_{max}$: reference strain, and
$\tau_{max} = c\cos\phi + \sigma'_m\sin\phi$: shear strength

Dilatancy model: $\varepsilon_{vd} = Ae^B + \dfrac{\int de}{C + D\int de}$ (2)

Here c is cohesion, ϕ is internal friction angle, G_{max} is elastic shear modulus, and α and β are parameters to define the s–e relationships. Volumetric strain caused by dilatancy ε_{vd} is determined by four parameters A, B, C and D.

Internal friction angle is set 35 degrees, cohesion is set 1 kPa, and $G_{max} = 4700\sigma_m'^{0.5}$. The small cohesion is considered in order to control the shear strength after the onset of liquefaction.

Liquefaction strength of the model is shown in Figure 3. Stress–strain relationships and stress paths are drawn in Figure 4 as an example.

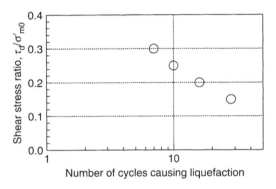

Figure 3. Liquefaction strength (initial confining stress = 100 kPa).

4 RESULT OF ANALYSIS AND DISCUSSION

A computer code STADAS (Yoshida, 1993) is employed to analyze the model. Reduced integration with an anti-hourglass stiffness is used in building an element stiffness matrix of a solid element. Therefore, bending behavior of a wall can be simulated well.

Both one-dimensional analysis and two-dimensional analysis are conducted. A horizontally layered deposit is assumed in the one-dimensional analysis.

4.1 One-dimensional analysis

A horizontally layered deposit is analyzed first in order to obtain the behavior of the bare ground without remediation. Peak responses are summarized in Figure 5, Excess porewater pressure time histories are shown in Figure 6 for all layers, and acceleration and displacement time histories at the ground surface are shown in Figure 7. Maximum displacement at the ground surface relative to the base layer is about 1 cm, and peak acceleration at the ground surface is about $2\,\mathrm{m/s^2}$.

In the excess porewater time history shown in Figure 6, lines with smaller excess porewater pressure belongs to the shallower location. From Figures 5 and 6, one can recognize that liquefaction occurred between GL-0.5 and 3.5 m. Liquefaction occurred at the shallow depth first, then gradually to the deep depths.

A maximum strain is about 0.3%. This value seems very small compared with the ordinary definition of liquefaction by shear strain (double shear strain amplitude = 7.5%), which will be discussed later. Excess porewater pressure generates quicker under random loading as can be seen in Figure 8. Maximum displacement occurs at about 12 seconds which is much later than main amplitude of ground shaking. This delay of displacement has been frequently seen

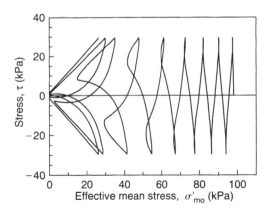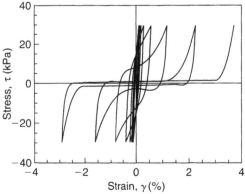

Figure 4. An example of stress path and stress–strain relationships.

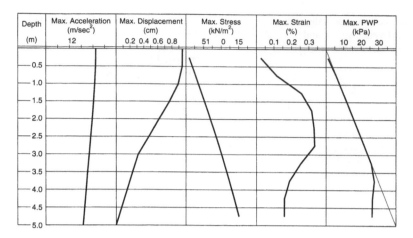

Figure 5. Maximum response by one-dimensional analysis.

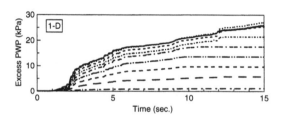

Figure 6. Excess porewater time history.

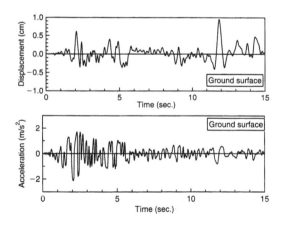

Figure 7. Acceleration and displacement time history at ground surface.

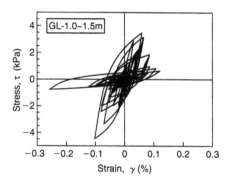

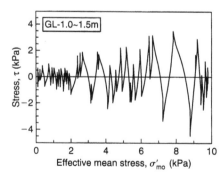

Figure 8. Example of behavior during liquefaction test (GL-1~1.5 m).

in past analyses using the same wave as input motion. Unfortunately, since liquefaction occurs from the element near the ground surface, phenomena that liquefaction do not occur near the ground if liquefaction occurs earlier in the deep depths cannot be seen.

4.2 Two-dimensional analysis

Two-dimensional analysis is conducted to investigate the behavior of the ground-underground wall system. Maximum displacements are shown for one-and two-dimensional analyses in Figure 9. Since the direction of the maximum displacement depends on nodes (they can take either positive or negative sign), absolute

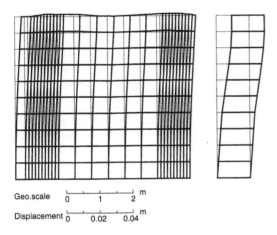

Geo.scale   0   1   2   m

Displacement   0   0.02   0.04   m

Figure 9. Absolute value of maximum displacement by two- and one-dimensional analyses.

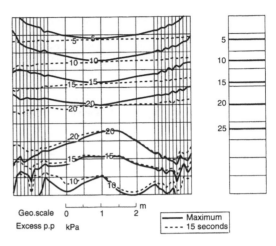

Geo.scale   0   1   2   m

Excess p.p   kPa

———— Maximum
---- 15 seconds

Figure 10. Contour of excess porewater pressure by two- and one-dimensional analyses.

values are used. Therefore, although the figure is drawn as if it is a deformed shape, only absolute values have meaning.

Three features can be easily seen in the figure.

At first, maximum horizontal displacement at the ground surface is about 1 cm in the one-dimensional analysis whereas it is about 0.4 cm in the two-dimensional analysis; displacement by the two-dimensional analysis is less than a half of that by the one-dimensional analysis. Therefore, the underground wall worked in order to reduce the horizontal displacement of the ground as expected. It is however noted that the displacement of 0.4 cm is very large compared with the displacement of the ground when average stiffness is used as shown later. In this sense, the underground wall does not work as expected.

Next, as seen in Figure 5, large deformation is observed at the middle depth and deformation near the ground surface is small in the one-dimensional analysis. On the other hand, shear strain seems to increase towards the ground surface in the two-dimensional analysis. Deformation constraint effect by underground wall also seems to appear.

Finally there appears vertical displacement, especially near the boundary of the underground wall. As discussed in the previous section, vertical displacement appears even in the horizontal shaking. Since vertical displacement near the center is nearly zero, inverse symmetric shapes of displacement is implied to occur in the whole ground. This deformation pattern agrees with the mechanism described in the previous section.

Contour lines of the excess porewater pressure are shown in Figure 10; both maximum values and residual values (excess porewater pressure at the end of shaking) are shown for two- and one-dimensional

analyses as dashed and solid lines. It is noted that excess porewater pressure increases nearly monotonically in the one-dimensional analysis as seen in Figure 6, maximum value and residual value are nearly the same. This indicates that cyclic mobility behavior did not occur in the one-dimensional analysis. Considering the nature of the input earthquake motion, waves with large amplitude comes at about 2 seconds, but stress point does not path phase transform region at this stage. Then amplitude becomes small. In such a situation, excess porewater pressure can generate without causing cyclic mobility behavior and without causing large strain. Similar behavior is seen in the test (e.g., Kusakabe, 1996).

Residual excess porewater pressures are nearly identical for both one- and two-dimensional analyses in the layers above GL-3 m. This fact clearly indicates that onset of liquefaction cannot be prevented by the underground wall.

On the other hand, excess porewater pressure generation is clearly smaller in the two-dimensional analysis than in the one-dimensional analysis at depths larger than 3 m. It is noted that vertical displacement becomes more difficult to occur at deep depth than at shallow depth, and vertical displacement is zero at the base. These facts prove that the mechanism shown in this paper is important.

The maximum excess porewater pressure is nearly the same with the residual excess porewater pressure along the centerline of the ground. Differences between the maximum excess porewater pressure and the residual excess porewater pressure become large towards the wall. As can be seen in Figure 5, region above GL-3 m liquefy in the one-dimensional analysis. Therefore, if excess porewater is larger than the one in the one-dimensional analysis as seen above, total overburden stress must be increased, which is possible only when vertical movement occurs.

This can also be recognized from the excess pore-water pressure time history in Figure 11, in which excess porewater pressures time history by the one-dimensional analysis are compared with the one at the elements adjacent to wall (shown as side in the figure) and at the center of the ground (shown as center)

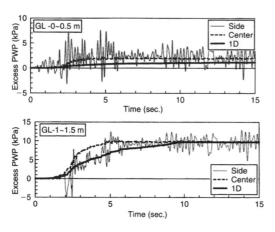

Figure 11. Comparison of excess porewater pressure time history.

by the two-dimensional analysis are shown at depths GL ± 0~0.5 m and GL-1~1.5 m.

Liquefaction does not occur at the ground surface in the one-dimensional analysis as shown in the left figure. Significant vibration is observed at the element near the wall in the two-dimensional analysis, but excess porewater pressure increases monotonically at the center. In spite of these difference, residual values at the end of the earthquake are nearly the same and they are nearly equal to initial overburden stress; element at the ground surface liquefy in two-dimensional analysis, which is different with the result of one-dimensional analysis. Same with the element at the surface, excess porewater pressure vibrates at the element adjacent to wall at depth between GL-1~1.5 m, but amplitude of vibration becomes much smaller. Excess porewater pressure monotonically increases same with the element at the ground surface at the center, too. Compared with the one-dimensional element, generation of excess porewater pressure occurs earlier in the two-dimensional analysis. These observations also imply that existence of underground wall does not reduce excess porewater pressure in the ground; it contributes excess porewater pressure.

Behavior of elements at the side and at the center is shown in Figure 12 at GL-1~1.5 m depth, which can be compared with the behavior at the same depth in one-dimensional analysis in Figure 8. General behaviors do

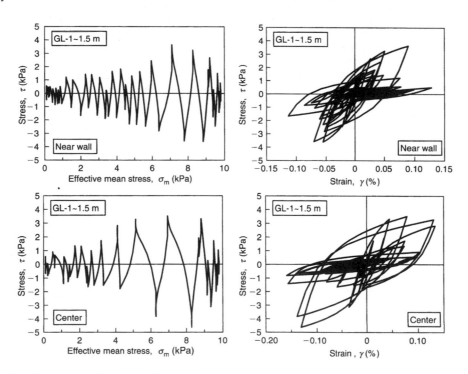

Figure 12. Behavior of soil element at GL-1~1.5 m.

not differ very much. It is noted that shear stresses and shear strains are the ones in horizontal plane. Therefore, shear strain is the largest in one-dimensional analysis, then at the center of ground and finally adjacent to wall. This again implies that deformation in horizontal direction is constrained. This also explains why excess porewater generates larger at the center element than at the element adjacent to wall as shown in Figure 11.

Figure 13 shows displacement time histories at the ground surface for both one- and two-dimensional

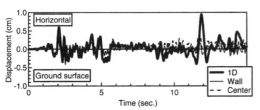

(a) Comparison with one-dimensional analysis (horizontal component)

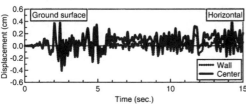

(b) Comparison between elements near the wall and at the center (horizontal component)

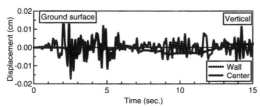

(c) Comparison between elements near the wall and at the center (vertical component)

Figure 13.   Comparison of displacements at the ground surface.

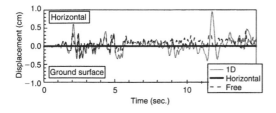

Figure 14.   Comparison of horizontal displacement at the ground surface.

analyses. As shown in Figure 13(a), displacement in the one-dimensional analysis is larger at about 2 second where the largest wave comes. It is also larger at about 12 second where displacement becomes the largest. Since behaviors in the two-dimensional analysis cannot be distinguished well in Figure 13(a), they are enlarged in Figure 13(b). Waveforms are nearly identical at the center and at the side of wall except that displacement at the center drifts at about 6 seconds. Elastic behavior of wall is supposed to cause the difference.

4.3   *Effect of vertical displacement constraint*

Through the preceding discussions, it becomes clear that vertical displacement accelerates excess porewater pressure generation. The idea to use average stiffness, however, does not consider vertical displacement presented in this paper. In order to investigate the effect of vertical displacement to the excess porewater pressure generation, two-dimensional analysis is conducted by constraining the vertical displacement; only horizontal displacement is allowed. Figure 14 compares displacements at the center of two-dimensional analyses and one-dimensional analysis. Solid line indicated "Horizontal" is the result of this case (no vertical displacement) whereas dashed line indicated "Free" is the result in the preceding (free of vertical displacement). Displacement under vertical displacement constraint is much smaller than other cases; Solid line seems horizontal line or no horizontal displacement. Related to this, excess porewater pressure hardly generates. When vertical displacement is constrained, earthquake load is distributed depending on shear stiffness. Shear stiffness of the underground wall is much larger than that of soil. Therefore almost all earthquake load is carried by the underground wall, which result in almost zero excess porewater pressure generation.

An analysis to constrain displacements of nodes that belong to underground wall is also conducted, whose result is not shown in this paper. Displacement is also very small compared with ordinary two-dimensional analysis even in this case although it is larger than the case that vertical displacement allowed. This fact also validates the mechanism proposed here.

5   CONCLUSION

Discussions are made in order to make the mechanism how remedial measures against soil liquefaction by shear strain constrained method works. Following conclusions are obtained from the case study using a simplified model.

The ground vibrates vertically by the shear stress transmission from the underground wall to the ground, which generates additional shear deformation in the ground, resulting in excess porewater pressure generation. Therefore, it cannot be said that existence of the rigid or stiffer underground wall constrains onset of soil liquefaction. A simple method to use average stiffness in evaluating the shear strain of the ground results in too small shear deformation, which is a critical side evaluation in design.

This kind of vertical deformation is easier to occur near the ground surface, and is difficult to occur at deep depth. Existence of nonliquefied layer above the liquefied layer will also work to make vertical deformation difficult although it is not proved in this paper.

It is, however, noted that horizontal displacement becomes much smaller by the existence of the underground wall. Since much damage to underground structures is caused by the ground displacement but not by the excess porewater pressure generation, remedial measures by shear deformation constraint method works because horizontal displacement of the ground is much smaller than the one of a bare ground.

It is possible to reduce shear stress acting on the wall by reducing the shear stress transmitted from the wall. One example is an existence of underground wall parallel to the place of shaking.

Inertia force from the superstructure also causes shear stress in the underground wall; therefore it will affect onset of soil liquefaction.

A mechanism that has not been known or discussed is made clear to accelerate excess porewater pressure generation. Since a simplified model is used in the case study, quantitative evaluation how underground wall works cannot be made in this paper. Establishment of the design method remains a future problem.

REFERENCES

Japanese Geotechnical Society (1993): Investigation, design and application of remedial measures against soil liquefaction.

Kusakabe, S. (1996): A study of dynamic behavior of ground by on-line earthquake response test, Theses submitted to Yamaguchi University for partial fulfillment of Doctor of Engineering.

Yoshida, N. and Tsujino, S. (1993): A simplified practical stress-strain model for the multi-dimensional analysis under repeated loading, Proc., The 28th Japan National Conference of Soil Mechanics and Foundation Engineering, pp. 1221–1224.

Fukutake, T. and Ohtsuki, A. (1993): Analysis by ALISS, Proc. Symposium on Remedial Measures against Liquefaction, JCSMFE, pp. 125–134.

Yoshida, N. (1993): STADAS, A computer program for static and dynamic analysis of ground and soil-structure interaction problems, Report, Soil Dynamics Group, The University of British Columbia, Vancouver, Canada.

Finite Elements in Civil Engineering Applications, Hendriks & Rots (eds.)
© 2002 Swets & Zeitlinger, Lisse, ISBN 90 5809 530 4

New constitutive models in DIANA for rubber-like materials

Z. Guo, P. Stroeven, L.J. Sluys
Delft University of Technology, Delft, The Netherlands

Y. Cheng
Northern Jiaotong University, Beijing P.R. China

ABSTRACT: New User-Supplied Subroutines in DIANA have been proposed for the elastic matrix of rubber-like materials in this paper. When the first and second derivatives of the strain energy function are known, other information can be generated automatically in DIANA through the new User-Supplied Subroutines. Knowles' and Gao's constitutive laws for rubber-like materials have been implemented in DIANA by this means. Singular problems are described successfully by those constitutive laws. We demonstrate two representative examples: uniaxial and biaxial tension analyses of a rubber cube show that the DIANA finite element package provides data which are consistent with the analytical results. When a new strain energy function is proposed in the future, the new constitutive law could be more conveniently inserted into DIANA than before. The new User-Supplied Subroutines in DIANA render new possibilities to study mechanics of rubber-like materials.

1 INTRODUCTION

Many strain energy functions have been proposed for rubber-like materials, for example by Moony, Knowles, Ogden, etc. These analytical expressions of strain energy functions differ from each other. In non-linear finite element packages, special routines are needed for each material model to form a tangential stiffness matrix. This complexity causes problems in analyzing different rubber-like materials.

Usually, in different finite element programs different strain energy functions are employed. It is difficult to cover all strain energy functions in one finite element program because of the aforementioned reason. Neo-Hookean, Modified Neo-Hookean and Ogden models have been implemented for hyperelastic behavior in the finite element program FEAP. Mooney-Rivlin, Besseling and Ogden models are available for hyperelastic behavior in finite element program DIANA. Evidently, available finite element programs are able to simulate certain aspects of mechanical behavior of rubber-like materials. The new model needs development of a special code.

Fortunately, DIANA offers the user-supplied subroutine mechanism to specify a nonlinear material behaviour. It offers an end-user the opportunity to supply Fortran source code for some predefined subroutines.

Therefore, we have developed a new User-Supplied Subroutine for rubber-like materials based on DIANA user supplied subroutine. The Improved User-Supplied Subroutine is for elastic matrix $\mathbf{D}$ in which $\dot{\sigma} = D\dot{\varepsilon}$, where $\dot{\sigma}$ is incremental Kichoff stress and $\dot{\varepsilon}$ is incremental Green strain. In this framework, the only code programmed by user is concerned with the first and second derivatives of the strain energy function with respect to strain invariants (W_i and W_{ij}). So, it makes approach more convenient than before when the constitutive law is inserted in user's supply in DIANA.

Gao proposed two strain energy functions for rubber-like materials in 1990 and 1997, respectively. Many phenomena in singular problems are described by the two laws. For instance, the expanding domain and shrinking domain around crack tip in Mode I, corner contact point, point of concentrated load, and so on, can be properly simulated. Therefore, Gao's strain energy function will be useful to study mechanics of rubber-like materials. So, we insert Gao's strain energy function into DIANA by means of the new User-Supplied Subroutine. This approach will be shown to work properly for a square element and a cube element.

2 FRAMEWORK ON MATRIX D

2.1 *Basic definition*

A three-dimensional domain of the material is considered. Let **P** and **Q** denote the position vectors of a material point before and after deformation, respectively, and x^i (i = 1, 2, 3) denote the Lagrangian coordinates of a point. Two sets of local triads are defined as follows,

$$P_i = \frac{\partial P}{\partial x^i}, \quad Q_i = \frac{\partial Q}{\partial x^i} \tag{1}$$

The displacement gradient tensor is

$$F = Q_i \otimes P^i \tag{2}$$

where P^i is the conjugate of P_i, $\otimes$ the dyadic symbol, and the summation rule is implied. The right and left Cauchy-Green strain tensors are

$$G = F^T \cdot F = \left(Q_i \cdot Q_j\right) P^i \otimes P^j$$
$$C = F \cdot F^T = \left(P^j \cdot P^j\right) Q_i \otimes Q_j \tag{3}$$

where superscript T indicates transposition. G and C possess the same invariants such as

$$I_1 = G:U, \quad I_2 = G:G, \quad I_3 = G^2:G \tag{4}$$

$$I_{-1} = G^{-1}:U, \quad I_{-1} = G^{-1}:G^{-1} \tag{5}$$

in which U denotes unit tensor and : denotes dual multiplication. Among these invariants, only three are independent. Further, we have commonly used invariant J,

$$J = \left[\frac{1}{6}\left(I_1^3 - I_1 I_2 + 2I_3\right)\right]^{1/2} \tag{6}$$

Let λ_i denote the values of principal strain, then

$$\begin{aligned} I_1 &= \lambda_1^2 + \lambda_2^2 + \lambda_3^2 \\ I_{-1} &= \lambda_1^{-2} + \lambda_2^{-2} + \lambda_3^{-2} \\ J &= \lambda_1 \lambda_2 \lambda_3 \end{aligned} \tag{7}$$

For isotropic material, the strain energy per unit undeformed volume can be expressed by three independent invariants, for example,

$$W = W(I_1, I_2, I_3) \tag{8}$$

The Kirchhoff stress is

$$\sigma = 2\frac{\partial W}{\partial G} \tag{9}$$

The Cauchy stress is

$$\tau = J^{-1} F \cdot \sigma \cdot F^T \tag{10}$$

2.2 *Framework on matrix D*

Before deformation, the base vectors along x^i direction at a point in curved coordinates are e_i (i = 1, 2, 3) where $\|e_i\| = 1$ and $e_i \cdot e_j = \delta_i^j$.

Let $\dot{\varepsilon} = \varepsilon_{ij} e_i \otimes e_j$ denote the Green strain tensor, and U be the unit tensor. Let

$$G = 2\dot{\varepsilon} + U \tag{11}$$

in which

$$G = g_{ij} e_i \otimes e_j \tag{12}$$

with

$$g_{ij} = 2\varepsilon_{ij} + \delta_{ij} \tag{13}$$

we have

$$\frac{\partial I_1}{\partial G} = U, \quad \frac{\partial I_2}{\partial G} = 2G, \quad \frac{\partial I_3}{\partial G} = 3G^2 \tag{14}$$

and

$$\begin{aligned} \frac{\partial I_1}{\partial G} &= 0 e_i \otimes e_j \otimes e_k \otimes e_l \\ \frac{\partial I_2}{\partial G} &= 2\delta_k^i \delta_l^j e_i \otimes e_j \otimes e_k \otimes e_l \\ \frac{\partial I_3}{\partial G} &= 3\left(\delta_k^i g_{lj} + g_{ik}\delta_l^j\right) e_i \otimes e_j \otimes e_k \otimes e_l \end{aligned} \tag{15}$$

Further, the Kirchoff stress is

$$\sigma = \frac{\partial W}{\partial \varepsilon} = \frac{\partial W}{\partial G} \cdot \frac{\partial G}{\partial \varepsilon} = 2\frac{\partial W}{\partial G} = 2W_i \frac{\partial I_i}{\partial G} \tag{16}$$

where $W_i = \partial W / \partial I_i$, so

$$\frac{\partial \sigma}{\partial \varepsilon} = \frac{\partial \sigma}{\partial G} \cdot \frac{\partial G}{\partial \varepsilon} = 2\frac{\partial \sigma}{\partial G}$$
$$= 2\frac{\partial}{\partial G}\left(2W_i \frac{\partial I_i}{\partial G}\right) = 4\frac{\partial}{\partial G}\left(W_i \frac{\partial I_i}{\partial G}\right) \quad (17)$$

It can easily be seen that

$$\frac{\partial}{\partial G}\left(W_i \frac{\partial I_i}{\partial G}\right) = \frac{\partial I_i}{\partial G} \otimes \frac{\partial W_i}{\partial G} + \frac{\partial}{\partial G}\left(\frac{\partial I_i}{\partial G}\right)W_i$$
$$= \frac{\partial I_i}{\partial G} \otimes \left(W_{ij}\frac{\partial I_j}{\partial G}\right) + \frac{\partial^2 I_i}{\partial G^2}W_i \quad (18)$$

in which $W_{ij} = \partial^2 W / \partial I_i \partial I_j$. Considering Equation 4, we have

$$\frac{\partial I_1}{\partial G} \otimes \left(W_{1j}\frac{\partial I_j}{\partial G}\right)$$
$$= U \otimes (W_{11}U + W_{12}2G + W_{13}3G^2)$$
$$\frac{\partial I_2}{\partial G} \otimes \left(W_{2j}\frac{\partial I_j}{\partial G}\right) \quad (19)$$
$$= 2G \otimes (W_{21}U + W_{22}2G + W_{23}3G^2)$$
$$\frac{\partial I_3}{\partial G} \otimes \left(W_{3j}\frac{\partial I_j}{\partial G}\right)$$
$$= 3G \otimes (W_{31}U + W_{32}2G + W_{33}3G^2)$$

Substituting Equation 12 into Equation 19 yields

$$\frac{\partial I_1}{\partial G} \otimes \left(W_{1j}\frac{\partial I_j}{\partial G}\right)$$
$$= (W_{11}\delta_j^i\delta_l^k + 2W_{12}\delta_j^i g_{kl} + 3W_{13}\delta_j^i g_{km}g_{ml})$$
$$\times e_i \otimes e_j \otimes e_k \otimes e_l$$
$$\frac{\partial I_2}{\partial G} \otimes \left(W_{2j}\frac{\partial I_j}{\partial G}\right)$$
$$= (2W_{21}g_{ij}\delta_l^k + 4W_{22}g_{ij}g_{kl} + 6W_{23}g_{ij}g_{km}g_{ml}) \quad (20)$$
$$\times e_i \otimes e_j \otimes e_k \otimes e_l$$
$$\frac{\partial I_3}{\partial G} \otimes \left(W_{3j}\frac{\partial I_j}{\partial G}\right)$$
$$= (3W_{31}g_{im}g_{mj}\delta_l^k + 6W_{32}g_{im}g_{mj}g_{kl}$$
$$+ 9W_{33}g_{im}g_{mj}g_{kn}g_{nl})e_i \otimes e_j \otimes e_k \otimes e_l$$

Let

$$4\frac{\partial}{\partial G}\left(W_i \frac{\partial I_i}{\partial G}\right) = D_{ijkl}\, e_i \otimes e_j \otimes e_k \otimes e_l \quad (21)$$

Substituting Equations 20 and 15 into Equation 18 and 17, and considering Equation 21, we have

$$D_{ijkl} = 4[W_{11}\delta_j^i\delta_l^k + 2W_{12}\delta_j^i g_{kl} + 3W_{13}\delta_j^i g_{km}g_{ml}$$
$$+2W_{21}g_{ij}\delta_l^k + 4W_{22}g_{ij}g_{kl} + 6W_{23}g_{ij}g_{km}g_{ml}$$
$$+3W_{31}g_{im}g_{mj}\delta_l^k + 6W_{32}g_{im}g_{mj}g_{kl}$$
$$+ 9W_{33}g_{im}g_{mj}g_{kn}g_{nl}$$
$$+2W_2\delta_k^i\delta_l^j + 3W_3(\delta_k^i g_{lj} + g_{ik}\,\delta_l^j)] \quad (22)$$

Equation 17 gives

$$d\sigma^{ij} = D(d\varepsilon_{11}, d\varepsilon_{22}, d\varepsilon_{33}, d\varepsilon_{12}, d\varepsilon_{23}, d\varepsilon_{31})^T \quad (23)$$

where

$$D = (D_{ij11}, D_{ij22}, D_{ij33}, D_{ij12} + D_{ij21}, D_{ij23}$$
$$+ D_{ij32}, D_{ij13} + D_{ij31}) \quad (24)$$

is the tangential stiffness matrix. The Kirchhoff stress is

$$\sigma = D \cdot \Delta\varepsilon + \sigma_0 \quad (25)$$

The tangential stiffness matrix D can be formed automatically in Equation 22 where the strain function $W = W(I_1, I_2, I_3)$ is given analytically and a subroutine, in which W_i and W_{ij} are generated, is given.

The main steps of new interface framework for D are indicated in Figure 1. Firstly, this new interface framework receives general data from main program, for example, ε_0, $\Delta\varepsilon$, USRMOD, USRVAL. ε_0 is strain vector at start of increment; $\Delta\varepsilon$ is total strain increment; USRMOD is the user-supplied model name from input table "MATERI". This name can be used as a switch to various material models coded in the Improved User-Supplied Subroutine. USRVAL are the user-supplied material parameters from input table "MATERI". Secondly, this framework provides upgraded tangential stiffness matrix D and total stress vector.

So, it is very convenient for a user to form matrix D for a new constitutive law.

3 THREE KINDS OF CONSTITUTIVE LAWS

3.1 Knowles and Sternberg's law

In 1973, Knowles and Sternberg proposed a constitutive law for rubber-like materials.

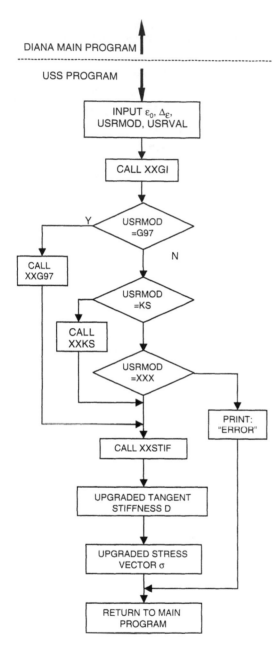

DIANA MAIN PROGRAM

USS PROGRAM

INPUT ε_0, Δ_ε, USRMOD, USRVAL

CALL XXGI

USRMOD =G97

CALL XXG97

USRMOD =KS

CALL XXKS

USRMOD =XXX

PRINT: "ERROR"

CALL XXSTIF

UPGRADED TANGENT STIFFNESS D

UPGRADED STRESS VECTOR σ

RETURN TO MAIN PROGRAM

Figure 1. Main steps of new interface framework of developed computer program.

The analytical expression of strain energy function is

$$W = (A_1 I_1 + A_2 J + A_3 I_1 J^{-2})^N \qquad (26)$$

where, A_1, A_2, A_3 and N are material constants, and J is mentioned before.

From Equation 26 we can get

$$\frac{\partial W}{\partial I_i} = N(A_1 I_1 + A_2 J + A_3 I_1 J^{-2})^{N-1}$$

$$\times \left[A_1 \delta_i^1 + A_2 \frac{\partial J}{\partial I_i} + A_3 \delta_i^1 J^{-2} - A_3 I_1 J^{-4} \frac{\partial J^2}{\partial I_i} \right] \qquad (27)$$

Further,

$$\frac{\partial^2 W}{\partial I_i \partial I_j} = N(N-1)(A_1 I_1 + A_2 J + A_3 I_1 J^{-2})^{N-2}$$

$$\times \left[A_1 \delta_j^1 + A_2 \frac{\partial J}{\partial I_j} + A_3 \delta_j^1 J^{-2} \right.$$

$$\left. - A_3 I_1 J^{-4} \frac{\partial J^2}{\partial I_j} \right]$$

$$\times \left[A_1 \delta_i^1 + A_2 \frac{\partial J}{\partial I_i} + A_3 \delta_i^1 J^{-2} \right.$$

$$\left. - A_3 I_1 J^{-4} \frac{\partial J^2}{\partial I_i} \right]$$

$$+ N(A_1 I_1 + A_2 J + A_3 I_1 J^{-2})^{N-1}$$

$$\times \left[A_2 \frac{\partial^2 J}{\partial I_i \partial I_j} - A_3 \delta_i^1 J^{-4} \frac{\partial J^2}{\partial I_j} \right.$$

$$- A_3 \left(\delta_j^1 J^{-4} \frac{\partial J^2}{\partial I_i} - 2 I_1 \left(J^2 \right)^{-3} \frac{\partial J^2}{\partial I_j} \frac{\partial J^2}{\partial I_i} \right.$$

$$\left. \left. + I_1 J^{-4} \frac{\partial^2 J^2}{\partial I_i \partial I_j} \right) \right] \qquad (28)$$

Upon substitution of Equation 6 into Equation 28, W_i and W_{ij} are obtained.

3.2 Gao's constitutive laws

From a physical point of view, the necessary conditions for a reasonable constitutive relation of a solid can be stated in two ways:

1. A material element should possess stiffness to resist both shape change and volume change.
2. A material element should possess stiffness to resist both extension and compression.

A strain energy function was proposed that encompassed only the two terms of Gao (1990) according to statement 1.

$$W = a \left[\frac{I_1}{J^{2/3}} \right]^n + b \left(J^2 - 1 \right)^m J^{-2q} \qquad (29)$$

where, a, b, m, n and l are positive constitutive parameters. m should be an even integer. The first term of Equation 29 reflects pure shape change, whereas the second term of Equation 29 reflects pure volume change.

An even simpler strain energy function was proposed by Gao (1997) (in accordance with statement 2). Hence,

$$W = a(I_1^n + I_{-1}^n) \qquad (30)$$

where I_1 is strain invariant, and I_{-1} is a strain invariant defined by

$$I_{-1} = \frac{3(I_1^2 - I_2)}{I_1^3 - 3I_1I_2 + 2I_3} \qquad (31)$$

It is obvious that Equations 29, 30 and 31 can be written as $W = W(I_1, I_2, I_3)$. So, we can get the first and second derivatives of the strain energy function with respect to strain invariants from the analytical expressions of strain energy function.

4 EXAMPLES

DIANA offers the user-supplied subroutine mechanism to specify a general nonlinear material behaviour. It offers an end-user the opportunity to supply Fortran source code for particular predefined subroutines. The framework of Equation 24 is inserted into DIANA. As an example, the user-supplied material model defined by Equations 26, 29 and 30 are coded.

4.1 Example 1: Uniaxial load

We consider a cubic material element with edges of unit length as shown in Figure 2 (left). Under the action of load P along x^1 direction, the edge lengths have become λ, μ and μ, respectively, but edges still remain perpendicular, as shown in Figure 2 (right). Let e_i ($i = 1, 2, 3$) denote the unit vectors along x^i direction, then according to Equations 2–7, it follows that:

$$F = \lambda e_1 \otimes e_1 + \mu(e_2 \otimes e_2 + e_3 \otimes e_3) \qquad (32)$$

$$C = \lambda^2 e_1 \otimes e_1 + \mu^2(e_2 \otimes e_2 + e_3 \otimes e_3) \qquad (33)$$

$$C^{-1} = \lambda^{-2} e_1 \otimes e_1 + \mu^{-2}(e_2 \otimes e_2 + e_3 \otimes e_3) \qquad (34)$$

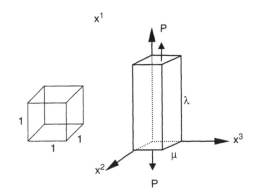

Figure 2. A rubber specimen before and after deformation.

$$I_1 = \lambda^2 + 2\mu^2, \quad I_{-1} = \lambda^{-2} + 2\mu^{-2}, \quad J = \lambda\mu^2 \quad (35)$$

Further, Equation 10 becomes

$$\tau = \frac{\rho}{\rho 0}\left(\frac{\partial W}{\partial I_1} \cdot \frac{\partial I_1}{\partial Q_i} + \frac{\partial W}{\partial J} \cdot \frac{\partial J}{\partial Q_i}\right) \otimes Q_i$$
$$= \left(2\frac{\partial W}{\partial I_1}C + J\frac{\partial W}{\partial J}U\right)J^{-1} \qquad (36)$$

Substituting Equation 33 into Equation 36 yields

$$\tau = \begin{bmatrix} \left(2\frac{\partial W}{\partial I_1}\lambda^2 + J\frac{\partial W}{\partial J}\right)e_1 \otimes e_1 \\ + \left(2\mu^2\frac{\partial W}{\partial I_1} + J\frac{\partial W}{\partial J}\right) \\ \times (e_2 \otimes e_2 + e_3 \otimes e_3) \end{bmatrix} \cdot J^{-1} \qquad (37)$$

Considering $\tau^{22} = \tau^{33} = 0$, Equations 37 and 26 give

$$2\mu^2\frac{\partial W}{\partial I_1} + J\frac{\partial W}{\partial J} = 0 \qquad (38)$$

$$\frac{\partial W}{\partial I_i} = N(A_1I_1 + A_2J + A_3I_1J^{-2})^{N-1}\left(A_1 + A_3J^{-2}\right) \qquad (39)$$

$$\frac{\partial W}{\partial J} = N(A_1I_1 + A_2J + A_3I_1J^{-2})^{N-1}$$
$$\times \left(A_2 - 2A_3I_1J^{-3}\right) \qquad (40)$$

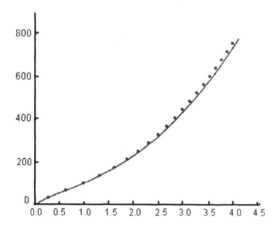

Figure 3.   Curve of load versus deformation in x^1 direction.

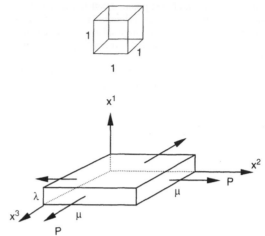

Figure 4.   A rubber specimen before and after deformation.

From Equation 38 we can obtain μ. Then, substituting μ into Equation 37, the total load acting on the element is

$$L = \mu^2 \tau^{11} = \frac{\mu^2}{J}\left(2\frac{\partial W}{\partial I_1}\lambda^2 + J\frac{\partial W}{\partial J}\right)$$

$$= \frac{1}{\lambda}\left(2\frac{\partial W}{\partial I_1}\lambda^2 + J\frac{\partial W}{\partial J}\right) \tag{41}$$

Substituting Equations 39 and 40 into Equation 41, we have

$$L = N(A_1 I_1 + A_2 J + A_3 I_1 J^{-2})^{N-1}$$
$$\times \left[2\lambda(A + A_3 J^{-2}) + \mu\left(A_2 - 2A_3 I_1 J^{-3}\right)\right] \tag{42}$$

Considering Equation 35, total load acting on element becomes

$$L = N\left[A_1\left(\lambda^2 + 2\mu^2\right) + A_2\lambda\mu^2 \right.$$
$$\left. + A_3\frac{\lambda^2 + 2\mu^2}{\left(\lambda\mu^2\right)^2}\right]^{N-1}\cdot\left[2\lambda\left(A_1 + \frac{A_3}{\left(\lambda\mu^2\right)^2}\right)\right.$$
$$\left. + \mu\left(A_2 - 2A_3\frac{\lambda^2 + 2\mu^2}{\left(\lambda\mu^2\right)^3}\right)\right] \tag{43}$$

The analytical solution and corresponding finite element solution with the new User-Supplied material model are presented in Figure 3.

4.2   Example 2: Biaxial load

Consider the same cubic material element with edges of unit length as shown in Figure 2 (left). Under the action of load P in x^2 and x^3 direction, the edge lengths have become λ, μ and μ, respectively, as shown in Figure 2 (right). Equations (32) to (37) are still valid but the relation of λ and μ must be given by $\tau^{11} = 0$, so

$$2\frac{\partial W}{\partial I_1}\lambda^2 + J\frac{\partial W}{\partial J} = 0 \tag{44}$$

From Equation 38 we can obtain μ. Then, substituting μ into Equation 37

$$\tau^{22} = \tau^{33} = J^{-1}\left(2\mu^2\frac{\partial W}{\partial I_1} + J\frac{\partial W}{\partial J}\right)$$
$$= 2\mu^2 J^{-1}\frac{\partial W}{\partial I_1} + \frac{\partial W}{\partial J} \tag{45}$$

The total load acting on the element is

$$L = \lambda\mu\tau^{22} = 2\lambda\mu^3 J^{-1}\frac{\partial W^2}{\partial I_1} + \lambda\mu\frac{\partial W}{\partial J} \tag{46}$$

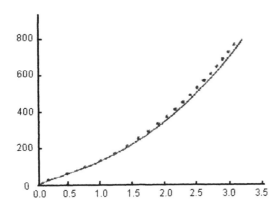

800

600

400

200

0

0.0   0.5   1.0   1.5   2.0   2.5   3.0   3.5

Figure 5.   Curve of load versus deformation in x^1 direction.

Hence

$$L = N(A_1 I_1 + A_2 J + A_3 I_1 J^{-2})^{N-1}$$
$$\times \left[2\mu(A + A_3 J^{-2}) + \lambda\mu\left(A_2 - 2A_3 I_1 J^{-3}\right) \right] \quad (47)$$

The analytical solution and corresponding finite element solution with the improved user-supplied material model are presented in Figure 5.

5   DISCUSSION

New User-Supplied Subroutines for the elastic matrix of rubber-like materials were developed in DIANA. Knowles' and Gao's constitutive laws for rubber-like materials have been implemented in DIANA in this way. They are available for a square element and a cube element.

We take Knowles' model as representative example. Uniaxial and biaxial tension analyses of rubber cube show that the DIANA finite element results are consistent with the analytical results.

When a new strain energy function is proposed in the future, the new constitutive law could be easily inserted into DIANA under the condition that the first and second derivatives of the strain energy function are known. The new User-Supplied Subroutine in DIANA provides a new way to study mechanics of rubber-like materials.

REFERENCES

Fu, Y.B. & Ogden, R.W. 2001. *Nonlinear Elasticity.* Cambridge: University Press.
Gao, Y.C. 1990. Elastostatic crack tip behaviour for a rubberlike material. *Theoretical and Applied Fracture Mechanics*, 14: 219–231.
Gao, Y.C. 1997. Large deformation field near a crack tip in a rubberlike material. *Theoretical and Applied Fracture Mechanics*, 26: 155–162.
Gao, Y.C. 1999. *Foundation of Solid Mechanics.* Beijing: Railway Publisher of China.
Guo, Z. 2000. A Framework for rubber materials in DIANA, *International. Symposium. On Modern concrete composites. Beijing, 30 Nov.–2 Dec. 2000.* Delft: University press.
Knowles, J.K. & Sternberg, Eli. 1973. An Asymptotic finite deformation analysis of the elastostatic field near the tip of a crack, *Journal of elasticity* 3: 67–107.
Mooney, M.A. 1940. A theory of large elastic deformation, *Journal of Applied Physics* 11: 582–592.

Finite Elements in Civil Engineering Applications, Hendriks & Rots (eds.)
© 2002 Swets & Zeitlinger, Lisse, ISBN 90 5809 530 4

Shape optimization of plane trusses

S. Šilih, S. Kravanja & B.S. Bedenik
University of Maribor, Faculty of Civil Engineering, Maribor, Slovenia

ABSTRACT: The paper presents the shape optimization of steel plane trusses, performed by the nonlinear programming (NLP) approach. Introduced is structural synthesis, where the optimization of the dimensions of cross-sections and the shape of trusses are carried out simultaneously. The objective function to be minimized defines the mass of the structure; it is subjected to the set of (in)equality constraints. The finite element equations are as the equality constraints defined for the calculation of internal forces and deflections of the structure. Constraints for the dimensioning of steel members are determined in accordance with Eurocode 3. Beside the theoretical basics, two examples of the shape optimization of trusses are presented at the end of the paper.

1 INTRODUCTION

The basic function of each civil engineering structure is to provide stability and safety of the building. Since each building serves to a specific purpose, the structure has to assure its functionalism. Nevertheless, the design of the building has also to satisfy conditions of the aesthetics and symbiosis with its environment. As these functions had the major influence on the design of buildings and structures in the near past, they are not sufficient in the modern time. Growing competition on the market in the last decades together with the technological progress have led to the problem how to achieve the maximal possible reduction of costs of structures. In this way, different optimization techniques, particularly various structural optimization methods, become very useful in the up-to-day engineering practice.

The term structural optimization may be defined as the calculation of a structural design that is the best within all possible designs in the direction of a prescribed objective function (e.g. mass minimization) and a given set of geometrical and/or behavioral limitations.

The classical force-stress-deflection calculation of the structure is called structural analysis. At structural analysis, the structural design is given in the beginning of the calculation and the result of the analysis is the structural response, e.g. whether the chosen structure does or does not satisfy all required conditions. Using structural optimization on the other hand, the input information are only some general data like load and span of the structure, while other parameters of the structure like cross-sections, shape, forces, stresses

and deflections appear as variables. Defined is the objective function which is to be minimized; it is subjected to the set of (in)equality constraints. The variables are calculated in the optimization process when the objective function is converged.

Considering the physical meaning of the optimization variables, the optimization problems are in general divided into three basic groups: *sizing optimization* (calculation of optimal cross-section dimensions of structural elements), *shape optimization* (beside optimal cross-sections also the calculation of optimal shape of the structure) and *topology optimization* (beside optimal cross-section and the shape also the calculation of optimal number and layout of structural elements). While sizing and shape optimizations are today mostly performed by the continuous nonlinear programming (NLP) approach, Šilih & Kravanja (2002), topology optimization requires the calculation in the discrete/continuous space, carried out by the mixed-integer nonlinear programming (MINLP), Kravanja et al. (1998a,b) or Genetic algorithm, Goldberg (1989).

The shape optimization of steel plane trusses is presented in the paper as an application of structural optimization. The objective of the optimization is to calculate the optimal shape of the truss and the optimal dimensions of all cross-sections so that the mass of the structure is minimal. The finite element equations are as the equality constraints defined for the calculation of internal forces and deflections of the structure. Constraints for the dimensioning of steel members are determined in accordance with Eurocode 3 (1992). The nonlinear programming approach is applied. Beside the theoretical basics, two examples of the shape optimization of trusses are presented at the end of the paper.

2 NONLINEAR PROGRAMMING FORMULATION

As problems in structural mechanics are in general nonlinear, the nonlinear programming (NLP) is the most commonly used structural optimization approach. The general form of the nonlinear optimization problem can be defined as follows:

$$
\begin{aligned}
\min \quad & Z = f(x) \\
\text{subjected to: } & h(x) = 0 \\
& g(x) \leq 0 \quad\quad\quad \text{(NLP)} \\
& x \in X = \{x | x \in \mathbf{R}^n, x^L \leq x \leq x^U\}
\end{aligned}
$$

where x is a vector of continuous variables, defined within the compact set X. Functions $f(x)$, $h(x)$ and $g(x)$ are nonlinear functions involved in the objective function Z, equality and inequality constraints, respectively. All functions $f(x)$, $h(x)$ and $g(x)$ must be continuous and differentiable.

In the context of structural optimization, variables include dimensions, cross-section characteristics, strains, materials, stresses, economic parameters, etc. Equality and inequality constraints and the bounds of the variables represent a rigorous system of the design, loading, stress, deflections and stability functions taken from structural analysis. The optimization of structures may include various objectives worthy of consideration. The most popular criterion used today is the minimization of mass.

There are several methods for solving the NLP problems successfully by the direct determination of points satisfying the Karush-Kuhn-Tucker optimality constraints. As the optimization problems are usually very large with a high number of variables and constraints, the effective solving is possible only with the appropriate computer programs. The shape optimization of steel trusses were carried out by the general reduced gradient method GRG, Abadie & Carpenter (1969) with the computer code CONOPT2, Drudd (1994).

3 SHAPE OPTIMIZATION OF PLANE TRUSSES

The classical way of searching the best structural design is structural analysis. Analyzing a high number of alternative structural designs may become in the case of trusses a comprehensive and time-consuming calculation process, particularly when trusses are composed from a high number of structural elements. Further, each change of the cross-section of

each individual member in the case of statically indeterminate structure, requires a new analysis of the whole structure, which leads to the calculation of new values of internal forces. When the search of the best shape is included in the analysis, the comprehension of the problem is additionally increased. The method can easily become a never-ending cyclic process.

Using structural optimization methods on the other hand, the dimensions of the cross-sections as well as the shape of the truss, represented by the nodal coordinates, appear as optimization variables. The structure is optimized in the direction of the minimizing the objective function. The obtained structure satisfies all design conditions, which are defined in the set of (in)equality constraints. In other words, the result of the structural optimization process is the best (cheapest, lightest) possible structure, which satisfies all required geometrical and behavioral conditions.

The optimization model TRUSSOPT was developed for the shape optimization of steel plane trusses. The considered trusses are composed from the top and bottom chords and from vertical and diagonal bracing members. The finite element equations are defined in the set of constraints for the calculation of the axial forces and the deflections, while other design constraints are determined in accordance to Eurocode 3. The members of the truss are checked for the tensional and the compressive/buckling resistance. In addition, the vertical deflections of the truss are also constrained.

The elements of the trusses are considered to be made of circular steel hollow sections. Using the nonlinear programing approach, the dimensions of the calculated optimal cross-sections can theoretically amount to any value lying within the prescribed lower and upper bounds. For this reason additional limitations, defined as the ratios between the diameter and thickness of the cross-sections, were included in the optimization to enforce the dimensions of the obtained optimal cross-sections to be calculated close to the sizes of standard circular hollow sections. The conditions of Annex K of Eurocode 3 for the hollow section lattice girder connections were also considered.

The objective function of the optimization model TRUSSOPT defines the mass of the structure (1). The main components of the optimization constraints set are the finite element equations (2), (3) and (4) for the calculation of internal forces and deflections; the tensional and the compressive/buckling resistance design conditions (5), (6), (7); as well as deflection conditions (8). The basic finite element notation for the definition of joints and elements of the truss is shown in Figure 1:

$$
\min MASS = \sum_{el(i,j)=1} s_{ij} \cdot A_{ij} \cdot \rho \qquad (1)
$$

$$\sum_{j:el(i,j)=1} \frac{A_{ij} \cdot E}{s_{ij}} \cdot \left(l_{ij}^2 \cdot \left(u_i - u_j \right) \right.$$
$$\left. + l_{ij} \cdot m_{ij} \cdot \left(v_i - v_j \right) \right) = F_{i,x} \quad (2)$$

$$\sum_{j:el(i,j)=1} \frac{A_{ij} \cdot E}{s_{ij}} \cdot \left(l_{ij} \cdot m_{ij} \cdot \left(u_i - u_j \right) \right.$$
$$\left. + m_{ij}^2 \cdot \left(v_i - v_j \right) \right) = F_{i,y} \quad (3)$$

$$F_{ij} = \frac{A_{ij} \cdot E}{s_{ij}} \cdot \left[l_{ij} \cdot \left(u_i - u_j \right) + m_{ij} \cdot \left(v_i - v_j \right) \right] \quad (4)$$

$$\gamma \cdot F_{ij} \le A_{ij} \cdot f_y / 1,1 \quad (5)$$

$$\gamma \cdot F_{ij} \ge -A_{ij} \cdot f_y / 1,1 \quad (6)$$

$$\gamma \cdot F_{ij} \ge -\chi_{ij} \cdot A_{ij} \cdot f_y / 1,1 \quad (7)$$

$$v_i \le L/250 \quad (8)$$

where:

i, j joint
$el(i,j)$ component of the element matrix $el(i,j) = 1$ if the element $i-j$ exists; otherwise $el(i,j) = 0$
A_{ij} area of cross-section of element $i-j$
E modulus of elasticity of steel
s_{ij} length of element $i-j$
l_{ij} cosine of the angle between global axis x and local axis x_e of element $i-j$
m_{ij} cosine of the angle between global axis y and local axis x_e of element $i-j$
$u_{i(j)}$ displacement of joint $i(j)$ in the direction of global axis x
$v_{i(j)}$ displacement of joint $i(j)$ in the direction of global axis y
$F_{i,x}$ force in joint i in the direction of global axis x
$F_{i,y}$ force in joint i in the direction of global axis y

F_{ij} axial force in element $i-j$
γ load safety factor
f_y yield strength of steel
χ_{ij} buckling reduction coefficient (Eurocode 3)
ρ unit mass of steel
L span between supports

The independent variables of the optimization model TRUSSOPT are the dimensions of the cross-sections of elements as well as the coordinates of the joints. The input data for the optimization are span, load, supports as well as the number and layout of the bracing elements.

The optimization model TRUSSOPT was developed using the General Algebraic Modelling System (GAMS), Brooke, Kendrick & Meeraus (1988). The optimizations were carried out by the general reduced gradient method with the computer code CONOPT2.

4 THE EXAMPLES

4.1 Simply supported truss

The first example is the shape optimization of a simply supported steel plane truss over the span of 30 m. The truss is subjected to vertical nodal forces of 100 kN, acting on the nodes of the bottom chord. Considering the topology with 12 panels (Figure 2) the number of elements of the truss is 45. The cross-sections of the chords is fixed constant along the entire span. Each structural element has two independent variables: diameter and thickness of tube. The overall number of independent variables for two chords and 21 bracing members representing the cross-sections is 46. It should be noted that both side diagonals (Figures 2 and 3) are considered to be parts of top chord.

The optimizations were performed for two different cases concerning the boundary conditions of the optimization problem. In the first case, the height of the truss is adopt to be constant along the span. As the independent variables of the problem are the cross-sections dimensions and the height, the total number of independent variables is therefore 47.

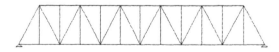

Figure 2. Optimal simply supported truss: constant height.

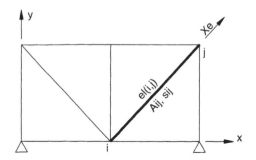

Figure 1. Joints and elements of the truss.

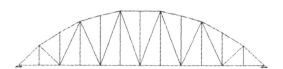

Figure 3. Optimal simply supported truss: variable height.

371

Table 1.	Optimal results – simply supported truss.		
	Case 1 Constant height	Case 2 Variable height	
Mass [kg]	4877.95	2213.21
Maximal height [cm]	472.69	648.09
Cross-section of the top chord [cm²]	72.44	63.18
Cross-section of the bottom chord [cm²]	58.63	6.25

Table 2.	Optimal results – continuous truss.		
	Case 1 Constant height	Case 2 Variable height	
Mass [kg]	16854.65	10049.05
Maximal height [cm]	414.11	787.49
Cross-section of the top chord [cm²]	102.52	62.06
Cross-section of the bottom chord [cm²]	102.52	62.06

Figure 4.   Optimal continuous truss: constant height.

Figure 5.   Optimal continuous truss: variable height.

In the second case, the height of the truss may vary along the span, i.e. the y coordinates can differ for each individual node of the top chord. In this case the total number of independent variables is 56. The obtained optimal trusses are presented for both cases in Figures 2 and 3. The results are presented in Table 1.

From the results it is evident, that there exists a significant 55% reduction of mass, obtained by allowing the height of the truss to vary along the span. The area of the cross-section of the top chord of the variable-height truss remains relatively large (due to required buckling resistance and larger buckling lengths in the case of variable height) with respect to the constant-height truss, while the area of the bottom chord is reduced drastically to almost 90%.

4.2 Continuous truss

The second example is the shape optimization of a continuous truss, defined over 3 spans. The overall span of the truss is 80 m long; the outer spans are 25 m long and the inner span is 30 m long. The truss is subjected to vertical nodal forces of 100 kN, acting on the nodes of the bottom chord. Considering the topology with 32 panels (Figure 4), the number of elements of the truss is 125. The number of independent variables representing the cross-sections is 126.

As in the case of simply supported truss, considered were two different cases: the first case with the constant height of the truss (127 independent optimization variables) and the second case with the variable height of the truss (157 independent optimization variables).

Both obtained optimal trusses are shown in Figures 4 and 5 respectively. The results are presented in Table 2.

The mass of the variable-height truss is about 40% lower than the mass of the constant-height truss. About the same reduction is observed by comparing the areas of cross-sections of chords for both considered cases. Comparing to the results of the simply supported truss, the reduction of mass is smaller, but still significant. The cross-sections of chords are reduced by the same amount, as the buckling resistance is required in both the top chord (mid-span areas) and the bottom chord (areas near the inner supports).

5   CONCLUSIONS

The paper presents the shape optimization of steel plane trusses, performed by the nonlinear programming (NLP) approach. The structures are optimized in the direction of the minimizing the objective function. The result of the structural optimization process is the optimal (the cheapest, lightest) possible structure, which satisfies all required geometrical and behavioral conditions.

The optimization model TRUSSOPT for shape optimization of steel plane trusses was developed using the General Algebraic Modelling System (GAMS). The objective function to be minimized defines the mass of the structure and is subjected to the set of (in)equality constraints. The finite element equations are defined for the calculation of internal forces and deflections of the truss. Together with the design conditions from Eurocode 3 for the dimensioning the steel members they form the set of optimization constraints. The optimization is carried out by the generalized reduced gradient method with the computer code CONOPT2.

In the paper are presented two practical examples of shape optimization: simply supported truss and continuous truss. In both examples the optimization was performed for two different cases. In the first case, the structural height was kept constant along the span, while in the second case we allowed variable height of the truss. From both examples it is evident that the variable height enables significant reduction of mass of the truss. In the case of simply supported beam, the mass reduction amounts to almost 55%, while in the case of continuous truss the reduction is smaller, but still significant: 40%.

The results of the presented examples have proved the superiority of the shape optimization when compared with the sizing optimization.

REFERENCES

Abadie, J. & Carpenter, J. 1969. Generalization of the Wolfe Reduced Gradient Method to the case of Nonlinear Constraints. In R. Fletcher (ed.), *Optimization*: 37–47, New York: Academic Press.

Brooke, A., Kendrick, D. & Meeraus, A. 1988. *GAMS (General Algebraic Modelling System), a User's Guide*. Redwood City: The Scientific Press.

Drudd, A.S. 1994. CONOPT – A Large-Scale GRG Code. *ORSA Journal on Computing* 6(2): 207–216.

Eurocode 3, Design of Steel Structures, Bruxelles, European Committee for Standardization, 1992.

Goldberg, D.E. 1989. Genetic Algorithms in Search, Optimization and Machine Learning, MA.: Addison-Wesley Publishing.

Kravanja, S., Kravanja, Z. & Bedenik, B.S. 1998a. The MINLP approach to structural synthesis. Part I: A general view on simultaneous topology and parameter optimization. *International Journal for Numerical Methods in Engineering* 43: 263–292.

Kravanja, S., Kravanja, Z. & Bedenik, B.S. 1998b. The MINLP approach to structural synthesis. Part II: Simultaneous topology, parameter and standard dimension optimization by the use of the Linked two-phase MINLP strategy. *International Journal for Numerical Methods in Engineering* 43: 293–328.

Šilih, S. & Kravanja, S. 2002. Comparison of composite floor systems. In C.A. Brebbia & W.P. deWilde (eds.), *High Performance Structures and Composites; Proc. intern. conf., Seville, 11–13 March 2002*. Southampton: WIT Press.

Finite Elements in Civil Engineering Applications, Hendriks & Rots (eds.)
© 2002 Swets & Zeitlinger, Lisse, ISBN 90 5809 530 4

Finite element modeling for the high damping rubber bearing

J. Yoshida
Ph.D, Assistant Professor, Yamanashi University, Yamanashi, Japan

M. Abe
Ph.D, Associate Professor, the University of Tokyo, Tokyo, Japan

Y. Fujino
Ph.D, Professor, the University of Tokyo, Tokyo, Japan

ABSTRACT: In this paper, a finite element model of high damping rubber bearings is developed. Firstly, an accurate constitutive model of high damping rubber material, which was proposed by the authors is briefly explained. Then, it is formulated into the form which is applicable to the finite element method. Secondly, a mixed finite element method that can introduce the proposed constitutive model is described. Thirdly, on the basis of the proposed constitutive model and the finite element method, a three-dimensional finite element model of the bearings is constructed. Comparing the simulation by the model with the experimental results, the model is found to be in good agreement with the experimental results and accurately predict the seismic response of structures under multi-axial loading excitations.

1   INTRODUCTION

A rubber material with damping property, which is called as High Damping Rubber material (HDR), was developed in Japan and has mainly been applied to the laminated rubber bearings for the base-isolation of bridges and buildings. This type of laminated rubber bearing is called as high damping rubber bearing.

The high damping rubber bearing not only supports the structure by restricting the bulge of the rubber layers with the reinforced steel plates, but also reduces the inertia force of the structure by extending the natural period and absorbs the earthquake energy by its damping property under the large deformation. Hence, slight compressibility, kinematic nonlinearity, and material nonlinearity must simultaneously be considered in its modeling.

So far, no accurate constitutive model of HDR has existed and hyper-elasticity which neglects the energy absorbing property is used as an approximated model (Seki, Fukahori, Iseda & Mtsunaga 1987, Hosam-Eddin & Abdel-Ghaffar 1995, Takayama, Tada & Tanaka 1990). Therefore, finite element formulations that have been mainly developed in the past are based on the hyper elasticity and a constitutive model with history dependency can not be applied to these

types of finite element methods. Since reliable results can not be obtained from numerical simulation due to the above-mentioned reasons, experimental investigation which is time and cost consuming work has been required in order to confirm the performance of each bearing in the actual design of the base-isolation.

According to these backgrounds, the authors developed an accurate constitutive model of HDR on the basis of the results of the material tests (Yoshida, Abe & Fujino 2002). In this study, in order to utilize the proposed constitutive model in the finite element analysis:

1. A formulation is shown, in which the proposed constitutive model is applied in the finite element method.
2. A mixed finite element method is developed in order to apply the proposed constitutive model.
3. A finite element model for the high damping rubber bearing is constructed and it is verified in comparison with the experimental results of the bearing.

At first, the constitutive models of HDR and steel material are briefly explained and the four-order constitutive tensor, which is needed in the application of the finite element method, is derived. Then, a projected mixed finite element method (Watanabe 1995) is extended into an updated Lagragian formulation for the application of the history dependent constitutive model. Utilizing the finite element method and the constitutive model, a three-dimensional finite element model of the high damping rubber bearings is constructed. Finally, the simulation by the model is compared with the results of multi-axial loading experiment and hybrid seismic response experiment (Abe, Yoshida & Fujino 2002) in order to show the validity of the model.

2 CONSTITUTIVE MODEL

High damping rubber bearing is a complex member of HDR and steel material. Hence, to construct an accurate finite element model, adaptive constitutive models of HDR and steel are needed. In this section, Constitutive models of these two materials used in our finite element model are described.

2.1 Constitutive model of high damping rubber

In this study, the constitutive model proposed by the authors (Yoshida, Abe & Fujino 2002) is used. In this model, hyper-elasticity is combined with elasto-plasticity in parallel as shown in Figure 1.

2.1.1 Hyper-elastic part
In the hyper-elastic part, the function $\bar{W}$ is used as a strain energy density function is as follows:

$$\bar{W} = gW_1 + hW_2 + \frac{\chi}{2}(W^V)^2 \tag{1a}$$

In the above equation, W_1, W_2 and W^V are the functions of the invariants of right Cauchy Green tensor $\mathbf{C}$ and detailed forms of them are:

$$W_1 = c_1(\bar{I}_C - 3) + c_2(\bar{II}_C - 3) \tag{1b}$$

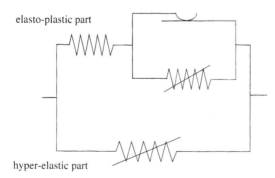

elasto-plastic part

hyper-elastic part

Figure 1.   Illustration of the constitutive model of HDR.

$$W_2 = \frac{c_3 c}{n+1}\left(\frac{\bar{I}_C - 3}{c}\right)^{n+1} \tag{1c}$$

$$W^V = 2\left(\sqrt{III_C} - 1\right) \tag{1d}$$

where I_C, II_C and III_C are the first, second and third invariant of $\mathbf{C}$, respectively. $\bar{I}_C$ and $\bar{II}_C$ are called reduced invariants and defined as $\bar{I}_C = I_C/III_C^{1/3}$ and $\bar{II}_C = II_C/III_C^{2/3}$, respectively. In equation (1a), g and h are damage functions and, in this study, the following forms are employed.

$$g(x) = \beta + (1 - \beta)\frac{1 - e^{-x/\alpha}}{x/\alpha} \tag{2a}$$

$$x(t) = \max_{s \in (-\infty, t]} \sqrt{2W_1(s)} \tag{2b}$$

$$h(y) = 1 - \frac{1}{1 + \exp\{-a_H(y - b_H)\}} \tag{2c}$$

$$y(t) = \max_{s \in (-\infty, t]} \left(\bar{I}_C(s) - 3\right) \tag{2d}$$

In equation (1) and (2), χ, c_1, c_2, c_3, c, n, α, β, a_H and b_H are parameters and t is the current time. Using the strain energy density function in equation (1a), the relation between second Piola-Kirchhoff stress tensor $\mathbf{S}$ and Green–Lagrange strain tensor $\mathbf{E}$ is obtained as:

$$\mathbf{S} = g\frac{\partial W_1}{\partial \mathbf{E}} + h\frac{\partial W_2}{\partial \mathbf{E}} + \chi W^V \frac{\partial W^V}{\partial \mathbf{E}} \tag{3}$$

Furthermore, equation (3) is transformed into a relation with Cauchy stress tensor $\mathbf{T}$:

$$\mathbf{T} = \frac{1}{J}\mathbf{F} \cdot \left(g\frac{\partial W_1}{\partial \mathbf{E}} + h\frac{\partial W_2}{\partial \mathbf{E}} + \chi W^V \frac{\partial W^V}{\partial \mathbf{E}}\right) \cdot \mathbf{F}^T \tag{4}$$

where $\mathbf{F}$ is the deformation gradient tensor and $J \equiv \det(\mathbf{F})$.

2.1.2 Elasto-plastic part
In the model, energy absorbing property of HDR is represented by the elasto-plasticity. Detailed equations of the elasto-plastic part are:

$$\mathring{\mathbf{T}}_{(J)} = \mathbf{C}^{(E)} : (\mathbf{D} - \mathbf{D}^P) \tag{5}$$

where

$$\mathbf{D}^p = (3K_2)^{1/2}(3J_2)^{(N-1)/2}\frac{\mathbf{T}'_{ij}}{\tau_y} \tag{6a}$$

$$K_2 = \frac{\mathbf{D}':\mathbf{D}'}{2} \tag{6b}$$

$$J_2 = \frac{\mathbf{T}':\mathbf{T}'}{2\tau_y^2} \tag{6c}$$

$$\tau_y = \tau_0\left\{1 + \left(\frac{\bar{I}_C - 3}{c}\right)^b\right\} \tag{6d}$$

In equation (6), $\mathbf{D}$, $\mathbf{T}$ and $\overset{\circ}{\mathbf{T}}_{(J)}$ are the deformation rate tensor, Cauchy stress tensor and Jaumann rate of Cauchy stress tensor, respectively and $\mathbf{A}:\mathbf{B}$ means $\mathrm{tr}(\mathbf{A}\cdot\mathbf{B}^T)$.

In equation (5), $\mathbf{C}^{(E)}$ is the elastic constitutive tensor based on the hyper elasticity as:

$$C^{(E)}_{pqrs} = \frac{1}{J}F_{pi}F_{qj}F_{rk}F_{sl}C^{(0)}_{ijkl} + \delta_{pr}T^{(h)}_{sp}$$
$$+ \delta_{qs}T^{(h)}_{pr} - \delta_{rs}T^{(h)}_{pq} \tag{7a}$$

$$\mathbf{C}^{(0)} = \frac{\partial^2 W_E}{\partial\mathbf{E}\partial\mathbf{E}} \tag{7b}$$

$$\mathbf{T}^{(h)} = \frac{1}{J}\mathbf{F}\cdot\frac{\partial W_E}{\partial\mathbf{E}}\cdot\mathbf{F}^T \tag{7c}$$

$$W_E = c_4(\bar{I}_C - 3) + c_5(\overline{II}_C - 3)$$
$$+ \frac{c_4 c}{m+1}\left(\frac{\bar{I}_C - 3}{c}\right)^{m+1} \tag{7d}$$

where c_4, c_5, m, b, τ_0 and N are parameters.

2.1.3 Constitutive tensor
In the application of the proposed model in the finite element method, the constitutive tensor is required. In the hyper-elastic part, the constitutive tensor is easily derived by taking the time derivatives of equation (3) (Simo 1987). In this subsection, the constitutive tensor of the elasto-plastic part is derived.

At first, $(3K_2)^{1/2}$ in equation (6a) is transformed into:

$$(3K_2)^{1/2} = \begin{cases} \dfrac{3K_2}{(3K_2)^{1/2}} & if \ K_2 \neq 0 \\ 0 & if \ K_2 = 0 \end{cases} \tag{8}$$

Then, K_2 is described:

$$K_2 = \frac{D'_{ij}D'_{ij}}{2} = \frac{D'_{ij}D_{ij}}{2} \tag{9}$$

Using equation (9), $\mathbf{D}^p$ in equation (6a) is expressed as:

$$D^p_{ij} = \begin{cases} \dfrac{D'_{kl}D_{kl}}{2(3K_2)^{1/2}}(3J_2)^{(N-1)/2}\dfrac{T'_{ij}}{\tau_y} & if \ K_2 \neq 0 \\ 0 & if \ K_2 = 0 \end{cases} \tag{10}$$

Hence, substituting equation (10) into equation (5), the elasto-plastic constitutive model is arranged as:

$$\overset{\circ}{T}_{(J)ij} = C^{ep}_{ijkl}D_{kl} \tag{11}$$

where

$$C^{ep}_{ijkl} = \begin{cases} C^{(E)}_{ijkl} - \dfrac{D'_{kl}(3J_2)^{(N-1)/2}}{2(3K_2)^{1/2}}\cdot\dfrac{C^{(E)}_{ijpq}T'_{pq}}{\tau_y} & if \ K_2 \neq 0 \\ C^{(E)}_{ijkl} & if \ K_2 = 0 \end{cases} \tag{12}$$

The constitutive tensor $\mathbf{C}^{ep}$ in equation (12) includes the value related to $\mathbf{D}$, which is unknown in assembling the tangent stiffness matrix in the finite element method. Hence, in the numerical computation, $\mathbf{D}$ at the last step is used as an initial value and, then, $\mathbf{D}$ is updated according to the iteration procedure in solving the nonlinear equations until convergence. It is noted that the tangent stiffness matrix based on the equation (11) is asymmetric due to the condition $C^{ep}_{ijkl} \neq C^{ep}_{klij}$.

2.2 Constitutive model of steel

As a constitutive model for steel, the elasto-plasticity proposed by Chaboche (1998) is extended to the model corresponding to large strain state. The original model by Chaboche possesses the advantages in that it can well represent the cyclic behavior, ratcheting effect of metals and can be introduced into the finite element method.

2.2.1 Extension to large strain state
Additive decomposition of the deformation rate tensor $\mathbf{D}$ is used for this extension (Hisada & Noguchi 1995). In the additive decomposition, infinitesimal strain in the small deformation theory is replaced by the deformation rate tensor.

Here, to obtain a symmetric tangent stiffness matrix, stress in the small deformation theory is replaced by the Jaumann rate of relative Kirchhoff stress $\hat{\mathbf{T}}_{(J)}$ with reference to the current time.

377

2.2.2 Yield criteria

Von Mises yield criteria shown below is used.

$$F = \bar{\sigma} - R - \sigma_y \geq 0 \qquad (13)$$

where

$$\bar{\sigma} = \left(\frac{3}{2}\tilde{\mathbf{T}}' : \tilde{\mathbf{T}}'\right)^{1/2} \qquad (14a)$$

$$\tilde{\mathbf{T}}' = \hat{\mathbf{T}} - \hat{\mathbf{b}} \qquad (14b)$$

$$\tilde{\mathbf{T}}' = \tilde{\mathbf{T}} - \frac{\mathrm{tr}(\tilde{\mathbf{T}})}{3}\mathbf{I} \qquad (14c)$$

In equation (13), σ_y is the initial yielding stress and R shows a variation of yielding stress due to the isotropic hardening law. In equation (14), $\mathbf{I}$ is the unit tensor and $\mathbf{b}$ is the back stress tensor, which is the same type of relative Kirchhoff stress tensor and physically shows the center of the yield surface.

2.2.3 Associated flow rule

Plastic flow occurs, when $F = 0$ and $(\partial F / \partial \hat{\mathbf{T}}) : \overset{\circ}{\hat{\mathbf{T}}}_{(J)} \geq 0$. According to the normality condition in plastic potential theory (Akhtar & Sujian 1995), the plastic component of deformation rate tensor is:

$$\mathbf{D}^p = \dot{p}\frac{\partial F}{\partial \hat{\mathbf{T}}} = \frac{3}{2}\dot{p}\frac{\tilde{\mathbf{T}}'}{\sigma_y + R} \qquad (15)$$

where $\dot{p}$ is the plastic coefficient and equal to the equivalent plastic strain rate in case of using Von Mises yield criteria.

$$\dot{p} = \left(\frac{2}{3}\mathbf{D}^p : \mathbf{D}^p\right)^{1/2} \qquad (16)$$

2.2.4 Kinematic hardening law

As the kinematic hardening of yield criteria, the following equation is used

$$\overset{\circ}{\mathbf{b}} = \frac{2}{3}c\mathbf{D}^p - \gamma\dot{p}\hat{\mathbf{b}} \qquad (17)$$

where c and γ are unknown constants, and $\overset{\circ}{\mathbf{b}}$ is the Jaumann rate of the back stress $\hat{\mathbf{b}}$. The formulation in equation (17) is called nonlinear kinematic hardening model and it represent the cyclic behavior or ratcheting effect of metals.

2.2.5 Isotropic hardening law

In equation (13), $R(t)$ grows according to the Evolution equation of yield surface as:

$$\dot{R} = \beta(q - R)\dot{p} \qquad (18)$$

where q and β are unknown constants. In the above equation, q is the limit value when R reaches infinite values and, in this case, yield stress becomes $q + \sigma_y$. Integrating the both side of equation (18) on time under the initial condition $R|_{t=0} = 0$, equation as shown below is obtained.

$$R = q(1 - e^{-\beta p}) \qquad (19)$$

2.2.6 Compatibility condition

During the plastic deformation, the yield condition $F = 0$ is satisfied and, therefore, $\dot{F} = 0$.

$$\dot{F} = \frac{\partial F}{\partial \tilde{\mathbf{T}}} : \overset{\circ}{\tilde{\mathbf{T}}}_{(J)} + \frac{\partial F}{\partial p}\dot{p} = 0 \qquad (20)$$

Equation (20) is called as compatibility condition and used for deriving the plastic coefficient.

2.2.7 Elasto-plastic constitutive tensor

In the elasto-plasticity, elastic part is directly connected to the plastic part and the constitutive model is described as:

$$\overset{\circ}{\tilde{\mathbf{T}}}_{(J)} = \mathbf{C}^{(e)} : (\mathbf{D} - \mathbf{D}^p) \qquad (21)$$

In equation (21), $\mathbf{C}^{(e)}$ is the elastic constitutive tensor and, in this case, linear elastic Hook's law is used.

$$C^{(e)}_{ijkl} = \lambda_0 \delta_{ij}\delta_{kl} + 2\mu_0 \delta_{ik}\delta_{jl} \qquad (22)$$

where λ_0 and μ_0 are Lame constants.

Arranging equation (20) by using equation (13), (15), (17), (19), plastic coefficient $\dot{p}$ is computed as follows.

$$\dot{p} = \frac{\varphi_{ij}D_{ij}}{\partial R/\partial p + \varphi_{ij}C_{ijkl}\varphi_{kl} + \varphi_{ij}M_{ij}}$$
$$= \frac{3\mu}{R + \sigma_y} \cdot \frac{\tilde{T}_{ij}D_{ij}}{\beta q e^{-\beta p} + 3\mu_0 + \varphi_{ij}M_{ij}} \qquad (23)$$

where

$$\varphi_{ij} = \frac{\partial F}{\partial \tilde{T}_{ij}} = \frac{3\tilde{T}'_{ij}}{2(R + \sigma_y)} \qquad (24a)$$

$$M_{ij} = \frac{2c}{3} \varphi_{ij} - \gamma \hat{b}_{ij} \qquad (24b)$$

Then, computing $\mathbf{D}^p$ from equation (23) and substituting $\mathbf{D}^p$ into (21), the constitutive model is expressed as follows.

$$\overset{\circ}{T}_{ij} = C_{ijkl}^{ep} D_{kl} \qquad (25)$$

where C_{ijkl}^{ep} is the elasto-plastic constitutive tensor shown below.

$$
\begin{aligned}
C_{ijkl}^{ep} &= C_{ijkl} - \frac{C_{ijab}\varphi_{ab}\varphi_{cd}C_{cdkl}}{\partial R/\partial p + \varphi_{ab}C_{abcd}\varphi_{cd} + \varphi_{ab}M_{ab}} \\
&= C_{ijkl} - \frac{9\mu_0^2}{\left(R + \sigma_y\right)^2} \cdot \frac{\tilde{T}_{ij}\tilde{T}_{ij}}{\beta q e^{-\beta p} + 3\mu_0 + \varphi_{ab}M_{ab}}
\end{aligned} \qquad (26)
$$

3 FINITE ELEMENT FORMULATION

Simulating a continuum with a material whose stress is decided only from the deformation from the reference configuration $\mathbf{X}$ such as elastic body, Total Lagragian method is usually applied, because the formulation becomes simple (Hisada & Noguchi 1995). Therefore, in modeling the rubber material that is mainly approximated by hyper-elastic body, total Lagragian formulation with incompressible condition has been employed in the past researches (Seki, Fukahori, Iseda & Matsunaga 1987, Hosam-Eddiin & Abdel-Ghaffar 1995).

On the other hand, total Lagrange method is inconvenient to a material with history dependency such as HDR or steel material, since this type of material needs the value defined at the current configuration.

In this section, a weak form and its time derivatives for the mixed finite element formulation of hyper-elastic material (Watanabe 1995) is described. This week form is superior in that the mixed finite element formulation based on it can treat the slight compressibility of HDR and is stable in the numerical computation.

Then, the week form and its time derivatives are transformed into the equations, which are described in the current configuration. Then, these equations are discretized by finite element method.

3.1 *Weak form for total Lagragian method*

Watanabe (1996) proposed the following weak form for the slightly compressible material with hyper-elastic body.

$$
\begin{aligned}
&\int_{\Omega} \left(\frac{\partial W}{\partial C_{ij}} + \lambda \frac{\partial W^V}{\partial C_{ij}} \right) \delta C_{ij}\, d\Omega \\
&= \int_{\partial\Omega} \mathbf{t} \cdot \delta\mathbf{u}\, d\Omega = \int_{\Omega} \rho_0 \mathbf{g} \cdot \delta\mathbf{u}\, d\Omega
\end{aligned} \qquad (27a)
$$

$$\int_{\Omega} \left(W^V - \frac{\lambda}{\chi} \right) \partial\lambda\, d\Omega = 0 \qquad (27b)$$

where W and W^V are the elastic strain energy density functions related to the non-volumetric deformation and volumetric deformation, respectively, $\mathbf{t}$ is the traction force vector, $\mathbf{g}$ is the body force, $\mathbf{u}$ is the displacement vector, λ is the hydro-static pressure, $\mathbf{C}$ is the right Cauchy-Green tensor, Ω and $\partial\Omega$ indicate the region of the continuum and its boundary respectively in the initial configuration, χ is a material constant which possesses the relation with the bulk modulus κ as $\kappa = 4\chi$.

Equation (27a) and (27b) are the weak forms, which become the basic equations for deriving the inner force vector by total Lagragian method.

Defining the external force vector R as

$$R = \int_{\partial\Omega} \mathbf{t} \cdot \mathbf{u}\, d\Omega + \int_{\Omega} \rho_0 \mathbf{g} \cdot \mathbf{u}\, d\Omega \qquad (28)$$

and taking the time derivative of the right hand side of (27a,b), the following equations are derived.

$$
\begin{aligned}
\delta\dot{R} = \int_{\Omega} \Bigg[\delta C_{ij} \Bigg\{ &\left(\frac{\partial^2 W}{\partial C_{ij}\partial C_{kl}} + \lambda \frac{\partial^2 W^V}{\partial C_{ij}\partial C_{kl}} \right) \dot{C}_{kl} \\
&+ \frac{\partial W^V}{\partial C_{kl}} \dot{\lambda} \Bigg\} + \delta F_{ki} \left(2\frac{\partial W}{\partial C_{ij}} + 2\lambda\frac{\partial W^V}{\partial C_{ij}} \right) \dot{F}_{kj} \Bigg] d\Omega
\end{aligned} \qquad (29a)
$$

$$\int_{\Omega} \delta\lambda \left(\frac{\partial W^V}{\partial C_{kl}} \dot{C}_{kl} - \frac{\dot{\lambda}}{\chi} \right) d\Omega = 0 \qquad (29b)$$

Equation (29a) and (29b) are the basic equations for deriving the tangent stiffness matrix in the total Lagragian formulation.

3.2 *Transformation of the weak form*

In order to utilize the constitutive model with rate form in the finite element formulation, equation (28) and (29) are transformed into the forms, which are described by the quantities of deformed configuration (Hisada & Noguchi 1995).

The final forms of equation (27) after transformation are as follows:

$$\delta R = \int_v T_{ij}\delta A_{ij}\, dv \qquad (30a)$$

$$\int_v \left(W^V - \frac{\lambda}{\chi} \right) \frac{\delta\lambda}{J}\, dv = 0 \qquad (30b)$$

where J is the determinant of deformation gradient tensor and $\delta\mathbf{A}$ is the variation of the linear part of Almansi strain tensor.

Furthermore, the time derivatives of the weak forms are also transformed into the following forms.

$$\delta\dot{R} = \int_v \left[\delta A_{ij} D^t_{ijkl} \dot{A}_{kl} + \delta F_t(t)_{ki} T_{ij} L_{kj} + \delta A_{ij} T^p_{ij} \dot{\lambda} \right] dv \tag{31a}$$

$$\int_v \delta\lambda \left(T^p_{kl} \dot{A}_{kl} - \frac{\dot{\lambda}}{\chi J} \right) dv = 0 \tag{31b}$$

where

$$D^t_{ijkl} = \frac{1}{J} F_{ip} F_{jq} F_{kr} F_{ls} D^0_{pqrs} \tag{32a}$$

$$D^0_{ijkl} = 4\frac{\partial^2 W}{\partial C_{ij} \partial C_{kl}} + 4\lambda \frac{\partial^2 W^V}{\partial C_{ij} \partial C_{kl}} \tag{32b}$$

$$L_{ij} = \frac{\partial \dot{u}_i}{\partial x_j} \tag{32c}$$

$$F_t(t)_{ij} = \frac{\partial \dot{u}_i}{\partial x_j} \tag{32d}$$

$$\mathbf{T}^p = \frac{1}{J} \mathbf{F} \cdot 2\frac{\partial W^V}{\partial C_{ij}} \cdot \mathbf{F}^T \tag{32e}$$

3.3 Discretization by the finite element method

Discretizing equation (30) and (31) by the mixed finite element, the following simultaneous equations are obtained in each element.

$$\begin{bmatrix} [K] & [H] \\ [H]^T & [G] \end{bmatrix} \begin{Bmatrix} \{\dot{u}\} \\ \{\dot{\lambda}\} \end{Bmatrix} = \begin{Bmatrix} \{F_{out}\} \\ \{0\} \end{Bmatrix} - \begin{Bmatrix} \{FD\} \\ \{FP\} \end{Bmatrix} \tag{33}$$

where $[K]$, $[H]$ and $[G]$ is the submatrices of tangent stiffness matrix that is derived from (31a,b), $\{FD\}$ and $\{FP\}$ are the inner force vectors which come from (30a) and (30b), respectively and $\{F_{out}\}$ is the external force vector that is acting on the nodes in the element. $\{\dot{u}\}$ and $\{\dot{\lambda}\}$ are the time derivatives of displacement and pressure at the nodes in the element.

It is noted that equation (33) includes the equations of the displacement method. Hence, this method can also be applied to the constitutive model without interpolation of the pressure. In this type of constitutive model, only the inner force vector $\{FD\}$ and tangent stiffness matrix $[K]$ are used.

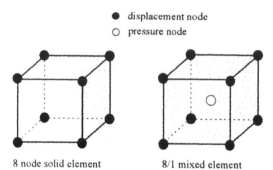

● displacement node
○ pressure node

8 node solid element   8/1 mixed element

Figure 2. Solid elements used in the model.

In the application to the constitutive model of HDR described in this paper, the inner force vector and the tangent stiffness matrix that is computed from elasto-plastic part are superposed on the $[K]$ and $\{FD\}$ in equation (33).

3.4 Numerical method

In the numerical computation, Lagragian eight nodes hexahedral with constant pressure mixed element is used for rubber material and Lagragian eights node hexahedral element is used for steel material. Figure 2 illustrates these elements. In the solution of the incremental equations, Newton-Raphson method (Bathe 1996) is employed. In the stress integration of elasto-plastic constitutive model, subincremental integration scheme (Crisfield 1991) with forward-Euler method is applied.

In the actual loading of laminated rubber bearings, vertical force is loaded with keeping its upper and lower plate horizontal. This boundary condition means that all nodes in these planes have the same displacement. However, the value of the displacement is unknown and only the sum of the distributed load acting on the nodes in the plane is known.

In order to include this boundary condition in the computation, Multi-point-constraint method is employed. In this method, all loads acting in the plane are gathered into one node by the manipulation of the tangent stiffness matrix and the inner force vector (Watanabe 1995).

4 COMPARISON WITH THE EXPERIMENTAL RESULTS

Using the afore-mentioned constitutive model and finite element method, a three-dimensional finite element model of high damping rubber bearings is developed. In this section, the developed model is proven to be valid by comparing the simulation with the results of the multi-axial loading experiments and the hybrid seismic response experiments which were conducted by the authors (Abe, Yoshida & Fujino 2002).

4.1 Bearing for modeling

Figure 3 illustrates the size of the finite element model of the high damping rubber bearing, which was used in the loading experiments (Abe, Yoshida & Fujino 2001). The parameters of the constitutive model of HDR used in this modeling are as follows (Yoshida, Abe & Fujino 2002).

$c_1 = 0.323[MPa]$, $c_2 = 0.127[MPa]$,
$c_3 = 0.0686[MPa]$, $\chi = 16.0[MPa]$, $c = 6.59$,
$n = 2.71$, $\alpha = 0.800$, $\beta = 0.302$, $a_H = 0.138$,
$b_H = 14.0$, $\tau_0 = 0.588[MPa]$, $N = 0.400$,
$c_4 = 0.206[MPa]$, $c_5 = 1.67$, $m = 1.43$, $b = 1.72$

4.2 Parameters of steel material

In the bearing, SS400 steel material is used for inner reinforced plates. In this study, material constants for the steel are decided using the result of the uni-axial cyclic experiment (Nishimura, Ono & Ikeuchi 1995). Parameters of the model are chosen to give the minimum difference of the simulation by the model and the experimental result. Identified parameters are shown below.

$E = 20.7[MPa]$, $\sigma_Y = 250[MPa]$, $Q = 108[MPa]$,
$b = 13.5$, $c = 5880[MPa]$, $\gamma = 55.0$, $\Phi_\infty = 0.800$,
$\nu = 0.3$

Figure 4 shows the simulation by the model with identified parameters in comparison with the experimental results. From Figure 4, it is understood that although the model can not accurately reproduce the initial yielding behavior, the other parts of the cyclic behaviors are well simulated by the model.

4.3 Comparison with multi-axial loading experiment

In order to show the validity of the developed finite element model, simulations by the model are compared with the results of multi-axial loading experiments. Detailed conditions of the loading experiments are described in Table 1. From Figure 5 to Figure 7, the deformation modes of the bearings in each experiment are illustrated. In these figures, the dot line shows the boundary of the symmetric deformation and only one of these symmetric parts is selected as the object of the modeling. The analytical conditions of the finite element model in each loading are summarized in Table 2.

Figure 8, Figure 9 and Figure 10 show the comparisons of the simulations and the results of the uni-axial, vertical and bi-axial loading experiments, respectively. From Figure 8 and Figure 10, the simulated results are found to be in good agreement with the experimental

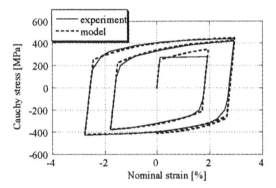

Figure 4. Comparison of the constitutive model with the result of material test of the steel SS400.

Table 1. Loading conditions of the experiment.

Type of loading	Uni-axial	Vertical	Bi-axial
Horizontal displacement	Sine wave	–	Eight-shaped path
Velocity [Hz]	0.01	0.01	0.01
Amplitude [%]	50–200	50–150	–
Average vertical pressure [MPa]	3.92	3.92	0–7.84

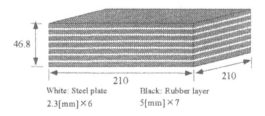

White: Steel plate
2.3[mm]×6
Black: Rubber layer
5[mm]×7

Figure 3. Illustration of the size of the high damping rubber bearing used in the modeling.

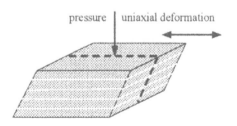

Figure 5. Loading conditions of the uni-axial loading experiment.

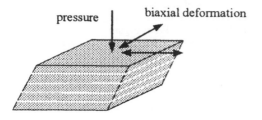

Figure 6. Loading conditions of the bi-axial loading experiment.

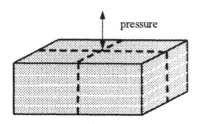

Figure 7. Loading conditions of the vertical loading experiment.

Table 2. Analytical conditions of the finite element models.

Type of loading	Uni-axial	Vertical	Bi-axial
Size of the model	1/2	1/4	1/1 (full)
NE* in plane	5 × 10	5 × 5	10 × 10
NE in vertical direction	41	41	41
	7[Rubber layer] × 5 + 6[steel layer] × 1		
Total NE	2050	1025	4100
Total number of nodes	4822	2537	9182

*NE means the Number of Element.

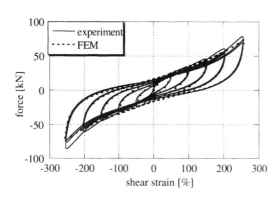

Figure 8. Comparison of the finite element model with the result of uni-axial loading experiment.

results. This indicates that the finite element model can accurately reproduce the basic characteristics of horizontal restoring forces of the bearing.

In Figure 9, on the other hand, a big difference is found in the residual displacement. This difference is considered to be the results of using the elastoplastic

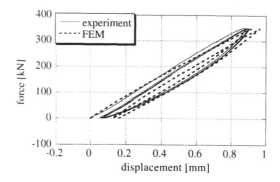

Figure 9. Comparison of the finite element model with the result of vertical loading experiment.

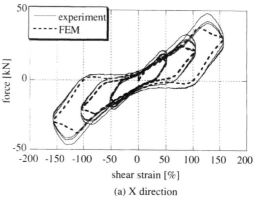

(a) X direction

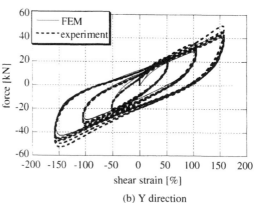

(b) Y direction

Figure 10. Comparison of the finite element model with the results of bi-axial loading experiment.

constitutive model for energy absorption of HDR. However, it is seen that the model averagely reproduces the load-displacement relation of the experiment.

4.4 Comparison with hybrid seismic response experiment

The same conditions of the hybrid seismic response experiment (Abe, Yoshida & Fujino 2002) are simulated by the finite element model in order to verify the

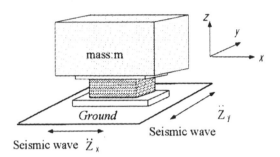

Figure 11. A structural system used in the hybrid seismic response experiment.

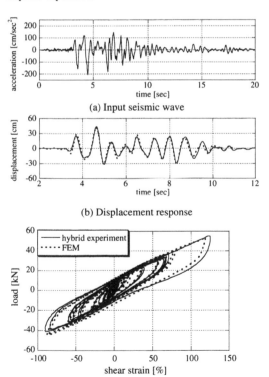

(a) Input seismic wave

(b) Displacement response

(c) Restoring forces

Figure 12. Comparison of the results of the uni-axial hybrid experiment with the simulation by FEM.

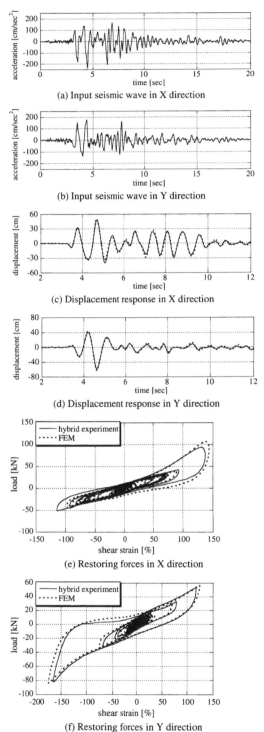

(a) Input seismic wave in X direction

(b) Input seismic wave in Y direction

(c) Displacement response in X direction

(d) Displacement response in Y direction

(e) Restoring forces in X direction

(f) Restoring forces in Y direction

Figure 13. Comparison of the results of the bi-axial hybrid experiment with simulation by FEM.

performance of predicting the seismic response by the model.

In the hybrid experiment, a mass supported by the laminated rubber bearing is considered to have the uni-axial or bi-axial seismic excitations. This is illustrated in Figure 11. In the actual experiments, acceleration records observed at Kobe Marin are scaled 1/3.5 times as input seismic waves in order to obtain the maximum shear deformation of the bearing as around 150 [%].

Figure 12 and Figure 13 show the comparison of the experimental results and the simulation in the case of the uni-axial and the bi-axial seismic excitation, respectively. From Figure 12 and Figure 13, it is understood that the experimental results are well simulated by the developed finite element model. This shows that the model can accurately predict the multi-axial seismic response of the base-isolated structure.

5 CONCLUSIONS

In this study, a three-dimensional finite element model of the high damping rubber bearing is developed. Major findings are summarized as follows.

1. The constitutive tensor of the proposed constitutive model of HDR is derived and it is understood that the model can be applied in the finite element analysis.
2. A mixed finite element method of the updated Lagragian formulation are developed. But this method, the rate form constitutive model with slight compressibility can be employed in the finite element analysis.
3. A finite element model of the high damping rubber bearing is constructed by using the developed finite element method and constitutive model. This finite element model is found to well reproduce the experimental results and accurately predict the multi-axial seismic responses of the structure.

ACKNOWLEDGEMENT

Dr. Hiroshi, Watanabe, Associate Professor in the University of Tokyo, gave helpful advices in the construction of the finite element code. Mr. Sadafumi, Uno and Mr. Yasuhisa, Hishigima of Kawaguchi Metal Industries Co., Ltd, Mr. Tiaki, Sudo and Mr. Yoji, Suitsu of Bridgestone, Mr. Kazuo, Endo of The Yokohama Rubber Co., Ltd and Mr. Haruo, Iseki and Mr. Hideaki Yokokawa of Oils Corporation provided rubber specimens and valuable comments during the study. Their assistances are very much appreciated. This study is supported by the Ministry of Education, Science and Culture, Japan and scholarship of Japanese Society of Promoting Science.

REFERENCES

Abe, M., Yoshida, J. & Fujino, Y. 2002. Uniaxial and Biaxial Property of Base-Isolation Bearings and its Modeling. *Journal of Structural and Earthquake Engineering, JSCE.* No. 696/I-58, pp. 125–144 (in Japanese).

Bath, K.J. 1996. *Finite Element Procedures.* Prentice-Hall.

Chaboche, L.J. 1989. Constitutive Equations for Cyclic Plasticity and Cyclic Viscoplasticity. *International Journal of Plasticity*, Vol. 5, pp. 247–301.

Crisfield, M.A. 1991. *Non-linear Finite Element Analysis of Solids and Structures.* Volume 1, John Wiley & Sons.

Hisada, T. & Noguchi, H. 1995. *Basis and Application of Nonlinear Finite Element Method.* Maruzen (in Japanese).

Hosam-Eddin, M.A. & Abdel-Ghaffar, A.M. 1995. Modeling of Rubber and Lead Passive Control Bearings for Seismic Analysis. *Journal of Structural Engineering*, ASCE, Vol. 121, No. 7, pp. 1134–1144.

Nishimura, S., Ono, K. & Ikeuchi, T. 1995. Constitutive Equation for Steel under Cyclic Loading Based on the Monotonic Loading. *Journal of Structural and Earthquake Engineering, JSCE*, No. 513/I-31, pp. 27–38 (in Japanese).

Seki, W., Fukahori, Y., Iseda, Y. & Matsunaga, T. 1987. A large deformation finite element analysis for multilayer elastomeric bearings. *Transaction of a meeting of the Rubber Division, American Chemical Society*, pp. 856–870.

Simo, J.C. 1987. On a Fully Three-Dimensional Finite-Strain Viscoelastic Damage Model: Formulation and Computational Aspect. *Computer Methods in Applied Mechanics and Engineering*, Vol. 60, pp. 153–173.

Akhtar, S.K. & Sujian, H. 1995. *Continuum Theory of Plasticity.* John Wiley & Sons.

Sussman, T. & Bathe, K.J. 1987. A Finite Element Formulation for Nonlinear Incompressible Elastic and Inelastic Analysis. *Computer & Structure*, Vol. 26, No. 1/2, pp. 357–409.

Takayama, M., Tada, H. & Tanaka, R. 1990. Finite Element Analysis of Laminated Rubber Bearing used in Base-Isolation System. *The meeting of the Rubber Division*, American Chemical Society, Washington D.C., pp. 46–62.

Watanabe, H. 1995. Mixed Finite Element Method for Incompressible Hyper Elastic Body. Ph.D. dissertation, Department of Mechanical Engineering, University of Tokyo (in Japanese).

Yoshida, J., Abe, M. & Fujino, Y. 2002. Constitutive Model of High Damping Rubber Material. *Journal of Engineering Mechanics,* ASCE (submitted).

Reinforced concrete structures

Finite Elements in Civil Engineering Applications, Hendriks & Rots (eds.)
© 2002 Swets & Zeitlinger, Lisse, ISBN 90 5809 530 4

Invited paper: Modal analysis versus time history analysis – concepts for the seismic design

Th. Baumann & J. Böhler
WALTER/DYWIDAG, Central Technics, Munich, Germany

ABSTRACT: The forces induced into a structure by earthquake ground motions are prescribed traditionally by elastic response spectra. They can be reduced by behaviour factors to account for the possibility of energy dissipation due to plastic deformations. Presupposition for the utilization of this favourable effect is a sufficient ductility. The consistent definition and verification of this ductility is an essential part of the structural design for earthquakes. In the present paper, the principal features of this task are discussed. In order not to persist in general terms, the governing effects are quantified within a case study of a 3-storey steel frame with composite beams, considering the following steps: Modal analysis, application of behaviour factors, usual assumptions for the required plastic deformations, push over-analysis with plastic rotation of hinges, connection of modal analysis and push over-analysis, nonlinear time history-analysis. Only the last step creates the basis for a reliable assessment of the relevant deformations of the plastifying frame, depending on the applied accelerogram and its peak value PGA.

1 BEHAVIOUR FACTOR, DUCTILITY AND MODAL ANALYSIS

The maximum seismic force of an elastic structure (F_{el}) is proportional to the peak value (PGA) of the ground acceleration $a_o(t)$. The force which causes yielding of the structure is denominated F_y, the corresponding deformation Δ_y and the acceleration $(PGA)_y$. If the structure has a sufficient ductility allowing plastic deformations, it may sustain also stronger earthquakes. In Eurocode 8 (2001) and other codes this fact is taken into account by a behaviour factor q, which describes the possible increase of PGA up to $q \cdot (PGA)_y$. This increase requires plastic deformations $\mu_\Delta \cdot \Delta_y$. For the definition of μ_Δ alternative assumptions have been proposed acc. to fig. 1 (Paulay et al. 1990): a) "same displacement" like an elastic structure with the same stiffness, and b) "same work". As pointed out already by Baumann & Böhler (1997), these assumptions are questionable. Especially the "same work"–formula has no rational base, because the areas shadowed in fig. 1 do not designate "works" or energies at all, which are characteristic for the different behaviour of elastic and non-elastic structures.

On the other hand, the realistic valuation of the plastic deformations dependent on the value of PGA is an essential part of the seismic design. The following report shows, how the informations which are necessary in

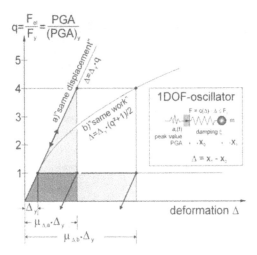

Figure 1. Plastic deformations as presupposition for the application of behaviour factors.

this respect can be found by various types of analyses, i.e. modal analysis (response spectrum analysis), push over-analysis and time history-analysis. In order not to remain in general terms, a steel frame with composite beams excited horizontally by earthquake is considered within a case study.

In areas with low seismicity, bracing of frames by diagonal rods acc. to fig. 2 is an economic solution (MGS 2001). However, this type of structure is not able to sustain plastic deformations and horizontal forces in reverse directions. Therefore a moment-resisting frame acc. to figs. 3 and 4 has been investigated. The frame distance of about 5 m is bridged by a concrete slab which is connected to the rolled steel girders by studs. In this way the stiffness of the composite beams is increased by the factor 3.5 compared to the steel girders only. The formation of plastic hinges adjacent to the columns is supported by omission of bond in this region.

Fig. 5 shows the first two eigenmodes for horizontal excitation. For $T_E = 0.437$ s (mode 1), the response spectrum of fig. 6 (acc. to Eurocode 8 (2001) for Soil B/Type 1) defines an elastic response value of 2.5 for a viscous damping of $\xi = 5\%$. For the lower damping of a steel or composite structure ($\xi = 2\%$) this response

is increased acc. EC 8 by the factor $\eta = 1.2$. For an assumed value of PGA = 0.12 g and a total weight of the structure of about $3 \cdot 35 \cdot 15 = 1575$ kN we get a total lateral force $H = 0.12 \cdot 2.5 \cdot 1.2 \cdot 1575 = 567$ kN. From a more detailed modal analysis considering 10 horizontal modes results only $H = 500$ kN.

For time history-analyses we need natural or artificially generated accelerograms. Fig. 7 shows the two accelerograms, which have been applied for this case study. They are based on PGA = $k \cdot S \cdot a_g = 1.0 \cdot 1.2 \cdot 0.10$ g = 0.12 g. The value $S = 1.2$ belongs to soil type B acc. to EC 8. The comparison of the target spectrum with the response spectra of the artificially generated accelerograms shows larger deviations of the spectral values $S_e(T)$, especially in the range of maximum amplification (0.1 s $< T < 0.7$ s) (fig. 6).

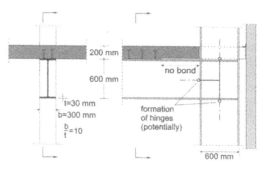

Figure 4.   Beam-column-connection.

Figure 2.   Bracing of steel frame with composite beams by diagonal rods.

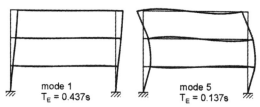

Figure 5.   Eigenmodes for horizontal excitation.

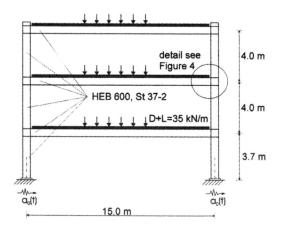

Figure 3.   Moment resisting frame with composite beams.

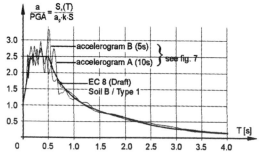

Figure 6.   Elastic response spectrum for $\xi = 5\%$.

The influence of this inaccuracy on the results of time history-analyses may be reduced only by probabilistic methods, i.e. using a set of histories and considering mean values of each result of interest. In this report – as its main aim is not the final dimensioning, but to show the principal concepts for a design – a number of two histories has been considered being sufficient.

Time history-analyses as described in section 4 yield for the elastic frame and PGA = 0.12 g extreme horizontal forces of totally H = +543/−455 kN (for accelerogram A: duration 10 s) and +482/−554 kN

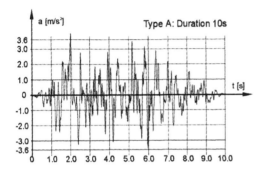

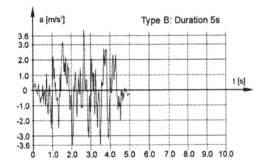

Figure 7. Accelerograms for PGA = $a_g \cdot k \cdot S$ = 0.36 g (fitting to response spectrum of fig. 6).

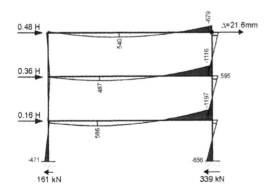

Figure 8. Bending moments due to the elastic response for PGA = 0.12 g (H = 500 kN).

(for accelerogram B: duration 5 s). The deviation from the result of the modal analysis (500 kN) is in agreement with the deviation of the response values of the 1-DOF-oscillator from the original response spectrum (fig. 6).

Fig. 8 shows the bending moments for D + L = 35 kN/m (fig. 3) and H = 500 kN (PGA = 0.12 g), this force being distributed to the 3 floors proportional to the horizontal displacement of mode 1. The top displacement amounts to Δ = 21.6 mm. The stresses in the steel sections remain below f_{yk}/γ_m = 240/1.15 = 209 N/mm².

The yield strength f_{yk} will be reached first for H_y = 745 kN or 952 kN (in hinge 1, see figs. 14 and 15; without or with overstrength, see fig. 10). The extrapolation from the results of the modal analysis yields a corresponding (PGA)$_y$ of 0.12 g · 745/500 ≈ 0.18 g or 0.12 g · 952/500 ≈ 0.22 g. The corresponding top displacements are Δ_y = 32 or 41 mm. This means, that the frame acc. to fig. 3 and 4 is able to sustain earthquakes up to (PGA)$_y$ = 0.18 g or 0.23 g (without or with overstrength) with small displacements and within the elastic range. The following investigations shall clarify to which level the value of PGA can be increased, if plastic deformations of the structure are taken into account.

2 PUSH OVER-ANALYSIS

Within a push over-analysis, the forces and deformations of the frame are evaluated for an increasing horizontal force H, considering the formation of plastic hinges, when the yield moment M_y is reached. For this purpose, hinges are provided acc. to fig. 9 at locations where a plastification is expected. The hinges are blocked by rotational springs with the characteristic acc. to fig. 10. In order to increase the moments and plastic rotations in the columns, an "overstrength of beams" is taken into account in the definition of the maximum moments in beams and columns. This corresponds to the requirements of the so-called capacity-design acc. to Paulay et al. (1990) and EC 8 (2001).

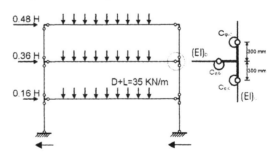

Figure 9. Statical systems with hinges blocked by rotational springs.

The plastic rotation requires plastic strains and curvatures acc. to fig. 11 within a plastic length l_p. Depending on the distribution of ΔK within the plastic length, the maximum curvature may be $\Delta K_p = \vartheta_{pl}/l_p$ or $3\Delta K_p = 3\vartheta_{pl}/l_p$. Spangemacher & Sedlacek (1992) demonstrated a method for the experimental and numerical assessment of the rotational capacity under monotonous loading up to failure.

Under earthquake-induced oscillations, stresses and plastic rotations occur in reverse directions. The

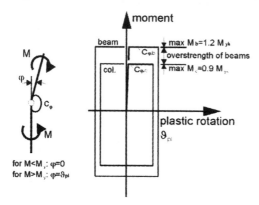

Figure 10. Characteristic of rotational springs (with over-strength of beams).

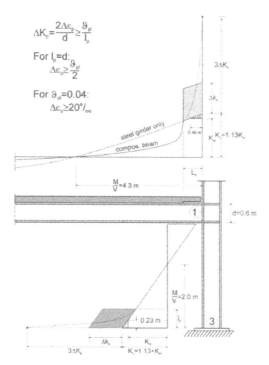

Figure 11. Plastic curvatures ΔK_p and strains $\Delta \varepsilon_p$ required for plastic rotations θ_{pl}.

behaviour under such low cycle fatigue-conditions has been studied by Vayas & Spiliopoulos (1999). With the static-cyclic tests acc. to fig. 12 (alternately load- and deformation controlled), the rotational capacity can be checked for histories of load, total deflection and plastic deflection, which are defined by time history-analyses for the considered structure and excitation. The test addresses directly ϑ_{pl}, thus it is not necessary to define the auxiliary values l_p, ΔK_p and $\Delta \varepsilon_p$.

The sequence of yielding of the individual hinges is shown in fig. 13, the horizontal deformation characteristic of the frame in fig. 14. The force which causes yielding of the first hinge, is denominated H_y, the maximum force producing a kinematic system H_u.

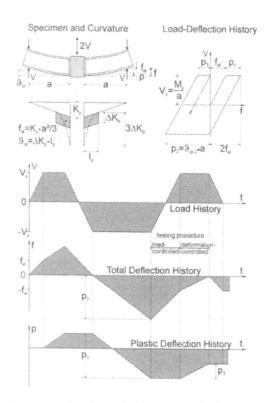

Figure 12. Test for verification of the plastic deformability θ_{pl}.

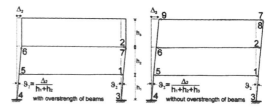

Figure 13. Sequence of yielding of hinges and kinematic deformations for $H = H_U$.

390

The corresponding top displacements of the frame are Δ_y and Δ_1. If additional displacements Δ_2 are induced by an earthquake, they cause plastic angles ϑ_2 on the kinematic system (fig. 13).

Fig. 15 shows the bending moments for $H_u = 1380$ kN and $\Delta_1 = 115$ mm (frame with overstrength of beams). It has to be pointed out, that the capacity of the steel section is exceeded at hinge 3 for the combined action of $M = 1383$ kN and $V = 694$ kN. However, the required strengthening measures are not subject of this report.

Now the seismic capacity of the frame (with overstrength) shall be defined, going back to the principles of fig 1. Assuming a moderate behaviour factor $q = 4$ (acc. to EC 8 factors up to $5 \cdot 1380/950 \approx 7$ would be possible), the frame can withstand an earthquake, which would produce an elastic response $H_{el} = q \cdot H_y = 4 \cdot 952 \approx 3800$ kN (fig. 16).

From the modal analysis with 0.12 g it can be concluded, that $H_{el} = 3800$ kN is reached for an earthquake with PGA $= 4 \cdot 0.22 \text{ g} \approx 0.9$ g. With the assumptions of fig. 1, the maximum top displacement required for the dissipation of the high seismic energy is either $\Delta_{el} = 4\Delta_y = 163$ mm ("same deformation") or $\mu_\Delta \cdot \Delta_y = 346$ mm and 502 mm, resp. ("same work").

It has to be checked first, if horizontal displacements in the above range are acceptable with regard to the serviceability of the building, e.g. with regard to cracks in partition walls, function of mechanical equipment, facade etc.

The next question is the amount of plastic rotation $(\vartheta_{pl} = \vartheta_1 + \vartheta_2)$ produced by the displacement $\Delta_1 + \Delta_2 = \Delta_{el}$ or $\mu_\Delta \cdot \Delta_y$ and its effect on the stability. To answer this question, the relations between horizontal force, top displacement and plastic rotations of hinges have to be evaluated as shown in fig. 17. From this results for $\Delta_{el} = 163$ mm a rotation of $\vartheta_{pl} = 0.0160$, for $\mu_\Delta \cdot \Delta_y = 346$ mm $\vartheta_{pl} = 0.0409$ in hinge 1.

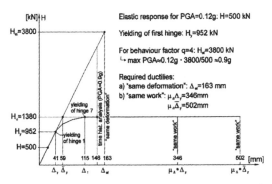

Figure 16. Seismic capacity and assumptions for the corresponding deformations of frame acc. to modular analysis.

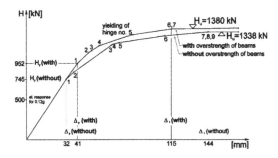

Figure 14. Horizontal deformation characteristic of frame.

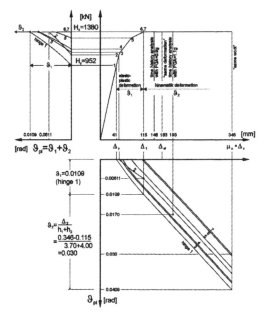

Figure 17. Plastic rotation of hinges required for ductile behaviour.

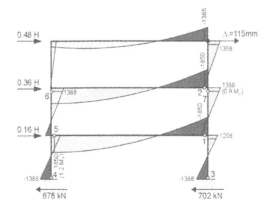

Figure 15. Bending moments for $H_u = 1380$ kN (with overstrength of beams).

The scatter of these deformations and rotations is quite large, depending on the arbitrary assumptions for Δ_y and μ_Δ. Besides it is necessary for a realistic assessment of their effect to know the number of load cycles. These informations can be found only by time history-analyses, delt with in the following sections.

For the frame considered in this study, the effect of non-conservative loading (theory of second order) is small. For a top displacement Δ of maximum 200 mm (fig. 21), the contribution $P \cdot \Delta$ to the sectional forces is less than 6% of those of first order theory and can therefore be neglected.

A further item to be mentioned is the soil-structure interaction. The discussion of this problem, however, would go beyond the scope of this paper. Therefore, a fully constrained foundation has been assumed.

3 TIME HISTORY-ANALYSIS OF 1-DOF-OSCILLATORS

Fig. 18 compares the assumptions a) and b) of fig. 1 with the results of time history-analyses of two 1-DOF-oscillators with $T_E = 0.4$ s and $T_E = 1.5$ s. The assumption b ("same work") yields to large displacements, whilst the assumption a ("same deforma-

tion") is closer to the results of time histories, at least for $q \leq 4$.

Fig. 19 demonstrates the results for an individual time history (PGA = 0.5 g, $F_y = 50$ kN corresponding to $q = 4$, $T_E = 0.4$ s). The points, which have been used as input in fig. 18, are marked by circles. They are not sufficient to describe the behaviour of this oscillator, which is characterized by the following items:

- 7 load cycles with plastic deformations at $F_y = \pm 50$ kN within the duration of 10 s
- increase of the effective period by plastic deformations to 0.5 s = 1.25 T_E
- remaining deformation of -22 mm (shift of zero-position)
- plastic deformations in alternate directions (relative to the shifted zero-position).

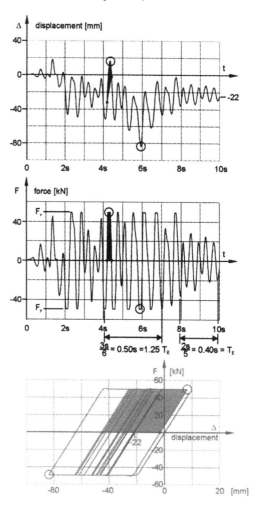

Figure 19.   History of force and displacement of 1 DOF-oscillator designed with q = 4 for accelerogram A (duration 10 s) and PGA = 0.5 g.

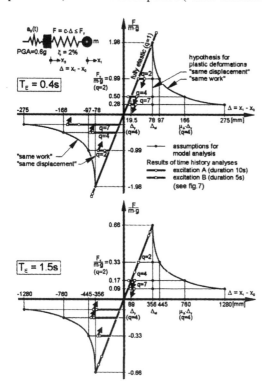

Figure 18.   Comparison of assumptions acc. Fig. 1 with the results of nonlinear time history-analyses.

As a further result of the performed time history-analyses, fig. 20 shows a typical half cycle of the plastifying oscillator (q = 4; F_y = 50 kN; detail from fig. 19) and of the fully elastic system (q = 1; F_y > 200 kN). From the F-Δ-diagram, two types of energy can be recognized: The elastic spring or kinetic energy and the dissipated energy (only for q = 4) corresponding to the plastic deformation p.

Not infrequently misleading definitions and interpretations of energies can be found in contradiction to the above simple scheme. The hysteretic behaviour of a plastifying spring can not be emulated e.g. by increased viscous damping as it is possible for high damping rubber bearings, which are used as seismic isolators. Fig. 20 underlines also the fact, that an assumption "same work" as cited in fig. 1 makes no sense with regard to the definition of the plastic deformations required for a certain value q.

The energies occurring during an earthquake have been discussed more generally by Baumann & Böhler (2001). They depend on the elastic and non-elastic properties of the structure as well as on the seismic input and can be defined only by time history-analyses for the individual case.

4 TIME HISTORY-ANALYSIS OF FRAME

Already for a 1-DOF-oscillator it is difficult to predict the plastic displacements required for a certain behaviour factor q without a time history-analysis. For frames, the connection of the results of push over-analyses and modal analyses as shown in figs. 16 and 17 requires even more a verification by time history-analyses. Such analyses are much more expensive for a frame than for a 1-DOF-oscillator especially because of the plastic hinges. The time steps have to be reduced from 10^{-2} s to 10^{-4} s. Thus, a nonlinear analysis requires a powerful software like DIANA 8.1 (2002) and takes nevertheless not seconds as for 1-DOF-oscillators, but hours.

Eurocode 8 recommends behaviour factors q = 5 · α_u/α_1 for moment resisting frames with sufficient

ductility. Therefore a time history-analysis has been performed with the assumption q = 5, i.e. for an accelerogram with PGA = 5 · $(PGA)_y \approx 1.1$ g, whereas $(PGA)_y = 0.22$ g is the value causing first

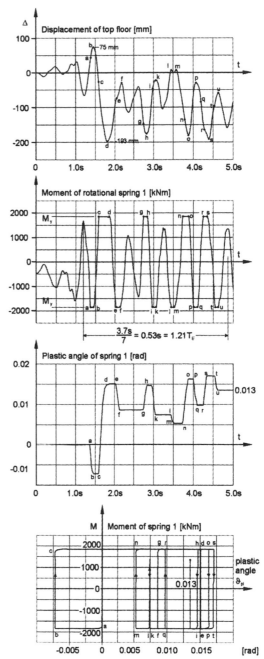

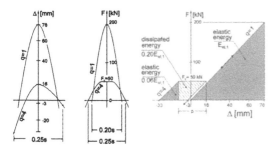

Figure 20. Displacements, forces and energies during one half-cycle (accelerogram A, PGA = 0.5 g, q = 1 and 4).

Figure 21. Relevant moments and deformations of frame for accelerogram B (duration 5 s) and PGA = 1.1 g.

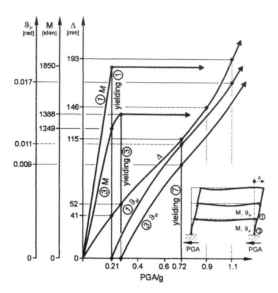

Figure 22. Extreme moments and deformations dependent on PGA (for accelerogram B, duration 5 s).

yielding of hinge 1 ($H_y = 952$ kN, fig. 14) acc. to the modal analysis.

The time history for the moments in hinge 1 and for the relevant deformations are given in fig. 21. The characteristic points are marked by letters. The following items are to be mentioned:

– 5 load cycles of hinge 1 with plastic deformations at $M_y = \pm 1850$ kN/m within the duration of 5 s
– increase of the effective period of the frame by plastic deformations to $0.53\,s = 1.21\,T_E$
– remaining plastic angle in hinge 1 of 0.013
– maximum top displacements $+75$ mm/-193 mm.

Fig. 22 shows the maximum moments and deformations dependent on PGA. Diagrams like this are the basis for a reliable assessment of the value of PGA which can be accepted with regard to deformations and stresses in the serviceability and ultimate limit state of a given structure.

The resistance of the plastic hinges under low cycle fatigue has to be considered with regard to the number and amplitude cycles of plastic rotation acc. to fig. 21. They can be used to define the program for pseudodynamic tests acc. to fig. 12.

In fig. 16, the results of the modal analysis for PGA = 0.12 g have been extrapolated for q = 4 to PGA $\approx$ 0.9 g. The assumptions for the required top displacement resulted $\Delta_{el} = 163$ mm ("same deformation") or $\mu_\Delta \cdot \Delta_y = 346$ mm and 502 mm, resp. ("same work"). The time history-analysis yields $\Delta = 146$ mm for PGA = 0.9 g, that means the

assumption "same deformation" is more realistic in this special case.

Nevertheless it has to be emphasized, that the degree of damage under seismic loads is governed not only by the extreme deformation values, but even more by the cycles of plastic rotations. These informations can be obtained only by time history-analyses.

5 CONCLUDING REMARKS

The seismic displacement, which is related to elastic and plastic strains, rotations and deformations addresses the seismic performance of a structure more directly than do response forces obtained by elastic or inelastic spectra. The survival capability of a building is more a function of its displacement capacity rather than its initial yield strength (Chandler et al. 2001). The ground motion, which can be sustained by a given structure with regard to the criteria of a SLS- or ULS-design, can be defined only on the base of a reliable assessment of the plastic deformations, including the extreme values as well as the number of plastic load cycles.

The complex and random nature of earthquake ground motions requires many approximations in the description of the excitation by various types of spectra or by accelerograms. Thus, the "accuracy" of a seismic design is always limited. The consecutive uncertainties can be quantified only by time history-analyses for a representative set of accelerograms.

The unavoidable scatter of the seismic input should not be mixed up with inconsistent methods of the structural analysis. The verification of a sufficient ductility, which is the main presupposition for the energy dissipation by plastic deformations, can and should be based on rational premises and calculations as discussed in this report rather than on arbitrary assumptions.

REFERENCES

Baumann, Th. & Böhler, J. 1997. Effect of nonlinear behaviour of structures under earthquake action. Finite elements in engineering and science. Proc. of the 2nd International DIANA conference. Amsterdam 4–6 June 1997. Rotterdam: Balkema.

Baumann, Th. & Böhler, J. 2001. Seismic design of Base Isolated LNG-Storage-Tanks. Structural Engineering International, Vol. 11, No. 2. Zürich: IABSE.

Chandler, A., Lam, N., Wilson, J. & Hutchinson, G. 2001. Review of modern concepts in the engineering interpretation of earthquake response spectra. Structures and buildings 146 Issue I. London: Proceedings of the Institution of Civil Engineers.

DIANA 8.1.2002. Computer software programs. Delft: Issue May 2002.

Eurocode 8: Design provisons for earthquake resistance of structures – Part 1.1: General rules 2001. ENV 1998-1-1. Draft December 2001.

Münchner Gesellschaft für Stadterneuerung mbH-MGS. 2001. Parkgarage München. Architekt: R. Schwaighofer, Innsbruck. Structural design: rb-BauPlanung GmbH.

Paulay, Th., Bachmann, H. & Moser, K. 1990. Erdbebensicherung von Stahlbetonhochbauten. Basel: Birkhäuser-Verlag.

Spangemacher, R. & Sedlacek, G. 1992. Zum Nachweis ausreichender Rotationsfähigkeit von Fließgelenken bei der Anwendung des Fließgelenkverfahrens. Stahlbau 61. Berlin: Ernst & Sohn.

Vayas, I., Ciutina, A. & Spiliopoulos, A. 1999. Low cycle fatigue – gestützter Erdbebennachweis von Rahmentragwerken aus Stahl. Bauingenieur 74. Düsseldorf: Springer-VDI-Verlag.

Finite Elements in Civil Engineering Applications, Hendriks & Rots (eds.)
© 2002 Swets & Zeitlinger, Lisse, ISBN 90 5809 530 4

Invited paper: Integrated structural performance assessment for reinforced concrete under coupled environment and seismic actions

K. Maekawa, T. Ishida & R. Mabrouk
University of Tokyo, Tokyo, Japan

ABSTRACT: The authors propose a so-called life-span simulator that can predict concrete structural behaviors under arbitrary external forces and environmental conditions. In order to realize such a technology, two computational systems have been developed; one is a thermo-hygro system that covers microscopic phenomena in C–S–H gel and capillary pores, and the other is a structural analysis system, which deal with macroscopic stress and deformational field. In this paper, the unification of mechanics and thermo-dynamics of materials and structures has been made, and the proposed multi-scale integrated system can be used for simultaneous overall evaluation of structural and material performances.

1 INTRODUCTION

Performances of concrete structure vary with time under various environmental actions during its life-cycle. In order to realize durable and reliable structures, a computational method that can predict the overall behavior of concrete structures is thought to be one of the most powerful tools for engineers. The objective of our research is to develop a so-called lifespan simulator capable of predicting structural behaviors for arbitrary conditions. Figure 1 shows the schematic representation of the lifespan simulator of material science and structural mechanics. Our research

group has been developing two numerical simulation tools. One is a thermo-hygro system named DuCOM (Maekawa et al. 1999), which covers the micro-scale phenomena governed by thermodynamics. This computational system can evaluate an early age development of cementitious materials and deterioration processes of hydrated products under long-term environmental actions. The other one is a nonlinear path-dependent structural analytical system named COM3 (Maekawa et al. 1999, Okamura et al. 1991). For arbitrary mechanical actions including temperature, shrinkage and creep effects, the structural response as well as mechanical states of constituent elements can

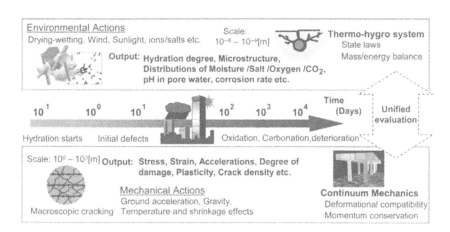

Figure 1. Lifespan simulation for materials and structures.

be predicted. In the following section, the overall scheme of these systems and their unification will be introduced.

2 THERMO-HYGRO PHYSICS FOR CONCRETE PERFORMANCE

In order to trace the early-age development of cementitious materials, it is necessary to consider the inter-relationships among the hydration, moisture transport and pore-structure development processes. The authors have been developing a 3D FE analysis program code-named DuCOM that simulates these phenomena. A detailed discussion of the material models and system dynamics for describing their interactions can be found in a published book, several papers, and DuCOM online web site (Maekawa, et al. 1999, Ishida, et al. 2000, 2001).

The overall computational scheme is shown in Figure 2. The constituent material models are formulated based on microphysical phenomena, and they take into account the inter-relationships in a natural way. The inputs required in the scheme are mix proportion, powder material characteristics, casting temperature, geometry of the target structure and the boundary conditions to which the structure will be exposed during its life-cycle.

The hydration of both constituent minerals of cement and pozzolans is traced by simultaneous differential equations based on the Arrhenius' law of chemical reaction (Kishi et al. 1996, 1997). The rate of hydration is mathematically specified in terms of temperature, free water content in capillary pores, degree of hydration and associated cluster thickness of C–S–H gel layers precipitated around non-reacted cement particles. Then, the chemical process and its interaction among minerals and additive pozzolans are considered by sharing common variables associated with pore solution, water and temperature.

During the hydration process, mass and heat energy conservations have to be satisfied with respect to moisture and temperature. At the same time, moisture migration in terms of vapor and liquid water and heat flux are incorporated in the conservation conditions of the second law of thermo-dynamics. The equilibrium conditions are simultaneously to be solved together, and the mass and energy transport resistance denoted by permeability and conductivity has to be formulated.

The permeability of vapor and liquid water is mathematically formulated based on the micro-pore size distribution as demonstrated in Figure 3 (Maekawa et al. 1999). The path of moisture in cement paste is thought to be assembly of small sized fictitious pipes and its integration results in the macroscopic permeability. Tortuosity on percolation and the thermo-dynamic activation of surface energy onto the micro-scale viscosity of pore water are taken into account. It is to be noted that a simple micro-mechanical modeling is applied without any variable fitting.

The equilibrium state of moisture is described by applying principal thermodynamic theories, such as, Gibbs energy balance and the Clausius-Clapeyron equation, to the moisture existing in pore structures. As the driving forces of moisture transport, temperature, vapor density and pore pressure differences are specified. These modeling enable us to evaluate moisture profiles varying with temperature in the system, although the generalized modeling capable of the precise prediction is still under development (Ishida et al. 2002).

As a natural way, the pore structure formation model, as illustrated in Figure 4, is added in the system dynamics of transient concrete performance modeling. The statistical approach to the micro pore structural geometry of hardened cement paste having interlayer, C–S–H gel and capillary pores is used. The porosity distribution of hydrated and non-hydrated compounds around referential cement particles is calculated and the surface area of micro-pores is estimated

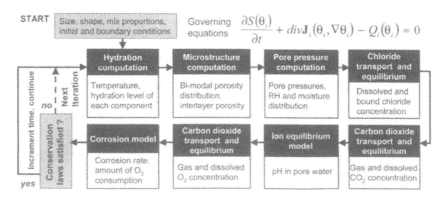

Figure 2. Framework of DuCOM thermo-hygro physics.

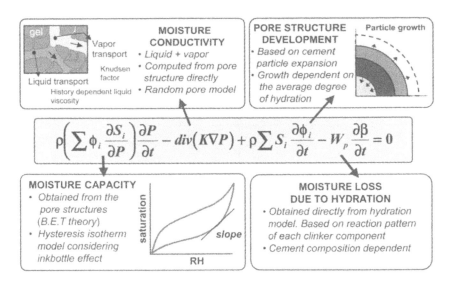

Figure 3. Schematic representation of moisture transport modeling in concrete.

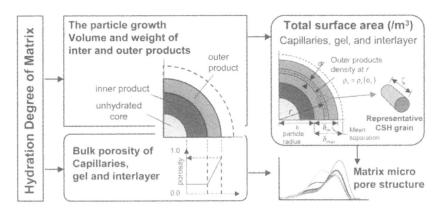

Figure 4. Outline of the pore structure development computation.

mathematically. By assuming statistical distribution function with regard to the pore sizes, the authors extend the geometrical description of micro pores. The connective mode of each pore volume is also defined with simple probability on the basis of which the path-dependency of isotherm of moisture is successfully described (Maekawa et al. 1999).

In addition to the above modeling related to an early age development phenomenon, the authors have been widening the application of DuCOM in order to cover the deterioration and resolution of cementitious materials and steel corrosion under long-term environmental actions (Ishida et al. 2000, 2001). For that purpose, as degrees of freedom, concentrations of chloride ion, oxide, and carbon dioxide are solved in the system such that the mass/energy balance equations are fully

satisfied (Fig. 2). Potential term $S(\theta)$, flux term $J(\theta)$ and sink term $Q(\theta)$ constituting the governing equations are formulated as nonlinear functions of variables θ_i based on thermodynamic theory. The obtained material properties are shared through the common variables beyond each sub-system.

For example, in the case of carbon dioxide (Fig. 5), the potential term can be obtained by applying Henry's equilibrium law under estimated liquid and vapor distribution in porous materials (Ishida et al. 2001). The pore structure development model gives geometric characteristics of pores. Based on these properties, the CO_2 transport in both phases of dissolved and gaseous carbon dioxide can be quantified by considering the effect of Knudsen diffusion, tortuosity and connectivity of pores on the diffusivity. Ionic concentrations,

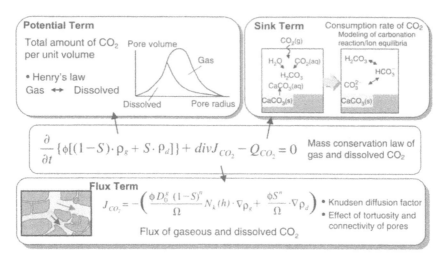

Figure 5. Modeling of CO_2 equilibrium and transport.

their reaction rate and pH value can be determined by modeling of ion equilibria. In the reaction, calcium hydroxide is consumed, and the total amount is balanced with the production rate estimated by the multi-component hydration heat model.

By applying these treatments, an interactive problem, such as corrosion due to the simultaneous attack of chloride ions and carbon dioxide, can be simulated in a natural way. Coupling these materials modeling, an early age development process and deterioration phenomenon during the service period can be evaluated for arbitrary materials, curing and environmental conditions.

3 UNIFICATION OF THERMO-PHYSICS OF MATERIALS AND MECHANICS OF STRUCTURES

It is a well-known fact that creep, shrinkage, and their coupling behaviors show strong dependency on moisture condition, pore structure, and hydration process of concrete. The authors have been developing an integrated computational system for thermo-physics and structural mechanics, which can deal with time-dependent behaviors of young aged concrete as well as mechanical behaviors of hardened concrete, such as deformation, ductility, fracture, and cracking. A system dynamics for micro-scale pore structure formation and macro-scale defects and structural deformation is shown in Figures 6–8.

The frame of structural mechanics, named COM3 (Maekawa et al. 1999, Okamura et al. 1991), has an inter-link with thermo-hygro physics in terms of mechanical performances of materials through the

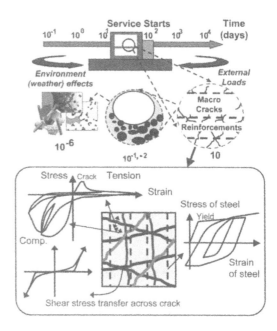

Figure 6. Macro-scale defects and micro-scale pore structures.

constitutive modeling in both space and time. Instantaneous stiffness, short-term strengths of concrete in tension and compression, free volumetric contraction rooted in coupled water loss and self-desiccation caused by varying pore sizes are considered in the creep constitutive modeling of liner convolution integral (Figs. 7, 8). In a unified solidification modeling for creep (Mabrouk et al. 1998), aggregates are idealized as suspended continuum media of perfect elasticity, whereas cement paste is treated as the solidified non-aging clusters having individual creep properties (Fig. 7). The aging process itself is represented by the solidification of these clusters. In the rheological model of cement paste layers, each cluster is assumed to have several components representing elastic, visco-elastic, visco-plastic and instantaneous plastic strain. The properties of constituting spring, dashpot and slider are directly linked with the motions of capillary, gel, and interlayer water.

The combined effect of external loads and pore water pressure is treated as the driving force for the

concrete deformation. Of course, the micro-pore size distribution and moisture balance of thermo-dynamic equilibrium are given from the code DuCOM at each time step. By means of this methodology, it is possible to unify the material science and structural mechanics in a natural way.

4 NUMERICAL SIMULATIONS

4.1 Mass transport and degradation phenomena under various environmental actions

Using the computational system, several simulations were carried out to demonstrate durability performances of concrete structures under environmental actions. Figure 9 shows the distribution of free and bound chlorides from the boundary surface. The size of mortar specimens was $5 \times 5 \times 10$ [cm] and the water to powder ratio was 50%. After 28 days of sealed curing, the specimens were exposed to cyclic alternate drying and wetting cycles. As shown in the analytical results, the distribution of bound and free chlorides can be reasonably simulated with advective transport due to the rapid suction of pore water under wetting phase.

Next, accelerated carbonation tests were verified (Fig. 10). After two days of sealed curing, specimens were kept in a controlled chamber where the concentrations of CO_2 gas (1.0% and 10%), temperature (20°), and relative humidity (55%RH) were kept constant. All of the input values in the analysis corresponded to the experimental conditions. Analytical results show the relationship between exposure time and the concrete depth at which the pore water pH falls below 10.5,

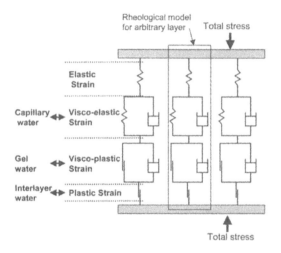

Figure 7.   Rheological model of cement paste layers.

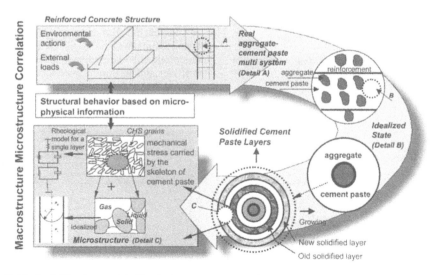

Figure 8.   Unified solidification model of hardening concrete composite.

401

which is the phenolphthalein indicator point. The simulations are able to roughly predict the progress of carbonation for different CO_2 concentrations and water-to-powder ratios.

Chloride content [wt% of cement]

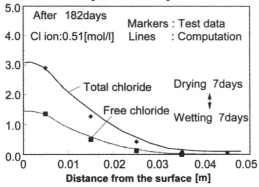

Figure 9. Chloride content profile in concrete exposed to cyclic wetting and drying.

Depth of carbonation[mm]

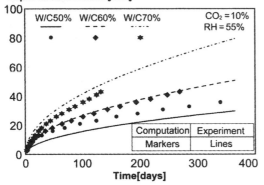

Figure 10. Carbonation phenomena for different CO_2 concentrations and water-to-cement ratio.

4.2 Moisture distribution in concrete damaged by internal stresses

As an example of coupling analysis of micro-scale thermo physics and mechanical behavior in macro-scale, a numerical simulation of moisture loss in cracked concrete is shown in Figure 11. The target structure in this analysis was a concrete slab, which has a 30% water-to-powder ratio using medium heat cement. The volume of aggregate was 70%. After 3 days of sealed curing, the specimen was exposed to 50% RH. Figure 11 shows the mesh layout and the restraint conditions used for this analysis. Moisture conductivity after cracking was formulated as a function of average strain of concrete. Figure 11 shows the cracked elements, the distribution of moisture, and the normalized tensile stress at each point from the boundary surface exposed to drying conditions. The moisture distribution calculated without stress analysis is also shown in Figure 11. As these results show, cracking begins from an element near the surface, and it is also clear that the moisture loss rises due to cracking.

4.3 Deformation of beam specimens exposed to loading and drying conditions

The phenomenon of the drying creep behavior has been well known as the Pickett's effect. In the conventional approach to studying the drying creep, total deformation is assumed to be the summation of the basic creep, shrinkage, and instantaneous elastic strain. In the proposed system, however, it is not necessary to distinguish each contribution. The Pickett's effect can be predicted as a natural outcome of the model computation, only by giving boundary conditions.

For verification the beams studied by Pickett (Pickett, 1942) are analyzed. The size of the beams is $5 \times 5 \times 86$ [cm]. Three different boundary conditions were studied; the first case was dried at the age of 7 days without any loading, the second was loaded at the same age but while submerged under water to prevent drying, and the third was subjected to both loading

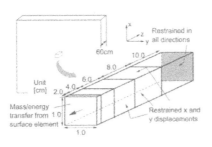

Water content[kg/m3]

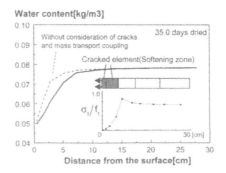

Figure 11. Moisture and internal stress distribution in concrete exposed to drying condition.

and drying at the age of 7 days. As shown in Figure 12, deformational behaviors of beam specimens can be reasonably predicted for each boundary condition.

4.4 Temperature effect on time-dependent behaviors of concrete

The specimens analyzed here are cylindrical specimens of radius 10 cm and height 20 cm. After 1 day of sealed curing, the specimens were subjected to both loading and drying (80%RH). The loading stress was 10 MPa, and a number of unloading-loading cycles was applied. The results of time-dependent mechanical behaviors under different temperature are shown in Figure 13. It can be seen that the behaviors of the specimens are strongly affected by ambient temperature. The specimen exposed to the temperature of 65 degrees Celsius shows larger deformation

with time, reflecting temperature-dependent moisture profile, hydration and micro-structure formation. In the model, visco-elastic and visco-plastic deformations are described by the viscosity of liquid water, moisture distribution and microstructure profiles. Moreover, capillary stress generated by drying and self-desiccation is taken into account as the driving force for the deformation of cement paste as well as applied external forces. Analytical results implicitly involve these influential factors on mechanical behaviors of concrete, once initial and boundary conditions are given to the system.

Here, it has to be noted that the temperature sensitivity of movement of interlayer water is neglected in this study. The interlayer water existing between tobermorite sheets of nano-meter scale would play a major role in plastic deformation especially at high temperature. The authors understand that it is necessary

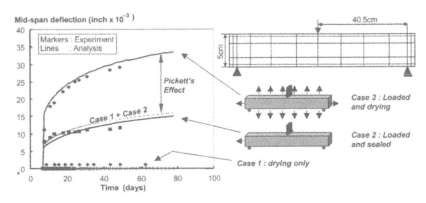

Figure 12.   Time-dependent behavior of beams exposed to external loading and drying.

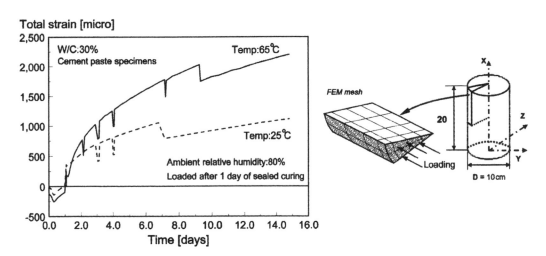

Figure 13.   Time-dependent mechanical behaviors of specimens subjected to both drying and loading under different ambient temperatures.

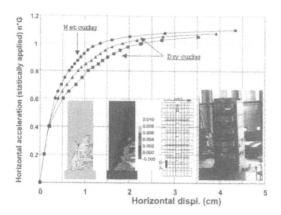

Figure 14. Load-displacement relation of shear wall exposed to drying.

to establish a more generalized treatment for interlayer part, but this enhancement remains for future study.

4.5 *Effect of moisture profile on macroscopic mechanical behavior*

Figure 14 shows the computed load displacement relation of the thin-wall specimen (CAMUS, 1998) subjected to drying and horizontal forces. Since the wall thickness is small, the effect of drying shrinkage is not negligible on the post-crack behaviors even under dynamic actions. The volume change induced by the loss of moisture accompanies self-equilibrated stress in concrete due to the presence of reinforcement, and the early cracking under loads tends to be experienced. The post-crack tension-stiffening continues to be influenced by the magnitude of drying and its effect can be comparatively seen in the wall-type structures because the surface area is larger. As can be seen in Figure 14, the drying shrinkage substantially makes the structural stiffness lower. The integrated system of moisture migration and structural mechanics with damage can capture the coupled response under loads and ambient conditions.

5 CONCLUSIONS

A numerical simulation system that can evaluate structural behaviors under coupled forces and environmental actions was presented in this paper. This system consists of two computational system, that is, one is a thermo-hygro system that covers microscopic

phenomena in C–S–H gel and capillary pores, and the other is structural analysis system, which deal with macroscopic stress and deformational field.

Macroscopic structural behaviors were linked with both the microphysical phenomenon and external load and restraint conditions. In this paper, the unification of mechanics and thermo-dynamics of materials and structures has been made. Though further progress and development is still needed for accomplishing entire system, the system dynamics of micro-scale pore structure formation and macro-scale defects and deformation of structures can be shown as a possible approach in this study.

REFERENCES

Maekawa, K., Kishi, T. and Chaube, R.P. 1999. *Modeling of Concrete Performance*, E&FN SPON. http://concrete.t.u-tokyo.ac.jp/en/demos/ducom/ index.html

Maekawa, K., Irawan, P. and Okamura, H. 1997. Path dependent three dimensional constitutive laws of reinforced concrete – Formation and experimental verifications, *Structural Engineering and Mechanics*: **15**, (6), 743–754.

Okamura H. and Maekawa, K. 1991. *Nonlinear Analysis and Constitutive Models of Reinforced Concrete*, Gihodo.

Kishi, T. and Maekawa, K. 1996. Multi-component model for hydration heating of portland cement, *Concrete Library of JSCE*: No. 28, 97–115.

Kishi, T. and Maekawa, K. 1997. Multi-component model for hydration heating of blended cement with blast furnace slag and fly ash, *Concrete Library of JSCE*: No. 30, 125–139.

Ishida T. and Maekawa, K. 2000. An integrated computational system for mass/energy generation, transport, and mechanics of materials and structures, *Concrete Library of JSCE*: No. 36, 129–144.

Ishida T. and Maekawa, K. 2001. Modeling of pH profile in pore water based on mass transport and chemical equilibrium theory, *Concrete Library of JSCE*: No. 37, 131–146.

Ishida T. and Maekawa, K. 2002. Solidified cementitious material-structure model with coupled heat and moisture transport under arbitrary ambient conditions: *fib* Congress in Osaka (to be published).

Mabrouk, R., Ishida, T. and Maekawa, K. 1998. Solidification model of hardening concrete composite for predicting autogenous and drying shrinkage, *Autogenous Shrinkage of Concrete Edited by Ei-ichi Tazawa*, E&FN SPON.

Pickett, G. 1942. The effect of change in moisture-content of the creep of concrete under a sustained load, *Journal of ACI*: Vol. 52, 333–355.

Commissariat a l'Energie Atomique. 1998. "CAMUS" INTERNATIONAL BENCHMARK – Experimental Results Synthesis of the participants' report – Organized by CEA and GEO, a French research network, part of the CAMUS Working Group under the auspices of the French Association of Earthquake Engineering (AFPS).

Finite Elements in Civil Engineering Applications, Hendriks & Rots (eds.)
© 2002 Swets & Zeitlinger, Lisse, ISBN 90 5809 530 4

Concrete excavation for patch repair: Non-linear modelling of propped and unpropped conditions

T.D.G. Canisius
BRE Ltd, Bucknalls Lane, Garston, Watford, WD25 9XX, England

ABSTRACT: DIANA offers the possibility to conduct phased analyses of construction processes. This paper presents an exercise at utilising this capability to model the excavation of concrete and the subsequent filling up of the void with new material during the patch repair of a reinforced concrete structure. A particular difficulty was that the model, which was not developed with the repair analysis in mind, consisted of embedded reinforcements which were to be exposed during excavation. As embedded reinforcements need mother elements, and would not function without them, it was necessary to devise a way to overcome this problem. This paper presents the methodology adopted and some analytical results from the conducted exercise. The analyses considered the non-linear behaviour of concrete, including cracking.

1 INTRODUCTION

The deterioration of reinforced concrete structures is a common phenomenon. Patch repair is the most common method of repair of concrete deterioration. In patch repair, the deteriorated concrete is excavated out, usually to a level below that of the reinforcement, and new concrete is placed as replacement. This may happen with or without the replacement or supplement of reinforcements dependent upon the existing reinforcement conditions.

The excavation of reinforced concrete for patch repair purposes, if not done properly, can lead to overstressing of a structure and to future problems. There have been experimental and analytical investigations of this aspect in the past by various researchers. Canisius and Waleed (2002a and 2002b) have referred to these and the importance of holistic considerations, including the modelling of the process of patch repair.

A method of considering the effects of patch repair excavations of a structure is to conduct an analysis of its behaviour under the processes involved. Although the phased analysis capability of DIANA offers a powerful method of doing, while including non-linear effects – examples of which the authors have not come across before. This paper presents a simulation of the excavation of a reinforced concrete slab (plate) near its column support.

When the original model for the analysis of the particular structure was developed, there was no intention of conducting a phased analysis of the repair

process. However, with the need for a phased analysis arising later, the embedded reinforcement concept used in the model seemed a significant obstacle for conducting it. This is because embedded reinforcing bars need mother elements to embed them – something not possible if the concrete around the bars are to be excavated. However, remodelling the structure was not seen as practical – both due to the time constraints and also the complexity due to various bars within the 3-dimensional structure modelled with solid brick elements. Thus, it became necessary to devise a way of conducting the analysis using the available model. This paper also presents the method the author devised to overcome the problem.

2 THE STRUCTURE

The structure analysed is shown in Figure 1. It is essentially a rectangular flat slab, connected to a column using some steel collar. The concrete slab had both top and bottom steel.

The column, which was considered as linear elastic, was modelled coarsely. The central part of the slab had a rectangular finite element grid in plan. This was changed to a trapezium grid in the outer regions so that fineness of the discretisation decreased towards the perimeter. There were seven finite elements in depth. However, they were not of equal depth as they were sized to consider other geometric parameters such as the column connection.

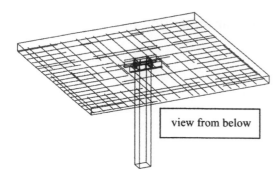

Figure 1. The analysed structure: flat plate on a linear elastic column. Also shown are the bottom reinforcement bars. Top bars, to be exposed by excavation not shown.

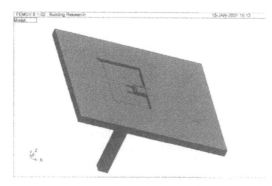

Figure 2. Region of excavation for patch repair.

The bottom of the column was treated as clamped. The perimeter of the slab was free, with concentrated forces that represented the bending moments, in-plane forces and transverse shear forces. To prevent local cracking due to these concentrated forces applied at nodes, the perimeter row of the slab finite elements and the row next to it were maintained as linear elastic.

3 OBJECTIVE

Figure 2 shows the region of repair considered in the analyses presented here. The repair was assumed to be conducted by first excavating the region to a depth that coincided with the three top most layers of finite elements in this 7-layer model. The top reinforcements lied within the excavated depth.

The intention of these analyses was to find out the different stress conditions that would occur in the original concrete and in the repair material if the structure is either not propped ("unpropped") or propped during a repair process.

4 METHODOLOGY

The methodology for achieving the objective was to conduct phased analysis of the different processes involved.

The analyses essentially consisted of three phases for each of the two alternatives of "propped" and "unpropped" repair. Propping was considered to be "complete", with all the finite element nodes at the slab soffit restrained against vertical movement – either downward or upward (with the latter assumption being slightly unrealistic but not significant).

5 THE PHASED ANALYSIS

The conducted analyses had three phases for each of the two repair scenarios under "Unpropped" and "Propped" conditions. The steps involved in each are described below. The final step of the third phase was the application of an additional load to study the response of the repaired structure.

PHASE 1: Analysis with self weight and boundary loads (from a whole structure analysis). Common to both cases.

PHASE 2: Propped Case:

- Apply vertical restraints to all the finite element nodes at the soffit of the slab.
- Remove concrete from the repair patch, except for those in contact with the reinforcements and the column. The concrete in contact with the reinforcements, and within the repair excavation, are called BARELS (Figure 3).
- Conduct an equilibrium analysis to obtain stresses. Note: The self weight is now smaller than in Phase 1 due to the elimination of some concrete elements.

Unpropped Case:

- Remove concrete from the repair patch, except for those in contact with the reinforcements and the column. The concrete in contact with the reinforcements, and within the repair excavation, are called BARELS, and are provided with a new set of properties.
- Conduct an equilibrium analysis to obtain stresses. Note: The self weight is now smaller than in Phase 1 due to the elimination of some concrete elements.

(The deflected shape of the floor slab after this unpropped excavation is shown in Figure 4.)

PHASE 3: Propped Case:

- Apply the repair patch elements with its own concrete properties. Further weaken the effect of BARELS by reducing its stiffness. *(The analyses in the third phase can be carried out even without BARELS. Figure 10, which compares the first principal stress distributions in the repair materials, contains results from an unpropped repair analysis that did not use BARELS elements in the 3rd phase.)*

- Remove the vertical restraints provided at the soffit of the slab.
- Conduct an equilibrium analysis to obtain stresses and the new position of the structure.
- Apply an additional load of 0.2 times the current loads and obtain the equilibrium stress state.

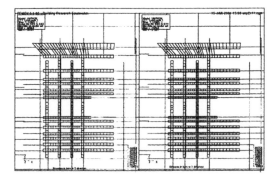

Figure 3.   The concrete elements (BARELS) in contact with the top reinforcements within the excavated region.

Unpropped Case:

- Apply the repair patch elements with its own concrete properties. Further weaken the effect of BARELS by reducing its stiffness.
- Conduct an equilibrium analysis to obtain stresses and the new position of the structure.
- Apply an additional load of 0.2 times the current loads and obtain the equilibrium stress state.

6   CONCRETE PROPERTIES USED IN PHASED ANALYSES

As mentioned in Section 3, the properties of the finite element group BARELS, within which the reinforcements in the excavated region are embedded, were progressively reduced during the phased analysis procedure. These and other properties used are described below. The properties for Phase 1, the original state, are provided in Table 1.

6.1   *Properties for Phase 2*

The material properties used during the second phase, under both propped and unpropped conditions, are provided in Table 1.

The element group BARELS was considered to be weak in tension, but with a relatively high Young's modulus. The low tensile strength helped to crack

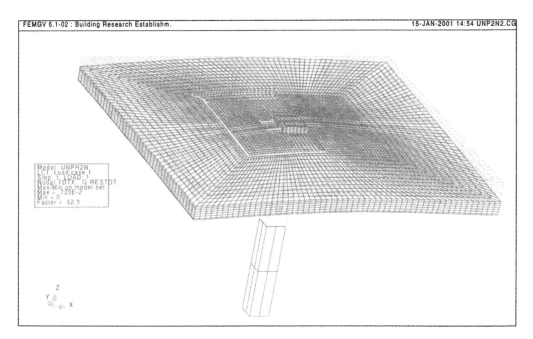

Figure 4.   The deflected shape of the unpropped excavated structure at the end of the second phase. The figure shows the finite element discretisation used for the floor slab and the column.

407

Table 1. Concrete properties used in the Phase 1 and Phase 2 Analyses.

Property	Original concrete	BARELS
Young's modulus (N/m^2)	2.05E10	2.05E08
Poisson ratio	0.2	0.2
Density (kg/m^3)	2400	0.01
Compressive strength (N/m^2)	21.0E06	0.02E06
Cracking criterion	Maximum principal stress	Maximum principal stress
Tensile strength (N/m^2)	2.1E06	0.0021E06
Shear stiffness criterion	Variable stiffness	Constant stiffness
Beta for shear criterion	0.99	0.99
Tension softening criterion	Linear	Linear
Strain at zero tensile strength	0.00125	0.000125

Table 2. Concrete properties used in the Phase 3.

Property	Repair material	BARELS
Young's modulus (N/m^2)	3.075E10	Unpropped 1.0E04 Propped 1.0E02
Poisson ratio	0.2	0.2
Density (kg/m^3)	2400	0.01
Compressive strength (N/m^2)	60.0E06	
Cracking criterion	Maximum principal stress	None (Linear elastic)
Tensile strength (N/m^2)	6.0E06	
Shear stiffness criterion	Variable stiffness	
Beta for shear criterion	0.99	
Tension softening	Linear	
Strain at zero tensile strength	0.00125	

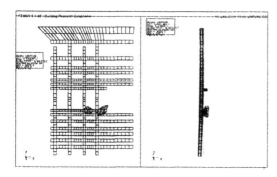

Figure 5. Unpropped repair analysis, Phase 2: The very much exaggerated views, in plan and in elevation, of the displacement of BARELS elements that embedded reinforcements within the excavated region.

these elements, preventing the occurrence of high direct stresses. The higher Young's modulus provided the stiffness necessary to generate the cracking stress. The shear stiffness after cracking was maintained at a constant value of 0.99 times the stiffness before the development of cracks. This was necessary to prevent transverse movement of the BARELS group.

Figure 5 shows the much exagerrated displacements of BARELS at the end of Phase 2 of unpropped repair analysis. (These displacements are so small as to not affect the subsequent analyses.)

The actual BARELS properties used were determined by trial and error with several analyses. They had to be such that the results (for example slab stresses, crack strains, BARELS lateral displacements) were not that sensitive to small changes in their values.

6.2 Properties for Phase 3

The material properties used during the third phase, for both propped and unpropped conditions, are provided in Table 2. Although the actual analyses were carried out with different values for BARELS properties, they were not sensitive to the values used. (In fact, as mentioned earlier, Figure 10 has been plotted for a propped repair analysis that contained no BARELS elements in Phase 3.)

7 COMPARISON OF PHASE 3 RESULTS FROM UNPROPPED AND PROPPED REPAIR IMPLEMENTATIONS

7.1 Concrete cracking

Figure 6 presents a comparison of first crack normal strain vectors for the two repair implementations subsequent to the application of the additional 20% of original the load. As can be seen the region of cracking is greater in the unpropped repair implementation.

A detail of these crack strains vectors are presented in Figure 7. The greater level of cracking in the repair region of the unpropped repair can be seen there. (The vectors in the two figures have not been plotted to the same scale.) The maximum normal crack strains are 0.0125 and 0.0106, respectively, for the unpropped and propped repairs.

7.2 Transverse shear stresses

As the particular flat slab structure was susceptible to shear failure near the column head, the transverse shear stresses that result from different patch repair implementations were studied.

Figures 8 and 9 present contours of these stresses, flowing out-of-plane in the plots, at a certain 1*D perimeter. This perimeter (and others of 2*D and 3*D considered in the study) were defined as that at one (or 2 and 3) slab effective depth distance from the column perimeter. Although the column was square, the considered perimeter was rectangular. This was because of a long angle iron that existed in the slab-column connection (Figure 10). Only the original concrete are shown in the figures.

From Figure 8 it can be seen how the transverse shear stress Syz is concentrated near the repair excavation. Similarly, from Figure 9, it can be seen that the transverse shear stress Szx is higher under unpropped repair implementation. In the Figure, the concentration of stress contours is around the two reinforcement bars that occur at those locations.

7.3 Participation of repair material in resisting load

Figure 11 provides plan views of the repair material (only). Plotted in them are the first principal stress contours from the two repair implementations discussed.

As to be seen, the repair material from the unpropped repair implementation (left window) is stressed much less than those of the propped implementation (right window). This can be appreciated from the higher density of contours in the latter of these figures plotted to the same contour ranges.

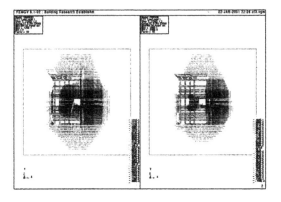

Figure 6.   Comparison of first crack normal strain vectors (in plan) at the end of Phase 3 from unpropped (left) and propped (right) repair analyses. Complete floor slab. As to be seen the cracking is more widespread for the unpropped repair case. Also see Figure 7.

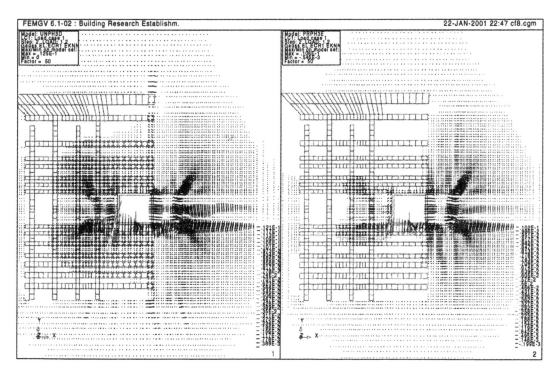

Figure 7.   Detail near repair region and column head. Comparison of first crack normal strain vectors (in plan) at the end of Phase 3 from unpropped (left) and propped (right) repair analyses.

409

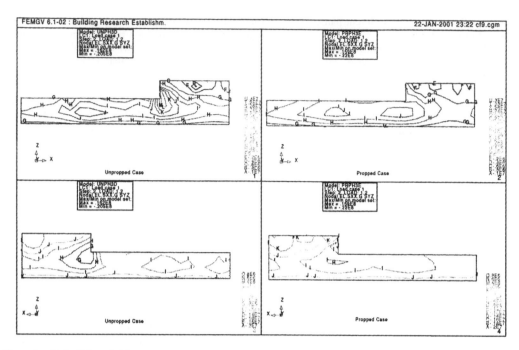

Figure 8. End of Phase 3. The transverse shear stress Syz (MPa) contours in the −Y (top windows) and +Y faces (bottom windows) at a defined 1*D perimeter block that contains the repair region and the column head. The unpropped repair case is in the left windows and the propped repair case is in the right windows.

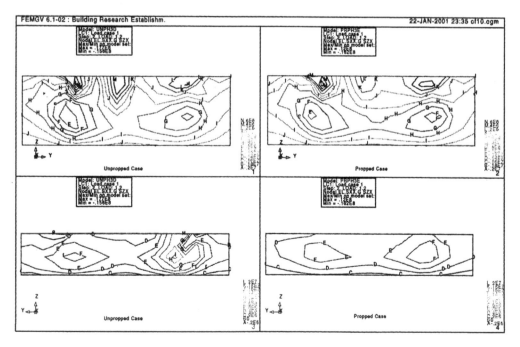

Figure 9. End of Phase 3. The transverse shear stress Szx (MPa) contours in the +X (top windows) and −X faces (bottom windows) of a defined 1*D perimeter block that contains the repair region and the column head. The unpropped repair case is in the left windows and the propped repair case is in the right windows.

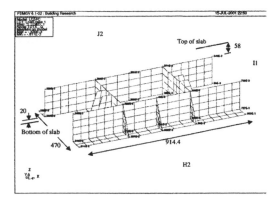

Figure 10. The steel collar that connected the floor slab to the column. The slab rested on the steel angles. Dimensions are in mm. Also to be seen are numerical values of some vertical deflections (in m) from an analysis.

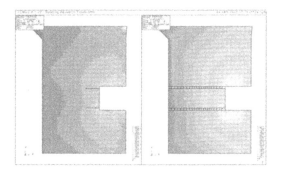

Figure 11. The first principal stress (MPa) on the surface of the repair concrete (at end of Phase 3). Unpropped repair at left. Propped repair at right.

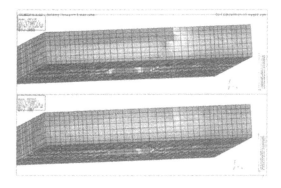

Figure 12. End of Phase 3. The transverse shear stress Syz in 1*D perimeter block. Unpropped repair at top, propped repair at bottom.

Figure 12 is a figure which shows the 1*D perimeter block of the repaired structure. The out-of-the-facing plane transverse shear stress (Syz) contours are plotted there. As can be seen from the figure in the top window for unpropped repair implementation, the shear stresses are nearly all carried by the original concrete. The repair material, seen on the top left corner of the block, is hardly stressed. In contrast, the stress contours in the figure in the lower window for propped repair implementation is nearly continuous. In this case, the repair material and the original concrete act together in resisting the loads.

8 DISCUSSION

This paper presents a study of the effects of propped and unpropped patch repair of a reinforced concrete flat slab. The during repair (not presented here) and post-repair stress states, including stress concentrations, and cracking were the issues of interest.

It was seen that the propping before repair excavation has resulted in a reduction of stresses and cracking within the original concrete. It also helped to make the repair material participate in later load carrying. However, an engineer needs to be careful as this stressing of the repair material (among other reasons) can cause cracking, leading to reinforcement corrosion. Reasons such as this has led to studies and disagreements related to the advantageous and disadvantageous of propping during structural patch repair of concrete (Canisius and Waleed, 2002a and 2002b). Thus, in the absence of general rules, it is prudent to conduct studies, such as that presented here, before excavation for patch repairs is undertaken. DIANA's phased analysis capability, combined with the extensive number of concrete constitutive models it has, is very useful in this.

This study was a part of a forensic investigation. As the report of this investigation has not yet been published, the author has refrained from commenting in detail on the issues involved.

9 CONCLUSIONS

- This paper presented a study of the effects of propped and unpropped patch repair of a reinforced concrete flat slab.
- The capabilities of DIANA in relation to phased analysis and reinforced concrete modelling were very useful in studying the repair process. This included the effects of different repair implementation.
- Embedded reinforcements need mother elements to contain them, and this is impossible if the concrete surrounding them is to be excavated during

repair. The paper presented a method of overcoming this problem.

- Behaviour of structures is complex both during and after structural patch repair. Its ultimate performance may depend on how the repair is carried out, for example whether propped or unpropped during excavation and if propped, to what level. Thus, consideration should be given to phased analysis of the repair process, as carried out here, for aiding repair engineers in their decision making.

ACKNOWLEDGEMENTS

The author wishes to acknowledge:

- The Health and Safety Executive of the UK for funding this work.
- Prof. Jonathan Wood for his support.
- Wilde & Partners for advice on phased analysis.

DISCLAIMER

The opinions expressed in this paper are of the author alone and do not in any way reflect or represent the position or opinion of the Health and Safety Executive.

REFERENCES

Canisius, T.D.G. and Waleed, N. Concrete Patch Repair: Is it time for a Risk-based strategy?. To be presented at FIB2002 Conference, Osaka, Japan, 2002.

Canisius T.D.G. and Waleed N. *The Behaviour of Concrete Repair Patches Under Propped and Unpropped Conditions*, FBE Report 3, CRC Press, March 2002.

Finite Elements in Civil Engineering Applications, Hendriks & Rots (eds.)
© 2002 Swets & Zeitlinger, Lisse, ISBN 90 5809 530 4

A 3D general-purpose dynamic analysis system for bridges considering pounding effects between girders – theory and implementation

P. Zhu
Research Institute of Science and Technology for Society, Japan Atomic Energy Research Institute, Tokyo, Japan

M. Abe & Y. Fujino
Department of Civil Engineering, The University of Tokyo, Tokyo, Japan

ABSTRACT: Bridges play an important role in transportation of modern societies. Under severe earthquakes, a bridge may lose its functionality even though the structure itself doesn't collapse. To precisely simulate seismic response of a bridge, 3D modeling of the total structure is needed. This paper introduces a general-purpose dynamic analysis program for bridges based on 3D modeling with highlighting modeling of pounding between girders. The paper presents theory and implementation of modeling for each component of bridges. A case study of a steel elevated bridge has also been conducted.

1 INTRODUCTION

During severe earthquakes, in addition to damage caused directly to bridge structures, a bridge may lose its functionality from the viewpoint of serviceability, even though the bridge itself has not collapsed. During the 1995 Kobe earthquake, the traffic service system in the emergency situation experienced substantial difficulties as considerable damage happened in elevated bridges. To precisely simulate seismic response of a bridge, 3D modeling of the total structure is needed. This paper introduces a general-purpose dynamic analysis program for bridges – DABS (Dynamic Analysis of Bridge Systems). DABS has been developed based on 3D modeling with highlighting modeling of pounding between girders.

The first part of this paper presents 3D modeling for each component of elevated bridges. Implementations of DABS are introduced in the second part. A case study with a three-span steel bridge is presented in the final part.

2 3D MODELING OF BRIDGES

Elevated bridges are generally composed of foundations, piers, abutments, girders/decks, bearing supports and expansion joints. To conduct precise analyses, detailed modeling for each of these components is given as follows. In addition, a 3D contact-friction model for pounding between girders/decks and models of restrainers and bumpers are given as well.

2.1 Fiber model for piers

The fiber model, known as a discretized-section model for nonlinear analysis (Li & Kubo 1998, Zhu 2001), is used to model piers. Figure 1 illustrates the concept of

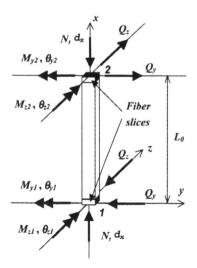

Figure 1. An illustration of the fiber model.

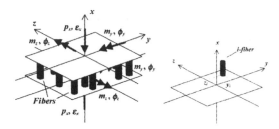

Figure 2. A fiber slice with forces and displacement in positive directions.

fiber model. Two fiber slices are picked from cross sections at two ends of a column. A column section is discretized into fibers with uniaxial nonlinear properties. All fibers in a slice are assumed to work together according to plane section assumption (Fig. 2).

For distributed nonlinearity, in cases of cantilever column or columns (or parts of a column) without inflexion point, linear distribution can be assumed. Upon assuming flexibility distribution along with the column, by calculating flexibilities of the two fiber slices, the end stiffness of the column is given as follows (Eqs 1–4).

$$\{F\} = [K]\{D\} \tag{1}$$

$$\{D\} = \left\{\theta_{y1}, \theta_{y2}, d_{z1}, d_{z2}, \theta_{z1}, \theta_{z2}, d_{y1}, d_{y2}, d_{x1}, d_{x2}\right\} \tag{2}$$

$$\{F\} = \left\{M_{y1}, M_{y2}, Q_{z1}, Q_{z2}, M_{z1}, M_{z2}, Q_{y1}, Q_{y2}, N_{x1}, N_{x2}\right\}^T \tag{3}$$

$$[K]_{10 \times 10} = [B]^T_{10 \times 5} [\delta]^{-1}_{5 \times 5} [B]_{5 \times 10} \tag{4}$$

A transformation matrix $[B]$ (Eq. 5) is employed to expend the local stiffness matrix $[\delta]^{-1}$:

$$[B]_{5 \times 10} = \begin{bmatrix} -1 & 0 & 1/L_0 & -1/L_0 & & & & & & \\ 0 & -1 & 1/L_0 & -1/L_0 & & & & & & \\ & & & & 1 & 0 & 1/L_0 & -1/L_0 & & \\ & & & & 0 & 1 & 1/L_0 & -1/L_0 & & \\ & & & & & & & & 1 & -1 \end{bmatrix} \tag{5}$$

The local flexibility matrix $[\delta]$ can be obtained by Equation 6:

$$[\delta]_{5 \times 5} = \frac{L_0}{12}\begin{bmatrix} 3f^1_{zz}+f^2_{zz} & -f^1_{zz}-f^2_{zz} & 3f^1_{yz}+f^2_{yz} & -f^1_{yz}-f^2_{yz} & -4f^1_z-2f^2_z \\ & f^1_{zz}+3f^2_{zz} & -f^1_{yz}-f^2_{yz} & f^1_{yz}+3f^2_{yz} & 2f^1_z+4f^2_z \\ & & 3f^1_{yy}+f^2_{yy} & -f^1_{yy}-f^2_{yy} & -4f^1_y-2f^2_y \\ & & & f^1_{yy}+3f^2_{yy} & 2f^1_y+4f^2_y \\ & Symm & & & 6f^1_x+6f^2_x \end{bmatrix} \tag{6}$$

Items in matrix $[\delta]$ are computed by stiffness items of fiber slices at each end (Eqs 7–10) (Fig. 2), where i and j represent the i-th fiber and slice number (1 or 2 in Fig. 1) respectively; and E_i, A_i and (y_i, z_i) are the area, Young's modulus and coordinate of the i-th fiber respectively:

$$[f^i] = \begin{bmatrix} f^j_{zz} & f^j_{yz} & f^j_z \\ & f^j_{yy} & f^j_y \\ Symm & & f^j_x \end{bmatrix} = [k^j_f]^{-1} \tag{7}$$

$$[k^j_f] = \begin{bmatrix} k^j_{zz} & k^j_{yz} & k^j_z \\ & k^j_{yy} & k^j_y \\ Symm & & k^j_x \end{bmatrix} \tag{8}$$

$$k^j_{zz} = \sum_i E^j_i A_i \cdot z_i^2, \quad k^j_{yz} = \sum_i E^j_i A_i \cdot y_i z_i,$$
$$k_z = \sum_i E^j_i A_i \cdot z_i \tag{9}$$
$$k^j_{yy} = \sum_i E^j_i A_i \cdot y_i^2, \quad k^j_y = \sum_i E^j_i A_i \cdot y_i, \quad k^j_x = \sum_i E^j_i A_i$$

$$E^j_i = E^j_i(\varepsilon^j_i) \tag{10}$$

2.2 Beam element for girders

A linear elastic straight beam element (Bathe 1996) is adopted for girders. Figure 3 shows a beam element located by three points (i, j, k) in 3D.

2.3 Foundations

Soil-structure interactions should also be taken into account. A soil-grouped pile model with simplifications has been adopted (Konagai 1999). As illustrated in Figure 4, the cap stiffness of the single equivalent beam is judged in frequency domain. The equation of motion in frequency domain can be presented in Equation 11, where $[S(\omega)]$, $\{u(\omega)\}$ and $\{P(\omega)\}$ are the stiffness, response and loads in frequency domain respectively. The stiffness in frequency domain is given in Equation 12 (Wolf 1985). After evaluating the cap stiffness of the equivalent beam in frequency domain, equivalent stiffness parameters, $[K]$, $[C]$ and

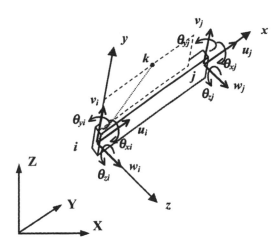

Figure 3. The coordinate system of a beam element.

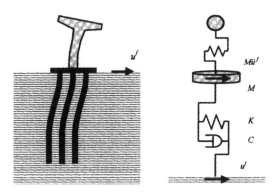

Figure 4. The equivalent model of foundation.

$[M]$, can be given and then can be used for modeling of foundations:

$$[S(\omega)]\{u(\omega)\} = \{P(\omega)\} \qquad (11)$$

$$[S(\omega)] = [K] + i\omega[C] - \omega^2[M] \qquad (12)$$

2.4 Bi-axial model of rubber bearings

A bi-axial model for rubber bearing by Yoshida et al (1999), which is capable at two horizontal directions, is used in the analysis (Fig. 5). The model is given by Equations 13a to 13g:

$$F_i = F_{i1} + F_{i2} \qquad (13a)$$

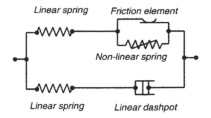

Figure 5. Modeling of rubber bearings.

$$\frac{\dot{F}_{i1}}{Y} = \frac{\dot{U}_i}{U_0} - \frac{\sqrt{\dot{U}_x^2 + \dot{U}_y^2}}{U_0}$$
$$\times \left| \frac{\sqrt{(F_{x1} - S_x)^2 + (F_{y1} - S_y)^2}}{Y} \right|^{n-1} \frac{F_{i1} - S_i}{Y} \qquad (13b)$$

$$F_{i2} = \eta \dot{U}_i^{dashpot} = k U_i^{spring} \qquad (13c)$$

$$U_i^{dashpot} + U_i^{spring} = U_i \qquad (13d)$$

$$\frac{S_i}{Y} = \left(\frac{U_i}{U_0} - \frac{F_i}{Y} \right) \left(\alpha_0 - \beta \left| \frac{U_{max}}{U_0} \right|^q \right) \qquad (13e)$$

$$Y = Y_0 \left\{ 1 + \gamma \left(\frac{\sqrt{U_x^2 + U_y^2}}{U_0} \right)^p \right\} \qquad (13f)$$

$$U_{max} = \left. \sqrt{U_x^2 + U_y^2} \right|_{max\_past} \qquad (13g)$$

where F_i, U_i, S_i are force, displacement and back force at i direction; Y, U_0 are yield load and displacement; α, β, γ, n, p, q, η are parameters; and $i = x, y$.

2.5 Modeling of pounding between girders/decks

A 3D contact-friction model developed by Zhu et al (2001) is illustrated in Figure 6. This model is based on point-surface contact with penetration. The target surface, named as $abcd$, is assumed as a rigid plane. Node k is the contactor node at the contactor body, which penetrates into the target surface during contact. Point p is the physical contact point at the target surface $abcd$.

Upon contact, a universal spring K_{cnt} between node k and point p is created to compute the force of contact. Two dashpots, C and C_t, are also applied to node k for simulating energy loss during contact. The contact force at node k, $\mathbf{F}_k$, can be computed as $\mathbf{F}_k = K_{cnt} \cdot \Delta_k$ and be

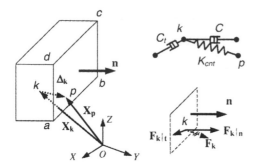

Figure 6.   Illustration of the 3D contact-friction model.

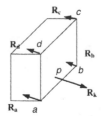

Figure 7.   Contact forces at contact node and target surface.

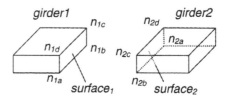

Figure 8.   An illustration of pounding modeling between two girders.

divided into normal and tangent components ($\mathbf{F}_k|_n$ and $\mathbf{F}_k|_t$ respectively), where vector $\mathbf{n}$ is the outer normal vector of the target surface and vector $\mathbf{t}$ is a projection vector of $\mathbf{F}_k$ to the target surface. During contact, status can be divided into stick contact and slide contact which can be decided by the ratio of tangent component of the contact force $|\mathbf{F}_k|_t|$ to the normal one $|\mathbf{F}_k|_n|$.

Contact status can be divided into stick and slide contact, which can be decided into stick and slide according to Equations 14a and 14b respectively:

$$\left|\mathbf{F}_k|_t\right| < \mu_s \left|\mathbf{F}_k|_n\right| \tag{14a}$$

$$\left|\mathbf{F}_k|_t\right| \geq \mu_s \left|\mathbf{F}_k|_n\right| \tag{14b}$$

where $\mathbf{F}_k|_n$, $\mathbf{F}_k|_t$ are normal and tangent components of $\mathbf{F}_k$ to the target surface respectively; and μ_s is static friction coefficient.

The contact force at the contactor node k can be calculated separately for stick and slide conditions, as given in Equations 15a and 15b respectively:

$$\mathbf{R}_k = \mathbf{F}_k + \mathbf{F}_c|_n + \mathbf{F}_c|_t \tag{15a}$$

$$\mathbf{R}_k = \mathbf{F}_k|_n + \mathbf{F}_c|_n + \mathbf{F}_f|_t \tag{15b}$$

Components of damping forces at the normal and tangent directions, $\mathbf{F}_c|_n$ and $\mathbf{F}_c|_t$, are given by Equations 15c and 15d respectively. Kinetic friction $\mathbf{F}_f|_t$ is given by Equation 15e:

$$\mathbf{F}_c\big|_n = -C \cdot \mathbf{V}_{kp}\big|_n \tag{15c}$$

$$\mathbf{F}_c\big|_t = -C_t \cdot \mathbf{V}_{kp}\big|_t \tag{15d}$$

$$\mathbf{F}_f\big|_t = -\mu_k \cdot \left|\mathbf{F}_k|_n\right| \frac{\mathbf{V}_{kp}\big|_t}{\left|\mathbf{V}_{kp}\big|_t\right|} \tag{15e}$$

where $\mathbf{V}_{kp}$ is the relative velocity of node k to point p; and μ_k is kinetic friction coefficient.

Contact forces at the target surface can be obtained by linear interpolation according to static equilibrium (Fig. 7). The total contact forces for a contact-pair are given in Equation 16:

$$\mathbf{R}_{cnt}\big|_{cnt\_pair} = \left\{ \begin{matrix} \mathbf{R}_k \\ \mathbf{R}_{target\_surface} \end{matrix} \right\} = \left\{ \begin{matrix} \mathbf{R}_k \\ \mathbf{R}_a \\ \mathbf{R}_b \\ \mathbf{R}_c \\ \mathbf{R}_d \end{matrix} \right\} \tag{16}$$

To simulate arbitrary pounding between two girders, contactor nodes and target surfaces are designated in both girders. As shown in Figure 8, contact pairs for girder1 to girder2, which means nodes at girder1 contact with target face on girder2, can be defined as (n_{1a}, surface2), (n_{1b}, surface2) etc. The same way is for contact pairs of girder2 to girder1, which as (n_{2a}, surface1), (n_{2b}, surface1) etc. This is for the simplest case of contact between two girders. In fact, a node may be involved into more than one contact pairs (for instance, during to refined modeling of girder's ends), which means the node may contact with more than one surfaces, in most cases, not simultaneously, according to real situations.

2.6 Restrainers and bumpers

Restrainers and bumpers are modeled in bilinear. As shown in Figure 9, a restrainer works in tension side as a linear spring with stiffness k_{0r} after an initial clearance

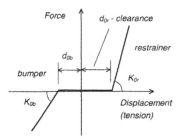

Figure 9. Modeling of restrainers and bumpers.

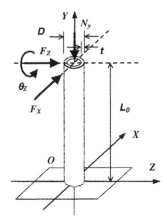

Figure 10. A test of the fiber model with a steel pier.

d_{0r}. Similarly, a bumper works in compression side with stiffness k_{0b} after an initial clearance d_{0b}.

2.7 Earthquake excitations

As elevated bridges are usually longitudinally lengthy structures, when a seismic wave travels, supports along a bridge may receive different excitations. The seismic wave propagation effect can be the dominant factor leading to poundings between bridge girders/ decks (Jankowski et al 1998). Therefore, a separate excitation for each foundation of an elevated bridge is considered. Accordingly, an absolute coordinate system is adopted as the coordinate system for analysis and seismic excitation.

2.8 Governing equation

The governing equation of motion is given in Equation 17. To simulate pounding between girders, a vector of contact forces, $\mathbf{R}_{cnt}$, is added into the equation:

$$
\begin{aligned}
\mathbf{M}_{ii}\ddot{\mathbf{u}}_i &+ \mathbf{C}_{ii}\dot{\mathbf{u}}_i + \mathbf{K}_{ii}\mathbf{u}_i \\
&= \mathbf{R}_i - \mathbf{M}_{ib}\ddot{\mathbf{u}}_b - \mathbf{C}_{ib}\dot{\mathbf{u}}_b - \mathbf{K}_{ib}\mathbf{u}_b + \mathbf{R}_{cnt}
\end{aligned}
\tag{17}
$$

where i and b represent inner and boundary nodes respectively.

3 IMPLEMENTATION OF ANALYSIS PROGRAM – DABS

A general-purpose dynamic analysis program for bridges, DABS (Dynamic Analysis of Bridge Systems), has been developed (Zhu 2001). DABS implements 3D models of bridge structures presented in this paper. Written in C++, DABS takes advantages of object-oriented programming to realize numerical models for bridge structures. A free-formatted text file input interface has been designed to model bridge structures and to give computing and output conditions.

Crosschecks and tests of non-linear models of DABS are presented as follows.

Table 1. Parameters of the steel pier.

Dimension	$D = 3\,\text{m}$, $t = 0.038\,\text{m}$, circular ring, $L_0 = 18.5\,\text{m}$
Material	$E_0 = 2.06 \times 10^5\,\text{N/mm}^2$, $\sigma_y = 235.2\,\text{N/mm}^2$, $\nu = 0.3$
Hysteresis model	bilinear model: $\beta = E_y/E_0 = 0.01$
Fiber	Number of fibers: 200

3.1 Crosschecks

To ensure the correctness of DABS implementation, crosschecks have been conducted with a trustable program, NEABS, which was initialed by Tseng and Penzien (1973). The Crosschecks include elastic time-history analysis for a single-pile system and computations of eigenvalues for a three-span steel bridge. Coincide results have been observed from these computations (Zhu 2001).

3.2 Test of the fiber model

A steel pier with circular ring section is chosen for the test, as shown in Figure 10. Parameters of the steel pier and features of the steel are given in Table 1.

The maximum capacity of this steel pier, by assuming full yielding of the section, is 7.84×10^4 kNm. Very good agreement can be seen from Figure 11. Results of two directions force driving are given in Figure 12, which show the bi-direction capacity of the fiber model.

3.3 Test of the foundation model

A structure of single pier with lumped mass at the top is chosen for the test (Fig. 13, Table 2). Takatori EW wave from the 1995 Kobe earthquake is used for earthquake

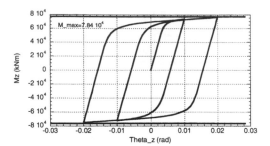

Figure 11.   Displacement driving.

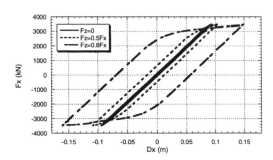

Figure 12.   Bi-direction force driving.

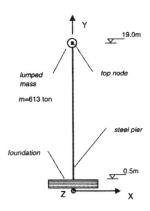

Figure 13.   Single pier model for foundation test.

Table 2.   Parameters.

Steel pier	As shown in table 1
Foundation	$M_x = 107.8$ ton, $I_z = 3371.2$ ton $\cdot$ m$^2 \cdot$ s^2/rad, $K_x = 4.8902 \times 10^5$ kN/m, $K_{zz} = 5.7624 \times 10^5$ kN $\cdot$ m/rad, $C_x = 3.675 \times 10^4$ kN s/m, $C_{zz} = 1.1172 \times 10^6$ kN $\cdot$ m $\cdot$ s/rad.

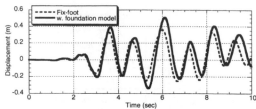

(a) Displacement comparison

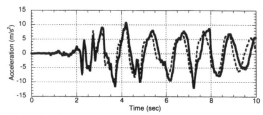

(b) Acceleration comparison

Figure 14.   Test of foundation model – response at the top node.

Table 3.   Parameters of the rubber bearing.

Capacity	5000 kN
Size	970×970 mm
Area (A)	9.409×10^5 mm^2
Thickness of a rubber bearing layer	26 mm
Number of rubber bearing layers	7
Total thickness of rubber layers (t_r)	182 mm
Parameters	$Y_0 = 0.0368$[kN/cm^2] $\times A$, $U_0 = 0.0386$ [strain] $\times t_r$, $\alpha_0 = 0.44, \beta = 0.39$, $\gamma = 8.00 \times 10^{-10}$, $n = 0.201, p = 4.91, q = 0.0168$, $k = 0.0201$ [(kN/cm^2)/strain] $\times A/t_r$, $\eta = 16.2$ [(kN/cm^2)/strain] $\times A/t_r$.

excitation in X direction. Computation conditions include cases of fix-foot and with foundation model.

Results are given in Figure 14. Enlargements of displacement response at the top of the pier, as effects of the foundation, can be observed.

3.4   Test of the bi-axial model for rubber bearing

A rubber bearing supporter, parameters given in Table 3, is chosen for the computation. Bi-direction displacement driving is adopted. Results are shown in Figure 15. The bi-axial capacity of this model can be seen from the results.

3.5 Test of the model of pounding

To illustrate the pounding model, a numerical test of adjacent rods with point masses on free vibration was selected (Fig. 16). In the test an initial displacement, which is given to point mass m_1, starts a free vibration on this two point mass system where energy is transferred in between as pure elastic impacts (Fig. 16(a)). A theoretical solution of the problem is shown in Figure 16(b). Parameters of the model were chosen as Figure 16(a) for simulation. Two time intervals were chosen for time-history analysis. Results in Figure 17 show that an accurate solution of displacement response can be obtained by choosing a smaller time interval for contact.

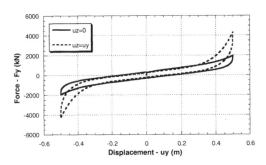

Figure 15. Test of rubber bearing – bi-direction displacement driving.

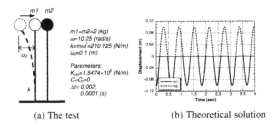

(a) The test   (b) Theoretical solution

Figure 16. A numerical test – two point masses on free vibration.

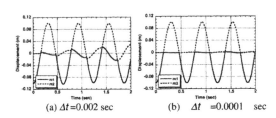

(a) Δt =0.002 sec   (b) Δt =0.0001 sec

Figure 17. Analytical results of the test.

Experimental verifications in more complicated 2D cases have also been conducted and good results have been obtained (Zhu et al 2001).

4 A CASE STUDY WITH A STEEL ELEVATED BRIDGE

A typical three-span steel bridge has been selected for a case study. As shown in Figure 18, fiber model is adopted at the first segment of each pier from foundation. Base-isolation rubber bearings are applied for each pier. For computation of pounding, a simple supported girder in each span is assumed. Restrainers are adopted as a countermeasure for pounding effects.

4.1 Computing conditions and parameters

Computations were conducted in three cases: (1) without pounding,)2) considering pounding and (3) considering pounding and applying restrainers. All parameters and the bridge structure model are the same for these three cases, except that in third case, restrainers were adopted. In the first case of without pounding, no material contacts were assumed, though girders may overlap each other in a real case. Takatori waves from the 1995 Kobe earthquake were used as earthquake excitations in three-dimensional. Parameters for time-history analysis are as follows:

Time interval = 0.001 sec, duration = 10 sec.

4.2 Results and analyses

One node, node A, was picked from the center of girders of the mid span. Displacements for all the cases and accelerations for case2 and case3 were selected to output in longitudinal, transversal and rotating directions. Rotating direction is in a plane parallel to the ground.

Results of computations are given in Figure 19 and Figure 20, which show comparisons of displacements and results of accelerations at node A. Responses of

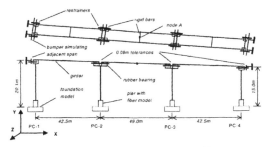

Figure 18. Modeling of a three-span steel bridge.

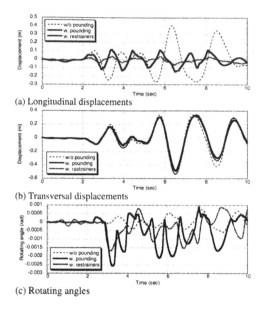

(a) Longitudinal displacements

(b) Transversal displacements

(c) Rotating angles

Figure 19. Displacement comparisons of the mid span at node A.

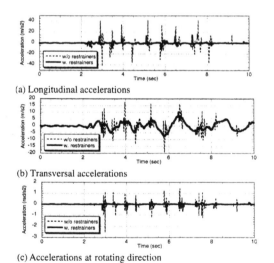

(a) Longitudinal accelerations

(b) Transversal accelerations

(c) Accelerations at rotating direction

Figure 20. Accelerations at mid span at node A for the pounding case.

pounding at the middle span and side span show same trends. Longitudinal displacements are reduced according to poundings, while displacements at transversal do not have much change, as there is no face-to-face pounding assumed in this direction. A remarkable increase of rotating angle of the girder can be seen as a result of pounding. Results of accelerations show that

the structure experiences strong reaction forces during pounding. Comparisons of displacement results for case2 and case3 show that the application of restrainers can reduce the pounding effecting dramatically, not only in the longitudinal direction, but also in the rotational direction. As a view of acceleration results, which refer to the forces induced by pounding, restrainers can also work effectively to reduce pounding forces.

5 CONCLUSIONS

This paper presented a practical way to establish precise 3D modeling for elevated bridges including modeling of pounding between girders in 3D. Upon implementing a general-purpose dynamic analysis program, DABS, tests of non-linear components of bridges were conducted to show their seismic features according to the models presented in this paper. A case study of a chosen three-span steel elevated bridge was conducted with discussions of pounding effects and pounding mitigation.

REFERENCES

Bathe. K.J. 1996. *Finite Element Procedures.* New Jersey: Prentice-Hall, Inc.
Li, K.N. & Kubo, T. 1998. Reviewing the multi-spring model and fiber model. *The 10th Symposium on Earthquake Engineering of Japan:* 2369–2374.
Jankowski. R., Wilde, K. & Fujino, Y. 1998. Pounding of superstructure segments in elevated bridges during earthquakes. *Earthquake Engineering and Structural Dynamics* 27: 487–502.
Konagai. K. 1999. Shaking table test allowing interpretation of damage to structure in terms of energy influx and efflux through soil-structure interface. *Report of Research Project 1999 Grant-in-aid for Scientific Research (B) (No. 10450174).* The Ministry of Education, Science, Sports and Culture, Japan.
Tseng. W.S. & Penzien, J. 1973. Analytical Investigations of the Seismic Response of Long Multiple Span Highway Bridges. *EERC 73–12,* Earthquake Engineering Research Center. University of California, Berkley.
Wolf, J.P. 1985. *Dynamic Soil-Structure Interaction.* New Jersey: Prentice-Hall, Inc.
Yoshida. J., Takesada, S., Abe, M. & Fujino, Y. 1999. A bi-axial restoring force model on rubber bearings considering two-direction horizontal excitations (in Japanese). *Proceedings of the 25th Conference of Research on Earthquake Engineering* 2: 741–744.
Zhu, P., Abe, M. & Fujino, Y. 2001. A 3D Contact-friction Model for Pounding at Bridges during Earthquakes. *Proceedings of The First M.I.T. Conference on Computational Fluid and Solid Mechanics:* 575–578.
Zhu, P. 2001. Seismic Analysis and Serviceability Evaluation of Elevated Bridges on 3D Modeling with Pounding Effects of Girders. *Ph. D. Dissertation,* The University of Tokyo.

Finite Elements in Civil Engineering Applications, Hendriks & Rots (eds.)
© *2002 Swets & Zeitlinger, Lisse, ISBN 90 5809 530 4*

3D finite element analysis of multi-beam box girder bridges – assessment of cross-sectional forces in joints

C.M. Frissen, M.A.N. Hendriks
TNO Building and Construction Research, Delft, The Netherlands

N. Kaptijn
Ministry of Public Works and Water Management, The Netherlands

ABSTRACT: Three finite element models have been developed to analyze the cross-sectional forces in the joints of multi-beam box girder bridges. A model based on shell elements is used in nonlinear analyses including two analysis phases. The analysis phases simulate the construction phases. For straight bridges current design methods based on orthotropic plate models give reasonable results compared to the results of the advanced model. For bridges with a skew or curved geometry plan orthotropic plate models give poor predictions for the forces and moments in the joints. Advanced modeling, as presented in this paper, is then to be preferred.

1 INTRODUCTION

In this introductory section we will present a design question related to multi-beam box girder bridges. This question has lead to an intensive numerical study of a multi-beam box girder bridge. We will start with a general description.

1.1 *Multi-beam box girder bridges*

As indicated by the name, multi-beam box girder bridges consist of several box shaped girders. The topsides of the girders compose the car deck. Figure 1

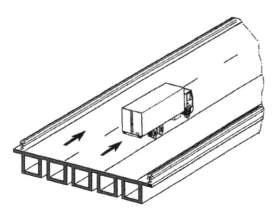

Figure 1. An illustrative sketch of a multi-beam box girder bridge.

is an illustrative sketch of a multi-beam box girder bridge. Prefabricated girders are frequently used for bridges crossing highways where you do not want not disturb the traffic during the construction phase. In the Netherlands multi-beam box girder bridges are not commonly applied. Because of the costs normally inverted prestressed T-beams are used. Multi-beam box girder bridges are regularly applied for spans up to 40 meters.

Concrete multi-beam box girder bridges are always prestressed in longitudinal direction. The curved prestressed tendons or strands in the girder webs of the boxes result in an upward load, in the opposite direction of the gravity load. The greater the curvature of the tendons, the greater this effect is. The maximal curvature of the tendons depends on the ratio between the height of the boxes and the length of the span.

1.2 *Current design practice*

The following loads have to be taken into account:

– the dead weight of the girders and the joints between the girders,
– the prestress of the tendons (or strands),
– distributed loads, of asphalt, rails, mobile loads,
– concentrated wheel loads.

The wheel loads should be located on the most unfavorable position.

The force transmission of the loads through a multi-beam box girder bridge is a complex issue. The complexity increases even more when, instead of a straight plan arrangement, bridges of curved or skew plan geometry have to be analyzed. A popular, finite element based, methodology makes use of 2D plate bending models. For multi-beam box girder bridges the overall bending stiffness in longitudinal direction is substantial higher than the bending stiffness in transverse direction. In transverse direction the relatively soft joints between the box girders dominate the bending stiffness. In 2D plate bending models this is modeled by applying plate models with geometrically orthotropic properties, where moment of inertia I_{xx} in longitudinal direction differs from the moment of inertia I_{yy} in transverse direction. The determination of I_{xx} and I_{yy} follows Timoshenko's theory for plates and shells (Timoshenko, 1970) and reflects the cross-sectional shape of the box girder bridge. Linear static analyses are adopted where reduced stiffness moduli account for cracking of the concrete.

1.3 Design considerations

The design considerations that form the basis of the numerical study as presented in this paper focuses on the cross-sectional forces in the joints between the girders. The girders are prefabricated boxes; the joints are casted after placement of the girders. These two construction phases cause the joints to be initially stress free, apart from the influence of the dead weight on the joints. After completing the bridge with asphalt and after introducing additional loads such as mobile loads and possible (wind) loads resulting from baffle boards, this situation has been changed. The design of the reinforcements in the joints requires a good insight in the (shear) forces and bending moments in the joints. Furthermore, the question could be raised whether it is necessary to apply post-tensioned tendons in the cross direction. It is believed that the current design procedure does not give sufficient insight in the forces and bending moments in the joints. For this, a more advanced analysis is in its place. More generally, such an advanced analysis could serve as an assessment of the design method based on orthotropic plate models.

1.4 Approach

In this study we have selected a relatively small bridge which we will describe in detail in section 2. The economic relevance of this case study is motivated by the popularity of such bridges. As a variation study a wind load on baffle board was investigated. As a geometric variation study a skew bridge was investigated. In each case the forces and moments in the joints are the primary concern.

An advanced analysis should include the construction phases as these have a crucial effect on the stresses in the joints. Also, such an analysis should include nonlinear effects, such as cracking of the concrete and plasticity of the reinforcements. Three models will be used: a full 3D model with solid brick elements, a degenerated full 3D model based on shell elements and a 2D plate bending model. The full 3D model is, from geometrical viewpoint, the most accurate model. However, for the nonlinear analyses the 3D solid model is not considered as a feasible option. For the nonlinear analyses the 3D shell model will be adopted. The 2D plate bending model will serve as a reference and represents the standard design practice.

Section 2 gives a more detailed description of the selected case. Section 3 and 4 then subsequently present linear and nonlinear analyses. The linear analyses are mainly meant to justify the three models. Section 5 concludes this paper by reverting to the design question.

2 CASE STUDY: BOX GIRDER BRIDGE HEULTSEDREEF

As a case study we selected viaduct "KW 14 Heultsedreef", a typical multi-beam box girder bridge, located near Boxtel in the south of the Netherlands, crossing the 2 × 2 lanes highway "A2". We will subsequently describe its geometry, the material properties and the design loads.

2.1 Geometry

The structure consists of five prestressed and prefabricated box girders. The bridge comprises two fields, each spanning two lanes. Because of symmetry we will consider one half of the structure with a span of 27,225 mm. Each box is 1180 mm wide and 900 mm high. The center to center distance between the boxes is 1320 mm. The top slabs are designed such that a strip of 230 mm wide and 160 mm thick will be casted on the site. We will refer to this strip as the "joint". The width of the deck is 6960 mm. Each girder is placed on four rubber supports, located 250 mm from the girder ends. The general configuration is illustrated in Figure 2. This figure also indicates the horizontal supports, as how these will be schematized in the analyses.

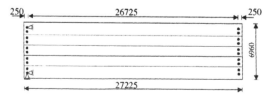

Figure 2. Plan of viaduct "Heultsedreef".

The shape of the prestressed strands is illustrated in Figure 3. The figure shows the strands in one of the ten webs. Each web includes 20 bended strands. Note that the course of the strands is such that they come down 440 mm from the girder end to the middle of the girder. Each of the bottom slabs includes 24 prestressed strands, which are straight.

2.2 *Materials*

The Tables 1 and 2 summarize the material properties of the reinforced concrete and the properties of the strands. The concrete quality of the joints is lower than the quality of the prefab girders. The rubber support blocks will be modeled as springs with a stiffness 227,000 N/mm for the supports in the left hand side and a stiffness of 289,000 N/mm for the supports on the right hand side.

2.3 *Design loads*

The design loads comprise (i) the dead weight of girders and joints, (ii) the prestress in the strands, (iii) distributed loads from the asphalt layer, rails, etc., (iv) a mobile load, as an equally distributed load (v) and wheel load. Figure 4 illustrates the distributed loads. Figure 5 illustrates the wheel load. The wheel load is

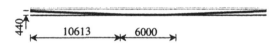

Figure 3.   Shape of the prestressed strands.

Table 1.   Material properties of concrete and reinforcements.

Box girders	Concrete class B55
Young's modulus	36,000 N/mm^2
Poisson ratio	0.2
Density (rebars incl.)	25 kN/m^3
Joints	**Concrete class B35**
Young's modulus	31,833 N/mm^2
Poisson ratio	0.2
Density (rebars incl.)	25 kN/m^3
Reinforcements	**FeB 500 HWL steel**
Young's modulus	210,000 N/mm^2

Table 2.   Properties of strands.

Strands	FeP 1860 steel
Young's modulus	200,000 N/mm^2
Area per strand	100 mm^2
Prestress (applied)	140.9 kN
Prestress (actual)	125 kN

composed of three axle loads. Each axle load is distributed over four wheels, resulting in 3×4 wheel prints. The critical position of the wheel load, in transverse as well as in longitudinal direction will be determined with an influence field analysis. This is described in section 3.

3   LINEAR ANALYSIS

In the next three subsections we will subsequently present the element meshes for the full 3D model, the shell model and the orthotropic plate model. Section 3.4 presents a comparison of the major results. Section 3.5 presents an influence field analysis to assess the most critical position of the wheel loads. For this analysis the shell model was adopted.

3.1 *Full 3D model*

The 3D model could be termed as a straightforward model, as it required the fewest geometrically simplifying assumptions. The Figures 6–8 illustrate the mesh that comprises five girders. The mesh consists of 20-node brick elements, embedded reinforcement bars for the strands and embedded reinforcement grids for the reinforcements.

3.2 *Shell model*

The shell model consists of 8-node quadrilateral elements and 6-node triangular elements. These elements could be envisaged as degenerated solid elements, assuming the shell hypotheses of straight normals and zero normal stress. Compared to the full 3D model the main point of concern is the geometric treatment of the corners in the model. As a guiding principle for the geometric discretization the bulk concrete volume bulk is preserved. Figure 9 shows a cross-section of the shell model. The embedding of strands and reinforcements is analogous to that of the full 3D model.

3.3 *Orthotropic plate model*

The orthotropic plate model consists of 8 node quadrilateral plate elements. The geometrically orthotropic properties of the 2D plate elements reflect the actual

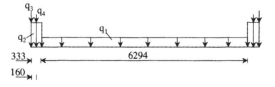

Figure 4.   Distributed loads on the top slabs, cross-sectional view.

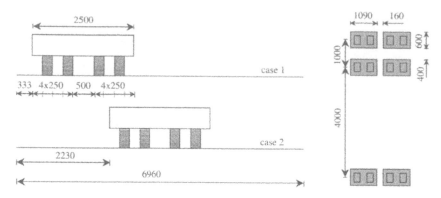

Figure 5. Wheel loads on the top slabs, cross-sectional view and top view.

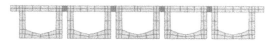

Figure 6. Full 3D model – cross section of the model.

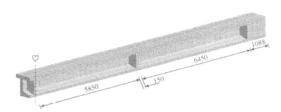

Figure 7. Full 3D model – mesh of halved girder.

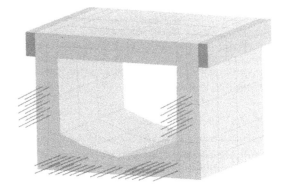

Figure 8. Full 3D model – locations of the strands.

geometry of the bridge, see e.g. Timoshenko, 1970. Two geometry groups are distinguished: one for the two edge girders and one for the three girders in between, see Figure 10. Table 3 presents the moment of inertia I_{xx} in longitudinal direction, the moment of inertia I_{yy} in transverse direction and the moments of

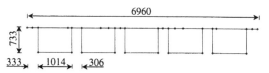

Figure 9. Shell model – cross-section of the model.

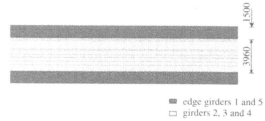

■ edge girders 1 and 5
□ girders 2, 3 and 4

Figure 10. 2D orthotropic plate model – moments of inertia.

Table 3. Geometrically orthotropic properties of the plate elements. Between parentheses: the properties for the two edge girders.

Property	[10^6 mm^4/mm]
I_{xx}	45.69 (42.80)
I_{yy}	00.76 (00.76)
$I_{xy} = I_{yx}$	35.06 (35.06)

inertia I_{xy} and I_{yx}. To estimate the moments $M_{xy,girder}$ in the girders and the moments $M_{xy,joints}$ in the joints the calculated moments M_{xy} resulting from the analysis should be multiplied with respectively 2.0 and 0.02 (Frissen, 2000).

3.4 Results and justification

The force transmission of the wheel loads could easily be demonstrated by plotting the reaction forces in the elastic supports. On each edge the bridge is supported

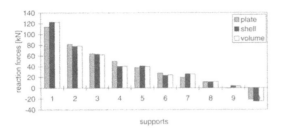

Figure 11. Reaction forces in the elastic supports – wheel load.

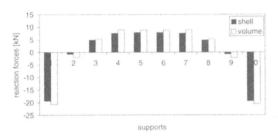

Figure 12. Reaction forces in the elastic supports – pre stress load.

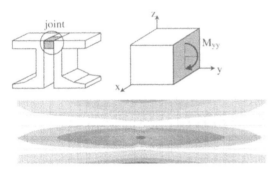

Figure 13. Influence field for cross moment M_{yy} in the joint between edge girder 2 and girder 3 in the middle of the span.

by ten elastic supports, two per girder. Figure 11 shows the reaction forces on one edge of the bridge as resulting from wheel load "case 1" in the middle of the span (cf. Fig. 5). Compared to the reaction forces as obtained with the volume model, the shell model gives a difference of about 2% and the plate model gives a difference of about 10%. Figure 12 shows a similar graph but then for a prestress load applied in the strands. Note that this load could not be analyzed with the plate model, as this model does not explicitly model the strands.

3.5 Influence fields

The shell model was adopted to calculate influence fields. Figure 13 shows the influence field of a wheel load. The colors indicate the cross moment M_{yy} in the joint between edge girder 2 and girder 3 at the middle of the span. Note that we selected a point in a joint as that is the focus of the present paper. Based on various influence field analyses for several points and for both bending moments and shear forces, wheel load "case 2", cf. Figure 5, was selected as the normative wheel load.

4 NONLINEAR ANALYSIS

The linear analyses revealed that the differences between the results of the three models are small. For the assessment of the cross-sectional forces in the joint we will proceed with nonlinear analyses, all performed with the shell model. In this section we will specify the loading history and analysis phases, the material modeling and results. The nonlinear results are compared with results obtained from the orthotropic plate model.

4.1 Analysis phases, loading and load history

The analysis comprises two phases. In the first phase the model consists of five girders, which are not yet connected with joints. The loads are the gravity and the prestress in the strands.

In the second analysis phase the joints are added to the model. The reasoning for distinguishing two phases is that the concrete in the joint is added free from strains, stresses and possible cracks. This also holds true for the reinforcements in the joints. As opposed to the joints, the initial state of the girders in the second phase equals the resulting state of the first analysis phase. The loads from the first phase are maintained in the second analysis phase. Next the design loads are applied including wheel load "case 2", which was the outcome of the influence field analysis.

A bridge modeled in such a way might be modeled too stiff. In reality, cracking due to the ongoing traffic loads might influence the transfer of loads. To simulate this in the analysis the bridge will be pre-damaged by applying and subsequently removing wheel loads at ten locations. Figure 14 illustrates the initial loading and unloading of wheel load "case 2", the subsequent loading and unloading of the 10 wheel loads and finally the loading of wheel load "case 2".

4.2 Material modeling

For the concrete Diana's multiple fixed crack model is adopted, with linear tension softening. Full tension stiffening is assumed to model the composite behavior of reinforced concrete. The compressive behavior is modeled with a Von Mises plasticity model, which is acceptable for membrane stress situations.

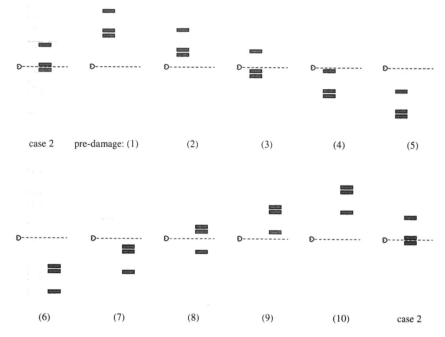

case 2   pre-damage: (1)   (2)   (3)   (4)   (5)

(6)   (7)   (8)   (9)   (10)   case 2

Figure 14.   Load history of the 12 wheel loads.

The reinforcements are modeled with ideally elasto-plasticity. For the strands hardening elasto-plasticity is adopted.

4.3   *Results for construction phase 1 and 2*

In this section some exemplary results for the ultimate limit state analysis are presented. The nonlinear analysis comprises two analysis phases. Figure 15 gives the bending of the topsides of the girders due to dead weight and prestress for analysis phase 1. Note that the elements representing the joints are not active in this analysis phase. In the middle of the span the displacement is 38 mm in upward direction. Figure 16 shows the crack pattern for construction phase 2. Longitudinal cracks arise in the joints, whereas cracks in transverse direction could be observed in the bottom sides of the girders. Figure 17 further illustrates the cross-sectional deformations at the middle of the span.

A typical comparison between the nonlinear shell model and the plate model is given in Figure 18. This graph presents the bending moment M_{yy}, in transverse direction, in joint 2 along the span. Bending moment M_{yy} is one of the many forces and moments that were compared in this study, but was considered as one of the most crucial ones for the joints. The lines distinguish between:

- shell models versus orthotropic plate models,

Figure 15.   Phase 1 – the placement of the girders: bending of the topsides of the girders due to dead weight and prestress.

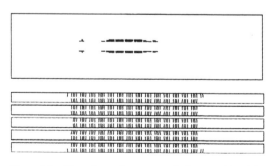

Figure 16.   Phase 2: crack pattern at the bottom side of the joints (top panel) and at the bottom side of the girders (bottom panel).

- reduced stiffness (red.), to account for cracking, and non-reduced stiffness (for the plate model),
- with (dam.) and without pre-damaging (for the shell model).

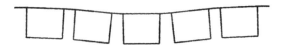

Figure 17. Phase 2: vertical displacements at the middle of the span.

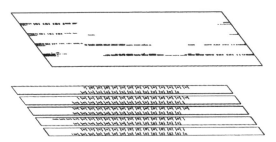

Figure 18. Phase 2: bending moment M_{yy} in joint 2.

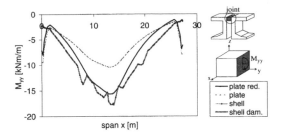

Figure 19. Skew viaduct, phase 2: crack pattern at the bottom side of the joints (top panel) and at the bottom side of the girders (bottom panel).

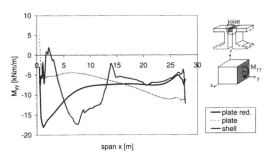

Figure 20. Phase 2: bending moment M_{yy} in joint 2 for the skew viaduct.

The lines show that the effect of pre-damaging the shell model could be neglected. Further, reducing the stiffness in the plate elements yields reasonable results compared to the results of the shell model. The reduced plate model underestimates M_{yy} up to 7%. Non-reduced plate models underestimate M_{yy} up to 38%. Variation studies with skew viaducts see Figure 19, revealed that also the reduced plate models fail to accurately predict M_{yy} in the joints. Figure 20 shows the same results as Figure 18 but now for the skew viaduct. Although, in absolute sense, the maximum value of M_{yy} for the plate model with reduced stiffness is comparable with the maximum value for the shell model, we observe that the course is completely different.

5 CONCLUSION

It is feasible to perform a nonlinear analysis of box girder bridges taking into account the two main construction phases. Based on the current variation studies, including the modeling of a skew viaduct, effect of the position of the wheel loads and the effect of wind loads on baffle boards, we concluded that nonlinear analyses have an added value above the orthotropic plate models. This holds especially true for the skew viaduct and will also hold true for bridges with a curved planar geometry. In this case plate models give bad predictions for the bending moments and forces in the joints. Nonlinear analysis is then to be preferred.

For the "Heultsedreef" bridge relative high bending moments M_{yy} in the joints occur but it was not necessary to apply post-tension strands in transverse direction.

ACKNOWLEDGEMENT

Apart from the authors, A. de Boer and H. Nosewicz, both from the Dutch Ministry of Public Works and Water Management, complemented the project team. The authors wish to thank them for the many stimulating discussions, advice and support in the course of this project.

REFERENCES

Frissen, C.M. 2000, *Numerieke Analyse Kokerbalkviaduct*. (in Dutch) TNO report 1999-MIT-NM-R0013, TNO Building and Construction research, Rijswijk.
Timoshenko, S.P. & Woinowsky-Krieger, S. 1970, *Theory of plates and Shells*, 2nd ed. McGraw-Hill.

Finite Elements in Civil Engineering Applications, Hendriks & Rots (eds.)
© 2002 Swets & Zeitlinger, Lisse, ISBN 90 5809 530 4

Numerical simulations of tests on a segmented tunnel lining

A.H.J.M. Vervuurt
TNO Building and Construction Research, The Netherlands

C. van der Veen
Delft University of Technology

F.B.J. Gijsbers
TNO Building and Construction Research, The Netherlands

J.A. den Uijl
Delft University of Technology

ABSTRACT: In order to gain a better understanding of the complex structural behavior in shield driven segmented tunnel linings for the specific Dutch soil conditions, a full scale test set-up has been realized in the Stevin Laboratory at Delft University of Technology in the Netherlands. The finite element program DIANA has been used for analyzing the experimental behavior of the tests. In earlier studies with the experimental facility it was shown that the stress distribution in the serviceability limit state is dominated by concentrated loads in the construction phase. In additional tests the behavior of the lining during construction is emphasized. In these tests initial gaps between the segments are prescribed in order to study the effect of concentrated load transfers. The data of the former tests and the intended tests will serve for validating the results of three dimensional FEM analyses. In this paper emphasis is placed upon the finite element model for analyzing the tests. Moreover, the experimental set-up is addressed as well as the measurements used for validating the numerical model. Finally the test results are compared to the results of the FEM model.

1   INTRODUCTION

From 1995 several projects involving shield driven segmented tunnels have been realized in the Netherlands. De first experiences have been gained with the Second Heijnenoord Tunnel. In the lining of this tunnel two rings have been instrumented and measurements have been performed with respect to the strain distribution in the lining and the loads acting upon the lining. The results showed that the stresses that develop during the construction phase, are considerably higher than expected. Therefore, contrarily to conventional design, the construction loads can not be neglected.

In the Second Heijnenoord Tunnel the stresses in the construction phase led to damage to the construction. In shield driven tunnel projects carried out after 1995 similar problems were faced. Again damage of the construction was observed during the construction phase. In order to examine the observed aspect in more detail a research project was started by the Delft Cluster. Delft Cluster is a corporation of five research institutes in which, among others, Delft University of Technology

and TNO participate. The research program was sponsored by the Delft Cluster, the Dutch Government and two project organizations which are involved in shield driven tunnel projects (HSL Zuid and Betuweroute).

The test model consists of a full scale tunnel lining containing three rings with 9.45 m external diameter and a segment thickness of 0.4 m (lining center radius $R = 4.525$ m).

In the first tests with the full scale equipement, which were performed in 1999, the ring behavior under several loading conditions was emphasized. The results obtained from the evaluation of these tests are described in Vliet et al. (2000) and yielded the following conclusions:

- For modelling the behaviour of the lining correctly, the behaviour of the joints between the segments and between the rings is crucial.
- Locally large differences were observed between the measured moments and forces and the numerically predicted values. No reasonable explanation could be given for this.

- From the experiments a significant effect was found of initial gaps in the joints during the construction phase.

The experiences in the first series of tests gave reason to perform a second series of tests with the full scale test facility. The numerical analyses performed on behalf of these additional tests are described in this paper. The tests focus upon the behavior in the final stage (service loads) as well as in the construction phase. The effect of inaccurate positioning of the segments leading to possible damage of the construction is studied by applying initial gaps in the ring joints.

In section 3 and 4 of this paper the numerical analyses carried out with the finite element program DIANA are outlined. Attention is focused upon the model and a comparison of the results with the experimental results. More specifically, attention has been paid to the modeling of the behavior of the segment joints. For this purpose a *user supplied subroutine* was adopted. In the following section (section 1) the full scale test facility which was used for performing the tests is outlined.

A summary of the conclusions and a preview of the final tests is given at the end of the paper.

2 EXPERIMENTS

2.1 Set-up

The full scale test facility adopted for the experiments is shown in Figure 1. It can be seen from the photograph that the three rings are mounted in vertical direction, in stead of horizontal direction, thereby denying the dead weight of the construction. The effect of the dead weight is significantly less than the effect of the active loads due to the grout and the Tunnel Boring Machine (TBM).

From Figure 1 it can be seen that the lining is surrounded by a steel reaction frame for balancing the

radial loads which are applied by 28 hydraulic actuators per ring. The axial forces representing the loads from the TBM are applied by 14 actuators placed at the top and connected by a closed frame to the bottom of the lining. In order to avoid friction at the bottom supports the bottom ring (ring 1) is mounted on Teflon layers. Moreover, the bottom ring is supported by four tangential active supports, whereas the reaction frame is prevented from rotating by four tangential pendulums.

2.2 Loading

As mentioned previously radial as well as axial loads can be applied to the test specimen. For each actuator a different load can be applied, with the condition that the radial actuators are diametrically coupled to ensure symmetry. The tested load scheme with respect to the radial hydraulic actuators is illustrated in Figure 2. Based on the radial loads shown in Figure 2a and Figure 2b the total radial load is given by:

$$p = q - \Delta q \cdot \cos(2\theta) \qquad (1)$$

The expected (radial) deformation due to the so called ovalisation load (Figure 2b) is illustrated in Figure 3. The shape of the deformed tunnel is characterised by a positive horizontal ovalisation and a negative vertical ovalisation. The rotational angle θ refers

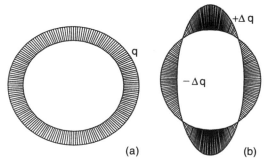

Figure 2.   Uniform radial loads (a) and ovalisation loads (b).

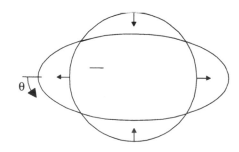

Figure 3.   Shape of the expected ovalisation of the lining.

Figure 1.   Overview of the full-scale test facility in the Stevin Laboratory of Delft University.

to the angle which is used for presenting the results (section 4).

The axial loads in the tests described in this paper are uniformly distributed along the circumference of the lining. In the set-up, however, also a global bending moment may be applied to the lining, simulating the actual load distribution from the TBM.

2.3 Measurements

During testing the loads are applied in a number of steps. After each load step several measurements are performed that are used for the evaluation and for comparison to the results of the finite element calculations. During testing the following data are recorded:

- Oil pressure of the (axial and radial) hydraulic actuators.
- Radial forces by means of load cells.
- Radial displacements of the lining by means of laser measurements as well as LVDTs.
- Radial displacements of the reaction frame.
- Axial, tangential and diagonal strains at the inner and outer side of the concrete segments. In total 312 strain gauges are used.
- Displacements at the (ring and segment) joints by means of 128 Linear Variable Displacement Transducers (LVDTs).

3 NUMERICAL ANALYSES

3.1 Finite element model

As mentioned earlier two series of tests are performed. A full description of the two series is given in Vervuurt et al. (2002a). The results of each test are analysed and compared to the results of calculations carried out with the finite element program DIANA.

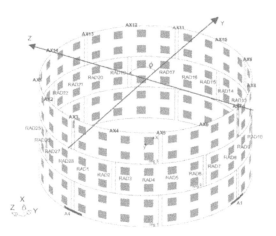

Figure 4. Schematic view of the outlines of the finite element model.

An overview of the outlines of the model is given in Figure 4. As can be seen from the figure a shell model was used for modelling the specimen, using eight-node quadrilateral isoparametric curved shell elements (type CQ40S) with 2 · 2 integration points. Because the material behavior of the elements is assumed linear elastic, three integration points over the height of the shell elements was adopted.

A detail of the element distribution of a single segment is given in Figure 5. Moreover, in the figure, the cross sections of the segments, including the positions of the segment and ring joints are given.

In the joints between the segments and one single rings, the segments are connected to each other by concrete tot concrete contact. In the ring joints plywood is adopted for transferring the axial loads and shear forces.

The principle for modeling the joint behavior is elucidated in detail in section 3.2. Moreover, in the following sections the material properties are described as well as the boundary conditions (section 3.3) and the analyzed load combinations (section 3.4).

3.2 Material properties

3.2.1 Concrete segments

The material properties of the concrete segments are summarized in Table 1. Because the joint behavior is dominant for the overall behavior of the lining and non linear effect are only foreseen in the joints, the segments have been assumed to behave linear elastically during testing.

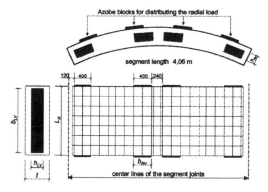

Figure 5. Detail of the element distribution of one single segment related to the actual geometry of the segment.

Table 1. Material properties of the concrete segments.

Modulus of elasticity	$40 \cdot 10^6$ kN/m^2
Poisson's ratio	0.2

3.2.2 *Joint behavior*

In general the behavior of the joints between the segments is characterized by contact behavior (no stiffness in tension and infinite stiffness in compression). Moreover, as can been seen from Figure 6a, contact is only present over a limited height of the segment thickness. Due to this, on the one hand, the behavior of a joint in compression is less stiff than a full cross section of a segment. On the other hand, because the joint is not able to transfer tensile stresses, the bending stiffness of the joint strongly depends on the normal force in the joint.

For modeling the joint behavior, a concept based on a model developed by Janßen (1983) has been adopted. In the model the reduced (normal and bending) stiffness is obtained by a beam element between the segments. The beam element has a limited length ($l = h$, see Figure 6b) and a reduced cross section. The normal dependant rotational stiffness is incorporated by non linear material behavior of the beam ($f_t = 0$).

A schematized detail of the joints in the finite element model is given in Figure 7. The interface element (NI6F) which is placed is series with the (non linear) beam element (L7BEN) provides the slip behavior of the joint. The slip behavior also depends on the normal stress in the element.

The rotational joint behavior is characterized by the M-N-κ-diagram. In Figure 8 the rotational joint behavior is illustrated for the geometry adopted in the tests and an arbitrary normal force. In the figure the theoretical solution according to Janßen (1983) is compared to the numerical solution when adopting L7BEN elements (11 integration points over the height of one beam). Moreover, in the figure, the situation where concrete under compression behaves linear elastically is addressed as well as the situation where the compressive strength of the concrete is taken into account. In the case of linear elastic compressive behavior the ultimate moment (M_u) is determined by the height of the beam ($M_u = 0.5 \cdot N \cdot h$). When non linearity's with respect to the compressive concrete behavior are taken into account the ultimate moment decreases about 20%.

Moreover, it can be seen from Figure 8 that a good prediction is given with the numerical (beam) model. Because of the limited number of integration points over the height (i.e. 11), a too stiff rotational stiffness is calculated for large curvatures (when all but one interfaces are loaded in tension). In case of non linear compressive behavior divergence of the numerical system is obtained for large curvatures, indicating failure of the joint.

When, during construction, the segments are mounted perfectly, the joints are closed from the beginning and no initial gaps are present. Modeling the interface joint behavior for this case can be achieved by using default material models available in DIANA (i.e. Mohr Coulomb). Nevertheless, in practice initial gaps in the joints have shown to be inevitable, resulting in local peak stresses and possible cracking of the segments.

Due to the relatively small initial gaps, contact elements have proven to be less efficient for modeling the joint behavior is such cases. For that reason, a specially

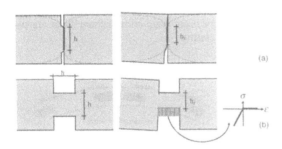

Figure 6. Physical joint behavior (a) and principles of modeling the joints (b), after Van Empel (1997).

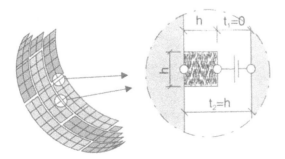

Figure 7. Element schematization of the joints.

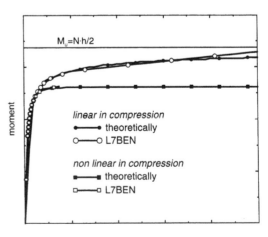

Figure 8. Comparison of the numerical and theoretical joint behavior, for both linear and non linear compression.

developed material model (*user supplied subroutine*) has been implemented for the interface elements.

The basics of the material model for joints with (and without) initial gaps are illustrated in Figure 9. In Figure 9b the relation between the normal stress and displacement is illustrated for three different positions at an oblique positioned segment is shown. Depending on the gap, the normal stiffness in compression is activated at a larger displacement. In Figure 9b this is illustrated by the constant stiffness for two regions (tension and compression). Because in tension also no shear can be transferred by the joint, the shear stiffness also depends on the normal displacement in the joint as well as on the initial gap distance.

The effect of the key stones on the behavior of the lining has been modeled by assigning specific material properties to one of the segment joints in the lining.

3.3 *Boundary conditions*

At the bottom of the model the lining is supported in vertical direction, circumferentially at the same angles as where the plywood (ring) joints are adopted in the model.

Because the loads are symmetrically in the YZ-plane, theoretically no supports are required in this plane. Nevertheless, as explained in section 2, at four positions at the two axes of symmetry at the bottom of the model, the tangential deformations are restricted (indicated by A1 to A4 in Figure 5).

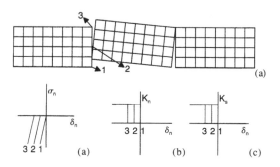

Figure 9. Basics of the implemented material model with respect to the joint behavior in case of initial gaps in the joints.

Table 2. Tested load combinations in series A (forces in kN/actuator).

Test number	Axial uniform	Radial uniform	Ovalisation
A1	800	225	20
A2	1600	450	20
A3	2400	675	30

3.4 *Performed analyses*

In the first test series the behavior of the construction under service loads has been studied (series A). The behavior due to the radial ovalisation load is significantly influenced by the magnitude of the uniform axial and radial loads. These loads determine the behavior of the ring and segment joints respectively. The behavior due to an ovalisation load is studied for three different combinations. The combinations which refer to a high, medium and low uniform load are given in Table 2. In the table the applied load per actuator is given. The areas at which the loads act, depend on the radius of the lining ($R = 4.525$ m), the height of the rings (H = 1.5 m) and the numbers of actuators (14 in axial direction and per ring 28 in radial direction).

In the second series of tests (series B) attention is paid to the effect of inaccurate positioning of the segments. The finite element model is used to study whether there is a relation between the width of gaps between segments and expected damage. In series B tests are foreseen with initial gaps between the second and third ring in the test specimen. At the moment this paper is written the tests as well as the numerical simulations with respect to these tests are being performed. The results of the second series of tests will be presented at the conference.

4 RESULTS

4.1 *Introduction*

In this section some results of the finite element analyses with respect to series A of the tests are presented. Moreover, some significant results are compared to the experimental measurements. Because the largest ovalisation is observed in test A3, in this section attention is focused mainly upon this particular test. A detailed description of all test results is given in Vervuurt et al. (2002a).

The results in the following sections are presented according to the polar coordinate system as illustrated in Figure 10. The view of the lining corresponds to the view of the lining from the inner side of the set-up. In the figure the positions of the key stones are marked by the gray areas and the dashed lines.

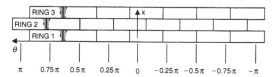

Figure 10. Overview of the lining seen from the inside of the set-up and the definition of the defined polar coordinate system.

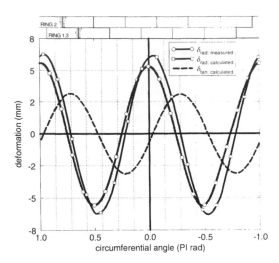

Figure 11. Calculated and measured average deformations (radial and tangential) in ring 2.

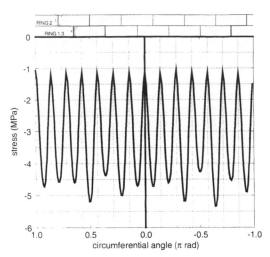

Figure 12. Calculated and measured average axial strain distribution in the middle ring of the test specimen (ring 2).

4.2 Deformations

The results with respect to the radial and tangential deformations in test A3 are given in Figure 11. In the figure the deformation is plotted on the vertical axis. A positive deformation refers to an increase of the diameter. On the horizontal axis the circumferential angle is given. The angle $\theta = 0$ refers to the axis between the AX10 and AX11 in Figure 5. The direction of rotation of the circumferential angle is shown in Figure 10 as well. The position of the segments, including the key segments is shown in the inset above the figure.

The deformations as shown in Figure 11 are dominated by the radial ovalisation loads (Figure 2b). Because this load is distributed uniformly over the height of the construction, it may be expected that almost no shear stresses develop in the ring joints, causing marginal differences between the deformations of the different rings. Therefore in Figure 11 only the results from the second ring are given. The solid line marked with the opaque circles indicates the measurements whereas the solid line with square markers refers to the finite element calculations. In the figure, moreover, the calculated tangential displacements (δ_{tan}) are given by means of the dotted line.

From the radial deformations it can be seen that the calculated and measured average ovalisation differ approximately 10% from each other. This difference is caused by either the assumed segment stiffness ($E = 40000\,N/mm^2$) or the segment joint stiffness. Since the effect of the segment stiffness is considerably less than the effect of the joint stiffness, it may be expected that the latter effect is dominant for the observed differences. In the calculation a higher segment joint stiffness should be applied. The strain

measurements presented hereafter confirm these findings.

When a monolithic lining (no joints) is considered the maximum ovalisation can be calculated from:

$$w_{max} = \frac{1}{9} \cdot \frac{\Delta q \cdot R^4}{EI} \qquad (2)$$

where Δq is according to Figure 2, R is the radius ($R = 4.525\,m$) and EI is the tangential bending stiffness of the segments. For a maximum amplitude of the load of $30\,kN/actuator$ it follows that $\Delta q = 19.7\,kN/m^2$ (Table 2) and $w_{max} = 4.3\,mm$. Considering the average measured amplitude in test A3 (5.5 mm) the presence of the segment joints cause a decrease of the bending stiffness of 22% compared to a monolithic lining.

Next to the radial deformations also the calculated tangential deformations are plotted in Figure 11. For a monolithic lining the maximum tangential deformation is half of the radial ovalisation. Moreover, the results are shifted $\pi/4$ rad. This is in correspondence to the calculations (Figure 11).

4.3 Axial strains and stresses

In the tests the axial loads due to the TBM forces have been assumed to be uniformly distributed along the circumference of the tunnel (2400 kN/actuator for test A3). Consequently an average axial stress of almost $3\,N/mm^2$ may be expected. This is confirmed by the stresses obtained from the calculations (Figure 12). Moreover it can be seen from this figure that the stresses are strongly influenced by the concentrated load transfer at the joints. Close to the working lines of

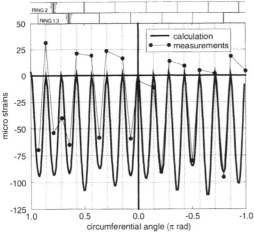

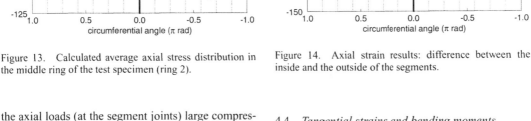

Figure 13. Calculated average axial stress distribution in the middle ring of the test specimen (ring 2).

Figure 14. Axial strain results: difference between the inside and the outside of the segments.

the axial loads (at the segment joints) large compressive stresses are observed (almost 5 N/mm^2), whereas in between the segment joints (at each quarter of a segment) a rather low compressive stress is found (about 1 N/mm^2).

With respect to the axial strains (Figure 13) it is found that the shape of the strain distribution equals the stress distribution (minimum at the joints and maximum at a quarter of each segment). The maximum strains calculated, however, are about zero, whereas a maximum stress of -1 N/mm^2 is found. The difference is attributed to the influence of lateral contraction due to the radial compressive loads. When the calculations and the experiments are compared it is found that in the calculations the maximum strain is limited to about 0, whereas in the experiments the maximum strain varies between 0 and $+25 \cdot 10^{-6}$.

In Figure 14 the difference in axial strain between the inner and the outer side of the segments is shown, indicating local axial bending in the segments (due to the tapered shape of the rings). In the figure the results of the calculations are indicated by the solid lines whereas the measurements are given by the same type of markers (yet filled) connected by the dashed lines.

The results in Figure 14 show that the measurements and the calculations correspond both qualitatively and quantitatively. The influence of the segment joints is shown by the large differences between the neighboring rings. Because both the axial and the radial loads are applied symmetrically it is expected that the observed differences are caused by contraction effects due to the radial loads. In the calculations the differences are larger than in the experiments, indicating a too low assumed stiffness of the segment joints.

4.4 Tangential strains and bending moments

In radial direction a uniform load is applied as well as an ovalisation load. In a monolithic lining a uniform radial load causes a uniform tangential stress, which can be calculated from:

$$\sigma_{\tan} = q \cdot \frac{R}{t} \qquad (3)$$

where q is according to Figure 2a, R is the radius and t is the thickness of the segments. For tests A3 (q = 443 kN/m^2, see Table 2) this leads to an expected uniform stress of 5 N/mm^2. The finite element calculation yields the same conclusions (Vervuurt et al, 2002b). Similar to the axial stress and strain distribution (Figure 12 and Figure 13 respectively), a considerable effect of the concentrated load introduction is found. In Vervuurt et al. (2002a) it is shown that the shape of the distribution is opposite to the distribution of the mean axial stress. This corresponds to the effect of lateral contraction in the segments.

The tangential bending moments are characterized by the tangential curvature in the segments. When the curvature is defined as $\Delta\varepsilon_{tan}/t$ (Figure 15) it can be seen that there are quite some differences found in the curvature between each neighbouring ring. In the experiments the same tendency is found, however, less pronounced.

The differences between the measurements and the calculations are mainly attributed to the (locally) low bending stiffness of the segment (and ring) joints. For infinite stiffness of the segment joints a uniform behavior may be expected, whereas for zero stiffness

$\Delta \varepsilon_{tan}$ yields to zero. Consequently, the large differences between the subsequent rings in the calculations indicate that the bending stiffness of the segment joints is assumed too low. This corresponds to the findings considering the axial strains (Figure 14).

Because the tangential curvature and the bending moment are related linearly to each other, the bending moment distribution is expected to be comparable to Figure 15. In Figure 16 the ring moments are plotted by means of the solid lines.

The dotted line in Figure 16 indicates the average ring moment. Because the average moment only depends on the externally applied radial loads, the

distribution should be similar to the moment distribution in a monolithic lining. For a monolithic lining the moment distribution can be calculated from:

$$M = \frac{1}{3} \cdot \Delta q \cdot R^2 \cdot \cos(2\theta) \tag{4}$$

For test A3 ($\Delta q = 19.7 \, \text{kN/m}^2$, see Table 2) the maximum moment equals 202 kNm/ring, which corresponds to the finite element calculations.

The effect of the low bending stiffness of the segment joints is that the moment in that particular ring decreases, whereas the moment in the neighboring rings increases. Therefore a considerable increase of the bending moment in ring 2 is found in the case of a segment joint in ring 1 and 3 (e.g. for $\theta = -0.5 \cdot \pi$ rad). When, on the other hand, a segment joint is present in ring 2 (e.g. for $\theta = +0.5 \cdot \pi$ rad), the bending moment in ring 1 and 3 increases, however, substantially less.

4.5 Principal strains

Because in series B tests are foreseen with initial gaps at the ring joint between the second and third ring of the test specimen, additional strain measurements have been performed in the third ring. By placing strain gauges at 45° with the tangential and axial strain gauges, the principal strains can be calculated from the results of one rosette containing three strain gauges.

The results with respect to the calculated principal strains are given in Figure 17. It can be seen that the experiments and calculations correspond well, however, the maximum principal strain in the calculations is

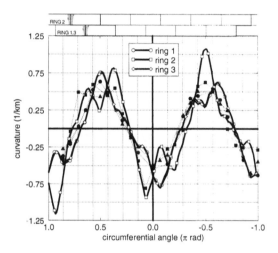

Figure 15. Measured and calculated tangential curvature in the segments ($\Delta \varepsilon_{tan}/t$).

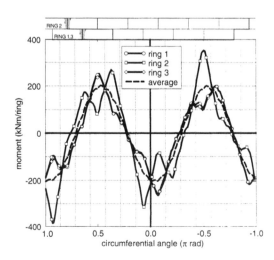

Figure 16. Calculated bending moments in the segments.

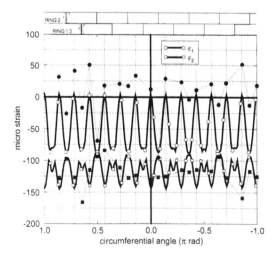

Figure 17. Measured and calculated average principal strains in ring 3.

436

generally smaller than in the experiments. The principal strains correspond approximately to the axial and tangential strains (ε_1 and ε_2 respectively).

5 CONCLUSIONS

Finite element calculations with respect to full scale experiments of segmented shield driven tunnels have shown that the lining behavior can be predicted quite well. Differences are mainly caused by the assumed stiffness of the segment joints. The stiffness of the ring joints is of less influence in the tests carried out. Moreover the study has shown that the comparison of the results should be carried out at the level of strains. Stresses can not be extracted directly from the strain measurements in the tests. Due to contraction effects the stress distribution differs significantly from the strain distribution.

In order to study the effect of inaccuracies in the positioning of the segments a second series of tests is foreseen. In these tests initial gaps are prescribed between the second and the third ring. At the time this paper is written the tests are about to be carried out.

The results of the numerical analyses with respect to these tests will be presented at the conference.

REFERENCES

Janβen, P. (1983). *Tragverhalten von Tunnelausbauten mit Gellenktübings*. PhD thesis TU Carolo-Wilhelmina, Braunschweig.

Van Empel, W.H.N.C. (1997). Liggerwerking van gesegmenteerde boortunnels. Masters thesis Delft University of Technology.

Vervuurt, A.H.J.M., Den Uijl, J.A., Gijsbers, F.B.J. and Van Der Veen, C. (2002a). *Aanvullende Proeven in de Tunnelproefopstelling: Constructiegedrag Onder Gebruiksbelastingen en het Effect van Plaatsingsonnauwkeurigheden. Deel 1: Opzet en Resultaten van Serie A,* **Delft Cluster Report DC 01.06.02-01** (in Dutch).

Vervuurt, A.H.J.M., Den Uijl, J.A., Gijsbers, F.B.J. and Van Der Veen, C. (2002b). Full scale tests on a segmented tunnel lining. *In proc.* of the First *fib* congress, October 13–19 2002, Osaka, Japan (submitted).

Vliet, C. v.d., Ros, P., en Zeilmaker, A. (2000). *Postdictierapport, Analyse van de 3-D rekenresultaten en vergelijking met de meetresultaten,* HSL-Zuid rapport **NOH\77506** (in Dutch).

Finite Elements in Civil Engineering Applications, Hendriks & Rots (eds.)
© 2002 Swets & Zeitlinger, Lisse, ISBN 90 5809 530 4

Advanced modeling of innovative bored tunnel design Amsterdam North-Southline

W.H.N.C. van Empel & F.J. Kaalberg
North/Southline Consultants & Witteveen + Bos, Amsterdam, Netherlands

ABSTRACT: The two bored tunnels that will be part of the North/South metroline under the city centre of Amsterdam are characterized by an innovative design of radial and circular joints. This innovative joint design calls for an innovative and advanced numerical model of the joints and the tunnel tube as a whole. While developing the 3-D phased model of the tunnel lining and its components the vast array of modeling facilities offered by DIANA was optimally used. An important aspect of the development process consists of the validation regarding the behavior of the numerical components of the tunnel structure.

1 INTRODUCTION

1.1 *The Amsterdam North-South metroline*

Under the historic city center of Amsterdam two bored tunnels will be constructed that are part of North-South metroline. These tunnels, bored with separate tunnel boring machines (TBMs) are approximately 3.6 km long and follow a route below existing streets. The bored tunnels pass through strongly varying soil layers, ranging in nature from stiff sand to soft Eem-clay. Along the proposed route of the bored

tunnels the TBM's have to be driven along narrow curvatures as low as 190 meter. The poor soil conditions at some stretches along the proposed route, combined with the locally narrow curvatures and an urban environment that is very sensitive to settlements make the Amsterdam North-Southline a challenging project (Kaalberg & Hentschel 1999).

1.2 *Innovative design of the bored tunnels*

Because of the project specific adverse conditions encountered along the proposed route it was acknowledged in an early stage that enhancement and optimization of conventional design concepts of bored tunnels was necessary in order to minimize and control the risks regarding the structural performance of the bored tunnels during the construction phase and the serviceability phase. After thorough investigation of the theoretical and practical performance of conventional lining concepts it was concluded that conventional lining concepts in practice often give rise to structural failure in the sense of crack formation and even breaking off of ring couplers. Given the sensitive and adverse environment in which the tunnels are to be constructed a lining concept has been developed which ensures a conditioned behavior of the lining, which can be managed by design and design calculations.

The innovative lining concept is characterized by a centrally placed elastomere profile in both the radial joints as the circular joints through which the normal forces at the joint are mainly transferred

Figure 1. Proposed route of the North-Southline.

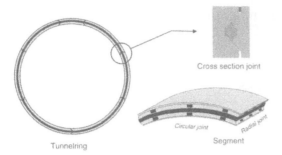

Figure 2.   Lining concept.

(see figure 2). The elastomere profile not only has a structural function but also ensures a watertight joint.

At several locations at the joints elastic plates are applied in order to transfer bending moments and shear forces at the joint.

1.3   *Numerical modeling of the behavior of the tunnellining and its components*

Creating a lining concept that results in a defined structural behavior of the lining and its components is the first step in the direction of managing and minimizing the risks regarding the structural performance of the tunnel lining.

The second important step comprises realistic and state of the art numerical modeling of the behavior of the tunnellining, its segments and its joints. Because of the fact that the detailed design (for example the stiffness of the joint plates) of the tunnellining and its components have to be determined in the same phase where the models are applied, the models not only serve a purely predictive goal but are also used to determine the optimal constellation of design parameters, which gives the design and calculation process an additional dimension.

This is a major difference with standard tunnel design practice where (numerical) models are mostly used to predict the behavior of a lining according to a given conventional concept and detailed design and therefore given but mainly assumed mechanical parameters of the tunnellining like joint stiffnesses. In the conventional design practice the behavior of the tunnellining solely depends on boundary conditions like soil pressures acting on the lining. In the conventional situation (numerical) models are therefore only used to determine the (given) loads on or force in the lining components in order to determine the strength of the components needed to withstand these loads and forces.

Because of the fact that the stiffnesses of the lining components in the North-Southline lining concept

are open and designable, parameters the distribution of the forces among these components can be manipulated by design which makes the design and calculation process regarding the North-Southline lining concept differ from a conventional design and calculation process applied in bored tunnel engineering.

The fact that the North-Southline lining concept hasn't been applied in a bored tunnel project before and therefore no empirical data is available calls for sophisticated and extensive modeling of all relevant mechanisms combined with an extensive test program, which also is an important aspect of the design process.

2   CONCEPTUAL NUMERICAL MODEL OF THE BORED TUNNEL

2.1   *General*

The behavior of a bored tunnel, consisting of several segments is strongly determined by the interaction of these segments at the circular and radial joints. In order to predict the global behavior of the tunnellining the individual behavior of these circular and radial joints has to be modeled in a realistic and accurate way. This joint behavior is determined by the geometrical layout of the joint and the mechanical properties of the joint components (elastomere profile and elastic plates). Globally the external loads on the tunnellining will be distributed among circular and radial joint according to their relative stiffness.

2.2   *Designing the numerical model*

Globally the tunnellining can be regarded as a thin walled structure. Modeling of the segments by means of curved shell elements is therefore appropriate.

In order to model the joint a spring system is used as presented in figure 3.

The centrally placed profile is modeled by means of an N6IF interface element, orientated along the normal axis of the joint. The plates in the joint are modeled by eccentrically placed N6IF interface elements with an orientation conform the inclination of the actual plates. The two nodes that are part of the N6IF element are both eccentrically tied to the respective nodes of the shell element.

The joint and its model have 3 degrees of freedom corresponding to 3 types of loads acting on the joint.

Degrees of freedom and forces acting on the joint are related by a stiffness matrix according to equation 1.

$$
\begin{bmatrix} M \\ N \\ Q \end{bmatrix} = \begin{bmatrix} k_{11} & k_{12} & k_{13} \\ k_{21} & k_{22} & k_{23} \\ k_{31} & k_{32} & k_{33} \end{bmatrix} \cdot \begin{bmatrix} \varphi \\ u_t \\ u_r \end{bmatrix} \qquad (1)
$$

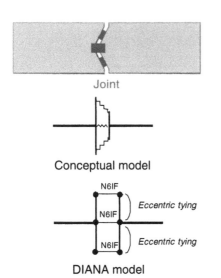

Joint

Conceptual model

N6IF

N6IF } Eccentric tying

N6IF } Eccentric tying

DIANA model

Figure 3.   Joint and numerical model.

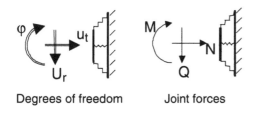

φ u_t

U_r

M N

Q

Degrees of freedom   Joint forces

Figure 4.   Degrees of freedom and forces acting on the joint.

If the joint would be modeled with separate springs for transfer of bending moment, normal force and shear force a model as shown in figure 4 would arise.

In case of this model the relationship between degrees of freedom and forces in the joint would be as described in equation 2.

$$\begin{bmatrix} M \\ N \\ Q \end{bmatrix} = \begin{bmatrix} k_{11} & 0 & 0 \\ 0 & k_{22} & 0 \\ 0 & 0 & k_{33} \end{bmatrix} \cdot \begin{bmatrix} \varphi \\ u_t \\ u_r \end{bmatrix} \qquad (2)$$

A joint model as shown in figure 4 would not take into account any interaction between the 3 different degrees of freedom of the joint. In reality this interaction exists, which is illustrated in figure 6. If the joint is subjected to a combined shear force and bending moment the rotation will differ from the case where the joint is purely subjected to a bending moment.

Based on the considerations above a joint model conform figure 2 is applied instead of the model, shown in figure 5.

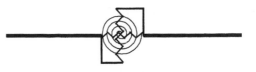

Figure 5.   Joint model with uncoupled springs.

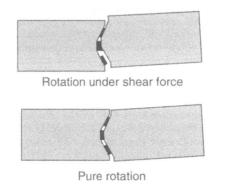

Rotation under shear force

Pure rotation

Figure 6.   Rotation with and without shear force.

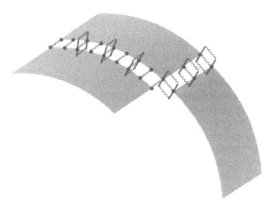

Figure 7.   3D-view at joint model.

In figure 7 the circular joint and radial joint are shown in a 3D perspective. The centrally placed elastomere profile at the joints is modeled by means of a row N6IF interface elements because no 3-D line interface elements are offered by DIANA at this moment. By means of lumping the individual stiffness of each interface element is assigned in such a way that the row of individual spring will act if it were a continuous line interface element.

At the locations of the joint plates the eccentrically placed N6IF elements are applied and orientated according to the individual orientation of the plates. The nodes of these interface elements are eccentrically tied to the corresponding nodes of the shell elements representing the segments.

441

3 VALIDATION OF THE NUMERICAL MODEL

3.1 General

The joint model to be applied in the macro model of the tunnelrings is to be validated in order to assure a correct prediction of the response of the tunnellining and its components under the governing boundary conditions.

Also the behavior of the macro model is to be validated by means of a number of benchmark studies based on simple loadcases leading to a response of the tunnellining that can be predicted analytically.

In order to validate the simplified spring joint model applied in the joints this simplified model is to be validated based on a higher order reference model (2D continuum model) which will be discussed later on. The behavior of the reference model can be validated on its turn by means of a full scale physical model of the joint to be tested in a laboratory. In the ultimate case the behavior of the macro model can be validated based on a full scale physical model consisting of several rings.

3.2 Validation of the simplified spring model based on a 2D continuum model of the joint

In order to validate the simplified spring model to be applied in the macro model of the tunnellining a 2-D continuum model has been used.

Compared to the reference model the simplified spring model has one limitation. The springs in the simplified model do not have dimensions and represent the net behavior of the corresponding plate or elastomere profile. In order to assess the implications of this simplification several benchmark calculations have been performed. During one calculation the normal compression of the joint was set at 10 mm and a transverse offset of 2 mm was applied. Under these conditions and realistic stiffnesses of the elastomere profile and joint plates the rotation φ of the joint was varied between $-30 \cdot 10^{-3}$ rad and $30 \cdot 10^{-3}$ rad and the M-φ, N-φ and Q-φ relationships were determined for both the reference model as the simplified spring model. In figure 8 the M-φ diagram is shown as an example. This particular benchmark calculation and also the other calculations showed an outstanding resemblance between the behavior of the reference model and the simplified spring model.

It can therefore be concluded that the simplified spring model is able to behave in a way that is equal to the behavior of a more complex 2-D continuum model.

3.3 Validation of the macro model of the lining based on simple loadcases

By applying a simple loadcase with regard to the macro model of the lining and predicting the distributions of

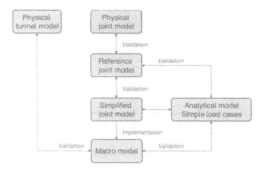

Figure 8.   Validation procedure.

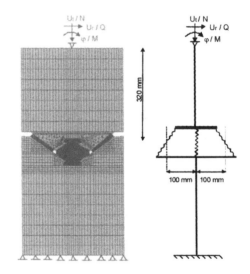

Figure 9.   Reference model and spring model.

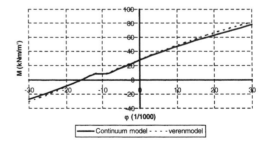

Figure 10.   Reference model and spring model.

stresses and forces in the circular and radial joints the global behavior of the macro model and the integral 3D joint behavior can be validated. In this paper the validation of the numerically predicted behavior of the circular joint will be discussed.

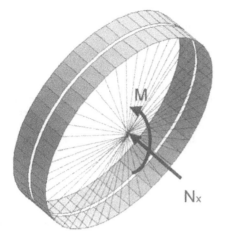

Figure 11.   Macro model, axially loaded.

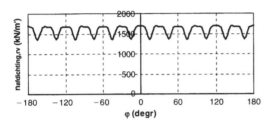

Figure 12.   Distributed normal force in elastomere profile.

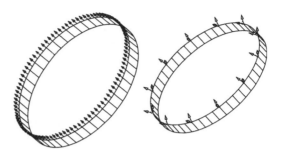

Figure 13.   Vector representation of forces in circular joint.

Figure 14.   Elastomere profile placed in a steel cavity.

The model used consisted of 2 tunnelrings, loaded by a resulting normal force acting in the direction of the tunnel axis (see figure 11). Also an additional global bending moment could be applied. In this paper only the loadcase comprising a pure axial normal force will be discussed.

In order to apply these global loads a master node was used. The nodes of the shell elements along the tunnelperimeter were eccentrically tied to this masternode. In this way the global loads were automatically distributed along the tunnelperimeter in a proper way.

For a given set of material properties of elastomere profile and joint plates the distribution of the normal force along the elastomere profile in the circular joint was determined. These material properties were chosen in such a way that the normal load was partially transferred by the joint plates.

In figure 12 the distribution of the normal distributed force along the elastomere profile is shown. The diagram shows that at the locations of the plates the normal load is partially taken over by the joint plates thus leading to a local reduction of the distributed force in the elastomere profile.

The way the global axial force is transferred in the circular joint is graphically shown in figure 13.

Qualitative and quantitative comparison of the output of the numerical model with the analytical predictions showed that the model predicts the global behavior of the lining in a proper way.

3.4   Validation of the simplified spring model based on results of a physical model

After determination of the optimal constellation of design parameters based on evaluation of the output of the applied (numerical) models, described in chapter 4 the abstract numerical elements and the according desired behavior have to be translated into physical elements. For example the desired stiffnesses of the plates (interface elements) in radial and circular joints are deduced from the modelresults resulting in functional specification for the real physical plates to be produced. Once these plates have been produced testing of the characteristics of these plates is to be performed in a laboratory.

At this moment a number of tests has been performed on the elastomere profile in order to determine the normal characteristic of this profile. Figure 14 shows the profile as it was tested in a laboratory.

Figure 15 shows the characteristic of the elastomere profile that was measured. This characteristic can be approximated with a multi-linear diagram that serves as material input for the used interface element.

The same procedure can and will be performed in the nearby future for the elastic plates thus resulting in input for interface elements representing these plates.

Once the properties of elastomere profile and elastic plates are known the constitutive input for the joint model is present. The final steps lies in validation of the numerically predicted kinematics of the joint as a whole based on a 1:1 test specimen of the joint.

Once the behavior of the numerical joint model is validated based on the physical joint model one can be sure that the behavior of the joints in the numerical macro model is accurately simulated.

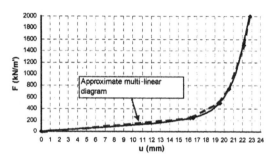

Figure 15.   Measured characteristic elastomere profile and according approximate multi-linear diagram.

4   APPLICATION OF THE MODEL IN PRACTICE

4.1   *Modeling of construction phase*

The construction phase during which the bored tunnel is being constructed by erecting individual segments and forming a tunnelring is commonly recognized as a phase to be more critical than the serviceability phase of the tunnelling.

During the construction phase the segments are loaded by strongly varying jack forces, applied by the TBM thrust jacks. Once a tunnelring has been completed the TBM will continue boring and gradually the tunnelring will leave the shelter of the TBM-shield and will be subjected to pressures, present in the grout suspension injected in the tailvoid.

The complex succession of strongly varying load cases calls for a sophisticated phased 3-D model commonly known as a 4-D model.

In figure 18 the used 4-D model is shown. At one side of the model the thrust jacks are modeled. Also the conical keystone has been modeled. The gradual construction of the tunnelling can be modeled by means of a ring-by-ring or even segment-by-segment phasing of the calculation.

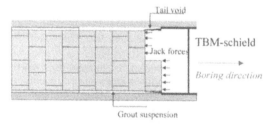

Figure 17.   Construction of a shield driven tunnel.

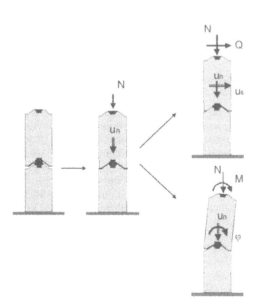

Figure 16.   1:1 test specimen as to be tested in a laboratory.

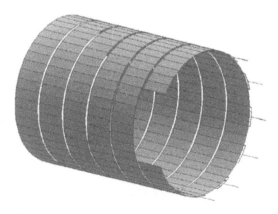

Figure 18.   Phased 3-D numerical model of the tunnellining.

444

4.2 Modeling of the serviceability phase

During the serviceability phase no significant changes in the boundary conditions regarding the tunnel occur within a limited distance along the tunnel axis. Therefore a semi 3D model is used, consisting of two adjacent tunnelrings (shell elements) and a continuum model of the surrounding layered soil.

The soil has been modeled with a Modified Mohr Coulomb material model. Before application this material model has been successfully validated based on laboratory results of specimen of the soil layers encountered.

4.3 Parameter study

Besides performing a parameter study in order to determine the most critical configuration of conditions or parameters which cannot be manipulated by design (boundary conditions like stiffnesses of soillayers, external loads etcetera) a separate parameter study is performed in order to determine the optimal individual stiffness of the elastic plates in radial and circular joint. The design and therefore the mechanical properties of these plates are not determined in advance but are to be determined based on a prediction of the behavior of the lining and it's components

as a function of these mechanical properties. Once the impact of the mechanical properties of the elastic joint plates on the response of the tunnellining and its components is quantified by means of a parameter study the optimal joint plate design can be selected. This procedure is illustrated by a case regarding two tunnelrings loaded by an external pressure distribution $p(\varphi)$ (soil/groutpressures).

The behavior of the tunnel as coupled rings is commonly referred to as ring action which results in an oval deformation or ovalisation of the tunnelrings.

In the case of ring action the radial joints globally act as rotational springs with an equivalent rotational stiffness depending on the stiffness of the plates applied and their eccentricity.

The plates in the circular joint will merely act as translational springs with an equivalent radial stiffness depending on the stiffness of the plates applied.

For a given distribution of the external pressure along the perimeters of the tunnelrings and a selected combination of rotational stiffness of the radial joints and a stiffness of the plates in the circular joints a calculation can be performed and the relevant loads on and deformation of the lining components determined.

As an illustration a qualitative distribution of bending moments in both tunnelrings is shown in figure 22.

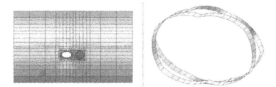

Figure 19. Continuum model layered soil and deformation of the tunnellining under soil pressures.

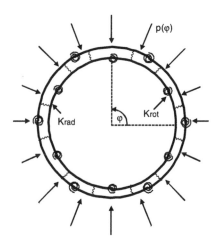

Figure 20. Tunnelring loaded by external pressure.

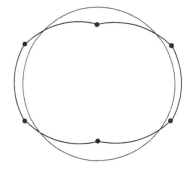

Figure 21. Ovalisation of a tunnelring.

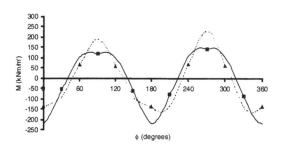

Figure 22. Illustrative distribution of bending moments in tunnelrings.

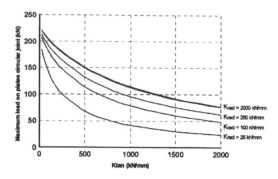

Figure 23. Maximum load on plates in circular joint as a function of (K_{rad}, K_{tan}) combination.

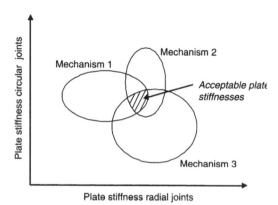

Figure 24. Determination of range of acceptable plate stiffnesses.

Depending on the ratio of rotational stiffness of the radial joints and radial stiffness of the plates in the circular joint the external load will be distributed among radial and circular joints. The stiffer the plates in radial joints are designed relative to the plates in the circular joints the higher the bending moments acting on the radial joints will become and the lower the forces acting on the plates in the circular joints will become. The loads acting on radial and circular joints can therefore be manipulated by manipulating the relative stiffnesses of these joints.

As an illustration the maximum force acting on the plates in the circular joint is qualitatively shown as a function of the radial stiffness K_{rad} of the plates in the circular joint and the tangential (along ring perimeter) stiffness K_{tan} of the plates in the radial joints in figure 23.

Figure 23 shows that the maximum load on the plates in the circular joint decreases with increasing K_{rad}/K_{tan} ratio.

The maximum load on the plates in the circular joint is only one design parameter to be taken into account. Other design parameters like deformation of the tunnelring, forces in the segments and loads on the plates in the radial joint also have to be taken into account. The optimal K_{rad}/K_{tan} ratio is that ratio which leads to minimal reinforcement needed for the transfer of internal forces in the tunnellining and its components combined with an acceptably small deformation or ovalisation of the lining.

Besides the ring action mechanism discussed in this paper other mechanisms also play a role of importance. Each mechanism results in a range of acceptable or preferred stiffnesses regarding the elastic plates in the radial and circular joints. Integral evaluation of all ranging will eventually result in one range of plate stiffnesses which will meet all design criteria.

5 DETAILED MODELING OF ELASTOMERE PROFILE (AND ELASTIC PLATES)

5.1 General

In the preceding chapters the implementation of the behavior of elastomere profile and elastic plates in the joint- and tunnelring model has been discussed. The input for the material models (multi-linear characteristic) is eventually obtained from laboratory test results. Before production and successive testing of elastomere profile and elastic plates these components have to be designed based on the defined specifications.

Practice shows that production and testing of custom made products like the needed joint components is costly and time consuming. Therefore DIANA also has been used to predict the behavior of the elastomere profile and elastic plates. In this paper the modeling of the elastomere profile will be briefly discussed.

5.2 Modeling of the elastomere profile

DIANA offers a number of hyperelastic material models, which can be used to model rubber-like materials. In order to model the behavior of the elastomere profile a Mooney-Rivlin material model has been used which makes use of 2 material constants C1 and C2 and an a bulkmodulus. Because of the fact that elastomere is almost incompressible the bulkmodulus was set at a high value. The material constants C1 and C2 were obtained from formulae that describe the relationship between these constants and the shore stiffness SH.

Because of the fact that strong compressions of the elastomere profile lead to numerical instability the numerical model was eventually used to semi-quantitatively assess the deformation mode of the profile under small compression. Figure 25 shows the deformation mode of a profile which was considered as a design candidate.

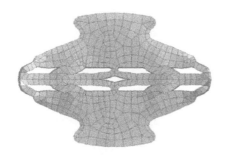

Figure 25. Deformation mode of an elastomere profile.

6 CONCLUSIONS

The complex action of radial and circular joints in the tunnellining has successfully been modeled by means of a simplified system of interface elements resulting in a macro tunnel model with minimal size and complexity.

Successful validation of these simplified joint models based on higher order models and test results guarantees a realistic prediction of the joint action in the macro model of the tunnellining and therefore a correct prediction of the response of the tunnellining itself.

REFERENCE

Kaalberg, F.J. & Hentschel, V. 1999. Tunneling in Soft Soil with a High Water Level and Pile Foundations: Towards the development of settlement-oriented and settlement-minimizing TBM operation. *Proceedings of ITA World Tunnel Congress, Oslo.*

Finite Elements in Civil Engineering Applications, Hendriks & Rots (eds.)
© 2002 Swets & Zeitlinger, Lisse, ISBN 90 5809 530 4

Application of DIANA in concrete building design

J.P. Straman
Delft University of Technology

ABSTRACT: In the past years different components of concrete buildings have been analysed using the finite element method computer package DIANA. Some of the results have been published. The first publication concerns the analyse of a prefabricated floor slab, consisting of prestressed hollow core elements, loaded in his plane. The second one describes the analysis of a prefabricated shear wall, especially the distribution of the shear stresses in the vertical joints between the elements. Both studies were parameter studies and carried out in order to develop simple rules to determine the dimensions of elements of a building in the preliminary design stage. This article gives a short summary of these studies and describes also the design of the bearing structure of a building of which the upper six stories cantilever over an existing old warehouse. One of the design options was the application of reinforced or prestressed concrete walls, providing cantilevers of about 13 m. To determine the stresses and deformation DIANA has been used.

1 INTRODUCTION

On a world scale the scope for high specification precast concrete in buildings is still very large. In the Netherlands e.g. about 50% of the utility buildings are prefabricated and this percentage is even higher concerning the horizontal components of buildings, such as prestressed floors.

So it is not surprising that a substantial part of our desk research concerns the schematization, load distribution, detailing and construction of prefabricated bearing structures of buildings, in order to optimize the elements, the connections and the whole structure.

To utilize the benefits of prefabrication one of the basic rules is to keep the connections as simple as possible e.g. pin-jointed connections. These connections can be made without the need for the additional time required to cure in-situ concrete joints.

However the connections are not moment-resisting, so that the stability of the structure must be provided by other means.

It is generally accepted that the most economic solution in terms of production and erection, is a pin-jointed braced frame, see Figure 1.

1.1 Stability

Stability is provided usually around staircases and and/or lift shafts, by precast concrete shear walls,

cores or shear boxes. Stability can also be provided by in-situ concrete shear cores or shear walls but that is not preferable because they do not utilize the benefits of prefabrication.

The two important elements to provide for stability in this structure are:

– The floor, which has to be sufficiently strong and rigid in the horizontal plane in order to transmit the lateral loads to the cores or walls. If the floor is prefabricated, as it is often the case, measures must be taken to ensure the proper diaphragm action.

Figure 1. Braced structure.

– Cores or walls to transmit the loads from the floors to the foundation. If they are prefabricated, an effort should be made to avoid tensile stresses. Tensile stresses can not be transferred from one precast element to another without additional provisions like reinforcing bars passing from one element to another, welding steel plates or prestressing.

Both, the prefabricated floor as well as the shear wall has been analysed. Together with a third construction, a reinforced concrete wall, providing cantilever of about 13 m, they will be discussed in this publication.

2 PREFABRICATED FLOOR

A hollow core floor slab of an office building, shown in Figure 2, spans from façade to façade, supported by beams extending in the longitudinal direction. In horizontal direction the floor is supported by a core, situated in the middle of the building.

To realise the horizontal diaphragm action the joints between the elements are filled with mortar and a continuous reinforced concrete tie is applied around the edges of the floor. To simulate all precast units, the longitudinal mortar joints and the transverse tie beams the finite element package DIANA has been used.

2.1 Modelling

For the benefit of the computer calculations the model of Figure 3 is used. This model is developed step by step during the different investigations in order to analyse the individual construction elements. On symmetry reasons half the floor plan is taken into consideration. On the symmetry line roller bearings are assumed. The core is schematised by a void in the floor plan, along which bearings are placed. At five places (marked by*) bars are protruding out from the core into the floor structure, modelled as extension and shear springs. Besides the present tension capacity also the shear capacity of the joint is increased by these bars, because the longitudinal deformation goes together with separation of the interfaces, by which tensile stresses in the reinforcement are introduced. Due to that the normal stress perpendicular to the

joint will increase and also the shear resistance. This phenomenon is called dowel action.

2.2 Some results

The deflection at the working load was about 1 mm, and the span-to-deflection ratio at failure was 1/4000. The results have shown that the floor diaphragm failed due to yielding of the ties in the edge beam with a factor of safety of 3.13 with respect to the design strength. The compressive stresses were formed in a triangular distribution and concentrated across 2.5 m of the floor, giving a lever arm factor of $z/\beta = 0.97$. This suggests that the proposed design lever arm factor of 0.8 in case of calculations by hand is very conservative.

The maximum compressive stress obtained in the modelling was $3.6\,N/mm^2$ and the maximum shear stress in the longitudinal joint was $0.4\,N/mm^2$. The final deformation pattern is shown in Figure 4 where

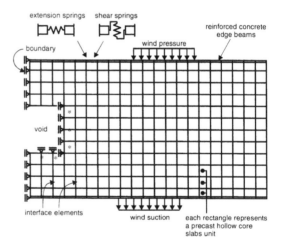

Figure 3. Final model.

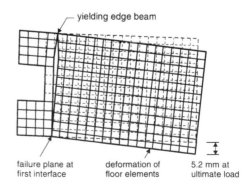

Figure 4. Floor diaphragm deformation.

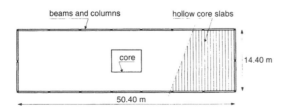

Figure 2. Prefabricated floor.

the critical longitudinal joint is not at the support but at the second interface between the slabs.

3 PREFABRICATED SHEARWALL

Prefabrication of shear walls or cores implies the prefabrication of elements of transportable dimensions, which are connected together at the building site.

Storey-high shear wall panels are connected in such a way that the total wall can function as a cantilevering unit and can be designed as deep beams, as shown in Figure 5.

However, in case of insufficient vertical loading tensile stresses will occur in the horizontal joints and a bridging of these joints by welding or a lap joint is necessary in order to transmit the tensile forces. This is in most cases not an economic solution.

It is than preferred to overpress the tensile stresses by means of vertical prestress. In the case of horizontal load application, displacements can occur in the vertical joints. There are several junction constructions which solve this problem and which are already investigated. By analytical as well as by numerical models these effects have been investigated.

3.1 *Analytical model*

Because the vertical joint possesses certain rigidity, a part of prestress load will be transferred through the vertical joint to the neighbouring elements. The shear rigidity of the joint between the elements determines the shear stress and the magnitude of the deformation in the vertical joint. The shear rigidity K is defined as

the ratio of the shear stress τ and the deformation δ of the junction, see Figure 6.

For the relationship between the prestressing force and the shear stresses distribution in the vertical joint differential equation has been deduced.

3.2 *Numerical analysis*

Modelling
Influence of the load spread at the top end of the wall on the shear stress distribution was determined by means of a numerical analysis.

The wall made up of precast parts is subdivided into:

– prefabricated wall parts;
– vertical joints (continuous joints or specific joints);
– horizontal joints;
– the prefabricated wall parts are modelled by plane stress element CQ16M with dimensions 0.5 × 0.25;
– the vertical joints are simulated by non-linear interface elements CL121;
– the horizontal joints are not modelled because they are subjected to pressure.

3.3 *Comparison of analytical and numerical calculations*

Comparison between analytical (a) and numerical (b) distribution of the shear stress indicated in Figure 7 shows good agreement over nearly the whole wall height. Only in the vicinity of the top end, there is a difference. The closer the load application point to the joint, the greater the accuracy of the analytical method is. With the numerical model, the maximum shear stress is reached at a distance from the top end, which results from the obliquity of the load spread at 45°.

Some conclusions are:

1. The shear stress in the vertical joint of coupled shear wall as a result of prestressing, concentrates at the upper end of the wall.

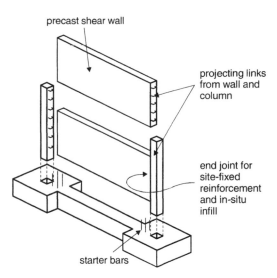

precast shear wall

projecting links from wall and column

end joint for site-fixed reinforcement and in-situ infill

starter bars

Figure 5. Precast shearwall.

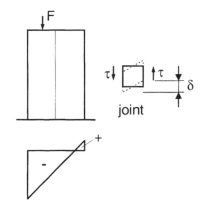

joint

Figure 6. Shear stress in joint.

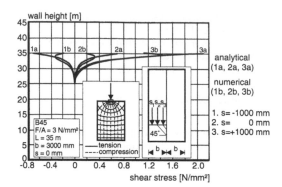

Figure 7. Effect of load eccentricity.

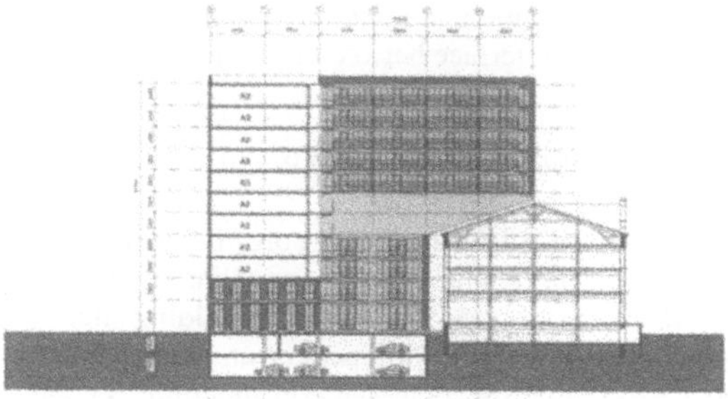

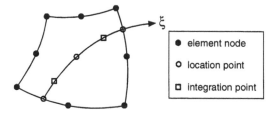

Figure 8. Cantilever wall.

2. The maximum shear stress and its distribution along the wall are dependent on the shear rigidity of the joint.
3. Eccentrically arranged prestressing tendons result in increased joint stresses.
4. The shear stresses are highest in the joint, which is nearest to the prestressing tendon.
5. In the design of joints, the residual stresses resulting from prestressing must be taken into account.
6. Normal compressive stresses in the lower end of the wall may be designed with a monolithical cross section only when the joint shear stress in that position is equal to zero.

BAR particle in plane stress element

Figure 9. Bar particle in plane stress element.

4 CANTILEVER WALL

Figure 8 shows a building of which the upper six floors cantilever over an existing old warehouse. As bearing structure different solutions are possible like steel truss structures, in-situ concrete reinforced or prefabricated pretensioned walls.

One of these solutions, the reinforced concrete wall, has been analysed.

To determine the stresses and deformation of the walls DIANA non-lin has been used.

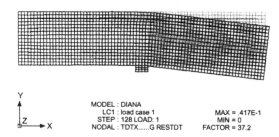

Figure 10. Deformation of wall.

4.1 Data

The dimensions of the walls are:

– total length: 26.4 m
– cantilever : 13.0 m
– height : 5.6 m
– width : 0.4 m

Other data:

– load on wall state: 450 kN/m' serviceability
– reinforcement : at the top : 0.022 m²
  at both sides: 0.003 m²/m'

4.2 Modelling

The wall is modelled by plane stress elements CQ16M.

To represent the reinforcement REINFO B elements have been used, coupled to the plane stress elements.
Figure 9 shows both elements.
The elements are about 350 mm square.

4.3 Some results

1. The failure load of the wall was 617 kN/m', resulting in a mean load factor of 1.37.

2. The maximum deformation of the wall service-ability limit state is 45,4 mm (Fig. 10).
3. The walls were analysed with different calculations methods. It was found that DIANA non-lin was a reliable method to analyse these kind of walls.

REFERENCES

1. Straman, J.P. 1999 Finite element Analysis of Prefabricated floors. Diana World 1999. Issue No. 1.
2. Straman, J.P. 2000. Finite Element analysis of Prefabricated Shear Walls. Diana World 2000. Issue No. 1.

Finite Elements in Civil Engineering Applications, Hendriks & Rots (eds.)
© 2002 Swets & Zeitlinger, Lisse, ISBN 90 5809 530 4

Two-dimensional finite element analysis on shear performance of RC interior beam-column joints reinforced by a new reinforcing method

Zhang Dachang
Graduate School Sc. & Tech., Chiba Univ., Japan

Noguchi Hiroshi & Kashiwazaki Takashi
Dept. of Design and Architecture, Chiba Univ., Japan

ABSTRACT: In order to study the seismic performance of RC interior beam-column joints reinforced by a new reinforcing method, two-dimensional finite element analysis was carried out, with the variations of reinforcing and bonding condition in a joint. Analytical results considering bonding slip between beam and column main bars and concrete in the joint showed a good agreement with the test results. Loading-deformation relations, stresses and strains in joint, the deformation ratio of the beam, column and joint, and deformation modes and failure modes were discussed. It was recognized that the new reinforcing method for the joint could improve bonding condition between beam main bars and joint concrete, and increase the joint shear capacity with changing the failure mode.

1 GENERAL INSTRUCTIONS

When the ultimate strength design method is adopted, the shear stress of RC interior beam-column joints will be increased, due to the increased amount of beam main bars avoiding beam or column's brittle failure. What is more, frame's relative weakness will move to the beam-column joint. Therefore, joint's ductility reduces, and the possibility of brittle failure increases with the increase of joint shear stress and beam main bars' stress. Many reinforcing methods have been investigated and developed, in order to increase the joint shear capacity. In this study, the two-dimensional FEM analyses are carried out about the RC interior joint specimens tested by Shiohara[1] in Tokyo University.

The purpose of the paper is to investigate the deformation performance and failure mode of the RC interior beam-column joint and the effectiveness of the new reinforcing method proposed by Shiohara, based on the macroscopic model. The loading-deformation relation, beam main bars strain distribution, and concrete stress distribution in the joint are also investigated.

2 ANALYTICAL SPECIMENS

2.1 *Failure modes proposed by Shiohara[2]*

The moment-resisting mechanisms of the RC interior beam-column joint suggested by Shiohara are shown

in Figure 1, which are two kinds of joint failure modes. It is B-mode whose beam main bars are pulled out on the flexural cracks and the flexural cracks open at the beam end, and the beam's rotational deformation becomes remarkable. On the other hand, X-shaped diagonal shear cracks occurs in the joint under cyclic loading, and the joint panel rotates around the center of the joint, as shown in Figure 1 (b), J-mode shear failure occurs. In this case, the joint shear deformation is outstanding. The deformations of B-mode and J-mode are mainly concentrated at the end of the beams, and the joint panel, respectively.

2.2 *Analytical specimens*

In this paper, the RC interior beam-column joint specimens S3, S4 tested by Shiohara are chosen as the analytical objects in order to investigate the effectiveness

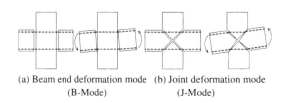

(a) Beam end deformation mode   (b) Joint deformation mode
        (B-Mode)   (J-Mode)

Figure 1.   Resisting mechanisms of the RC interior beam-column joint.

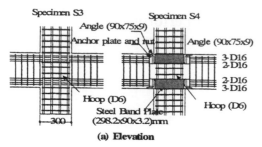

(a) Elevation

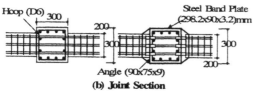

(b) Joint Section

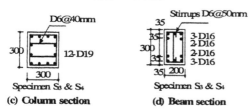

(c) Column section   **(d) Beam section**

Figure 2. Geometry of beam-column joint and steel bars arrangement.

Table 1. Mechanical properties for steel bars and plates.

Reinforcing bars and plates	Nominal sectional area (cm^2)	Yield strength f_y (MPa)	Tensile strength f_u (MPa)	Young's modulus E_S (GPa)
D6	0.32	390	580	185
D16	1.99	470	660	194
D19	2.87	450	680	208
Steel Plate	2.88	353	475	205

of the reinforcing method considering above-mentioned failure modes. The dimensions of the specimens and the reinforcing details of the joints are shown in Figure 2. Specimen S3 that had conventional reinforcing details was designed as the joint shear failure type before the beam bars' yielding. The joint S4 has the same dimensions, reinforcing bars and material strength as the specimen S3. It had an additional new joint reinforcing detail in the joint core. Four steel angles (90 × 75 × 9) perpendicular to beam axis were welded to the ends of the steel bands as shown in Figure 2. The mean compressive strength for concrete was 28.0 Mpa. The material mechanical properties of the steel bars and tie plates are shown in Table 1.

During the test, statically cyclic lateral load was applied maintaining constant axial load at the top of the column.

Figure 2 shows the comparison of the story shear-story deflection angle for specimens S3 and S4. Attained maximum story shears were 128.4 kN and 164.8 kN for specimens S3 and S4, respectively. The capacity of the specimen S4 in term of the story shear was 28% larger than that of S3, and the ductility of the specimen S4 was better than that of specimen S3.

3 ANALYTICAL PROGRAM

In this study, FEM analytical computer program DIANA was used, and the subroutine for additional material constitutive models was developed based on the subroutine function.[3]

3.1 Finite elements and material constitutive models

(1) Concrete

For concrete, the four-node quadrilateral isoparameteric plane stress element was applied. The equivalent strain relation proposed by Darwin, Pecknold[4] was used. When referred to the orthotropic principal axes, the form of the incremental constitutive relation for four-node quadrilateral isoparameteric plane element can be written as equation (1).

$$\begin{Bmatrix} d\varepsilon_1 \\ d\varepsilon_2 \\ d\gamma_{12} \end{Bmatrix} = \begin{bmatrix} E_1^{-1} & -v_{12}E_2^{-1} & 0 \\ -v_{21}E_1^{-1} & E_1^{-1} & 0 \\ 0 & 0 & G_{12}^{-1} \end{bmatrix} \begin{Bmatrix} d\sigma_1 \\ d\sigma_2 \\ d\tau_{12} \end{Bmatrix} \quad (1)$$

Being inversed, equation (1) becomes

$$\{d\sigma\} = [C]\{d\varepsilon\} \quad (2)$$

in which $\{d\sigma\}$ and $\{d\varepsilon\}$ = the vectors of incremental stresses and strains appearing in the Eq. 1, the constitutive matrix $[C]$ is as follows:

$$[C] = \frac{1}{1-u_{12}^2} \begin{bmatrix} E_1 & \sqrt{E_1 E_2}\,u_{12} & 0 \\ & E_2 & 0 \\ sys & & (1-u_{12}^2)G_{12} \end{bmatrix} \quad (3)$$

and $u_{12}^2 = v_{12}v_{21}$

$$G_{12} = \frac{1}{4(1-u_{12}^2)}\left[E_1 + E_2 - 2u_{12}\sqrt{E_1 E_2}\right]$$

Let Eq. 2 be written as follows

$$\begin{Bmatrix} d\sigma_1 \\ d\sigma_2 \\ d\tau_{12} \end{Bmatrix} = \begin{bmatrix} E_1 B_{11} & E_1 B_{12} & 0 \\ E_2 B_{21} & E_2 B_{22} & 0 \\ 0 & 0 & G_{12} \end{bmatrix} \begin{Bmatrix} d\varepsilon_1 \\ d\varepsilon_2 \\ d\gamma_{12} \end{Bmatrix} \qquad (4)$$

Carrying out the multiplication in Eq. 4 yields

$$\begin{aligned} d\sigma_1 &= E_1(B_{11}d\varepsilon_1 + B_{12}d\varepsilon_2) \\ d\sigma_2 &= E_2(B_{21}d\varepsilon_1 + B_{22}d\varepsilon_2) \\ d\tau_{12} &= G_{12}d\gamma_{12} \end{aligned}$$

which can be written, in the matrix form, as follows:

$$\begin{Bmatrix} d\sigma_1 \\ d\sigma_2 \\ d\tau_{12} \end{Bmatrix} = \begin{bmatrix} E_1 & 0 & 0 \\ 0 & E_2 & 0 \\ 0 & 0 & G_{12} \end{bmatrix} \begin{Bmatrix} d\varepsilon_{1u} \\ d\varepsilon_{2u} \\ d\gamma_{12} \end{Bmatrix} \qquad (5)$$

The vector on the right-hand side of Eq. 5 may be defined as the vector of "equivalent incremental uniaxial strains", whose components are defined in axial strains", whose components are defined in the terms of actual incremental strains as follows:

$$d\varepsilon_{iu} = B_{i1}d\varepsilon_1 + B_{i2}d\varepsilon_2; \quad i = 1,2 \qquad (6)$$

The incremental equivalent uniaxial strain is evaluated from Eq. 5 as follows:

$$d\varepsilon_{iu} = \frac{d\sigma_i}{E_i} \qquad (7)$$

and the total equivalent uniaxial strain is determined by integrating Eq. 7 over the load path as follows:

$$\varepsilon_{iu} = \int \frac{d\sigma_i}{E_i} \qquad (8)$$

The uniaxial compressive stress-strain relationship of Saenz[5] is used for the ascending branch of the stress-equivalent strain curve, and the Modified Kent-Park model[6] is used on the peak strength and descending branch as follows:

$$\sigma_i = \frac{E_0 \varepsilon_{iu}}{1 + \left(\dfrac{E_0}{E_s} - 2\right)\dfrac{\varepsilon_{iu}}{\varepsilon_{ic}} + \left(\dfrac{\varepsilon_{iu}}{\varepsilon_{ic}}\right)^2} \qquad (9)$$

The tension stiffening is considered by using the Shirai model.[7] That can be written as follows:

$$\sigma_{t,eq} = \left(a_0 + a_1 X + a_2 X^2 + a_3 X^3\right)f_t \qquad (10)$$

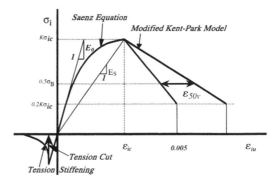

Figure 3. Equivalent uniaxial stress-strain curves for compression and tension.

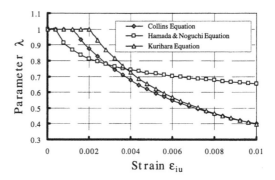

Figure 4. Reducing parameter for concrete compressive strength due to cracking.

where, $X = (\varepsilon_{av} - \varepsilon_{cr})/(\varepsilon_{Bu} - \varepsilon_{cr})$, f_t is the tension strength of concrete, ε_{av} is the mean tension strain, ε_{cr} is the cracking strain, ε_{Bu} is the bond strain, and a_1, a_2, a_3 are constant parameters 1.0, -2.748, 2.654, -0.906, respectively.

The above equations of stress-strain relations for concrete are illustrated in Figure 3.

The Darwin model is used as the failure based on the Kupfer's failure criteria.[8] What is more, the Hamada and Noguchi equation[9] is used to consider the reducing effectiveness on the concrete compressive strength due to cracking in the perpendicular direction shown in Figure 4, and reducing parameter can be written as follows:

$$\lambda = \frac{1}{0.27 + 0.96\left(\varepsilon_{1u}/\varepsilon_0\right)^{0.167}} \qquad (11)$$

where, ε_{1u} is the mean tensile strain perpendicular to the crack's direction, ε_0 is the strain of the uniaxially compressive strength of concrete.

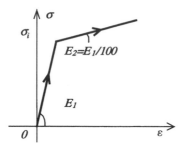

Figure 5. Steel bar stress-strain relation.

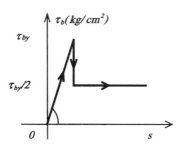

Figure 6. Bond stress-slip relation between steel and concrete.

(2) Steel bars

The 2-node truss element is used for beam and column main bars. The stress-strain relation is approximated by bilinear shown in Figure 5. The second module is $E_2 = E_1/100$. The layer element is used for the beam and column hoops, anchor plates and steel band plates. The Von-Mises condition is used as the yield criteria of the layer steel.

(3) Interface element

The interface element is used to consider the bond slip between the beam and column bars and the concrete. Bond stress-slip relation is approximated by bilinear model shown in Figure 6 according to various test results. The bond strength is calculated based on the Morita and Kaku's tests and theoretical equations on the bond strength.[10][11][12]

Besides these, the discrete crack model is used through the interface elements along the X-shaped preset diagonal shear cracks in the joint panel shown in Figure 7. The strength of the interface element for discrete crack is the tensile strength of concrete, and the failure energy is $0.05\,N/mm^2$. Concrete shear stiffness between the cracks is reduced to 10% of the former after cracking.[3]

3.2 Finite element model for analysis of RC interior joint specimens

In Figure 8, the finite element mesh for the analysis of interior joint specimens is shown. The bond between

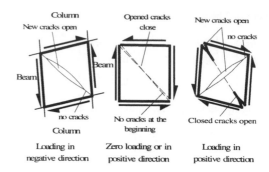

Figure 7. Discrete cracks model for diagonal shear cracks in the joint panel.

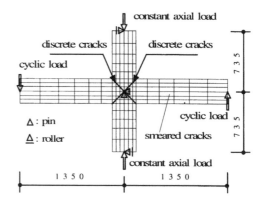

Figure 8. Mesh of the specimen and the boundary condition.

beam and column bars and concrete is considered by two kinds of conditions: perfect bond and bond slip by the interface element. The crack models include smeared cracks that are used in the whole specimen, and discrete cracks that are used in joint diagonal shear cracks. In the test, cyclic lateral load was used at the top of the column while maintained constant axial load on the column. But there are some difficulties on the FEM analysis of RC interior joints under cyclic lateral load, the joint analyses are carried out with monotonic load. It is thought to be possible that the X-shaped preset diagonal discrete cracks could illustrate the effects of cyclic loading on the cracks' opening and closing even by monotonic loading analysis.

The steel band plates of Specimen S4 are modeled by the layer element. The beam bars are welded to the anchor angles, and are supposed to be unable to slide at the panel and have perfect bond with concrete.

3.3 Loading and boundary condition

As shown in Figure 8, a 100 kN axial load is firstly loaded at the top of the column and maintains constant.

458

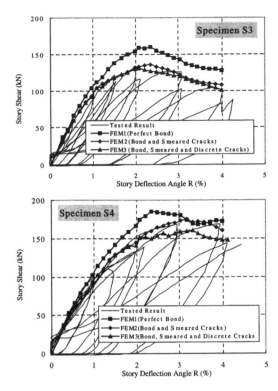

Figure 9. Story shear-story deflection angle relation.

After this, a lateral load is applied at the end of the beam and increased step by step, which left is downward and right is upward. The boundary conditions are that center of the joint panel is fixed by a pin, and that rollers are applied at the ends of the column in x direction. According to the calculation, the joint deformation from the joint P-δ effectiveness is only 1/100 that of the loads, therefore, the P-δ effectiveness is not considered during the analysis.

4 ANALYTICAL RESULTS

4.1 *Story shear-story deflection relations*

The relations of the story shear-story deflection angle from the test and analysis are compared in Figure 9.

In Figure 9, the result FEM1 presents the result of perfect bond, and the result FEM2 presents the result of considering the bond slip between beam and column main bars and concrete, when only smeared cracks model is applied in the whole specimen. The result FEM3 is obtained from the analysis that smeared cracks model is applied in the whole specimen and discrete cracks model is applied in the X-shaped preset diagonal cracks of the joint panel.

The shear strength of specimen S3 and S4 for perfect bond from the analysis are 24.8%, 20.0% higher than the test results, respectively. The shear strength of the specimen S3 obviously reduces after the maximum strength, but specimen S4 has not this obvious reduction like specimen S3. The analytical results have a good agreement with the tested results, in which the bonding slip between the main bars and concrete is considered and the smeared cracks is applied. Besides these, the results FEM3 that the discrete cracks model is applied can be consistent with the test results.

It can be concluded that the bond condition of beam and column main bars with concrete has a great effect on the story shear strength, and the new reinforcing method is effective on the joint performance of loading-deflection.

4.2 *Stress and strain performance of the joint*

4.2.1 *Joint concrete*

(1) *Compressive principal stress distribution*
The compressive principal stress of the concrete in the joint is shown in Figure 10 for three kinds of bonding conditions and two kinds of crack model, when the story deflection angle R is 1%, 2%, 3% and 4%, respectively.

For specimen S3, seen from the above stress distribution from four phases, it can be known that the stress magnitude changes, and the compressive flow from the right above to the left bottom generates at the central zone of the joint, which is called a compressive strut. When the bond between main bars and concrete is considered as perfect bond (FEM1), the shape of the compressive strut expands with the increase of the load. But when the bond is considered through bond slip (FEM2), the shape of the compressive strut does not expand with load increase, and the stress is higher than that of the perfect bond (FEM1), and the distribution of the stress flow is a little concentrated on the diagonal line.

For specimen S4 that was reinforced by a new reinforcing method, the shapes of the compressive strut of the perfect bond and bond slip, and the stress distribution and stress magnitude have not obvious difference like that of specimen S3, because the beam bars are welded with the anchor plate and can hardly slide in the joint panel.

For FEM3 results with discrete crack and smeared cracks model, the compressive strut flow were generated with the increase of the load, and the stress magnitude and distribution have some differences from those of FEM2 and FEM1, especially for specimen S3, because the X-shaped diagonal discrete cracks open and close during the loading and some parts of concrete around the discrete cracks fail.

What is more, the compressive strut concrete of the specimen S3, fails by compression and the compressive

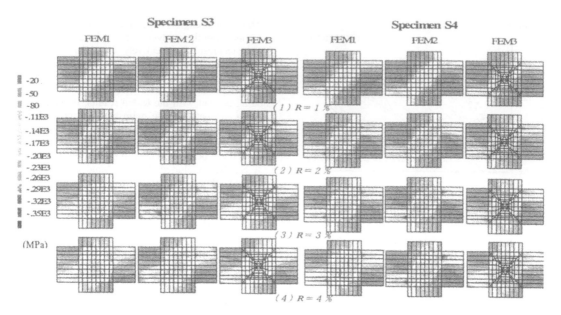

Figure 10. Compressive principal stress distribution in joint panel (FEM1: perfect bond, FEM2: bond slip and smeared cracks, FEM3: bond slip, smeared and discrete cracks).

strut becomes unobvious and goes to vanish, because the load for the perfect bond condition is larger than that of bond slip condition on the same story deflection angle, after the story deflection angle R = 2%, joint reaches the shear strength. But for specimen S4 reinforced by a new reinforcing method, the compressive strut is maintained until the story deflection angle R reaches 4%.

It can be concluded that the bond condition of the beam and column bars, and the reinforcing condition have effect on the change of the shape of the concrete compressive strut flow, and stress distribution in the panel and the shear strength of the joint.

(2) *Compressive principal stress-equivalent uniaxial strain relation of concrete at the middle point in joint panel*

It is necessary to investigate the stress condition of concrete combining with the strain condition in details, though the stress condition of concrete can be estimated through the distribution of compressive principal stress shown in some levels in Figure 11. Besides this, the strength of the concrete increases under biaxial compressive stress condition, but the compressive strength of concrete is reduced with the increase of tensile strain according to Hamada & Noguchi equation, when cracks occur in the perpendicular direction of compression.

In Figure 11, concrete compressive principal stress-equivalent uniaxial strain relation is shown at the middle point in joint panel. The curve of the strength increase of confined concrete is the curve "Modified Kent Park", according to the Saenz equation. In figure, the strength increase of concrete is shown in solid line, and the reduced actual strength of concrete is shown in dashed line. The stress condition of concrete is discussed in two conditions, perfect bond and bond slip of beam and column bars with concrete. Seen from the figure, and confined condition can increase concrete strength, and the concrete strength of specimen S4 is larger than that of specimen S3 because of the new additional reinforcing method applied in joint core.

For specimen S3, in the same strain, the strength increase of concrete of perfect bond is a little quicker than that of considering bond slip, and the reduced actual strength of concrete of perfect bond is higher than that of considering bond slip. So, it can be known that the concrete damage for RC interior joint in general reinforcement on considering bond slip is a little quicker than that of perfect bond. But for specimen S4, the strength increase of concrete of perfect bond and bonding slip is almost the same, and the reduced actual concrete strength in two kinds conditions is almost the same too, because the beam and column bars are welded with the anchor plates and hardly slide in the joint panel.

Based on concrete compressive principal stress – equivalent uniaxial strain relationship, confining and bond condition has a valid effect on the concrete strength and damage, and the new reinforcing method can improve concrete strength and damage in joint.

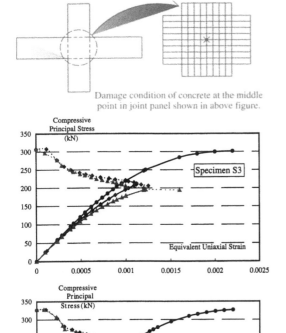

Damage condition of concrete at the middle point in joint panel shown in above figure.

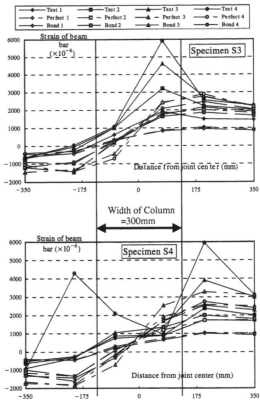

Figure 11. Concrete compressive principal stress–equivalent uniaxial strain relation. (In figure, the strength increase of concrete is shown in solid line, and the reduced actual strength of concrete is shown in dashed line.)

Figure 12. Strain distribution along the bottom middle longitudinal beam bars.

4.2.2 Strain distribution of beam bars at the range of the joint

Strain distribution of bottom bars at the range of the joint from the analysis and test is shown in Figure 12, when the story deflection angles are 1%, 2%, 3%, and 4%, respectively. " Test 1, Test 2, Test 3, Test 4" indicate the test results on story deflection angles are 1%, 2%, 3%, and 4%. "Perfect 1, Perfect 2, Perfect 3 and Perfect 4" indicate the analytical results of perfect bond and "Bond 1, Bond 2, Bond 3, Bond 4 indicate the analytical results on considering bond slip.

There are some differences between the tested results and the analytical results, but the strain distribution form the analysis along the bottom middle longitudinal bar can indicate the strain change along beam bars as the test.

As in usual cases of flexural failure of the beam, the yielding of the bottom middle longitudinal bar starts at the beam column interface at 2% story drift angle and extends to outside of the face of the column with the increase of the load.

Comparison of the strain distribution along the bottom middle longitudinal bars through the joint core for specimen S3, S4 are shown in Figure 12 respectively, which are obtained from the test and analysis. Apparently, the strain in compressive bar is kept in compression for specimen S4 until story deflection angle R = 3%. This is the result of combined effect of steel plate confining the volume expansion of the joint and the anchor angle plates fixed by nuts preventing the bond slip in the joint panel. However, when the story drift angle increases to 4%, it experiences tensile strain, this comes as the result of yielding of steel plates which lose its function of preventing bond slip. Moreover, the level of strain inside the joint is kept lower significantly than the yielding strain in specimen S4 contrary

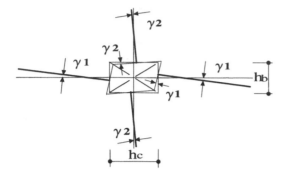

(a) Lateral story displacement due to shear distortion

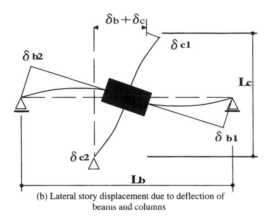

(b) Lateral story displacement due to deflection of beams and columns

Figure 13.   Decomposition of inter story displacement.

to specimen S3 in which large inelastic strain occurs. Comparing the strain level for 3% story deflection angle, it is evident that the effectiveness of increasing the moment capacity of joint relative beam in changing the mode of failure from joint failure which is observed in specimen S3 to favorable flexural beam failure of specimen S4.

4.3   Deformation decomposition and failure mode of RC interior joint

4.3.1   Decomposition ratio of the deformation of beam, column and joint

In general, the imposed story displacement of joint is composed of elastic and inelastic deformation of the structural elements in the test specimen. The structural elements of the joint specimen consists of two beams, columns and a joint panel, and the story displacement includes the component of joint deformation in Figure 13 (a), beams deformation and columns deformation in Figure 13 (b). The imposed story displacement of the

test specimen comprises several components, and can be written as the sum of each component as follows:

$$\Delta = \delta_b + \delta_c + \delta_j \qquad (12)$$

where, δ_b is contribution of beams including flexural and shear deformation, δ_c is contribution of columns including flexural and shear deformation, δ_j is contribution of joint shear deformation.

In Figure 14, deformation ratios for beam, column and joint panel are shown. From figures, it is evident that the deformation ratios of each structural element from the analyses changes with the increase of the load, and all have the tendency that the deformation ratio of beam decreases and the deformation ratio of the column has no great change, while the deformation ratio of the joint increases. According to the analytical results (a) and (c) of test specimen S3, the deformation ratio of each structural element in perfect bond is nearly the same as that of the bonding slip before R = 0.5%. But after this, the deformation ratio of beam in bond slip is lager than that of perfect slip until R = 2%, because the slip of beam bars occurs and the beam deformation quickly develops contrary to that of the perfect bond. At last, the deformation ratio of the joint with perfect bond is larger than that of bond slip, but the deformation ratio of the beam of perfect bond is smaller than that of bond slip. What is more, the deformation ratio of the joint increases quickly after the story deflection angle R = 2%.

For specimen S4, there are some differences for deformation ratio between perfect bond and bond slip, but the deformation ratio of two kinds of condition is nearly the same.

Seen from the deformation ratio of specimen S3 and S4, the deformation ratio of beam for S3 and S4 increases with the increase of the load and the joint deformation decreases. But the deformation ratio of the joint for specimen S3 increases quickly, and at last is larger than that of specimen S4. The deformation ratio of the joint for specimen S4 increases too, and it is slowly than that of specimen S3. At last, the deformation ratio of joint is smaller than that of beam. Based on those, it is indicated that specimen S3 fails in joint at last, but specimen S4 fails in beam.

At the same time, in the test research it was measured that the deformation in diagonal of the joint panel for specimen S3 was nearly the same to that of specimen S4 before the story deformation angle R = 1%. But after that, the deformation in diagonal of the joint panel for specimen S3 was larger than that of specimen S4. At last, the diagonal deformation of the joint panel for specimen S3 increased quickly, but that of specimen S4 increased a little slowly than that of S3.

Based on the analytical analysis and test research, it is concluded that specimen S3 fails in joint shear failure and specimen S4 fails in beam flexural failure,

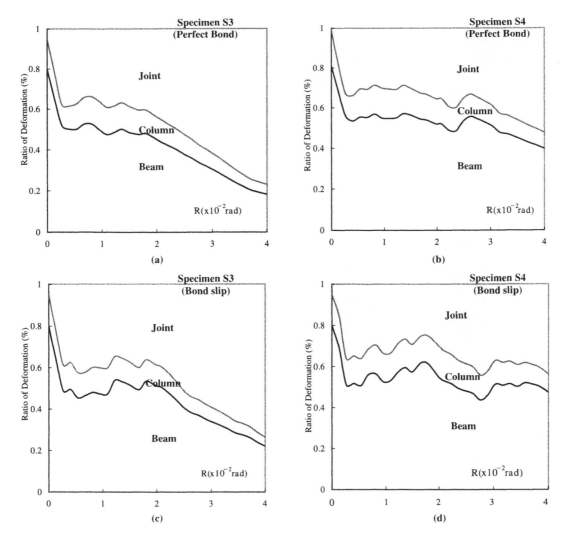

Figure 14.   Ratio of deformation of beam, column and joint (%).

because of the effectiveness of new reinforcing method on specimen S4.

4.3.2 *Deformation mode of joint*

During the analysis, the joint deformation in every step can be seen, and the characteristic of joint deformation can be evaluated based on the analytical results.

Before the story deformation angle R = 1%, the joint deformation of perfect bond and smeared cracks is nearly the same to that of bond slip for beam bars and smeared cracks, and even to that of bond slip and smeared cracks and discrete cracks. After story deformation angle R = 1%, compared with the deformation of perfect bond with the deformation of bonding slip,

the joint panel deformation of perfect bond is quickly increased than that of bong slip considered.

It is evident that the joint deformation of specimen S3 is larger than that of specimen S4, while the beam deformation of specimen S3 is smaller than that of S4, because the beam main bars cannot slide in the joint panel and decrease the beam deformation from bars' pulling out.

Seen the Figure 15(b) and (c), it can be known that the joint deformations of bond slip with smeared cracks for specimen S3 is nearly the same to that of bind slip and smeared cracks and discrete cracks, especially for specimen S4. And from the opening and closing of the discrete cracks in Figure (c) of specimen S3, it can be

Specimen S3   Specimen S4

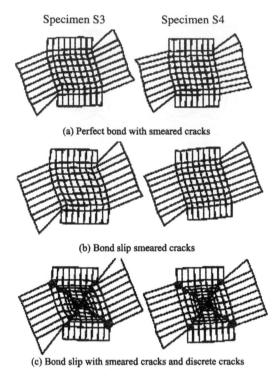

(a) Perfect bond with smeared cracks

(b) Bond slip smeared cracks

(c) Bond slip with smeared cracks and discrete cracks

Figure 15. Deformation Mode of Joint (R = 3%, Scale = 20).

known that the joint failure as J-model for specimen S3 occurs. And the discrete cracks in specimen S4 hardly opens, and the joint deformation is smaller than specimen S3. But the beam deformation of specimen S4 increases slowly, at last the beam flexural failure occurs as B-mode and the joint does not fail, because of the confined effect of the steel plates and steel angles.

4.3.3 Failure mode for joint

On one hand, specimen S3 reaches the story shear strength 128 kN on the story deflection angle R = 2%, and the deformation of the joint increases. Joint shear failure occurs at last with diagonal crack occurred opening. On the other hand, specimen S4 reaches the story shear strength 165 kN until the story deflection angle R = 3%, and the deformation of the joint increases slowly, and beam flexural failure occurs with flexural cracks opening at the end of the beam. These can be conformed based on the deformation ratio of each structural element (Figure 14), and on the measured deformation about diagonal deformation during the test. As shown in Figure 15, the deformation mode of J-mode failure can be reproduced by the analysis when the discrete cracks are used along the diagonal

line of the joint, at the same time, the smeared cracks are used in the whole model.

It was conformed in the test research by Shiohara that the new reinforcing method makes the failure mode and the shear strength change, and it is conformed through the analysis of the deformation ratio and deformation mode, too. Specimen S3 fails in J-mode, because the concrete crushes and fails around the discrete crack after shear strength on R = 2%. As for specimen S4, the concrete failure does not occur around the discrete crack in joint without crack opening and closing, but beam flexural failure occurs in compressive zone.

At last, the specimen S3 fails in joint shear failure of J-mode when the discrete cracks opens and closes. Specimen S4 fails in beam flexural failure of B-mode, and opening of the discrete shear cracks in the joint are prevented.

5 CONCLUSIONS

(1) Based on test research, it is analytically conformed to be effective that the new joint reinforcing detail method suggested by Shiohara can increase the story shear strength, and makes the joint failure from joint shear failure to the beam flexural failure that the failure mode changed J-mode to B-mode.

(2) Using smeared crack and considering bond between beam main bars and concrete, the analytical result has a good agreement with the test result.

(3) When the smeared crack is used in whole model and the discrete crack is applied in the diagonal of joint core, not only has the analytical result a good agreement with the test result, but also the failure mode for joint suggested by Shiohara could be reproduced, through the discrete crack opening and closing. That is to say, the joint failure J-mode could be reproduced in analysis.

(4) The bond characteristic of beam main bars has a great effect on the joint shear capacity, and the joint failure mode. And improving the bond condition of the beam main bars can improve the performance of the joint.

REFERENCES

1. Safaa Zaid, Hitoshi Shiohara, Ahunsuke Otani: Test of Joint Detail Improving Joint Capacity of R/C Interior Beam-column Joint, The 1st Japan-Korea Joint Seminar on Earthquake Engineering for Building Structures, Oct.29-30, 1999, Faculty Club House, Seoul National University, Seoul, Korea.
2. Shiohara etc.: An Analysis on Two Failure Modes for Reinforced Concrete Beam-Column Joints, Summaries

of Technical Papers of Annual Meeting Architectural Institute of Japan, Structure II, pp 797–799, 2000.

3. DIANA 7.2, User's Manual: Nonlinear Analysis, Nov 19, 1999.

4. Darwin, D., and Pecknold D.A.W.: Inelastic Model for Cyclic Biaxial Loading of Reinforced Concrete, A Report on A Research Project Sponsored by the NSF, University of Illinios, July 1974.

5. Chen W.F.: Plasticity in Reinforced Concrete, McGraw-Hill International Book Company, New York, U.S.A., 1982.

6. Robert Park, M. ASCE, M. J. Nigel Priestley and Wayne D. Gill: Ductility of Square-Confined Concrete Column, April 1982.

7. Toshio Sato, Nobuaki Shirai: A Study on the Elastic and Inelastic Performance of RC Seismic Wall, Summaries of Technical Papers of Annual Meeting Architectural Institute of Japan, Structure II, September 1978.

8. Kupfer, H.B., Hilsdorf, H.K., and Rusch, H.: Behavior of Concrete under Biaxial Stresses, Journal of the American Concrete Institute, Vol. 66, No.8 pp 656–666, 1969.

9. Satoshi Hamda, Hiroshi Noguchi: Basic Experiment about compressive Characteristic Damage of Cracked Concrete, Summaries of Technical Papers of Annual Meeting Architectural Institute of Japan, Structure II, pp 397–399, 1988.10.

10. Shiro Morita, Tetuzo Kado: Study on the Bonding between Bar and Concrete under Cyclic Loading, Summaries of Technical Papers Architectural Institute of Japan, pp 15–24, March, 1975.

11. Ei Fujii, Shiro Morota: Study on the Bonding Crack Strength of Deformed Bar, Report 1: Test Result, Summaries of Technical Papers Architectural Institute of Japan, pp 47–54, September 1982.

12. Ei Fujii, Shiro Morota: Study on the Bonding Crack Strength of Deformed Bar, Report 2: Proposal about Calculating Formula, Summaries of Technical Papers Architectural Institute of Japan, pp 45–52, February 1983.

Finite Elements in Civil Engineering Applications, Hendriks & Rots (eds.)
© *2002 Swets & Zeitlinger, Lisse, ISBN 90 5809 530 4*

Finite element analysis of prestressed concrete containment vessel subjected to internal pressure

H.-W. Song, B. Shim & K.-J. Byun
Department of Civil Engineering, Yonsei University, Seoul, Korea

ABSTRACT: In this paper, finite element analysis is carried out for failure analysis of prestressed concrete containment vessel (PCCV) subjected to accidental internal pressure. For an 1 : 4 model of prestressed concrete containment vessel, layered shell finite element modeling technique along with prestressing tendon modeling is applied. For verification of the modeling, global and local behaviors of the PCCV obtained from the analysis are compared with results of the so-called both limit state test and structural failure mechanism test at the Sandia National Laboratories in USA. As a new analysis technique for the PCCV subjected to internal pressure, the so-called volume control technique is also introduced. Layered shell elements with the pressure node attached to the shell elements are applied to modeling of the PCCV and in-plane reinforced concrete constitutive models are implemented. Formulation of the technique and comparisons of analytical results with experimental data are presented. The comparison shows that modeling technique including the volume control technique in this paper can be effectively used for the prediction of the ultimate pressure capacity of the prestressed concrete reactor containment structures subjected to internal pressure.

1 INTRODUCTION

Prestressed concrete reactor containment vessel (PCCV) is a typical PSC shell structure which has been increasingly built recently. Finite element modeling technique to obtain the failure behaviors of PSC shell structures including ultimate load and post-peak behavior have been attempted rigorously in the field of computational mechanics.

For design of prestressed concrete reactor containment vessels considering extremely high internal pressure due to fatal accidents, it is necessary to develop analytical methods to predict ultimate pressure capacity of the vessels subjected to accidental internal pressure due to the fatal accidents. The ultimate capacity of a PCCV subjected to internal pressure can be determined by two failure modes, which are structural failure mode due to mainly primary reinforcement bar yielding of the wall of the vessel and local failure mode due to tearing of inner steel-liner and subsequent leakage (Dunham et al. 1985).

Recently, a round robin test co-sponsored by Nuclear Power Engineering Corporation of Japan and US Nuclear Regulatory Commission (NRC) has been performed for a prestressed concrete containment vessel (PCCV) 1 : 4 model at the Sandia National Laboratories (SNL) in USA to find out the ultimate behavior of the PCCV model experimentally (Hessheimer 2000). In

this paper, experimental results of the test along with test set-up are explained and a finite element modeling technique of the PCCV model is described and a new numerical technique for the PCCV model subjected to pressure is also introduced. Analytical techniques are verified by comparisons with the experimental results.

2 PCCV 1 : 4 MODEL

A test on a PCCV model was carried out at New Mexico, USA under the supervision of the SNL. The model is an 1 : 4 scale of prestressed concrete reactor containment vessel. Figure 1 shows layout and dimension of the PCCV 1 : 4 model. The model has 1.6 mm thick steel inner liner, equipment hatch, and personal airlock. There are 90 and 18 hoop post-tensioning systems in wall, and in dome, respectively. And, there are 90 hairpin type meridional post-tensioning systems. Each tendon consists of 3 strands whose diameter is 13.7 mm. The PCCV model mainly consists of cylindrical wall and spherical dome to resist membrane tension force of the wall due to internal pressure. Additional thickening or reinforcing is provided to strengthen junctions between wall and base and transition areas near equipment hatch of the wall. A previous study on an RCCV (Amin et al. 1993) showed that behavior of general area like wall and dome other than

467

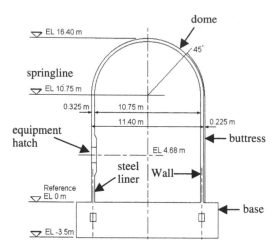

Figure 1. Layout and dimension of PCCV 1 : 4 model.

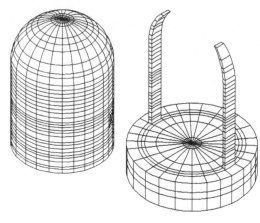

Figure 2. Finite element discretization of PCCV 1:4 model.

the junctions and the transition area governs the failure behavior of RCCV. Thus, failure behavior of the general areas of the PCCV model is a main concern in the analysis. Strengths of concrete are 44 MPa and 30 MPa for wall and base, respectively. The design pressure P_d of the PCCV 1 : 4 model is 0.39 MPa.

3 FINITE ELEMENT MODELLING OF PCCV

In this study, the finite element program DIANA 7.1 (DIANA 1998) is used for finite element analysis of the PCCV model. Figure 2 shows finite element discretization of the PCCV model using shell and solid elements. The shell elements are used for discretization of wall and dome, and isoparametric solid elements for discretization of buttress and base.

As shown in Figure 3, layered shell modeling is used for shell elements, i.e. 8 nodes quadrilateral isoparametric layered curved shell elements are used for wall and dome, and 6 nodes quadrilateral isoparametric layered curved shell elements for the apex of dome, respectively. Figure 3 also shows a section of wall and the layered shell element, which consists of 8 layers; a steel liner layer, a concrete layer, an interior meridional rebar layer, an interior hoop rebar layer, a concrete layer, an exterior meridional rebar layer, an exterior hoop rebar layer, and an outer concrete layer in order from the inner surface. Thickness of each layer is calculated from each area ratio of each material section, and rebar layers have only one direction stiffness according to hoop or meridional directions.

The rebars in the solid element for buttress and base as shown in Figure 2 are modeled by embedded grid reinforcement element. Tendons in wall of the PCCV

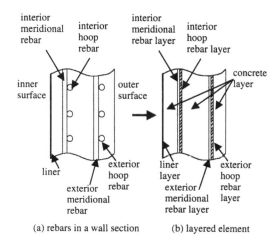

(a) rebars in a wall section   (b) layered element

Figure 3. Layered shell modeling of wall and dome.

model are modeled by embedded reinforcement bar as shown in Figure 4. By assuming perfect bond between the reinforcement and surrounding material, the strains of rebars are computed from the displacement field of mother elements using the BOND option in the DIANA (Witte & Feenstra 1998). Since strains of tendons do not vary with loading in the analysis with NOBOND option, the NOBOND option in the DIANA is not used for the unbonded tendon. Note that bond slip modeling in the DIANA needs interface element as for the modeling of interface element in complicate 3D modeling.

For entire discretization of the PCCV model, 1060 layered shell elements are used for wall and dome, 296 isoparametric solid elements for base and buttress, 198 embedded reinforcement bar elements for

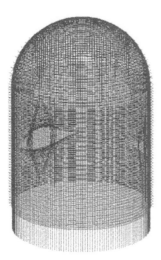

Figure 4. Tendon modeling.

tendon modeling, and 128 embedded reinforcement grid for rebars in solid elements, respectively.

4 MATERIAL MODELING OF PCCV

4.1 Modeling of concrete

Table 1 shows material properties of concrete used in the PCCV model. The properties are obtained from average values of concrete specimens after standard curing and field curing. An ideal stress-strain curve, suggested by Hognestad (Park & Paulay 1975), is used for modeling of compression in concrete. For plastic modeling in concrete, a plastic model based on the failure criteria by Drucker Prager is applied. It is assumed that tension of concrete behaves as linear up to tensile strength and linear softening for post peak and smeared crack model is employed.

In the smeared crack model, it is assumed that a first crack occurs when a principal tensile stress reaches the tensile strength at an integration point and then the second crack occurs when another principal stress in other direction reaches tensile strength of concrete and the difference in degree between the direction of the first crack and the direction of the second crack becomes larger than 60°. In order to consider effect of decreased shear strength due to interlocking of aggregates between crack faces, shear retention parameter 0.2 is applied.

4.2 Modeling of steel

Stress-strain relation of reinforcing steel (rebar) is assumed as multi-linear hardening with four modified straight lines, and initial Young's modulus

Table 1. Material properties of concrete.

	Compressive strength (MPa)	Tensile strength (MPa)	Young's modulus (MPa)	Poisson's ratio	Density (ton/cm3)
Base	46.54	3.65	28.49	0.19	2.23
Wall and dome	54.53	3.83	29.47	0.19	2.23

Table 2. Multi-linear plastic models of rebar (D19).

	Strain (%)	Stress (MPa)
Yield point	0.344	491.9
Second point	1.474	488.1
Third point	5.068	609.1
Ultimate point	9.434	630.4

Table 3. Multi-linear plastic models of steel liner.

	Strain (%)	Stress (MPa)
Yield point	0.18	382.6
Second point	5.08	465.9
Third point	33.2	589.5

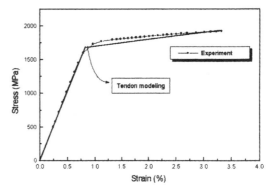

Figure 5. Material modeling of tendon.

210,000 MPa is applied according to experimental results. The yield points of D19 rebar are shown in Table 2. Stress-strain relation of steel liner is also assumed as elasto-plastic multi-linear hardening with three modified lines based on experimental results. Young's modulus and Poisson's ratio are 210,000 MPa and 0.3, respectively. Three yield points of the steel liner are shown in Table 3.

Stress-strain relationship of prestressing tendon is obtained as bilinear elasto-plastic behavior from experimental results (Hessheimer 2000) as shown in Figure 5. Wobble factor and friction loss are considered for calculation of prestress loss and setting loss resulting from sliding of the wedge of anchors is also considered.

Since the tension behaviors of steel rebars, liners, and tendons are same in compression behaviors when there is no buckling, von-Mises yield condition and work-hardening hypothesis are used for plastic modeling of those steel materials.

5 TEST AND NONLINEAR ANALYSIS OF PCCV MODEL

5.1 Limit state test (LST)

The so-called limit state test (LST) on the PCCV 1 : 4 model is carried out by pressurizing the model by nitrogen gas (Hesseimer 2000). For the continuous measurement of deformation and strain of the model at each pressurized stage, strain gages, pressure gages, displacement transducers and load cells are installed. 542 gages are installed for liner, 391 gages for rebars, 193 gages for tendons, 94 gages for concrete, respectively. Additionally, 101 displacement transducers, 68 load cells, 99 temperature and pressure sensors are used for extra measurements. Total 1493 instruments are used for the LST. Figure 6 shows the layout for the installed instruments (Hesseimer 2000). After structural soundness test and leakage test was performed, the PCCV model was pressurized until no additional pressurizing is possible due to leakage. At several pressure steps, which are 1.5 P_d, 2.0 P_d, 2.5 P_d and 3.3 P_d, leakage rate was calculated by measuring pressure and temperature during several hours without any additional pressure.

Even structural failure was not occurred, functional failure due to leakage was occurred at the pressure 3.3 P_d due to tearing of liner. In order to observe the structural failure, the SNL carried out an additional test which is called structural failure mechanism test (SFMT) for the same PCCV model.

5.2 Structural failure mechanism test (SFMT)

In order to obtain failure behavior of the PCCV model, the SFMT was carried out after the LST. Inner surface of the PCCV model was firstly sealed by the elastomeric liner and then 97% of inner volume of the model was filled by water for pressurizing. Figure 7 shows an instrumentation layout of the SFMT. Air pipe inside the model as shown in Figure 7 helps rapid pressurizing. The SFMT takes only about an hour to pressurize the model from beginning to failure.

For the SFMT, additional 17 radial and 3 vertical displacement transducers, 18 exterior liner strain gages and 82 rebar gages are installed. All surviving tendon stress gages and all load cells after the LST are reused.

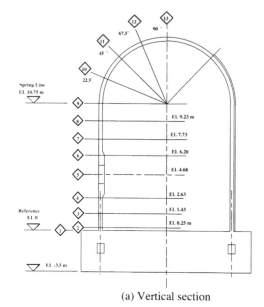

(a) Vertical section

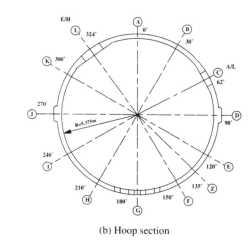

(b) Hoop section

Figure 6. Instrumentation layout.

5.3 Comparison between analysis and test results

Figure 8 shows magnified (×100) deformation profiles at 135°, at each pressurizing stage; 1.0 P_d (0.39 MPa), 1.5 P_d, 2.0 P_d, 2.5 P_d, 3.0 P_d and 3.3 P_d (1.29 MPa). Figure 8 shows that analysis was comparably well predicted experimental results. For higher pressure, analysis predict larger deformation than those of experiment.

More detail comparisons for the deformation at several parts of the wall of PCCV are shown in Figures 9–12. The SFMT results show generally more flexible behavior than LST results, but displacements

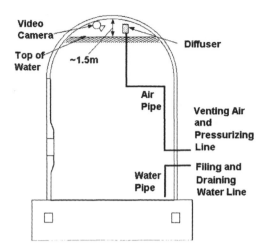

Figure 7. SFMT instrumentation layout.

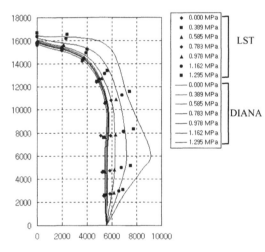

Figure 8. Deformation (×100) at 135°.

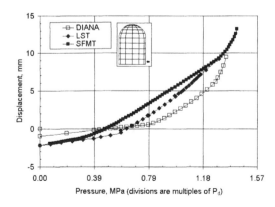

Figure 9. Radial displacement at base (EL 2.63 m).

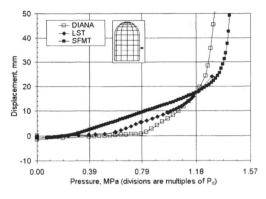

Figure 10. Radial displacement at midheight (EL 6.20 m).

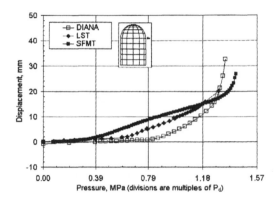

Figure 11. Radial displacement at springline (EL 10.75 m).

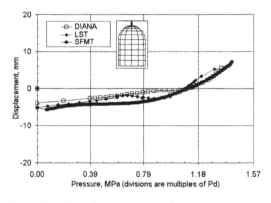

Figure 12. Vertical displacement at dome apex.

obtained from the SFMT become close to those from the LST as internal pressure becomes large at higher pressure. In the LST, rapid displacement increase was observed at 1.5 P_d, which is due to initiation of concrete cracks detected by the acoustic sensors in the

471

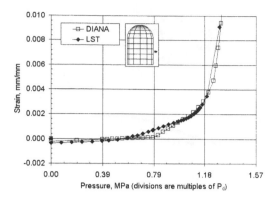

Figure 13. Outer rebar hoop strain at midheight (EL 6.20 m).

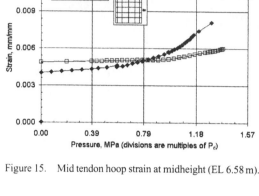

Figure 15. Mid tendon hoop strain at midheight (EL 6.58 m).

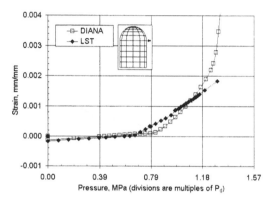

Figure 14. Outer rebar hoop strain at stringline (El 10.75).

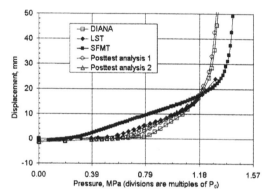

Figure 16. Comparison of displacement at midheight.

tests. But DIANA results show that the increase is occurred generally at 2.0 P_d. It is estimated that the difference was due to the difference of concrete tensile strength and the unbonded tendon modeling used in the analysis. 2.25 MPa of tensile strength of concrete obtained from direct tensile test (Lenke & Gerstle 2001), which is smaller than reported strength in Table 1, was used in the posttest analysis.

Figures 13 and 14 show comparisons for rebar hoop strain at dome apex and spring-line. It shows similar trends as those in displacements.

Figure 15 shows that analytically obtained hoop strain of a tendon at midheight of the model nearly remains constant up to 2.5 P_d, but the LST shows larger strain due to the unbonded tendon modeling. Note that DIANA can not predict distributed tendon strain of unbonded tendon.

In order to see the effect of tensile strength in the posttest analysis, two different concrete tensile strengths as a controlling factor are used. Tensile strength of 2.25 MPa obtained from the test (Lenke & Gerstle 2001) was used for posttest analysis 1 and

1.00 MPa for posttest analysis 2. Figure 16 shows that use of smaller tensile strength in the analysis give close result to that of the LST.

For the case of full-scale PCCV modeling, an improved result was obtained by attaching 100 mm steel plate to anchorage surface in the modeling for anchorage area.

6   VOLUME CONTROL TECHNIQUE FOR ANALYSIS OF PCCV

For concrete structures showing quasi-brittle behaviors after limit load or peak load, displacement control method can be used more effectively than load control method. On the other hand, when a specific displacement can not be easily selected to represent global behavior of shell structures under specific loadings, such as structures like prestressed concrete reactor containment vessels subjected to internal pressure or external pressure, so-called volume-control method, that can control volume-change of structures and predict load-change due to the volume-change, can be used more effectively than the displacement control method (Song et al. 2002a, b). In this paper, the volume-control method is applied to analysis of the PCCV 1 : 4 model, and then the validity of volume-control technique is verified with experimental results.

6.1 Volume-control technique

The so-called volume control method can be utilized with a pressure node (Song & Tassoulas 1993) which is a single degree of freedom, namely, the uniform change of pressure on the finite shell element, denoted by Δp. The change in volume is defined by

$$\Delta V = \int_{b^e} N^T \cdot \Delta u \, db^e \tag{1}$$

Since the traction increment Δt on element boundary b^e due to Δp is approximated by $-\Delta pn$, the traction term in the element load vector is given by

$$\int_{b^e} N^T (t + \Delta t) \, db^e = -(p + \Delta p) \int_{b^e} N^T n \, db^e \tag{2}$$

where n is a unit vector normal to element boundary and N is shape function. Using Equation (2), the equilibrium equation in the element is given by

$$K_e \Delta U = -(p + \Delta p) \int_{b^e} N^T n \, db^e + F_e \tag{3}$$

where K_e is the element stiffness matrix and F_e is element internal load vector.

Absorbing the term involving Δp into the stiffness matrix in Equation (3) and inserting one additional equation of Equation (1) for the change in volume, one finds that the element equilibrium equation with matrix form is now given by

$$\begin{bmatrix} K^e & \int_{b^e} N^T n \, db^e \\ \int_{b^e} n^T N \, db^e & 0 \end{bmatrix} \begin{bmatrix} \Delta U \\ \Delta p \end{bmatrix} = \begin{bmatrix} -P\int_{b^e} N^T n \, db^e + F^e \\ \Delta V \end{bmatrix} \tag{4}$$

where, the last row and column in modified stiffness matrix correspond to the additional degree of freedom Δp, while the load vector now contains an additional component ΔV. The presence of ΔV in the load vector indicates that one can prescribe the change in volume enclosed by the PCCV model and determine the required change in pressure. Thus, by controlling the volume in nonlinear iterative analysis, one can solve for the required increase or decrease in pressure along equilibrium paths around the ultimate point of pressure. The technique can overcome the instability problem

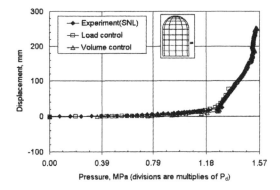

Figure 17. Radial displacement at midheight (EL 6.20 m).

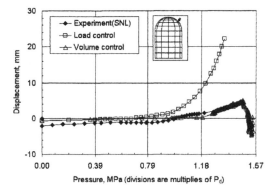

Figure 18. Radial displacement at dome 45°.

of the conventional force-control method as well as the difficulty of selecting a local characteristic point to control the local displacement which governs the global behavior of shell structures in the conventional displacement-control method.

For the failure analysis of the PCCV model subjected to internal pressure, the wall of PCCV is modeled by layered shell elements implemented with the in-plane reinforced concrete constitutive models (Song et al. 2002c) and the volume control technique using the pressure node attached to the shell elements is applied. The shell element in this study is an isoparametric degenerated shell element with six degrees of freedom including a drilling degree of freedom in each node. In the layered element formulation (Song et al. 2002b), shell is divided into several layers of panel where 2-D constitutive models are applied to take into account material non-linearity.

6.2 Evaluation of volume-control technique

Radial displacement at midheight and dome 45° are shown in Figures 17 and 18, respectively. The

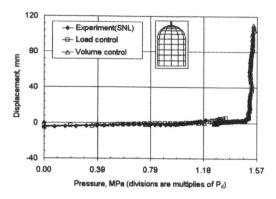

Figure 19.   Vertical displacement at dome apex.

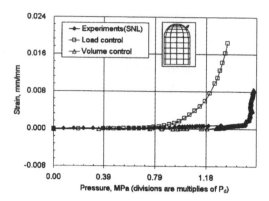

Figure 21.   Outer rebar hoop strain at dome 45°.

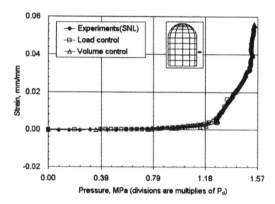

Figure 20.   Outer rebar hoop strain at midheight (EL 6.20 m).

experiment was also stopped when additional pressurizing was not possible due to tearing of liner. The analysis by the volume control method is stopped when the strain of a hoop tendon is reached 3%, which is assumed to be ultimate state. After that, displacement increased sharply with very small increment of internal pressure.

It can be seen that the difference between analytical results by volume control method and experimental results is very small. Note that the influence of boundary condition of foundation of the PCCV model is ignored in volume control analysis.

Figure 19 shows that there is a little difference between analytical result and experimental result on vertical displacement at dome apex after 3.0 P_d. Note that volume control method can give extensive results near limit pressure.

The comparison for local behavior of PCCV was carried out by comparing the strains of hoop outer rebar at midheight of wall (Fig. 20) and at doom 45° (Fig. 21), respectively. Figures 20 and 21 show that local behavior of steel strain has a similar trend with the behavior of radial displacement on the model. In

early stage, the behaviors obtained from both experiment and analysis are very similar, but a little difference occurs after the limit point where tendons of the model start to yield.

7   CONCLUSIONS

For the prediction of ultimate capacity of the prestressed concrete reactor containment vessel subjected to internal pressure, a finite element modeling technique is applied and the so-called volume control method is introduced.

It shows that finite element modeling technique in this paper can predict effectively the ultimate behavior of a PCCV 1 : 4 model subjected to internal pressure. Test layout and test results obtained from both limit state test and structural failure mechanism test are explained. Analytical techniques in this paper are verified by comparison of analytical results with test results. Comparison also shows that improvement of unbonded tendon modeling in DIANA code and use of proper tensile strength of concrete is necessary for better prediction of the ultimate capacity.

REFERENCES

Amin, M., Eberhardt, A. C. & Erler, B. A. 1993. Design consideration for concrete containments under severe accident loads, *Nuclear Engineering and Design* 145:223–331.
DIANA 1998. DIANA 7.1, TNO Building and Consruction Research, Delft.
Dunham, R. S., Rashid, Y. R., Yuan, K. A. & Lu, Y. M., 1985. *Methods for ultimate load analysis of concrete containments*, NP-4046, Anatech International Co., CA.
Hessheimer, M. F. 2000. *Pretest round robin analysis of a prestressed concrete containment vessel model*, NUREG/CR-6678, Sandia National Labs, Albuquerque, NM.
Lenke, L. R. & Gerstle, W. 2001. *Mechanical property evaluation of concrete used in the NUPEC/NRC prestressed*

concrete containment vessel (PCCV): prestress and limit state test results, ATR Institute, University of New Mexico, Albuquerque, NM.

Park, R. & Paulay, T. 1975. *Reinforced concrete structures*, New York, John Wiley & Sons.

Song, H.-W., Bang, J.-Y., Byun, K.-J. & Choi, K.-R. 2002a. Finite element analysis of reinforced concrete containment vessel subjected to internal pressure, *Journal of Nuclear Engineering and Design* (accepted).

Song, H.-W., Shim, S. H., Byun, K. J. & Maekawa, K. 2002b. Failure analysis of RC shell structures using layered shell element with pressure node, *Journal of the Structural Engineering* 128(5):655–664.

Song, H.-W., You, D.-W., Byun, K.-J. & Maekawa, K. 2002c. Finite element failure analysis of reinforced concrete T-girder bridges, *Engineering Structures* 24(2):151–162.

Song, H.-W. & Tassoulas, J. L. 1993. Finite element analysis of propagating buckles, *International Journal of Numerical Methods in Engineering* 36(20):3529–3552.

Witte, F. C. & Feenstra, P. H. 1998. *DIANA user's manaul nonlinear analysis*, Delft, TNO Building and Construction Research.

Finite Elements in Civil Engineering Applications, Hendriks & Rots (eds.)
© 2002 Swets & Zeitlinger, Lisse, ISBN 90 5809 530 4

FE guided structural rehabilitation of cooling towers

P. Dalmagioni
QUALITALIA-CONTROLLO TECNICO, Milan, Italy

R. Pellegrini
ENEL. HYDRO – B.U. ISMES, Seriate, Italy

ABSTRACT: Surveys to ascertain the conditions and possible deterioration of the materials of cooling towers made of reinforced concrete have been carried out for many Italian geo-thermal power plants. Physically and geometrically non-linear FE models have been developed for the structural analyses of the towers under static loads and analysed with DIANA v.7.1.

The towers build in the 40s–70s, are now approaching their service life and deterioration is visible in the reduction of the nominal thickness and also decreased material performance. Using material non-linearity and modelling the real conditions of the structures, all the ultimate resources have been taken into account, allowing determining the safety coefficient of the towers.

The FE models have been used also to design the most effective technique for the structural rehabilitation. For the largest tower a fibre reinforced tie has been designed. The structural models have been used also to guide the design in the demolition-reconstruction phases.

1 INTRODUCTION

Typically cooling towers of geo-thermoelectric plants built up to the seventies are in reinforced concrete. Air under the physical status of vapour is conveyed through them by natural convection. Along this path vapour cool down and loose water, which is collected at the bottom in a concrete pool, together with the cooled down cycle water.

Vapours of geothermal water are rich in many chemical species that generally fasten the natural degradation of concrete exposed to humidity and temperature gradients.

Moreover, these structures have now approached their service life and require a general safety reassessment with a view to their future use. In fact, while some are meant for continuing service, others are kept standing as testimonials of the industrial cultural heritage. Decommissioning is seldom followed by dismantling, also in consideration to the environmental impact due to the disposal of important amounts of potentially contaminated concrete.

2 THE STRUCTURE

Cooling towers investigated in this study are thin hyperboloid shells of concrete, reinforced by two layers of reinforcement bars, which coincide with the straight generating lines of the hyperboloid.

At the bottom, the cooling pool has a circular plate mat 40–50 m diameter, and a retaining circular wall 0.4–0.5 m thick, over which a system of inclined circular columns (diameter 0.35–0.5 m) bear the tower

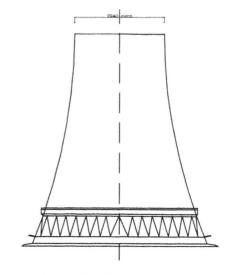

Figure 1. Layout of a cooling tower.

477

shell. The columns system is designed as to allow for fresh air entering into the chimney.

The shell is 60–70 m high and its diameter is 25–30 m at its top. Thickness varies from 0.5 m at the bottom to 0.12 m at the top, where a circular beam is present to provide extra stiffness against adverse ovalization effects.

3 OBJECTIVES AND INTERVENTION STRATEGIES

Enel Green Power, which produces most of renewable geothermal energy in Italy, has contracted to Enel. Hydro-ISMES an extensive campaign to ascertain the current conditions of their cooling towers standing in plants of the Larderello area (Tuscany), with a view to identify major rehabilitation interventions as well as a maintenance plan, and to allow for the planning of maintenance protocols of small impact on the production activities.

The diagnostic approach is that of collecting information about the current conditions of the structure, of their materials, to represent them in a mathematical model by which the current safety margins are calculated, and, finally, the type and extension of rehabilitation actions are reproduced.

As to arrive to the "as built" description of the structure the following investigations were made:

– Visual survey (from the exterior and the interior of the shell) and mapping of degradation under the form of cracks, concrete consumption, spalling, scabbing, presence of humid paths, absence of the internal protection layer (in shotcrete).
– Measure of concrete strength and its potential for durability (carbonation, oxidation, chloride attack, thin section SEM observations). These assessments were made by laboratory tests on samples taken from the structure.
– Measure of column concrete soundness (in terms of stiffness degradation) by vibration *in-situ* tests.
– *In-situ* rip tests purposely designed to understand the behaviour of the system (concrete and bars) in special locations (through cracks) under bending.

A finite element model has been then set-up where all evidences of deterioration have been included, i.e. cracks, thickness changes either due to absence of concrete, or shotcrete, or to highly deteriorated concrete.

The constitutive models of reinforced concrete were selected as to describe the full range of response up to cracking and crushing. Calibration took into consideration the actual stiffness and strengths measured in laboratory tests.

With the above set of data, finite element analyses were run with DIANA Version 7.1 to capture the ultimate response of the structure in order to have a

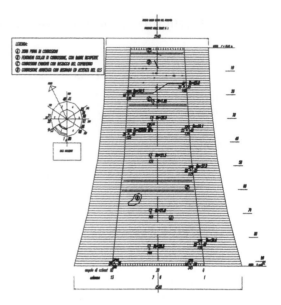

Figure 2. Test positions and damaged areas (light grey) for the Serrazzano tower.

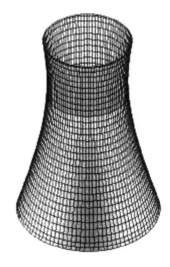

Figure 3. The finite element model of the tower shell. The top portion requiring stiffening is visible in darker grey.

measure of its current safety margin against the loads used for its dimensioning.

The results have provided a priority ranking for interventions in the towers investigated, allowing identifying maintenance protocols for each. For one it could be demonstrated that the rehabilitation intervention foreseen by the owner was justified. In this case the numerical analysis has guided the selection and validation of the rehabilitation, which has introduced state-of-the-art technologies such as the use of Fibre Reinforced Carbon (FRC).

4 STRUCTURAL ANALYSES

The analyses have been carried out with the finite element code DIANA Version 7.1, reproducing the tower as a fully three-dimensional structure: 8 node layered shell elements have been used in meshing the shell, the hot water adduction channel and the cooling pool. Three nodes beam elements have been used to model the beams connecting the later to the tower (rectangular $0.15 \cdot 0.15$ m cross section), the balustrade of the adduction channel (trapezoidal $0.60 \cdot (0.10-0.15)$ m cross section) and, finally, the columns supporting the chimney (circular 0.30 m diameter section). The finite element model resulted of about 10000 nodes and 60000 degrees of freedom.

Damage, concentrated in the chimney structure, under the form of reduced concrete thickness and/or cracks has been directly introduced in the model. Cracks have been modelled by disconnecting the elements' faces. To reproduce the rehabilitation interventions, crack faces have been re-connected.

4.1.1 Reinforcement
The reinforcement layout and dimensions have been taken from the original design drawings since they were found from the test campaign substantially unaltered. The bottom part of the chimney structure, up to one third of the chimney's height, is reinforced with a net of two inclined bars ($\alpha = 15-20°$ respect to the vertical direction) and by one in the hoop direction. The concrete cover of the rebar spanned 3 cm.

The upper part of the tower structure is reinforced with two single inner and outer layers of helicoidally nets inclined again 15–20°, which ideally cross to form a lozenge. The concrete cover of the rebar spans 5 cm.

Hoop and vertical bars are the reinforcement of the adduction channel.

The reinforcement has been modelled explicitly in all structural members, with the only exception of the stirrups embedded in columns and beams.

4.1.2 Constraints
The structure has been fully constrained at its base along all available d.o.f. This representation corresponds to the design assumptions, which were confirmed during the inspection by the absence of any trace of differential settlement.

4.1.3 Material models
As to determine the safety margin of the structure, non-linear material modelling has been selected for reinforced concrete.

Concrete response in compression has been described via a Mohr-Coulomb criterion, and by a Galileo-Rankine in the tensile range. An isotropic hardening law rules both criteria. The uni-axial behaviour of concrete is then linear elastic up to the tensile and compressive strength. In traction a linear softening law has been selected characterised by a crack band of 0.6 m, which is the typical dimension of the finite elements of the cooling tower, and by fracture energy, G_f, of 105 $Nm^{-3/2}$. Unloading in traction is modelled by a damage of the elastic stiffness.

An elastic-perfectly plastic model, adopting the Von Mises plasticity criterion, models reinforcement steel.

The plastic threshold is set to 375 MPa. The elastic Young modulus is set to 2.1 GPa.

Material constants have been calibrated keeping into consideration the results of the *in situ* and laboratory tests and the general evidence of damage extension over the actual structure.

As for the concrete, the following values have been selected throughout the shell:

- Volume mass, $\rho = 2400-2450$ kg/m³, which includes also the weight of the steel mass.
- Young modulus, $E_c = 25000-28000$ MPa
- Poisson coefficient $\nu = 0.18$
- Compression design strength, $f_{cd} = 10-15$ MPa
- Tensile design strength = $1.2-1.5$ MPa.

As for beams and columns, they have been assumed to behave elastically. Their elastic modulus is calibrated from the vibration tests results, and it is the same as that of the shell.

4.2 Results

The load combination is that suggested by the design guidelines [1,6], considering, in this case, the specific target operation condition of the tower (some are operating, other are decommissioned), with a return period of loads of fifty years:

$$\sum_{j \geq i} \gamma_{Gi} \cdot G_{kj} + \gamma_p \cdot P_k + \gamma_{Ql} \cdot Q_{kl} + \sum_{j \geq i} \gamma_{Qj} \cdot \varphi_{0i} \cdot Q_{kj}$$

where
γ_G = 1.0 for self weight (PP)
γ_P = 1.35 for the thermal gradient (DT)
γ_{Ql} = 1.5 for radiation and wind load (I)
φ_0 = 0.6 additional for wind load (V)

Assuming the tower in operation, the above load combination is the following:

$$C = PP + 0.9 \cdot V + 1.5 \cdot I + 1.35 \cdot DT$$

The draft effect, which increases the severity of the load combination, has been included.

479

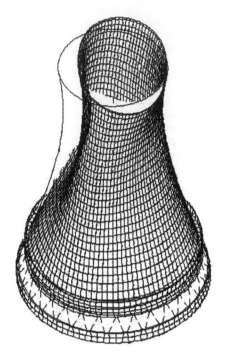

Figure 4.   Deformed shape highlighting the ovalization effect.

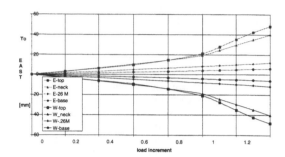

Figure 5.   Displacement in the East direction of Serrazzano tower points belonging to the East and West meridians.

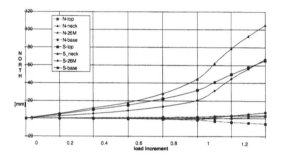

Figure 6.   Displacement in the North direction of the Serrazzano tower points belonging to the North and South meridians.

The finite element analysis allows to verify the structure response under operational loads and to determine the safety margins of such conditions, by factorising operational loads up to collapse.

The analysis has been then carried out by increasing the load combination C by steps of 0.1.

Collapse has been observed at C = 1.1–1.2 in towers with severer damage conditions and at 1.4–1.5 in case of better/localised damage conditions.

According to limit state approach, the structure is safe when:

– Concrete: $\sigma_c < f_{cd}$
– Steel rebar: $\sigma_s < f_{sd}$

The results of the finite element analyses, which took into consideration all the margins offered by the non-linear response of the concrete, have demonstrated the vulnerability of the cooling towers shells to wind and temperature loads.

The severer effect stands in the ovalization of the tower that adds an adverse flexural state, which the chimney membrane is progressively less able to respond to. This condition is worsened by the presence of typically longitudinal cracks. In this case maximum displacements can be metric at the top at ultimate loads.

4.2.1   Tensile and deformation behaviour

To catch the behaviour of the tower from service conditions up to the ultimate load (C factor from 0 to 1.4) the horizontal displacements of four points of the tower (nodes of the mesh) were monitored at different elevation:

– Four at the base, where the tower shell is connected with the supporting beams;
– Four at one third of the tower height, from bottom to top;
– Four at one third, from top (neck of the structure);
– Four at the top.

These displacements are collected as an example for the case of the Serrazzano tower after the intervention (new structure) in the following figures.

It may be observed that the global structural behaviour is substantially linear up to the service condition, and beyond that a moderate non-linear behaviour follows, which is due to diffused cracking of concrete in traction. Beyond C = 1.2 ovalization occurs in the top part of the structure: displacements along the North South direction before rehabilitation reached in the Serrazzano tower 1–2 m. The former is markedly non-linear. The rehabilitation intervention is studied to substantially reduce this effect.

The worst stress condition is originated in the shell by anomalous flexural conditions in the structure,

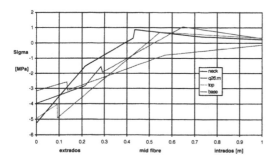

Figure 7. Hoop stress distribution in operating conditions of the Serrazzano tower. Negative sign indicates compression.

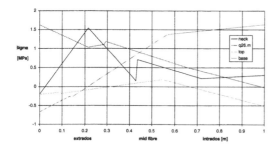

Figure 8. Vertical stress distribution in operating conditions of the Serrazzano tower. Negative sign indicates compression.

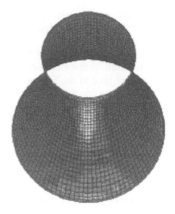

Figure 9. Von Mises equivalent stresses (MPa) in service conditions. The area where the highest stresses occur is indicated in lighter gray.

which is designed to respond as a membrane. These are due essentially to wind and temperature. For the case of the Serrazzano tower, after rehabilitation, hoop, vertical stress distribution within the thickness of the shell are given at the four levels previously defined in correspondence of the service conditions.

It is here possible to observe that a portion of the shell has reached the tensile strength in its inner face, while the outer is compressed. Further, discontinuities in the stress distribution reveal the presence of stiffness changes due to material properties change and/or damage/cracks.

Compressive stresses reach in the outer Southern face values of some MPa: 7 in the rehabilitated structure for the load combination increase of C = 1.3. Tensile stresses keep lowest than the tensile strength, since cracking is produced. Compressive stresses, while are near to the compressive strength of pre-existent concrete, are well beyond the compressive strength of new materials used for rehabilitiation (mortars reaches 28 MPa).

As for the reinforcement bars, Fig. 9, the highest Mises stresses are detected in the inner part of the structure along the Southern meridian, while otherwise they keep smaller than 70 MPa. The reinforcement bars work well within the elastic range ($F_{sd} = 200$ MPa).

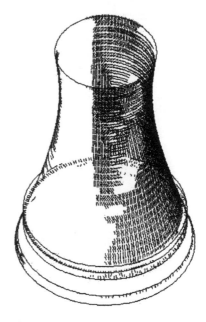

Figure 10. Crack pattern in service conditions (before rehabilitation).

4.2.2 Crack pattern

Extensive micro-cracking is produced by the temperature. It encourages stress redistribution towards the inner part of the structure, which becomes more compressed.

As for the Serrazzano tower, cracks in service conditions (C = 1) are spread over the inner part of the Southern face along both the vertical and the hoop direction. Cracks are present also in the Northern outer face. Vertical cracks are detected over the Eastern and Western meridians.

Upon increasing loads, cracking of the outer face spread all over the shell. Some horizontal cracks are generated in the outer face of the Southern meridian. These cracks are responsible for the increase in displacements at load factors ≥1.2 and promote the tower collapse.

5 DESIGN OF THE REHABILITATION INTERVENTION

The design intervention, where required after the safety assessment, complies with the rationale of keeping unaltered the overall mass and the mass distribution of the structure, therefore avoiding unnecessary extra-actions on the supporting elements, while providing more-stiffness.

In the inner part of the shell of the Serrazzano Tower, the main intervention is that of recovering the pristine waterproofing 1.5 cm thick shotcrete layer and the concrete cover of the rebar. The original shotcrete layer has been previously removed together with a relevant thickness of the concrete cover by water-scarification. A layer of high strength and waterproof mortar has been then set, producing a final thicker concrete cover (by 4–5 cm).

In the following figure the contour plot of the shear stress exchanged along the contact between the new mortar layer and the "old" concrete is displayed.

The values are well below the rip strength of the mortar used (2.5 MPa) and of the tensile strength of concrete (2 MPa).

A millimetre film of epoxy-polyurethane resin guarantees the waterproofing action.

In the outer part of the shell, a film of acrylic paint assures the protection against the aggressive action of the vapours conveyed out by the tower.

In the outer top portion (top 15 m) a stiffening support has been provided by laminated carbon fibres 40 cm high, laid in the hoop and vertical direction to form a square net of 3 m edge. This reinforcement ring net reduces wind-induced ovalization of the tower's neck to one fifth, to few centimetres.

Accordingly the crack pattern anticipated for the repaired structure in service conditions is substantially more constrained, in value and extension (Fig. 12 compared to Fig. 9).

6 CONCLUSIONS

This study has proved to the Authors and to the Owner of the plant, the fully mature capabilities of non-linear finite element modeling in the rehabilitation process

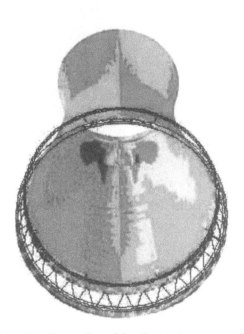

Figure 11. Contour lines of the adhesion shear stress at the contact between new mortar and old concrete in the rehabilitated Serrazzano tower.

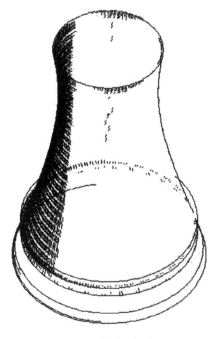

Figure 12. Crack pattern distribution in service conditions after rehabilitation.

of old structures, where it is essential to provide a realistic representation of the current conditions of the structure and accurate modeling of rehabilitation measures.

REFERENCES

Syndacat National du Beton Armé et des Techniques Industrialiées (SNBATI), "Regles professionnelles applicables a la costruction des refrigerants atmospheriques en beton arme" texte provisoire, 1986. In French.

Circolare n.252 AA.GG./S.T.C. del 15 ottobre 1996, Instructions for the application of the "Norme tecniche per il calcolo, l'esecuzione e il collaudo delle strutture in cemento armato, normale e precompresso e per le strutture metalliche" di cui al decreto ministeriale 9 gennaio 1996. In Italian.

D.M. 16 gennaio 1996, Technical rules concerning the "Criteri generali per la verifica di sicurezza delle costruzioni e dei carichi e sovraccarichi", in Italian.

ENV 1992-1-1 Eurocodice 2 (Dec. 1991), "Progettazione delle strutture in calcestruzzo", Part 1-1: Regole generali e regole per gli edifici (in Italian).

SIDEROCEMENTO S.p.A.: "ENEL – FIRENZE – Centrale geo-termoelettrica di Serrazzano: Refrigerante iperbolico da 8000 m^3/h – Progetto esecutivo":
  – Relazione tecnica,
  – Certificati dei materiali,
  – Disegni n. 17507, 512, 513, 514, 516, 576, 583, 586, 587.
  – Disegno ENEL n.072-027-07231.

Prog. STR/9526, Doc. RAT-STR-2623/97, "Programma di riqualifica strutturale di refrigeranti atmosferici – Guida di progetto per la riqualifica strutturale delle torri di raffreddamento a tiraggio naturale. – Validazione teorica dei metodi proposti. – Software di progettazione", per conto ENEL – VDPT AGI/UPG di Pisa, 20-10-97. In Italian.

ELECTRICITE DE FRANCE – Refrigerants atmospheriques humides a tirage naturel Directives d'utilisation du CST et de reaction du CCTP, 2/6/1983. In French.

DIANA v.7.1 User's manuals.

A.M. Neville: "Le proprietà del calcestruzzo". Sansoni, 1980. Firenze.

Finite Elements in Civil Engineering Applications, Hendriks & Rots (eds.)
© 2002 Swets & Zeitlinger, Lisse, ISBN 90 5809 530 4

FEM-models applied for unreinforced underwater concrete

C. van der Veen
Delft University of Technology, Delft, the Netherlands

A. de Boer
Ministry of Transport, Public Works and Water Management, Utrecht, the Netherlands

ABSTRACT: In the Netherlands many deep building pits, created by sheet-pile walls, tensile piles and underwater concrete, have been used. An underwater concrete floor is necessary to prevent water leakage. Up till now only limited flexural tensile stresses are allowed in the concrete floor. However, a new design code for underwater concrete allowed cracking in the floor. Consequently, the floor can be made less thick.

In order to investigate the flow of forces three different FEM-models have been used. The models consist of beam, shell or volume elements. The differences in results based on the models were very small with respect to the linear elastic analysis. However, the non-linear analysis showed larger differences. Not only the load factor was lower but also the failure mode was different in the FEM-model based on the volume elements compared to the other two models.

1 INTRODUCTION

In the Netherlands a deep building pit can be created by sheet-pile walls, tensile piles and underwater concrete. Creating a building site with underwater concrete is necessary because of the high ground water level; otherwise an excessive amount of water must be drained. Furthermore, the permitted drainage of water is limited in the Netherlands. The building of such a site can be explained as follows. It starts with the ramming of steel sheet-pile walls into the ground. These walls consist of steel sheet piles of the Hoesch type, or consist of combi-walls. The type of wall used depends on the depth of excavation. After the ramming of the sheet-pile walls, the excavation and anchoring starts. For the anchoring screw-injections and strand-anchors are used. After the excavation the ramming of the foundation piles starts and mostly separate building pits are created by ramming partitions. When the ramming of the foundation piles is finished, underwater concrete is poured into each building pit. As soon as the underwater concrete has hardened, the water in the building pit is pumped away and the concrete work starts an example of a building pit is shown in Figure 1.

1.1 *Flow of forces*

The foundation, which consists of the underwater concrete floor and the piles, will be loaded by the hydraulic water pressure (buoyancy of the building pit), resulting in tensile forces in the piles.

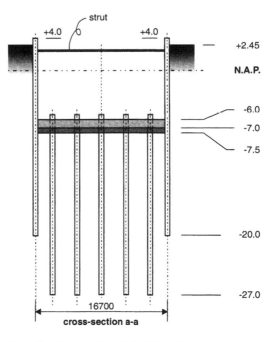

Figure 1. Cross section of a building pit.

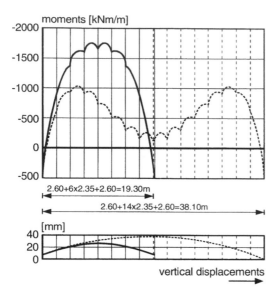

Figure 2. Distribution of the displacements and moments.

In the horizontal direction the underwater concrete floor acts as a compression strut between the opposite steel sheet walls. The sheet walls are loaded by the water pressure and the effective soil pressure. As a result of the tensile forces in the piles and sheet walls the underwater concrete floor lifts upwards. However, the axial stiffness of the piles and sheer walls differs. Consequently, a different elongation occurs which affects the moment distribution in the underwater concrete floor, see Figure 2. Two different building pits are shown in this figure; a small one 19.3 m and a wide building pit. Clamping moments develop in the connection concrete floor-sheet wall, because the sheet wall has a higher stiffness than the piles.

The moment distribution is comparable to a floor on an elastic foundation. In the small building pit the clamping moments enlarged the field moments. However, in the wider building pit, represented by the dotted line in Figure 2, no effect on the moment was noticed in the midspan. The bottom panel of Figure 2 shows the displacements of both building pits. Up till now cracking of the concrete floor was not allowed and only limited flexural tensile stresses were possible. The distribution of forces was based on a linear elastic distribution. Recently, a new design code for underwater concrete was presented, which allowed cracking in the unreinforced concrete floor. Thus, the carrying capacity of the concrete floor made use of the compressive membrane force, which acts in the floor.

In order to investigate in more detail the flow of forces different FEM-models have been used. In the models, which consist of beam, shell or solid elements,

respectively, special attention has been paid to the interaction between the steel sheet piles and concrete floor. With the aid of a physical non-linear analysis features cracking and plasticity have been included in the calculations.

2 CASE STUDY

The different models used will be explained. We start with the most simple model. First the beam model has been used. The use of the models are demonstrated by a case.

We investigate a concrete floor of 16.7 m wide and 40 meter long. The point of departure is a beam of 1 m wide which represents a strip of the underwater concrete floor. It has a length of 16.7 m (this is the short direction of the building pit) and a depth of 1.0 m. The floor is supported by foundation piles with a $450 \times 450 \, mm^2$ cross-section. A concrete strength for the underwater concrete and the piles of C25 and C55 respectively was used. The floor was loaded by 7-meter water pressure.

The supports of the concrete floor from the sheet wall and piles are translated into translation spring elements. A spring constant has been calculated:

Sheet wall : K_{vert} = 60,000 kN/m for 1 m sheet wall
K_{hor} = 4,000 kN/m for 1 m sheet wall
Piles : $K_{p.vert}$ = 130,000 kN/m for each pile

3 LINEAR ELASTIC CALCULATIONS

3.1 Beam model

The compressive force in the concrete floor is calculated with a load factor 1 and amounts 550 kN/m'.

As a result a clamping moment between the sheet wall and concrete floor is generated. This moment equals 137.5 kN/m', and is applied on the model by an external constant moment (load case). After calculation the following moment distribution has been found, see Figure 3. It is shown from the figure that the clamping moment is 138 kNm. We can check the sum of moments between the pile distance which is 3.1 m. Thus $1/8 \cdot 47 \cdot (3.1)^2 = 107.5$ kNm and this equals the found moment values.

In order to simulate the real structure better an advanced beam model has been used. Firstly, the piles are no longer modelled with an axial spring but with line interface elements (CL12I).

Consequently, the wideness of the pile gets its own dimension. Thus a surface (cross-section of the pile) can be modelled. Secondly, the sheet wall will be modelled as a surface in the horizontal direction. Again a line interface element (CL12I) has been used

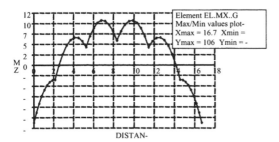

Figure 3. Moment distribution Mz over the length of the beam model.

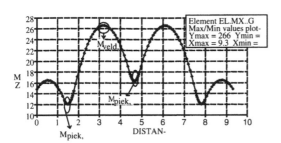

Figure 4. Detail of moment distribution M_z.

for this purpose. In this way the concrete floor is supported fully over the depth. For the concrete floor 3-dimensional beam elements has been used, (CL18B). By using the option "zone" the wideness of the concrete floor is increased to 2.5 m, (distance of the piles). The calculation has been made again. Some details about the results of the moment distribution in the short direction of the building pit are shown in Figure 4.

It is clearly shown that the sharp changes in moment above the piles are now smoother. The moments above the piles are about 10 to 15% larger than in the first beam model. The moments in the midspans between the piles are lower. This phenomenon is caused by the thickness of the piles. It is expected that the advanced model describes reality more appropriate than the first beam model.

3.2 Advanced shell model

In this model the concrete floor has been modelled with 2 dimensional shell-elements, (CQ40S). The intersections with the piles and concrete floor are modelled with a 3 dimensional volume interface elements, (CQ48I). Volume interface elements (CQ48I) have been used for the sheet walls both in horizontal and vertical direction. Consequently, the external moments should be translated to an equivalent surface loading, see Figure 5.

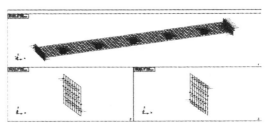

Figure 5. Moment translated to a surface loading.

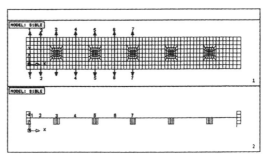

Figure 6. Cross-sections 1 to 7 in the shell-model.

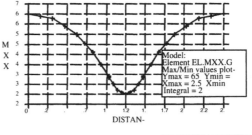

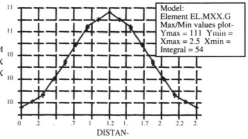

Figure 7. Effect of load eccentricity.

Some results will be explained. Because the moments and displacements vary along the wideness of the model results in different cross-sections are explained, see Figure 6.

The distribution of the moments over the wideness of the model above and between the piles respectively is shown in Figure 7. Based on these results an

487

Table 1. Load factor at first cracking and ultimate load.

	Element type		
	Beam	Shell	Solid
Load factor first Cracking	2.5	2.5	1.4
Position crack	Midspan	Midspan	Around pile
Ultimate load factor	7.4	6.0	4.5
Crack depth (m)	0.67	0.67	0.75

effective wideness of 1.55 m can be found. This is less than the pile distance of 2.5 m. When we compare the moment distribution in the different cross sections 1 to 7 with the results of the advanced shell model, only small (2%) differences were found.

3.2 3-D model

Three-dimensional solid elements have been used for the last model. The concrete floor has been modelled with solid elements (CHX60). When we compare the results of the 3-D model with the advanced shell model hardly any differences have been found.

4 NON-LINEAR ANALYSIS

In all models the options of cracking and plasticity have been added. The cracking and ultimate load are compared and discussed. The different load factors are summarized in Table 1.

The differences in the beam and shell models are small. Much lower load factors are found with the 3-Dimensional solid model, 1.4 and 4.5 against 2.5 and 7.4 (6.0). Furthermore, the position of first cracking differs with the 3-D models. In the model cracking started around the pile at the underside of the concrete floor. The cracks run to the top of the floor. Despite the fact that these cracks are micro cracks, problems could occur with the bond between the pile and concrete

floor. This is the topic for a new research in the future. Further, water tightness of the concrete floor could be a problem. In general a concrete compressive zone of 200 mm insured a watertight floor. Because the crack depth at the ultimate load is 0.75 m enough compression depth is left ($1.00 - 0.75 = 0.25$ m).

5 CONCLUSIONS

1. The results of the three different models, beam, shell and solid, are close to each other, if the structure is analysed linear elastic.
2. A non-linear analysis gives comparable results for the beam and shell model. The results of the 3-D model differ. Not only the ultimate load factor is much lower but also the load factor at first cracking is lower.
3. In a 3-D analysis the position of the cracks are different from the 1-D and 2-D analysis.

More research is needed to control the connection between the pile and the concrete floor. We have to be sure that no punching shear will occur before failure in bending of the concrete floor occurs.

ACKNOWLEDGEMENT

The research reported was carried out in cooperation with the members of CUR-committee VC-61. Computational work was done by Mr. G. Kwaaitaal. The practical help of Mr. A. Zeilmaker was very much appreciated.

REFERENCES

1. Kwaaitaal. G.J.J. Flow of forces and cracking in underwater concrete floors, graduation report 2001 (in Dutch), TU Delft.
2. Design rules for unreinforced underwater concrete floors (in Dutch). CUR Recommendation 77, 2001.

Finite Elements in Civil Engineering Applications, Hendriks & Rots (eds.)
© 2002 Swets & Zeitlinger, Lisse, ISBN 90 5809 530 4

3-D finite element deformation analysis of concrete-filled tubular steel column

T. Matsumura
Chubu University, Kasugai-City, Aichi, Japan (The Takigami Steel Construction Co. Ltd.)

E. Mizuno
Chubu University, Kasugai-City, Aichi, Japan

ABSTRACT: This paper deals with the 3-D finite element deformation analysis of concrete-filled tubular (CFT) steel columns subjected to monotonic loading. The finite element package program, DIANA is used to carry out the verification analysis. As for the constitutive model of materials, the Drucker-Prager model in DIANA is applied as the concrete material. This model has the features such as the linear tension cut off in tension range, the compressive softening behavior in compression range and the confining effect on concrete strength. The multi-linear constitutive material model is used as the steel material, and the interface element is used to represent the gap and contact behavior between steel and concrete. The 3-D FEM analysis on monotonic behavior of concrete-filled tubular (CFT) steel columns is investigated by comparing the analytical results with experimental results.

1   INTRODUCTION

The concrete-filled tubular (CFT) steel columns are used for composite structure like hybrid pier as well as composite girder in Japan. Using its advantage of tensile strength region on steel material and compressive strength region on concrete material, CFT structure makes a good performance on durability compared to a single material structure. Although a lot of experimental study and 2D analytical verifications has conducted recently, few 3D analytical study has progressed due to its difficulty of physical and material nonlinear modeling. That is, in spite of the local buckling of the steel, cracking and confining effect of the concrete has been simultaneously occurred in the real structure at the axial loading and the bend loading state, it is difficult to express these behavior without using interface elements in the analysis.

This study treats 3D non-linear deformation analysis of CFT steel column, under constant axial loading and lateral deformed monotonic loading, considering such steel, concrete and interface behavior mentioned above.

2   3-D DEFORMATION ANALYSYS OF CFT

2.1   *Analytical modeling of CFT*

Experimental model is shown in Fig. 1, from the experimental data. In the model, constant axial loading

(6860 kN) and lateral deformed loading is carried out, that is, monotonic loading of CFT steel column. From the experimental data, analytical model is constructed, as shown in Fig. 2, which relatively fine mesh

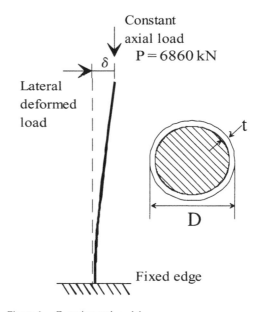

Figure 1.   Experimental model.

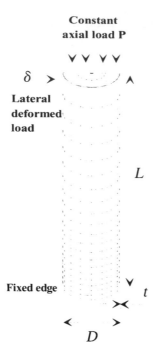

Figure 2. Analytical model.

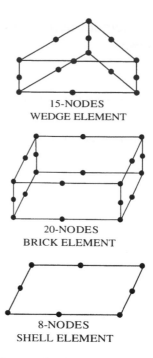

Figure 3. Elements of analytical model.

Table 1. Material Property.

Diameter D (mm)	Thickness (steel) t (mm)	Length L (mm)	Quality (steel)
700	6	2,920	SM490Y
Yield point σy (MPa)	Young's modulus (steel) Es (MPa)	Uniaxial strength (concrete) fc (MPa)	Young's modulus (concrete) Ec (MPa)
431.2	205,800	42.73	29,498

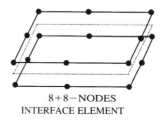

Figure 4. Interface element.

from the fixed edge to the height of 2D length and gradually coarse mesh from the above, is adapted. Here, D is outer diameter of CFT. Physical and material property is shown in Table 1.

For the steel column, 8-nodes isoparametric curved shell elements are used, for the mortal concrete, 15-nodes isoparametric solid wedge elements and 20-nodes isoparametric solid brick elements with quadric interpolation and Gauss integration scheme are used in the package program DIANA as shown in Fig. 3.

As for structural interface between steel column and mortar concrete, 8 + 8 nodes quadrilateral plane interface elements are adapted to describe relation between traction in the tangential plane and perpendicular to the tangential behavior as shown in Fig.4.

In order to express the diameter of the concrete and steel tubular exactly, virtual nodes, which consists the interface element, are settled between solid element and shell element. These virtual nodes also play a role of steel tubular with connected nodes of shell elements, using equal tying, which constrained the displacement degrees of freedom. Considering above assumption, the boundary connection is constructed, shown as Fig. 5.

Thickness of interface element is assumed to thin, or, t/100 is adopted so as not to affect the analysis.

2.2 Constitutive law of concrete material

As for the constitutive model of materials, the Drucker-Prager model in DIANA as shown in Fig. 6, is applied as the concrete material. This model has the features such as the linear tension cut off in tension range, the compressive softening behavior in

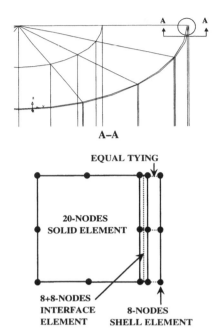

Figure 5. Boundary connection.

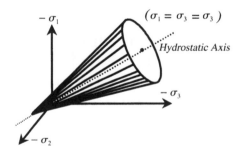

Figure 6. Drucker-Prager constitutive model.

compression range as shown in Fig. 7 and the confining effect on concrete strength.

2.3   *Linear tension cut off and crack criterion*

The initiation of cracks is governed by tension cut-off criterion as shown in Fig. 8a and a threshold angle between consecutive cracks. Smeared crack constitutive model is adopted. The following two criteria, which satisfied simultaneously, must be occurred, when crack arises.

1. The principal tensile stress violates the maximum stress condition.

    Here, the maximum stress is assumed to tensile stress (MPa), which is $1/10\,f_t$ of uniaxial compressive strength in the analysis.

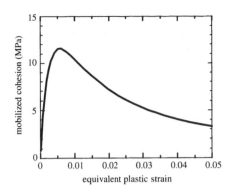

Figure 7.   Softening behavior on compression.

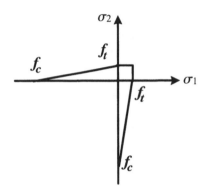

Figure 8a.   Linear tension cut off.

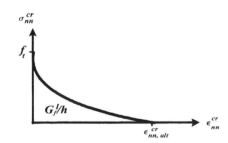

Figure 8b.   Nonlinear tension softening.

2. The angle between the existing crack and the principal tensile stress exceeds the value of a threshold angle.

    Here, α_{TD} is the angle between existing crack and the tensile principal stress direction. The 60 degrees in default value is considered.

After crack arises, nonlinear tension softening behavior is estimated as shown in Fig. 8b. Here, G_f^I is fracture energy, h is equivalent length, $\varepsilon_{nn,ult}^{cr}$ is ultimate crack strain, respectively.

2.4 Constitutive law of material steel

The multi-linear constitutive material model is used in material steel, in order to express characteristic of the yield plateau and strain hardening behavior of the steel material as shown in Fig. 9.

2.5 Modeling of interface behavior

If the interface element experienced tension, gap arises between tubular steel and concrete, that is, no contact between them. On the other hand, when the faces are in contact, normal forces are developed between the two materials as shown in Fig. 10. Normal forces are estimated from concrete Young's modulus Ec, because interface elements behave as a concrete when transmit the normal forces between two materials. No frictional forces are considered in tangential plane for interface element, so as to evaluate the buckling behavior of steel and concrete behavior, separately.

3 ANALYTICAL RESULTS

3.1 Buckling mode of steel tubular

Local buckling behavior of steel tubular is observed at the 2nd layer (layer 2) mesh near to the bottom of CFT as shown in Fig. 10. This behavior caused by the assumptions that touch and detachment perpendicular to the tangential plane and no frictional forces are considered in the tangential plane in the interface element. For simplicity, interface elements are eliminated in Fig. 11.

3.2 Load–displacement relation of CFT

Load–displacement relation at the tip of the CFT is shown in Fig. 12. The analytical result lowers than the experimental result, particularly in the post peak

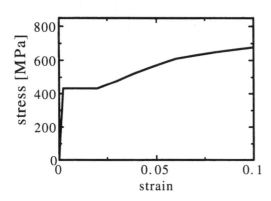

Figure 9.   Multi-linear constitutive model.

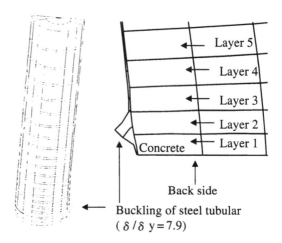

Figure 11.   Local buckling of steel tubular.

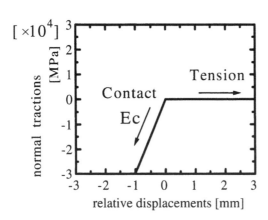

Figure 10.   Interface behavior in normal direction.

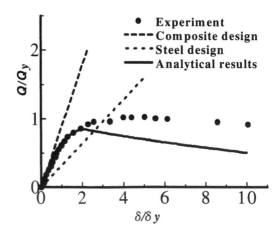

Figure 12.   Load–displacement relation.

behavior. The reasons why the analytical result is lower than experimental results is assumed that no frictional interface behavior in the tangential plane is considered in analysis. In addition, different constitutive law of concrete also affects the post peak behavior. Although those behaviors must be considered in order to fit the experimental result with analytical result, each material behavior to be grasped is important for the composite structure in the first place.

3.3 Stress–strain relationship of the concrete material

In order to verify the stress distribution, confined effect provided by steel tubular, and tensile behavior of concrete material, stress–strain relationship is investigated.

As it can be assumed that the base of cantilever beam governs the total behavior of CFT, axial stress distribution of fixed edge is investigated.

3.4 Stress distribution in fixed edge

Axial stress distribution of σzz, which is the vertical component of the total stress, at the fixed edge has been changed as the deformed load made progress. Although compression range is larger than tensile range in $\delta/\delta y = 0.8$, tensile range gradually spread and compression range decreased, but compressive strength reached approximately 2.5 times of uniaxial compressive strength fc in $\delta/\delta y = 7.9$ as shown in Fig. 13.

The neutral axis shifted toward center of the concrete core into deformed direction as the deformed load progress. Moreover, confined effect of concrete core can be seen only in the tip area of compression side.

3.5 Confined effect in longitudinal direction

So as to check the behavior, elements and integration point is chosen as shown in Fig. 14. Tip of the brick elements in arrow on deformed side and integration point number 6 are adopted. In the longitudinal direction, Layer 1 to 10 is selected to compare with the uniaxial stress–strain relationship, as the elastic behavior can be seen in the above Layers.

At the bottom Layer 1 shows approximately 2.2 times of uniaxial strength (92.6 MPa) and confined effect is gradually decreased onto the above Layers shown in Fig. 15. At the Layer 2, buckling of the steel tubular occurred to outside of the concrete core, however, no particular behavior can be seen in the axial stress-strain relationship.

3.6 Stress invariant space

On the other hand, stress invariant space is calculated at the same integration point, number 6 of brick element, in the above investigation as shown in Fig. 16.

All integration points reach failure surface and back to within the surface, as it can be seen in the uniaxial loading state, softening behavior on compression is occurred. Particularly, highly triaxial stress state can be observed in Layer 1 rather than other Layers, as the stress slope is gentle. It is simply because the bottom element Layer 1 suffer the highly damage.

3.7 Cracking behavior

In tension range of the concrete, cracking behavior is observed as shown in Fig. 17.

As the constant axial loading of CFT initiated,

1. Loading behavior is started on compression side in concrete material.

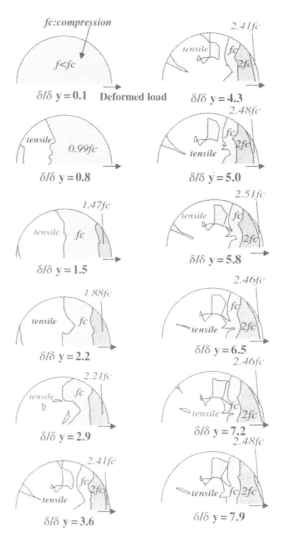

Figure 13. Axial stress distribution at the fixed edge.

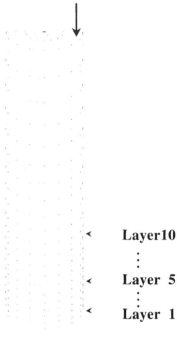

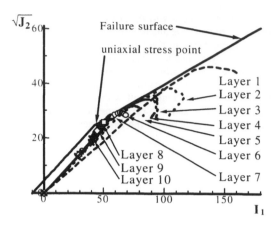

Figure 16.   Stress invariant space.

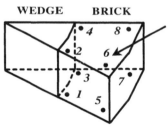

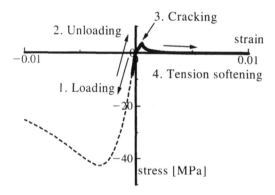

Figure 17.   Cracking behavior in tension range.

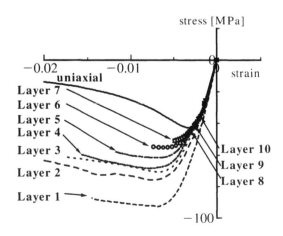

Figure 14.   Element and integration point.

Figure 15.   Axial stress–strain relationship.

After the constant axial loading, the deformed loading of CFT initiated.
2. Unloading started.
    And if it violates the crack criteria,
3. Crack arises.
    And finally,
4. Tension-softening behavior started.

3.8   *Development of crack strain*

Crack strain normal to the crack plane grows as shown in Fig. 18, corresponds to displacement of CFT. Crack arises at the surface of steel tubular in tension range. Initial crack strain gradually grows at the vertical direction from the bottom side to the top, and also multiplied into the concrete core from outer radius to center axis of CFT. At the large displacement loading state, for instance $\delta/\delta y = 7.9$, bottom of concrete core of CFT in tension range is highly suffered from cracking behavior.

It is also observed that cracking behavior of tensile range gradually spread and compression range

Deformed load

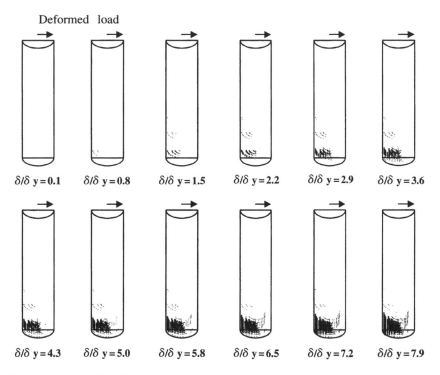

| $\delta/\delta\,y = 0.1$ | $\delta/\delta\,y = 0.8$ | $\delta/\delta\,y = 1.5$ | $\delta/\delta\,y = 2.2$ | $\delta/\delta\,y = 2.9$ | $\delta/\delta\,y = 3.6$ |

| $\delta/\delta\,y = 4.3$ | $\delta/\delta\,y = 5.0$ | $\delta/\delta\,y = 5.8$ | $\delta/\delta\,y = 6.5$ | $\delta/\delta\,y = 7.2$ | $\delta/\delta\,y = 7.9$ |

Figure 18. Development of crack strain.

decreased at the bottom region of CFT as is discussed in the stress distribution in fixed edge.

4 CONCLUSIONS

This paper presents 3D analytical study of CFT in monotonic loading state. At the process of modeling, constitutive law of Drucker-Prager tri-axial stress state of concrete material and multi-linear biaxial stress state of steel material are considered.

Cracking behavior of linear tension cut off and nonlinear tension softening are also considered in the concrete constitutive modeling. As well as interface behavior of normal direction and no frictional forces in tangential plane to express the connection between steel tubular and concrete core are considered.

The buckling behavior of steel tubular can be observed in the analytical results under composite action between steel tubular and concrete.

As for concrete behavior, confined effect of the tri-axial stress state enclosed by steel tubular and cracking behavior of loading, unloading and tension softening behavior can be observed in the analytical results. Still more, it is observed that crack strain normal to the crack plane developed at the vertical direction and also multiplied into the concrete core from outer radius to *center axis of CFT.*

Due to the elaborate modeling, such composite action between steel tubular and concrete core of CFT can be observed in the 3D finite element deformation analysis.

REFERENCES

Stephen P. Schneider (1998), "Axially loaded concrete-filled steel tubes", Journal of Structural Engineering, 1125–1138.
Ishigure T. (1997), "On cyclic deformation characteristics of concrete-filled tubular steel columns, bachelor thesis of Nagoya University, 6–9.
M. Shames, M.A. Saadeghvaziri (1999), "Nonlinear Response of Concrete-Filled Steel Tubular Columns under Axial Loading", ACI Structural Journal, 1009–1017.
Mizuno E., Hatanaka S. (1991), "Strain-Space Plasticity Modeling for Compresive Softening Behavior of Concrete Materials", Concrete Research and Technology, vol. 2, No.2, 85–95.
Oshita H., Oda T., Kawamura T. (2001), "Modeling of Deformational Behavior on Joint Surface between Steel and concrete and Its Applicability", Concrete Research and Technology, vol. 12, No.3, 1–13.
Diana-7 User's Manual (1998), "Nonlinear Analysis", 8.3, 299–301.

Finite Elements in Civil Engineering Applications, Hendriks & Rots (eds.)
© 2002 Swets & Zeitlinger, Lisse, ISBN 90 5809 530 4

Analysis and design of concrete shells using ultra-high performance fiber reinforced concrete

E.M.R. Fairbairn, R.C. Battista, R.D. Tolêdo Filho
COPPE/Universidade Federal do Rio de Janeiro, RJ, Brazil

J.H. Brandão
Universidade Federal de Mato Grosso, MT, Brazil

ABSTRACT: This paper presents the analysis of a concrete shell designed with an ultra-high performance fiber reinforced concrete: Ductal®, produced by Lafarge. It pertains to a new generation of cementitious materials conceived on a multi-scale reinforcement concept that allows for an improved damage resistance. The analysis carried out in this study is performed in two steps. Firstly, an inverse analysis is carried out to find out the post-cracking constitutive relation of the material. For this aim the experimental results of three-point bending tests are fitted by the numerical results obtained by the finite element simulation of these tests. Secondly, this stress-displacement curve is used to analyze a folded plate roof. With the aid of the obtained numerical results, together with the high performance characteristics of new materials such as Ductal®, it is argued that lightweight thin concrete shells could have renewed interest to designers and a promising structural alternative to many applications.

1 INTRODUCTION

Concrete shell structures had their golden age during the 1950s and 1960s, when they were very frequently designed and constructed throughout the world (Haas & Bouma 1961). Since their stability depend mainly on their configuration and ability to resist the applied loads through membrane action/strength, a multitude of extreme shell forms had been proposed resulting in the design and construction of elegant thin shell and shell-like structures.

In the late sixties, space structures of steel, timber and fabrics appeared as a new form of thin elements construction and concrete shells became less frequent (Falter 2000). Several reasons have been given to explain the disuse of concrete shells, such as the high costs of scaffolding, the big share of labor work during construction and the complexity of concrete shells design and construction (Ketchun 2002).

At present time, some new scientific and technological advances can give a new impetus to concrete shell engineering: finite element computer modeling have provided the designer with more accurate and easier design tools, and the advent of the new high performance and ultra-high performance cementitious materials have introduced a new paradigm to the discipline of shell design.

This paper explores the use of an ultra-high performance fiber reinforced concrete in the design of a thin concrete shell structure. This new material is Ductal®, produced by Lafarge. The simulation of the shell behavior under load, including its post cracking behavior, is performed with DIANA Finite Element code. This computer program is also used to carry out an inverse analysis to find out the post cracking constitutive relation of this fiber reinforced material.

The results presented in this paper show that very thin shells can be designed using this new generation of materials. Besides their light weight, it is noticeable the absence of reinforcing bars that greatly reduces the complexity of the analysis, design and construction of these structures. Thus, it can be expected that a new generation of concrete shells will emerge in the close future.

2 MAIN CHARACTERISTICS OF DUCTAL®

Ductal® is a new fiber reinforced cementitious material designed on a multi-scale reinforcement concept. It has very high compressive strength, ductility, toughness, and excellent durability. Its main characteristics (taken from (Acker 1999) and (Orange et al. 2000)) are listed in the following tables (Tables 1–3).

Table 1. Mechanical characteristics.

Compressive strength	180–230 MPa
Flexural strength	40–50 MPa
Young's modulus	50–60 GPa
Cracking toughness Gc	20–30 J/m^2
Fracture energy Gf	20000–30000 J/m^2
Creep coefficient	0.2–0.5
Shrinkage after curing	$<10^{-5}$
Density	2.45–2.55

Table 2. Rheological characteristics.

Abrahms cone	50–70 cm
ASTM shock table	250 mm

Table 3. Durability characteristics.

Chloride ion diffusion	$0.02 \cdot 10^{-12}$ m^2/s
Carbonation (depth of penetration)	<0.5 mm
Abrasion (relative volume loss)	1.2 index 1
Freeze-thaw (residual E-mod after 300 cycles)	100%
Capillary porosity ($>10\,\mu$m)	$<1\%$
Total porosity	2–6%

The highly ductile behavior of Ductal®, both in tension and flexure is close to the behavior of elastic-plastic materials. This deformation performance level results from an improvement in the micro-structural properties of the mineral matrix (micro-scale reinforcement) and the control of the link between matrix and fibers (macro-scale reinforcement) (Acker 1999). Natural inorganic micro-fibers might be employed as micro-scale reinforcement. In this way, mica flakes, or wollastonite micro-fibers can be incorporated by partial sand substitution (Orange et al. 2000). Macro-fibers, made of high-grade steel, or organic materials are between 3 and 50 times longest than the largest particle, and have a small cross-section to ensure an adequate bond.

Besides its ability to be virtually self-placing or dry-cast, the improved physical characteristics of Ductal® eliminates the need for reinforcing steel bars. Also, it can be produced with customary industrial tools by casting, injection, or extrusion (Perry 2001).

Performance measured in terms of ductility, mechanical strength and permeability of the material, combined with its easy and quick application, indicates that Ductal® can be a good choice for the construction of thin concrete shells.

3 POST CRACKING CONSTITUTIVE LAW

In what follows an inverse analysis is performed to find out the post-cracking constitutive relation of Ductal®.

Table 4. Typical concrete composition (Ductal®-1)

cement	sand	silica flour	silica fume	micro-fiber	steel fiber	sp	w/c
1	1.25	0.1	0.3	0.2	0.22	0.016	0.21

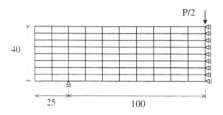

Figure 1. Finite element mesh for the simulation of 3 point bending test (all dimensions in mm).

For this aim, Ductal®-1 as defined by (Orange et al. 2000) is used as the construction material. The typical concrete composition is given in Table 4.

The three points bending test, for a 40 × 40 × 250 mm (beam span: 200 mm) specimen (Orange et al. 2000) was simulated with DIANA computer code. The element used was Q8MEM, a 4 nodes quadrilateral isoparametric plane stress element (DIANA 1998). The mesh is depicted in Figure 1.

The smeared crack model with strain decomposition was used. Within the framework of this model, the total strain is decomposed into an elastic strain and a crack strain:

$$\varepsilon = \varepsilon^e + \varepsilon^{cr} \qquad (1)$$

In the directions normal (n) and transversal (t) to the crack plane, the relationship between stress and crack strain, for a two dimensional analysis, is given by:

$$\begin{Bmatrix} \sigma_{nn}^{cr} \\ \tau_{nt}^{cr} \end{Bmatrix} = \begin{bmatrix} D_{secant}^I & 0 \\ 0 & \dfrac{\beta E}{2(1 + \nu)} \end{bmatrix} \begin{Bmatrix} \varepsilon_{nn}^{cr} \\ \gamma_{nt}^{cr} \end{Bmatrix} \qquad (2)$$

where β is the shear retention factor, taken, in the present analysis, as dependent on the crack width:

$$\beta = \frac{1}{1 + 4447\varepsilon_{nn}^{cr}} \qquad (3)$$

and D_{secant}^I can be computed if the $\sigma_{nn} - \varepsilon_{nn}^{cr}$ curve is known (see Figure 2). Here, ε_{nn}^{cr} is the crack mouth opening displacement, divided by the crack bandwidth length, as defined in equation (4), and Gf is the fracture energy.

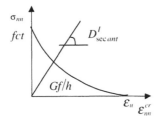

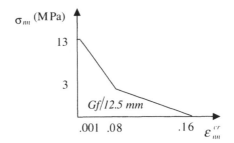

Figure 2. Stress-crack strain curve in the direction normal to crack.

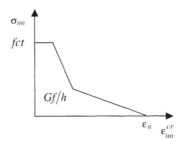

Figure 5. Post cracking constitutive relation obtained by inverse analysis.

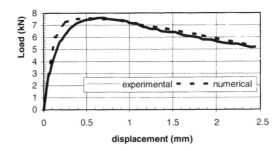

Figure 3. Shape of the cracking constitutive relation as a basis for inverse analysis.

Figure 6. Load-displacement curves for numerical and experimental 3 point bending tests.

The experimental and numerical load-displacement curves (the displacement was measured at the point where the load was applied) for the 3 point bending test are shown in Figure 6.

The results displayed in Figure 6 indicate that the constitutive relation of Figure 5 can be used to simulate the load-displacement behavior of the beam submitted to 3 point bending test. It is capable to reproduce the ductile behavior of the material, as well as the flexural strength of the beam. This constitutive relation will then be used in the analysis presented in the next paragraph.

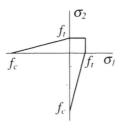

Figure 4. Linear tension cut-off criterium.

$$\varepsilon_{nn}^{cr} = w_{nn}^{cr}/h \qquad (4)$$

Using concepts presented on previous research works by Aveston et al. (1971) and Li (1992), the shape for the $\sigma_{nn} - \varepsilon_{nn}^{cr}$ curve depicted in Figure 3 was used a starting point for the inverse analysis. Then, DIANA's multi-linear tension softening model TENSIO 2 was used to model the smeared cracking behavior.

Linear tension cut-off, as displayed in Figure 4, was used with $f_c = 180$ MPa and $f_t = 13$ MPa.

Concrete in compression was modeled with associated plasticity using the Mohr-Coulomb yield condition with cohesion $c = 52$ MPa and friction angle $\phi = 30^0$.

The inverse analysis found the constitutive relation depicted in Figure 5.

4 FINITE ELEMENT ANALYSIS OF A FOLDED PLATE

The analyzed structure is a 10 mm thick folded plate used as a bus station roof. A perspective of the structure is shown in Figure 7 and a frontal view is shown in Figure 8.

The roof is designed without steel reinforcing or prestressing bars. The folded plate was discretized with CQ40S, quadrilateral eight-node isoparametric curved shell elements (DIANA 1998). The mesh and the boundary conditions are displayed in Figure 9.

The nonlinear behavior of concrete under tension (smeared crack), for the shell element, derived from

499

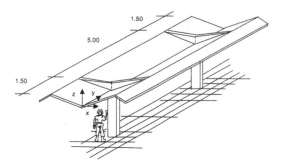

Figure 7. Perspective of the folded plate roof (all dimensions in meters).

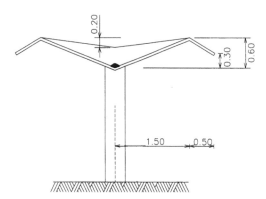

Figure 8. Frontal view of the folded plate roof (all dimensions in meters).

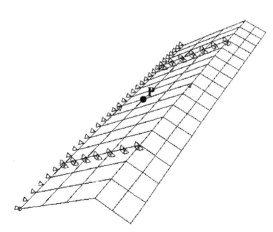

Figure 9. Mesh and boundary conditions.

the constitutive relation found by inverse analysis (see section 3) is depicted in Figure 10.

Uniformly distributed vertical load was applied through an incremental scheme corresponding to the Riks arc-length method as supplied by DIANA (1998).

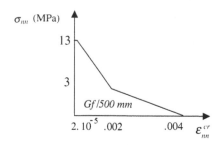

Figure 10. Post cracking constitutive relation for the shell element.

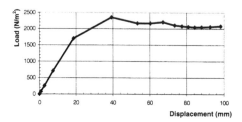

Figure 11. Load-displacement curve.

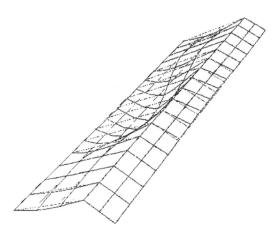

Figure 12. Deformed configuration.

The load-displacement curve for the point P marked in Figure 9 is displayed in Figure 11. This equilibrium path indicates that this shell-like structure has a great capacity to deform and that it can support a maximum load corresponding to approximately nine times its dead weight of $250 \, \text{N/m}^2$.

The deformed configuration for a load close to the ultimate load (displacement of 60 mm in Figure 11) is shown in Figure 12.

The cracking pattern for the same load-displacement configuration is shown in Figure 13.

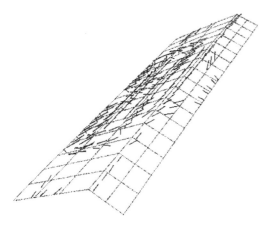

Figure 13.   Crack pattern.

5   CONCLUDING REMARKS

Concrete shells are nowadays in disuse for several reasons. Among them, it stands out the complexity of the design, analysis and construction, originating high costs.

The use of new ultra high performance cementitious materials, such as Ductal®, allows for the design and construction of highly lightweight thin shells. For the case study presented in this paper, the folded plate was as thick as 10 mm and presented an excellent structural performance from the point of view of both load carrying capacity and deformability.

Besides the lightness of the structure, the use of such new cementitious material, as well as the structural analysis and design, are considerably simplified through the elimination of reinforcing steel bars and the facility for placing the material.

It should also be pointed out that the analysis and design of concrete shells has also been made easier by the fabulous development of numerical analysis that provides the design engineer with accurate and user friendly computer codes.

It is also noticeable that the low porosity of the material improves the impermeability and durability of the structure, what corresponds to an improved environmental performance. Moreover, lightweight structures demand low quantities of cementitious materials, contributing to reduce the consumption of non-renewable resources and the emissions of greenhouse gases.

ACKNOWLEDGEMENTS

The authors would acknowledge FAPERJ, the Science Foundation of the state of Rio de Janeiro for granting the research study presented in this paper.

REFERENCES

Acker, P. 1999. *Ultra-high-performance concretes – properties and applications*, New concrete products, Lafarge Group, Paris.

Aveston, J., Cooper, G. A. & Kelly, A. 1971. Single and multiple fracture. *In Proc. of NPL Conf. on the Properties of Fiber Composites, National Physical Laboratory, London*, 15–26.

DIANA. 1998. *User's Manual*, TNO Building and Construction Research.

Falter, H. 2000. Conference Report: International Association of Shell and Spatial Structures (IASS), *Nexus Network Journal* 1 (4), http://www.nexusjournal.com/.

Haas, A.M. & Bouma, A.L. 1961. *Shell research, Proc. RILEM symp. on shell research, Delft, Aug 30-September 2, 1961*. North Holand.

Ketchun, M. 2002. *The home page of Milo Ketchum on Concrete Shell Structures*, http://www.ketchum.org/-milo/.

Li, V.C. 1992. Postcrack scaling relations for fiber reinforced cementitious composites, *ASCE Journal of Materials in Civil Engineering* 4 (1): 41–57.

Orange, G., Dugat, J. & Acker, P. 2000. Ductal®: New ultra high performance concretes. Damage resistance and micromechanical analysis. In P. Rossi & G. Chanvillard (eds) *Fibre-Reinforced Concretes (FRC) BEFIB'2000 Proc. of the 5th Int. RILEM Symp., Lyon, 13–15 September 2000*. RILEM Publ.

Perry, V. 2001. What is reactive powder concrete? Bridge Views, issue No 16, July-August 2001, http://hpc.fhwa.dot.gov.

Finite Elements in Civil Engineering Applications, Hendriks & Rots (eds.)
© 2002 Swets & Zeitlinger, Lisse, ISBN 90 5809 530 4

Non-linear behavior of low ductility RC frame under base excitations

S.T. Quek & C. Bian
Department of Civil Engineering, National University of Singapore, Singapore

ABSTRACT: Non-linear time-history analysis of low-ductility reinforced concrete (RC) frames subjected to base excitations were performed numerically using DIANA and compared with experimental results. The crack and inelastic behavior of concrete as well as the plasticity of steel reinforcements were included in the numerical model. The capacity, ductility and crack pattern of the RC frames from the FE analysis matched the experimental results reasonably well, giving confidence to the model and the subsequent study of the failure mechanism of the structure. The results also confirmed that such RC frames exhibit limited ductility and hence not suitable against sizeable base excitations.

1 INTRODUCTION

The behavior of RC structure under ground vibration has been widely studied for earthquake design (CEB 1996). The approach for seismic-resistant design is to provide capacity in terms of strength, stiffness, deformability, stability and energy dissipation greater than demanded by some specific design earthquake. For instance, the ductility of moment-resisting frame is supposedly assured if the reinforcement details specified in codes such as UBC (1970) are followed.

However, buildings in seismically inactive regions, such as Singapore, are generally not designed specifically to withstand sizeable ground-transmitted excitations. Furthermore, ground excitations can be induced by non-earthquake activities, such as quarry blasting, pile driving, the construction of underground transit systems and underground explosions. The effect of sizeable accidental underground explosion can be severe if the source is shallow, close to buildings and of significant charge weight.

For future planning purposes, such as the siting of underground ammunition chambers, which the minimum safe distance from the source is a critical consideration as it has both economic and safety consequences, the effect of blast-induced ground shock on non-seismic designed surface structure needs to be understood.

Considerable experience and research work on the behavior of RC structures subjected to earthquake-induced excitations has been documented (CEB 1996). However, these findings may not be entirely relevant for blast-induced ground shock in view of the different excitation characteristics, such as frequency range and duration, where the latter for blast is much shorter, typically less than 2 seconds (Cording, et al. 1975). Research work based on blast-induced ground excitation is scarce, even more so for its effect on low ductility RC structures, designed to take predominantly gravity type loads.

The objective of this paper is to study the non-linear behavior of low ductility RC frames subjected to base excitation. Half-scale RC frames were subjected to sinusoidal base excitations using the shake table at the Research Institute of Engineering Structures, Tongji University in China (Quek et al. 2002). Finite element analysis was performed using the input data experimentally obtained. The FE analysis software DIANA version 7.1 was used to simulate the non-linear behavior of the RC frame. The collapse mechanism and lateral-load capacity were also studied.

2 FINITE ELEMENT MODEL

2.1 *Half-scale RC frame subjected to base excitations*

Half-scale two-bay, two-story test frames were designed to satisfy similitude with regard to parameters related to geometry, loading and material properties, except for mass density. To satisfy the similitude requirement governing the mass density, and the applied dead and live loads from each floor, additional mass blocks were attached on top of the beams and columns. All the attached mass blocks were designed to be non-structural

Figure 1. Half-scale RC frame with mass blocks on shake table.

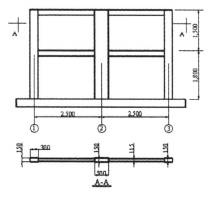

Figure 2. Half-scale RC frame for shake table test.

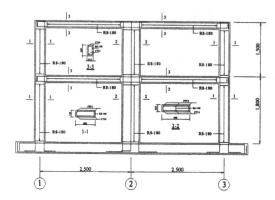

Figure 3. Reinforcement detailing of the specimen.

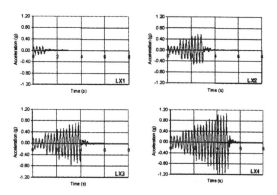

Figure 4. Time histories of input base excitations.

to minimize the effect on the structural stiffness and strength of the bare frame. Figure 1 shows a specimen on the shake table including the mass blocks.

The dimensions and layout of the half-scale models, which satisfy geometrically the laws of similitude, are shown in Figure 2. The detailing of the reinforcement (shown in Figure 3) was designed according to local practice based on the British Standard BS 8110 code (Mosley & Bungey 1990). The concrete used was Grade 35, consistent with the typical mix in local construction practice.

During the test, base excitations with frequencies close to the fundamental resonant frequency of the frames were applied. A typical time history of the input base excitation is shown as in Figure 4. In total, there were 4 excitations applied with different amplitude and durations, namely as LX1 to LX4, respectively, as shown in Figure 4.

3 FINITE ELEMENT MODEL

The above-mentioned RC frame was modeled by finite element using DIANA. The mesh generated is shown in Figure 5. The concrete beams and columns were modeled with 8-nodes quadrilateral plane stress element CQ16M with embedded rebar for longitudinal reinforcements and links. The size of the elements is about 50 mm × 50 mm (totaled >2200 elements). The mass blocks were modeled with point mass at the corresponding positions. The non-linear material properties were modeled as below.

3.1 Concrete

The non-linear behavior of concrete is complex under tension and compression and various options are available in DIANA. In the current study, the inelastic behavior of concrete in the compression zone was modeled using a Drucker-Prager description of the yield surface. Smeared crack model was used to simulate cracking behavior, as shown in Figure 6. DIANA

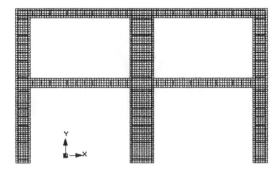

Figure 5. FE model for the half-scale RC frame.

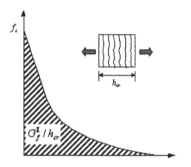

Figure 6. Smear crack model for concrete.

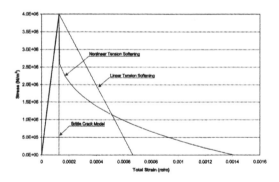

Figure 7. Models for tension softening behaviour.

provides various types of tension softening and shear retention models, as shown in Figures 7 and 8. In the current study, linear tension softening and variable shear retention models were adopted.

3.2 Reinforcement

Elastic–plastic with failure condition described by Von-Mises yield criterion was used for reinforcement bars, as described in Figure 9.

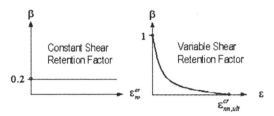

Figure 8. Shear retention models available from DIANA.

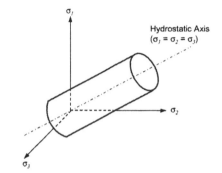

Figure 9. Von-Mises yield criterion.

4 DISCUSSION OF THE RESULTS

The experiment results were obtained by various instrumentations. Accelerometers were used to record the actual input base acceleration, which would be used as input for FE analysis. The experimental results and numerical predictions were compared and discussed as below.

4.1 Mode shape and natural frequencies

Eigenvalue analysis was first performed to compare the natural frequencies of the structure. During the test, white noise sweep was employed to identify these natural frequencies. The validity of the FE model that was constructed can be partially determined by comparing the numerical and experimental results. The frequencies of modes 1 and 2 are listed in Table 1.

The result showed that the numerical predictions of the natural frequencies are about 1.07 times of the experimental results. Such difference may be due to the fabrication process and inherent variability of the material properties. It is difficult to estimate the actual weight and stiffness distribution due to factors such as reinforcement overlaps, etc. The discrepancy of material properties could affect the results, as the properties used for numerical analysis were mean values obtained from laboratory tests. The mode shapes from the eigenvalue analysis of the first 2 modes are shown in Figure 10.

Table 1. Comparison of natural frequencies obtained experimentally and numerically.

	Mode 1	Mode 2
Experimental (Hz)	5.1	32.4
Numerical (Hz)	5.5	35.0

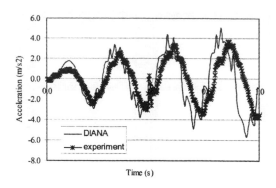

Figure 11. Comparison of experimental and numerical time histories of roof acceleration.

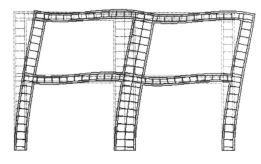

(a) Mode 1

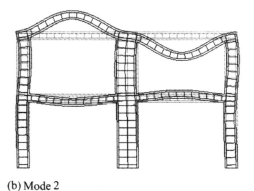

(b) Mode 2

Figure 10. Mode shapes from eigenvalue analysis.

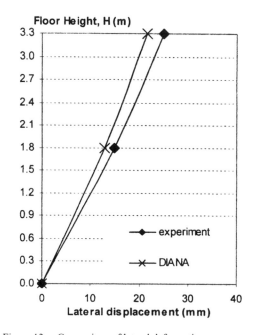

Figure 12. Comparison of lateral deformation.

4.2 Structural response under base excitation

Figure 11 compares the time history of floor acceleration under the first excitation (LX1) predicted from DIANA with those measured from the test. The base excitation recorded from the test was used as the input load. The results agree well considering the variability in the material properties and fabrication process. The numerical results have a slightly higher value of peak acceleration and shorter period, which could be due to the reasons discussed in the forgoing section.

Typical predicted lateral deformation profiles, corresponding to the instant that the roof displacement reached a maximum value, is given in Figure 12, together with the corresponding experimental results. The DIANA prediction of lateral displacements is about 10% smaller than those obtained from the test

(consistent with the numerical higher frequency obtained), and it gives the same profile of the deformation under the base excitation.

4.3 Crack pattern and ductility behaviour of the structure

During the test, low-frequency excitations close to the fundamental frequency of the structure were applied to induce severe damage. It was observed that flexural cracks started at the end of the beams. Further excitation would cause more severe cracks along the whole beams as well as cracks penetrating into the column. A sketch of the crack pattern is shown in Figure 13.

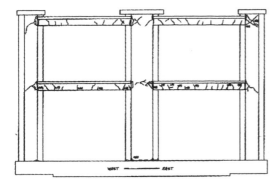

Figure 13. Crack pattern of the structure given by the test.

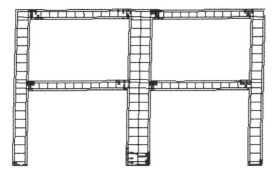

Figure 14. DIANA prediction of crack distribution (shaded) when the lateral deformation is maximum.

The distribution of initial cracks of the structure under base excitation predicted by DIANA is shown in Figure 14. The figure shows the crack locations after the second cycle of the applied base excitation LX1. It can be seen that cracks appeared at the end of the beams and bottom of the columns. Because of the strong column, yielding would occur in the beam first.

Figure 15 gives a magnified view of the cracks near the beam-column joint and column base. Photographs of actual cracks after the test are shown in Figure 15(a), while the corresponding DIANA predictions are given in Figure 15(b). Both the experimental and numerical results showed that cracks would occur near the bottom of columns due to insufficient links in reinforcement detailing.

In order to study the lateral collapse mechanism for the non-seismic designed frame, the lateral capacity is of interest. A static non-linear FE analysis was performed with lateral static force applied. The result is compared with the experimental one, as shown in Figure 16.

The results in Figure 16 showed that DIANA predicted a smaller deformation with higher yield force

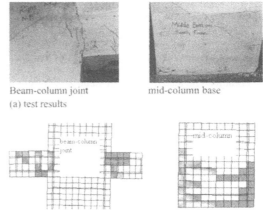

Beam-column joint   mid-column base

(a) test results

(b) DIANA prediction of the crack location

Figure 15. Crack near the beam-column joint and column base.

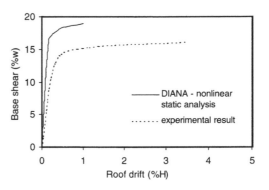

Figure 16. Comparison of lateral load-deformation relationship.

compared to the experimental results. The numerical prediction shows that the structure will most likely yield at a lateral force of 16%w (w = self-weight of the structure), while from the test, the structure yielded at base shear of 13%w, with a maximum lateral load capacity of 16%w. The difference could be caused due to the following reasons:

(1) Discrepancy in material properties. Mean values of the material properties were used for FE analysis, which could be slight different from the values of the true model.
(2) Discepancy due to the fabrication. Additional mass and support may affect the total weight and stiffness of the specimen and could be difficult to estimate accurately.
(3) Computational approximation. The base shear force in Figure 16 is obtained by taking the product

of the measured floor acceleration and the corresponding mass at the each floor since more than 85% of the total mass is concentrated at the floor levels. However, neglecting the term with the distribution of mass (e.g. columns) other than the floor level multiplied by the corresponding acceleration could lead to a smaller value in the results.

5 SUMMARY AND CONCLUSIONS

The FE analysis software DIANA was used to simulate a half-scale RC frame subjected to horizontal base excitation. The result showed that:

(1) By selecting proper models, the non-linear response of the structure under base excitation can be predicted reliably by FE simulation using DIANA.
(2) The non-seismic designed structure showed some degree of ductility, although it is not adequate for sizeable low-frequency base excitations. Structural damage initiatee at a drift ratio of 0.005-0.01 H (H equals to the height of the frame) for the frame considered.
(3) It would be advisable to enhance the structure by proper reinforcement detailing to prevent highly undesirable damage, such as shear, in the event that unintended ground excitations occur.

ACKNOWLEDGMENTS

The experimental work was performed jointly by the National University of Singapore and the Research Institute of Engineering Structure at Tongji University. The financial support provided by the DSTA (Singapore) is greatly appreciated.

REFERENCES

Comite Euro-International Du Beton. 1996. *RC Frames under Earthquake Loading: State of the Art Report* T. Telford, London, UK,.

Cording E.J., Hendron A.J., Hansmire W.H., Macpherson H., Jones R.A. & O'Rourke T.D. Method for geotechnical observations and instrumentation in tunnelling, *The National Science Foundation* 1975; **23**, Grant GZ 33644x.

Mosley W.H. & Bungey J.H. 1990. *Reinforced Concrete Design*, Macmillan Education Ltd., UK.

Quek, S.T., Bian, C., Lu, X., Lu, W. & Xiong, H. 2002. Tests on low-ductility RC frames under high- and low-frequency excitations. *Earthquake engineering and structural dynamics* 2002; 31: 459–474.

Uniform Building Code, International Conference of Building Officials, Pasadena, California, 1970.

Finite Elements in Civil Engineering Applications, Hendriks & Rots (eds.)
© 2002 Swets & Zeitlinger, Lisse, ISBN 90 5809 530 4

Finite element analysis in natural draught cooling towers

V.N. Heggade
Gammon India Limited, India

ABSTRACT: The column based hyperbolic reinforced concrete cooling towers are ecstatic in aesthetics, gigantic in nature and complex in geometry. The evolution of "Design and Constructions" of these towers have come long way in India and as on date, 141.0 m height Panipat Natural Draught Cooling Towers are the tallest commissioned towers in the country. In the past, before the computer revolution, the analysis used to be done only for membrane forces in shell elements. However, as the towers grew in stature beyond 100.0 m height clubbed with infamous episode of Ferry Bridge cooling tower in UK emphasized the relevance of bending analysis coupled with membrane forces. Invariably it is found that the structural criticality in NDCTs is due to the wind induced dynamic forces, which compounds the complexity of the above analysis. The following aspects,

- the peculiar geometry,
- the real boundary conditions,
- stress, temperature and time dependent behaviour of constituent materials including composites,
- space-time dependent loads including random excitations,
- different types of behaviors, such as linear, dynamic stability, etc., of structure to be examined,

make the Finite Element Analysis apt and natural choice for the NDCTs.

In the paper followed, the complexity of wind induced forces and rationale behind the applications, the progression from membrane analysis to bending analysis, the simple practice of descretising Finite Element Model for NDCTs and finally the illustrations and discussions on FEA for Panipat NDCT, the tallest tower in India as on date is discussed in length.

1 INTRODUCTION

In any plant, turbo-generations are driven by treated steam, generating the electric energy. The steam is condensed in order to create an effective heat sink behind the turbine and recycled into the boiler. This requires large amount of cooling water where during the condensation process, heat energy is transferred into cooling water media and disposed of to atmosphere.

Earlier days, water used to be drawn from natural water resources like sea, river, etc. and after condensing the steam; the cooling water carried the absorbed heat energy and discharged back to sea or river from where it was drawn. However, thermal pollution of natural resources, which has many adversarial environmental consequences, forced the technologists to resort to the cooling towers where the cooling water resource is recycled and reused.

Among the cooling towers, Natural Draft Cooling Towers do not need any external cooling devices driven by prime movers, consuming some sort of energy and hence are able to balance environmental factors and operating costs of the power plants.

The Gammon India Limited has been the pioneer in India in the field of NDCTs having hyperbolic profile in elevation. This shape provides the most beneficial heat exchange efficiency besides structural stability.

The very first such construction in India in 1934 measuring 34.0 m dia and 38.0 m height has now crystallized in structures of 120.0 m dia and 141.0 m high – huge ones by any standard (fig. 1). The structural design of this highly complicated parabolic shell has been perfected over the last 50 years along with the sophisticated constructional systems for realizing in space the three dimensional accuracy of the thin shell of minimum thickness 200 mm. The internal prestressed louvers for the heat exchange stack are the only ones of its type in the world and have been adopted in India to a large extent for NDCTs.

During the evolutionary process of structural analysis and design of this peculiar structure, one had to deal with either one or combination of the following features:

- Unconventional geometry/form.
- Real boundary conditions.
- Stress, temperature and time dependent linear/non linear behaviour of constituent materials including composites.
- Space-time dependent loads including random excitation.
- Different types of behaviour – linear, nonlinear dynamic stability – of structures to be examined.

At the nascent stage of evolution, the classical methods of structural mechanics were resorted to. These methods were also available for one, two and three-dimesional models representing the behavior and analysis of structural components and systems. These classical methods in general may be classified into two categories. In the first category materials and compatibility conditions are constituted in the modeling and the differential equation of equilibrium of the fields are solved for the

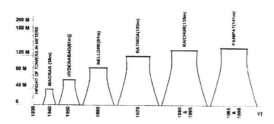

Figure 1.   Evolution of NDCTs in India.

given boundary condition. In the second method the principle of minimum potential energy is employed to arrive at the deformation of the structure and then the stresses on the structure are evaluated. Normally, Finite difference and Rayleigh-Ritz methods are the numerical methods employed for the first and second category respectively. In case of continuum structures when these numerical methods are used, the solution of linear and nonlinear simultaneous algebraic equations warrant enormous amount of labor and time.

The computer revolution encouraged the structural engineers to develop unified approaches for analysis using matrix method suited for programming, giving an impetus for the development of Finite Element Method. This method is well documented by now and many commercial soft wares are available for structural analysis. However the efficiency and reliability of the structural analysis solution depends upon expertise and knowledge of the engineer. The suitability of the model also has the direct bearing on the above.

In the succeeding paper, the principle behind the classical numerical method used till eighties and the simple, effective and reliable Finite Element Model beyond eighties, which can be analyzed using any commercially available package is discussed in detail while taking up "Panipat NDCT" for illustration.

2   PANIPAT "NDCT"

The tower consists essentially of an outside hyperbolic shell of reinforced concrete, the principal function of

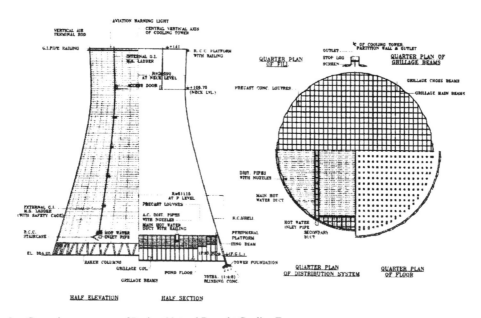

Figure 2.   General arrangement of Panipat Natural Draught Cooling Tower.

510

which is to create draught of air in a similar way to a chimney. The other water-cooling and collecting components of the tower such as internal grillage structure, pond floor, fill; water distribution system is structurally independent of hyperbolic shell and is housed inside the tower. Shell diameter at the base and throat, shell height and air inlet-height is governed by thermic design consideration. Figure 2 shows typical cross section of the cooling tower (Fig. 2 is given at the end).

The tower is 141.0 m high above basin sill level with a base diameter of 122.977 m and a top diameter of 78.709 m. Diameter of the throat located at 103.635 m above the basin sill level is 73.20 m. Shell thickness of the tower varies from 540 mm at the top to 100 mm at the bottom with a minimum shell thickness of 215 mm at the throat level. The shell is supported on 52 pairs ("V" type) of raker columns of 800 mm diameter which transverse a 6.45 m high air inlet space below the shell and merge at the lower ends with reinforced concrete pedestals, which are cast integral with the RC basin wall. Foundation for the tower consists of 1092 cast-in-situ, driven RC raker piles, diametering 400-Φ and load capacity of 550 KN each. The piles, which are raked outwards at 1 horizontal to 6 vertical, are distributed uniformly along four concentric rows and topped by 4.30 m wide RC pile cap.

The natural draught cooling towers are special structures, in view of the hyperbolic shape and large size combined with very small shell thickness, appreciably less in proportion than that of an egg shell, and sensitiveness to horizontal forces. These towers are one of the largest civil engineering structures where wind forms the major applied loading.

3 SHELL ANALYSIS

The shape of the tower structure represents a hyperboloid of revolution (fig. 3) and the element in the shell assumes the curvatures in two way of Negative Guaussian value $\{1/r_1 * 1/r_2\}$.

Up to eighties it was assumed that for doubly curved shells, except for the edge zones, the bending stresses developed are negligible because of the large vertical loads. Thus the membrane analysis, which does not involve complex analytical process with bending, was considered to be adequate.

As shown in the fig. 4, the loads operate in 3 coordinate directions and the following general equations of equilibrium are considered to solve for the co-planar stresses:

$$1 \quad \frac{\partial}{\partial \phi}(rN_{\theta\phi}) - \frac{\partial}{\partial_\theta}(RN_\theta) + R\cos\phi \, N_{\theta\phi}$$
$$- rRP_\theta = 0$$

$$2 \quad \frac{\partial}{\partial \phi}(rN_\phi) - \frac{\partial}{\partial_\theta}(RN_{\theta\phi}) - R\cos\phi \, N_\theta + rRP_\phi = 0$$

$$3 \quad R\sin\phi N_\vartheta + rN_\phi + rRPz = 0$$

After having applied the necessary boundary conditions, the "Stress Function" is induced to bring down the three variables ϕ, θ & $\phi\,\theta$ into one.

Though the tower behaves predominantly as a membrane structure, the wind tunnel studies carried out at various international laboratories revealed that the corrections are required near boundaries especially at the cornice and ring beam locations (where raker columns meet the shell) due to the edge perturbations. As the tower sizes increased beyond 120.0 m height, it was imperative to predict the bending behaviour of the structure.

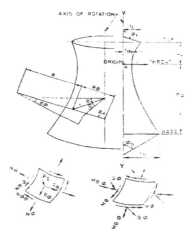

Figure 3. Hyperboloid of revolution with 2-way curvature.

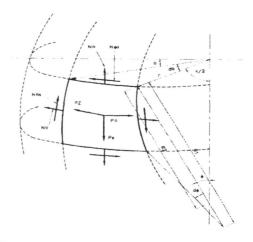

Figure 4. Forces and Loads on shell element.

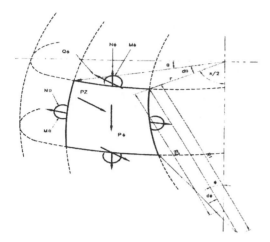

Figure 5. Stress resultants and Loads on shell element.

If the shell element is subject to the external loadings in three directions as shown in the fig. 5, in the shells of revolutions, the internal forces developed will be:

$N\phi$, $N\theta$

$M\phi$, $M\theta$

$S \cdot Mt$

The following general equations of equilibrium in terms of the above six quantities can be written:

$$1 \quad \frac{\partial}{\partial \phi}(rs) - \frac{\partial}{\partial \theta}(RN_\theta) + R\cos\phi\, S + \frac{\sin\phi}{r}$$
$$\times \left\{ 2\frac{\partial}{\partial \phi}(rMt) - \frac{\partial}{\partial \theta}(RM_\theta) \right\}$$
$$+ 2Mt\cos\phi - rRP_\theta = 0$$

$$2 \quad \frac{\partial}{\partial \phi}(rN_\theta) - \frac{\partial}{\partial \theta}(RS) - R\cos\phi N_\theta$$
$$+ \frac{1}{R}\left\{ \frac{\partial}{\partial \phi}(rM_\phi) - R\cos\phi\, M_\theta - 2\frac{\partial}{\partial \theta}(RMt) \right\}$$
$$+ rRP_\phi = 0$$

$$3 \quad R\sin\phi\, N_\theta + rN_\phi$$
$$+ \frac{1}{r}\frac{\partial}{\partial \theta}\left\{ \frac{\partial}{\partial \phi}(rMt) - \frac{\partial}{\partial \theta}(RM_\theta) + R\cos\phi\, Mt \right\}$$
$$- \frac{\partial}{\partial \phi}\frac{1}{R}\left\{ \frac{\partial}{\partial \phi}(rM_\phi) - \frac{\partial}{\partial \theta}(RMt) - R\cos\phi\, M_\phi \right\}$$
$$+ rRPz = 0$$

where

$$Mt = \frac{1}{2}(M_{\theta\phi} + M_{\phi\theta}) \quad \text{and}$$
$$S = \left(N_{\theta\phi} - \frac{M_{\phi\theta}}{R} \right) = \left(N_{\phi\theta} - \frac{M_{\theta\phi}\sin\phi}{r} \right)$$

However, since there are six unknowns, additional three equations are required to make the problem determinate which are obtained by compatibility relations between the strain and curvature parameters of the middle surface of the shell.

$$4 \quad -\frac{R\,\partial k\phi}{\partial \theta} - \frac{r\,\partial T}{\partial \phi} - 2R\cos\phi\, T + w\cos\phi$$
$$+ \frac{\sin\phi}{r}\left\{ r\frac{\partial w}{\partial \phi} + R\cos\phi\, w + R\frac{\partial e\phi}{\partial \theta} \right\} = 0$$

$$5 \quad \frac{rk_\theta}{\partial \phi} + R\cos\phi\,(K_\theta - K\phi) + \frac{R\,\partial T}{\partial \theta}$$
$$- \frac{1}{R}\left\{ R\frac{\partial w}{\partial \theta} + r\frac{\partial e_\theta}{\partial \phi} + R\cos\phi\,(e_\theta - e\phi) \right\} = 0$$

$$6 \quad rk_\theta + R\sin\phi K\phi + \frac{1}{r}$$
$$- \frac{\partial}{\partial \phi}\left\{ R\frac{\partial e\phi}{\partial \theta} + \frac{r}{2}\frac{\partial w}{\partial \phi} R\cos\phi w \right\}$$
$$+ \frac{\partial}{\partial \phi}\frac{1}{R}\left\{ r\frac{\partial e_\theta}{\partial \phi} + R\cos\phi\,(e_\theta - e\phi) + \frac{R}{2}\frac{\partial w}{\partial \theta} \right\} = 0$$

where

$$e_\theta = \frac{1}{Ecd}(Ne - \gamma N\phi)$$

$$e_\phi = \frac{1}{Ecd}(N\phi - \gamma N\theta)$$

$$w = \frac{2(1 + \gamma)}{Ecd} S$$

$$K\theta = \frac{12}{Ecd^3}(M\theta - \gamma M\phi)$$

$$K\phi = \frac{12}{Ecd^3}(M\phi - \gamma M\theta)$$

$$T = \frac{12(1 + \gamma)}{Ecd^3} Mt$$

The above classical bending analysis by numerical method, apart from being very complex, does not take into consideration the soil structure interaction at the foundation levels, which adds further complexities.

4 FINITE ELEMENT ANALYSIS

As explained earlier, since there are many commercial soft wares available and the method can be used to any structure with complicated boundary conditions, the Finite Element Method is extensively used in the analysis of NDCTs, beyond eighties. This method was used by "Hill and Collin" for the first time for cooling towers. In this method, it is possible to include variation in thickness, arbitrary loading, the prestressing effect at the edge beam and the idealization of edge beam (ring beam) as the part of the shell, which are difficult to include in other methods.

Also for an accurate determination of frequencies of free vibration of cooling towers, the flexibility of shell supports and foundations should be included in the analysis. The bottom and top stiffening will also have a significant effect on the frequency of free vibration. Perhaps Finite Element Method is the only convenient tool when allowance must be made for such distinct structural features.

The standard Finite element Model, which has been evolved over a period of time and employed in Panipat NDCT (fig. 6) is taken for illustrating the discretisation.

The base of the natural draft cooling tower is supported on 52 equi-spaced pairs of inclined columns to facilitate air intake. The raker of the shell supporting columns (fig. 7) and also the angle of foundations is matched with the base of hyperbola, to subject the column and foundation to maximum axial forces, avoiding the kinks and consequential flexural stress formations. However, the column supports produce concentrated reactions along the bottom edge of the shell, while between the columns shell edge remains stress free.

These edge reactions in turn, produce stress concentrations and give rise to bending moments and transverse shears in the shell, which is otherwise predominantly membrane shell. Thus the accurate determination of stresses at the junction of columns and shell elements is of significant structural importance. The presence of columns adds further complications by destroying axi-symmetric nature of shell geometry, which would have been simple, idealization wise.

In the case under the illustration, most of the cooling tower shell is modeled using plane stress quadrilateral element with orthotropic material properties. Each plane stress or shell element can cater for different thickness and can be located in an arbitrary plane with respect to the three dimensional coordinate system. These elements are based on iso parametric formulations.

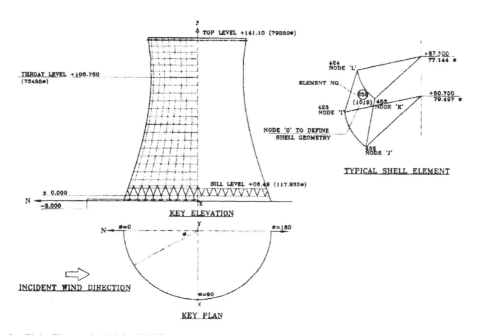

Figure 6.   Finite Element Model for NDCTs.

513

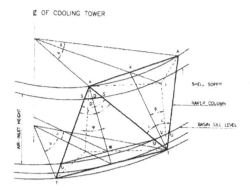

₵ OF COOLING TOWER

SHELL SOFFIT

RAKER COLUMN

BASIN SILL LEVEL

AIR INLET HEIGHT

Figure 7.   Schematic view of raker columns.

For the accurate determination of the concentration of stresses and also to make the bottom edge (soffit) of the tower as the part of shell, a combination of quadrilateral and triangular shell elements are employed in the layer at the top of the raker columns.

The raker columns are idealized as three-dimensional solid beam elements, which consider torsional bending about two axes, axial and shearing deformation, and are prismatic.

Barring the cases where fixed and free conditions are specified, both 4 noded, shell elements and two noded beam elements have 6 nodal degree of freedom, hence where beam element is attached to shell element at a node, the problem free compatibility is ensured.

The top thickness, i.e. the cornice of the tower and the portion of the soffit between the raker columns are specified to be free which shall yield zero stress resultants.

It has been shown by "Abasing and Martin" by considering four different idealized boundary conditions that base conditions do not significantly affect stress resultant, except near the base and the fixed conditions give an upper bound solution. Also, since the base edge beam (ring beam) is 1/8 to nearly 1/10 of overall dimension of the tower, it is realistic to consider the base as fixed.

Since the loading of the tower is symmetrical about Z-axis, only half of the tower is considered for analysis (However, 1/4th shown on the fig. 6). As the displacements in X-direction and the rotation in "Y" & "Z" axis of point of symmetry are zero, these nodal points at θ = 0° and θ = 180° are considered to be fixed against displacements in X-direction and rotation about "Y" & "Z" directions.

5   LOADINGS AND DESIGN OF ELEMENTS

The various load combinations studied before inferring the critical stress resultants/forces for the design are as below:

- Dead load + seismic load for shell and fill structures
- Dead load + construction load for shell structure only
- Dead load + wind load
- Dead load + seismic load + temp. Gradient + sun's radiation
- Dead load + wind load + temp. Gradient + sun's radiation

The wind loading forms the major part of the external loading in the design of the shell and supporting components. This has a large steady component and a significant random component because of the air turbulence. The latter dynamic components can be calculated in frequency domain by natural frequency analysis, which is established to be contributing 50% at the total peak responses. However, this involves very complex analysis both in meridonial and circumferential directions at different level, for tensile, compressive forces and bending moments as separate combinations. The above has necessitated the codal provisions to translate the external forces and structural responses to "quasi-static" analysis by application of "Gust-factor" (G). The gust factor depends on the natural frequency in the fundamental mode, with speed, terrain and size of structure and the peak response occurring in a time interval of 1-hour duration. In the case of Panipat NDCT the fundamental frequency calculated by the spectral analysis using commercial package is 1.002 HZ and the gust factor which is worked out using the formula given in IS: 875 (Part 3)-1987 is 1.74. Normally "G" value fluctuates between 1.6 to 2.2 being in ascendancy with smaller tower height and rough terrain.

The wind speed of 47.0 m/sec has been considered at 10.0 m height and wind pressures are based on peak wind speed of 3-second gust with a return period of 50 years. The design wind pressures in the meridonial direction depend upon factors related to probable life of structure, terrain, and local topography, size of the structure for both peak factor and gust factor methods. The combined effect of these factors deduced by multiplication is shown in fig. 8. It may be interesting to note that up to 40.0 m height, the peak factor method attains criticality.

The variation across a particular elevation is ascertained by normalizing the values of equal angle increments from the windward direction and is represented by Fourier Series, H = □ An cosN□ The pressure coefficients "An" as per IS: 11504-1985 and BS: 4485-1996 which is used for NDCT is given in following table.

The above coefficients include internal pressure of 0.4; the effect of internal negative pressure increases the circumferential compressive forces by around 40% and correspondingly decreases the circumferential tensile forces. However, it does not have any impact

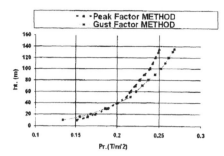

Figure 8.   Wind effects on cooling tower.

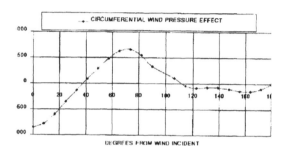

Figure 9.   Pattern showing wind effects circumferentially.

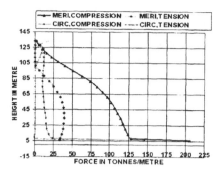

Figure 10.   Membrane forces on the shell.

modulus of rupture, as has been the assumption in the state of the art analysis.

Ha	An
0	−0.00071
1	0.24611
2	0.62296
3	0.48833
4	0.10756
5	−0.09579
6	0.01142
7	0.04551

on meridonial forces. The fig. 9 depicts the effect of Fourier series coefficients due to wind at foundation nodal points.

When the NDCTs are grouped together as in the case of said towers, the wind pressures are augmented because of the aerodynamic interference effects and generally on the basis of wind tunnel testing, the enhancement factors are considered in the design. On the basis of wind tunnel experiments conducted by Dev and Fiddler, the IS and BS give simplified empirical formula for the critical wind pressure to cater for global buckling behaviour of the tower. Since the formula does not account for the influences of bottom and top edge stiffening, vertical variation of wind pressure, curvature effect and aspect ratio of tower, the factor of safety used is as high as 5. However, the latest BS: 4485-96 recommends local buckling criteria, which is a function of both meridonial and circumferential stress resultants at a particular elevation. This is called Buckling Stress State (BSS) approach and developed by Mungun of Ruhr University Bochum.

After having deduced the critical stress resultants, the elements are designed for membrane forces and bending moments both meridonially and circumferentially, restricting the tensile forces in concrete to

The following inferences can be made from the membrane forces graph (fig. 10), in case of Panipat NDCT, which are general trends in all NDCTs. The meridonial compression is very high at the lintel level, which warrants additional thickening, and same goes on reducing along the height of the tower. The same is the case with circumferential compression, though in this case, the force remains almost constant above the throat level to bottom of top thickening level.

The meridonial tension gradually reduces and vanishes above throat level where as the circumferential tension is high where the top and bottom stiffening of the shell is there and in fact starts building up above throat level perhaps because of which the latest BS: 4485-1996 recommends higher minimum percent steel at top 1/3 height of tower.

The accompanying B.M. variation along the height is shown in fig. 11. Though the graph looks too much scattered, the keen observation reveals, the B.M. mainly fluctuates between 0.1 TM/M to 1.0 TM/M, and the fluctuation is mainly because of local effects. Also quite pronounced B.M. effect can be seen at top and bottom of shell, which can be attributed to stiffening effect, where thickening is necessary to cater for large membrane forces at top and bottom.

The following observations highlight the general non-consensus among the international community pertaining to design for temperature loading.

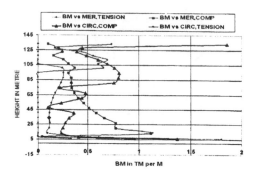

Figure 11. Bending moment variation on shell.

The German VGB guidelines specifies $I_{eff} = 1/2$ (I gross + I crack), while BS & IS only state that consideration shall be given in the analysis for the strain resulting from temperature gradient, without mentioning anything about stiffness criteria. IASS and also ACI are silent about the stiffness factors to be considered for temperature stresses. However, IS: 4998, which deals with calculations of thermal stresses, recommends the use of working stress method, using cracked section stiffness. Thus it is inevitable to take recourse into companion Indian Code IS: 4998 and use "Superposition" method for calculating temperature stresses, though the "Combined Simultaneous" method gives slightly larger stresses. However the latest BS: 4485 deems to cater for thermal gradient loading by provision of minimum 0.3% steel in the shell.

Unlike other stack like structures, in NDCTs, the maximum meridional tension occurs at windward 0° and compression occurs at around 72° instead of leeward 180°. Thus the raker columns and pedestals have to be checked for maximum axial tension and compressive forces with the corresponding bending moments. Also it has to be ensured that the foundation is not uplifted under maximum axial tensile forces.

6 CONCLUSION

The use of the quadrilateral plane stress elements for the shell and 3D prismatic beam elements for raker columns are found to be economical and fairly accurate and is recommended in latest BS: 4885-1996 in the form of substructure for periodically rotational structures. Using any of the commercially available softwares such as GT STRUDL, SAP2000, ANSYS, NISA, etc can very conveniently employ the typical finite element model presented in the illustration.

However, it has to be kept in mind that the modelling problems of structural analysis have a significant influence on the accuracy, reliability and economy with which the results can be achieved. Therefore, in-depth understanding of the various aspects of modelling and their effects is imperative for the designer. In view of this, though the commercially available packages can be conveniently used for the analysis of NDCTs, the user of such packages is advised to thoroughly study the background of finite element library and facilities available for modelling before making specific choices for the given requirements.

REFERENCES

Hill D.W., Collin G.K., "Stress and Deflection in Cooling Tower Shells due to Wind Loading" – Bulletin of IASS, No. 35. Sept. 1968, pp 43–51.

Albasing. E.L., Martin D.W., "Bending and Membrane Equilibrium in Cooling Towers" – Journal of Engineering Mechanics Division of ASCE, Vol. 93, June: 1967, pp 1–19.

IS: 11504-1985, "Criteria for Structural Design of Reinforced Concrete Natural Draught Cooling Towers" – Indian Standards Institute, 1986, pp 15–18.

V.N. Heggade, "Raichur Natural Draught Cooling Towers – Units 5 & 6" – Journal of Indian Concrete Institute, vol. 1, July-Sept. 2000, No. 2 pp 7–18.

BS: 4485 Part 4: 1996 "Code of practice for Structural Design and Construction.

Short papers

Finite Elements in Civil Engineering Applications, Hendriks & Rots (eds.)
© 2002 Swets & Zeitlinger, Lisse, ISBN 90 5809 530 4

Constitutive models for reinforced concrete: Challenges and proposals for future research

E. Fehling & T. Bullo
Department of Structural Concrete, University of Kassel, Germany

ABSTRACT: A concise state-of-the art review of the constitutive models formulated in the past using the concept of classical local macro-continuum mechanics and used for the finite element analysis of reinforced concrete structures is presented. The complex composite material behaviour of reinforced concrete under different loading conditions is discussed. A direction for future research and some suggestions on the techniques that may be applied to improve the existing models are highlighted.

1 CONSTUTITIVE MODELING: A CONCISE STATE-OF-THE-ART REVIEW

Over the past nearly four decades, several material models were developed to simulate the experimentally observed behaviours of reinforced or/and prestressed cement-based composite structural elements subjected to both static and dynamic loading (Chen 1991, Bazant 1985, Ishikawa 1985). Even though it is not the objective of this succinct paper to reassess the vast constitutive models developed in the past; an endeavour is made to briefly summarize the material models available so far and give some opinions, remarks and suggestions for their improvements. Based on the information obtained from the experimental investigations hitherto, the material behaviour of the composite material reinforced concrete is mainly characterised by:

- Non-linear and non-elastic material behaviours under multiaxial states of stress (strain);
- Damage induced degradation of the linear elastic material constants (stiffness) of the matrix (concrete) in the pre-peak stress region;
- Stress- or strain-induced anisotropy due to matrix cracking, which requires a material stiffness capable of modelling the coupling effect of normal stresses and shear strains;
- Non-uniqueness of the post-peak compressive (tensile) stress-strain behaviour for specimens of different size tested; due to multidirectional and non-uniform strain distribution induced by the progressive fracturing of the heterogeneous matrix;

- The resistance of the cracked composite to applied shear stress as a result of the aggregate interlock mechanism along the crack surface and the dowel action of rebars;
- Interfacial bond stress and slip (or debonding) between rebars and concrete;
- Tension stiffening due to the tensile stress-carrying capacity of the uncracked concrete between two adjacent cracks;
- The difference between the yield strength of reinforcing bars embedded in the concrete matrix and that of bare rebars and
- Time-dependent behaviours such as creep, thermal and shrinkage of concrete, and the relaxation of prestressing steel.

The incorporation of the enumerated behaviours into the material model to be developed helps to simulate at best the structural responses of reinforced concrete elements subject to multidirectional loading. Thus, the quite accurate prediction of the element behaviour leads to a relatively exact numerical simulation of the responses of reinforced concrete composite structures such as ultimate strength, load deformation history, crack patterns, modes of failure, above all the likelihood stress and strain distributions within the structure that can be adopted for design purposes. Some of the main constitutive models available to date and that are adopted in the past for the simulation of the structural responses of reinforced concrete elements under different states of stress (strain) are summarised in Tables 1a and 1b.

Table 1a.　Capability of the existing material models.

No.	Capability to model	Material model	
		Elasticity based models	Elasto-plasticity based models
1	Path dependency (dependency on strain history)	Possible using formulation based on hypoelasticity	Using yield (loading) surfaces, hardening/softening curves and flow rules
2	Stress/strain induced anisotropy (coupling of hydrostatic and deviatoric responses)	Possible using hypoelastic (hyperelastic) models	Not possible using isotropic loading surfaces; but using anisotropic loading surfaces
3	Post-peak material behaviour	Possible mainly using secant stiffness	Possible using strain based loading surfaces or hardening/softening curves
4	Damage induced degradation of elastic constants	Generally possible	Not possible using classical theory of plasticity
5	Unloading and reloading behaviour	Generally possible	Only elastic unloading is possible
6	Cracked matrix behaviour	Possible using hyper- or hypoelasticity	Possible, but for cracked behaviour it should be based on anisotropic plasticity
7	Experimentally observed behaviours	Primarily for uniaxial and biaxial states of stress (strain)	Mainly for ductile materials and to some extent for quasi-brittle materials
8	Simplicity and practicability for use	Simple and easy to use	Requires deep knowledge

Table 1b.　Capability of the existing material models.

No.	Capability to model	Material model	
		Continuum damage mechanics	Fracture mechanics
1	Path dependency (dependency on stress-strain history)	Possible if combined with a plasticity model	It is based on energy release rate
2	Stress/strain induced anisotropy (coupling of hydrostatic and deviatoric responses)	Possible for distributed cracks	Not yet formulated
3	Post-peak material behaviour	Not yet well developed	Well formulated for tensile behaviour
4	Damage induced degradation of elastic constants	Adequately formulated for smeared cracks	Not yet formulated for a coupled tensile-compressive loading
5	Unloading and reloading behaviour	Not yet exclusively formulated	Only elastic unloading
6	Cracked matrix behaviour	Only for well distributed cracks	Mainly for discrete cracks; applicable using the smeared crack version for well distributed cracks
7	Experimentally observed behaviours	For well distributed cracks	For discrete cracks
8	Simplicity and practicability for use	Requires deep knowledge	Requires deep knowledge

2　REMARKS ON THE EXISTING MODELS AND PROPOSED IMPROVEMENTS

Based on the available and published research reports on the constitutive modelling of reinforced concrete, it can be said that research results founded on or supplemented by experimental investigations regarding the post-peak stress martial properties are limited only to uniaxial tensile and compressive behaviours. Besides, there exist few investigations on the non-uniform strain (stress) fields as a result of the matrix fracturing (i.e., splitting- sliding- and mixed-type fracture) and the local mortar-coarse aggregate interface behaviour at micro level; which require the so-called homogenisation technique. Moreover, the material anisotropy and the continuous degradation of the material stiffness of the matrix due to crack-induced damage that is common for quasi-brittle materials were not adequately investigated under multiaxial states of stress (strain) both at experimental and theoretical levels.

It can generally be said more remains to be done before a stage is reached, when realistic, rational, reliable and practical constitutive models are available to be adopted for the prediction of the non-uniform stress (strain) fields, failure modes, and design of reinforced concrete structures.

REFERENCES

Chen,Wai-Fah et al. 1991. Constitutive Models. J. Isenberg (ed.), Finite Element Analysis of Reinforced Concrete Structures II, Proc. of the International Workshop, New York, 2–5 June 1991. 36–117. New York: ASCE.

Bazant, Z.P. 1985. Fracture Mechanics and Strain-Softening of Concrete. C. Meyer & H. Okamura (eds), Finite Element Analysis of Reinforced Concrete Structures, Proc. Paper, Tokyo, 21–24 May 1985. 121–150. New York: ASCE.

Ishikawa, M. et al. 1985. The Constitutive Model in term of Damage Tensor. C. Meyer & H. Okamura (eds), Finite Element Analysis of Reinforced Concrete Structures, Proc. Paper, Tokyo, 21–24 May 1985. 104. New York: ASCE.

Finite Elements in Civil Engineering Applications, Hendriks & Rots (eds.)
© 2002 Swets & Zeitlinger, Lisse, ISBN 90 5809 530 4

Degree of hydration concept for early age concrete using DIANA

G. De Schutter

Magnel Laboratory for Concrete Research, Ghent University, Belgium

ABSTRACT: Thermal and mechanical simulations of hardening concrete elements have to take account of the evolving material properties. This can be realized in a fundamental way using the degree of hydration concept. In this concept, the material properties are related with the microstructural development during cement hydration. The implementation of the degree of hydration concept within the DIANA environment is evaluated. The accuracy of the results is studied in comparison with other simulation tools.

1   INTRODUCTION

During former research, thermal and mechanical material models have been developed for hardening concrete based on the degree of hydration as a fundamental parameter (De Schutter 1996). For the heat production during hardening, a new general hydration model, also valid for blast furnace slag cement, was developed (De Schutter & Taerwe 1995). Furthermore, evolving material laws were developed, describing the evolution of thermal and mechanical properties during hardening (De Schutter 1996).

In this paper, the implementation of de degree of hydration based hydration model and material laws into the finite element program DIANA is evaluated. The accuracy of the results is studied in comparison with other simulation tools.

2   THERMAL SIMULATION

2.1   *Specific heat and thermal conductivity*

The specific heat and the thermal conductivity of the hardening concrete can be introduced in DIANA as degree of hydration dependent parameters. A piece-wise linear relation can be defined for both parameters, using the instructions CAPREA resp. CONREA.

2.2   *Hydration model*

Within the DIANA environment, a degree of hydration depending heat production during hardening can be introduced. In DIANA, the degree of hydration is defined as the fraction of the heat of hydration already released. This definition is similar to the one considered by De Schutter & Taerwe (1995). De different model parameters occurring within the general hydration model developed by De Schutter & Taerwe (1995) are not always defined as such within DIANA. Some of the model parameters have to be combined to one parameter within DIANA. Moreover, DIANA does not have the possibility to consider the twofold character of the general hydration model, i.e. one reaction for Portland clinker and another reaction for blast furnace slag. In order to be able to make simulations for concrete elements based on blast furnace slag cement (or other blended cements), the hydration process has to be simplified by means of one reaction. Within these constraints, the hydration model developed by De Schutter & Taerwe (1995) can be implemented in DIANA as follows:

- The Arrhenius constant C_A (in °C) in DIANA should be considered as the ratio of the apparent activation energy E (in kJ/mole) and the universal gas constant R (in kJ/mole·K).
- The heat development rate as a function of the degree of hydration can be implemented directly in DIANA, as a piecewise linear approximation of the real function developed by De Schutter & Taerwe (1995), using the instructions REACTI and PRDKAR.
- The parameter α (expressed in W/m³, using the instruction ALPHA), can be calculated from the maximum heat production rate (in J/g·h) at 20°C, by means of the following expression:

$$\alpha = q_{max,20} \cdot e^{\frac{E}{R} \cdot \frac{1}{293}} \cdot C \cdot \frac{1000}{3600} \qquad (1)$$

with C the cement content of the concrete, expressed in kg/m³. The values of 1000 resp. 3600 represent the conversion from kg to g resp. from hours to seconds.

– The parameter Q (in J/m³, using the instruction MAXPRD), giving the total heat content of the concrete can be calculated as follows:

$$Q = Q_{max} \cdot 1000 \cdot C \qquad (2)$$

with Q_{max} the total cumulated heat production of the cement, expressed in J/g, C the cement content of the concrete, and where 1000 represents the conversion from kg to g.

The initial value of the degree of hydration should be specified within the command file, at the start of the transient analysis.

The instructions given above are valid for the case of one reaction. When a twofold reaction has to be considered, like for the case of blast furnace slag cement, a simplification has to be made when using DIANA. A twofold reaction cannot be introduced. The simplification toward one reaction can be obtained by means of the following conversions:

$$Q_{max} = Q_P + Q_S \qquad (3)$$

$$q_{max,20} = q_{p,max,20} \cdot \frac{Q_P + Q_s}{Q_p} \qquad (4)$$

where the subscripts P resp. S stands for the Portland reaction resp. slag reaction. By means of the approximation the heat production of the blended cement is not simulated in a perfect way. The obtained simulation results however are acceptable, as illustrated in the sequel.

2.3 Initial and boundary conditions

The convective boundary conditions can be implemented by the application of boundary elements with an attached conduction coefficient (in W/m²·K). Time dependent boundary conditions can be implemented using the instructions TIME and CONVTT.

The initial concrete temperature has to be introduced using the instructions INIVAR and POTENT. The temperature of the environment has to be specified by means of BOUNDA and EXPOT. A time dependent environmental temperature should be introduced using TIMEBO.

3 EVALUATION OF THE ACCURACY

3.1 Adiabatic temperature rise

As the heat of hydration is the driving force for thermo-mechanical problems in hardening concrete, it is first

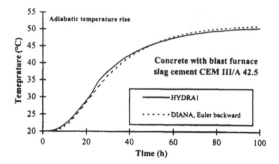

Figure 1. Adiabatic temperature rise.

evaluated whether DIANA enables an accurate simulation of the heat production rate. Especially for the case of blended cements, for which a simplification into one reaction has to be realized as mentioned before, a verification of the accuracy is needed.

For adiabatic conditions, the temperature rise of the concrete is simulated using DIANA, and the results are verified in comparison with the scientific program HYDRA1, developed by De Schutter (1996). The obtained results are very accurate, even in the case of blast furnace slag cement, as illustrated in Figure 1.

For cements with a higher slag content, the deviations due to the simplification into one hydration reaction might become more pronounced.

3.2 Convective problem with variable environment

The situation of a hardening massive concrete wall within a variable environment (daily temperature cycle) is simulated using DIANA. The results are evaluated in comparison with the results obtained by means of the scientific program HYDRA1. Figure 2 shows a very agreement between the results obtained by both methods.

The obtained deviations can mainly be attributed to the simplification of the twofold character of the blended cement into one reaction within the DIANA environment.

3.3 Three-dimensional convective problem

Although the simulations reported above have been realized by means of volume elements, the heat flow was mainly one-dimensional. A further evaluation has been done for real three-dimensional problems. In this case, the results were evaluated by means of the program CALIX, developed at Ghent University.

Figure 3 shows the simulation results for the case of a hardening concrete cube with side length 2 meter.

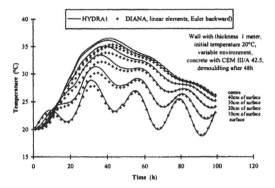

Figure 2. Temperature evolution in hardening wall in environment with variable temperature (daily cycle).

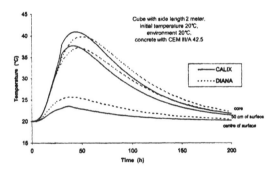

Figure 3. Temperature evolution in hardening cube.

Within DIANA, the simulation has been realized using 704 linear elements, of which 192 boundary elements. Only one eight of the cube was considered, due to symmetry. Similar calculations have also been realized using quadratic elements (112 elements, of which 46 boundary elements), yielding the same results, although with less calculation time.

4 MECHANICAL SIMULATION

For the mechanical calculation also a degree of hydration based approach can be used within DIANA. The degree of hydration based mechanical material laws, as developed by De Schutter (1996), can be implemented in DIANA. The visco-elastic behavior is implemented by means of a Kelvin chain, with spring and dash properties depending on the degree of hydration, as reported by De Schutter (1999).

The cracking behavior is simulated using a smeared cracking concept, with material properties also depending on the degree of hydration. A simple cracking criterion is used, stating that cracking occurs when the highest principal tensile stress reaches the actual and local tensile strength. A non-linear tension softening behavior is applied. In this softening model the tensile strength and the specific fracture energy depend on the degree of hydration, as described by De Schutter & Taerwe (1997).

A variable shear retention factor for the cracked concrete is also applied within the finite element simulation, depending on the normal crack strain. The Newton-Raphson method is used for the iterative solution of the resulting set of nonlinear equations which ensues after discretization of the nonlinear boundary value problem. A verification of the obtained results has been realized by means of literature results and by means of own experiments (De Schutter 2000).

5 PRACTICAL APPLICATIONS

The degree of hydration concept for hardening concrete using DIANA as a simulation tool has already been used for practical cases in Belgium, like the massive concrete armour units on the breakwaters in Zeebrugge and the quay walls of the Verrebroekdok in Antwerp. The obtained simulation results have shown to be in good agreement with reality.

6 CONCLUSIONS

A realistic three-dimensional simulation of early age thermal cracking in hardening massive concrete elements using the DIANA environment can be carried out considering the degree of hydration as a fundamental parameter. This parameter, which is closely related to the hydration reaction of the cement, enables a simple though fundamental description of most material properties during hardening. The fundamental material laws can be implemented in DIANA, although sometimes some simplification has to be made. The obtained results have shown to be in good agreement with the results of other numerical tools as well as with experimental results and real structural behavior.

REFERENCES

De Schutter, G. 1996. Fundamental and practical study of thermal stresses in hardening massive concrete elements. Doctoral thesis, Ghent University.
De Schutter, G. 1999. Degree of hydration based Kelvin model for the basic creep of early age concrete. Materials and Structures, Vol. 32, May, 260–265.
De Schutter, G. & Taerwe, L. 1995. General hydration model for portland cement and blast furnace slag cement. Cem. Con. Res., Vol. 25, No. 3, 593–604.

De Schutter, G. & Taerwe, L. 1997. Fracture energy of concrete at early ages. Materials and Structures, Vol. 30, March, 67–71.

De Schutter, G. 2000. Simulation of thermal cracking in massive hardening concrete elements using an internal parameter related to the hydration reaction. In *Computational Techniques for Materials, Composites and Composite Structures* (ed. B. Topping), Civil-Comp. Press, Edinburgh, Sept. 2000, pp. 7–14.

Finite Elements in Civil Engineering Applications, Hendriks & Rots (eds.)
© 2002 Swets & Zeitlinger, Lisse, ISBN 90 5809 530 4

Development of a procedure for crack width evaluation in continuous composite bridges

Q.U.Z. Khan, T. Honda & Y. Okui
Department of Civil and Environmental Engineering, Saitama University, Saitama, Japan

N. Masatsugu & I. Eiji
Department of Civil Engineering, Nagaoka University of Technology, Nagaoka, Niigata, Japan

ABSTRACT: A procedure for crack width evaluation in continuous steel-concrete composite bridges, which accounts for shear-lag effect, is proposed. The proposed procedure employs smeared crack finite element analysis and a governing differential equation for bond-slip behaviour. The numerical results are compared with the results of tests performed on beams subjected to negative bending.

1 INTRODUCTION

In continuous steel-concrete composite girder bridges, concrete slabs near internal supports are affected by negative bending moments and thus undergo cracking. The crack width control procedure in Eurocode is based on Hanswille theory (Hanswille 1986), while that in Japanese code (Expressway Technology Center 1997) is based on experimental results. Both procedures employ the beam theory with the effective width concept to take account of the shear-lag effect. The applicability of the effective width formulae in design codes to crack width evaluation has not been confirmed yet. In this paper, we propose a numerical procedure to evaluate the crack width, which accounts for the shear-lag effect.

2 METHOD OF CRACK WIDTH EVALUATION

2.1 *FE modeling*

The proposed method of crack width evaluation is combination of FE analysis and Hanswille theory. The smeared crack model is used in the FE analysis. A constant cut-off criterion was used as a cracking criterion for modeling of concrete. Perfect bonding between concrete and embedded reinforcement is assumed in this modeling. Bond-slip effect was taken into account by using an average stress-strain relationship of reinforced concrete including tension stiffening effect. The stress-strain relationship of concrete is modeled as the

multilinear tension softening relation derived from average stress-strain relation using the bond-slip differential equation proposed by Hanswille. The steel girder and concrete slab were modeled with four-node isoparametric shell elements and eight-node solid ones, respectively.

2.2 *Crack width evaluation*

For a RC member subjected to axial tensile force, two cracking states can be distinguished, namely initial cracking state and stabilized cracking state. In the initial cracking state, since there is perfect bond region between two cracks, there is no interaction between two adjacent cracks. On the other hand, in the stabilized cracking state, the bond slip regions of two adjacent cracks overlap each other.

Using the same constitutive relation for bond-slip behaviour as used in Hanswille's theory and the bond-slip differential equation solution, the crack widths expressions for the two cracking states were derived. The crack width w_R in the initial cracking state is given by (Hanswille 1986)

$$w_R = 2 \left[\frac{1+N}{A\beta_w} \frac{d_s}{8} \frac{\Delta\sigma_{s,r}}{E_s} \left(\sigma_{s,r} - E_s\varepsilon_0 \right) \right]^{\frac{1}{1+N}} \quad (1)$$

where β_z is the tensile strength of concrete; E_s and d_s are Young's modulus and the diameter of a reinforcement steel bar, respectively. A and N are parameters for the bond-slip constitutive model and ε_0 denotes the strain

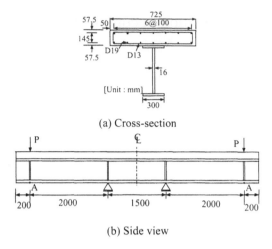

(a) Cross-section

(b) Side view

Figure 1.   Test specimen of composite girder.

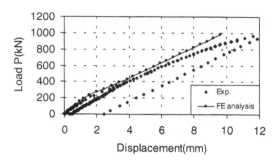

Figure 2.   Comparison of load-displacement curves between experimental data and numerical results.

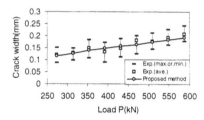

Figure 3.   Comparison of crack width-load curves between experimental data and numerical results.

of concrete due to shrinkage (contraction is negative). σ_{sr} is the steel stress at the crack in the case of first cracking and $\Delta\sigma_{s,r} = \beta_z/\mu$ where μ is reinforcement ratio.

In the stabilized cracking stage, crack width is given by

$$ w = w_R \alpha^{\frac{2}{1-N}} \left[1 - \frac{(2\alpha - \eta_m)^2}{\alpha - N(\alpha - \eta_m)} \frac{1-N}{4\alpha} \right] \qquad (2) $$

where η_m is the mean nondimensional crack spacing and

$$ \alpha = \left(\frac{\sigma_{sII} - \varepsilon_0 E_s}{\sigma_{s,r} - \varepsilon_0 E_s} \right)^{\frac{1-N}{1+N}} \qquad (3) $$

with σ_{sII} is the stress in steel at a crack.

In the proposed procedure, first 3D finite element analysis, including shear-lag effect of composite girders, is employed to evaluate average concrete and steel stresses. Then α and σ_{sr} are calculated from these average stresses from FE analysis.

3   COMPARISON WITH EXPERIMENT DATA

3.1   *Specimens and test setup*

To verify the proposed method we compared the numerical results with experimental data (Nagao 2001). The cross-sectional dimensions, span and loading details for a specimen are shown in Fig. 1. The vertical

displacement was measured at point A in Fig. 1b. The FE analysis was carried out using DIANA.

3.2   *Comparison*

A comparison of the vertical displacements obtained from the FE analysis with the experimental results is presented in Fig. 2. The agreement between the numerical results and the experimental data is fairly good.

The average crack widths were calculated at different load levels and its comparison with experimental crack widths is shown in Fig. 3. In addition, the maximum and minimum crack widths observed in the tests are plotted. The numerical results are in good agreement with the experimental mean values.

4   CONCLUDING REMARKS

In this paper, a procedure is developed to evaluate the crack width using the smeared crack finite element analysis and a differential equation for bond-slip behaviour. The theoretical prediction is closer to the mean crack widths. Since the maximum crack width is more important from a durability point of view, it is required to convert the mean values into maximum crack width. After this conversion, the proposed method can be used to propose the formula for effective width to evaluate crack width.

REFERENCES

Expressway Technology Center, 1997, Report on Optimization of 2-I Girder Composite Bridges with PC Slabs, (in Japanese).

Hanswille, G. Zur Riβbreitenbeschränkung bei Verbundträgern, Technisch-Weissenschaftliche Mitteilungen, Institut für Konstruktiven Ingenieurbau Ruhr- Universität Bochum, Mittelilung Nr. 86-1, 1986.

Nagao, T. 2001, Study on Cracking in Composite Girders under Negative Bending Moment, Master thesis, Nagaoka University of Technology, (in Japanese).

Finite Elements in Civil Engineering Applications, Hendriks & Rots (eds.)
© 2002 Swets & Zeitlinger, Lisse, ISBN 90 5809 530 4

Lateral reaction modulus for single piles – Simplified modelling by linear finite elements

A. Bouafia, L. Belmouloud & A. Henniche
Department of Civil Engineering University of Blida, Blida, Algeria

ABSTRACT: Pile foundations often behave under lateral forces. The complexity of soil-pile interaction is due to many geotechnical and geometrical factors as well as to the 3D nature of the phenomenon. This paper is aimed to present the results of a simplified 3D modelling of soil/pile system by linear solid finite elements. After validation of the model, a parametric study was carried out. The lateral subgrade reaction modulus was derived from a process of fitting and differentiation of shear forces curves along the pile, and a simplified formulation in homogeneous soil as well as in Gibson's soil was proposed. At last, an assessment of this formulation was carried out by predicting the deflections of experimental piles tested at full scale.

1   INTRODUCTION

It is recognised that lateral load-deflection behaviour is an important aspect of soil/pile interaction analysis.

Many slender structures built on pile foundations transmit to these latter important lateral forces. Resulting pile deflections should be within permissible values allowed by the structure.

Numerous deflection analysis methods are available in practice. They may be classified into three main categories, namely the elastic methods, the subgrade reaction methods or P-Y curves-based methods and the finite elements methods. Soil surrounding the pile is modelled in the last category as an isotropic elastic semi-infinite medium characterised by an elastic modulus E and a Poisson's coefficient ν.

The aim of this paper is to present the results of a simplified 3D analysis by finite elements of the lateral response of a single pile embedded in a homogeneous soil or in a Gibson's soil whose soil modulus linearly increases with depth. It is focused in this study on the subgrade reaction modulus E_s in relation with soil/pile interaction parameters as shown later. Figure 1 illustrates the relevant parameters in such a problem.

2   PRESENTATION OF THE MODEL

The 3D nature of the pile/soil response necessitates a laborious cylindrical finite elements mesh. In order to facilitate the parametric study, simplified regular rectangular mesh, as illustrated in Figure 2, was considered with rectangular cross section for the pile. It has been however checked by comparison with the results of some cases with irregular cylindrical mesh that the size effect on pile top displacements is negligible for semi-rigid and rigid piles (Belmoulod & Henniche, 1996). The advantage of such an approach is to analyse many cases with the same mesh.

The program SAP was used with 8-node brick elements based upon an isoparametric formulation including nine optional incompatible bending modes

2 $\times$ 2 $\times$ 2 numerical integration scheme is used for the solid elements (Wilson & Habibullah, 1989). It is to be noted that stresses are evaluated by the program

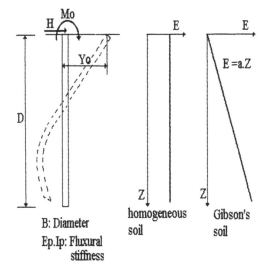

Figure 1.   Notations used in the problem.

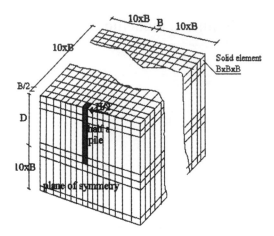

Figure 2. Scheme of regular FEM mesh.

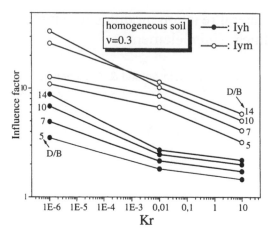

Figure 3. Influence factors in homogeneous soil.

at the centre of each element and averaged at the joints of the element. The study was then limited to the interpretation of forces and displacements in joints.

The mesh was dimensioned by increasing the size of the mesh until top displacements of a typical pile do not depend on the mesh size. Dimensions retained are shown in Figure 2.

3   PARAMETRIC STUDY

The relative soil/stiffness K_r is defined according to Figure 1 as follows:

$$K_r = \frac{E_p \cdot I_p}{E(D) \cdot D^4} \qquad (1)$$

Dimensional analysis leads to link the surface deflection Y_0 to the slenderness ratio D/B, K_r and ν as follows:

$$f(I_{YH}, I_{YM}, D/B, \nu, \nu_p, K_r) = 0 \qquad (2)$$

where I_{YH} and I_{YM} are the influence factors due to the lateral force H and the moment M_0 respectively.

$$I_{YH} = \frac{Y_0^h \cdot E(D) \cdot D}{H} \qquad (3)$$

$$I_{MH} = \frac{Y_0^m \cdot E(D) \cdot D^2}{M_0} \qquad (4)$$

Y_0^h and Y_0^m are the deflections due to H and M_0 respectively.

Influence factors I_{YH} and I_{YM} for different values of D/B, are shown in Figures 3 and 4 in bi-logarithmic

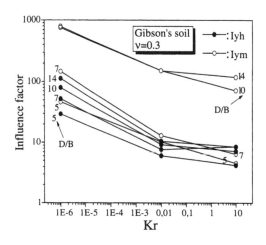

Figure 4. Influence factors in Gibson's soil.

scales with a considerable reduction of pile deflection with the fluxural soil/pile rigidity K_r. Deflections of short and rigid piles are therefore less than those of long and flexible piles.

The linear shape of the curves suggests a power variation of I_{YH}-K_r and I_{YM}-K_r relationships. Values of D/B of 5 and 7 rather correspond to piers, and those of 10 and 14 to long piles.

4   LATERAL SOIL REACTION MODULUS

Shear forces profile T(Z) was fitted by a cubic polynomial function then differentiated to obtain the lateral

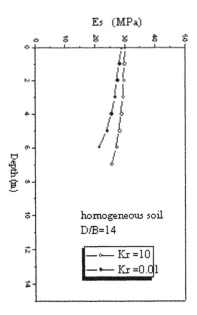

Figure 5.　Lateral reaction modulus profile.

soil reaction $P(Z)$. Lateral soil modulus E_s is computed as follows:

$$E_s(Z) = \frac{P(Z)}{Y(Z)} \qquad (5)$$

Figure 5 illustrates a typical profile of E_s in a homogeneous soil. In the neighbouring zone of zero deflection (centre of rotation), equation 5 provides disturbed values of E_s. The analysis was then limited to the upper part of pile shaft.

For flexible piles ($K_r < 10^{-6}$), shear forces at pile top do not correspond to the lateral load H, which leads to inaccuracies in values of E_s determined according to the above procedure. The analysis of lateral soil reaction modulus was then limited to semi-rigid to rigid piles ($K_r > 10^{-2}$).

It is remarkable that the ratios E_s/E in homogeneous soil and N_H/a in Gibson's soil slightly depend on K_r. N_H is the rate of linear increase of E_s with depth.

Moreover, as shown in Figure 6, logarithmic function fits well the data obtained, as follows:

$$\frac{E_s}{E} = 2 - \frac{Ln\left(\dfrac{D}{B}\right)}{4} \qquad (6)$$

$$\frac{N_H}{a} = 2.65 - 0.42x\,Ln\left(\frac{D}{B}\right) \qquad (7)$$

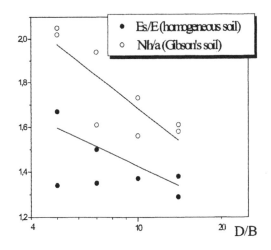

Figure 6.　Variation of E_s/E and N_H/a with D/B.

These simple relationships may be used for the evaluation of pile deflections with the lateral reaction modulus-based methods.

5 PREDICTION OF DEFLECTIONS

Comparative study was undertaken in order to assess the quality of prediction of pile deflection on the basis of lateral piles loading tests at full-scale (Bouafia, 1990).

The first site is located in Salledes (France). Steel pipe pile long of 6.14 m with an outside diameter of 0.91 m was installed into a homogeneous normally consolidated marl characterised by a pressuremeter modulus of 5.4 MPa. Fluxural stiffness $E_p \cdot I_p$ is equal to 1070 MN·m². Measured deflection under H = 100 kN and M_0 = 50 kN·m is 4.16 mm.

From equation 6, E_s = 12.34 MPa. The elastic length L_0 was found equal to 4.30 m and D/L_0 = 1.42. The pile should be considered as semi-rigid one. Computed deflection by the lateral reaction theory is equal to 6.1 mm.

The second site is located in Plancoet (France) and composed of a quite homogeneous clayey layer with occasional presence of sand. Average pressuremeter modulus was estimated to 2.24 MPa and the clay was classified as soft clay.

Test pile is 6.10 m long and has an outside diameter of 0.28 m and $E_p \cdot I_p$ = 28.71 MN·m². For a lateral load of 4.6 kN and M_0 = 2.07 kN·m, measured lateral deflection is 2.13 mm.

According to equation 6, E_s = 5.51 MPa. L_0 = 2.14 m and D/L_0 = 2.85. Computed deflection is 1.0 mm.

As a finding, one can say that the simple formulation proposed for the lateral reaction modulus leads to a reasonable prediction of lateral deflections and may be used for a preliminary estimation of small lateral deflections of piles. The elastic modulus E may be estimated from the pressuremeter test or by correlation with usual in-situ tests parameters.

6 CONCLUSIONS

Simplified 3D regular mesh was built for finite element modelling of lateral load-deflection response in homogeneous soil or in Gibson's soil.

Simple formulation of the lateral reaction modulus in relation with the elastic soil modulus was proposed for semi-rigid to rigid piles. Comparison of predicted deflections to the ones measured during full-scale lateral loading tests has shown a reasonable agreement, which may be subsequently improved by refining the modelling and studying other values of D/B.

REFERENCES

Bouafia, A. 1990. Modelling of laterally loaded piles in centrifuge (in French), Ph.D thesis, University of Nantes, Ecole Centrale de Nantes\LCPC, 267 p.
Henniche, A. & Belmouloud, L. 1996. Behaviour of deep foundations neara sloppy ground – Finite element modelling (in French), B.SC of Civil Engineering dissertation, Departyement of Civ. Engg, University of Blida, 157 p.
Wilson, E.L. & Habibullah, A. 1989. SAP90 – A series of computer programs for the static and dynamic finite element analysis of structures, CSI, 188 p.

Finite Elements in Civil Engineering Applications, Hendriks & Rots (eds.)
© 2002 Swets & Zeitlinger, Lisse, ISBN 90 5809 530 4

A concrete beam finite element accounting for shear effects

F. Kaiser
S.A. Ronveaux, Ciney, Belgium

H. Degée
M&S Department, University of Liège, Belgium

ABSTRACT: This paper aims at presenting a new beam finite element allowing to simulate the behaviour of structures made of concrete beam structural elements where shear can have a significant effect. An incremental MCFT-based constitutive law (Modified Compression Field Theory) describes behaviour of concrete. Results obtained by using this element are compared with membrane models and tests results.

1 INTRODUCTION

Beam finite elements are very attractive for civil engineers because of the possibility of an easy definition of the geometry of structures and of great facilities in interpreting computation results. However, quite few of the great amount of available beam elements can account for shear deformations.

If classical elements without shear deformations are normally accurate to study steel structures, this accuracy is not so obvious for concrete structures. Indeed, concrete beams are generally more massive than steel beams, leading to a greater influence of shear energy. Furthermore, if steel can be considered as an isotropic and homogeneous material, the disposition of longitudinal and transversal reinforcements in concrete beams can induce critical cross sections very sensitive to shear and therefore lead to an early collapse of the structure.

In this perspective, we develop a beam element that allows accounting for shear deformation in general, and for shear cracking in concrete in particular.

2 INCREMENTAL CONSTITUTIVE LAW FOR REINFORCED CONCRETE

The constitutive law is an incremental version of the Modified Compression Field Theory. This theory, elaborated by Collins and Vecchio (1986) for cross section verification is based on the following basic principles:

– Stresses and strains are managed in terms of average effective values (cracks are not explicitly modeled);
– Equilibrium of stresses and compatibility of strains in concrete and steel are ensured;
– Principal stresses and principal strains in concrete are assumed to be parallel;
– Uniaxial stress-strain relations in principal directions are used for both concrete and reinforcement steel.

The organization of the concrete incremental law can be summarized as follows (see also Fig. 1):

1. Computation of ε_{xx}, ε_{yy} and γ_{xy} from the FE displacements;
2. Computation of principal strains ε_1 and ε_2, and of their orientation θ, which gives also crack orientation. Computation of the principal strain increments $\Delta\varepsilon_1$ and $\Delta\varepsilon_2$;
3. Actualisation of the damage parameter β, function of the principal tensile strain ε_2;

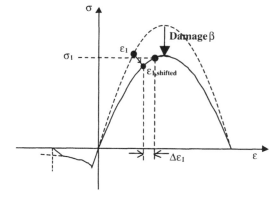

Figure 1. Incremental constitutive law.

4. Shifting to the actualized curve, according to the new value of β ($\varepsilon_1 \rightarrow \varepsilon_{1,\text{shifted}}$);
5. Computation of the new principal stresses σ_1 and σ_2;
6. Computation of σ_{xx}, σ_{yy} and τ_{xy} stresses;

This procedure has first been validated in a membrane finite element.

3 BEAM ELEMENT INCLUDING SHEAR DEFORMATIONS

The new finite element is based on the element developed by De Ville (1986) in his Ph.D. thesis. This element is a curved beam element with 3 nodes and 7 degrees of freedom in Total Corotational Lagrangian formulation, allowing dealing with both geometrical and material non-linearities. The element is implemented in home-made non-linear finite element code Finelg.

Displacement field of the plane beam element is given by Eq. (1.1) and (1.2) where the rotation $\theta(x)$ is assumed to remain small (see also Fig. 2):

$$u(x, y) = U(x) - y\,\theta(x) \tag{1.1}$$

$$v(x, y) = V(x) \tag{1.2}$$

The most simple finite element formulation of (1.1) and (1.2) should be a independent discretisation of $V(x)$ and $\theta(x)$ with linear shape functions (Hencky). However a due examination of equilibrium equations of the beam shows that transverse displacements and rotations are dependent and related together by Eq. (2). Furthermore, if the beam is subjected to a uniformly distributed load q, Eq. (3) must also be verified.

By integrating (2) and (3), we come to expressions (4.1) and (4.2) of $V(x)$ and $\theta(x)$. The four constants B, C, D and F can then be expressed in terms of nodal connectors V_1, V_2, θ_1 and θ_2, leading to the discretised formulation of the element.

$$EI\,\theta''(x) + GA^{sh}[v'(x) - \theta(x)] \tag{2}$$

$$EI\,\theta''(x) = q \tag{3}$$

$$\theta(x) = \frac{1}{EI}\left(q\frac{x^3}{6} + B\frac{x^2}{2} + Cx + D\right) \tag{4.1}$$

$$v(x) = \frac{1}{EI}\left(q\frac{x^4}{24} + B\frac{x^3}{6} + C\frac{x^2}{2} + Dx\right)$$
$$- \frac{1}{GA^{sh}}\left(q\frac{x^2}{2} + Bx\right) + F \tag{4.2}$$

Such a formulation exhibits the great advantages to avoid any shear locking and to provide the exact solution of the linear elastic problem independently of the number of elements. This leads therefore to a very good convergence in the non-linear range.

4 APPLICATION

The reinforced concrete beams tested by Bresler and Scordelis (1961) were designed to be shear critical and are now often used as benchmarks. We present in Fig. 3 a comparison between tests results, membrane elements with MCFT-based law (with and without possibility of transverse reinforcements yielding) and beam elements (20 FE over the beam length).

5 CONCLUSION

We propose in this paper an incremental law based on MCFT. This approach has firstly been validated with a membrane model, then implemented in an efficient shear-deformable beam element. In its current version, this beam model is accurate if no yielding of transverse reinforcements is observed.

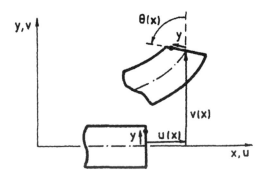

Figure 2. Displacement field of a plane beam.

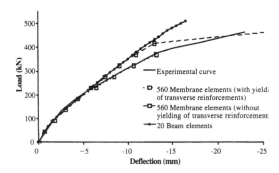

Figure 3. Comparison of test and models for beam A2 of Bresler and Scordelis.

REFERENCES

Bresler B. & Scordelis A.C. 1963. Shear strength of reinforced concrete beams. ACI Journal, Vol. 60 n°1.

De Ville de Goyet V. 1986. L'analyse statique non linéaire par la méthode des éléments finis des structures spatiales formées de poutres à section non symétrique. Ph.D. Thesis. University of Liège.

Kaiser F. 2000. Internal Research Reports n°236, 240 & 241. University of Liège, M&S Dpt.

Vecchio F.J. & Collins M.P. 1986. The modified compression field theory for reinforced concrete beam elements subjected to shear. ACI Journal (March–April 1986).

FINELG User's Manual (2002). Version 8.5. Greisch Info – M&S Department (University of Liège).

Finite Elements in Civil Engineering Applications, Hendriks & Rots (eds.)
© 2002 Swets & Zeitlinger, Lisse, ISBN 90 5809 530 4

DIANA test procedures

J.C.M. Jansen

TNO Building and Construction Research, Delft, Netherlands

ABSTRACT: The quality of the results of a finite element calculation depends highly on the quality of the finite element package used. In DIANA, testing is an integral part of the software development process. It can be divided in static and dynamic testing, two methods that complement each other. Both methods are described with their pros and cons.

1 INTRODUCTION

The quality of a finite element calculation depends on a lot of factors. Transforming the geometry to a correct finite element model with boundary conditions, using the correct material model, choosing the iteration procedures, and interpreting the results is mainly the responsibility of the engineer. However, if this is done carefully, a bad program can still lead to wrong results. Although an experienced engineer will recognize these incorrect answers, generally we do not accept that the program delivers incorrect results. To avoid errors due to program failures, DIANA is being developed under a lot of Quality Assurance rules. Several kinds of testing are embedded in a lot of places of these QA rules. Except from the acceptation tests for software that we develop on request of a specific client, static and dynamic tests are carried out to detect failures as early as possible.

2 STATIC TESTING

To detect bugs, even before running, the code can be tested just by looking at it. The QA rules require that one or more peers inspect all new or modified code. In earlier stages of the software development process, like requirement specification and design, reviews are performed as well. On these stages review is even more important to avoid notification of design errors afterwards, when all the programming is done. We aim to detect defects as soon as possible, because fewer costs are involved when a repair is done in an early stage of the development process.

During the source code review, the code is inspected for correctness, style and performance. Besides the peer reviews we developed a tool for static code checking. Originally this tool was designed for debugging

purposes, but it evolved into a QA tool. Besides the checking of new code, this tool is also used regularly to check the complete code on consistency.

The advantage of static checking is that it is quite easy to inspect the complete code. The disadvantage is that it is hard to follow all the possible paths through the software.

3 DYNAMIC TESTING

The opposite of static testing is dynamic testing. With this kind of testing the application is being executed and the analysis results are being verified against the expected results. These expected results may come from a known analytical solution, from other programs, from literature, like the NAFEMS benchmark journal or even from experiments. If these results are not available, for instance for more complex geometrical models or combined material models, the reference results can be constructed from a first run by an engineer. This guarantees that the behavior of the program, while being developed, does not change without notification.

The results of a test are not only related to the source code, but also to the used hardware and third party software (compilers, external libraries, etc.). When one of the components is changed, the tests must be carried out again. Due to the daily changes of the source code, the testing process must be performed automatically. At present we have a test suite that contains over 1800 tests. The nature of these tests varies. There are tests which only test a subroutine (-set) which will be used as black box in the rest of the code. Other tests contain small finite element jobs (1 to 10 elements) that test a specific feature. There are more complicated tests to test the combination of certain features. Finally there are large finite element jobs that run for a longer time to test a real life situation. Some tests of the last group are stored in a performance test suite. Also the examples from the "DIANA example manual" are tested this way.

The test suite of DIANA automatically runs every night on several platforms with different versions of operating systems and compilers. Before the test-suite is started, the application is updated with the last modification of the sources on each target machine. The test suite checks whether all the existing functionality still works and whether the newly added tests give machine independent results. In other words, the new development is being checked for portability and absence of unwanted side effects. When one or more of the tests give results that differ from the expected results, an e-mail is send to the responsible developer. Normally this failure is caused by a source code change in the last 24 hours. Therefore, the cause of the new failure should be easy to trace.

In contradiction to the static testing, it is quite hard to test all of the source code with dynamic testing. At the moment 82 percent of the source code lines in DIANA is covered with the automatic test suite. For new parts even over 90 percent. About four times a year we check the line coverage by running the test suite with instrumented code. From this coverage analysis we inspect the untested lines and try to extend the test suite for more coverage. This way we try to minimize the number of untested lines. 100% coverage seems to be the ideal situation, but coverage analysis has its limitations. Indeed, you can see that a line is tested, but not under which condition.

It is still possible that a tested line contains bugs, for instance a division by a value that is derived from user input. Although a particular line of code has been reached in one or more tests in the test suite, this method does not guarantee that the user input is checked in a way a division by zero is prevented. Another limitation of the line coverage analysis is, that it gives no insight in the tested combination of features. If, for instance, certain material models interfere which each other, their influence should be tested.

4 TEST SUITE ACCESSIBILITY

The complete test suite is stored in a tree-structured file system. Tests are grouped by application and it is easy to partial run the test suite. In fact the DIANA developers frequently run parts of the test suite themselves to verify their changes. Another way of accessing the test suite is via the dtest program. Most of the tests have keywords that indicate the features that are tested. The dtest program can locate a test that has a specified combination of keywords. For instance, it is easy to find a test that models a face load on a flow element.

The dtest program, together with most of the test suite, is delivered within the DIANA product distribution. This way the customers have a clear view about the testing that has been done at TNO. They also can verify that the DIANA test suite covers a test with a certain combination of features. Of course, in an ideal situation, all appropriate combinations of features are available in the test suite. Unfortunately, in practice this cannot be achieved. At the moment we have over 500 keywords that result in an almost unlimited amount of combinations. At the moment all keywords are covered in the DIANA test suite. We have been thinking about an automated keyword coverage analysis that gave some insight in the coverage of combination of keywords. The major problem at this point is that not all combinations make sense, and that it is rather arbitrary to define the combinations that do make sense.

5 TEST SUITE MAINTENANCE

We have spent a lot of effort to build the test suite as we have it now. To build a test, verify the significant results, and make it completely portable takes from about 4 to 8 hours. This means that the effort to setup such a test suite is over five man-years. While we are still expanding the test suite, the existing part needs to be maintained.

All the tests must be kept working and input syntax changes should be done in a consistent way over all the tests. To avoid the risk of mistakes involved by these syntax conversions, and limit the amount of work, we always try to write conversion programs that can modify the input data and commands from an old version to a new version. These conversion programs are kept between releases and delivered to the customers to help them with the syntax changes between two releases (d8com and d8dat).

Another concern is the output format. When something is changed in the layout, we do not want to interpret all differences and adapt the reference files. To avoid this updating effort, we have developed a lot of filters to pickup relevant results in a machine independent way for the output devices.

6 CONCLUSION

During a software development process a lot of human resources (and money) are used for testing. The global idea is that this investment leads to a higher quality of the software. To reach a constant level of quality it is essential that the testing is described in the software development QA rules. These QA rules can help the development team to find defects as early as possible. To improve the testing processes, it is essential that the process is being evaluated regularly. To make testing more efficient, it is profitable to automate the routine jobs.

Finite Elements in Civil Engineering Applications, Hendriks & Rots (eds.)
© 2002 Swets & Zeitlinger, Lisse, ISBN 90 5809 530 4

A steel beam element accounting for the cross section deformation

H. Degée
M&S Department, University of Liège, Belgium

ABSTRACT: Classical beam elements are based on the assumption that the cross section of the element remains unchanged during the entire loading. However, this can be too simple for modeling the behavior of thin-walled steel profiles, where local and distortional buckling are likely to occur. In this perspective, a special element taking into account the deformation of the cross section is presented and some examples of computations of critical bifurcation loads are presented.

1 INTRODUCTION

In the field of civil engineering, a frequent use is made of structural elements for which one dimension (the length) is significantly greater than the two others (the width and the depth). In order to study the behavior of such so-called beam elements, some classical assumptions are made, in order to reduce the behavior of the whole element to the behavior of its axis. Among these assumptions, the conservation of the shape of the cross section can be pointed out.

Such an approach leads in most cases to an accurate model of the member's real behavior. However, this assumption can be too strong, particularly in the frame on thin-walled structures. Indeed some phenomena including a deformation of the cross-section can occur in thin-walled members, such as for example local or distortional buckling.

If we have a look at the numerical methods available to study such profiles with deformable cross-section, and if we except the finite strip method, very efficient but essentially restricted to the study of isolated prismatic members, we can see a duality between plate finite elements on one hand, and special beam elements on the other hand. Plate models are convenient for almost any geometry, load and support condition, but they can be very heavy for what regards computation time and storage space. At the contrary, beam models are much faster and economic, but their use is more limited.

In this paper, we present such a special plane beam element allowing to account for local deformation of the cross section. This element is implemented in our home-made FE code FINELG.

2 DISPLACEMENT FIELD OF THE ELEMENT

The displacement field of the element is composed of two part (Eq. 1). The first term is representing the global displacement of the beam and is described by a Bernoulli formulation (Eq. 2.1 and 2.2).

For what regards the local deformation of the cross section (second term of Eq. 1), the displacement is described by a superposition of sectional modes chosen a priori and depending on the main global sollicitation of the structural element. For example, Fig. 1 shows the sectional modes used for an I-section subjected to pure compression.

So if $w_{L,i}^k(\eta_k)$ represents the local displacement of the wall k in the sectional mode i, the additional local displacement of wall k is given by Eq. 3. In this equation, η_k is the local coordinate along the transverse edge of wall k while $f_i(x)$ functions are representing the variation of the magnitude of mode i along the longitudinal axis. These functions are polynomial.

$$\mathbf{u}(x,y,z) = \mathbf{u^{bm}}(x,y) + \mathbf{u^{pl}}(x,y,z) \tag{1}$$

$$u^{bm}(x,y) = u^{bm}(x) - y\frac{dv^{bm}(x)}{dx} \tag{2.1}$$

$$v^{bm}(x,y) = v^{bm}(x) \tag{2.2}$$

$$w_L^{pl}(x,\eta_k) = \sum_i f_i(x)w_{L,i}^k(\eta_k) \tag{3}$$

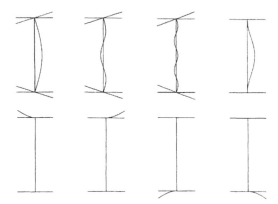

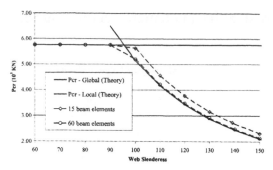

Figure 2. Influence of the web slenderness on the buckling load of an I-section profile under compression.

Figure 1. Sectional modes for an I-section under compression.

3 STRAINS AND STRESSES

Strains can easily be derived from the displacements through Eq.3.1 to 3.3 (example given for the case of a wall with its normal oriented along the global z-direction, for which η_k is identical to y).

Stresses can then be obtained by using a constitutive plane stress state law for each wall of the profile.

$$E_{xx} = \frac{du^{bm}}{dx} - y\frac{d^2v^{bm}}{dx^2} - z\frac{\partial^2 w_L^{pl}}{\partial x^2} + \frac{1}{2}\left(\frac{\partial w_L^{pl}}{\partial x}\right)^2 \tag{3.1}$$

$$E_{yy} = -\nu\frac{du^{bm}}{dx} + \nu y\frac{d^2v^{bm}}{dx^2} - z\frac{\partial^2 w_L^{pl}}{\partial y^2} - \frac{\nu}{2}\left(\frac{\partial w_L^{pl}}{\partial x}\right)^2 \tag{3.2}$$

$$2E_{xy} = -z\frac{\partial^2 w_L^{pl}}{\partial x \partial y} \tag{3.3}$$

4 APPLICATIONS

We present two examples of computation of critical bifurcation load. The first case is plotted at Fig. 2. On this figure, we compare the evolution of the critical buckling load of an I profile subjected to pure compression when the web slenderness increases. The computation is carried out with the new element and

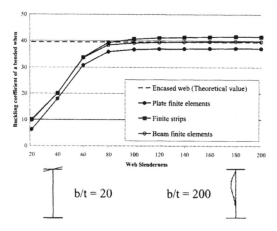

Figure 3. Influence of the web slenderness on the buckling load of an I-section profile subjected to pure bending.

compared to theoretical values. We can see that the buckling is shifting from global to local, and that the local behavior is very well estimated with the new element if a sufficient number of beam elements are used over the whole length of the profile.

The second example (Fig. 3) presents the evolution of the buckling coefficient of the web of an I-section subjected to pure bending when the web slenderness increases. Computation is carried out with plate finite elements, beam finite elements and finite strips. We can observe that the behavior of the cross section is shifting from the local buckling of the compressed flange to the local buckling of the bended web considered as clamped in the flanges. We can see that our adapted beam element seems to be the more accurate among the three compared models.

Some other examples dealing with both critical and non linear elastic analysis have also been carried out. Results are presented by Degée (2001).

5 CONCLUSION

We propose in this paper an approach allowing enriching efficiently the displacement field of a thin-walled beam element in order to account for the cross-section deformation. In its current version, this element can deal with box-, channel- and I-profiles subjected to in-plane global sollicitations.

REFERENCES

Degée H. 2000. Contribution à la prise en compte de la deformablité de la section droite dans un élément fini de poutre. Ph.D. thesis. University of Liège.
Degée H. 2001, 2002. Internal research reports n°248 & 6.

Finite Elements in Civil Engineering Applications, Hendriks & Rots (eds.)
© 2002 Swets & Zeitlinger, Lisse, ISBN 90 5809 530 4

Author index